ADVANCED MATHEMATICS

BOOK 2

Martin and Patricia Perkins

Bell & Hyman

First published 1983 by
BELL & HYMAN
An imprint of Unwin Hyman Limited
Denmark House
37–39 Queen Elizabeth Street
London SE1 2QB

Reprinted 1983, 1984, 1986, 1987

British Library Cataloguing in Publication Data

Perkins, Martin
Advanced mathematics.
Book 2.
1. Mathematics–1961–
I. Title II. Perkins, Pat
510 QA39.2

ISBN 0 7135 1322 5

Typeset by Polyglot Pte Ltd, Singapore
Printed in Great Britain at the
University Press, Cambridge

Contents

Preface

This is the second of two books designed to cover most Advanced Level single subject Mathematics syllabuses. The two volumes together provide a thorough treatment of the Pure Mathematics and Mechanics required at this level, together with a choice of further options including probability and statistics, numerical methods, matrices and group theory. We hope that the inclusion of these additional topics will enable students and teachers to plan a course of study to suit individual syllabus requirements. In allowing this opportunity for choice we have covered various topics which appear on Further Mathematics syllabuses. This means that Book 2 could be used as a starting point for some of this more advanced work.

We are most grateful for permission to use questions from past G.C.E. Advanced Level examinations. These are acknowledged as follows:

University of London University Entrance and Schools Examination Council (L)
The Associated Examining Board (AEB)
Joint Matriculation Board (JMB)
University of Cambridge Local Examinations Syndicate (C)
Oxford and Cambridge Schools Examination Board (O&C)
Oxford Delegacy of Local Examinations (O)
Southern Universities' Joint Board for School Examinations (SU)
Welsh Joint Education Committee (W)

We must again thank the staff of Messrs. Bell and Hyman, especially Caroline Paines and Antonia Murphy, for both their enthusiasm and their patience. We have also been helped and encouraged by many friends and colleagues.

M.L.P.
P.P.

List of symbols

$\mathbb{Z}$	set of integers
$\mathbb{Z}^+$	set of positive integers
$\mathbb{Z}_n$	set of integers modulo n, $\{0, 1, 2, \ldots, n-1\}$
$\mathbb{Q}$	set of rational numbers
$\mathbb{R}$	set of real numbers
$\mathbb{R}^+$	set of positive real numbers
$\mathbb{R}_0^+$	set of positive real numbers and zero
$\mathbb{R}^2$	set of ordered pairs of real numbers
$\mathbb{C}$	set of complex numbers
$[a, b]$	the closed interval $\{x \in \mathbb{R}: a \leqslant x \leqslant b\}$
(a, b)	the open interval $\{x \in \mathbb{R}: a < x < b\}$
$p \Rightarrow q$	p implies q; if p then q
$p \Leftarrow q$	p is implied by q; if q then p
$p \Leftrightarrow q$	p is equivalent to q
$\sim p$	not p
$\sum_{r=1}^{n} f(r)$	$f(1) + f(2) + \ldots + f(n)$
$\binom{n}{r}$	the binomial coefficient: ${}_nC_r = \dfrac{n!}{r!(n-r)!}$ for $n \in \mathbb{Z}^+$, $\dfrac{n(n-1)\ldots(n-r+1)}{r!}$ for $n \in \mathbb{Q}$
$\lvert x \rvert$	the modulus of x
$[x]$	the integral part of x
e^x, $\exp x$	exponential function of x
$\log_a x$	logarithm to the base a of x
$\ln x$	the natural logarithm of x, $\log_e x$
$\lg x$	the common logarithm of x, $\log_{10} x$

$\sin^{-1}$	inverse function of sin with range $[-\frac{1}{2}\pi, \frac{1}{2}\pi]$
$\cos^{-1}$	inverse function of cos with range $[0, \pi]$
$\tan^{-1}$	inverse function of tan with range $(-\frac{1}{2}\pi, \frac{1}{2}\pi)$
δx	small increment of x
$f'(x), f''(x), f'''(x)$	first, second and third derivatives of $f(x)$ with respect to x
$f^{(r)}(x)$	rth derivative of $f(x)$ with respect to x for $r > 3$
$\dot{x}, \ddot{x}, \ldots$	first, second, ... derivatives of x with respect to time
i	square root of -1
z	complex number, $z = x + iy = r(\cos\theta + i\sin\theta)$
$\text{Re}(z)$	real part of z, $\text{Re}(z) = x = r\cos\theta$
$\text{Im}(z)$	imaginary part of z, $\text{Im}(z) = y = r\sin\theta$
z^*	complex conjugate of z, $z^* = x - iy$
$\lvert z\rvert$	modulus of z, $\lvert z\rvert = \sqrt{(x^2 + y^2)} = r$
$\arg z$	principal argument of z, $\arg z = \theta$, where $\sin\theta = x/r$, $\cos\theta = y/r$ and $-\pi < \theta \leqslant \pi$
Δ	determinant of a matrix
$\det \mathbf{A}$	determinant of the matrix $\mathbf{A}$
$\mathbf{A}^{-1}$	inverse of the matrix $\mathbf{A}$
$\mathbf{A}^T$	transpose of the matrix $\mathbf{A}$
$\text{adj}\,\mathbf{A}$	adjoint of the matrix $\mathbf{A}$
$(G, *)$	group G with binary operation $*$
$+_n, \times_n$	addition and multiplication modulo n
$\cong$	is isomorphic to
$\mathbf{PQ}$	vector represented by line segment $\overrightarrow{PQ}$
$\lvert\mathbf{PQ}\rvert, PQ$	magnitude of vector $\mathbf{PQ}$
$\lvert\mathbf{a}\rvert, a$	magnitude of vector $\mathbf{a}$
$\hat{\mathbf{a}}$	unit vector in the direction of $\mathbf{a}$
$\mathbf{i}, \mathbf{j}, \mathbf{k}$	unit vectors in the directions of the x-, y- and z-axes
$\mathbf{a}\,.\,\mathbf{b}$	scalar product of $\mathbf{a}$ and $\mathbf{b}$
g	acceleration due to gravity (taken as $9{\cdot}8\ \text{m s}^{-2}$ unless otherwise stated)
$P(A)$	probability of event A
$P(A \mid B)$	probability of event A given event B
$X, Y, R, \ldots$	random variables
$x, y, r, \ldots$	values of the random variables $X, Y, R, \ldots$
$p(x)$	value of the probability function $P(X = x)$ of the discrete random variable X
$p_1, p_2, \ldots$	probabilities of the values $x_1, x_2, \ldots$ of the discrete random variable X
$f(x), g(x), \ldots$	value of the probability density function of the continuous random variable X
$F(x), G(x), \ldots$	value of the (cumulative) distribution function $P(X \leqslant x)$ of the continuous random variable X

$E(X), \mu$	expectation (or mean) of the random variable X
$\mathrm{Var}(X), \sigma^2$	variance of the random variable X
$G(t)$	probability generating function for a random variable which takes the values $0, 1, 2, \ldots$
$B(n, p)$	binomial distribution, parameters n and p
$N(\mu, \sigma^2)$	normal distribution, mean μ and variance σ^2
Z	the standardised normal variable with distribution $N(0, 1)$
ϕ	probability density function for $N(0, 1)$
Φ	cumulative distribution function for $N(0, 1)$
$x_1, x_2, \ldots$	observations
$f_1, f_2, \ldots$	frequencies with which the observations $x_1, x_2, \ldots$ occur
n	sample size
$\bar{x}$	sample mean
$\mathrm{var}(x)$	sample variance, $\mathrm{var}(x) = \frac{1}{n}\sum(x - \bar{x})^2$
s^2	unbiased estimate of population variance from a sample, $s^2 = \frac{1}{n-1}\sum(x - \bar{x})^2$
$\bar{X}$	random variable representing the sample mean
V	random variable representing the sample variance
μ	population mean
σ^2	population variance
σ	population standard deviation
H_0	null hypothesis
H_1	alternative hypothesis
$\mathrm{cov}(x, y)$	covariance of observations $(x_1, y_1), (x_2, y_2), \ldots$
$\mathrm{Cov}(X, Y)$	covariance of random variables X and Y
r	sample (product moment) correlation coefficient
ρ	population (product moment) correlation coefficient
r_S	Spearman's rank correlation coefficient
r_K	Kendall's rank correlation coefficient
α, β	population parameters for a regression line
a, b	least squares estimates of α, β

1 Mathematical reasoning

1.1 Deductive and inductive processes

Pure mathematics is the study of pattern and structure in systems involving sets of objects such as numbers or points in space. The theory of the subject is developed by logical argument based on precise definitions of the terms used. In general, the type of reasoning used in scientific work is different from that used in mathematics. A scientist observes some natural phenomenon, then uses the data he has collected to formulate laws governing the behaviour of the objects concerned. If results obtained from later experiments also satisfy these laws, they may be accepted by other scientists and used as the basis of further work. This process, by which particular items of data lead to a general theory, is called *inductive* reasoning. Scientific laws obtained in this way cannot be regarded as absolute truths. New evidence may be found which shows that the theory is untenable or that it needs modification.

A mathematician may use inductive processes to produce *conjectures* which he then attempts to prove. One such conjecture is that there are no positive integers a, b and c such that $a^n + b^n = c^n$, where n is any positive integer greater than 2. It is believed that the mathematician Fermat[†] may have proved this, but he left no record of his method of proof. The conjecture has since been proved for certain values of n, but the general statement has yet to be proved or disproved. Another famous mathematical conjecture, known as the "four colour problem", has only recently been proved. This is the statement that any geographical map (or division of a plane into regions) can be coloured using only four different colours, so that no two countries (or regions) with a common boundary have the same colour.

To *prove* a mathematical statement, we must establish its truth by logical reasoning. A mathematical *proof* is a chain of reasoning in which a statement or theorem is shown to follow from previously proved statements or from initial assumptions called premises, axioms or postulates. This process, in which one

[†] *Fermat, Pierre de* (1601–1665) French mathematician. He made important contributions to probability theory and the theory of numbers. He wrote his famous conjecture in the margin of his copy of the *Arithmetica* written by the Greek mathematician Diophantus in about A.D. 275.

statement follows logically from another, is called *deductive* reasoning or *deduction*. We now give a simple example to illustrate the use of the process.

Example 1 What can be deduced from the following three statements?

(a) John always enjoys a boiled egg for breakfast.
(b) There was no bacon left for breakfast today.
(c) If there is no bacon, John boils an egg for breakfast.

If we form these statements into a chain, taking (b) then (c) then (a), we may deduce that John enjoyed his breakfast today.

The next example shows that an apparently logical argument may be invalid.

Example 2 Decide whether the conclusion drawn from the three given statements is valid.

(a) There is a robin in Mrs. Brown's garden.
(b) No other bird in Mrs. Brown's garden has a red breast.
(c) Mrs. Brown can see a bird with a red breast.
Conclusion: Mrs. Brown can see a robin.

The argument is invalid for various reasons. Robins are not the only birds with red breasts. Mrs. Brown may not be watching the birds in her own garden. She may, for instance, be looking at a bullfinch in a neighbour's garden.

We now demonstrate an invalid mathematical argument.

Consider the equation: $x + \frac{1}{x} = 1$ (1)

Multiply (1) by x, $x^2 + 1 = x$ (2)

Multiply (2) by x, $x^3 + x = x^2$ (3)

Rearranging (3), $x^3 = x^2 - x$

Rearranging (2), $x^2 - x = -1$

$\therefore$ $x^3 = -1$ (4)

Hence $x = -1$

Clearly $x = -1$ does not satisfy the original equation. The flaw in the argument becomes apparent when we consider the steps in more detail.

Rearranging (2), $x^2 - x + 1 = 0$

Rearranging (3), $x(x^2 - x + 1) = 0$

Adding, $(x + 1)(x^2 - x + 1) = 0$

i.e. $x^3 + 1 = 0$

We can now see that combining equations (2) and (3) to produce equation (4) results in the introduction of the factor $(x + 1)$. It is this extra factor which leads to the invalid solution $x = -1$.

[It can be shown that the equation $x + \frac{1}{x} = 1$ has no real roots.]

The way in which deductive reasoning is used in various types of mathematical proof is discussed later in the chapter. However, although deduction forms the basis of mathematical reasoning, in such fields as problem solving and original research less reliable processes such as intuition or "trial and error" may be involved.

In applied mathematics both inductive and deductive processes are used. An applied mathematician states in mathematical form laws obtained from experimental data. He may be interested in such things as the performance of an aircraft, the growth of a population of bacteria or the current in an electrical circuit. Using the laws governing the system under consideration, he sets up a mathematical model, often in the form of a set of equations. Predictions about the behaviour of the system can now be made using this model. In general, setting up a mathematical model is an inductive process, but using the model to make predictions involves mainly deductive reasoning.

Exercise 1.1

1. Decide what conclusion can be deduced from the following sets of statements.

(a) Most people disapprove of immorality.
Stealing is immoral.
The Knave of Hearts stole some tarts.

(b) N is an even number greater than 2.
A prime number has only itself and 1 as factors.
Any even number is divisible by 2.

(c) Boycott was the vice-captain of the England cricket team.
It is impossible to play cricket with a broken arm.
If the captain cannot play, the vice-captain takes over.
The England cricket captain broke his arm.

2. Decide whether the conclusions drawn from the following sets of statements are valid. Give reasons for your answer in each case.

(a) Swallow-tail butterflies feed on Milk Parsley.
Milk Parsley grows in the Norfolk fens.
Conclusion: There are swallow-tail butterflies in Norfolk.

(b) Mathematicians are highly intelligent.
People who are not absent-minded never wear odd socks.
Highly intelligent people are thought to be absent-minded.
Conclusion: Mathematicians are thought to wear odd socks.

(c) Opposite angles of a parallelogram are equal.
A rhombus is a parallelogram whose diagonals cut at right angles.
$ABCD$ has two pairs of equal angles.
The diagonals of $ABCD$ cut at right angles.
Conclusion: $ABCD$ is a rhombus.

3. Find the flaw in the following argument:

Let	$x = y$
then	$x^2 = xy$
and	$x^2 - y^2 = xy - y^2$
Factorising,	$(x + y)(x - y) = y(x - y)$
$\therefore$	$x + y = y$
Since $x = y$,	$2x = x$
Hence	$2 = 1$

4. Read the short story given below, then consider the statements which follow it. In each case decide whether the statement is (a) true, (b) false or (c) could be either true or false.

A pedestrian had just stepped off the pavement when a vehicle, travelling at high speed, came round the corner. The car driver quickly applied his brakes and the man on the zebra crossing narrowly escaped serious injury. The police were called to the scene of the incident and later the guilty party was prosecuted for dangerous driving.

(1) At the beginning of the story a man stepped off the pavement.
(2) The car was travelling at high speed.
(3) The man on the zebra crossing was not injured.
(4) After the incident the police were called to the scene.
(5) A driver was found guilty of dangerous driving.
(6) A man escaped serious injury because the car driver braked quickly.
(7) A vehicle came round the corner just after a pedestrian had stepped off the pavement.
(8) A man was crossing the road at the time of the incident.
(9) The car stopped on or near the zebra crossing.
(10) The car stopped.
(11) The car driver was prosecuted for dangerous driving.
(12) The incident involved at least one driver.
(13) The zebra crossing was near a bend in the road.
(14) The events described in the story could not have taken place on a dual carriageway.
(15) None of the people referred to in the story could have been riding a bicycle.
(16) It is possible that the incident involved two vehicles.
(17) The story states that a vehicle being driven at high speed knocked down a man on a zebra crossing, injuring him only slightly.
(18) Three people are mentioned in the story, apart from the police.

1.2 Implication and equivalence

It is sometimes convenient to use letters and symbols to represent statements and relations between them.

For two statements denoted by p and q,

$p \Rightarrow q$ means "p implies q" or "if p, then q",

$p \Leftarrow q$ means "p is implied by q" or "if q, then p",

$p \Leftrightarrow q$ means "p implies q and q implies p" i.e. "the statements p and q are equivalent".

Example 1 Use the symbols $\Rightarrow$ and $\Leftrightarrow$ to connect the following statements:

a: $\triangle ABC$ is isosceles b: In $\triangle ABC$, $\angle B = \angle C$
c: $\triangle ABC$ is equilateral d: In $\triangle ABC$, $AB = AC$

We see that $b \Rightarrow a, b \Rightarrow d; c \Rightarrow a, c \Rightarrow b, c \Rightarrow d; d \Rightarrow a, d \Rightarrow b$.
Since $b \Rightarrow d$ and $d \Rightarrow b$, we have $b \Leftrightarrow d$
i.e. the statements b and d are equivalent.

The symbols $\not\Rightarrow$ "does not imply" and $\not\Leftrightarrow$ "is not equivalent to" may also be used.

Example 2 Use the symbols $\not\Rightarrow$ and $\not\Leftrightarrow$ to connect statements a, b, c and d.

By considering an isosceles triangle ABC in which $AB = BC$, we see that $a \not\Rightarrow b$ and $a \not\Rightarrow d$. We also find that $a \not\Rightarrow c, b \not\Rightarrow c$ and $d \not\Rightarrow c$. Hence $a \not\Leftrightarrow b$, $a \not\Leftrightarrow c, a \not\Leftrightarrow d, b \not\Leftrightarrow c, c \not\Leftrightarrow d$.

The *negation* of a statement p is denoted by $\sim p$ (or sometimes p') and read "not p". The negations of a, b, c and d in *Example 1* are as follows:

$\sim a$: $\triangle ABC$ is not isosceles $\sim b$: In $\triangle ABC$, $\angle B, \neq \angle C$
$\sim c$: $\triangle ABC$ is not equilateral $\sim d$: In $\triangle ABC$, $AB \neq AC$

Here are some of the connections between these new statements:

$$\sim a \Rightarrow \sim c, \quad \sim b \Leftrightarrow \sim d, \quad \sim d \Rightarrow \sim c, \quad \sim d \not\Rightarrow \sim a.$$

These examples illustrate some basic properties of negation and implication.
(1) If $p \Rightarrow q$, then $\sim q \Rightarrow \sim p$.
(2) If $p \Leftrightarrow q$, then $\sim p \Leftrightarrow \sim q$.
It is also important to notice that writing $p \not\Rightarrow q$ is *not* equivalent to writing $p \Rightarrow \sim q$. For instance in *Example 2*, the statement $a \not\Rightarrow d$ means: "if $\triangle ABC$ is isosceles, then it does not necessarily follow that $AB = AC$". However, writing $a \Rightarrow \sim d$ would mean: "if $\triangle ABC$ is isosceles, then $AB \neq AC$", which is clearly false.

Other phrases which are used in mathematical statements, such as "if and only if" or "necessary and sufficient condition", can be replaced by implication symbols. The table below shows some sets of equivalent statements.

$p \Rightarrow q$	p only if q	q is a necessary condition that p
$p \Leftarrow q$	p if q	q is a sufficient condition that p
$p \Leftrightarrow q$	p if and only if q	q is a necessary and sufficient condition that p

Consider now the following statements concerning a positive integer n.

a: n is the square of an even number
b: n is a multiple of 4
c: n is an even number

We see that $a \Rightarrow b \Rightarrow c$ and that, using the phrases in the given table:

the statement b is true if a is true, but b is true only if c is true;
the truth of c is a necessary condition for b to be true, but a is a sufficient condition for b to be true.

Clearly implication signs provide the most economical way of expressing connections between statements. The process of deduction is based on a *law of implication* which we can now state in the form:

$$\text{if} \quad a \Rightarrow b \quad \text{and} \quad b \Rightarrow c, \quad \text{then} \quad a \Rightarrow c.$$

A deductive proof can always be reduced to a series of these steps or closely related ones such as, "if $a \Rightarrow b$ and $a \Rightarrow c$, where b and c together imply d, then $a \Rightarrow d$."

Exercise 1.2

1. Use the symbols $\Rightarrow$ and $\Leftrightarrow$ to connect the statements p, q and r.

(a) p: $ABCD$ is a square
  q: $ABCD$ has four equal sides
  r: $ABCD$ is a parallelogram

(b) p: Ivor is Welsh
  q: Ivor lives in Cardiff
  r: Ivor is British

(c) p: $KLMN$ is a rectangle 3 cm by 4 cm
  q: The area of rectangle $KLMN$ is $12\,\text{cm}^2$
  r: $KLMN$ is a rectangle with diagonals 5 cm long

(d) p: The integer n is an even multiple of 5
  q: The integer n has final digit 0
  r: The integer n has a pair of prime factors which differ by 3.

2. Use the symbol $\nRightarrow$ to connect the statements given in question 1.

3. Use the symbols $\Rightarrow$ and $\Leftrightarrow$ to connect the negations $\sim p$, $\sim q$ and $\sim r$ of the statements given in question 1.

4. Write down all possible connections between the statements p, q, $\sim p$ and $\sim q$ using the symbols $\Rightarrow$ and $\Leftrightarrow$.

(a) p: In quadrilateral $ABCD$, $AC = BD$
  q: $ABCD$ is a rectangle

(b) p: In $\triangle ABC$, $\angle A$ is obtuse
  q: In $\triangle ABC$, $\angle B = 90°$

(c) p: For integers x and y, $x - y$ is odd
  q: For integers x and y, $x^2 - y^2$ is divisible by 4

(d) p: C lies on the circle with AB as diameter
q: In $\triangle ABC$, $\angle C$ is not a right angle.

5. Using only the terms and symbols $p, q, \sim p, \sim q, \Rightarrow, \Leftrightarrow, \not\Rightarrow$ and $\not\Leftrightarrow$, write down a statement equivalent to each of the following:

(a) $\sim p \Rightarrow \sim q$, (b) $p \Leftrightarrow \sim q$, (c) $\sim p \not\Rightarrow q$,
(d) $p \not\Leftrightarrow q$, (e) $p \Rightarrow \sim q$, (f) $\sim p \Leftrightarrow q$,

6. Use the symbol $\Rightarrow$ to state the connection between the statements p and q in each of the following cases.
(a) q is true only if p is true,
(b) p is a necessary condition that q,
(c) q is a sufficient condition that p,
(d) p is true if q is true,
(e) p is a sufficient condition that q,
(f) p is true only if q is true.

7. Use the terms "if", "only if", "necessary condition" and "sufficient condition" to express the connection between the statements, "$\triangle ABC$ is equilateral" and "$AB = AC$" in four different ways.

1.3 Types of proof

We first give an example of an elementary deductive proof.

Example 1 Prove that $a \times 0 = 0$ for all real values of a.

[You may assume the following properties of real numbers:
(i) $a + (b + c) = (a + b) + c$ associative law for addition
(ii) $a(b + c) = ab + ac$ distributive law
(iii) $a + 0 = a = 0 + a$ for all real a
(iv) any real number a has an additive inverse $(-a)$, such that $a + (-a) = 0 = (-a) + a$.]

$$
\begin{array}{rll}
 & b + 0 = b & \text{property (iii)} \\
\Rightarrow & a(b + 0) = ab & \text{both sides multiplied by } a \\
\Rightarrow & ab + a \times 0 = ab & \text{distributive law} \\
\Rightarrow & -ab + (ab + a \times 0) = -ab + ab & -ab \text{ added to both sides} \\
\Rightarrow & (-ab + ab) + a \times 0 = -ab + ab & \text{associative law for } + \\
\Rightarrow & 0 + a \times 0 = 0 & \text{property (iv)} \\
\Rightarrow & a \times 0 = 0 & \text{property (iii)}
\end{array}
$$

Another important type of proof is called *proof by contradiction* (or "reductio ad absurdum".) To prove a theorem by this method we assume that the theorem does not hold, then show that this assumption leads to a contradiction. In its simplest form this process is equivalent to proving a statement p true by showing $\sim p$ to be false. For instance, the standard proof of the statement "$\sqrt{2}$ is irrational" begins with the assumption that $\sqrt{2}$ is rational. In the example given here a statement of the form $p \Rightarrow q$ is proved by showing that $\sim q \Rightarrow \sim p$.

Example 2 Prove that in $\triangle ABC$, if $AB^2 + BC^2 = AC^2$, then $\angle ABC = 90°$.

[You may assume Pythagoras' theorem.]

Let us assume that in $\triangle ABC$, $\angle ABC \neq 90°$.

(1) $\angle ABC < 90°$

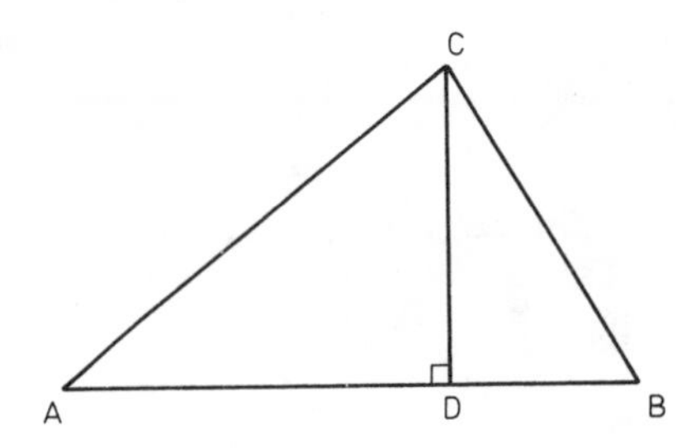

(2) $\angle ABC > 90°$

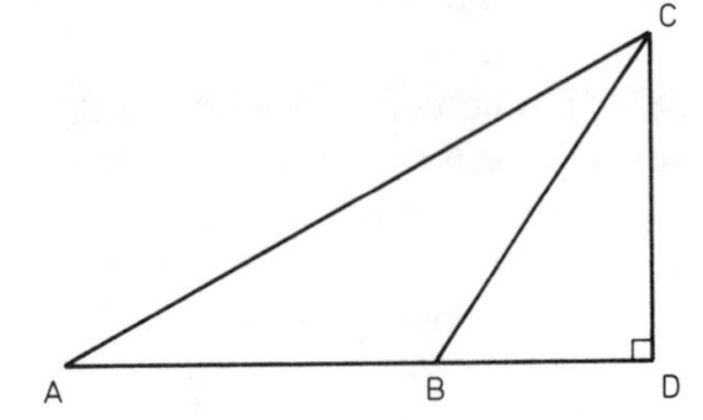

Case (1): By Pythagoras' theorem in $\triangle ADC$, $AC^2 = AD^2 + DC^2$

But $AD < AB$ and $DC < BC$, $\therefore\ AC^2 < AB^2 + BC^2$.

Case (2): By Pythagoras' theorem in $\triangle ADC$, $AC^2 = AD^2 + DC^2$

$\therefore\ AC^2 = (AB + BD)^2 + DC^2 = AB^2 + 2AB.BD + BD^2 + DC^2$

By Pythagoras' theorem in $\triangle BDC$, $BD^2 + DC^2 = BC^2$

$\therefore\ AC^2 = AB^2 + 2AB.BD + BC^2 > AB^2 + BC^2$.

Thus, in both cases, if $AB^2 + BC^2 = AC^2$, the assumption that $\angle ABC \neq 90°$ leads to a contradiction.
Hence, if $AB^2 + BC^2 = AC^2$, then $\angle ABC = 90°$.

This result is, of course, the converse of Pythagoras' theorem. In general, the *converse* of the statement $p \Rightarrow q$ is the statement $q \Rightarrow p$.

As in the case of Fermat's famous conjecture, mentioned in §1.1, it is sometimes difficult to establish the truth of a general statement about such objects as real numbers or sets. However, a statement can be proved false by producing just one *counter-example*.

Example 3 Discuss the statement that if n is a positive integer, then $n^2 - n + 17$ is prime.

[To prove that this statement is true it would be necessary to show that it holds for all positive integral values of n. However, the statement can be proved false by finding one value of n, such that $n^2 - n + 17$ is not prime.]

If we test the statement by writing $n = 1, 2, 3, 4, 5 \ldots$, we obtain the prime numbers $17, 19, 23, 29, 37, \ldots$ This suggests that the statement could be true. However, if $n = 17$, $n^2 - n + 17 = 17^2$, which is not prime. Hence the statement is false.

Exercise 1.3

1. Using the given diagrams, show that if $\angle ABC = \angle CAT$, the assumption that AT cuts the circle again at a point D leads to a contradiction. What conclusion can be drawn from this?

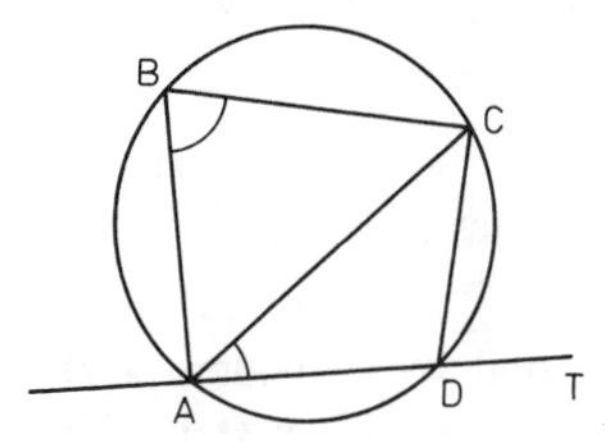

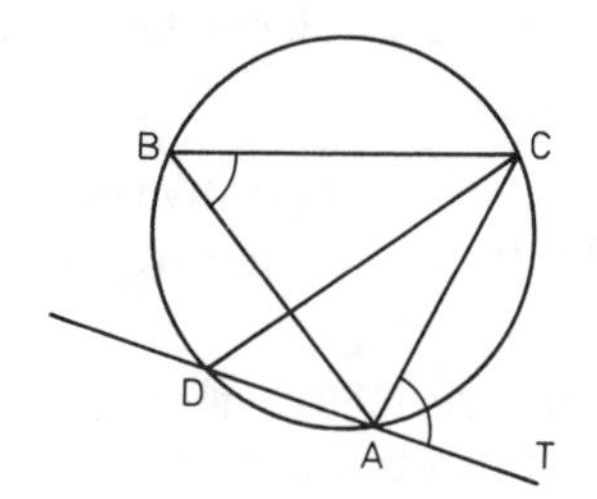

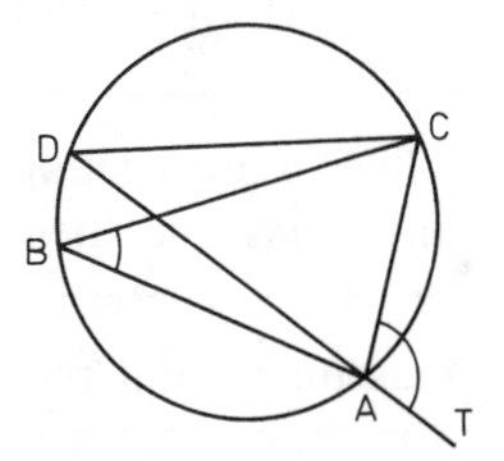

2. Show that if $p_1, p_2, p_3, \ldots, p_n$ are prime numbers, then the number $p_1p_2p_3 \ldots p_n + 1$ is either a prime not already listed or a multiple of such a prime. Hence prove by contradiction that there are infinitely many prime numbers.

3. Write down the converses of the following statements. In each case decide whether (i) the statement, (ii) its converse are true or false.
(a) If a diameter of a circle bisects a chord, it is perpendicular to that chord.
(b) If $2x - 3 = 0$, then $2x^2 - x - 3 = 0$.
(c) If a pentagon is regular, it must be equilateral.
(d) If x is greater than y, then x^2 is greater than y^2.
(e) If a positive integer N is divisible by 9, then the sum of its digits is divisible by 9.
(f) If any set of six socks taken from a drawer contains a pair, then there must be five pairs of socks in the drawer.

In questions 4, 5 and 6 decide whether the given statements are true or false. If a statement is true, prove it; if it is false, give a counter-example.

4. (a) The square of any odd number is of the form $4k + 1$, where k is a non-negative integer.
(b) Any number of the form $4k + 1$ is the square of an odd number.

5. (a) If $0 < p < q$, then $\log p < \log q$.

(b) If $0 < p < q$, then $\int_0^p f(x)dx < \int_0^q f(x)dx$.

6. (a) If $n^3 - n$ is divisible by 4, then n must be divisible by 4.
(b) If n is a positive integer, then $n^3 - n$ is divisible by 3.

1.4 Proof by induction

The elementary deductive methods discussed so far are sometimes found to be inadequate when attempting to prove that a proposition, such as "the number $4^n + 5$ is divisible by 3", holds for all positive integral values of n. The difficulty is that this general statements is equivalent to the infinite sequence of propositions "3 is a factor of $4^1 + 5, 4^2 + 5, 4^3 + 5, 4^4 + 5, \ldots$" However, propositions of this kind can often be proved using *mathematical induction.* Although the method is related to the type of inductive reasoning used in science, it is nevertheless a rigorous procedure in which conclusions are reached by sound deductive reasoning.

The principle of a proof by induction can be stated formally as follows:

> Given a proposition $P(n)$ involving a positive integer n, if (i) $P(k) \Rightarrow P(k+1)$ and (ii) $P(1)$ is true, then $P(n)$ is true for all positive integral values of n.

A useful mental picture of a proof by induction is created by considering the proposition that a man can climb a certain uniform staircase with n steps. The statement will be proved if we can show

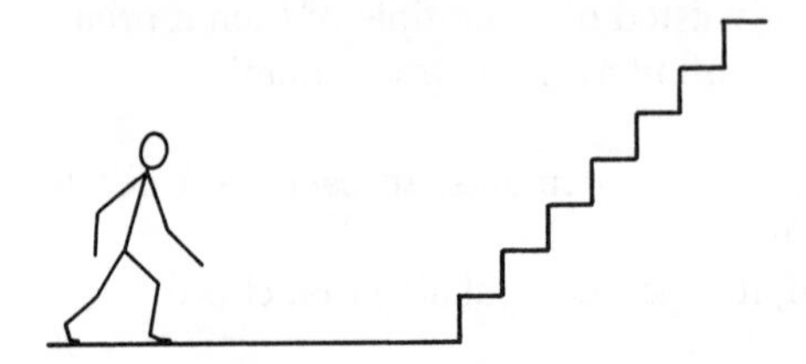

(i) that the man is capable of climbing from any step to the one above and
(ii) that the man can reach the foot of the staircase.

Thus it may be helpful to think of a general proposition $P(n)$ as a staircase to be climbed. The steps of the staircase are the propositions $P(1), P(2), P(3), \ldots$. It is possible to proceed from one "step" to the next if we can prove that for any k, $P(k+1)$ may be deduced from $P(k)$. If, in addition, it is shown that $P(1)$ is true (i.e. that it is possible to set foot on the first step of the staircase), then it follows that $P(2)$ is true, then $P(3)$ is true and so on for all positive integral values of n.

Example 1 Prove that the number, $a_n = 4^n + 5$, is divisible by 3 for all positive integral values of n.

Consider the numbers $a_k = 4^k + 5$ and $a_{k+1} = 4^{k+1} + 5$.

$$a_{k+1} - a_k = (4^{k+1} + 5) - (4^k + 5)$$
$$= 4^{k+1} - 4^k = 4^k(4 - 1) = 4^k \times 3$$
$$\therefore \quad a_{k+1} = a_k + 4^k \times 3$$

Hence, if a_k is divisible by 3, so is a_{k+1}.
$\therefore$ if the proposition is true for $n = k$, it is also true for $n = k + 1$.
$a_1 = 4^1 + 5 = 9$, which is divisible by 3,
$\therefore$ the proposition is true for $n = 1$.
Hence, by induction, the number $4^n + 5$ must be divisible by 3 for all positive integral values of n.

When dealing with a finite series, it is sometimes possible to suggest a formula for the sum to n terms by examining the first few sums, $S_1, S_2, S_3, \ldots$. Such a result can often be proved by the method of induction.

Example 2 If S_n is the sum of the first n terms of the series

$$\frac{1}{1 \times 2} + \frac{1}{2 \times 3} + \frac{1}{3 \times 4} + \ldots + \frac{1}{n(n+1)} + \ldots,$$

find S_1, S_2 and S_3, then suggest a formula for S_n. Prove that the result is correct by the method of induction.

$$S_1 = \frac{1}{1 \times 2} = \frac{1}{2}, \qquad S_2 = S_1 + \frac{1}{2 \times 3} = \frac{1}{2} + \frac{1}{6} = \frac{2}{3},$$

$$S_3 = S_2 + \frac{1}{3 \times 4} = \frac{2}{3} + \frac{1}{12} = \frac{3}{4}.$$

These results suggests that $S_n = \dfrac{n}{n+1}$

Assuming that this result is true for $n = k$, $S_k = \dfrac{k}{k+1}$

$$\therefore \quad S_{k+1} = S_k + \frac{1}{(k+1)(k+2)}$$

$$= \frac{k}{k+1} + \frac{1}{(k+1)(k+2)}$$

$$= \frac{k(k+2)+1}{(k+1)(k+2)}$$

$$= \frac{k^2+2k+1}{(k+1)(k+2)} = \frac{(k+1)^2}{(k+1)(k+2)} = \frac{k+1}{k+2}$$

$\therefore$ if the result holds for $n = k$, it also holds for $n = k + 1$.

When $n = 1$, $S_n = S_1 = \dfrac{1}{2}$ and $\dfrac{n}{n+1} = \dfrac{1}{1+1} = \dfrac{1}{2}$

$\therefore$ the result holds for $n = 1$.
Hence, by induction, it is true that $S_n = n/(n+1)$ for all positive integral values of n.

The method of induction can also be used to prove the binomial theorem:

$$(a+b)^n = a^n + {}_nC_1 a^{n-1}b + {}_nC_2 a^{n-2}b^2 + \ldots$$
$$+ {}_nC_r a^{n-r}b^r + \ldots + {}_nC_{n-1} ab^{n-1} + b^n.$$

Assuming that the theorem is true for $n = k$,

$$(a+b)^k = a^k + {}_kC_1 a^{k-1}b + {}_kC_2 a^{k-2}b^2 + \ldots$$
$$+ {}_kC_{r-1} a^{k-r+1}b^{r-1} + {}_kC_r a^{k-r}b^r + \ldots + b^k,$$

then, multiplying by $(a + b)$,

$$(a+b)^{k+1} = a^{k+1} + {}_kC_1 a^k b + {}_kC_2 a^{k-1}b^2 + \ldots + {}_kC_r a^{k-r+1}b^r \quad + \ldots$$
$$+ a^k b + {}_kC_1 a^{k-1}b^2 + \ldots + {}_kC_{r-1} a^{k-r+1}b^r + \ldots + b^{k+1}.$$

But ${}_kC_1 + 1 = k + 1 = {}_{k+1}C_1$, and

$$
\begin{aligned}
{}_kC_r + {}_kC_{r-1} &= \frac{k!}{r!(k-r)!} + \frac{k!}{(r-1)!(k-\{r-1\})!} \\
&= \frac{k!}{r!(k-r+1)!}\{(k-r+1)+r\} \\
&= \frac{(k+1)!}{r!(k+1-r)!} = {}_{k+1}C_r
\end{aligned}
$$

$$
\therefore \quad (a+b)^{k+1} = a^{k+1} + {}_{k+1}C_1 a^k b + {}_{k+1}C_2 a^{k-1}b^2 + \dots \\
\dots + {}_{k+1}C_r a^{k+1-r}b^r + \dots + b^{k+1}.
$$

$\therefore$ if the theorem is true for $n = k$ it is also true for $n = k + 1$.
Since $(a + b)^1 = a^1 + b^1$, the theorem is true for $n = 1$. Hence, by induction, the binomial theorem is true for all positive integral values of n.

[Note that the basic principle of induction can be extended in various ways. Suppose, for instance, that for some proposition $P(n)$, where n is a positive integer, it can be shown that $P(k) \Rightarrow P(k+1)$, but that the proposition $P(1)$ is false or meaningless. Clearly $P(n)$ cannot be proved true for all positive integral values of n. However, provided that there is some positive integer m such that $P(m)$ is true, then by induction, $P(n)$ holds for all integral values of n greater than or equal to m.]

Exercise 1.4

In questions 1 to 8 prove, by induction, that the given statements are true for all positive integral values of n.

1. The sum of the first n terms of the series
$(1 \times 3) + (2 \times 4) + (3 \times 5) + \dots + r(r+2) + \dots$ is $\frac{1}{6}n(n+1)(2n+7)$.

2. The sum of the first n terms of the series
$\dfrac{1}{1 \times 3} + \dfrac{1}{3 \times 5} + \dfrac{1}{5 \times 7} + \dots + \dfrac{1}{(2r-1)(2r+1)} + \dots$ is $\dfrac{n}{2n+1}$.

3. $\sum_{r=1}^{n} r(r+1) = \frac{1}{3}n(n+1)(n+2)$.

4. $\sum_{r=1}^{n} \dfrac{r}{2^r} = 2 - \dfrac{n+2}{2^n}$.

5. $n^3 + 3n^2 - 10n$ is divisible by 3.

6. $3^{2n} - 1$ is a multiple of 8.

7. $7^n + 4^n + 1^n$ is divisible by 6.

8. If $\mathbf{M} = \begin{pmatrix} 1 & 0 \\ 3 & 1 \end{pmatrix}$, then $\mathbf{M}^n = \begin{pmatrix} 1 & 0 \\ 3n & 1 \end{pmatrix}$.

9. Prove that $\sum_{r=1}^{n} r^2 = \frac{1}{6}n(n+1)(2n+1)$ for all positive integers n. Hence find the following sums:
(a) $1^2 + 2^2 + 3^2 + \ldots + 20^2$,
(b) $21^2 + 22^2 + 23^2 + \ldots + 40^2$,
(c) $2^2 + 4^2 + 6^2 + \ldots + 40^2$,
(d) $1^2 + 3^2 + 5^2 + \ldots + 39^2$.

10. Prove that the sum to n terms of the series

$$\frac{1}{1 \times 2 \times 3} + \frac{1}{2 \times 3 \times 4} + \frac{1}{3 \times 4 \times 5} + \ldots \text{ is } \frac{1}{4} - \frac{1}{2(n+1)(n+2)}.$$

Hence find the sum to infinity of the series.

11. If $\mathbf{A} = \begin{pmatrix} 3 & -1 \\ 4 & -1 \end{pmatrix}$, prove that $\mathbf{A}^n = \begin{pmatrix} 2n+1 & -n \\ 4n & 1-2n \end{pmatrix}$, where n is a positive integer. Given that the transformation represented by $\mathbf{A}$ is applied to the points of the x, y plane n times in succession, find the image of
(a) the point $(0, -1)$,
(b) the line $y = x$.

12. Given that $f(x) = x/(x+1)$, find $f^2(x)$ i.e. $f(f(x))$. Find also $f^3(x)$ and suggest a possible form for $f^n(x)$. Prove that your result is correct by mathematical induction.

Exercise 1.5 (miscellaneous)

In questions 1 to 6 explain why the given chain of reasoning is unsound.

1.
$$\begin{aligned} & -6x + 9 = 2x + 1 \\ \Rightarrow \quad & x^2 - 6x + 9 = x^2 + 2x + 1 \\ \Rightarrow \quad & (x-3)^2 = (x+1)^2 \\ \Rightarrow \quad & x - 3 = x + 1 \\ \Rightarrow \quad & -3 = 1 \end{aligned}$$

2.
$$\begin{aligned} & 2 > 1 \\ \Rightarrow \quad & 2\lg(0{\cdot}2) > \lg(0{\cdot}2) \\ \Rightarrow \quad & \lg(0{\cdot}2)^2 > \lg(0{\cdot}2) \\ \Rightarrow \quad & \lg(0{\cdot}04) > \lg(0{\cdot}2) \\ \Rightarrow \quad & 0{\cdot}04 > 0{\cdot}2 \end{aligned}$$

3.
$$\begin{aligned} & (2x+3)(3x+1) = (6x+5)(x+1) \\ \Rightarrow \quad & 6x^2 + 11x + 3 = 6x^2 + 11x + 5 \\ \Rightarrow \quad & 3 = 5 \end{aligned}$$

4.
$$\begin{aligned} & a + b > 0 \\ \Rightarrow \quad & a^2 + ab > 0 \\ \Rightarrow \quad & a^2 + 2ab + b^2 > ab + b^2 \\ \Rightarrow \quad & (a + b)^2 > (a + b)b \\ \Rightarrow \quad & a + b > b \end{aligned}$$

Hence if $a + b > 0$, then $a + b > b$.

5.

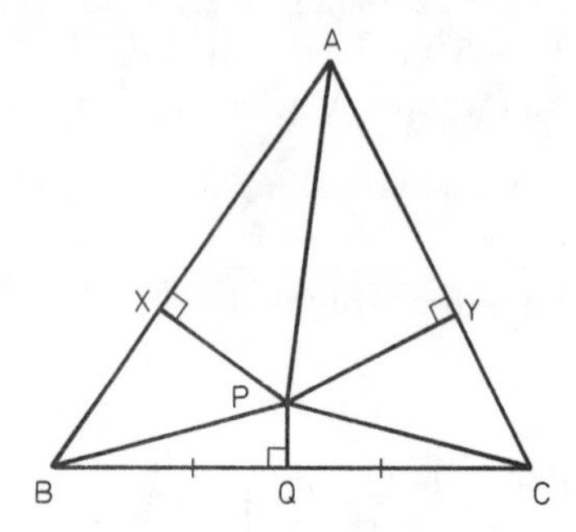

In $\triangle ABC$ the point P is constructed such that AP bisects $\angle A$ and PQ is the perpendicular bisector of BC. PX and PY are drawn perpendicular to AB and AC respectively.

In $\triangle$s APX and APY, $\angle AXP = \angle AYP$, $\angle PAX = \angle PAY$ and AP is common,
$\therefore$ $\triangle$s APX and APY are congruent.
Hence $PX = PY$ and $AX = AY$.
In $\triangle$s PXB and PYC, $\angle BXP = \angle CYP = 90°$, $PX = PY$ and $PB = PC$.
$\therefore$ $\triangle$s PXB and PYC are congruent and hence $XB = YC$.
Thus $AX + XB = AY + YC$ i.e. $AB = AC$.
Hence any triangle is isosceles.

6. To prove that $a_n = 7^n - 11$ is divisible by 6, where n is an integer greater than 1.

Proof: $a_{k+1} - a_k = (7^{k+1} - 11) - (7^k - 11)$
$= 7^{k+1} - 7^k = 7^k(7 - 1) = 7^k \times 6,$

$\therefore$ if a_k is divisible by 6, so is a_{k+1}.
Hence, by induction, a_n is divisible by 6 for all integral values of n greater than 1.

7. In proving a statement of the form $p \Rightarrow q$, it has been shown that p implies either r or s. What further steps are required to complete the proof?

8. In proving that the proposition $P(n)$ is true for all positive integral values of n, it has been shown that $P(k) \Rightarrow P(k + 2)$. What further steps are required to complete the proof?

9. Prove that if n is any positive integer,
$1 \times 2 \times 3 + 2 \times 3 \times 4 + \ldots. + n(n + 1)(n + 2) = \frac{1}{4}n(n + 1)(n + 2)(n + 3).$

10. Prove that $\dfrac{d}{dx}(x^n) = nx^{n-1}$ for all positive integral values of n.

11. Prove that $\displaystyle\sum_{r=2}^{n} \frac{1}{r^2 - 1} = \frac{3}{4} - \frac{2n + 1}{2n(n + 1)}.$

12. Given that $\mathbf{A} = \begin{pmatrix} 1+\cos\theta & \sin\theta \\ \sin\theta & 1-\cos\theta \end{pmatrix}$, prove that $\mathbf{A}^{n+1} = 2^n \times \mathbf{A}$, where n is a positive integer.

13. If $u_{n+1} = 2u_n + 1$ where n is a positive integer and $u_1 = 5$, prove by induction that $u_n = 3 \times 2^n - 1$.

14. If $\mathbf{A} = \begin{pmatrix} 1 & 1 \\ 0 & 2 \end{pmatrix}$, find $\mathbf{A}^2$, $\mathbf{A}^3$ and conjecture a form for $\mathbf{A}^n$ where n is any positive integer. Prove the truth of your conjecture by mathematical induction. (O & C)

15. Prove by induction that the sum of the first n terms of the series

$$\frac{1}{1\times 2\times 3} + \frac{4}{2\times 3\times 4} + \frac{7}{3\times 4\times 5} + \ldots \text{ is } \frac{n^2}{(n+1)(n+2)}.$$

Find the limit of this sum as n tends to infinity.

16. If $S_n = 1.n + 2(n-1) + 3(n-2) + \ldots + (n-1).2 + n.1$, where n is a positive integer, prove that $S_{n+1} - S_n = \frac{1}{2}(n+1)(n+2)$. By induction, or otherwise, prove that $S_n = \frac{1}{6}n(n+1)(n+2)$. (O)

17. Prove that, if n is a positive integer greater than one, $n^5 - n$ is a multiple of 5. Hence, or otherwise, prove that all square numbers are of the form $5r$ or $5r \pm 1$ where r is an integer. (JMB)

18. Let A, B and C be subsets of some universal set $\mathscr{E}$. For the three statements below, prove those that are true and, by taking $\mathscr{E}$ to be the set of all positive inte-ers, provide counter-examples to demonstrate the falsity of the others:
(i) $(A \cap B = A \cap C) \Rightarrow (B = C)$, (ii) $(A \cap B = A \cup B) \Rightarrow (A = B)$,
(iii) $(A \cap B) \cup C = (A \cup C) \cap B$.

[In this question Venn diagrams may be used to examine the plausibility of a statement but will not be accepted as proofs of truth or falsity.] (C)

19. The real function f, defined for all $x \in \mathbb{R}$, is said to be multiplicative if, for all $x \in \mathbb{R}$, $y \in \mathbb{R}$, $f(xy) = f(x)f(y)$. [$\mathbb{R}$ denotes the set of real numbers.] Prove that if f is a multiplicative function then
(i) either $f(0) = 0$ or $f(x) \equiv 1$,
(ii) either $f(1) = 1$ or $f(x) \equiv 0$,
(iii) $f(x^n) = \{f(x)\}^n$ for all positive integers n.
Give an example of a non-constant multiplicative function. (C)

20. The matrix $\mathbf{X} = \begin{pmatrix} p & q \\ r & s \end{pmatrix}$ has real non-zero elements; the matrix $\mathbf{P}$ is of the form $\begin{pmatrix} a & 0 \\ 0 & b \end{pmatrix}$ where $a \neq b$, and $\mathbf{M} = \begin{pmatrix} 4 & 2 \\ -3 & -1 \end{pmatrix}$. Given that $\mathbf{XM} = \mathbf{PX}$, show

that $\dfrac{q}{p} = \dfrac{4-a}{3} = \dfrac{2}{1+a}$ and find a possible matrix **P** and a possible corresponding matrix **X**. Show that, if n is a positive integer, $\mathbf{M}^n = \mathbf{X}^{-1}\mathbf{P}^n\mathbf{X}$. Deduce the elements of $\mathbf{M}^n$ in terms of n. (JMB)

21. Prove that if n is a positive integer, $10^n - 1$ is divisible by 9. Hence prove that a necessary and sufficient condition for a positive integer to be divisible by 9 is that the sum of its digits is divisible by 9.

2 Some functions and their properties

2.1 Symmetry, continuity and differentiability

A *function* f with *domain* X and *codomain* Y is a rule which assigns to each element $x \in X$ exactly one *image* $y \in Y$. The set containing the images under f of all the elements of the domain is called the *range* (or image set). In this chapter we will be considering functions with domain the set of real numbers $\mathbb{R}$ (or some subset of $\mathbb{R}$), and codomain $\mathbb{R}$. The notation used to define such functions will take one of the following forms: $f(x) = x^2 + 1$, $y = x^2 + 1$ or $f: x \to x^2 + 1$. Unless otherwise stated, the domain of a function f is assumed to be the set of all real numbers for which an image is defined. For instance, the domain of the function $f(x) = x/(x^2 - 1)$ is taken to be the set $\mathbb{R} - \{-1, 1\}$ i.e. all real numbers except -1 and $+1$. We now examine some of the properties possessed by various functions defined in $\mathbb{R}$.

A function $f(x)$ is said to be *odd* if $f(-x) = -f(x)$. The graph of an odd function is symmetrical about the origin, i.e. if (a, b) lies on the graph so does $(-a, -b)$. The function $f(x)$ is said to be *even* if $f(-x) = f(x)$. The graph of an even function is symmetrical about the y-axis i.e. if (a, b) lies on the graph so does $(-a, b)$.

Example 1 Decide whether each of the functions f, g and h is odd, even or neither, given that

$$f(x) = x + 1/x,\ x \in \mathbb{R},\ x \neq 0,$$
$$g(x) = 2\cos x - x,\ x \in \mathbb{R},$$
$$h(x) = \sqrt{(4 - x^2)},\ x \in \mathbb{R},\ |x| \leqslant 2.$$

$$f(-x) = (-x) + 1/(-x) = -x - 1/x = -f(x)$$

$\therefore$ f is an odd function.

$$g(-x) = 2\cos(-x) - (-x) = 2\cos x + x$$

$\therefore$ g is neither odd nor even.

$$h(-x) = \sqrt{(4 - \{-x\}^2)} = \sqrt{(4 - x^2)} = h(x)$$

$\therefore$ h is an even function.

The sketches below show the graphs of f, g and h.

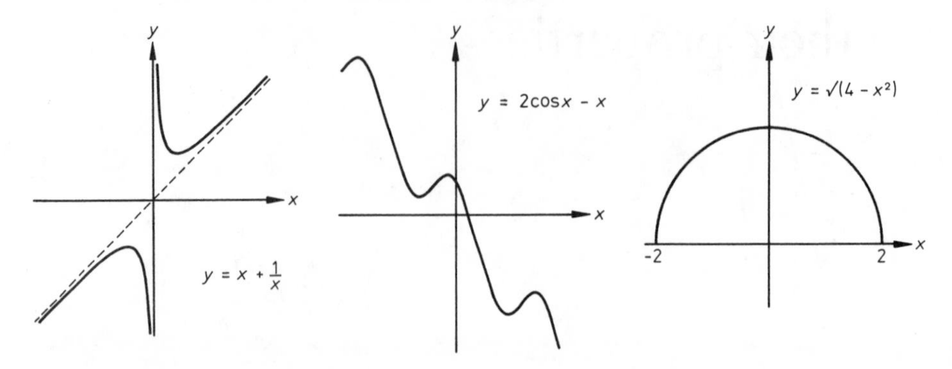

A function $f(x)$ is said to be *periodic* if there is a constant k such that $f(x + k) = f(x)$ for all values of x. The graph of the function forms a repeating pattern such that any translation parallel to the x-axis through k units leaves the pattern unchanged. The *period* of the function is the least positive constant a such that $f(x + a) = f(x)$ for all values of x. The trigonometric functions $\sin x$ and $\cos x$ are periodic and have period 2π. The function $\tan x$ is also periodic, but has period π.

Example 2 Sketch the graph of the function $f(x) = x - [x]$ and decide whether it is periodic. [Note that $[x]$ denotes the greatest integer less than or equal to x.]

When x is an integer, $[x] = x$, $\therefore \; f(x) = x - x = 0$
When $0 \leqslant x < 1$, $[x] = 0$, $\therefore \; f(x) = x$
For all values of x,
$$\begin{aligned} f(x+1) &= (x+1) - [x+1] \\ &= x + 1 - ([x] + 1) \\ &= x - [x] = f(x) \end{aligned}$$

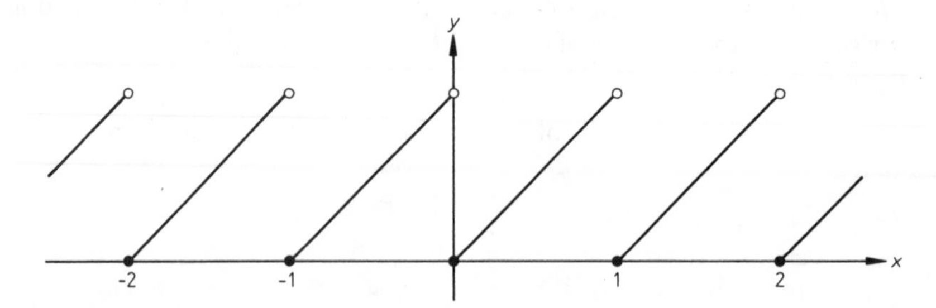

The sketch shows that $f(x)$ is periodic with period 1.
[Open and closed circles are used to show that, for instance, the value of $f(1)$ is 0 rather than 1.]

Example 3 Given that f is a periodic function such that for $0 \leqslant x \leqslant 4$, $f(x) = x(4 - x)$, sketch the graph $y = f(x)$ in each of the following cases,

(a) the period of f is 4,
(b) f is an odd function and has period 8.

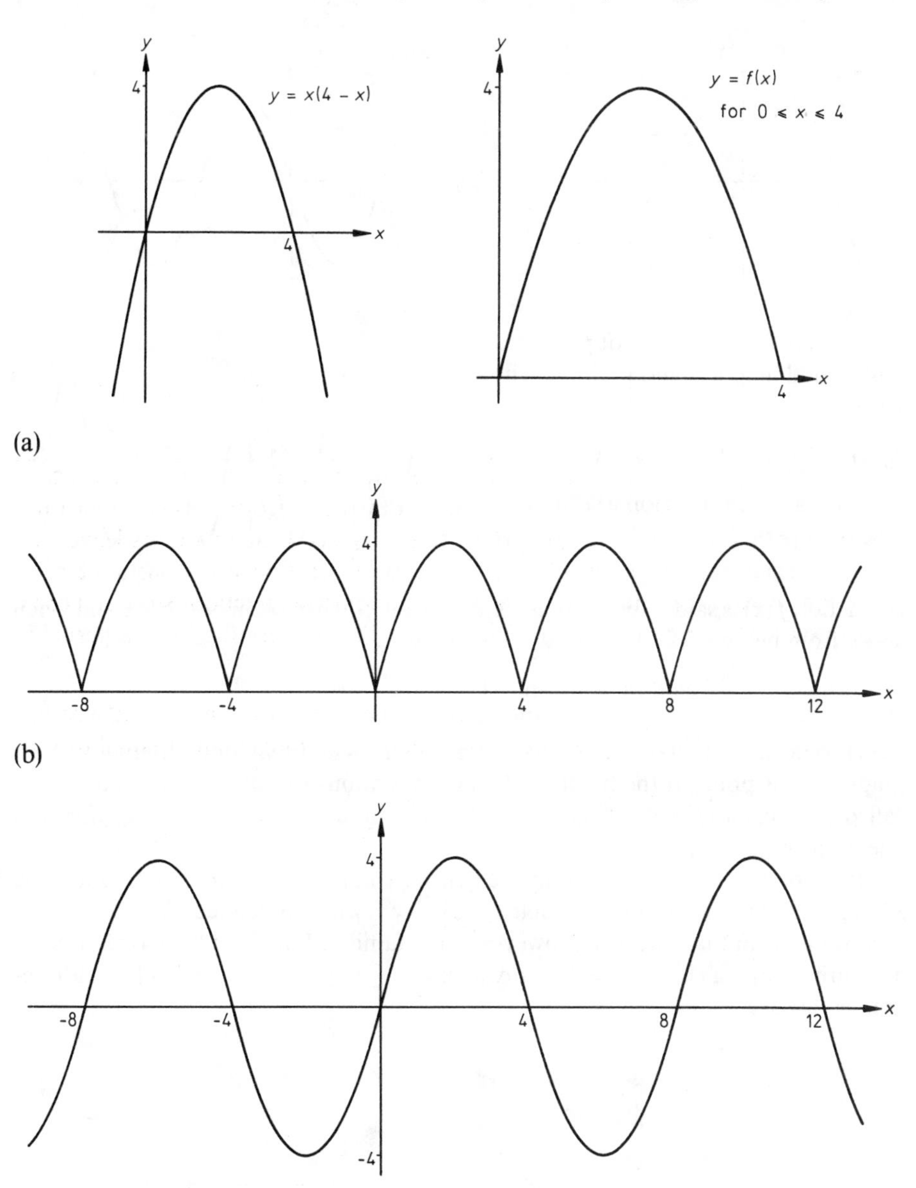

If f and g are functions defined on $\mathbb{R}$, then the *composite function* fg (also written $f \circ g$) is defined by $fg(x) = f(g(x))$.

Example 4 The functions f and g are defined by $f: x \to 1 - 2x^2$ and $g: x \to \sin x$. Sketch the graphs of the functions f, g and fg. State whether each of these functions is

(a) odd, even or neither,

(b) periodic, and if so give the period.

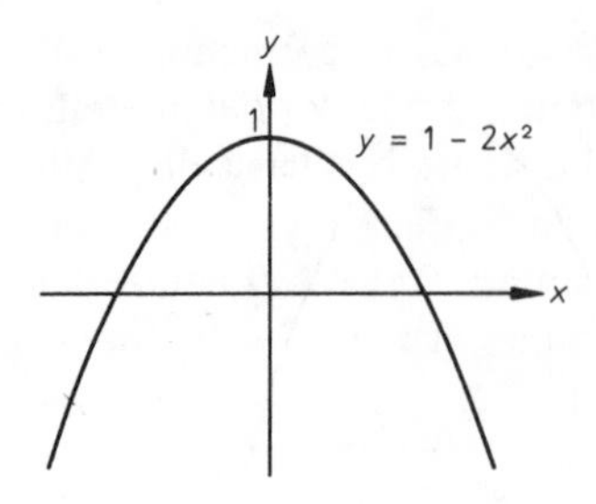

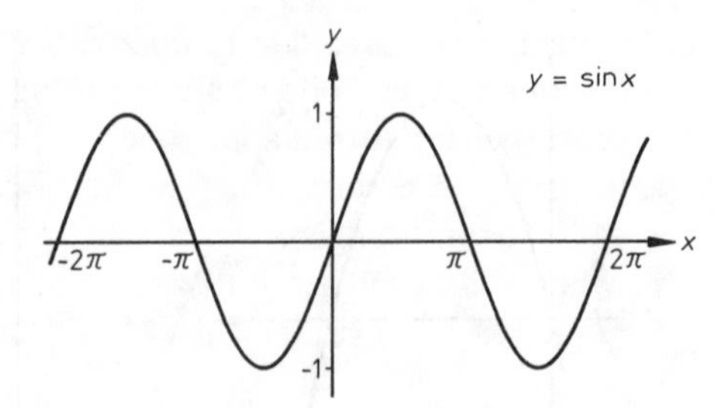

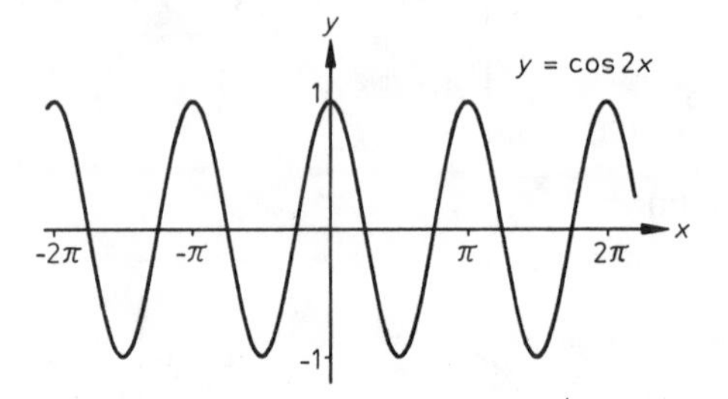

f is an even function but not periodic.
g is an odd function and periodic with period 2π.

$fg(x) = 1 - 2\sin^2 x = \cos 2x$

$\therefore$ fg is an even function and periodic with period π.

A function $f(x)$ is said to be *continuous* at $x = a$ if, when $x \to a$ from above and when $x \to a$ from below, $f(x) \to f(a)$

i.e. if
$$\lim_{x \to a} f(x) = f(a).$$

Less formally, $f(x)$ is continuous at $x = a$, if there is no break or *discontinuity* in its graph at that point. If the function $f(x)$ is continuous for all real values of x, it is called a *continuous function*. The graph of a continuous function is a single unbroken line or curve.

Many of the functions met in elementary mathematics are continuous e.g. $x^3 - x + 3$, $\sin x$, $|x|$. A function such as $\sqrt{(4 - x^2)}$, which is defined for $-2 \leqslant x \leqslant 2$, is continuous in that interval. However, other familiar functions have one or more discontinuities e.g. $1/x$ at $x = 0$; $\tan x$ at $x = \pm\frac{1}{2}\pi, \pm\frac{3}{2}\pi, \pm\frac{5}{2}\pi, \ldots$; $[x]$ at all integral values of x.

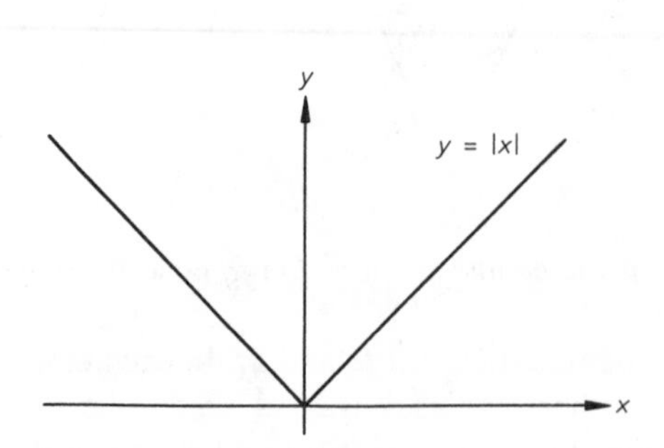

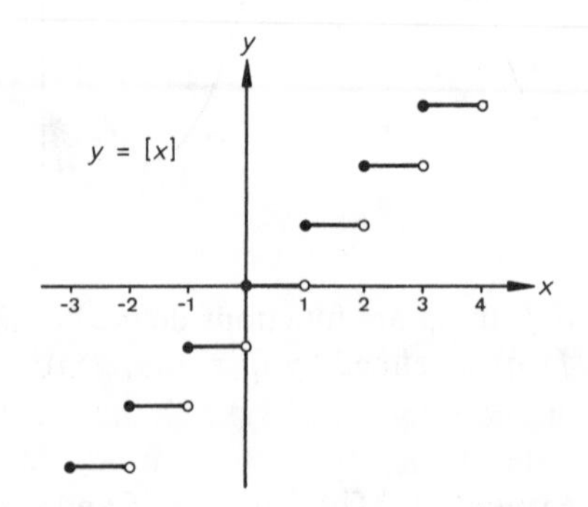

The *derivative* of a function $f(x)$ at $x = a$ is defined as

$$f'(a) = \lim_{h \to 0} \frac{f(a + h) - f(a)}{h}.$$

If this limit exists $f(x)$ is said to be *differentiable* at $x = a$. Given a function $f(x)$ the *derived function* $f'(x)$ can often be found using the standard rules for differentiation. However, any general formula for $f'(x)$ will be invalid where $f(x)$ has a discontinuity and possibly at certain other points. For instance, if $f(x) = 1/x$, the result $f'(x) = -1/x^2$ is meaningless when $x = 0$. The function $f(x) = [x]$ is not differentiable when x is an integer, but for non-integral values of x, $f'(x) = 0$. If we consider the function $f(x) = |x|$ near $x = 0$, we find that

for $h > 0$, $$\frac{f(0+h) - f(0)}{h} = \frac{h - 0}{h} = 1,$$

but for $h < 0$, $$\frac{f(0+h) - f(0)}{h} = \frac{-h - 0}{h} = -1.$$

Hence the expression $\dfrac{f(0+h) - f(0)}{h}$ approaches no single value as $h \to 0$. Thus although $|x|$ is a continuous function it is not differentiable at $x = 0$.

Exercise 2.1

In questions 1 to 4 decide whether the given functions are odd, even or neither.

1. (a) $x^4 - 6$, (b) $x^3 + 5$, (c) $x^3 + 5x$.

2. (a) $1 + \dfrac{1}{x^2}$, (b) $\dfrac{1}{1 + x^2}$, (c) $\dfrac{x}{1 + x^2}$.

3. (a) $1 - \cos x$, (b) $x^2 \sin x$, (c) $\sec x + \tan x$.

4. (a) $[x]$, (b) $[x] + \frac{1}{2}$, (c) $[x^2]$.

In questions 5 to 8 decide whether the given functions are periodic and if so state the period.

5. (a) $\sin 3x$, (b) $1 + \tan 2x$, (c) $\cos \frac{1}{4}(x + \pi)$.

6. (a) $x + \cos x$, (b) $\sin x + \cos x$, (c) $\sin x \cos x$.

7. (a) $\sin 2x - \sin 3x$, (b) $x \sin 3x$, (c) $\sin 3x \cos 2x$.

8. (a) $|\cos x|$, (b) $\sin |x|$, (c) $[\tan x]$.

In questions 9 to 12 sketch the graphs of the functions f, g, fg and gf. In each case state whether the function is odd, even or neither, give its range and, if periodic, its period.

9. $f(x) = x^2 - 1$, $g(x) = 2x + 1$.

10. $f(x) = \frac{1}{2}(\pi - x)$, $g(x) = \sin x$.

11. $f(x) = \cos x$, $g(x) = x^2$.

12. $f(x) = |x|$, $g(x) = x(x^2 - 3)$.

In questions 13 to 16 sketch the graphs of the given periodic functions.

13. $f(x) = x^2$ for $-2 < x \leqslant 2$, f has period 4.

14. $f(x) = |x|$ for $-1 < x \leqslant 1$, f has period 2.

15. $f(x) = x^3$ for $0 \leqslant x \leqslant 1$,
$f(x) = -x$ for $-1 < x < 0$, f has period 2.

16. $f(x) = x$ for $0 \leqslant x \leqslant k$,
$f(x) = k$ for $k < x < 2k$, f has period $2k$.

In questions 17 to 20 sketch the graphs of the given functions. State any values of x at which
(a) there is a discontinuity,
(b) the function is not differentiable.

17. $1/(x^2 - 1)$.

18. $|x^2 - 1|$.

19. $x + [x]$.

20. $f(x) = x$ for $|x| \leqslant 1$, $f(x) = x^2 - 2$ for $|x| > 1$.

21. Given that f is a periodic function such that for $0 \leqslant x \leqslant 1$, $f(x) = x^2 - 2x$, sketch the graph of the function in each of the following cases
(a) f has period 1,
(b) f is an even function and has period 2,
(c) f is an odd function and has period 2.

22. A function f is defined by $f(x) = x^2 + 3x$ for $x < 0$, $f(x) = kx$ for $x \geqslant 0$. Show that $f(x)$ is differentiable at $x = 0$ for only one value of k and state this value.

2.2 Some graphs with related equations

In this section we consider some graphs with equations of the form $y = f(x)$ together with graphs which have related equations such as $y = |f(x)|$, $y = 1/f(x)$, $y = \{f(x)\}^2$, $y^2 = f(x)$.

However, we first digress a little to note that not all graphs represent functions. There are many ways in which two real numbers x and y may be related, e.g. "$x < y$", "$y^2 = x$" or perhaps "$x - y$ is an integer". Each of these statements is said to define a *relation* on the set of real numbers. The graph of the relation is the set of points (x, y) such that x and y are related in the stated manner.

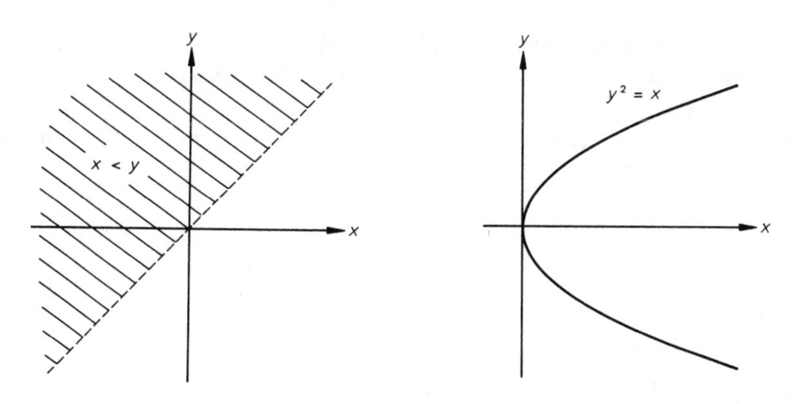

Thus, in general, an equation (or inequality) connecting x and y defines a relation on $\mathbb{R}$. Such an equation defines a function of x with domain $\mathbb{R}$ (or a subset of $\mathbb{R}$) only if there is at most one value of y corresponding to any given value of x. For instance the equation $y^2 = x$ specifies a relation, but not a function of the form $f:x \to y$, whereas the equation $y = \sqrt{x}$ defines a function with domain $\{x \in \mathbb{R}: x \geqslant 0\}$.

The equation $y = |f(x)|$

When $f(x) \geqslant 0$, the graph of $y = |f(x)|$ coincides with the graph of $y = f(x)$. When $f(x) < 0$, the graph of $y = |f(x)|$ is the reflection in the x-axis of the graph of $y = f(x)$.

[In the following sketches broken lines are used for the graph of $y = f(x)$ and unbroken lines for the graph of $y = |f(x)|$.]

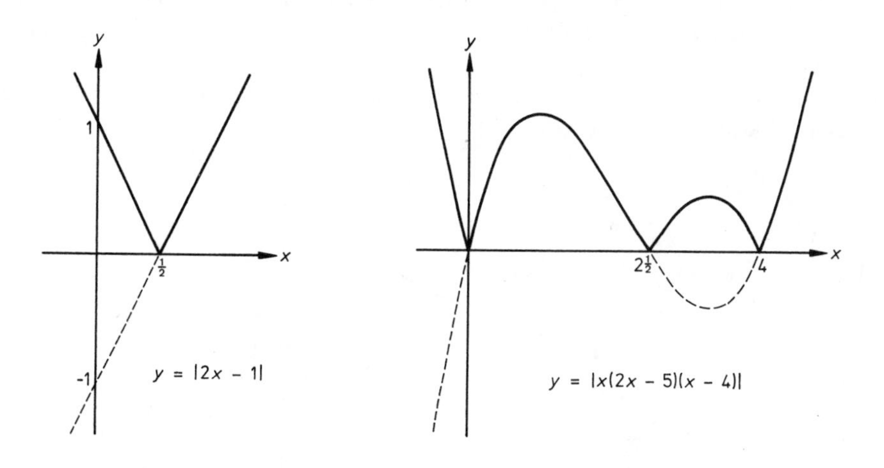

The equation $y = 1/f(x)$

If the graph of $y = f(x)$ meets the x-axis at $x = a$, then the line $x = a$ is an asymptote to the graph of $y = 1/f(x)$ and vice versa.

If there is a maximum (minimum) point on the graph of $y = f(x)$ when $x = k$, then provided that $f(k) \neq 0$, there is a minimum (maximum) point on the graph of $y = 1/f(x)$ when $x = k$.

The graphs of $y = f(x)$ and $y = 1/f(x)$ always lie on the same side of the x-axis and intersect when $f(x) = \pm 1$.

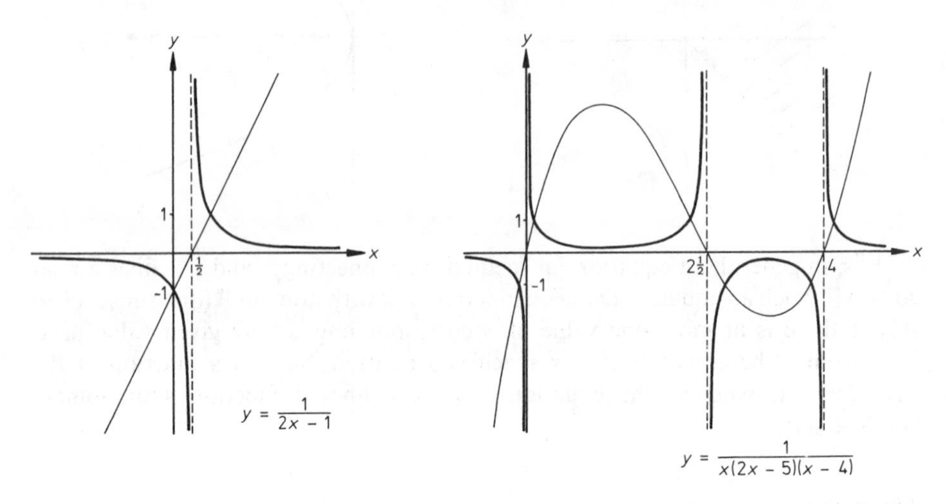

The equation $y = \{f(x)\}^2$

The graph of $y = \{f(x)\}^2$ lies entirely on or above the x-axis and cuts the graph of $y = f(x)$ when $f(x) = 0$ or 1.

$$y = \{f(x)\}^2 \Rightarrow \frac{dy}{dx} = 2f(x)f'(x)$$

Thus, since $dy/dx = 0$ when $f'(x) = 0$, there is a stationary point on the graph of $y = \{f(x)\}^2$ at any value of x for which there is a stationary point on the graph of $y = f(x)$.

Since $dy/dx = 0$ when $f(x) = 0$, any point at which the graph of $y = f(x)$ cuts the x-axis is a stationary point on the graph of $y = \{f(x)\}^2$.

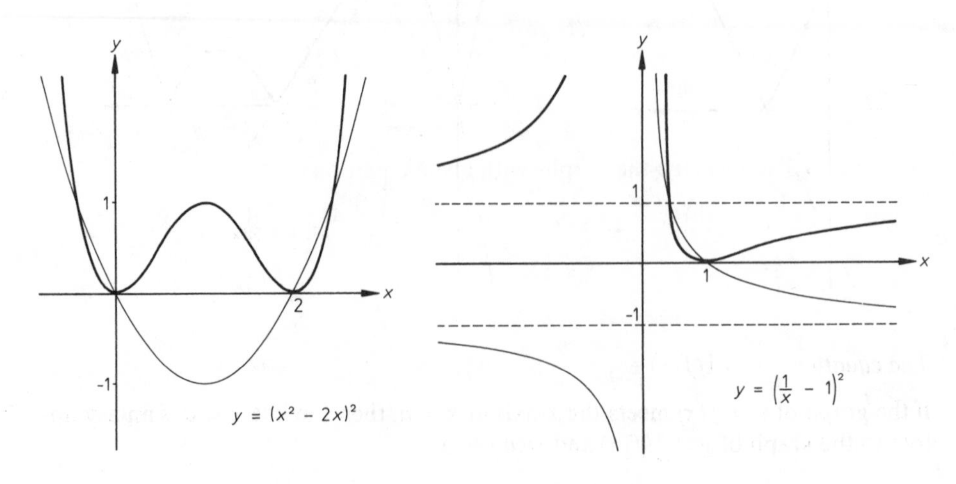

The equation $y^2 = f(x)$

At values of x such that $f(x) < 0$, there are no points on the graph of $y^2 = f(x)$.

At values of x such that $f(x) > 0$, there are two possible values of y, namely $\pm\sqrt{\{f(x)\}}$. Hence the graph of $y^2 = f(x)$ is symmetrical about the x-axis.

It is also useful to note that if $0 < f(x) < 1$, then $\sqrt{\{f(x)\}} > f(x)$ and if $f(x) > 1$, then $\sqrt{\{f(x)\}} < f(x)$.

$$y^2 = f(x) \Rightarrow 2y\frac{dy}{dx} = f'(x)$$

$$\therefore \quad \text{if } f'(x) = 0, \quad \text{and} \quad f(x) > 0, \quad \text{then} \quad \frac{dy}{dx} = 0.$$

Thus, any stationary point above the x-axis on the curve $y = f(x)$ gives rise to a pair of stationary points on the curve $y^2 = f(x)$.

To determine how the graph of $y^2 = f(x)$ behaves near the x-axis, the expression for dy/dx must be considered further. For instance, suppose that $f(x) = x(x-3)^2$.

If
$$y^2 = x(x-3)^2 = x^3 - 6x^2 + 9x$$

then differentiating with respect to x,

$$2y\frac{dy}{dx} = 3x^2 - 12x + 9 = 3(x-1)(x-3)$$

But $y = \pm(x-3)\sqrt{x}$

$\therefore$ provided that $x \neq 3$,

$$\frac{dy}{dx} = \pm\frac{3(x-1)}{2\sqrt{x}}$$

As $x \to 0$, $\dfrac{dy}{dx} \to \pm\infty$.

As $x \to 3$, $\dfrac{dy}{dx} \to \pm\sqrt{3}$.

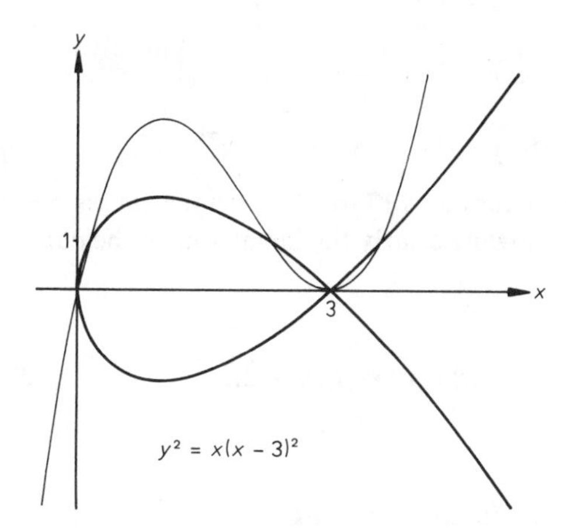

Exercise 2.2

In questions 1 to 6 sketch the graphs with given equations

1. (a) $y = x^2$, (b) $y = \frac{1}{2}x^2$, (c) $y = x^2 - 4$, (d) $y = (x+2)^2$.

2. (a) $y = \sin x$, (b) $y = 2\sin x$, (c) $y = \sin 2x$, (d) $y = \sin(x + \frac{1}{2}\pi)$.

3. (a) $y = \dfrac{1}{x}$, (b) $y = \dfrac{4}{x}$, (c) $y = \dfrac{1}{x+1}$, (d) $y = \dfrac{x+2}{x+1}$.

4. (a) $y = |3 - 2x|$, (b) $y = |x^2 - 4|$, (c) $y = |x^3 - 3x^2|$.

5. (a) $y = |\cos x|$, (b) $y = |\cos x| - 1$, (c) $y = |2\cos x + 1|$.

6. (a) $y = \dfrac{1}{x-2}$, (b) $y = \dfrac{1}{2-x}$, (c) $y = \dfrac{1}{|x-2|}$.

In questions 7 to 16 sketch the given pair of curves in the same diagram. Label clearly any stationary points and asymptotes.

7. $y = x^2 + 2x$, $y = 1/(x^2 + 2x)$.

8. $y = x^2 + 2x + 1$, $y = 1/(x^2 + 2x + 1)$.

9. $y = x^2 + 2x + 2$, $y = 1/(x^2 + 2x + 2)$.

10. $y = x^3 + 2x^2 - 15x$, $y = 1/(x^3 - 2x^2 - 15x)$.

11. $y = (2x - 1)/(x + 2)$, $y = (x + 2)/(2x - 1)$.

12. $y = 2x - 3$, $y = (2x - 3)^2$.

13. $y = 4 - x^2$, $y = (4 - x^2)^2$.

14. $y = \sin x$, $y = \sin^2 x$.

15. $y = 2x/(1 - x)$, $y = 4x^2/(1 - x)^2$.

16. $y = 3x^2 - x^3$, $y^2 = 3x^2 - x^3$.

In questions 17 to 20 sketch the curves $y = f(x)$ and $y^2 = f(x)$ in the same diagram, showing clearly the behaviour of the curves near the x-axis.

17. $f(x) = x^3$.

18. $f(x) = x(x + 3)^2$.

19. $f(x) = (x + 1)(x - 2)^2$.

20. $f(x) = x^2(4 - x^2)$.

2.3 Inverse functions

Consider a function f with domain X and range Y. If every element of Y is the image of one and only one element of X, then f is said to be *one-one*. For any one-one function $f: X \to Y$ we define an *inverse function* $f^{-1}: Y \to X$ such that if $y = f(x)$ then $f^{-1}(y) = x$.

It follows from the definition that

$$f^{-1}f(x) = f^{-1}(y) = x \quad \text{and} \quad ff^{-1}(y) = f(x) = y$$

i.e. the composite functions $f^{-1}f$ and ff^{-1} are the identity mappings on the sets X and Y respectively.

Example 1 One-one functions f and g are defined on $\mathbb{R}$ by $f: x \to 3x - 1$ and $g: x \to \lg(1 - x)$, $x < 1$. Find expressions for f^{-1} and g^{-1}.

Let $y = f^{-1}(x)$ then $f(y) = x$

$$\therefore \quad 3y - 1 = x$$
$$y = \tfrac{1}{3}(x + 1)$$

Hence $f^{-1}: x \to \tfrac{1}{3}(x + 1)$.
Let $y = g^{-1}(x)$ then $g(y) = x$

$$\therefore \quad \lg(1 - y) = x$$
$$1 - y = 10^x$$
$$y = 1 - 10^x$$

Hence $g^{-1}: x \to 1 - 10^x$.

Example 2 A one-one function f is defined by $f: x \to 1/\sqrt{(x - 1)}$ where $x \in \mathbb{R}$, $x > 1$.
Find an expression for f^{-1}, stating its domain. Verify that $f^{-1}f(x) = x$ where $x \in \mathbb{R}$, $x > 1$.

$$\begin{aligned} y = f^{-1}(x) \quad &\Rightarrow \quad f(y) = x \\ &\Rightarrow \quad 1/\sqrt{(y - 1)} = x \\ &\Rightarrow \quad 1/(y - 1) = x^2 \\ &\Rightarrow \quad 1 = x^2(y - 1) \\ &\Rightarrow \quad y - 1 = 1/x^2 \\ &\Rightarrow \quad y = 1 + 1/x^2 \end{aligned}$$

$$\therefore \quad f^{-1}: x \to 1 + 1/x^2.$$

The domain of f^{-1} is the same as the range of f
i.e. $\mathbb{R}^+$, the set of positive real numbers.

$$f^{-1}f(x) = f^{-1}(u), \quad \text{where} \quad u = f(x) = 1/\sqrt{(x - 1)}$$
$$\therefore \quad f^{-1}f(x) = 1 + 1/u^2 = 1 + (x - 1) = x.$$

Example 3 A function f is defined on $\mathbb{R}$ by $f: x \to x^2 - 2x$. Explain why f^{-1} does not exist. Suggest a restricted domain for which f^{-1} does exist.

$f(0) = f(2) = 0$ i.e. 0 is the image under f of both 0 and 2,
$\therefore$ the function f defined over $\mathbb{R}$ is not one-one and has no inverse.

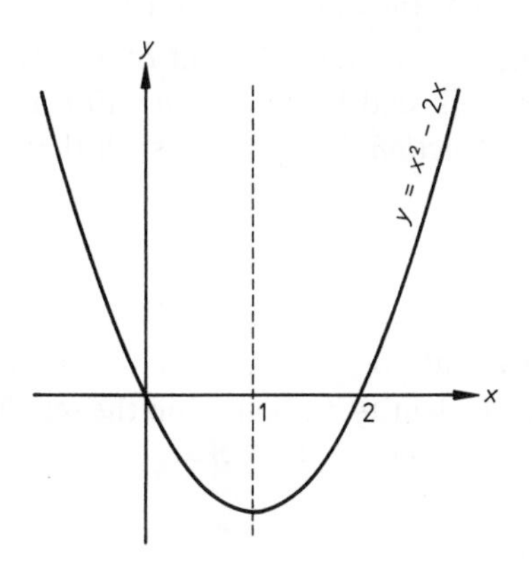

The sketch of the graph $y = x^2 - 2x$ shows that for $x \geqslant 1$ there is only one value of x corresponding to each value of y, $\therefore$ over the domain $\{x \in \mathbb{R}: x \geqslant 1\}$ the function f is one-one and f^{-1} does exist. [Note that f^{-1} also exists for the restricted domain $\{x \in \mathbb{R}: x \leqslant 1\}$.]

In general, a function f has an inverse if and only if any line drawn parallel to the x-axis cuts the graph $y = f(x)$ exactly once. Clearly a function whose graph has one or more turning points does not have an inverse.

If a point (a, b) lies on the graph $y = f(x)$ then the point (b, a) lies on the graph $y = f^{-1}(x)$. Hence each graph is the reflection of the other in the line $y = x$.

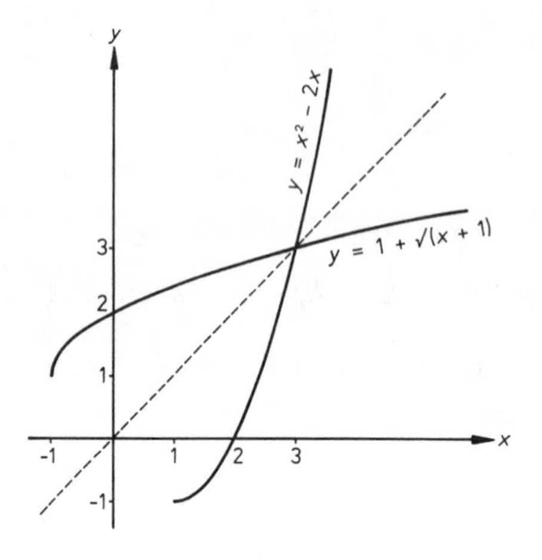

The diagram illustrates this relationship for the function

$$f: x \to x^2 - 2x, \quad x \geqslant 1.$$

The equation of the graph of f^{-1} is $y^2 - 2y = x, y \geqslant 1$, which may be rearranged as

$$(y-1)^2 = x + 1, \; y \geqslant 1$$

i.e. $$y - 1 = \sqrt{(x+1)}$$

i.e. $$y = 1 + \sqrt{(x+1)}.$$

In some cases it is possible to show that a function f has an inverse, but difficult to find an expression for $f^{-1}(x)$. Consider, for example, the function $f: x \to x + \sin x$.

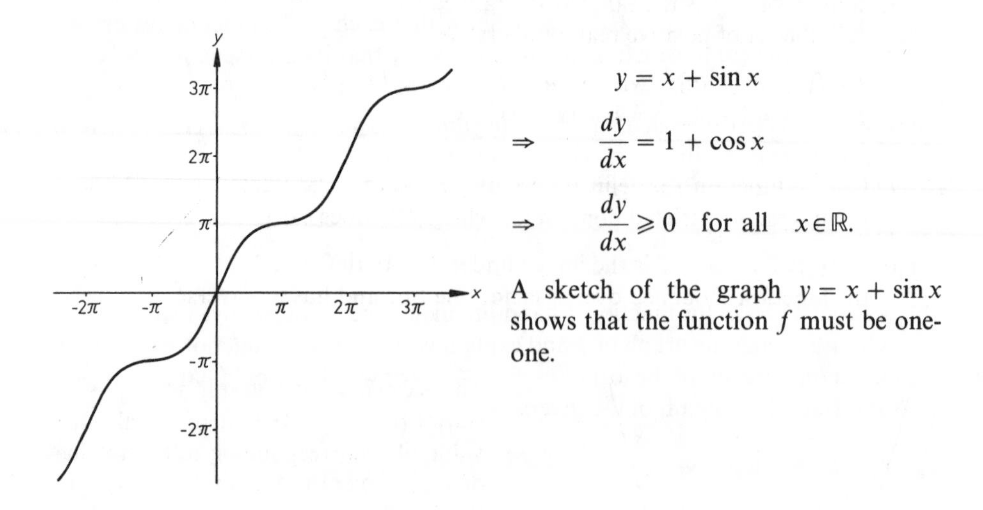

$$y = x + \sin x$$

$$\Rightarrow \quad \frac{dy}{dx} = 1 + \cos x$$

$$\Rightarrow \quad \frac{dy}{dx} \geqslant 0 \quad \text{for all} \quad x \in \mathbb{R}.$$

A sketch of the graph $y = x + \sin x$ shows that the function f must be one-one.

Hence the inverse f^{-1} exists and its graph has equation $y + \sin y = x$. However, it is impossible to rearrange this equation in the form $y = f^{-1}(x)$. Thus although f^{-1} exists, in practice it is difficult to find $f^{-1}(x)$ for many values of x.

Exercise 2.3

In questions 1 to 14 state whether or not the function f is one-one. If the function f is one-one, find an expression for f^{-1}, stating its domain. For any function that is not one-one, give two distinct values of x which have the same image.

1. $f:x \to 2x + 1.$

2. $f:x \to 5 - x.$

3. $f:x \to x(x + 2).$

4. $f:x \to x^2 - 2, x \geqslant 0.$

5. $f:x \to 1 + \dfrac{1}{x}, x \neq 0.$

6. $f:x \to \dfrac{1}{9 - x}, x \neq 9.$

7. $f:x \to \sqrt{(3 - x)}, x \leqslant 3.$

8. $f:x \to \sqrt{(1 - x^2)}, -1 \leqslant x \leqslant 1.$

9. $f:x \to 3^x.$

10. $f:x \to \lg(x + 2), x > -2.$

11. $f:x \to \dfrac{1}{x^2 + 4}.$

12. $f:x \to \dfrac{2x}{x - 1}, x \neq 1.$

13. $f:x \to \cos 3x.$

14. $f:x \to \sin^{-1} x, -1 \leqslant x \leqslant 1.$

15. One-one functions f and g are defined on $\mathbb{R}$ by $f:x \to 5 - 4x$ and $g:x \to 1 - x^2, x \geqslant 0$. Given that h is the composite function fg, find expressions for $h(x)$ and $h^{-1}(x)$. Verify that $h^{-1} = g^{-1}f^{-1}$.

16. Functions f, g and h are defined on $\mathbb{R}$ by $f:x \to 3 - x$, $g:x \to 3x/(x - 3)$, $x \neq 3$ and $h:x \to \sqrt{(4 - x^2)}, 0 \leqslant x \leqslant 2$. Show that each of the functions is its own inverse. Find the composite function gf and verify that its inverse is fg.

17. Find the inverse of each of the following functions defined on $\mathbb{R}$. Verify that, over the appropriate domains, $f^{-1}f(x) = x$ and $ff^{-1}(x) = x$.
(a) $f:x \to (x - 2)^2, x \geqslant 2,$
(b) $f:x \to 1 + \lg(x - 1), x > 1,$
(c) $f:x \to \dfrac{1 - x}{1 + x}, x \neq -1,$
(d) $f:x \to \dfrac{1}{\sqrt{(x^2 - 1)}}, x > 1.$

In questions 18 to 23 the function f has domain $\mathbb{R}^+$, the set of positive real numbers. In each case sketch the graph of f and explain why no inverse function exists. Find the maximum domain of the form $\{x \in \mathbb{R} : x \geqslant k\}$ for which f would have an inverse f^{-1} and state the domain of the inverse.

18. $f:x \to x^2 - 4x + 8.$

19. $f:x \to 6x - x^2.$

20. $f:x \to |x - 4|.$

21. $f:x \to 2x^3 - 3x^2.$

22. $f:x \to x + \dfrac{1}{x}.$

23. $f:x \to x - \sqrt{x}.$

24. Decide whether the following functions have inverses, giving brief reasons for your answers.
(a) $f:x \to x^3 + 3x^2, x \in \mathbb{R}$,
(b) $f:x \to x^3 + 3x, x \in \mathbb{R}$,
(c) $f:x \to 2x + \cos x, x \in \mathbb{R}$,
(d) $f:x \to x + 2\sin x, x \in \mathbb{R}$,
(e) $f:x \to |\lg x|, x \in \mathbb{R}^+$,
(f) $f:x \to |x| + 2x, x \in \mathbb{R}$.

25. Differentiable functions f and g are defined on $\mathbb{R}$ such that f is the inverse of g. Prove that, if $f(a) = b$, then $f'(a)g'(b) = 1$.

2.4 Polynomials; remainders and factors

If, when a polynomial $P(x)$ is divided by $(x - a)$, the quotient is $Q(x)$ and the remainder is R, then

$$P(x) \equiv (x - a)Q(x) + R.$$

This identity leads to two familiar results.

The remainder theorem: When $P(x)$ is divided by $(x - a)$ the remainder is $P(a)$.

The factor theorem: If $P(a) = 0$, then $(x - a)$ is a factor of $P(x)$.

This approach can be extended to the division of a polynomial $f(x)$ by a polynomial $g(x)$ of degree less than or equal to the degree of $f(x)$. If the division gives quotient $Q(x)$ and remainder $R(x)$ then

$$f(x) \equiv g(x)Q(x) + R(x)$$

where $R(x)$ is of lower degree than $g(x)$.

[In particular, if $g(x)$ is a quadratic function then $R(x)$ is of the form $Ax + B$.]

Example 1 A polynomial $P(x)$ is a multiple of $x - 3$ and the remainder when $P(x)$ is divided by $x + 3$ is 12. Find the remainder when $P(x)$ is divided by $x^2 - 9$.

When $P(x)$ is divided by $x^2 - 9$, let the quotient be $Q(x)$ and the remainder $Ax + B$, then

$$P(x) \equiv (x^2 - 9)Q(x) + Ax + B.$$

Since $x - 3$ is a factor of $P(x)$, by the factor theorem,

$$P(3) = 0 \qquad \therefore \quad 3A + B = 0 \qquad (1)$$

Since when $P(x)$ is divided by $x + 3$ the remainder is 12, by the remainder theorem,

$$P(-3) = 12 \qquad \therefore \quad -3A + B = 12 \qquad (2)$$

Using (1) and (2), $A = -2$, $B = 6$.

Hence the remainder when $P(x)$ is divided by $x^2 - 9$ is $-2x + 6$.

Suppose now that a polynomial $f(x)$ has a repeated factor $(x - a)$,

so that $\qquad f(x) \equiv (x-a)^2 g(x)$

Differentiating, $\qquad f'(x) = (x-a)^2 g'(x) + 2(x-a)g(x)$

$$= (x-a)\{(x-a)g'(x) + 2g(x)\}$$

Hence, if $f(x)$ has a repeated factor $(x - a)$, then $(x - a)$ is also a factor of $f'(x)$.

It can further be shown, by the method of question 21, Exercise 2.4, that $(x - a)^2$ is a factor of a polynomial $f(x)$ if and only if $f(a) = f'(a) = 0$.

Example 2 Given that the polynomial $f(x) = x^3 + 3x^2 - 9x + k$ has a repeated linear factor, find the possible values of k.

Differentiating, $\quad f'(x) = 3x^2 + 6x - 9$

$$= 3(x^2 + 2x - 3) = 3(x-1)(x+3)$$

$\therefore$ the repeated factor of $f(x)$ is $(x - 1)$ or $(x + 3)$.
If $(x - 1)$ is a factor of $f(x)$, then $f(1) = 0$
i.e. $1 + 3 - 9 + k = 0$ giving $k = 5$.
If $(x + 3)$ is a factor of $f(x)$, then $f(-3) = 0$
i.e. $-27 + 27 + 27 + k = 0$ giving $k = -27$.
Thus the possible values of k are 5 and -27.

When dealing with polynomial equations it may be helpful to use the relationships between roots and coefficients. In the case of the cubic equation $ax^3 + bx^2 + cx + d = 0$ with roots α, β and γ,

$$\alpha + \beta + \gamma = -\frac{b}{a}, \quad \alpha\beta + \beta\gamma + \gamma\alpha = \frac{c}{a}, \quad \alpha\beta\gamma = -\frac{d}{a}.$$

Similar results can be obtained for equations of higher degree.

Example 3 Find the roots of the equation $4x^3 + 12x^2 - 15x + 4 = 0$ given that it has a repeated root.

Let $\quad f(x) = 4x^3 + 12x^2 - 15x + 4$

then $\quad f'(x) = 12x^2 + 24x - 15$

$$= 3(4x^2 + 8x - 5) = 3(2x-1)(2x+5)$$

Any repeated root of the equation $f(x) = 0$ is also a root of the equation $f'(x) = 0$
$\therefore$ the repeated root must be either $-5/2$ or $1/2$.
[It is possible to decide which of these values is the required root by evaluating $f(-5/2)$ and $f(1/2)$. However, the method given below may be quicker.]

The product of the roots of $f(x) = 0$ is $-4/4$ i.e. -1.
Thus, if the repeated root is $-5/2$, the remaining root is $-1 \div (-5/2)^2$ i.e. $-4/25$.
If the repeated root is $1/2$, the remaining root is $-1 \div (1/2)^2$ i.e. -4.
Since the sum of the roots of $f(x) = 0$ is $-12/4$ i.e. -3, the roots must be $1/2$, $1/2$, -4.

Let us consider two polynomials $P(x) = a_0x^n + a_1x^{n-1} + \ldots + a_n$ and

$$Q(x) = b_0x^n + b_1x^{n-1} + \ldots + b_n.$$

It can be shown that if $P(x) \equiv Q(x)$ i.e. if $P(x)$ and $Q(x)$ are equal for all values of x, then $a_0 = b_0, a_1 = b_1, \ldots, a_n = b_n$.

One way of proving this result uses the fact that if $P(x) \equiv Q(x)$, then $P'(x) \equiv Q'(x)$, $P''(x) \equiv Q''(x), \ldots, P^{(n)}(x) \equiv Q^{(n)}(x)$.

Substituting $x = 0$, we find that $P(0) = Q(0) \Rightarrow a_n = b_n$,
$P'(0) = Q'(0) \Rightarrow a_{n-1} = b_{n-1}, \ldots, P^{(n)}(0) = Q^{(n)}(0) \Rightarrow a_0 = b_0$.

We deduce that if two polynomials are identically equal, then corresponding coefficients are equal. Problems involving such identities are solved by *comparing coefficients*, by substituting suitable values of x, or by a combination of these methods.

Example 4 Given that $x^2 - 3x + 7 \equiv a(x-1)^2 + b(x-1) + c$, find the values of a, b and c.

Comparing coefficients of x^2: $1 = a$ $\quad\therefore\ a = 1$
Putting $x = 1$: $1 - 3 + 7 = 0 + 0 + c$ $\quad\therefore\ c = 5$
Putting $x = 0$: $7 = a - b + c$ $\quad\therefore\ b = -1$
Hence $x^2 - 3x + 7 \equiv (x-1)^2 - (x-1) + 5$.

[Note that putting $x = 0$ is equivalent to comparing constant terms.]

Example 5 If $x^2 + 1$ is a factor of $3x^4 + x^3 - 4x^2 + px + q$, find the values of p and q.

Let $3x^4 + x^3 - 4x^2 + px + q \equiv (x^2 + 1)(ax^2 + bx + c)$.

Comparing coefficients of x^4: $3 = a$
x^3: $1 = b$
x^2: $-4 = a + c$
x: $p = b$
Comparing constant terms: $q = c$
Hence $p = 1$ and $q = -7$.

Exercise 2.4

1. Find the remainder when
(a) $x^3 - 4x^2 + 3x - 7$ is divided by $x - 4$,
(b) $x^4 + 2x^3 + 5x - 8$ is divided by $x + 3$,
(c) $2x^3 - 5x^2 + 3x + 1$ is divided by $2x + 1$,
(d) $x^3 - 2x^2 - 5$ is divided by $(x+1)(x-2)$,
(e) $x^4 + 3x^3 - 12x - 16$ is divided by $x^2 - 4$,
(f) $2x^4 - 3x^3 - 9x^2 - 7$ is divided by $x^2 - x - 6$.

2. When a polynomial $f(x)$ is divided by $(x - 2)$ the remainder is -2, and when it is divided by $(x - 3)$ the remainder is 3. Given that the remainder when $f(x)$ is divided by $(x - 2)(x - 3)$ is $px + q$, find the values of p and q.

3. When a polynomial is divided by $(x + 1)$ the remainder is 7, and when it is divided by $(x - 4)$ the remainder is -8. Find the remainder when the polynomial is divided by $(x + 1)(x - 4)$.

4. When the polynomial $x^3 + px^2 + qx + 7$ is divided by $x^2 - 3x + 2$, the remainder is $5x - 3$. Find the values of p and q.

5. When $P(x) = x^3 + ax^2 + bx + c$ is divided by $x^2 - 4$ the remainder is $2x + 11$. Given that $x + 1$ is a factor of $P(x)$, find the values of a, b and c.

6. Show, by putting $a = -b$, that $a + b$ is a factor of $(a + b + c)^3 - (a^3 + b^3 + c^3)$. Factorise the expression completely.

7. Show, by putting $x = y$, that $x - y$ is a factor of $(x - y)^3 + (y - z)^3 + (z - x)^3$. Factorise the expression completely.

8. Given that the polynomial $f(x) = 2x^3 - 3x^2 - 12x + k$ has a repeated linear factor, find the possible values of k and factorise $f(x)$ in each case.

9. Factorise the polynomial $3x^4 + 8x^3 + 16$ given that it has a repeated linear factor.

10. Find the values of a and b and factorise the polynomial

$$f(x) = x^3 + 4ax^2 + bx + 3a$$

given that it is divisible by $(x - 1)^2$.

11. Find the roots of the equation $x^3 - 6x^2 - 63x - 108 = 0$ given that it has a repeated root.

12. Given that $P(x) = 8x^3 - 12x^2 - 18x + k$, find the values of k such that the equation $P(x) = 0$ has a repeated root. Give the roots of the equation in each case.

13. The roots of the equation $2x^3 + 3x^2 + kx - 6 = 0$ are in arithmetic progression. By letting the roots be $\alpha - \lambda, \alpha, \alpha + \lambda$, or otherwise, find the value of k and solve the equation.

14. Given that the roots of the equation $3x^3 - 7x^2 + px + 24 = 0$ are in geometric progression, find the value of p and solve the equation.

15. If $ax^3 + 2x^2 + bx + 2b - c \equiv x^3 + 2x^2 + (c + 2)x + b + c$, find the values of a, b and c.

16. If $x^3 - 2x + 7 \equiv a(x-1)^3 + b(x-1)^2 + c(x-1) + d$, find the values of a, b, c and d.

17. Find constants a, b, c and d such that

$$n^3 \equiv an(n+1)(n+2) + bn(n+1) + cn + d.$$

18. Show that no constants a and b can be found such that

$$n^2 \equiv a(n+1)^2 + b(n+1).$$

19. Given that $x^4 - 6x^3 + 10x^2 + ax + b$ is a perfect square, find the values of a and b.

20. Given that $x^2 - x + 1$ is a factor of $2x^4 - 3x^3 + px^2 + qx - 3$, find the values of p and q.

21. If $f(x)$ and $g(x)$ are polynomials and $f(x) \equiv (x-a)^2 g(x) + Ax + B$, find $f'(x)$ and hence find A and B in terms of $f(a)$ and $f'(a)$. Deduce that $(x-a)$ is a repeated factor of $f(x)$ if and only if $f(a) = f'(a) = 0$.

Find all real values of k for which the equation $x^3 - 3kx^2 + 2k + 2 = 0$ has repeated roots and, for each such k, solve the equation completely. (O & C)

2.5 Graphs of rational functions

A rational function is a function of the form $P(x)/Q(x)$, where $P(x)$ and $Q(x)$ are polynomials. The graph of this function cuts the x-axis where $P(x) = 0$ and has asymptotes parallel to the y-axis where $Q(x) = 0$.

One of the simplest types of rational function takes the form $(ax+b)/(cx+d)$. In general, the graph of this function is a curve called a *rectangular hyperbola* with asymptotes parallel to the x- and y-axes.

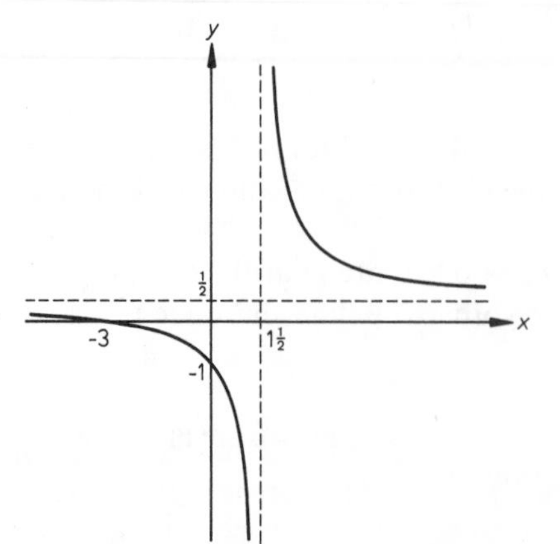

The diagram shows the graph of $y = (x+3)/(2x-3)$.
The curve cuts the axes at $(-3, 0)$ and $(0, -1)$.
As $x \to 1\frac{1}{2}$, $y \to \pm\infty$.

$$y = \frac{x+3}{2x-3} = \frac{1 + 3/x}{2 - 3/x}$$

$\therefore$ as $x \to \pm\infty$, $y \to \frac{1}{2}$.

Hence the lines $x = 1\frac{1}{2}$ and $y = \frac{1}{2}$ are asymptotes to the curve.

[If necessary, we can verify that such curves have no turning points by finding dy/dx.

In this case

$$\frac{dy}{dx} = \frac{(2x-3)1-(x+3)2}{(2x-3)^2} = -\frac{9}{(2x-3)^2}.$$

Thus the gradient of the curve is always negative.]

Another important set of rational functions are defined by expressions of the form $y = (ax+b)/(px^2+qx+r)$. When sketching the graphs of these functions it is helpful to find the range of values taken by y and also to note where changes in the sign of y occur.

We illustrate this approach by considering the graph of $y = 4x/(x^2+1)$.
The graph cuts the axes only at the origin $(0,0)$.
Since (x^2+1) is always positive, there can be no asymptotes parallel to the y-axis and the sign of y will always be the same as the sign of x.
As $x \to \infty$, $y \to 0$ from above and as $x \to -\infty$, $y \to 0$ from below,
$\therefore$ the x-axis is an asymptote to the curve.
To find the range of the function $y = 4x/(x^2+1)$, we rearrange the equation as a quadratic equation in x:

$$y(x^2+1) = 4x \quad \text{i.e.} \quad yx^2 - 4x + y = 0.$$

Since we are concerned only with values of y given by real values of x, the range of values taken by y is the set of values for which this quadratic equation has real roots.
The equation $ax^2+bx+c=0$ has real roots if $b^2-4ac \geqslant 0$,

$$\therefore \quad \text{for real values of } x, \quad (-4)^2 - 4y \,.\, y \geqslant 0$$
$$16 - 4y^2 \geqslant 0$$
$$y^2 \leqslant 4$$

i.e. for real values of x, $\quad -2 \leqslant y \leqslant 2$.

Thus all points of the graph lie on or between the lines $y = \pm 2$.

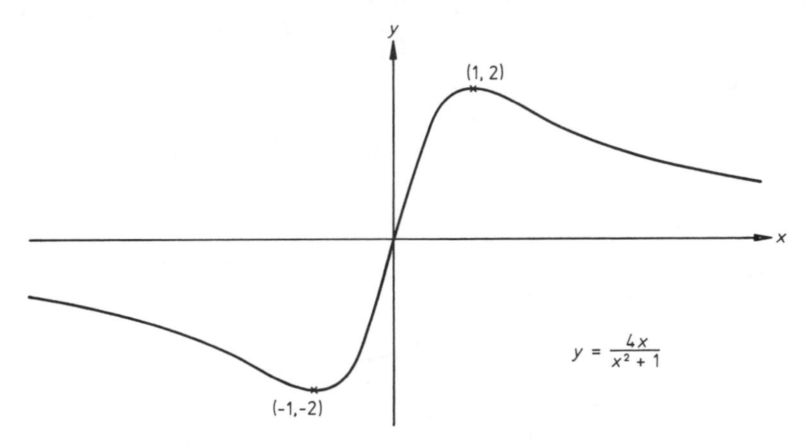

When $y = -2$ and when $y = 2$ the quadratic equation in x has equal roots, $x = -1$ and $x = 1$ respectively. Hence the line $y = -2$ is a tangent to the curve at the point $(-1,-2)$ and the line $y = 2$ is a tangent to the curve at the point $(1,2)$. It follows that $(-1,-2)$ must be a minimum point on the curve and $(1,2)$ a maximum point.

For every non-zero value of y in the interval $-2 < y < 2$ the quadratic equation in x has two distinct real roots. Hence there are no other turning points on the curve. [Note that a graph of similar form will be obtained for any function $(ax + b)/(px^2 + qx + r)$ if the denominator $(px^2 + qx + r)$ is never zero.]

Example 1 Sketch the curve $y = \dfrac{2x-3}{x^2+2x-3}$.

$x = 0 \Rightarrow y = 1$ $\quad\therefore$ the curve cuts the y-axis at the point (0, 1).

$y = 0 \Leftrightarrow 2x - 3 = 0 \Leftrightarrow x = 1\frac{1}{2}$

$\therefore$ the curve cuts the x-axis at the point $(1\frac{1}{2}, 0)$.

$x^2 + 2x - 3 = 0 \Leftrightarrow (x+3)(x-1) = 0 \Leftrightarrow x = -3$ or $x = 1$

$\therefore$ the lines $x = -3$ and $x = 1$ are asymptotes to the curve.

As $x \to \pm\infty, y \to 0$ $\quad\therefore$ the x-axis is an asymptote to the curve.

The following table shows the changes in the sign of y.

	$x < -3$	$-3 < x < 1$	$1 < x < 1\frac{1}{2}$	$x > 1\frac{1}{2}$
$x + 3$	$-$	$+$	$+$	$+$
$x - 1$	$-$	$-$	$+$	$+$
$2x - 3$	$-$	$-$	$-$	$+$
y	$-$	$+$	$-$	$+$

Rearranging the equation as a quadratic in x:

$$y(x^2 + 2x - 3) = 2x - 3 \quad \text{i.e.} \quad yx^2 + (2y-2)x - 3y + 3 = 0$$

$\therefore$ for real values of x, $(2y-2)^2 - 4y(-3y+3) \geqslant 0$

$$4(y-1)^2 + 12y(y-1) \geqslant 0$$

$$4(y-1)\{(y-1) + 3y\} \geqslant 0$$

$$(y-1)(4y-1) \geqslant 0$$

$\therefore$ for real values of x, $y \leqslant \frac{1}{4}$ or $y \geqslant 1$.

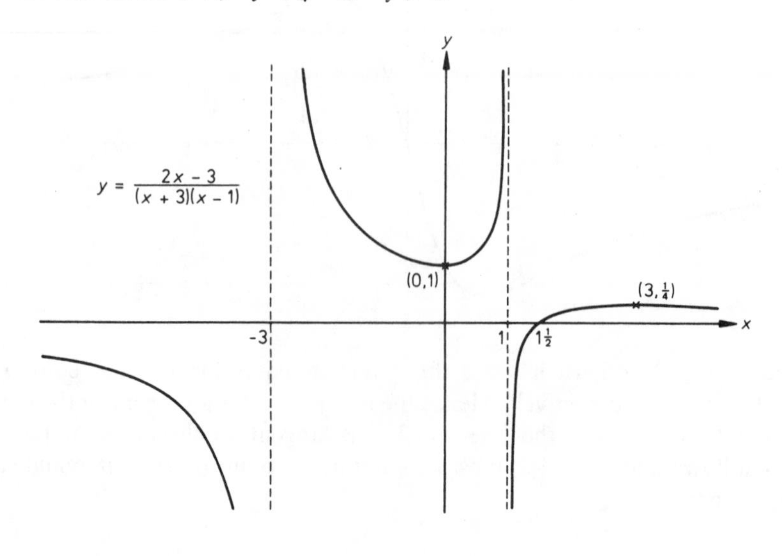

Hence there are no points on the curve between the lines $y = \frac{1}{4}$ and $y = 1$.
There is a stationary point on the curve whenever the quadratic equation in x has repeated roots, i.e. when $y = \frac{1}{4}$, $x = 3$ and when $y = 1, x = 0$. Since y cannot take values between $\frac{1}{4}$ and 1, $(3, \frac{1}{4})$ is a maximum point and $(0, 1)$ a minimum point.

There is one other form that the graph of $y = (ax + b)/(px^2 + qx + r)$ can take. Consider, for example, the curve $y = x/(x + 2)(x - 1)$.
This curve cuts the axes only at the origin $(0, 0)$.
The asymptotes to the curve are the lines $x = -2$, $x = 1$ and the x-axis.
Rearranging the equation as a quadratic in x:

$$y(x^2 + x - 2) = x \quad \text{i.e.} \quad yx^2 + (y - 1)x - 2y = 0.$$

The nature of the roots of this quadratic equation depends on the sign of the expression

$$(y - 1)^2 - 4y(-2y) \quad \text{i.e.} \quad (y - 1)^2 + 8y^2.$$

Since this expression is always positive, the equation must have two real distinct roots for all non-zero values of y. Hence there are no stationary points on the curve and no restrictions on the values of y.

Again it is useful to determine the sign of y throughout the domain of the function.

	$x < -2$	$-2 < x < 0$	$0 < x < 1$	$x > 1$
$x + 2$	$-$	$+$	$+$	$+$
x	$-$	$-$	$+$	$+$
$x - 1$	$-$	$-$	$-$	$+$
y	$-$	$+$	$-$	$+$

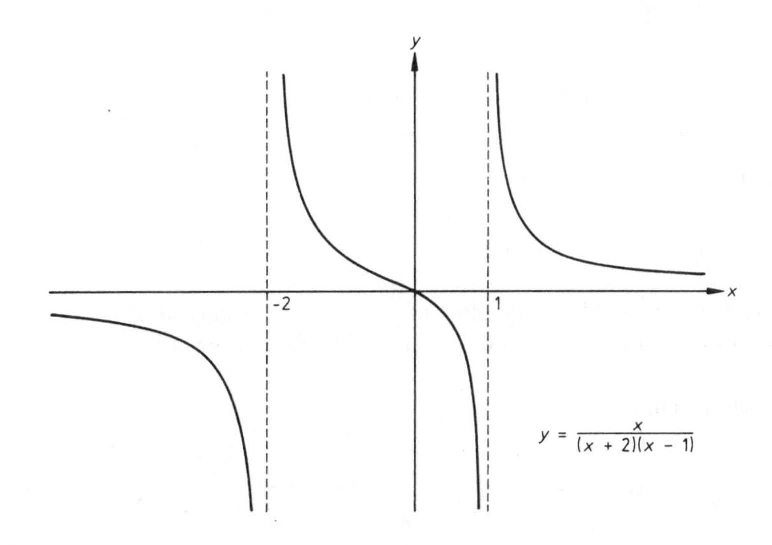

[These methods can be extended to functions of the form $P(x)/Q(x)$ where both $P(x)$ and $Q(x)$ are quadratic functions.]

Exercise 2.5

In questions 1 to 6 sketch the given curve.

1. $y = \dfrac{1}{x-1}$ 2. $y = \dfrac{x}{x-1}$ 3. $y = \dfrac{x-1}{x}$

4. $y = \dfrac{3-x}{2+x}$ 5. $y = \dfrac{2x+1}{x-2}$ 6. $y = \dfrac{x}{3x+4}$

In questions 7 to 16 find the range of possible values of y. Sketch each curve showing clearly any asymptotes and turning points.

7. $y = \dfrac{5x-25}{x^2+3x-4}$ 8. $y = \dfrac{2x+1}{x^2+2}$

9. $y = \dfrac{x+1}{x(x+3)}$ 10. $y = \dfrac{x-1}{x(x+3)}$

11. $y = \dfrac{2-3x}{x^2-3x+2}$ 12. $y = \dfrac{15-10x}{4+x^2}$

13. $y = \dfrac{1-x}{1-x+x^2}$ 14. $y = \dfrac{x+6}{x^2+7x+10}$

15. $y = \dfrac{1}{x-1} + \dfrac{1}{4-x}$ 16. $y = \dfrac{1}{x-1} + \dfrac{1}{x-4}$

17. If a, b and c are all positive, sketch the curve $y = \dfrac{x-a}{(x-b)(x-c)}$

when

(i) $a < b < c$,
(ii) $b < a < c$,
(iii) $b < c < a$.

18. By considering the limiting value of y as $x \to \pm\infty$, show that the line $y = 1$ is an asymptote to the curve $y = (x^2+1)/(x^2-1)$. Sketch the curve.

19. Sketch the curves

(a) $y = \dfrac{3x^2}{x^2+x-2}$, (b) $y = \dfrac{3x^2}{x^2+x+2}$.

20. Sketch the curve $y = \dfrac{x^2 - 3x + 4}{x - 3}$ showing clearly any asymptotes and turning points.

2.6 Partial fractions

A sum of two fractions can be expressed as a single fraction by a method such as the following:

$$\frac{3}{3x-2} - \frac{x}{x^2+1} \equiv \frac{3(x^2+1) - x(3x-2)}{(3x-2)(x^2+1)} \equiv \frac{2x+3}{(3x-2)(x^2+1)}.$$

In this section we consider the reverse process by which a rational function is expressed as a sum of two or more simpler fractions called *partial fractions.*

We will deal first with *proper fractions* i.e. fractions in which the degree of the numerator is less than the degree of the denominator. The methods used in the examples which follow are based on the assumption that a proper fraction can be expressed as a sum of partial fractions which are also proper, such as

$$\frac{A}{x-a}, \quad \frac{Ax+B}{ax^2+bx+c}.$$

[A rigorous proof of this statement is beyond the scope of this course.]

Example 1 Express $\dfrac{2x-5}{(x+2)(x-1)}$ in partial fractions.

Let $$\frac{2x-5}{(x+2)(x-1)} \equiv \frac{A}{x+2} + \frac{B}{x-1}$$

then, multiplying both sides of the identity by $(x+2)(x-1)$,

$$2x - 5 \equiv A(x-1) + B(x+2)$$

Putting $x = 1$: $\quad 2 - 5 = 3B \quad \therefore \quad B = -1$

Putting $x = -2$: $\quad -4 - 5 = -3A \quad \therefore \quad A = 3$

$$\therefore \quad \frac{2x-5}{(x+2)(x-1)} \equiv \frac{3}{x+2} - \frac{1}{x-1}.$$

Example 2 Express $\dfrac{x+1}{2x^3 - 5x^2 + 2x}$ in partial fractions.

$$2x^3 - 5x^2 + 2x \equiv x(2x^2 - 5x + 2) \equiv x(x-2)(2x-1)$$

$$\therefore \text{ let } \quad \frac{x+1}{2x^3 - 5x^2 + 2x} \equiv \frac{A}{x} + \frac{B}{x-2} + \frac{C}{2x-1}$$

then $$x + 1 \equiv A(x-2)(2x-1) + Bx(2x-1) + Cx(x-2)$$

Putting $x = 0$: $1 = 2A$ $\quad\therefore\quad A = \frac{1}{2}$

Putting $x = 2$: $3 = 6B$ $\quad\therefore\quad B = \frac{1}{2}$

Putting $x = \frac{1}{2}$: $\frac{3}{2} = C \times \frac{1}{2} \times (-\frac{3}{2})$ $\quad\therefore\quad C = -2$

$$\therefore \quad \frac{x+1}{2x^3 - 5x^2 + 2x} \equiv \frac{1}{2x} + \frac{1}{2(x-2)} - \frac{2}{2x-1}.$$

So far we have determined unknown constants by substituting suitable values of x. In examples involving quadratic factors some constants can be obtained more quickly by comparing coefficients of powers of x.

Example 3 Express $\dfrac{4x^2}{(x-3)(x^2+3)}$ in partial fractions.

Let $$\frac{4x^2}{(x-3)(x^2+3)} \equiv \frac{A}{x-3} + \frac{Bx+C}{x^2+3}$$

then $$4x^2 \equiv A(x^2+3) + (Bx+C)(x-3)$$

Putting $x = 3$: $36 = 12A$ $\quad\therefore\quad A = 3$

Putting $x = 0$: $0 = 3A - 3C$ $\quad\therefore\quad C = 3$

Comparing coefficients of x^2:

$$4 = A + B \qquad \therefore \quad B = 1$$

$$\therefore \quad \frac{4x^2}{(x-3)(x^2+3)} \equiv \frac{3}{x-3} + \frac{x+3}{x^2+3}$$

In the next example we consider a fraction with the repeated factor $(x + 1)^2$ in the denominator. Using the "proper fraction" rule, there will be a corresponding partial fraction of the form $(Ax + K)/(x + 1)^2$. However, we obtained simpler partial fractions by writing $K = A + B$:

$$\frac{Ax+K}{(x+1)^2} \equiv \frac{Ax+A+B}{(x+1)^2} \equiv \frac{A(x+1)+B}{(x+1)^2} \equiv \frac{A}{x+1} + \frac{B}{(x+1)^2}.$$

Example 4 Express $\dfrac{x-5}{(x+1)^2(x-1)}$ in partial fractions.

Let $$\frac{x-5}{(x+1)^2(x-1)} \equiv \frac{A}{x+1} + \frac{B}{(x+1)^2} + \frac{C}{x-1}$$

then $$x - 5 \equiv A(x+1)(x-1) + B(x-1) + C(x+1)^2$$

Putting $x = 1$: $-4 = 4C$ $\quad\therefore\quad C = -1$

Putting $x = -1$: $-6 = -2B$ $\quad\therefore\quad B = 3$

Comparing coefficients of x^2:

$$0 = A + C \qquad \therefore \quad A = 1$$

$$\therefore \quad \frac{x-5}{(x+1)^2(x-1)} \equiv \frac{1}{x+1} + \frac{3}{(x+1)^2} - \frac{1}{x-1}$$

A fraction in which the degrees of the numerator and denominator are equal can be expressed as the sum of a constant and a proper fraction,

e.g. $$\frac{x^2+1}{x^2-1} \equiv 1 + \frac{2}{x^2-1}.$$

We use this fact in the next example.

Example 5 Express $\dfrac{x^2+2}{(x-1)(2x+1)}$ in partial fractions.

Let $$\frac{x^2+2}{(x-1)(2x+1)} \equiv A + \frac{B}{x-1} + \frac{C}{2x+1}$$

then $$x^2+2 \equiv A(x-1)(2x+1) + B(2x+1) + C(x-1)$$

Putting $x = 1$: $3 = 3B$ $\quad \therefore \quad B = 1$

Putting $x = -\frac{1}{2}$: $\frac{9}{4} = -\frac{3}{2}C$ $\quad \therefore \quad C = -\frac{3}{2}$

Comparing coefficients of x^2:

$$1 = 2A \qquad \therefore \quad A = \tfrac{1}{2}$$

$$\therefore \quad \frac{x^2+2}{(x-1)(2x+1)} \equiv \frac{1}{2} + \frac{1}{x-1} - \frac{3}{2(2x+1)}.$$

When dealing with fractions in which the degree of the numerator is greater than the degree of the denominator, we may either use long division to express the fraction as the sum of a polynomial and a proper fraction or extend the method of Example 5. If the degree of the numerator exceeds that of the denominator by n, then the expression in partial fractions must include a polynomial of degree n.

Example 6 Express $\dfrac{x^3-3x^2+1}{x^2-x-2}$ in partial fractions.

Method 1

$$\begin{array}{r|l} & x - 2 \\ \hline x^2-x-2 & x^3 - 3x^2 \qquad + 1 \\ & x^3 - x^2 - 2x \\ \hline & \quad -2x^2 + 2x + 1 \\ & \quad -2x^2 + 2x + 4 \\ \hline & \qquad\qquad -3 \end{array}$$

$$\therefore \quad \frac{x^3-3x^2+1}{x^2-x-2} \equiv x - 2 - \frac{3}{x^2-x-2}$$

Let $$\frac{-3}{x^2-x-2} \equiv \frac{A}{x+1} + \frac{B}{x-2}$$

then $\quad -3 \equiv A(x-2) + B(x+1)$

Putting $x = -1$: $\quad -3 = -3A \quad \therefore \quad A = 1$

Putting $x = 2$: $\quad -3 = 3B \quad \therefore \quad B = -1$

$$\therefore \quad \frac{x^3 - 3x^2 + 1}{x^2 - x - 2} \equiv x - 2 + \frac{1}{x+1} - \frac{1}{x-2}$$

Method 2

Let $$\frac{x^3 - 3x^2 + 1}{x^2 - x - 2} \equiv Ax + B + \frac{C}{x+1} + \frac{D}{x-2}$$

then $$x^3 - 3x^2 + 1 \equiv (Ax + B)(x+1)(x-2) + C(x-2) + D(x+1)$$

Putting $x = -1$: $\quad -3 = -3C \quad \therefore \quad C = 1$

Putting $x = 2$: $\quad -3 = 3D \quad \therefore \quad D = -1$

Putting $x = 0$: $\quad 1 = -2B - 2C + D \quad \therefore \quad B = -2$

Comparing coefficients of x^3:

$$1 = A.$$

$$\therefore \quad \frac{x^3 - 3x^2 + 1}{x^2 - x - 2} \equiv x - 2 + \frac{1}{x+1} - \frac{1}{x-2}.$$

We now summarise the rules for expressing a rational function in partial fractions.

(1) To a linear factor $(x - a)$ in the denominator, there corresponds a fraction of the form $A/(x - a)$.

(2) To a quadratic factor $(ax^2 + bx + c)$, there corresponds a fraction of the form $(Ax + B)/(ax^2 + bx + c)$.

(3) To a repeated factor $(x - a)^n$, there correspond n fractions of the form $A_1/(x - a), A_2/(x - a)^2, \ldots, A_n/(x - a)^n$.

(4) To a repeated factor $(ax^2 + bx + c)^n$, there correspond n fractions

$$\frac{A_1x + B_1}{(ax^2 + bx + c)}, \frac{A_2x + B_2}{(ax^2 + bx + c)^2}, \ldots, \frac{A_nx + B_n}{(ax^2 + bx + c)^n}.$$

[Note that there are many ways of expressing a rational function as a sum of two or more fractions e.g. $\dfrac{1}{(x+1)(x+2)} \equiv \dfrac{x+1}{x+2} - \dfrac{x}{x+1} \equiv \dfrac{x^2}{x+1} - \dfrac{x^2 + x - 1}{x+2}$.

However, the simplest and most useful form is the expression in partial fractions according to rules (1) to (4) above.]

Exercise 2.6

Express in partial fractions

1. $\dfrac{1}{(x+1)(x+2)}$

2. $\dfrac{4 - x}{(x-1)(x+2)}$

3. $\dfrac{3x-10}{(x-2)(x-4)}$

4. $\dfrac{4x-9}{x(x-3)}$

5. $\dfrac{x+7}{(x-2)(x+1)}$

6. $\dfrac{5x+3}{(x+1)(2x+1)}$

7. $\dfrac{x}{x^2-1}$

8. $\dfrac{5}{6-x-x^2}$

9. $\dfrac{10x-1}{(2x+1)(4x-1)}$

10. $\dfrac{4-2x}{(2x-1)(2x+5)}$

11. $\dfrac{17x+11}{(x-2)(x+3)(x+1)}$

12. $\dfrac{5x^2-12x-5}{(x^2-1)(x-2)}$

13. $\dfrac{3x+1}{(x+1)(x^2+1)}$

14. $\dfrac{x+2}{(2x-1)(x^2+1)}$

15. $\dfrac{3x+1}{x(2x^2+1)}$

16. $\dfrac{x^2-10}{(x^2+3)(2x-1)}$

17. $\dfrac{x^2-13}{(x-1)^2(x+2)}$

18. $\dfrac{3x^2+7x+1}{x^3+2x^2+x}$

19. $\dfrac{2x^2-3x-2}{x^3-x^2}$

20. $\dfrac{x^2+23}{(x+1)^3(x-2)}$

21. $\dfrac{2x^2-5x-5}{(2x^2+5)(4x-5)}$

22. $\dfrac{x+2}{(x-2)(x^2-x+2)}$

23. $\dfrac{x^2+1}{x^2-1}$

24. $\dfrac{x^2}{x^2-x-2}$

25. $\dfrac{x(x-2)}{(3x-1)(x-1)}$

26. $\dfrac{x^3}{x^2-4}$

27. $\dfrac{x^2-x}{(x^2+3)(x^2+2)}$

28. $\dfrac{3x^3+2x^2+2x-3}{(x^2+2)(x+1)^2}$

29. $\dfrac{2x^4-2x^3+x}{(2x-1)^2(x-2)}$

30. $\dfrac{x^6-x^5-4x^2+x}{x^4+3x^2+2}$

31. Express $f(x)=\dfrac{4x+5}{(x+1)(2x+3)}$ in partial fractions. Hence find $f'(x)$ and $f''(x)$.

32. Find the coordinates of the points of inflexion on the curves

(a) $y = \dfrac{2x}{1-x^2}$, (b) $y = \dfrac{7(x+1)}{x(x-7)}$.

Exercise 2.7 (miscellaneous)

1. The functions f, g, h and k are defined by

$$\begin{aligned} &f: x \to \sin^2 3x, && x \in \mathbb{R}, \\ &g: x \to 1/(1-x), && x \in \mathbb{R},\ x \neq 1, \\ &h: x \to \sqrt{(1-x^2)}, && x \in \mathbb{R},\ |x| \leqslant 1, \\ &k: x \to \operatorname{cosec} \tfrac{1}{2}x, && x \in \mathbb{R},\ x \neq 2n\pi (n \in \mathbb{Z}). \end{aligned}$$

For each function, state whether it is even, odd or neither. State also whether or not it is periodic, giving the period where appropriate.

2. Sketch the graphs of the following functions. State any values of x at which
(a) there is a discontinuity,
(b) the function is not differentiable.

$$\begin{aligned} &f(x) = |x-2|, && g(x) = [\tfrac{1}{2}x], \\ &h(x) = 1/(x^2 - x - 2), && k(x) = |x^4 - x^3|. \end{aligned}$$

3. The functions f and g are defined by $f: x \to \sin 2x, x \in \mathbb{R}$, $g: x \to \cot x$, $x \in \mathbb{R}$, $x \neq k\pi (k \in \mathbb{Z})$. State the periods of f and g. Find the period of the function $f.g$. On separate axes, sketch the graph of f, g and $f.g$ for the interval $\{x: -\pi < x < \pi, x \neq 0\}$. Find the range of the function $f.g$. (JMB)

4. Determine whether each of the three functions, f, g, h defined in $\mathbb{R}$ is an even function, an odd function or neither, given that $f(x) = \pi/2 - \sin(x/2)$, $g(x) = \pi/2 - \cos(x/3)$, $h(x) = \cos(\sin x)$. Determine the period, if it exists, of each of the two functions $G(x) = g(x) - f(x)$, $H(x) = g(x) - h(x) + \sin[f(2x)]$. (L)

5. Find the ranges of the given functions f, g, h and k. State whether f^{-1}, g^{-1}, h^{-1}, k^{-1} exist, giving brief reasons for each answer.

$$\begin{aligned} &f: x \to x^2 + 4, && x \in \mathbb{R},\ 0 \leqslant x \leqslant 4; \\ &g: x \to x^2 - 4x, && x \in \mathbb{R},\ 0 \leqslant x \leqslant 4; \\ &h: x \to 3\cos x + 4\sin x, && x \in \mathbb{R}; \\ &k: x \to 1 - \lg(x+1), && x \in \mathbb{R},\ x \geqslant 0. \end{aligned}$$

6. Sketch the graph of the function f defined by $f(x) = 4x/(1+x^2)$, x real. Let g denote the function f restricted to the domain $[-a, a]$, so that $g(x) = 4x/(1+x^2)$, $-a \leqslant x \leqslant a$, where a is the largest positive number such that the function g is one-one. What is the value of a? Obtain an expression for the inverse function g^{-1}, and state the domain and range of g^{-1}. (W)

7. The function ϕ is defined by $\phi(x) = x^3 + 2x - 1$, and the inverse function ϕ^{-1} is denoted by ψ. Find the values of $\psi(2)$ and $\psi'(2)$. (W)

8. The domain of the function f is the set $D = \{x : x \in \mathbb{R}, -2 < x < 3\}$. The function $f : D \to \mathbb{R}$ is defined by

$$f(x) = \begin{cases} 2x - 1 & \text{for} \quad -2 < x \leqslant 1, \\ x^2 & \text{for} \quad 1 < x \leqslant 2, \\ 10 - 3x & \text{for} \quad 2 < x < 3. \end{cases}$$

Find the range of this function and sketch its graph. Explain why there is no inverse function to f. Suggest an interval such that f, restricted to this interval, will have an inverse function. Give an expression for the inverse function in this case. (L)

9. Explain what is meant by the statement "f is a function from a set A to a set B." A function $f : \mathbb{R} \to \mathbb{R}$ is defined by

$$\begin{cases} f(x) = \sin 2x, & 0 \leqslant x \leqslant \pi/4, \\ f(x) = 2 - 4x/\pi, & \pi/4 < x \leqslant \pi/2, \\ f \text{ is an odd function, } f \text{ is periodic with period } \pi. \end{cases}$$

Sketch the graph of $y = f(x)$ for $-\pi \leqslant x \leqslant \pi$. Find two different intervals, each of length $\pi/2$, such that f, restricted to either interval, will have an inverse. In each case, give an expression for the inverse function. (L)

10. (i) f_1 is a periodic odd function with a period 2π and f_2 is a periodic even function with a period 3π. Functions F and G are defined by $F(x) = f_2(x) . f_1(x)$, $G(x) = f_2[f_1(x)]$. State whether the functions F and G are
(a) odd, even or neither,
(b) periodic, and if periodic state a period.
Functions f_3 and f_4 are defined by $f_3(x) = x \cos x$, $f_4(x) = \cos(\sin 3x)$. State whether f_3 and f_4 are odd, even or neither and if periodic state a period.
(ii) A mapping $f : A \to \mathbb{R}$, where $A = \{x : x \in \mathbb{R} \text{ and } -\pi/4 \leqslant x \leqslant \pi/4\}$, is defined by $f(x) = \cos x + \sin x$. Find $f^{-1}(1)$. (L)

11. A function f is defined on $\mathbb{R}$ by $f : x \to |x + [x]|$ where $[x]$ indicates the greatest integer less than or equal to x, e.g. $[3] = 3$, $[2{\cdot}4] = 2$, $[-3{\cdot}6] = -4$. Sketch the graph of the function for $-3 \leqslant x \leqslant 3$. What is the range of f? Is the mapping one-one?
The function g is defined by $g : x \to |x + [x]|, x \in \mathbb{R}^+, x \notin \mathbb{Z}^+$. Find the rule and domain of the inverse function g^{-1}. (C)

12. Sketch the curve $y = |x^2 - 5x|$. Find the ranges of values of x for which $|x^2 - 5x| < 6$.

13. Sketch the curves $y = x^2(3 - x)$ and $y^2 = x^2(3 - x)$, showing clearly the behaviour of the curves near the x-axis.

14. Using a sketch graph, or otherwise, solve the inequality $\left|\dfrac{1}{1+2x}\right| < 1$.

15. Prove that the curves $y = \dfrac{x}{x+1}$ and $y = \left(\dfrac{x}{x+1}\right)^2$ intersect at only one point.

Sketch these curves in the same diagram. Label your sketch so that it is immediately clear which curve is which. (O&C)

16. Given that $f(x) \equiv 6x^2 + x - 12$, find the minimum value of $f(x)$ and the values of x for which $f(x) = 0$. Using the same axes, sketch the curves $y = f(x)$ and $y = 1/f(x)$, labelling each clearly. Deduce that there are four values of x for which $[f(x)]^2 = 1$. Find these values, each to two decimal places. (L)

17. The polynomial $P(x)$ leaves a remainder of 2 when divided by $(x - 1)$ and a remainder of 3 when divided by $(x - 2)$. The remainder when $P(x)$ is divided by $(x - 1)(x - 2)$ is $ax + b$. By writing $P(x) \equiv (x - 1)(x - 2)Q(x) + ax + b$, find the values of a and b. Given also that $P(x)$ is a cubic polynomial with coefficient of x^3 equal to unity, and that -1 is a root of the equation $P(x) = 0$, obtain $P(x)$. Show that the equation $P(x) = 0$ has no other real roots. (JMB)

18. Find the remainder when the polynomial $P(x) = x^3 - 5x + 2$ is divided by
(a) $x^2 + 1$,
(b) $(x - 1)^2$.

19. Using the remainder theorem, or otherwise, show that $x + a + b + c$ is a factor of

$$(x + a)(x + b)(x + c) + (b + c)(c + a)(a + b).$$

Hence, or otherwise, solve the equation

$$(x + 2)(x - 3)(x - 1) + 4 = 0.$$ (C)

20. Factorise completely the expressions
(a) $x(y^3 - z^3) + y(z^3 - x^3) + z(x^3 - y^3)$,
(b) $12abc + 4(a^3 + b^3 + c^3) - (a + b + c)^3$.

21. If $f(x)$ is a polynomial in x, show that when $f(x)$ is divided by $x - a$ the remainder is $f(a)$.
When $x^3 + ax^2 + bx + c$ is divided by $x + 3$ the remainder is -26 and when divided by $x^2 - x - 2$ the remainder is 14. Find the values of a, b and c.
(AEB 1978)

22. Show that if a polynomial $f(x)$ is divided by $(x - a)(x - b)$, where $a \neq b$, then the remainder is $\left\{\dfrac{f(a) - f(b)}{a - b}\right\}x + \left\{\dfrac{af(b) - bf(a)}{a - b}\right\}$.

23. Solve the equation $x^4 - 4x^3 - 20x^2 + 32 = 0$ given that it has a repeated root.

24. Given that the polynomial $f(x) = x^4 - 8x^3 + 10x^2 + p$ has a repeated linear factor, find the possible values of p.

25. The equation $x^3 - px^2 + qx - r = 0$ is such that its roots are in geometric progression. Show that
(a) one root is $\sqrt[3]{r}$,
(b) $p^3r = q^3$.

26. If the expression $2x^2 - xy - y^2 + 2x + ky - 4$ is the product of two linear factors, find the possible values of k.

27. (i) Given that $x + 1$ is a factor of $x^3 + 2x^2 + 3x + c$, find c.
(ii) In the algebraic operation of polynomial division, one polynomial $f(x)$ is divided by another polynomial $g(x)$ of order equal to or lower than that of $f(x)$. Polynomials $Q(x)$ and $R(x)$ are found such that $f(x) = g(x)Q(x) + R(x)$ and $R(x)$ is of lower order than $g(x)$. Carry out the process described in the flow chart when $A(x) = x^3 + 2x^2 + 3x + 2$ and $B(x) = x^2 - x - 2$.

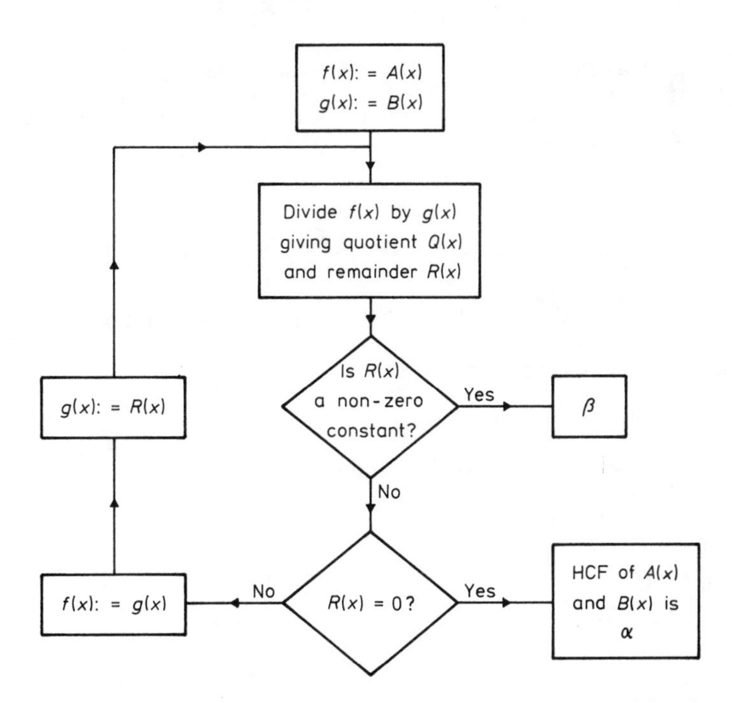

State what should be written in the spaces α and β for the flow chart to be applicable to any two polynomials of suitable degree. (AEB 1980)

28. By considering the equation $yx^2 + (y - 1)x + (y - 1) = 0$, where $y \neq 0$, as a quadratic equation in x, find the range of values of y for which this equation has real roots. State the value of x which satisfies this equation when $y = 0$. Using these results, or otherwise, determine the range of the function $f: x \rightarrow \dfrac{x + 1}{x^2 + x + 1}$, where x is real. Sketch the graph of $f(x)$ giving the coordinates of any turning points. Explain why f has no inverse function. (JMB)

29. In each of the following cases find the range of the function f. Sketch the graph of f, showing clearly any asymptotes and turning points.

(a) $f(x) = \dfrac{x - 1}{2x + 1}$, (b) $f(x) = \dfrac{4}{(1 - x)(3 + x)}$,

(c) $f(x) = \dfrac{3(x + 1)}{x(x - 3)}$, (d) $f(x) = \dfrac{2x - 1}{x^2 - 4}$.

30. Sketch each of the following curves, showing any stationary points and indicating clearly the form of the curve when $|x|$ becomes large.

(a) $y = x - \dfrac{1}{x + 1}$. (b) $y = \dfrac{x^2}{x - 1}$,

(c) $y = \dfrac{x^2}{1 + x^2}$, (d) $y = \dfrac{x - 1}{(x + 1)^2}$.

31. State the maximum domain and determine the range of the function $f(x) = \dfrac{x^2 + 2x + 3}{x^2 - 2x - 3}$. Sketch the curve $y = \dfrac{x^2 + 2x + 3}{x^2 - 2x - 3}$ indicating clearly any asymptotes. (O)

32. (a) The function f is defined by $f(x) = \dfrac{x}{1 + x^2}$. Show that $f'(x) = \dfrac{1 - x^2}{(1 + x^2)^2}$ and $f''(x) = \dfrac{-6x + 2x^3}{(1 + x^2)^3}$. Find the local maxima, minima and points of inflexion of f, and sketch the graph of f.

(b) The function g is defined by $g(x) = \dfrac{x^2}{1 - x^2}$. Show that $g(x)$ cannot take values between -1 and 0. Sketch the graph of g. (W)

Express in partial fractions:

33. $\dfrac{3x + 11}{(3x - 2)(2x + 3)}$ 34. $\dfrac{3x - 2}{(x - 1)^2(2x - 1)}$

35. $\dfrac{x}{x^3 - x^2 + x - 1}$ 36. $\dfrac{2}{x^3 - 2x^2 - x + 2}$

37. $\dfrac{5 + x^2}{3 + 5x - 2x^2}$ 38. $\dfrac{3x}{1 - x^3}$

39. $\dfrac{x(x + 1)(x - 2)}{(x - 1)(x + 2)}$ 40. $\dfrac{7x^2 - 2x + 3}{(x^2 - 1)^2}$

3 Exponential and logarithmic functions

3.1 Exponential functions

Exponent is another word for index. A function such as 2^x, in which the index is variable, is called *an exponential function*.

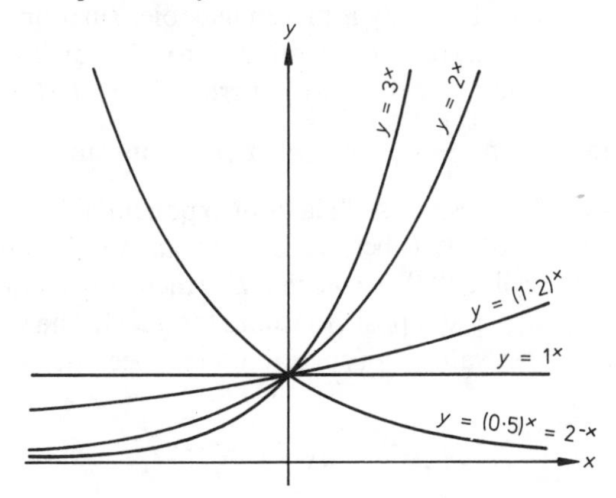

The diagram shows the graphs of $y = a^x$ for various positive values of a. We see that each graph lies entirely above the x-axis and cuts the y-axis at the point (0, 1).

We now consider how to differentiate the function a^x.

Let $$f(x) = a^x, \quad \text{then} \quad f(x+h) = a^{x+h}$$

$$\therefore \quad f(x+h) - f(x) = a^{x+h} - a^x = a^x(a^h - 1)$$

$$\therefore \quad f'(x) = \lim_{h\to 0} \frac{f(x+h) - f(x)}{h}$$

$$= \lim_{h\to 0} \frac{a^x(a^h - 1)}{h} = a^x \lim_{h\to 0}\left(\frac{a^h - 1}{h}\right).$$

In particular, $f'(0) = a^0 \lim_{h\to 0}\left(\dfrac{a^h - 1}{h}\right) = \lim_{h\to 0}\left(\dfrac{a^h - 1}{h}\right)$. Since $f'(0)$ is the gradient of the curve $y = a^x$ at the point (0, 1), it seems reasonable to assume that this limit exists and is equal to some constant k.

Thus, if $f(x) = a^x$, then $f'(x) = ka^x$, where the value of k depends on the value of a.

It follows that if $y = a^x$, then $\dfrac{dy}{dx} = ky$, i.e. the rate of change of y is proportional to y.

Because of this property the theory of exponential functions has many important applications. There are various scientific laws which state that the rate of change of a certain quantity is proportional to the quantity itself. We give here three simple examples of such laws.

(1) *Law of growth.* If a population of simple organisms such as bacteria is allowed to increase without restriction, then the rate of growth is proportional to the size of the population. If n is the number of organisms present at time t, then $\dfrac{dn}{dt} = kn$, where k is constant.

(2) *Law of radio-active decay.* The rate at which a radio-active substance decays is proportional to the quantity of the substance present. If a mass m is present at time t, then $\dfrac{dm}{dt} = -km$, where k is constant.

(3) *Newton's law of cooling.* If a body is placed in cooler surroundings then its temperature decreases at a rate proportional to the difference between the temperature of the body and that of its surroundings. If T is the temperature difference at time t, then $\dfrac{dT}{dt} = -kT$, where k is constant.

We arrive at a way of expressing such laws in exponential form by looking again at graphs of the form $y = a^x$. If, as before, the constant k represents the gradient of a particular curve at the point (0, 1), we see that as a increases, k also increases, taking negative values for $0 < a < 1$ and positive values for $a > 1$. The value of a for which $k = 1$ is denoted by e. This value is of special interest because when

$$y = e^x, \quad \frac{dy}{dx} = e^x = y.$$

It can be shown that e is an irrational number approximately equal to 2·718.

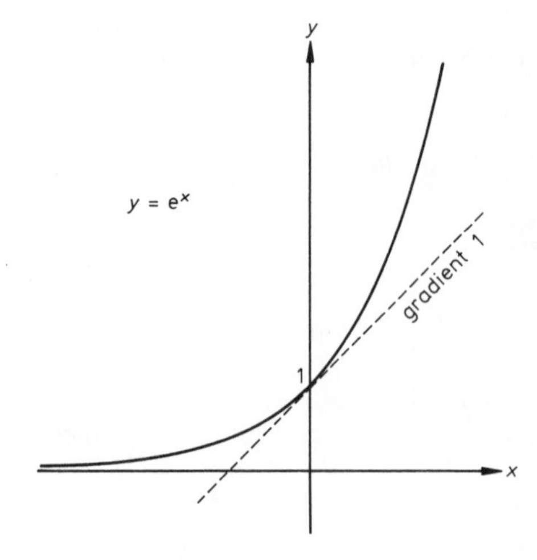

The function e^x is called *the exponential function* and is sometimes written exp x. [In numerical work values of e^x are found using a scientific calculator or mathematical tables.]

The result $\dfrac{d}{dx}(e^x) = e^x$ can be extended as follows.

Let $\quad y = e^u, \quad \text{where} \quad u = kx$

then
$$\frac{dy}{du} = e^u \quad \text{and} \quad \frac{du}{dx} = k$$
$$\frac{dy}{dx} = \frac{dy}{du} \times \frac{du}{dx} = ke^u.$$

Hence, if $y = e^{kx}$, then $\dfrac{dy}{dx} = ke^{kx} = ky$.

It is now possible to express the scientific laws mentioned earlier in exponential form:

(1) $n = n_0 e^{kt}$, (2) $m = m_0 e^{-kt}$, (3) $T = T_0 e^{-kt}$,

where n_0, m_0 and T_0 are the values of the variables when $t = 0$.

Some useful results to remember:

$$\frac{d}{dx}(e^x) = e^x, \qquad \int e^x\,dx = e^x + c$$
$$\frac{d}{dx}(e^{kx}) = ke^{kx}, \qquad \int e^{kx}\,dx = \frac{1}{k}e^{kx} + c.$$

Example 1 Differentiate e^{x^2+1} with respect to x.

Let $\quad y = e^u, \quad \text{where} \quad u = x^2 + 1$

$$\frac{dy}{du} = e^u, \qquad \frac{du}{dx} = 2x$$
$$\frac{dy}{dx} = \frac{dy}{du} \times \frac{du}{dx} = e^u \times 2x = 2xe^{x^2+1}.$$

[Students thoroughly familiar with the differentiation of composite functions may write:

$$\frac{d}{dx}(e^{x^2+1}) = e^{x^2+1}\frac{d}{dx}(x^2 + 1) = 2xe^{x^2+1}.]$$

Example 2 Find $\displaystyle\int \sin x\, e^{\cos x}\,dx$.

$$\frac{d}{dx}(e^{\cos x}) = e^{\cos x}\frac{d}{dx}(\cos x) = -\sin x\, e^{\cos x}$$

$$\therefore \quad \int \sin x\, e^{\cos x}\,dx = -e^{\cos x} + c.$$

Example 3 Sketch the curve $y = x^2e^{-x}$.

$x = 0 \Leftrightarrow x^2e^{-x} = 0 \Leftrightarrow y = 0$

$\therefore$ the curve cuts the x- and y-axes at the origin.
Differentiating (using the product rule),

$$\frac{dy}{dx} = x^2(-e^{-x}) + 2xe^{-x} = (-x^2 + 2x)e^{-x} = x(2 - x)e^{-x}$$

$$\therefore \quad \frac{dy}{dx} = 0 \Leftrightarrow x(2-x)e^{-x} = 0 \Leftrightarrow x = 0 \quad \text{or} \quad x = 2.$$

	$x < 0$	$0 < x < 2$	$x > 2$
x	$-$	$+$	$+$
$2 - x$	$+$	$+$	$-$
e^{-x}	$+$	$+$	$+$
$\frac{dy}{dx}$	$-$	$+$	$-$

$\therefore$ when $x = 0$, there is a minimum point $(0, 0)$,
when $x = 2$, there is a maximum point $(2, 4e^{-2})$.

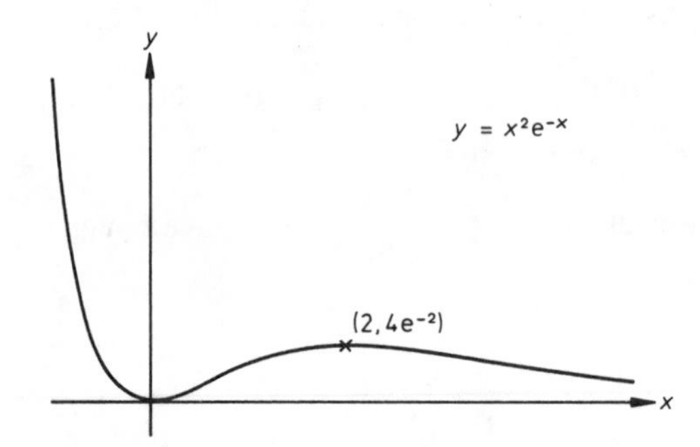

As $x \to -\infty$, $y \to \infty$.
As x increases through positive values, e^x increases at a much greater rate than x^2,
$\therefore$ as $x \to \infty$, $y \to 0$.

[The behaviour of the function x^ne^{-x} as $x \to \infty$ is also considered in *Exercise 9.4*, question 12.]

Exercise 3.1

In questions 1 to 7 differentiate the given functions with respect to x.

1. (a) e^{2x} (b) $5e^{-x}$ (c) e^{3x+5}

2. (a) e^{2x^3} (b) $e^{\sqrt{x}}$ (c) $e^{-1/x}$

3. (a) $e^{\sin x}$ (b) $e^{\cos 2x}$ (c) $e^{4 \tan x}$

4. (a) xe^{x^2} (b) x^2e^{5x} (c) $e^{x \cos x}$

5. (a) $e^{2x}\cos 3x$ (b) $e^{-x^2}\sin x$ (c) $(x+1)^3 e^{x/2}$

6. (a) $\sqrt{(1-e^{4x})}$ (b) $e^{(e^x)}$ (c) $\sin(e^x)$

7. (a) $\frac{1}{2}(e^x - e^{-x})^2$ (b) $\dfrac{e^x}{1+e^{-x}}$ (c) $\dfrac{e^{\sin^2 x}}{e^{-\cos^2 x}}$

8. Find (a) $\int e^{3x}\,dx$, (b) $\int e^{-x}\,dx$, (c) $\int e^{x+4}\,dx$.

9. Evaluate, leaving your answers in terms of e,

(a) $\int_0^1 6e^{2x}\,dx$ (b) $\int_{-2}^0 e^{(-x/2)}\,dx$, (c) $\int_{-1}^1 \{3/e^{3(x-1)}\}\,dx$.

10. Find dy/dx in terms of x when y is defined implicitly by:
(a) $e^y = x$, (b) $xe^y = 1$, (c) $e^{x+y} = x^2$.

11. If $y = e^x \sin x$, show that $\dfrac{d^2y}{dx^2} - 2\dfrac{dy}{dx} + 2y = 0$.

12. Find the equation of the tangent to the curve $y = e^{3x-5}$ at the point where $x = 2$.

13. Find in terms of e the area enclosed by the curve $y = e^{1-x}$, the axes and the line $x = 1$.

14. By considering the graphs of $y = e^x$ and $y = e^{-x}$, or otherwise, make rough sketches of:
(a) $y = e^x + e^{-x}$, (b) $y = e^x - e^{-x}$, (c) $y = e^x - x$.

In questions 15 to 20 sketch the given curves. Give the coordinates of any turning points and show clearly the behaviour of the curve when $|x|$ is large.

15. $y = xe^{-x}$. 16. $y = (x+1)e^x$. 17. $y = xe^{-x^2/2}$.

18. $y = (x-1)^2 e^{2x}$. 19. $y = e^x/x$. 20. $y = e^{1-\cos x}$.

21. A particle moves in a straight line so that after t seconds its distance from a fixed point O is s metres, where $s = t^2 e^{2-t}$. Find the distance of the particle from O when it first comes to rest and its acceleration at that point.

22. Find the values of x for which the function $(x^2 - 2x - 1)e^{2x}$ has maximum or minimum values, distinguishing between them.

23. Find the values of x between 0 and 2π for which the function $e^x \cos x$ has maximum or minimum values, distinguishing between them.

24. Find the area of the finite region bounded by the curves $y = e^{2x}$, $y = e^{-x}$ and the line $x = 1$. If this region is rotated completely about the x-axis, find the volume of the solid formed. (AEB 1976)

3.2 Logarithmic functions

If a and x are positive real numbers such that $x = a^p$, then p is the logarithm of x to the base a,

i.e. $$x = a^p \Leftrightarrow \log_a x = p.$$

The laws of logarithms can be summarised as follows:

$$\log_a xy = \log_a x + \log_a y$$
$$\log_a x/y = \log_a x - \log_a y$$
$$\log_a x^n = n \log_a x$$
$$(\log_a b)(\log_b c) = \log_a c$$

The relationship between logarithms to different bases leads to two useful formulae:

$$\log_a x = \frac{1}{\log_x a}, \quad \log_a x = \frac{\log_b x}{\log_b a}.$$

Now that the exponential function e^x has been defined, we can introduce logarithms to the base e. These are called *natural logarithms* or *Napierian logarithms*†. The natural logarithm of x may be written $\log_e x$, but is more often denoted by $\ln x$.

Thus $$x = e^p \Leftrightarrow \ln x = p.$$

In numerical work values of $\ln x$ are found using a calculator or mathematical tables.

Example 1 Solve the equation $e^x - 2 - 3e^{-x} = 0$.

Letting $y = e^x$,
$$\begin{aligned} & e^x - 2 - 3e^{-x} = 0. \\ \Rightarrow\ & y - 2 - 3/y = 0 \\ \Rightarrow\ & y^2 - 2y - 3 = 0 \\ \Rightarrow\ & (y-3)(y+1) = 0 \\ \Rightarrow\ & y = 3 \quad \text{or} \quad y = -1 \\ \Rightarrow\ & e^x = 3, \quad \text{since} \quad e^x > 0 \\ \Rightarrow\ & x = \ln 3 \approx 1{\cdot}1 \end{aligned}$$

To investigate the properties of logarithmic functions we use our knowledge of exponential functions.

If $f(x) = a^x$, then $y = f^{-1}(x) \Rightarrow f(y) = x \Rightarrow a^y = x \Rightarrow y = \log_a x$.

†*Napier, John* (1550–1617) Scottish mathematician. He was the inventor of logarithms and spent 25 years producing tables. He also invented the calculating apparatus known as "Napier's Bones".

Hence the function $\log_a x$ is the inverse of the function a^x.

Thus $$a^{\log_a x} = x \quad \text{and} \quad \log_a(a^x) = x.$$

It also follows that the graph of $y = \log_a x$ is the reflection in the line $y = x$ of the graph of $y = a^x$.

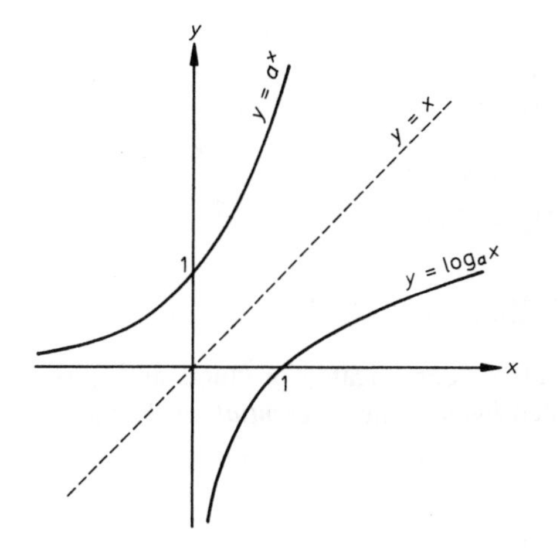

Since $\log_a x$ is defined only when $x > 0$, there are no points on the graph of $y = \log_a x$ for $x \leqslant 0$.
When $x = 1, y = \log_a 1 = 0$
$\therefore$ the graph of $y = \log_a x$ passes through the point $(1, 0)$ for all values of a.

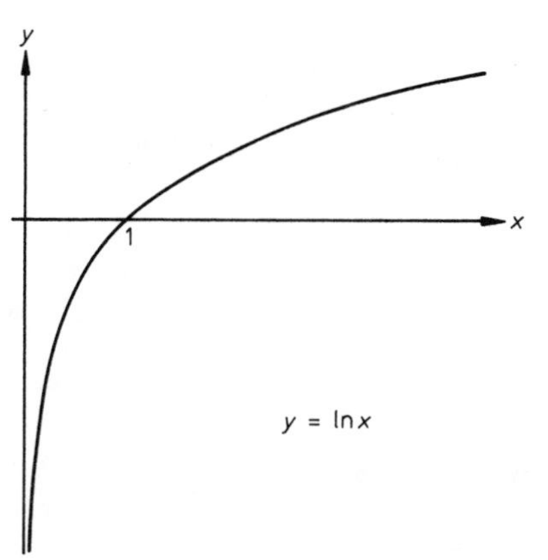

If $y = \ln x$, then $e^y = x$.

Differentiating with respect to x,

$$\frac{d}{dx}(e^y) = 1$$

But $$\frac{d}{dx}(e^y) = \frac{d}{dy}(e^y)\frac{dy}{dx}$$

$$= e^y\frac{dy}{dx} = x\frac{dy}{dx}$$

$$\therefore \quad x\frac{dy}{dx} = 1 \quad \text{i.e.} \quad \frac{dy}{dx} = \frac{1}{x}.$$

Hence, $$\boxed{\text{for } x > 0, \ \frac{d}{dx}(\ln x) = \frac{1}{x} \ \text{ and } \ \int \frac{1}{x}\,dx = \ln x + c.}$$

Example 2 Differentiate $\ln(x^2 + 1)$ with respect to x.

Let $y = \ln u$, where $u = x^2 + 1$

$$\frac{dy}{du} = \frac{1}{u} \qquad \frac{du}{dx} = 2x$$

$$\therefore \quad \frac{dy}{dx} = \frac{dy}{du} \times \frac{du}{dx} = \frac{1}{u} \times 2x = \frac{2x}{x^2+1}.$$

Some problems may be simplified by using the laws of logarithms.

Example 3 If $y = \ln\left(\frac{2x+1}{1-3x}\right)$, find $\frac{dy}{dx}$.

$$y = \ln\left(\frac{2x+1}{1-3x}\right) = \ln(2x+1) - \ln(1-3x)$$

$$\therefore \quad \frac{dy}{dx} = \frac{2}{2x+1} - \frac{(-3)}{1-3x} = \frac{2(1-3x)+3(2x+1)}{(2x+1)(1-3x)}$$

$$= \frac{5}{(2x+1)(1-3x)}$$

Some functions involving variable exponents or complicated products and quotients are differentiated using a process called *logarithmic differentiation.* The next two examples illustrate the method.

Example 4 Differentiate 2^x with respect to x.

Let $y = 2^x$, then $\ln y = \ln(2^x) = x \ln 2$.

Differentiating, $\frac{1}{y}\frac{dy}{dx} = \ln 2$

$$\therefore \quad \frac{dy}{dx} = y \ln 2 = 2^x \ln 2.$$

Example 5 If $y = \frac{x}{\sqrt{(x^2-2)}}$, find $\frac{dy}{dx}$.

$$\ln y = \ln x - \ln\sqrt{(x^2-2)} = \ln x - \tfrac{1}{2}\ln(x^2-2)$$

$$\therefore \quad \frac{1}{y}\frac{dy}{dx} = \frac{1}{x} - \frac{1}{2}\left(\frac{2x}{x^2-2}\right) = \frac{x^2-2-x^2}{x(x^2-2)} = \frac{-2}{x(x^2-2)}$$

$$\therefore \quad \frac{dy}{dx} = y\left\{\frac{-2}{x(x^2-2)}\right\} = \frac{x}{\sqrt{(x^2-2)}}\left\{\frac{-2}{x(x^2-2)}\right\} = -\frac{2}{(x^2-2)^{3/2}}$$

Example 6 Sketch the curve $y = \frac{1}{x}\ln x$.

$\ln x$ is not defined for $x \leqslant 0$,

$\therefore$ the curve lies entirely to the right of the y-axis.

$$y = 0 \Rightarrow \frac{1}{x}\ln x = 0 \Rightarrow x = 1$$

$\therefore$ the curve cuts the x-axis at the point $(1, 0)$.
Differentiating (using the product rule),

$$\frac{dy}{dx} = \frac{1}{x} \times \frac{1}{x} + \left(-\frac{1}{x^2}\right) \ln x = \frac{1}{x^2}(1 - \ln x)$$

$$\therefore \quad \frac{dy}{dx} = 0 \Leftrightarrow 1 - \ln x = 0 \Leftrightarrow x = e$$

When $x < e, \dfrac{dy}{dx} > 0$ and when $x > e, \dfrac{dy}{dx} < 0$,
$\therefore$ the point $(e, 1/e)$ is a maximum point.

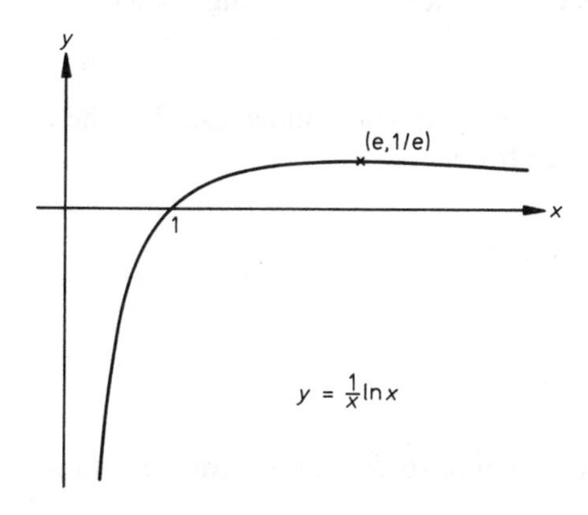

As $x \to 0, 1/x \to \infty$ and $\ln x \to -\infty$
$\therefore \quad y \to -\infty$.
When x is large, the rate of increase of $\ln x$ is very small
$\therefore$ as $x \to \infty$, $y \to 0$.
[The behaviour of the function $\dfrac{1}{x}\ln x$ as $x \to \infty$ is also considered in *Exercise 9.4*, question 12.]

Exercise 3.2

In questions 1 to 7 differentiate the given functions with respect to x.

1. (a) $\ln 3x$, (b) $\ln(x + 3)$, (c) $\ln(2x - 1)$.

2. (a) $\ln(x^3 + 4)$, (b) $\ln(\sin 2x)$, (c) $\ln(\sec x)$.

3. (a) $x^2 \ln x$, (b) $(\ln x)/x^2$, (c) $\ln(\ln x)$.

4. (a) $\ln(x^3)$, (b) $\ln\sqrt{(4x + 5)}$, (c) $x \ln(1/x)$.

5. (a) $\ln(\sec x + \tan x)$, (b) $\ln(x^2 + 4)^2$, (c) $\log_{10} x$.

6. (a) $\ln\left(\dfrac{x^2}{3x - 2}\right)$, (b) $\ln\sqrt{\left(\dfrac{x}{1 + x}\right)}$, (c) $\ln\dfrac{\cos x}{\sqrt{(1 - x^2)}}$.

7. (a) $e^{2 \ln x}$, (b) $\ln(e^{\tan x})$, (c) $\ln(3x^2 e^{-x})$.

8. Find the equation of the tangent to the curve $y = \ln(3x - 5)$ at the point where $x = 2$.

9. If $x = 1 + \ln t$ and $y = \ln(1 + t)$, find dy/dx in terms of t.

10. By considering the graph of $y = \ln x$, or otherwise, sketch the following:
(a) $y = \ln(-x)$, (b) $y = \ln(1/x)$, (c) $y = \ln(1 + x)$.

11. Given that $\ln 2 = p$ and $\ln 3 = q$, express the following in terms of p and q.
(a) $\ln 12e$, (b) $\log_3 2$, (c) $\log_6 e$, (d) 6.

12. Solve the following equations
(a) $e^{3t} + 2e^t = 3e^{2t}$, (b) $2e^x + 5 = 3e^{-x}$.

13. Sketch the following curves, giving the coordinates of any turning points.
(a) $y = |\ln x|$, (b) $y = 1/\ln x$, (c) $y = (\ln x)^2$.

14. Show that the curve $y = x - \ln x$ has one turning point. Show also that there are no points of inflexion on the curve. Sketch the curve.

15. If $y = \ln(x^2 - 5)$, show that $\dfrac{d^2y}{dx^2} + \left(\dfrac{dy}{dx}\right)^2 = 2e^{-y}$.

16. If $y = \sin 2x \ln(\tan x)$, show that $\dfrac{d^2y}{dx^2} + 4y = 4\cot 2x$.

In questions 17 to 20 use logarithmic differentiation to find the derivatives of the given functions.

17. (a) 3^x, (b) 5^{2x}, (c) $10^{\sin x}$.

18. (a) x^x, (b) $x^{\ln x}$, (c) $(\ln x)^x$.

19. (a) $\dfrac{x^2 + 1}{(2x + 1)^2}$, (b) $\dfrac{x^3}{\sqrt{(1 - x)}}$, (c) $\dfrac{(x - 1)^2 e^{4x}}{(x + 1)^2}$

20. (a) $\sqrt{\left\{\dfrac{(x + 2)^3}{1 - 5x}\right\}}$, (b) $\sqrt[3]{\left\{\dfrac{x^2 - 2}{x^2 + 4}\right\}}$, (c) $\dfrac{\sqrt{(\sin 2x)}}{\cos^3 x}$.

3.3 ln x defined in integral form

In this section we consider a more rigorous approach to the theory of exponential and logarithmic functions. In §3.1 we assumed with little justification that a^x is a continuous differentiable function defined for all real values of x. However, any elementary definition of a^x is valid only for rational values of x. Similarly, the laws of indices and the rule for differentiating x^n can be proved by elementary methods only for rational values of the exponents. We avoid these difficulties by using definitions of $\ln x$ and $\exp x$ which are valid for both rational and irrational values of x.

We define $\ln x$ as a definite integral by writing:

$$\ln x = \int_1^x \frac{1}{t}\,dt, \quad \text{where } x \text{ is real and positive.}$$

It follows that when $x = 1$, $\ln x = 0$. Other values of $\ln x$ are related to areas under the curve $y = 1/t$ as shown in the diagrams below.

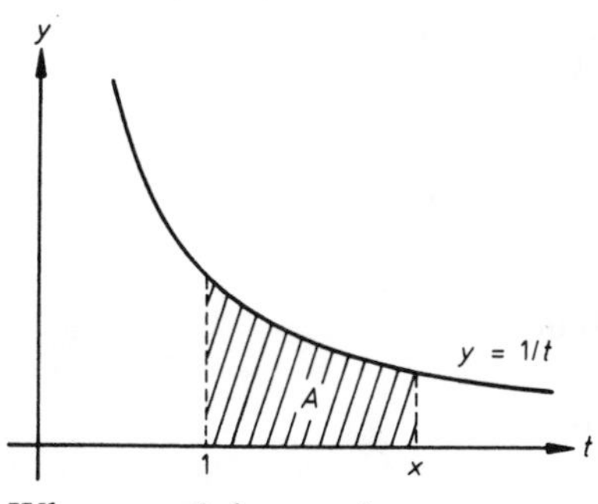

When $x > 1$, $\ln x = A$.

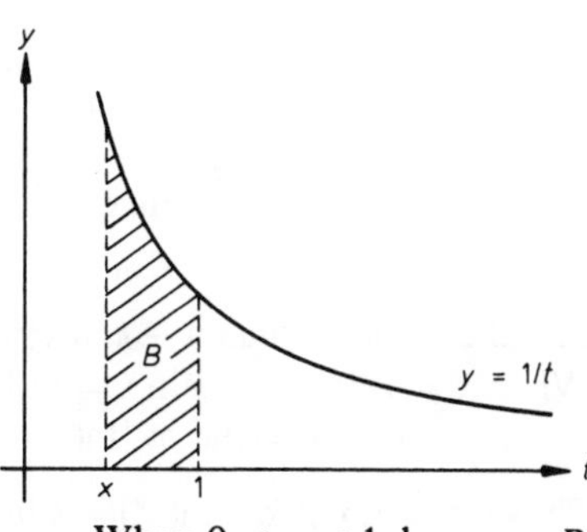

When $0 < x < 1$, $\ln x = -B$.

Thus the domain of the function $\ln x$ is the set of positive real numbers and its range is the set of all real numbers.

We can now define $\exp x$ for all real values of x by letting the function $\exp x$ be the inverse of the function $\ln x$, so that

$$y = \exp x \Leftrightarrow \ln y = x$$

To differentiate the function $\ln x$, let $f(x) = \ln x$, then

$$f(x+h) - f(x) = \ln(x+h) - \ln x$$
$$= \int_1^{x+h} \frac{1}{t}\,dt - \int_1^x \frac{1}{t}\,dt = \int_x^{x+h} \frac{1}{t}\,dt.$$

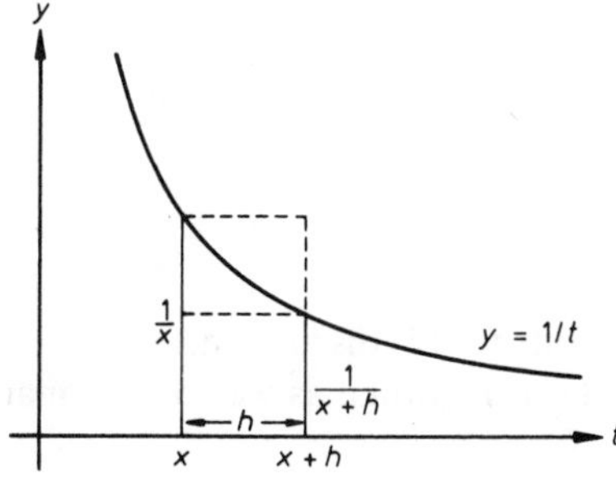

From the sketch we see that

$$h \times \frac{1}{x+h} < \int_x^{x+h} \frac{1}{t}\,dt < h \times \frac{1}{x}$$

$$\therefore \quad \frac{1}{x+h} < \frac{f(x+h) - f(x)}{h} < \frac{1}{x}$$

As $h \to 0$, $\dfrac{1}{x+h} \to \dfrac{1}{x}$,

$$\therefore \quad f'(x) = \lim_{h\to 0} \frac{f(x+h) - f(x)}{h} = \frac{1}{x}$$

i.e.
$$\frac{d}{dx}(\ln x) = \frac{1}{x}$$

To differentiate the function $\exp x$, let $y = \exp x$, then

$$\ln y = x \;\Rightarrow\; \frac{1}{y}\frac{dy}{dx} = 1 \;\Rightarrow\; \frac{dy}{dx} = y$$

i.e.

$$\frac{d}{dx}(\exp x) = \exp x$$

We develop the theory further by considering $\ln(x^k)$, where k is any rational number ($k \in \mathbb{Q}$).

$$\frac{d}{dx}\{\ln(x^k)\} = \frac{1}{x^k} \times kx^{k-1} = \frac{k}{x}$$

$$\therefore \quad \frac{d}{dx}\{\ln(x^k) - k\ln x\} = \frac{k}{x} - k \times \frac{1}{x} = 0$$

$\therefore$ the value of $\{\ln(x^k) - k\ln x\}$ must be constant.
When $x = 1$, $\ln(x^k) - k\ln x = \ln 1 - k\ln 1 = 0$,
$\therefore$ the constant value of $\{\ln(x^k) - k\ln x\}$ is zero.

Thus for rational k, $\quad \ln(x^k) = k\ln x$
and hence $\quad x^k = \exp(k\ln x) \qquad (1)$

The constant e is now defined as the real number such that $\ln e = 1$. Substituting $x = e$ in equation (1) we have

$$e^k = \exp(k\ln e) = \exp k, \quad \text{for rational } k.$$

Thus it seems reasonable to give a meaning to the expression e^x when x is irrational by letting $e^x = \exp x$ for all real x.
It follows from this definition that $\log_e x = \ln x$.
More generally, it is consistent with equation (1) to define a^x by writing

$$\boxed{a^x = \exp(x\ln a) \quad \text{for all real } x}$$

It is a consequence of this definition that

$$y = \log_a x \;\Rightarrow\; a^y = x \;\Rightarrow\; \exp(y\ln a) = x \;\Rightarrow\; y\ln a = \ln x$$

$$\therefore \quad \boxed{\log_a x = (\ln x)/(\ln a).}$$

We conclude by summarising the important points in this discussion.
We needed definitions of e^x and $a^x (a > 0)$ valid for irrational as well as rational values of x. We found that using the definitions

$$\ln x = \int_1^x \frac{1}{t}\,dt, \quad y = \exp x \;\Leftrightarrow\; \ln y = x,$$

it followed that, for rational values of k,

$$e^k = \exp k, \quad a^k = \exp(k\ln a).$$

Thus suitable definitions of e^x and a^x are

$$e^x = \exp x, \quad a^x = \exp(x \ln a).$$

We deduce from these definitions that

$$\log_e x = \ln x, \quad \log_a x = (\ln x)/(\ln a).$$

These results provide a firmer basis for the work of §3.1 and §3.2.

[Note that the theory of exponential and logarithmic functions defined in terms of $\exp x$ and $\ln x$ is extended in *Exercise 3.3* and in *Exercise 4.6*, question 11.]

Exercise 3.3

1. (a) Use the result $\dfrac{d}{dx}(\ln x) = \dfrac{1}{x}$ to prove that

$$\frac{d}{dx}(\ln xy - \ln x - \ln y) = 0.$$

(b) Deduce that $\ln xy = \ln x + \ln y$.
(c) By substituting $u = \ln x$, $v = \ln y$, show that

$$(\exp u)(\exp v) = \exp(u + v).$$

2. Use the methods of question 1 to show that
(a) $\ln(x/y) = \ln x - \ln y$.
(b) $(\exp u)/(\exp v) = \exp(u - v)$.

3. Use the definition $x^p = \exp(p \ln x)$ or the equivalent statement $\ln(x^p) = p \ln x$ to prove that, for all real values of p,

$$y = x^p \quad \Rightarrow \quad \frac{dy}{dx} = px^{p-1}.$$

3.4 Integration using logarithms

By differentiating logarithmic functions we can obtain various useful integration results.

For $x > 0$, $\dfrac{d}{dx}(\ln x) = \dfrac{1}{x}$,

$\therefore$ for $x > 0$,

$$\int \frac{1}{x}\,dx = \ln x + c. \qquad (1)$$

For $x < 0$, $\dfrac{d}{dx}\{\ln(-x)\} = \dfrac{1}{(-x)} \times (-1) = \dfrac{1}{x}$

$\therefore$ for $x < 0$,

$$\int \frac{1}{x}\,dx = \ln(-x) + c. \qquad (2)$$

For $ax + b > 0$, $\frac{d}{dx}\{\ln(ax+b)\} = \frac{1}{ax+b} \times a = \frac{a}{ax+b}$

$\therefore$ for $ax + b > 0$, $\int \frac{1}{ax+b}\,dx = \frac{1}{a}\ln(ax+b) + c.$

Example 1 Find $\int \frac{1}{3x-2}\,dx$ when (a) $3x - 2 > 0$, (b) $3x - 2 < 0$.

(a) $\int \frac{1}{3x-2}\,dx = \frac{1}{3}\ln(3x-2) + c, \quad \text{for} \quad 3x - 2 > 0.$

(b) $\int \frac{1}{3x-2}\,dx = -\int \frac{1}{2-3x}\,dx = -\frac{1}{(-3)}\ln(2-3x) + c$

$$= \frac{1}{3}\ln(2-3x) + c, \quad \text{for} \quad 3x - 2 < 0.$$

Example 2 Evaluate (a) $\int_5^9 \frac{1}{x-3}\,dx$, (b) $\int_1^2 \frac{1}{x-3}\,dx$.

(a) Within the limits of integration $x - 3 > 0$,

$$\therefore \quad \int_5^9 \frac{1}{x-3}\,dx = \Big[\ln(x-3)\Big]_5^9 = \ln 6 - \ln 2 = \ln(6/2) = \ln 3.$$

(b) Within the limits of integration $x - 3 < 0$,

$$\therefore \quad \int_1^2 \frac{1}{x-3}\,dx = -\int_1^2 \frac{1}{3-x}\,dx = \Big[\ln(3-x)\Big]_1^2 = \ln 1 - \ln 2 = -\ln 2.$$

When there are no stated restrictions on the values of x, results (1) and (2) can be combined in the statement:

$$\int \frac{1}{x}\,dx = \ln|x| + c$$

Similarly

$$\boxed{\int \frac{1}{ax+b}\,dx = \frac{1}{a}\ln|ax+b| + c.}$$

We stress, however, that it should not be necessary to use modulus signs when evaluating definite integrals.

Example 3 Find $\int \frac{x-1}{x+1}\,dx$.

$$\int \frac{x-1}{x+1}\,dx = \int \left\{1 - \frac{2}{x+1}\right\}dx = x - 2\ln|x+1| + c.$$

Example 4 Find $\int_1^3 \frac{2x^2+3x}{2x-1}\,dx$.

$$\begin{array}{r|l} & x+2 \\ 2x-1 & 2x^2+3x \\ & 2x^2-x \\ \hline & 4x \\ & 4x-2 \\ \hline & 2 \end{array}$$

By division $\frac{2x^2+3x}{2x-1} = x+2+\frac{2}{2x-1}$

Noting that within the limits of integration $2x-1>0$,

$$\int_1^3 \frac{2x^2+3x}{2x-1}\,dx = \int_1^3 \left\{x+2+\frac{2}{2x-1}\right\}dx$$

$$= \left[\tfrac{1}{2}x^2+2x+\ln(2x-1)\right]_1^3$$

$$= (\tfrac{9}{2}+6+\ln 5)-(\tfrac{1}{2}+2+\ln 1) = 8+\ln 5.$$

To obtain a general integration formula, let $y=\ln\{f(x)\}$, then

$$\frac{dy}{dx} = \frac{1}{f(x)} \times \frac{d}{dx}\{f(x)\} = \frac{f'(x)}{f(x)}$$

$\therefore$ for $f(x)>0$, $\int \frac{f'(x)}{f(x)}\,dx = \ln\{f(x)\}+c.$

When the sign of $f(x)$ is unknown we may write

$$\boxed{\int \frac{f'(x)}{f(x)}\,dx = \ln|f(x)|+c.}$$

Example 5 Find $\int \frac{x}{3x^2+5}\,dx$.

If $f(x)=3x^2+5$, then $f'(x)=6x$,
$\therefore$ noting that $3x^2+5>0$ for all values of x,

$$\int \frac{x}{3x^2+5}\,dx = \frac{1}{6}\int \frac{6x}{3x^2+5}\,dx = \frac{1}{6}\ln(3x^2+5)+c.$$

Example 6 Find $\int \tan x\,dx$.

If $f(x)=\cos x$, then $f'(x)=-\sin x$,

$$\therefore \quad \int \tan x\,dx = -\int \frac{(-\sin x)}{\cos x}\,dx = -\ln|\cos x|+c$$

$$= \ln\frac{1}{|\cos x|}+c = \ln|\sec x|+c.$$

Exercise 3.4

Find the following indefinite integrals.

1. (a) $\int \frac{1}{2x+7}\,dx$ for $2x + 7 > 0$, (b) $\int \frac{1}{5x-2}\,dx$ for $5x - 2 > 0$.

2. (a) $\int \frac{1}{1-x}\,dx$ for $1 - x > 0$, (b) $\int \frac{1}{2x-1}\,dx$ for $2x - 1 < 0$.

3. (a) $\int \frac{3}{3x-2}\,dx$ for $3x < 2$, (b) $\int \frac{8}{3-4x}\,dx$ for $4x < 3$.

4. (a) $\int \frac{1}{x+3}\,dx$, (b) $\int \frac{3}{3-x}\,dx$, (c) $\int \frac{2}{4x+5}\,dx$.

5. (a) $\int \frac{x}{x-1}\,dx$, (b) $\int \frac{x-2}{x-4}\,dx$, (c) $\int \frac{x+1}{2x+1}\,dx$.

6. (a) $\int \frac{x^2-1}{x}\,dx$, (b) $\int \frac{(1-x)^2}{1+x^2}\,dx$, (c) $\int \frac{1-x+x^2}{1-x}\,dx$.

7. (a) $\int \frac{x}{1-x^2}\,dx$, (b) $\int \frac{x^2}{x^3+8}\,dx$, (c) $\int \frac{x+1}{2x^2+4x-1}\,dx$.

8. (a) $\int \frac{\cos x}{1+\sin x}\,dx$, (b) $\int \cot x\,dx$, (c) $\int (1 + \tan 2x)\,dx$.

9. (a) $\int \frac{e^x}{1+e^x}\,dx$, (b) $\int \frac{1}{x \ln x}\,dx$, (c) $\int \frac{1}{(1+\sqrt{x})\sqrt{x}}\,dx$.

Evaluate the following definite integrals.

10. (a) $\int_2^4 \frac{5}{x}\,dx$, (b) $\int_3^9 \frac{1}{x-1}\,dx$, (c) $\int_{-4}^0 \frac{1}{1-2x}\,dx$.

11. (a) $\int_0^1 \frac{1}{3x+2}\,dx$, (b) $\int_0^3 \frac{1}{x-4}\,dx$, (c) $\int_3^5 \frac{x}{x+1}\,dx$.

12. (a) $\int_0^{\pi/2} \tan \tfrac{1}{2}x\,dx$, (b) $\int_3^4 \frac{2x-3}{x^2-3x+1}\,dx$, (c) $\int_0^{\ln 2} \frac{e^{-x}}{1+e^{-x}}\,dx$.

13. Evaluate, giving answers correct to 3 significant figures.

(a) $\int_2^4 \frac{x+2}{x^2+4x-7}\,dx$, (b) $\int_{\pi/6}^{\pi/3} \frac{\sec^2 x}{\tan x}\,dx$.

14. Express $\dfrac{1}{2x-1} - \dfrac{2}{3-x}$ as a single fraction and hence evaluate, correct to 3 decimal places, $\displaystyle\int_1^2 \frac{1-x}{(2x-1)(3-x)}\,dx$.

15. Find the area enclosed by the curve $y = 1/(x-2)$ and the line $y = 7 - 2x$.

16. Find the volume of the solid formed by rotating about the x-axis the area bounded by the curve $y = \sqrt{x} + 1/\sqrt{x}$, the x-axis and the lines $x = 1$, $x = 3$.

3.5 Introduction to hyperbolic functions

Hyperbolic sine and cosine are defined as follows:

$$\sinh x = \tfrac{1}{2}(e^x - e^{-x}), \quad \cosh x = \tfrac{1}{2}(e^x + e^{-x})$$

The remaining hyperbolic functions can be expressed in terms of $\sinh x$ and $\cosh x$.

$$\tanh x = \frac{\sinh x}{\cosh x}, \quad \coth x = \frac{\cosh x}{\sinh x}, \quad \operatorname{sech} x = \frac{1}{\cosh x}, \quad \operatorname{cosech} x = \frac{1}{\sinh x}$$

[Note that cosh, sech, cosech, coth are pronounced as they are written, tanh is pronounced 'tanch' and sinh as 'shine' or sometimes 'sinch'.]

The graphs of $y = \sinh x$ and $y = \cosh x$ may be sketched from the graphs of $y = e^x$ and $y = e^{-x}$.

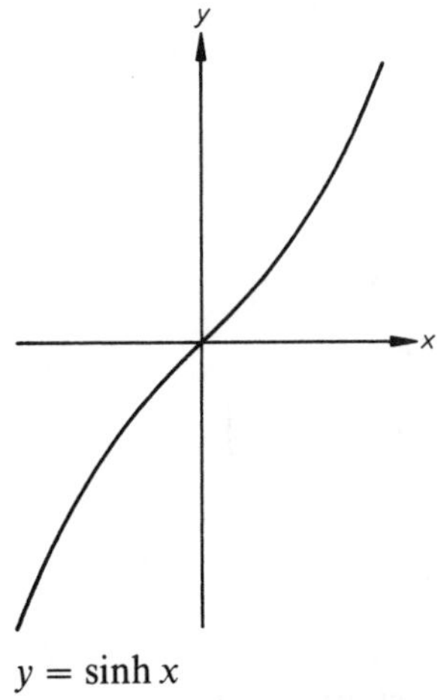

$y = \sinh x$

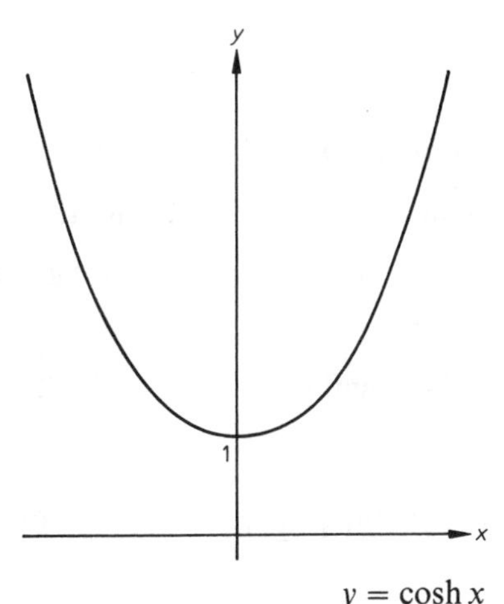

$y = \cosh x$

$$y = \sinh x = \tfrac{1}{2}(e^x - e^{-x}) \Rightarrow \frac{dy}{dx} = \tfrac{1}{2}(e^x + e^{-x}) = \cosh x$$

$$y = \cosh x = \tfrac{1}{2}(e^x + e^{-x}) \Rightarrow \frac{dy}{dx} = \tfrac{1}{2}(e^x - e^{-x}) = \sinh x$$

$$\therefore \qquad \frac{d}{dx}(\sinh x) = \cosh x, \quad \frac{d}{dx}(\cosh x) = \sinh x.$$

We now establish one of the most useful identities connecting hyperbolic functions.

$$\begin{aligned}\cosh^2 x - \sinh^2 x &= \{\tfrac{1}{2}(e^x + e^{-x})\}^2 - \{\tfrac{1}{2}(e^x - e^{-x})\}^2 \\ &= \tfrac{1}{4}\{(e^{2x} + 2 + e^{-2x}) - (e^{2x} - 2 + e^{-2x})\} \\ &= \tfrac{1}{4} \times 4 = 1\end{aligned}$$

i.e. $$\cosh^2 x - \sinh^2 x = 1.$$

Exercise 3.5

1. Differentiate with respect to x
(a) $\sinh 2x$, (b) $\cosh 3x$, (c) $\sinh^2 x$, (d) $\sqrt{(\cosh x)}$.

2. Find (a) $\int \sinh x\, dx$, (b) $\int \cosh 2x\, dx$, (c) $\int \tanh x\, dx$.

3. Differentiate with respect to x
(a) $\tanh x$, (b) $\coth x$, (c) $\operatorname{sech} x$, (d) $\operatorname{cosech} x$.

4. Use the definitions of $\sinh x$ and $\cosh x$ to prove
(a) $\sinh 2x = 2 \sinh x \cosh x$, (b) $\cosh 2x = \cosh^2 x + \sinh^2 x$.

5. By substituting $y = e^x$, solve the equations
(a) $3 \cosh x - 2 \sinh x = 3$, (b) $5 \cosh x + 7 \sinh x = 1$.

Exercise 3.6 (miscellaneous)

In questions 1 to 3 differentiate the given functions.

1. (a) xe^{3x}, (b) $e^{2x} \ln 4x$, (c) $a^x\, (a > 0)$.

2. (a) $\sin(\ln x)$, (b) $(1 - x^2)\ln x$, (c) $\ln\left(\tan \frac{1}{2}x\right)$.

3. (a) $\ln\left(\frac{x-3}{3x-1}\right)$, (b) $\ln\left(\frac{1+\sin x}{1-\sin x}\right)$, (c) $\ln\{(x^3)\sqrt{(1-x^2)}\}$.

4. Sketch the given curves, giving the coordinates of any stationary points.
(a) $y = e^x + 4e^{-x}$, (b) $y = x^3 e^{-x}$.

5. Find the value of x for which the curve $y = x^{1/x}\, (x > 0)$ has a stationary point. State the nature of this point.

6. Show that if $y = e^{4x} \cos 3x$, then d^2y/dx^2 can be expressed in the form $25e^{4x} \cos(3x + \alpha)$. Give the value of $\tan \alpha$.

7. Sketch the curves

(a) $y = \dfrac{x+1}{x-1}$, (b) $y = \left|\dfrac{x+1}{x-1}\right|$, (c) $y = \ln\left|\dfrac{x+1}{x-1}\right|$.

8. Given that $x \ln x \to 0$ as $x \to 0$, sketch the curves

(a) $y = x \ln x$, (b) $y = \dfrac{1}{x} + \ln x$.

9. Find the coordinates of any turning points on the curve $y = e^{-x/2}(x^2 + x - 2)$. Sketch this curve and the curve $y = x^2 + x - 2$ on the same axes.

10. Evaluate (a) $\displaystyle\int_2^6 \frac{1}{2x-3}\,dx$, (b) $\displaystyle\int_0^5 \frac{x}{x^2+5}\,dx$, (c) $\displaystyle\int_{\pi/2}^{\pi} \cot\frac{1}{3}x\,dx$.

11. Find (a) $\displaystyle\int \frac{x}{3x+1}\,dx$, (b) $\displaystyle\int \frac{(x-2)(x+2)}{x^3}\,dx$, (c) $\displaystyle\int \frac{2x^2+1}{x^2+x+1}\,dx$

12. Simplify $\log_e\left[\dfrac{(1+x)e^{-2x}}{1-x}\right]^{1/2}$ and show that its derivative is $x^2/(1-x^2)$. Hence, or otherwise, evaluate dy/dx at $x = 0$ for the function $y = \left[\dfrac{(1+x)e^{-2x}}{1-x}\right]^{1/2}$. (JMB)

13. If $\tan y = \log_e x^2$, show that $x\dfrac{dy}{dx} = 2\cos^2 y$. Hence show that

$$x^2\frac{d^2y}{dx^2} + 2(1 + 2\sin 2y)\cos^2 y = 0.$$

(AEB 1980)

14. If $y = (x - 0{\cdot}5)e^{2x}$, find dy/dx and hence or otherwise calculate, correct to one decimal place, the mean value of xe^{2x} in the interval $0 \leqslant x \leqslant 2{\cdot}5$. (AEB 1980)

15. The coordinates of a point on a curve are given in terms of a parameter θ by $x = a\{\ln(\cot\frac{1}{2}\theta) - \cos\theta\}$, $y = a\sin\theta$, where a is a positive constant and $0 < \theta \leqslant \frac{1}{2}\pi$. Show that, if $\theta \neq \frac{1}{2}\pi$, $\dfrac{dy}{dx} = -\tan\theta$.

The tangent at any point P of the curve meets the x-axis at T. Prove that the length of PT is equal to a. (C)

16. Sketch the region in the x-y plane within which all three of the following inequalities are satisfied:

(i) $y < x + 1$, (ii) $y > (x-1)^2$, (iii) $xy < 2$.

Determine the area of this region. (JMB)

17. Sketch the curve $y = 1 + 2e^{-x}$, showing clearly the behaviour of the curve as $x \to +\infty$. Find the area of the finite region enclosed by the curve and the lines $x = 0$, $x = 1$, $y = 1$. Find also the volume formed when this region is rotated completely about the line $y = 1$. (L)

18. Sketch the curve whose equation is $y + 3 = \dfrac{6}{x-1}$. Find the coordinates of the points where the line $y + 3x = 9$ intersects the curve and show that the area of the region enclosed between the curve and the line is $\frac{3}{2}(3 - 4\log_e 2)$. Determine the equations of the two tangents to the curve which are parallel to the line. (JMB)

19. Sketch the curve $y = x/(1 + x^2)$, finding its turning points, showing that the origin is a point of inflexion, and indicating the behaviour of y when $|x|$ is large. Calculate the area of the region defined by the inequalities

$$0 \leqslant x \leqslant a, 0 \leqslant y \leqslant x/(1 + x^2),$$

and show that this area has the value 1 when $a = \sqrt{(e^2 - 1)}$. (L)

20. A curve has the equation $y = a/(b + e^{-cx})$, $(a \neq 0, b > 0, c > 0)$. Show that it has one point of inflexion, and that the value of y at the point of inflexion is half the limiting value y as $x \to \infty$. (O)

21. Sketch the curve $y = (x^2 - 7x + 10)/(x - 6)$, indicating clearly any vertical asymptotes and turning values. Determine the finite area bounded by the x-axis and that part of the curve between the points where it crosses the x-axis. (O)

22. Sketch the curve $y = \log_e(x - 2)$. The inner surface of a bowl is of the shape formed by rotating completely about the y-axis that part of the x-axis between $x = 0$ and $x = 3$ and that part of the curve $y = \log_e(x - 2)$ between $y = 0$ and $y = 2$. The bowl is placed with its axis vertical and water is poured in. Calculate the volume of water in the bowl when the bowl is filled to a depth $h(<2)$. If water is poured into the bowl at a rate of 50 cubic units per second, find the rate at which the water level is rising when the depth of the water is 1·5 units. (AEB 1979)

23. By means of a diagram, or otherwise, show that

$$\frac{1}{r+1} < \int_r^{r+1} \frac{1}{x}\,dx < \frac{1}{r}.$$

Deduce that the sum to N terms of the series $1 + \frac{1}{2} + \frac{1}{3} + \ldots + \frac{1}{r} + \ldots$ lies between $\log_e(1 + N)$ and $1 + \log_e N$, where $N > 1$. Show that the sum of one hundred million terms of this series lies between 18·4 and 19·5. (JMB)

4 Methods of integration

4.1 Standard forms

We first list some important integration results.

$$\int x^n\,dx = \frac{1}{n+1}x^{n+1} + c, \quad (n \neq -1)$$

$$\int \sin x\,dx = -\cos x + c, \qquad \int \cos x\,dx = \sin x + c$$

$$\int \sec^2 x\,dx = \tan x + c, \qquad \int \operatorname{cosec}^2 x\,dx = -\cot x + c$$

$$\int e^x\,dx = e^x + c, \qquad \int \frac{1}{x}\,dx = \ln x + c, \quad (x > 0)$$

When using these standard forms it is worth noting that if we can integrate a function $f(x)$, we should also be able to integrate $f(ax + b)$. For instance, remembering that $\int \cos x\,dx = \sin x + c$, to find $\int \cos(2x - 3)\,dx$ we try differentiating $\sin(2x - 3)$:

$$y = \sin u, \quad \text{where} \quad u = 2x - 3$$

$$\Rightarrow \quad \frac{dy}{du} = \cos u \qquad \frac{du}{dx} = 2$$

$$\Rightarrow \quad \frac{dy}{dx} = \frac{dy}{du} \times \frac{du}{dx} = 2\cos u = 2\cos(2x - 3)$$

Hence $$\int \cos(2x - 3)\,dx = \tfrac{1}{2}\sin(2x - 3) + c.$$

Example 1 Find (a) $\int e^{x/2}\,dx$, (b) $\int (1-2x)^3\,dx$, (c) $\int \frac{1}{(3x+4)^4}\,dx$.

(a) $\frac{d}{dx}\{e^{x/2}\} = e^{x/2} \times \frac{1}{2} = \frac{1}{2}e^{x/2}$

$$\therefore \qquad \int e^{x/2}\,dx = 2e^{x/2} + c.$$

(b) $\frac{d}{dx}\{(1-2x)^4\} = 4(1-2x)^3 \times (-2) = -8(1-2x)^3$

$$\therefore \qquad \int (1-2x)^3\,dx = -\tfrac{1}{8}(1-2x)^4 + c.$$

(c) $\frac{d}{dx}\{(3x+4)^{-3}\} = -3(3x+4)^{-4} \times 3 = -\frac{9}{(3x+4)^4}$

$$\therefore \qquad \int \frac{1}{(3x+4)^4}\,dx = -\frac{1}{9(3x+4)^3} + c.$$

Integrals involving trigonometric functions can often be found using basic trigonometric identities.

$$\int \tan^2 x\,dx = \int (\sec^2 x - 1)\,dx = \tan x - x + c.$$

$$\int \cos^2 x\,dx = \int \tfrac{1}{2}(1 + \cos 2x)\,dx = \tfrac{1}{2}x + \tfrac{1}{4}\sin 2x + c.$$

$$\int \sin 3x \cos 2x\,dx = \int \tfrac{1}{2}(\sin 5x + \sin x)\,dx = -\tfrac{1}{10}\cos 5x - \tfrac{1}{2}\cos x + c.$$

In the next two examples we consider integrals of the form

$$\int \{f(x)\}^n f'(x)\,dx.$$

Example 2 Find $\int \sin^2 x \cos x\,dx$ and hence $\int \cos^3 x\,dx$.

[We note that $f(x) = \sin x \Rightarrow f'(x) = \cos x$ and therefore try differentiating $\sin^3 x$.]

$$\frac{d}{dx}(\sin^3 x) = 3\sin^2 x \cos x$$

$$\therefore \qquad \int \sin^2 x \cos x\,dx = \tfrac{1}{3}\sin^3 x + c.$$

$$\text{Hence} \qquad \int \cos^3 x\,dx = \int (\cos^2 x)\cos x\,dx$$

$$= \int (1 - \sin^2 x)\cos x\,dx$$

$$= \int (\cos x - \sin^2 x \cos x)\,dx$$

$$= \sin x - \tfrac{1}{3}\sin^3 x + c.$$

Example 3 Find $\int 2x(1 + x^2)^4\,dx$.

[We note that $f(x) = 1 + x^2 \Rightarrow f'(x) = 2x$ and therefore try differentiating $(1 + x^2)^5$.]

$$\frac{d}{dx}\{(1 + x^2)^5\} = 5(1 + x^2)^4 \times 2x$$

$$\therefore \quad \int 2x(1 + x^2)^4\,dx = \tfrac{1}{5}(1 + x^2)^5 + c.$$

Two general results involving logarithms were derived in §3.4:

$$\int \frac{1}{ax + b}\,dx = \frac{1}{a}\ln(ax + b) + c, \quad \text{for} \quad ax + b > 0$$

$$\int \frac{f'(x)}{f(x)}\,dx = \ln\{f(x)\} + c, \quad \text{for} \quad f(x) > 0$$

or

$$\int \frac{f'(x)}{f(x)}\,dx = \ln|f(x)| + c.$$

These were used to integrate functions such as:

$$\frac{1}{3x - 2}, \quad \frac{x - 1}{x + 1}, \quad \frac{2x^2 + 3x}{2x - 1}, \quad \frac{x}{3x^2 + 5} \quad \text{and} \quad \tan x.$$

Finally we note that when obtaining an indefinite integral it may be possible to produce several different answers.

(i) $\int (2x + 2)\,dx = x^2 + 2x + c$

or $\int (2x + 2)\,dx = \int 2(x + 1)\,dx = (x + 1)^2 + c.$

(ii) $\int 2\sin x \cos x\,dx = \sin^2 x + c$ or $\int 2\sin x \cos x\,dx = -\cos^2 x + c$

or $\int 2\sin x \cos x\,dx = \int \sin 2x\,dx = -\tfrac{1}{2}\cos 2x + c.$

However, apparently dissimilar results turn out to differ only by a constant. For example, in (ii)

$$\sin^2 x = -\cos^2 x + 1 = -\tfrac{1}{2}\cos 2x + \tfrac{1}{2}.$$

Exercise 4.1

Find the given indefinite integrals.

1. $\displaystyle\int \frac{x-3}{x^3}\,dx$ 2. $\displaystyle\int (x^3-1)^2\,dx$ 3. $\displaystyle\int (x+1)\sqrt{x}\,dx$

4. $\displaystyle\int \cos 3x\,dx$ 5. $\displaystyle\int e^{4x-3}\,dx$ 6. $\displaystyle\int \sin(1-2x)\,dx$

7. $\displaystyle\int (1-x)^4\,dx$ 8. $\displaystyle\int \frac{1}{(x+4)^2}\,dx$ 9. $\displaystyle\int \frac{1}{(2x-5)^3}\,dx$

10. $\displaystyle\int \frac{1}{2x-5}\,dx$ 11. $\displaystyle\int \sec^2 5x\,dx$ 12. $\displaystyle\int \cot\tfrac{1}{2}x\,dx$

13. $\displaystyle\int \sin^2 x\,dx$ 14. $\displaystyle\int \sin x\cos 3x\,dx$ 15. $\displaystyle\int \cot^2 x\,dx$

16. $\displaystyle\int \sin x\cos^2 x\,dx$ 17. $\displaystyle\int \frac{\sin x}{\cos^2 x}\,dx$ 18. $\displaystyle\int \sin^3 x\,dx$

19. $\displaystyle\int \frac{x+1}{x+2}\,dx$ 20. $\displaystyle\int \frac{x^2}{x-1}\,dx$ 21. $\displaystyle\int \frac{x^3-x}{x^2+5}\,dx$

22. $\displaystyle\int \frac{1}{\sqrt{(5x+3)}}\,dx$ 23. $\displaystyle\int \frac{x^3}{1+x^4}\,dx$ 24. $\displaystyle\int x^3(1+x^4)^3\,dx$

Evaluate the given definite integrals.

25. $\displaystyle\int_0^1 e^{1-x}\,dx$ 26. $\displaystyle\int_{-2/3}^{1/3} (2+3x)^5\,dx$ 27. $\displaystyle\int_0^{\pi/4} \cos^3 x\sin x\,dx$

28. $\displaystyle\int_0^{\pi/2} \sin 2x\sin 5x\,dx$ 29. $\displaystyle\int_0^{\pi/12} \cos^2 3x\,dx$ 30. $\displaystyle\int_0^{\pi/2} \sin^3 x\cos^2 x\,dx$

31. $\displaystyle\int_0^{\pi/3} \tan x\sec^2 x\,dx$ 32. $\displaystyle\int_0^{\pi/4} \cos 3x\cos x\,dx$ 33. $\displaystyle\int_0^{\pi/2} \tan^2\left(x-\frac{\pi}{4}\right)dx$

34. $\displaystyle\int_0^{\pi/2} \frac{\cos x}{1+\sin x}\,dx$ 35. $\displaystyle\int_1^2 \frac{6}{(1-2x)^2}\,dx$ 36. $\displaystyle\int_0^{1/2} \frac{x}{1-x^2}\,dx$

37. Show that $\int_0^{\pi/4} (1 + \tan x)^2\, dx = 1 + \ln 2$.

38. Find $\frac{d}{dx}\{\ln(\sec x + \tan x)\}$ and hence evaluate $\int_0^{\pi/6} \sec x\, dx$.

39. By expressing $\cos x$ in terms of $\cos \frac{1}{2}x$, or otherwise, find

$$\int \frac{1}{1 + \cos x}\, dx.$$

40. Prove that $\sin^4 x = \frac{1}{8}(\cos 4x - 4\cos 2x + 3)$. Hence evaluate

$$\int_0^{\pi} \sin^4 x\, dx.$$

4.2 Use of partial fractions

Most rational functions of the form $P(x)/Q(x)$, where $P(x)$ and $Q(x)$ are polynomials, are more easily integrated when expressed in partial fractions.

As shown in §2.6, a linear factor of the form $(ax + b)$ in the denominator of a rational function leads to a partial fraction $A/(ax + b)$. Such fractions are integrated using the rule:

$$\int \frac{1}{ax + b}\, dx = \frac{1}{a}\ln(ax + b) + c, \quad \text{for} \quad ax + b > 0.$$

Example 1 Find $\int \frac{1}{(x - 2)(x + 1)}\, dx$ for $x > 2$.

Let $$\frac{1}{(x - 2)(x + 1)} \equiv \frac{A}{x - 2} + \frac{B}{x + 1}$$

then $1 \equiv A(x + 1) + B(x - 2)$

Putting $x = 2$: $1 = 3A \quad \therefore \quad A = \frac{1}{3}$

Putting $x = -1$: $1 = -3B \quad \therefore \quad B = -\frac{1}{3}$

$$\therefore \quad \int \frac{1}{(x - 2)(x + 1)}\, dx = \int \left\{ \frac{1}{3(x - 2)} - \frac{1}{3(x + 1)} \right\} dx$$

$$= \tfrac{1}{3}\ln(x - 2) - \tfrac{1}{3}\ln(x + 1) + c$$

$$= \tfrac{1}{3}\ln\left(\frac{x - 2}{x + 1}\right) + c, \quad \text{for} \quad x > 2.$$

Partial fractions are not needed when integrating rational functions of the form $f'(x)/f(x)$.

Example 2 Find $\displaystyle\int \frac{2x-1}{(x-2)(x+1)}\,dx.$

If $f(x) = (x-2)(x+1) = x^2 - x - 2$, then $f'(x) = 2x - 1$

$$\therefore \quad \int \frac{2x-1}{(x-2)(x+1)}\,dx = \ln|(x-2)(x+1)| + c.$$

A quadratic factor in the denominator of a rational function may give rise to a partial fraction of the form $f'(x)/f(x)$.

Example 3 Find $\displaystyle\int \frac{1-x^2}{x(x^2+1)}\,dx$, for $x > 0$.

$$\text{Let} \qquad \frac{1-x^2}{x(x^2+1)} \equiv \frac{A}{x} + \frac{Bx+C}{x^2+1}$$

then $\quad 1 - x^2 \equiv A(x^2+1) + (Bx+C)x$

Putting $x = 0$ $\quad 1 = A \quad \therefore \quad A = 1$

Comparing coefficients, x^2: $\quad -1 = A + B \quad \therefore \quad B = -2$

x: $\quad 0 = C \quad \therefore \quad C = 0$

$$\therefore \quad \int \frac{1-x^2}{x(x^2+1)}\,dx = \int \left\{\frac{1}{x} - \frac{2x}{x^2+1}\right\} dx$$

$$= \ln x - \ln(x^2+1) + c$$

$$= \ln\left(\frac{x}{x^2+1}\right) + c, \quad \text{for} \quad x > 0.$$

Example 4 Evaluate $\displaystyle\int_0^1 \frac{1}{(x+1)(x^2+2x+2)}\,dx.$

$$\text{Let} \qquad \frac{1}{(x+1)(x^2+2x+2)} \equiv \frac{A}{x+1} + \frac{Bx+C}{x^2+2x+2}$$

then $\quad 1 \equiv A(x^2+2x+2) + (Bx+C)(x+1)$

Putting $x = -1$: $\quad 1 = A \quad \therefore \quad A = 1$

Putting $x = 0$: $\quad 1 = 2A + C \quad \therefore \quad C = -1$

Comparing coefficients of x^2: $\quad 0 = A + B \quad \therefore \quad B = -1$

$$\therefore \quad \int_0^1 \frac{1}{(x+1)(x^2+2x+2)}\,dx = \int_0^1 \left\{\frac{1}{x+1} - \frac{x+1}{x^2+2x+2}\right\} dx$$

$$= \int_0^1 \left\{\frac{1}{x+1} - \frac{1}{2}\left(\frac{2x+2}{x^2+2x+2}\right)\right\} dx$$

$$= \left[\ln(x+1) - \tfrac{1}{2}\ln(x^2+2x+2)\right]_0^1$$

$$= (\ln 2 - \tfrac{1}{2}\ln 5) - (\ln 1 - \tfrac{1}{2}\ln 2)$$

$$= \tfrac{3}{2}\ln 2 - \tfrac{1}{2}\ln 5$$

$$= \tfrac{1}{2}(3\ln 2 - \ln 5) = \tfrac{1}{2}\ln(8/5).$$

Example 5 Find $\int \frac{x-4}{(x-1)^2(2x+1)}\,dx.$

Let
$$\frac{x-4}{(x-1)^2(2x+1)} \equiv \frac{A}{x-1} + \frac{B}{(x-1)^2} + \frac{C}{(2x+1)}$$

then
$$x - 4 \equiv A(x-1)(2x+1) + B(2x+1) + C(x-1)^2$$

Putting $x = 1$: $\quad -3 = 3B \quad \therefore \quad B = -1$

Putting $x = -\frac{1}{2}$: $\quad -\frac{9}{2} = \frac{9}{4}C \quad \therefore \quad C = -2$

Comparing coefficients of x^2: $\quad 0 = 2A + C \quad \therefore \quad A = 1$

$$\therefore \quad \int \frac{x-4}{(x-1)^2(2x+1)}\,dx = \int \left\{ \frac{1}{x-1} - \frac{1}{(x-1)^2} - \frac{2}{2x+1} \right\} dx$$

$$= \ln|x-1| + \frac{1}{x-1} - \ln|2x+1| + c$$

$$= \frac{1}{x-1} + \ln\left|\frac{x-1}{2x+1}\right| + c.$$

Exercise 4.2

Find the following indefinite integrals.

1. $\int \frac{3}{(x-1)(x+2)}\,dx$

2. $\int \frac{1}{1-x^2}\,dx$

3. $\int \frac{1}{x(x-3)}\,dx$

4. $\int \frac{x}{x^2-4}\,dx$

5. $\int \frac{4x}{x^2-2x-3}\,dx$

6. $\int \frac{2x-5}{(x-2)(x-3)}\,dx$

7. $\int \frac{3x+5}{(x+1)(x+3)}\,dx$

8. $\int \frac{x-1}{x(2x-1)}\,dx$

9. $\int \frac{2x-11}{x^2-5x+4}\,dx$

10. $\int \frac{x^2}{1-x^2}\,dx$

11. $\int \frac{x-2}{x^2-4x-8}\,dx$

12. $\int \frac{x^3}{x^2-4}\,dx$

13. $\int \frac{2x^2 + 1}{(x + 1)^2(x - 2)} dx$

14. $\int \frac{x + 2}{x^2(x - 1)} dx$

15. $\int \frac{x}{(x + 3)^2} dx$

16. $\int \frac{x^2 + 2x - 1}{(x + 1)(x^2 + 1)} dx$

17. $\int \frac{x + 1}{(x^2 + 1)(x - 1)} dx$

18. $\int \frac{x}{(x^2 + 3)(x^2 + 5)} dx$

19. $\int \frac{4x^2 + x + 4}{(2x + 1)(x^2 + 2)} dx$

20. $\int \frac{x^3 - 3x - 4}{x^3 - 1} dx$

Evaluate the following definite integrals.

21. $\int_1^2 \frac{x}{(x + 1)(x + 2)} dx$

22. $\int_3^4 \frac{5}{x^2 + x - 6} dx$

23. $\int_4^5 \frac{2x}{x^2 - 4x + 3} dx$

24. $\int_0^{1/3} \frac{3x}{1 - x^2} dx$

25. $\int_0^{1/3} \frac{3 + x}{(1 - x)(1 + 3x)} dx$

26. $\int_2^3 \frac{1}{x(x - 1)} dx$

27. $\int_1^2 \frac{x - 3}{x(x + 1)(x + 3)} dx$

28. $\int_0^2 \frac{x^3 + 4x^2 + 3x - 2}{x^2 + 4x + 3} dx$

29. $\int_1^2 \frac{1}{x^2(x + 1)} dx$

30. $\int_3^4 \frac{x^2 + x - 3}{(x - 2)^2(x - 1)} dx$

31. $\int_3^4 \frac{x^3}{(x - 1)(x - 2)} dx$

32. $\int_4^5 \frac{2x^2 + 10x + 40}{(x - 3)^2(x + 1)} dx$

33. $\int_3^4 \frac{2x + 1}{(x - 2)(x^2 + 1)} dx$

34. $\int_1^2 \frac{2(x + 2)^2}{x^2(x^2 + 4)} dx$

35. $\int_0^1 \frac{(x - 2)^2}{x^3 + 1} dx$

36. $\int_2^3 \frac{x + 3}{(x - 1)(x^2 - x + 2)} dx$

37. Find $\displaystyle\int \frac{4}{(x-1)(x+3)}\,dx$ for (a) $x > 1$, (b) $-3 < x < 1$, (c) $x < -3$.

38. Find $\displaystyle\int \frac{x-4}{2x^2-x-3}\,dx$ for (a) $x > 1\frac{1}{2}$, (b) $x < 1\frac{1}{2}$.

4.3 Integration by substitution

The integration methods considered so far rely on the use of differentiation results in reverse. Integrating by substitution is a more systematic procedure, roughly equivalent to reversing the "chain rule" method for differentiating composite functions.

Suppose that as a result of applying the chain rule we have:

$$\frac{dy}{dx} = 10(2x+3)^4$$

We can work "backwards" to an expression for y as follows:

$$\frac{dy}{dx} = \frac{dy}{du} \times \frac{du}{dx}$$

$\therefore$ substituting $u = 2x + 3$ gives $\dfrac{dy}{dx} = 10u^4$, $\dfrac{du}{dx} = 2$ and thus

$$10u^4 = \frac{dy}{du} \times 2 \quad \text{i.e.} \quad \frac{dy}{du} = 5u^4.$$

Hence $\qquad y = u^5 + c = (2x+3)^5 + c.$

The above method can be modified to produce a useful general result.

Let $\qquad \dfrac{dy}{dx} = F(x), \quad$ so that $\quad y = \displaystyle\int F(x)\,dx.$

If, in terms of another variable u, $F(x) = G(u)$,

then $\qquad \dfrac{dy}{du} = \dfrac{dy}{dx} \times \dfrac{dx}{du} = F(x)\dfrac{dx}{du} = G(u)\dfrac{dx}{du}$

$\therefore$

$$y = \int G(u)\frac{dx}{du}\,du$$

i.e. $\qquad$ if $F(x) = G(u)$, then $\displaystyle\int F(x)\,dx = \int G(u)\frac{dx}{du}\,du.$

Hence, when simplifying an integral $\displaystyle\int F(x)\,dx$ by a change to a new variable u, we must express $F(x)$ in terms of u and replace dx by $\frac{dx}{du}du$.

[Note that in this context a statement such as $dx = \frac{dx}{du}du$ is often used as shorthand for "dx may be replaced by $\frac{dx}{du}du$".]

There are three different ways of expressing suitable substitutions.

(i) If $f(x) = u$, then $f'(x) = \frac{du}{dx}$,

and $$dx = \frac{dx}{du}du = \frac{1}{f'(x)}du$$

$\therefore$ we write $$f'(x)\,dx = du.$$

(ii) If $x = g(u)$, then $\frac{dx}{du} = g'(u)$,

$\therefore$ we write $$dx = g'(u)du.$$

(iii) If $f(x) = g(u)$, then $f'(x)\frac{dx}{du} = g'(u)$,

$\therefore$ we write $$f'(x)\,dx = g'(u)\,du.$$

Example 1 Find $\int 10(2x+3)^4\,dx$.

Let $2x + 3 = u$, then $2dx = du$.

$$\therefore \quad \int 10(2x+3)^4\,dx = \int 5(2x+3)^4 \times 2\,dx$$
$$= \int 5u^4 du$$
$$= u^5 + c = (2x+3)^5 + c.$$

Example 2 Find $\int \sin^3 2x \cos 2x\,dx$.

Let $\sin 2x = u$, then $2\cos 2x\,dx = du$.

$$\therefore \quad \int \sin^3 2x \cos 2x\,dx = \int \tfrac{1}{2}\sin^3 2x \times 2\cos 2x\,dx$$
$$= \int \tfrac{1}{2}u^3\,du$$
$$= \tfrac{1}{8}u^4 + c = \tfrac{1}{8}\sin^4 2x + c.$$

Example 3 Find $\int x\sqrt{(2x-1)}\,dx$.

Let	$2x-1=u^2,$	$2x=u^2+1$
then	$2dx=2u\,du,$	$x=\frac{1}{2}(u^2+1)$
$\therefore$	$dx=u\,du$	

$$\int x\sqrt{(2x-1)}\,dx=\int \tfrac{1}{2}(u^2+1)\times u\times u\,du$$
$$=\frac{1}{2}\int(u^4+u^2)\,du$$
$$=\tfrac{1}{2}(\tfrac{1}{5}u^5+\tfrac{1}{3}u^3)+c$$
$$=\tfrac{1}{30}u^3(3u^2+5)+c$$
$$\therefore\quad \int x\sqrt{(2x-1)}\,dx=\tfrac{1}{30}(2x-1)^{3/2}(3\{2x-1\}+5)+c$$
$$=\tfrac{1}{15}(2x-1)^{3/2}(3x+1)+c.$$

When the method of substitution is applied to a definite integral, the limits of integration must be changed as well as the variable.

Example 4 Evaluate $\int_0^1 4x(3x^2-1)^3\,dx$.

Let	$3x^2-1=u,$	$x=1\Rightarrow u=2$
then	$6x\,dx=du$	$x=0\Rightarrow u=-1$

$$\int_0^1 4x(3x^2-1)^3\,dx=\int_0^1 \tfrac{2}{3}(3x^2-1)^3\times 6x\,dx$$
$$=\int_{-1}^{2}\tfrac{2}{3}u^3\,du$$
$$=\left[\tfrac{1}{6}u^4\right]_{-1}^{2}$$
$$=\tfrac{1}{6}\{2^4-(-1)^4\}=\tfrac{15}{6}=2\tfrac{1}{2}.$$

Example 5 Evaluate $\int_0^2 \sqrt{(4-x^2)}\,dx$.

$$\int_0^2 \sqrt{(4-x^2)}\,dx$$

$$= \int_0^{\pi/2} 2\cos\theta \times 2\cos\theta\,d\theta$$

$$= \int_0^{\pi/2} 4\cos^2\theta\,d\theta$$

$$= \int_0^{\pi/2} (2+2\cos 2\theta)\,d\theta$$

$$= \Big[2\theta + \sin 2\theta\Big]_0^{\pi/2}$$

$$= \pi$$

Let $x = 2\sin\theta$
then $dx = 2\cos\theta\,d\theta$

$$4 - x^2 = 4 - 4\sin^2\theta = 4(1-\sin^2\theta) = 4\cos^2\theta$$

$\therefore\ \sqrt{(4-x^2)} = 2\cos\theta$

$x = 2 \Rightarrow \theta = \frac{1}{2}\pi$
$x = 0 \Rightarrow \theta = 0$

When integrating trigonometric functions one useful substitution is $\tan\frac{1}{2}\theta = t$, where

$$\sin\theta = \frac{2t}{1+t^2},\quad \cos\theta = \frac{1-t^2}{1+t^2},\quad \tan\theta = \frac{2t}{1-t^2}$$

Example 6 Evaluate $\displaystyle\int_0^{\pi/2} \frac{4}{3+5\cos\theta}\,d\theta$.

$$\int_0^{\pi/2} \frac{4}{3+5\cos\theta}\,d\theta$$

$$= \int_0^1 \frac{4(1+t^2)}{8-2t^2} \times \frac{2}{1+t^2}\,dt$$

$$= \int_0^1 \frac{4}{4-t^2}\,dt$$

$$= \int_0^1 \left\{\frac{1}{2+t} + \frac{1}{2-t}\right\}dt$$

$$= \Big[\ln(2+t) - \ln(2-t)\Big]_0^1$$

$$= (\ln 3 - \ln 1) - (\ln 2 - \ln 2)$$

$$= \ln 3$$

Let $\tan\frac{1}{2}\theta = t$
then $\frac{1}{2}\sec^2\frac{1}{2}\theta\,d\theta = dt$
$\therefore\ (1+\tan^2\frac{1}{2}\theta)\,d\theta = 2dt$
$\therefore\ d\theta = \dfrac{2}{1+t^2}\,dt$

$$3 + 5\cos\theta = 3 + 5\left(\frac{1-t^2}{1+t^2}\right) = \frac{3(1+t^2)+5(1-t^2)}{1+t^2} = \frac{8-2t^2}{1+t^2}$$

$\theta = \frac{1}{2}\pi \Rightarrow t = 1$
$\theta = 0 \Rightarrow t = 0$

A change of variable may be helpful when finding areas and volumes by integration. For instance, to evaluate $\int_a^b y\,dx$ it may be simpler to work in terms of y rather than x. If x and y are given in parametric form it will usually be best to use the parameter as variable in any integrations.

Example 7 Find the area bounded by the curve $x = t^3 - 1$, $y = 5t^2 - 4$, the x- and y-axes and the line $x = 7$.

As shown in the sketch below, values of x from 0 to 7 are given by values of t from 1 to 2.
Since $x = t^3 - 1$, we have $dx = 3t^2\,dt$.

$$\text{Required area} = \int_0^7 y\,dx$$
$$= \int_1^2 (5t^2 - 4) \times 3t^2\,dt$$
$$= \int_1^2 (15t^4 - 12t^2)\,dt$$
$$= \Big[3t^5 - 4t^3\Big]_1^2$$
$$= (96 - 32) - (3 - 4) = 65.$$

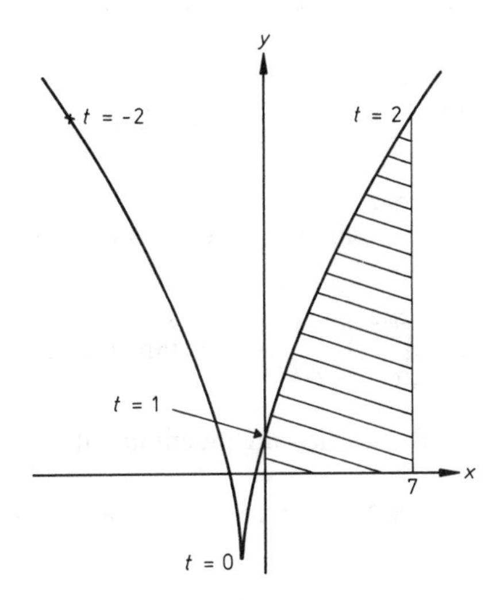

Exercise 4.3

Use the given substitutions to find the following indefinite integrals.

1. $\int (3x - 2)^3\,dx$, $3x - 2 = u$.
2. $\int (1 - 2x)^4\,dx$, $1 - 2x = u$.
3. $\int x(1 - x)^5\,dx$, $1 - x = u$.
4. $\int 2x^2(x^3 - 1)^3\,dx$, $x^3 - 1 = u$.
5. $\int \tan^2 x \sec^2 x\,dx$, $\tan x = u$.
6. $\int \sin x \cos^5 x\,dx$, $\cos x = u$.
7. $\int x\sqrt{(x^2 + 4)}\,dx$, $x^2 + 4 = t^2$.
8. $\int x\sqrt{(1 - x)}\,dx$, $1 - x = t^2$.
9. $\int \frac{x}{\sqrt{(2x + 1)}}\,dx$, $2x + 1 = t^2$.
10. $\int \frac{e^x}{1 + e^x}\,dx$, $e^x = t$.
11. $\int \cos^3\theta\,d\theta$, $\sin\theta = x$.
12. $\int \sin^3\theta \cos^2\theta\,d\theta$, $\cos\theta = x$.
13. $\int \frac{\sin\theta}{1 + \cos 2\theta}\,d\theta$, $\cos\theta = x$.
14. $\int \frac{\sec^2(\sqrt{x})}{\sqrt{x}}\,dx$, $\sqrt{x} = \theta$.

Use the given substitutions to find the following definite integrals.

15. $\int_4^5 (x-1)\sqrt{(x-4)}\,dx,\ x-4=u^2.$

16. $\int_0^1 x^2(1-x)^{1/2}\,dx,\ 1-x=u^2.$

17. $\int_e^{e^2} \frac{1}{x\ln x}\,dx,\ \ln x = t.$

18. $\int_0^1 \frac{e^x}{(1+e^x)^2}\,dx,\ e^x = t.$

19. $\int_0^2 \frac{1}{\sqrt{(4-x^2)}}\,dx,\ x = 2\sin\theta.$

20. $\int_0^{\sqrt{3}} \frac{1}{9+x^2}\,dx,\ x = 3\tan\theta.$

21. $\int_0^{\pi/2} \sin 2x\sqrt{(\sin x)}\,dx,\ \sin x = u^2.$

22. $\int_0^{1/2} (1-x^2)^{1/2}\,dx,\ x = \sin u.$

23. $\int_0^{\pi/2} \frac{1}{5\cos\theta+4}\,d\theta,\ \tan\frac{1}{2}\theta = t.$

24. $\int_0^{\pi/4} \frac{1}{5\cos^2\theta - 1}\,d\theta,\ \tan\theta = t.$

Find the following indefinite integrals.

25. $\int x(2x+3)^4\,dx$

26. $\int x^3(x^2-1)^3\,dx$

27. $\int \sqrt{(1+3x)}\,dx$

28. $\int \frac{x}{\sqrt{(3+x)}}\,dx$

29. $\int \frac{x^2}{(x^3-1)^2}\,dx$

30. $\int \frac{x+1}{(x-1)^3}\,dx$

31. $\int xe^{x^2}\,dx$

32. $\int \frac{1}{x\sqrt{(\ln x)}}\,dx$

33. $\int \frac{e^x}{4-e^{2x}}\,dx$

Find the following definite integrals

34. $\int_{-1}^2 x^2\sqrt{(x^3+1)}\,dx$

35. $\int_0^3 x\sqrt{(9-x^2)}\,dx$

36. $\int_0^3 \sqrt{(9-x^2)}\,dx$

37. $\int_1^2 \frac{x}{\sqrt{(4-x^2)}}\,dx$

38. $\int_1^2 \frac{x^2}{\sqrt{(4-x^2)}}\,dx$

39. $\int_0^{\ln 2} \frac{1}{1+e^x}\,dx$

40. Find the area between the x-axis and the curve $x = 2t^3 - 1$, $y = t^2 - 1$ for $1 \leqslant t \leqslant 2$.

41. Find the area of the finite region bounded by the curve $x = 5t^2$, $y = 2t^3$ and the line $x = 5$. Find also the volume of the solid formed when this region is rotated through π radians about the x-axis.

42. Find the area bounded by the x-axis and the curve $x = \theta - \sin\theta$, $y = 1 - \cos\theta$ for $0 \leqslant \theta \leqslant 2\pi$.

43. Using the substitution $\tan\frac{1}{2}x = t$, or otherwise, show that

$$\int_{\pi/3}^{\pi/2} \frac{1}{1 + \sin x + \cos x}\,dx = 0{\cdot}237$$

correct to 3 significant figures.

44. Using the substitution $x = 2\cos^2\theta + 5\sin^2\theta$, or otherwise, evaluate

$$\int_2^{7/2} \frac{1}{\sqrt{\{(x-2)(5-x)\}}}\,dx.$$

45. Using the substitution $\tan x = t$, or otherwise, show that

$$\int \frac{1}{1 - 2\sin^2 x}\,dx = \tfrac{1}{2}\ln\{\tan(x + \tfrac{1}{4}\pi)\} + c.$$

4.4 Integration by parts

A further general method of integration is obtained by considering the formula for differentiating a product. If u and v are functions of x, then

$$\frac{d}{dx}(uv) = u\frac{dv}{dx} + v\frac{du}{dx}$$

Integrating,

$$uv = \int u\frac{dv}{dx}\,dx + \int v\frac{du}{dx}\,dx$$

Rearranging,

$$\boxed{\int u\frac{dv}{dx}\,dx = uv - \int v\frac{du}{dx}\,dx} \qquad (1)$$

This rule provides a way of expressing the integral of a product in terms of a second integral, which may be easier to find. The method is called *integration by parts.*

Example 1 Find $\int x\cos x\,dx$.

In rule (1), let $u = x, \quad \dfrac{dv}{dx} = \cos x,$

$$\frac{du}{dx} = 1, \qquad v = \sin x,$$

$$\therefore \quad \int x\cos x\,dx = x\sin x - \int(\sin x \times 1)\,dx$$

$$= x\sin x - \int \sin x\,dx = x\sin x + \cos x + c.$$

[Note that it is not necessary to introduce an arbitrary constant when integrating $\frac{dv}{dx}$ to give v, since rule (1) is valid for any function v such that $\frac{dv}{dx} = \cos x$.]

Example 2 Find $\int x \ln x \, dx$.

In rule (1), let $\quad u = \ln x, \quad \frac{dv}{dx} = x,$

$$\frac{du}{dx} = \frac{1}{x}, \qquad v = \tfrac{1}{2}x^2,$$

$$\therefore \quad \int x \ln x \, dx = \tfrac{1}{2}x^2 \ln x - \int \left(\tfrac{1}{2}x^2 \times \frac{1}{x}\right) dx$$

$$= \tfrac{1}{2}x^2 \ln x - \int \tfrac{1}{2}x \, dx = \tfrac{1}{2}x^2 \ln x - \tfrac{1}{4}x^2 + c.$$

For definite integrals the rule for integration by parts becomes

$$\int_a^b u \frac{dv}{dx} dx = \Big[uv \Big]_a^b - \int_a^b v \frac{du}{dx} dx \qquad (2)$$

Example 3 Evaluate $\int_0^1 x^2 e^x \, dx$.

In rule (2), let $\quad u = x^2, \quad \frac{dv}{dx} = e^x,$

$$\frac{du}{dx} = 2x, \qquad v = e^x,$$

$$\therefore \quad \int_0^1 x^2 e^x \, dx = \Big[x^2 e^x \Big]_0^1 - \int_0^1 (e^x \times 2x) \, dx$$

$$= e - 2\int_0^1 xe^x \, dx.$$

In rule (2), let $\quad u = x, \quad \frac{dv}{dx} = e^x,$

$$\frac{du}{dx} = 1, \qquad v = e^x,$$

$$\therefore \quad \int_0^1 xe^x \, dx = \Big[xe^x \Big]_0^1 - \int_0^1 e^x \, dx$$

$$= e - \Big[e^x \Big]_0^1 = e - (e - 1) = 1.$$

Hence $\int_0^1 x^2e^x\,dx = e - 2\int_0^1 xe^x\,dx = e - 2.$

Integration by parts is also used to integrate functions which are not products.

Example 4 Find $\int \ln x\,dx$.

In rule (1), let $u = \ln x,\quad \dfrac{dv}{dx} = 1,$

$$\frac{du}{dx} = \frac{1}{x}, \qquad v = x,$$

$$\therefore \quad \int \ln x\,dx = x\ln x - \int\left(x \times \frac{1}{x}\right)dx$$

$$= x\ln x - \int dx = x\ln x - x + c.$$

The next example illustrates a further way in which integration by parts may be used.

Example 5 Find $\int e^x \cos x\,dx$.

Integrating by parts,

$$\int e^x\cos x\,dx = e^x\sin x - \int e^x \sin x\,dx \tag{i}$$

Integrating by parts,

$$\int e^x\sin x\,dx = e^x(-\cos x) - \int(-\cos x)e^x\,dx$$

$$= -e^x\cos x + \int e^x\cos x\,dx$$

Substituting this expression for $\int e^x\sin x\,dx$ in (i),

$$\int e^x\cos x\,dx = e^x\sin x - \left(-e^x\cos x + \int e^x\cos x\,dx\right)$$

$$= e^x\sin x + e^x\cos x - \int e^x\cos x\,dx$$

$$\therefore \quad 2\int e^x\cos x\,dx = e^x\sin x + e^x\cos x + k$$

Hence $\int e^x\cos x\,dx = \tfrac{1}{2}e^x(\sin x + \cos x) + c.$

Exercise 4.4

Find the following indefinite integrals.

1. $\int x \sin x \, dx$
2. $\int x \cos \frac{1}{2}x \, dx$
3. $\int x e^{-x} \, dx$
4. $\int x \ln 2x \, dx$
5. $\int x e^{2x} \, dx$
6. $\int x \sec^2 x \, dx$
7. $\int x \sin x \cos x \, dx$
8. $\int x \tan^2 x \, dx$
9. $\int x \cos^2 x \, dx$
10. $\int x^2 \cos x \, dx$
11. $\int x^3 e^x dx$
12. $\int x^3 \ln x \, dx$
13. $\int x(\ln x)^2 \, dx$
14. $\int \ln(x - 1) \, dx$
15. $\int e^{-x} \sin x \, dx$

Find the following definite integrals.

16. $\int_0^{\pi} x \cos x \, dx$
17. $\int_{-1}^{1} x e^x \, dx$
18. $\int_0^{\pi/3} x \sin 3x \, dx$
19. $\int_1^e \ln x \, dx$
20. $\int_0^{\pi} x \sin^2 x \, dx$
21. $\int_1^2 x^2 \ln x \, dx$
22. $\int_2^3 x^2 e^{-x} \, dx$
23. $\int_1^e \frac{1}{x^3} \ln x \, dx$
24. $\int_0^{\pi/2} x^2 \sin x \, dx$
25. $\int_0^{\pi} e^x \sin x \, dx$
26. $\int_1^2 x^2 e^{2x} \, dx$
27. $\int_e^{e^2} \ln(\sqrt{x}) \, dx$

28. Find $\int x(x + 1)^4 \, dx$ using (a) integration by parts, (b) the substitution $x + 1 = t$.

29. Find, using integration by parts or otherwise,

(a) $\int x\sqrt{(x - 1)} \, dx$, (b) $\int \frac{x}{(x + 2)^3} \, dx$.

30. Use integration by parts to show that

(a) $2\int \cos^2 x \, dx = \cos x \sin x + x + c$,

(b) $3\int \cos^3 x \, dx = \cos^2 x \sin x + 2 \sin x + c$.

4.5 Use of inverse trigonometric functions

The inverse trigonometric functions $\sin^{-1}x$ and $\tan^{-1}x$ are defined as follows:

$$y = \sin^{-1} x \;\Rightarrow\; \sin y = x \quad \text{and} \quad -\tfrac{1}{2}\pi \leqslant y \leqslant \tfrac{1}{2}\pi$$
$$y = \tan^{-1} x \;\Rightarrow\; \tan y = x \quad \text{and} \quad -\tfrac{1}{2}\pi < y < \tfrac{1}{2}\pi$$

In Book 1 two important integration results were obtained by differentiating $\sin^{-1}\dfrac{x}{a}$ and $\tan^{-1}\dfrac{x}{a}$.

$$\int \frac{1}{\sqrt{(a^2 - x^2)}}\,dx = \sin^{-1}\frac{x}{a} + c, \quad \int \frac{1}{a^2 + x^2}\,dx = \frac{1}{a}\tan^{-1}\frac{x}{a} + c.$$

[These integrals can also be found by substituting $x = a\sin\theta$ and $x = a\tan\theta$ respectively.]

Example 1 Evaluate (a) $\displaystyle\int_0^1 \frac{1}{\sqrt{(4 - x^2)}}\,dx$, (b) $\displaystyle\int_0^2 \frac{1}{4 + x^2}\,dx$.

(a) $$\int_0^1 \frac{1}{\sqrt{(4 - x^2)}}\,dx = \left[\sin^{-1}\frac{x}{2}\right]_0^1 = \sin^{-1}\frac{1}{2} - \sin^{-1}0 = \frac{\pi}{6}.$$

(b) $$\int_0^2 \frac{1}{4 + x^2}\,dx = \left[\frac{1}{2}\tan^{-1}\frac{x}{2}\right]_0^2 = \frac{1}{2}\tan^{-1}1 - \frac{1}{2}\tan^{-1}0 = \frac{\pi}{8}.$$

Example 2 Find (a) $\displaystyle\int \frac{1}{\sqrt{(9 - 4x^2)}}\,dx$, (b) $\displaystyle\int \frac{1}{25 + 16x^2}\,dx$.

(a) Let $2x = u$, then $4x^2 = u^2$ and $2dx = du$,

$$\therefore \quad \int \frac{1}{\sqrt{(9 - 4x^2)}}\,dx = \frac{1}{2}\int \frac{1}{\sqrt{(9 - u^2)}}\,du$$
$$= \frac{1}{2}\sin^{-1}\frac{u}{3} + c.$$
$$= \frac{1}{2}\sin^{-1}\frac{2x}{3} + c.$$

(b) Let $4x = u$, then $16x^2 = u^2$ and $4dx = du$,

$$\therefore \quad \int \frac{1}{25 + 16x^2}\,dx = \frac{1}{4}\int \frac{1}{25 + u^2}\,du$$
$$= \frac{1}{4} \times \frac{1}{5}\tan^{-1}\frac{u}{5} + c$$
$$= \frac{1}{20}\tan^{-1}\frac{4x}{5} + c.$$

The method of Example 2 can be extended to integrals of the form

$$\int \frac{1}{\sqrt{(ax^2+bx+c)}}\,dx, \quad \int \frac{1}{ax^2+bx+c}\,dx.$$

Example 3 Find $\int \frac{1}{x^2-2x+2}\,dx.$

$$x^2-2x+2 = (x^2-2x+1)+1 = (x-1)^2+1$$

$$\therefore \quad \int \frac{1}{x^2-2x+2}\,dx = \int \frac{1}{(x-1)^2+1}\,dx$$

$$= \int \frac{1}{u^2+1}\,du \qquad \text{letting } x-1=u,\ dx=du$$

$$= \tan^{-1}u + c$$

$$= \tan^{-1}(x-1)+c.$$

The remaining examples illustrate further uses of inverse trigonometric functions.

Example 4 Find $\int \frac{x+3}{x^2+9}\,dx.$

$$\int \frac{x+3}{x^2+9}\,dx = \int \left\{\frac{x}{x^2+9}+\frac{3}{x^2+9}\right\}dx = \frac{1}{2}\int \frac{2x}{x^2+9}\,dx + 3\int \frac{1}{x^2+9}\,dx$$

$$= \frac{1}{2}\ln(x^2+9)+\tan^{-1}\frac{x}{3}+c.$$

Example 5 Evaluate $\int_0^1 \frac{3-x}{(x+1)(x^2+1)}\,dx.$

Let $\frac{3-x}{(x+1)(x^2+1)} \equiv \frac{A}{x+1}+\frac{Bx+C}{x^2+1}$

then $\quad 3-x \equiv A(x^2+1)+(Bx+C)(x+1)$

Putting $x=-1$: $\quad 4=2A \quad \therefore \quad A=2$

Putting $x=0$: $\quad 3=A+C \quad \therefore \quad C=1$

Comparing coefficients of x^2:

$$0=A+B \quad \therefore \quad B=-2$$

$$\therefore \quad \int_0^1 \frac{3-x}{(x+1)(x^2+1)}\,dx = \int_0^1 \left\{\frac{2}{x+1}+\frac{(-2x+1)}{x^2+1}\right\}dx$$

$$= \int_0^1 \left\{\frac{2}{x+1}-\frac{2x}{x^2+1}+\frac{1}{x^2+1}\right\}dx$$

$$= \Big[2\ln(x+1)-\ln(x^2+1)+\tan^{-1}x\Big]_0^1$$

$$= (2\ln 2-\ln 2+\tan^{-1}1)-(2\ln 1-\ln 1+\tan^{-1}0)$$

$$= \ln 2+\tfrac{1}{4}\pi$$

Example 6 Find $\int \frac{2x^2}{\sqrt{(1-x^2)}}\,dx$.

Let $x = \sin\theta$
then $dx = \cos\theta\,d\theta$

$$1 - x^2 = 1 - \sin^2\theta = \cos^2\theta$$

$$\therefore \quad \sqrt{(1-x^2)} = \cos\theta$$

$$\begin{aligned}\int \frac{2x^2}{\sqrt{(1-x^2)}}\,dx &= \int \frac{2\sin^2\theta}{\cos\theta} \times \cos\theta\,d\theta \\ &= \int (1 - \cos 2\theta)\,d\theta \\ &= \theta - \tfrac{1}{2}\sin 2\theta + c \\ &= \theta - \sin\theta\cos\theta + c \\ &= \sin^{-1}x - x\sqrt{(1-x^2)} + c.\end{aligned}$$

Exercise 4.5

Find the following indefinite integrals.

1. $\int \frac{1}{\sqrt{(16-x^2)}}\,dx$
2. $\int \frac{1}{x^2+25}\,dx$
3. $\int \frac{1}{\sqrt{(1-4x^2)}}\,dx$
4. $\int \frac{x+1}{x^2+1}\,dx$
5. $\int \frac{1}{\sqrt{(4-25x^2)}}\,dx$
6. $\int \frac{1}{4+9x^2}\,dx$
7. $\int \frac{x^2-x}{(x+1)(x^2+1)}\,dx$
8. $\int \frac{4}{(1-x)(1+x^2)}\,dx$
9. $\int \frac{6(x-3)}{x(x^2+9)}\,dx$
10. $\int \frac{5x^2+4x-20}{(x+2)(x^2+4)}\,dx$
11. $\int \frac{1}{x^2+2x+5}\,dx$
12. $\int \frac{1}{\sqrt{(5+4x-x^2)}}\,dx$

Evaluate the following definite integrals.

13. $\int_0^{\sqrt{3}} \frac{1}{x^2+9}\,dx$
14. $\int_0^5 \frac{1}{\sqrt{(25-x^2)}}\,dx$
15. $\int_0^{1/2} \frac{1}{4x^2+1}\,dx$
16. $\int_{-2}^{2} \frac{x+2}{x^2+4}\,dx$
17. $\int_0^1 \frac{4}{(x+1)(x^2+1)}\,dx$
18. $\int_3^4 \frac{1}{x^2-6x+10}\,dx$
19. $\int_{-1/2}^{1/2} \frac{1}{\sqrt{(3-4x-4x^2)}}\,dx$

Use appropriate substitutions to find the following indefinite integrals.

20. $\displaystyle\int \frac{e^x + e^{2x}}{1 + e^{2x}}\,dx$

21. $\displaystyle\int \frac{1}{e^x + e^{-x}}\,dx$

22. $\displaystyle\int \sqrt{(1 - x^2)}\,dx$

23. $\displaystyle\int \frac{1 - x}{\sqrt{(1 - x^2)}}\,dx$

Use the given substitutions to find the following indefinite integrals.

24. $\displaystyle\int \frac{2x}{x^2 - 2x + 10}\,dx,\ x = u + 1.$

25. $\displaystyle\int \frac{x}{x^4 + 2x^2 + 2}\,dx,\ x^2 = u - 1.$

26. $\displaystyle\int \frac{1}{x\sqrt{(x^2 - 1)}}\,dx,\ x = \frac{1}{u}.$

27. Evaluate correct to 3 significant figures

(a) $\displaystyle\int_1^3 \frac{4 - 3x}{(2x - 1)(x^2 + 1)}\,dx,$ (b) $\displaystyle\int_0^1 \frac{7x + 2}{(x + 1)^2(x^2 + 4)}\,dx$

28. Use integration by parts to show that

$$\int \sin^{-1} x\,dx = x \sin^{-1} x + \sqrt{(1 - x^2)} + c.$$

4.6 Properties of the definite integral

The definite integral is defined formally as the limit of a sum by considering a function $f(x)$ in the interval $a \leqslant x \leqslant b$.

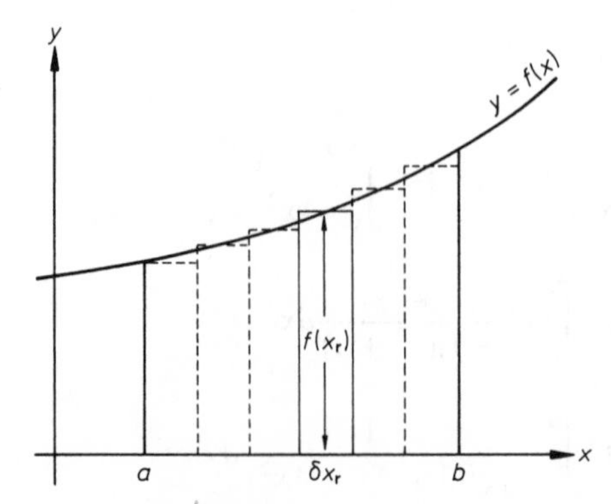

Suppose that this interval is divided into n subintervals of lengths $\delta x_1, \delta x_2, \ldots, \delta x_n$. If n is increased in such a way that the values of $\delta x_1, \delta x_2, \ldots, \delta x_n$ tend to zero, then

$$\int_a^b f(x)\,dx = \lim_{n\to\infty} \left\{ \sum_{r=1}^{n} f(x_r)\,\delta x_r \right\},$$

where x_r lies in the interval of length δx_r.

This definition may be expressed more simply as

$$\int_a^b y\,dx = \lim_{\delta x \to 0} \sum y\,\delta x.$$

In general the value of this integral represents the area bounded by the curve $y = f(x)$, the x-axis, the lines $x = a$ and $x = b$. We must remember, however, that areas above the x-axis are positive and areas below the x-axis are negative.

The most important properties of the definite integral are as follows:

$$\int_a^b f(x)\,dx + \int_a^b g(x)\,dx = \int_a^b \{f(x) + g(x)\}\,dx.$$

$$\int_a^b kf(x)\,dx = k\int_a^b f(x)\,dx, \quad \text{where } k \text{ is a constant.}$$

$$\int_b^a f(x)\,dx = -\int_a^b f(x)\,dx,$$

$$\int_a^b f(x)\,dx + \int_b^c f(x)\,dx = \int_a^c f(x)\,dx.$$

If $f(x)$ is an *even function* then the curve $y = f(x)$ will be symmetrical about the y-axis. Hence the area under the curve between $x = -a$ and $x = 0$ is equal to the area between $x = 0$ and $x = a$.

$\therefore$ if $f(x)$ is an even function $\displaystyle\int_{-a}^a f(x)\,dx = 2\int_0^a f(x)\,dx.$

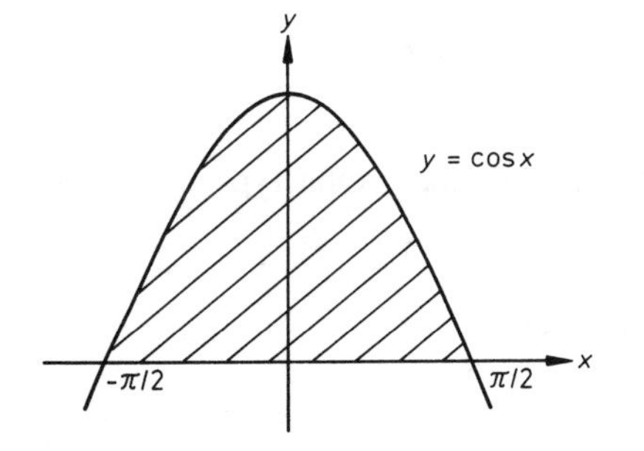

For instance, if $f(x) = \cos x$,
$f(-x) = \cos(-x) = \cos x = f(x)$

$$\therefore \int_{-\pi/2}^{\pi/2} \cos x\,dx = 2\int_0^{\pi/2} \cos x\,dx$$

$$= 2\Big[\sin x\Big]_0^{\pi/2} = 2.$$

If $f(x)$ is an *odd function*, the areas bounded by the curve $y = f(x)$ and the x-axis between $x = -a$ and $x = 0$ and between $x = 0$ and $x = a$ are numerically equal but opposite in sign,

$\therefore$ if $f(x)$ is an odd function $\displaystyle\int_{-a}^a f(x)\,dx = 0.$

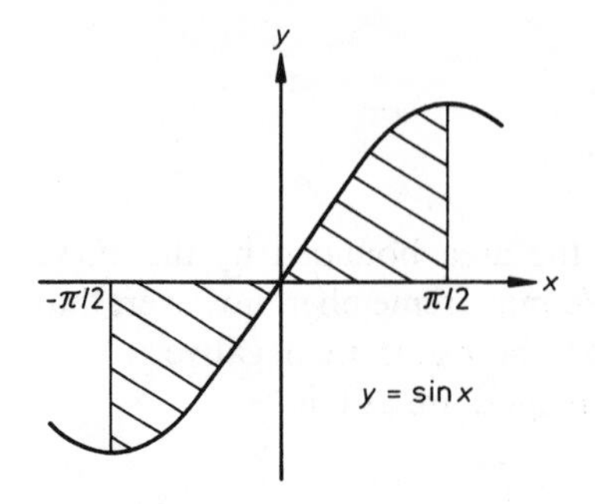

For instance, if $f(x) = \sin x$,
$f(-x) = \sin(-x) = -\sin x = -f(x)$

$$\therefore \quad \int_{-\pi/2}^{\pi/2} \sin x \, dx = 0.$$

The symmetries of the sine function can also be used when integrating functions of the form $xf(\sin x)$. We illustrate the method by a simple example.

Example 1 Evaluate $I = \displaystyle\int_0^{\pi} x \sin x \, dx.$

$$I = -\int_{\pi}^{0} (\pi - u)\sin(\pi - u) \, du$$

Let $x = \pi - u$
then $dx = -du$

$x = \pi \Rightarrow u = 0$
$x = 0 \Rightarrow u = \pi$

$$= \int_0^{\pi} (\pi - u)\sin u \, du$$

$$= \int_0^{\pi} \pi \sin u \, du - \int_0^{\pi} u \sin u \, du$$

$$= \pi \int_0^{\pi} \sin u \, du - I$$

$$\therefore \quad I = \tfrac{1}{2}\pi \int_0^{\pi} \sin u \, du = \tfrac{1}{2}\pi \Big[-\cos u \Big]_0^{\pi} = \tfrac{1}{2}\pi\{1 - (-1)\} = \pi.$$

Finally we consider definite integrals with *infinite limits*. One type of infinite integral is defined as follows:

$$\int_0^{\infty} f(x) \, dx = \lim_{t \to \infty} \int_0^{t} f(x) \, dx,$$

provided that this limit exists.

Example 2 Evaluate $\displaystyle\int_0^{\infty} e^{-x} \, dx.$

$$\int_0^{t} e^{-x} \, dx = \Big[-e^{-x} \Big]_0^{t} = -e^{-t} + 1$$

$$\therefore \quad \int_0^{\infty} e^{-x} \, dx = \lim_{t \to \infty} (-e^{-t} + 1) = 1.$$

Exercise 4.6

1. Evaluate the following integrals using the properties of odd and even functions.

(a) $\int_{-1}^{1} x^4\,dx$, (b) $\int_{-3}^{3} x(x^2-1)\,dx$, (c) $\int_{-\pi}^{\pi} \sin^2 x\,dx$,

(d) $\int_{-\pi/3}^{\pi/3} \frac{1}{1+\cos 2x}\,dx$, (e) $\int_{-4}^{4} \frac{x}{x^2-25}\,dx$, (f) $\int_{-1}^{1} \frac{x\cos x}{1+\tan^2 x}\,dx$.

2. Sketch the curve $y=(x^2-1)(x-3)$. Hence evaluate $\int_{-7}^{9} (x^2-1)(x-3)\,dx$.

3. Using sketch graphs, or otherwise, evaluate

(a) $\int_{-2}^{2} [x]\,dx$, (b) $\int_{-2}^{2} |x|\,dx$, (c) $\int_{-2}^{2} (x+|x|)\,dx$.

4. Evaluate (a) $\int_0^3 (x-2)\,dx$, (b) $\left|\int_0^3 (x-2)\,dx\right|$, (c) $\int_0^3 |x-2|\,dx$.

5. Evaluate (a) $\int_0^3 x(x-2)\,dx$, (b) $\int_0^3 |x(x-2)|\,dx$.

6. Prove that $\int_0^a f(x)\,dx = \int_0^a f(a-x)\,dx$. Hence, or otherwise, show that

(a) $\int_0^{\pi} \frac{x\sin x}{4-\cos^2 x}\,dx = \frac{1}{4}\pi\ln 3$, (b) $\int_0^{\pi} \frac{x\sin x}{1+\cos^2 x}\,dx = \frac{1}{4}\pi^2$.

7. Evaluate $\int_0^1 e^x\,dx$. Hence show that

$$\lim_{n\to\infty} \frac{1}{n}\left\{e^{1/n}+e^{2/n}+e^{3/n}+\ldots+e\right\} = e-1.$$

8. Evaluate $\int_0^1 \frac{x\,dx}{\sqrt{(1+x)}}$. By considering integration as the limit of a sum, or otherwise, show that

$$\lim_{n\to\infty} \frac{1}{n\sqrt{n}}\left\{\frac{1}{\sqrt{(n+1)}}+\frac{2}{\sqrt{(n+2)}}+\ldots+\frac{n}{\sqrt{(2n)}}\right\} = \frac{2}{3}(2-\sqrt{2}). \quad \text{(JMB)}$$

9. Explain what is wrong with the following mathematical arguments and evaluate the integrals I and J correctly:

(i) Let $I = \int_0^2 \frac{1}{x^2-2x+2}\,dx$. Then putting $x-1=\cot\theta$ we have

$$I = \int_{\cot^{-1}(-1)}^{\cot^{-1}(1)} \frac{-\operatorname{cosec}^2\theta}{\cot^2\theta+1}\,d\theta = -\int_{-\pi/4}^{\pi/4} d\theta = -\tfrac{1}{2}\pi.$$

(ii) Let $J = \int_0^{\pi} e^{|\sin x|}|\cos x|\,dx$. Then putting $u = |\sin x|$ we have $du = |\cos x|\,dx$ and

so $J = \int_0^0 e^u\,du = 0.$ (C)

10. Evaluate (a) $\int_0^{\infty} e^{-x/2}\,dx$, (b) $\int_1^{\infty} \frac{1}{x^2}\,dx$, (c) $\int_{-\infty}^{\infty} \frac{1}{1+x^2}\,dx$.

11. Use appropriate substitutions to show that

(i) $\int_1^{a^k} \frac{1}{t}\,dt = k\int_1^a \frac{1}{u}\,du$, (ii) $\int_a^{ab} \frac{1}{t}\,dt = \int_1^b \frac{1}{u}\,du$.

Defining $\ln x$ as $\int_1^x \frac{1}{t}\,dt$, deduce that

(iii) $\ln(a^k) = k\ln a$, (iv) $\ln(ab) = \ln a + \ln b$.

Exercise 4.7 *(miscellaneous)*

In questions 1 to 15 evaluate the given definite integrals.

1. $\int_0^4 \frac{x}{2x+1}\,dx$
2. $\int_{-1/2}^0 \frac{x}{(x+1)^2}\,dx$
3. $\int_1^2 \frac{1}{4x^2-1}\,dx$
4. $\int_{-1}^4 \frac{1}{\sqrt{(3x+4)}}\,dx$
5. $\int_2^3 \frac{x^2+1}{x(x-1)}\,dx$
6. $\int_0^2 (4-x^2)\sqrt{x}\,dx$
7. $\int_0^2 x\sqrt{(4-x^2)}\,dx$
8. $\int_0^{\pi} (\sin x + \cos x)^2\,dx$
9. $\int_0^{\pi/3} 2\sin 2x\cos x\,dx$
10. $\int_0^{\pi/4} \tan^3 x\,dx$
11. $\int_0^{\pi/4} \sin^3 x\cos^3 x\,dx$
12. $\int_{\ln 2}^{\ln 5} \frac{2}{e^x - e^{-x}}\,dx$
13. $\int_0^{\pi/4} \frac{\sin 2x}{1+\cos^2 x}\,dx$
14. $\int_0^1 x^2\sqrt{(1-x^2)}\,dx$
15. $\int_0^1 2^x\,dx$

16. Find $\int \frac{2x-1}{(x+2)(x-3)^2}\,dx$, given that $x > 3$.

17. Show that $\int_0^3 \frac{x+3}{x^2+3}\,dx = \ln 2 + \frac{\pi\sqrt{3}}{3}$.

In questions 18 to 21 use the given substitutions to evaluate the definite integrals.

18. $\int_{\sqrt{2}}^{2} \frac{1}{x\sqrt{(x^2-1)}}\,dx$, $x = \sec\theta$.

19. $\int_0^{1/2} \sqrt{\left(\frac{x}{1-x}\right)}\,dx$, $x = \sin^2\theta$.

20. $\int_0^1 \frac{1}{(1+x^2)}\,dx$, $x = \tan\theta$.

21. $\int_0^{\pi/6} \sec x\,dx$, $\sin x = u$.

22. Use integration by parts and the result of question 21 to show that

$$\int_0^{\pi/6} \sec^3 x\,dx = \tfrac{1}{3} + \tfrac{1}{4}\ln 3.$$

23. Show that the curve $y = \dfrac{3}{(2x+1)(1-x)}$ has only one turning point. Find the coordinates of this point and determine its nature. Sketch the curve. Find the area of the region enclosed by the curve and the line $y = 3$. (JMB)

24. Find the coordinates of the turning point of the curve $y = (x^2-4)/(x+1)^2$ and ascertain the nature of this turning point. Calculate the area of the region enclosed by the curve, the y-axis and the x-axis between $x = 0$ and $x = 2$. (AEB 1978)

25. If $f(x) = \dfrac{x^2+x+2}{(x^2+1)(x-1)}$ express $f(x)$ in partial fractions and hence evaluate $\int_2^3 f(x)\,dx$ and $\int_{-3}^{-2} f(x)\,dx$ each correct to 2 places of decimals. (SU)

26. The region enclosed by the loop of the curve $y^2(3-x) = x^2(3+x)$ is rotated about the x-axis through four right angles. Show that the volume of the solid of revolution is $18\pi(3\ln 2 - 2)$. (C)

27. (i) Given that $0 < a < b$, sketch the graph of $y = |x-a|$ for $-b \leqslant x \leqslant b$. Hence, or otherwise, find $\int_{-b}^{b} |x-a|\,dx$.

(ii) Find the value of $\int_1^2 \log_{10} x\,dx$ leaving your answer in terms of logarithms.

(iii) Using the substitution $x = \tfrac{1}{2}\sin\theta$, or otherwise, evaluate $\int_0^{1/4} \sqrt{(1-4x^2)}\,dx$.

(L)

28. Sketch the curve given in terms of a parameter t by $x = 2a + at^2$, $y = 2at$ where a is a positive constant. Find the equation of the tangent and of the normal to the curve at the point where $t = 1$. Determine the area of the finite region bounded by the curve and the straight line joining the points at which $t = \pm 1$. (L)

29. A curve is defined parametrically by the equations $x = \cos t$, $y = \sin^3 t$, $-\pi \leqslant t < \pi$. Show that the curve is symmetrical about each of the coordinate axes. Find dy/dx in terms of t and deduce the equation of the tangent to the curve at the point with parameter t_1. Sketch the curve. Show that the area, A, of the region enclosed by the curve is given by $A = 4\int_0^{\pi/2} \sin^4 t\, dt$ and use the relation $2\sin^2\theta = 1 - \cos 2\theta$ to deduce that $A = 3\pi/4$. (L)

30. The curve whose equation is $y = (1 - x)e^x$ meets the x-axis at A and the y-axis at B. The region bounded by the arc AB of the curve and the line segments OA and OB, where O is the origin, is rotated through a complete revolution about the x-axis. Show that the volume swept out is $\frac{1}{4}\pi(e^2 - 5)$. (JMB)

31. Obtain $\int x^2 \cos 2x\, dx$ and hence prove that $\int_0^{\pi} x^2 \cos 2x\, dx = \frac{1}{2}\pi$. Find the area of the region enclosed by the curve $y = x \sin x$, for $0 \leqslant x \leqslant \pi$, and the x-axis. The region is rotated through 2π radians about the x-axis. Find the volume of the solid of revolution thus generated. From your results, or otherwise, find the volume of the solid of revolution generated by rotation of the same region through 2π radians about the line $y = -\frac{1}{8}$. (O&C)

32. Express $y = \dfrac{2x^2 + 3x + 5}{(x + 1)(x^2 + 3)}$ in partial fractions and hence show that $\dfrac{dy}{dx} = -\dfrac{2}{3}$ when $x = 0$. Evaluate $\int_0^{\sqrt{3}} y\, dx$ and state the mean value of y in the interval $0 \leqslant x \leqslant \sqrt{3}$. (AEB 1979)

33. Determine the values of p and q for which $x^2 - 4x + 5 \equiv (x - p)^2 + q$ and hence evaluate $\displaystyle\int_2^3 \frac{1}{x^2 - 4x + 5}\, dx$. Calculate also $\displaystyle\int_2^3 \frac{2x - 4}{x^2 - 4x + 5}\, dx$ and deduce, or find otherwise, the value of $\displaystyle\int_2^3 \frac{2x}{x^2 - 4x + 5}\, dx$. (JMB)

34. (a) Using the formula for $\cos(A + B)$, express $\cos\theta$ in terms of $t = \tan\frac{1}{2}\theta$.

(b) Express $\dfrac{10(1 + t^2)}{(1 - t^2)(1 + 9t^2)}$ in partial fractions.

(c) With the help of (a) and (b), and using the substitution $t = \tan\frac{1}{2}\theta$, show that $\displaystyle\int_0^{\pi/3} \frac{5\, d\theta}{5\cos\theta - 4\cos^2\theta} = \log_e(2 + \sqrt{3}) + \frac{8}{9}\pi$. (W)

35. (a) Using integration by parts, or otherwise, evaluate

$$\int \log_e x \, dx \quad \text{and} \quad \int (\log_e x)^2 \, dx.$$

(b) Sketch the graph of the function $\log_e x$. Let A and C denote the points $(1, 0)$ and $(e, 1)$ on the graph, and let B denote the point $(e, 0)$. The region bounded by the graph and the lines, AB, BC is rotated about the x-axis; find the volume of the solid so formed. Show that the volume of the solid formed by rotating the same region about the y-axis is $\frac{1}{2}\pi(e^2 + 1)$.

(W)

36. Evaluate (i) $\displaystyle\int_0^{\pi/2} \frac{\sin^3\theta}{2 - \sin^2\theta} d\theta$, (ii) $\displaystyle\int_0^{\sqrt{2}} xe^{-x^2} dx$.

37. Let $I = \displaystyle\int_0^{\pi} \frac{x\,dx}{1 + \sin x}$. Show, by means of the substitution $y = \pi - x$ that $I = \pi \displaystyle\int_0^{\pi/2} \frac{dx}{1 + \sin x}$. Hence, by means of the substitution $t = \tan\frac{1}{2}x$, or otherwise, evaluate I.

(JMB)

38. Draw the circle $x^2 + y^2 = a^2$ and, on the same diagram, the line $x = a/2$. Find by integration the smaller area between the line and the circumference of the circle and the volume generated when this area is rotated through 4 right-angles about the x-axis.

(SU)

39. The area A in the first quadrant bounded by the curve $(2 - x)y^2 = x$, the x-axis and the line $x = 1$ is rotated completely about the x-axis to form a solid of volume V. Find: (i) the volume V, (ii) the area A.

(AEB 1976)

40. Find the area of the region enclosed between the curves $y = x^2/2a$ and $y = a - x^4/2a^3$ $(a > 0)$. Find also the volume of the solid obtained by rotating this region through two right angles about the y-axis.

(O)

41. Sketch the curve $y^2 = x^2(4 - x^2)$. Find the area enclosed by one loop of the curve. Find also the volume of the solid generated when this loop is rotated through two right angles about the x-axis.

42. The domain of the function defined by $f(x) = e^{-x}\sin x$ is the set of real numbers in the interval $0 \leqslant x \leqslant 4\pi$. Show that stationary values of this function occur for values of x for which $\tan x = 1$. Determine the nature of these stationary values. Sketch the graph of the function. By integrating by parts twice, or otherwise, evaluate $\displaystyle\int_0^{\pi} f(x)\,dx$. Show that the ratio of $f(a + 2\pi)$ to $f(a)$ is independent of the value of a, and hence write down the value of $\displaystyle\int_{2\pi}^{3\pi} f(x)\,dx$.

(JMB)

43. By interpreting $\int_a^b f(x)\,dx$ as an area, show that

(i) $\int_0^\pi \sin^5 x \cos^5 x\,dx = 0,$ (ii) $\int_0^\pi e^{2x} \cos x\,dx < 0,$

(iii) for $n > 1, \dfrac{1}{n+1}\left(\dfrac{\pi}{4}\right)^{n+1} < \int_0^{\pi/4} \tan^n x\,dx < \dfrac{\pi}{8}.$ (JMB)

44. (a) By considering $\int x^{1/2}\,dx$ between suitable limits, or otherwise, prove that $\frac{2}{3}n^{3/2} < \sqrt{1} + \sqrt{2} + \cdots + \sqrt{n} < \frac{2}{3}\{(n+1)^{3/2} - 1\}.$

(b) By considering an appropriate integral, prove that

$$\frac{1}{k}\left\{1 - \frac{1}{(n+1)^k}\right\} < \sum_{r=1}^{n} \frac{1}{r^{k+1}} < \frac{1}{k}\left\{k + 1 - \frac{1}{n^k}\right\}, \text{ where } k > 0.$$

(O&C)

5 Coplanar forces; centre of gravity

5.1 Parallel forces, moments and couples

In Book 1 we considered the action of forces on bodies small enough to be treated as particles and represented by single points in space. Applying Newton's laws of motion, we found that if the forces acting on such a body have a non-zero resultant then the effect produced is an acceleration in the direction of the resultant. However, if the vector sum of the forces on the body is zero, the body either remains at rest or continues to move with unchanged velocity.

There are many situations in which this elementary approach is inadequate. From experience we know that the forces on a body may tend to produce rotation as well as the types of motion discussed in earlier work. A projectile, such as a cricket ball, may spin as it moves. A block placed on an inclined plane may topple rather than slide. Thus, when solving a problem in which the rotational effects of forces could be important, we must use a more realistic representation of the bodies involved. We will then find that the effect produced by a set of forces depends not only on the magnitude and direction of the resultant, but also on its line of action.

We now develop these ideas by considering the effect of coplanar forces on a variety of rigid bodies i.e. bodies whose shapes are not changed by the forces acting on them.

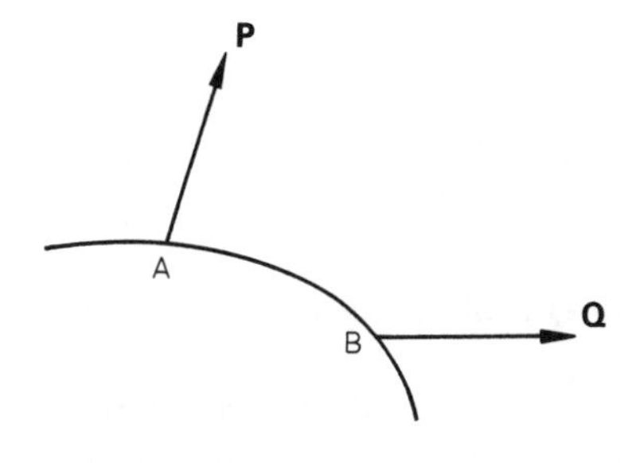

Let **P** and **Q** be *non-parallel* coplanar forces acting at points A and B of a rigid body. The magnitude and direction of their resultant **R** can be found using the parallelogram law. If the lines of action of **P** and **Q** intersect at the point C, then the line of action of **R** also passes through C.

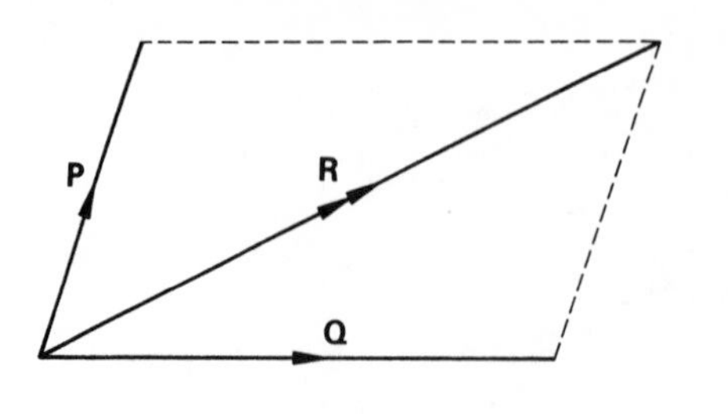

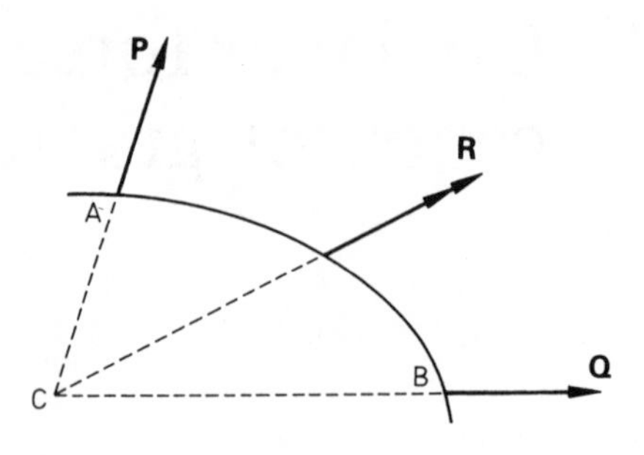

To find the line of action of the resultant of *parallel* forces **P** and **Q** acting at points A and B of a rigid body, we introduce forces **F** and $-$**F** acting along AB. The application of these equal and opposite forces will have no effect on the body and leave the line of action of the resultant unaltered.

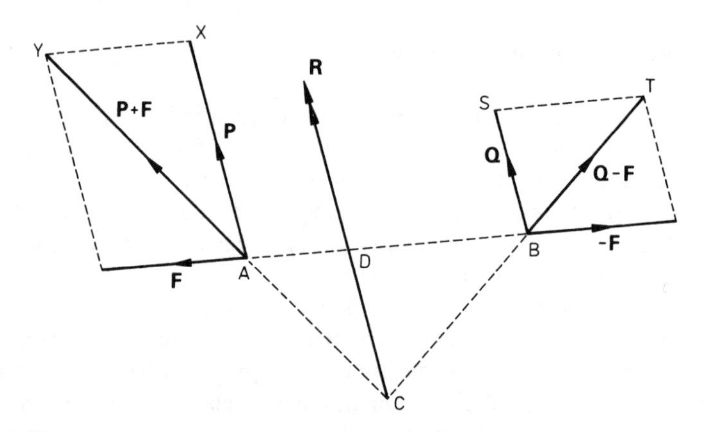

The diagram shows how the parallelogram law is used to construct a point C on the line of action of the resultant **R**, when **P** and **Q** are *like parallel forces* i.e. forces in the same sense.

In similar triangles ADC, YXA: $\quad \dfrac{AD}{DC} = \dfrac{YX}{XA} = \dfrac{F}{P}$

In similar triangles DBC, STB: $\quad \dfrac{DC}{DB} = \dfrac{SB}{ST} = \dfrac{Q}{F}$

$$\therefore \quad \frac{AD}{DB} = \frac{AD}{DC} \times \frac{DC}{DB} = \frac{F}{P} \times \frac{Q}{F} = \frac{Q}{P}$$

By vector addition, the magnitude of **R** is $P + Q$.
Hence the resultant of the like parallel forces **P** and **Q** is of magnitude $P + Q$ and divides AB internally in the ratio $Q:P$.

Using similar arguments it can be shown that the resultant of *unlike parallel forces* **P** and **Q**, where $P > Q$, is of magnitude $P - Q$ and divides AB externally in the ratio $Q:P$.

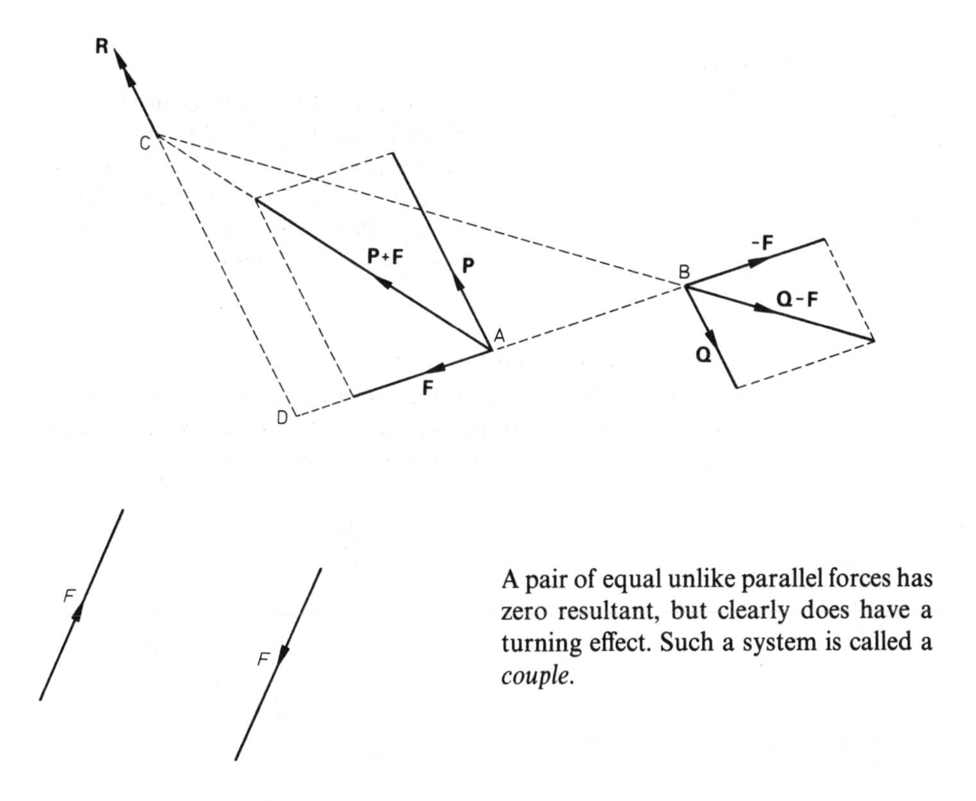

A pair of equal unlike parallel forces has zero resultant, but clearly does have a turning effect. Such a system is called a *couple.*

A quantity called the *moment of a force* or *torque* is used as a measure of "turning effect". The moment of a force about a given point is defined as the product of the magnitude of the force and its distance from the point. If the line of action of a force $\mathbf{F}$ is at a perpendicular distance d from a point O, then the moment of $\mathbf{F}$ about O is Fd.

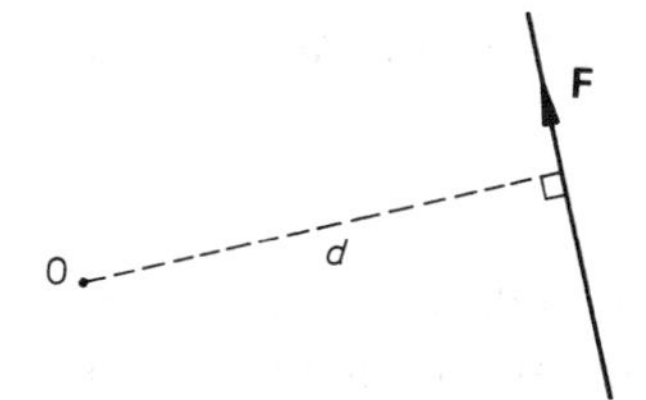

[In three-dimensional work the quantity Fd is more accurately described as the moment or torque about an axis through O perpendicular to the plane containing O and the force $\mathbf{F}$.]

The SI unit of torque is the newton metre (N m). Thus, if a force of magnitude 40 N acts along a line whose perpendicular distance from a point A is 2 m, then the moment of the force about A is 80 N m.

When finding the sum of the moments of a set of coplanar forces about a point, the moment of each force is taken to be positive or negative according to the sense of the rotation it would produce. The moment of a force about any point on its line of action is zero.

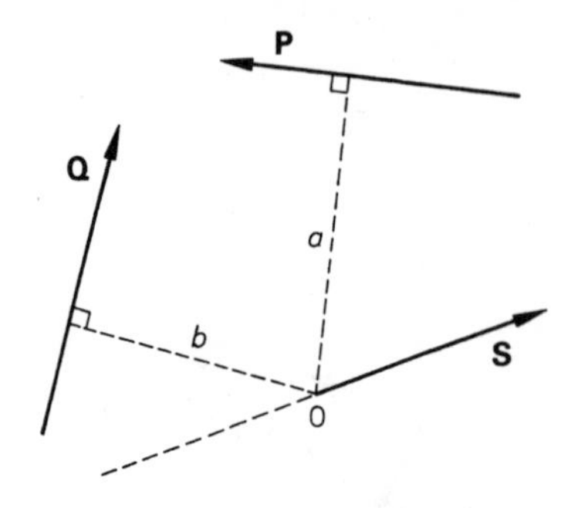

For instance, if forces **P**, **Q** and **S** act as shown in the diagram, then the sum of the anti-clockwise moments of **P**, **Q** and **S** about O is $Pa - Qb$, i.e. the combined turning effect of the system about O is $Pa - Qb$ anti-clockwise.

We now show that the sum of the moments of two coplanar forces about any point in their plane is equal to the moment of their resultant about the same point.

Suppose that **P** and **Q** are forces whose lines of action intersect at a point C. Let **R** be their resultant and let O be a point in the same plane such that $OC = d$.

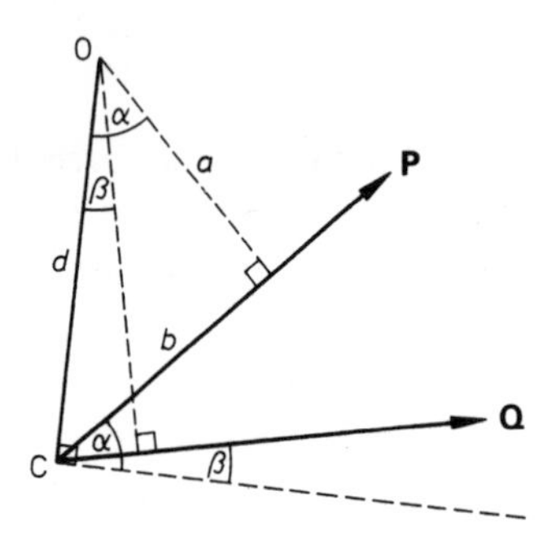

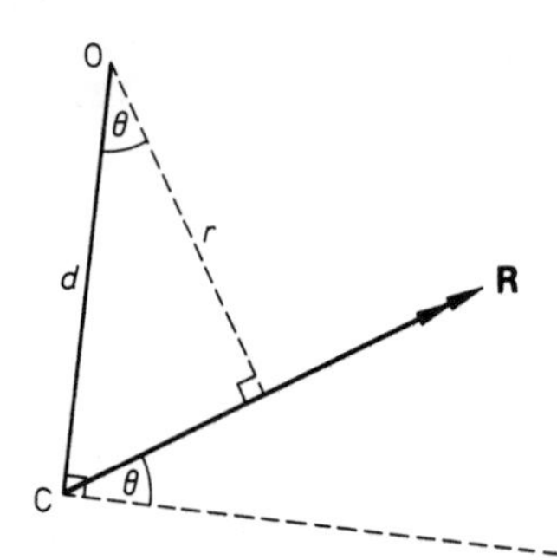

Using the notation of the above diagrams, the sum of the moments of **P** and **Q** about O↺

$$= Pa + Qb = Pd\cos\alpha + Qd\cos\beta = d(P\cos\alpha + Q\cos\beta)$$

The moment of **R** about O↺

$$= Rr = Rd\cos\theta = d(R\cos\theta).$$

Considering the components of **P**, **Q** and **R** in a direction perpendicular to OC, we have

$$P\cos\alpha + Q\cos\beta = R\cos\theta.$$

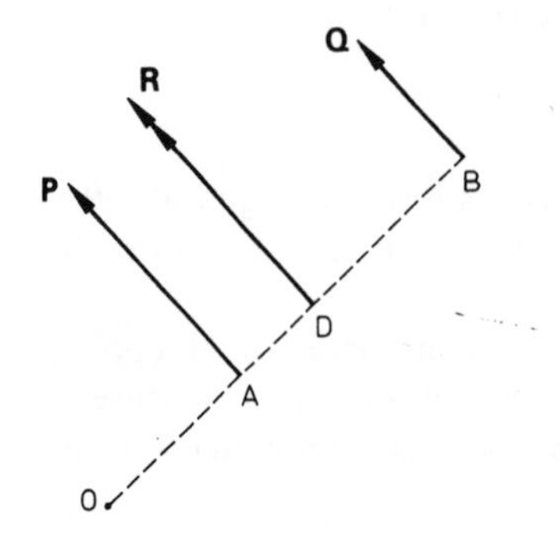

Hence the sum of the moments about O of **P** and **Q** is equal to the moment about O of **R**.

If **P** and **Q** are like parallel forces with resultant **R**, as shown in the diagram, then

$$\frac{AD}{DB} = \frac{Q}{P}$$

$$\therefore \qquad P \times AD = Q \times DB$$

Thus the sum of the moments of **P** and **Q** about O

$= P \times OA + Q \times OB$
$= P(OD - AD) + Q(OD + DB)$
$= (P + Q)OD - P \times AD + Q \times DB$
$= R \times OD$
$=$ the moment of **R** about O↺.

By extending these arguments to any set of coplanar forces with a non-zero resultant it is possible to establish the following general result called the *principle of moments*.

> The sum of the moments of a system of coplanar forces about any point in their plane is equal to the moment of their resultant about the same point

A couple has zero resultant but non-zero moment. Consider a couple consisting of a pair of equal and opposite forces of magnitude F with lines of action at a distance d apart. Suppose that O is any point in the plane of this couple.

(a)

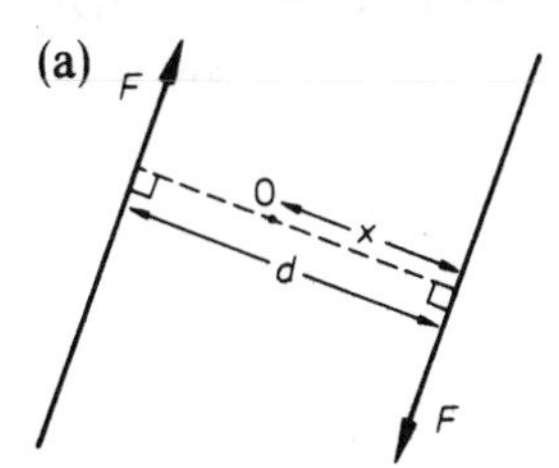

(b)

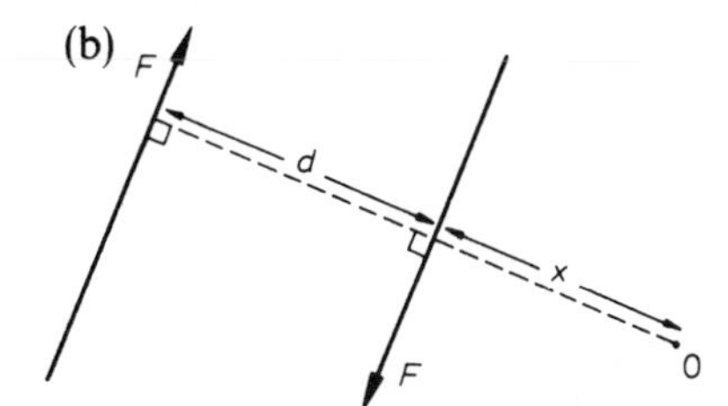

In diagram (a) the sum of the moments about O↻

$$= F(d - x) + Fx = Fd - Fx + Fx = Fd.$$

In diagram (b) the sum of the moments about O↻

$$= F(d + x) - Fx = Fd + Fx - Fx = Fd.$$

Since both results are independent of x, the couple has the same moment, i.e. Fd clockwise, for all positions of O.

Hence the *moment of a couple* about any point in its plane is the product of the magnitude of one of the forces and the perpendicular distance between the forces.

Exercise 5.1

1.

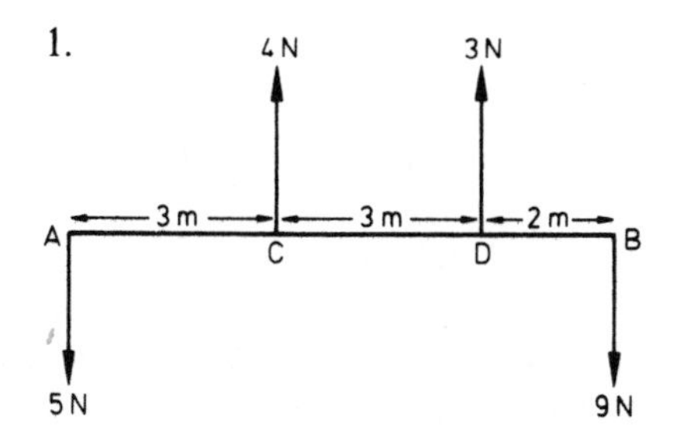

Forces act on a rod AB of length 8 m as shown in the diagram. Find the sum of the anti-clockwise moments of these forces about each of the points A, B, C and D.

2.

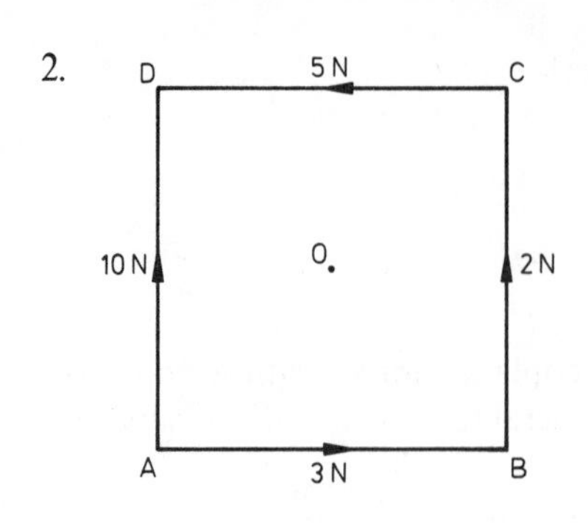

ABCD is a square of side 6 m with centre *O*. Forces act along the sides *AB*, *BC*, *CD*, *AD* as shown in the diagram. Find the sum of the anti-clockwise moments of these forces about each of the points *A*, *B*, *C*, *D* and *O*.

3.

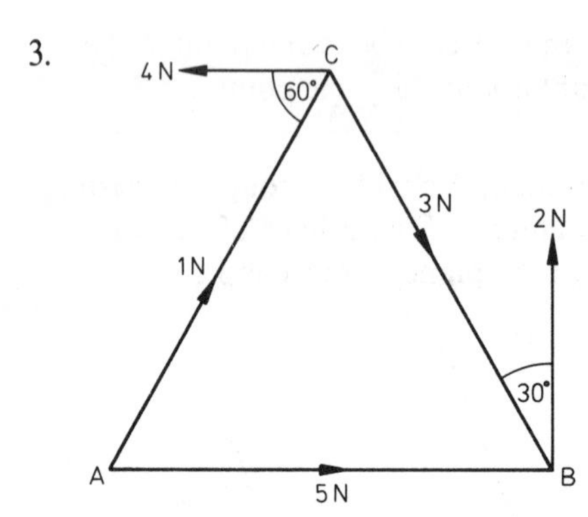

ABC is an equilateral triangle of side 2 m. Find the sum of the anti-clockwise moments of the forces shown in the diagram about each of the points *A*, *B* and *C*. [You may leave your answers in surd form.]

In questions 4 to 9 forces **F**, **G**, **P** and **Q** act along the sides of the rectangle *ABCD* as shown in the diagram.

4.

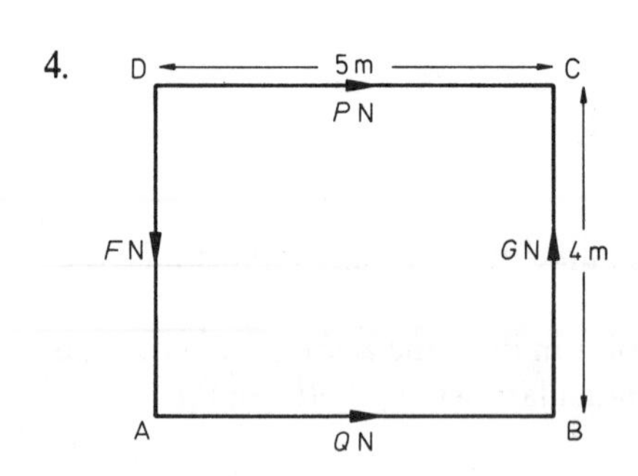

The resultant of **P** and **Q** acts through a point *X* on *BC*. Find the magnitude of the resultant and the length of *BX* if (a) $P = Q = 3$, (b) $P = 6, Q = 2$.

5. The resultant of **F** and **G** acts through a point *Y* on *AB* produced. Find the magnitude of the resultant and the length of *AY* if
(a) $F = 1, G = 2$, (b) $F = 2, G = 3$.

6. Find the moment of the couple formed by forces **F** and **G** if $F = G = 10$.

7. If the forces **F** and **G** form a couple of moment 20 N m, find *F* and *G*.

8. The resultant of **F** and **P** acts through a point *S* on *AB*. Find the magnitude of the resultant and the length of *AS* if (a) $F = P = 5$, (b) $F = 12, P = 9$.

9. The resultant of **G** and **Q** acts through a point T on DA produced. Find the magnitude of the resultant and the length of AT if (a) $G = Q = 4$, (b) $G = 6, Q = 2.5$.

10. Four forces each of magnitude 3 N act along the sides BA, BC, CD, AD of a square $ABCD$, in the directions indicated by the order of the letters. Find the magnitude and the line of action of their resultant.

11. Three forces each of magnitude 10 N act along the sides AB, DE, AF of a regular hexagon $ABCDEF$ of side 2 m. Find the sum of the moments of these forces about each of the points A, B, C, D, E and F. Find also the magnitude and the line of action of the resultant of the three forces.

5.2 Systems of coplanar forces

We consider first a system of parallel coplanar forces with non-zero resultant. The magnitude and direction of the resultant is found by vector addition. Its line of action can be determined using the principle of moments.

Example 1 Find the magnitude and line of action of the resultant of the system of parallel forces shown in the diagram.

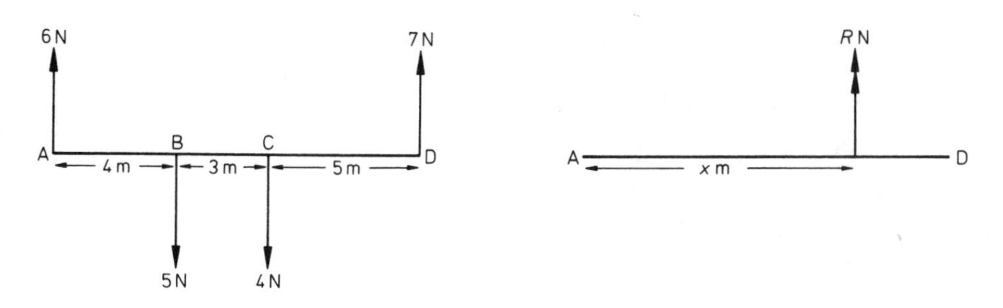

Let the resultant of the given system of forces be of magnitude R N acting at a distance x m from A.
Comparing magnitudes and moments about $A\circlearrowleft$

$\uparrow \quad R = 6 - 5 - 4 + 7 = 4$
$A\circlearrowleft \quad Rx = 0 - 5 \times 4 - 4 \times 7 + 7 \times 12 = 36$
$\therefore \quad R = 4 \quad \text{and} \quad x = 9.$

Hence the resultant has magnitude 4 N and its line of action cuts AD at a point 9 m from A.

Similar methods can be applied to any set of coplanar forces with a non-zero resultant. The magnitude and direction of the resultant is found by means of a vector polygon or by resolving in two perpendicular directions. As before the line of action of the resultant is determined using the principle of moments.

Example 2

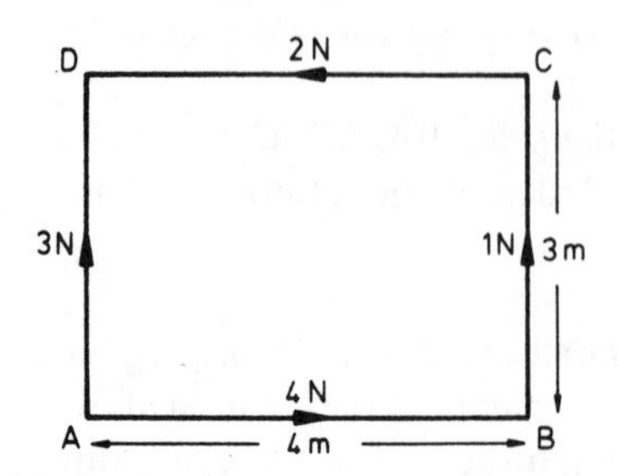

The resultant of the system of forces shown in the diagram is a force **R** acting through a point T on AB. Find the magnitude and direction of **R** and the length of AT.

Let X and Y be the components of **R** parallel and perpendicular to AB.

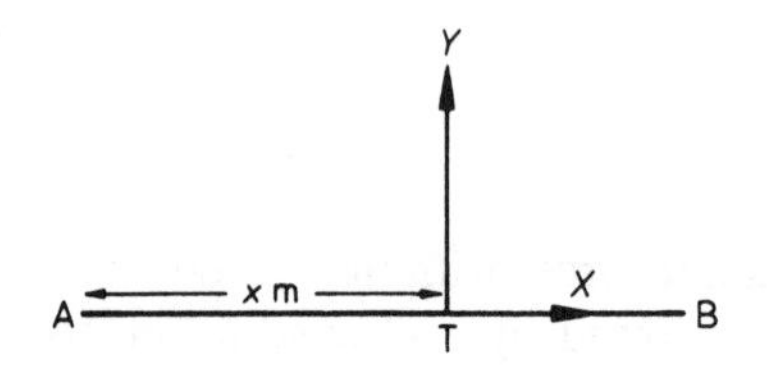

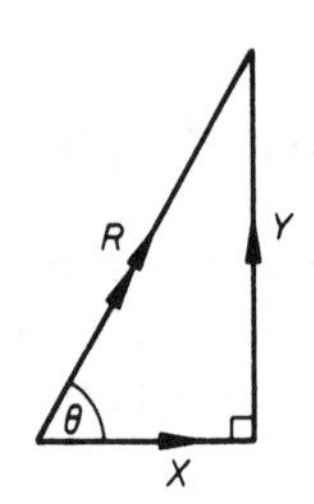

Resolving parallel and perpendicular to AB

$$\rightarrow \qquad X = 4 - 2 \qquad \therefore \quad X = 2$$
$$\uparrow \qquad Y = 3 + 1 \qquad \therefore \quad Y = 4$$

Taking moments about A ↺

$$Yx = 1 \times 4 + 2 \times 3 = 10 \qquad \therefore \quad x = 2{\cdot}5$$

$\therefore \quad R = \sqrt{(X^2 + Y^2)} = \sqrt{(4 + 16)} = \sqrt{20} = 2\sqrt{5}$ and

$$\tan\theta = \frac{Y}{X} = \frac{4}{2} = 2, \quad \theta \approx 63{\cdot}4°.$$

Hence the resultant has magnitude $2\sqrt{5}$ N and acts at 63·4° to AB. The length of AT is 2·5 m.

Example 3

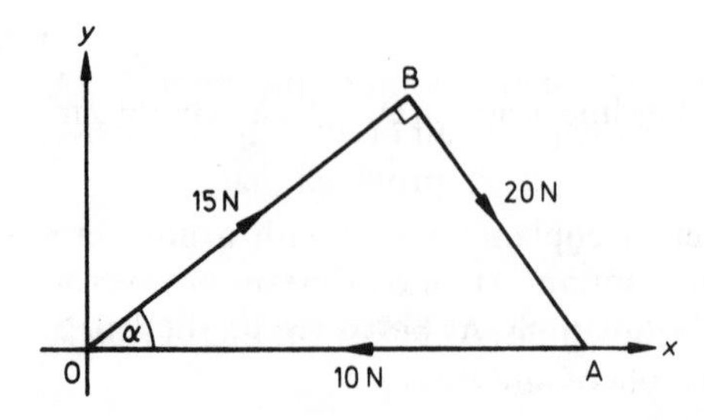

Find the equation of the line of action of the resultant of the forces shown in the diagram, given that $\tan\alpha = \frac{3}{4}$ and that A is the point $(7, 0)$.

Let X and Y be the components of the resultant in the directions of the x- and y-axes respectively. Let the line of action cut the y-axis at the point $(0, c)$.

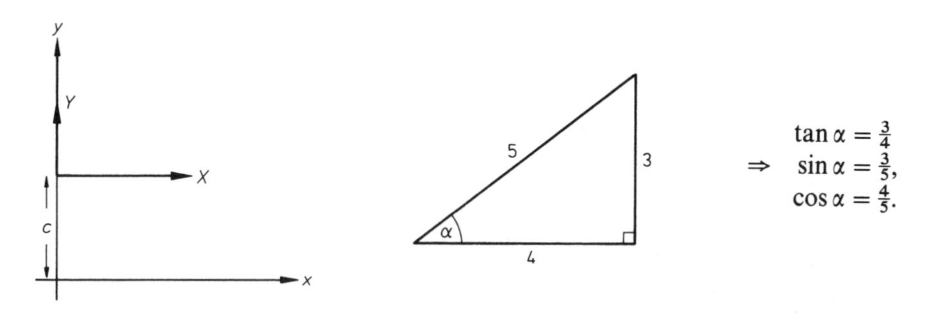

Resolving in the directions of the x- and y- axes,

$$\rightarrow \qquad X = 15\cos\alpha + 20\sin\alpha - 10$$
$$\uparrow \qquad Y = 15\sin\alpha - 20\cos\alpha$$

$\therefore \quad X = 12 + 12 - 10 = 14, \quad Y = 9 - 16 = -7.$

Taking moments about O ↺: $\quad Xc = 20 \times 7\cos\alpha$

$$\therefore \quad c = \frac{20 \times 7\cos\alpha}{X} = \frac{20 \times 7 \times 4}{14 \times 5} = 8$$

Hence the line of action of the resultant has gradient $Y/X = -7/14 = -1/2$ and intercept on the y-axis $c = 8$,

$\therefore$ the equation of the line of action is

$$y = -\tfrac{1}{2}x + 8 \quad \text{i.e.} \quad x + 2y = 16.$$

Two systems of forces are said to be *equivalent* if their effects are the same. In particular, two sets of coplanar forces are equivalent if

(i) the sums of the components of the forces in any given direction are equal and

(ii) the sums of the moments of the forces about any given point are equal.

Thus, the *resultant* of a system of forces is the single force equivalent to the whole system.

It also follows from conditions (i) and (ii) that any two coplanar *couples* with the same moment are equivalent. Similarly a set of coplanar couples with moments $G_1, G_2, G_3, \ldots, G_n$ is equivalent to a single couple with moment

$$G_1 + G_2 + G_3 + \ldots + G_n.$$

We next consider the resultant of a force $\mathbf{P}$ acting at a point O and a couple of moment G in the same plane. The couple is equivalent to a pair of equal and opposite forces of magnitude P and perpendicular distance d apart, provided that

$$Pd = G.$$

Therefore the force and the couple are equivalent to the three forces shown in diagram (a) and hence to the single force shown in diagram (b).

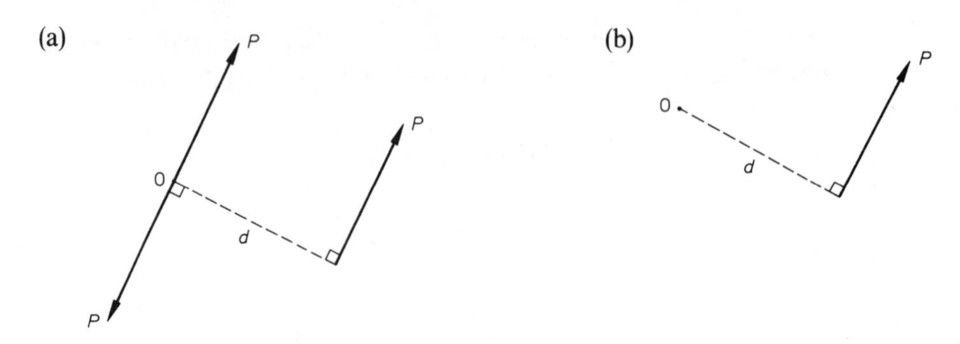

Thus the resultant of a force **P** acting at O and a couple of moment G is a force **P** acting at a distance d from O, where $d = G/P$.

Reversing the argument we may also deduce that a force **P** acting at a distance d from a point O is equivalent to a force **P** acting at O together with a couple of moment Pd.

This result can be used to find a simple equivalent system for any set of coplanar forces. Let $\mathbf{P}_1, \mathbf{P}_2, \ldots, \mathbf{P}_n$ be a system of coplanar forces acting at distances $d_1, d_2, \ldots, d_n$ respectively from a fixed point O in their plane. A typical force $\mathbf{P}_r$ acting at a distance d_r from O is equivalent to a force $\mathbf{P}_r$ acting at O together with a couple of moment $P_r d_r$. Applying this reasoning to all the forces in the system we obtain an equivalent system of forces $\mathbf{P}_1, \mathbf{P}_2, \ldots, \mathbf{P}_n$ acting at 0 together with couples of moments $P_1 d_1, P_2 d_2, \ldots, P_n d_n$.

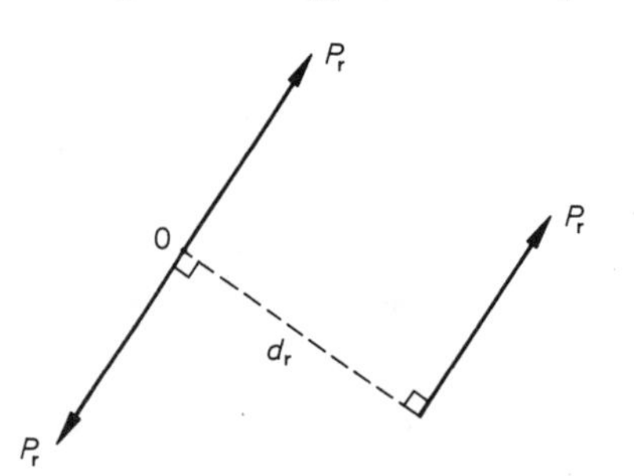

If **R** is the vector sum of the forces and G is the sum of the moments of the original forces about O, then this new system can be replaced by a single force **R** acting at O and a couple of moment G.

> Thus, in general, any system of coplanar forces is equivalent to a single force acting through a given point together with a couple.

We consider further four possible types of system.

(1) $R = 0$, $G = 0$. The system is in equilibrium.
(2) $R = 0$, $G \neq 0$. The system is equivalent to a couple of moment G.
(3) $R \neq 0$, $G = 0$. The system is equivalent to a single force **R** acting through O.
(4) $R \neq 0$, $G \neq 0$. The system is equivalent to a single force **R** acting at a distance G/R from O.

> Hence any system of coplanar forces which is not in equilibrium may be reduced either to a single force or to a couple.

Example 4 A square $ABCD$ of side 3 m has forces of magnitude 8, 3, 3, 4 and $2\sqrt{2}$ newtons acting along AB, CB, CD, AD and BD respectively. If the system is reduced to a force acting through A together with a couple, find the magnitude and direction of the force and the moment of the couple. If the system is reduced to a single force acting through a point E on AB at a distance d metres from A, find d.

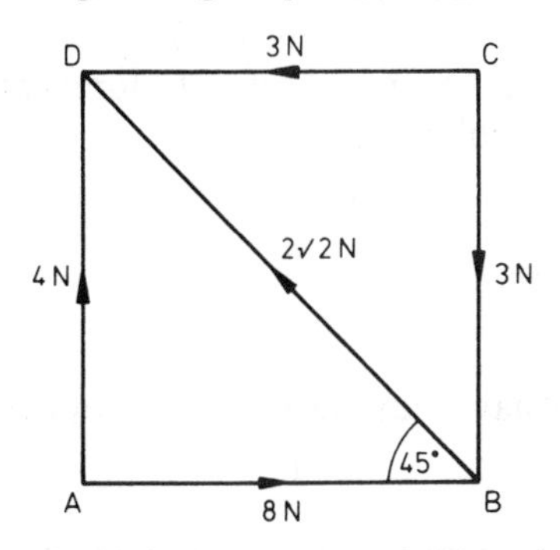

Suppose that the given system is equivalent to forces X and Y acting along AB and AD respectively, together with a couple of moment G in the sense $ABCD$.

Resolving in the directions of AB and AD:

$$\rightarrow \qquad X = 8 - 3 - 2\sqrt{2}\cos 45^\circ = 3$$
$$\uparrow \qquad Y = 4 - 3 + 2\sqrt{2}\sin 45^\circ = 3$$

Taking moments about A↺:

$$G = 3 \times 3 - 3 \times 3 + 2\sqrt{2} \times 3\sin 45^\circ = 6$$

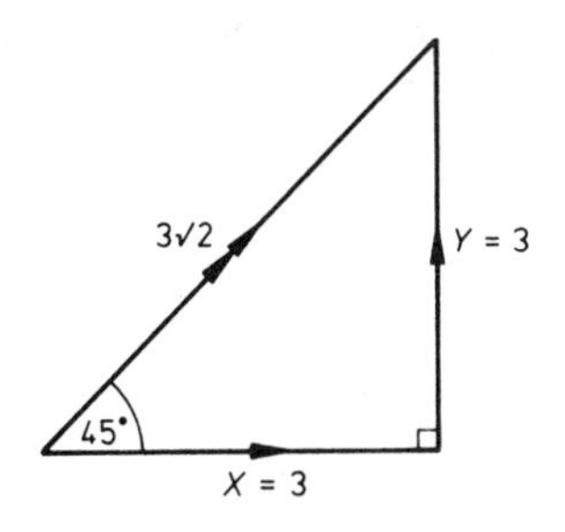

Hence the system is equivalent to a force of $3\sqrt{2}$ N acting along AC together with a couple of moment 6 N m in the sense $ABCD$.

The single equivalent force through E has moment G about A and components X and Y in the directions of AB and AD respectively.

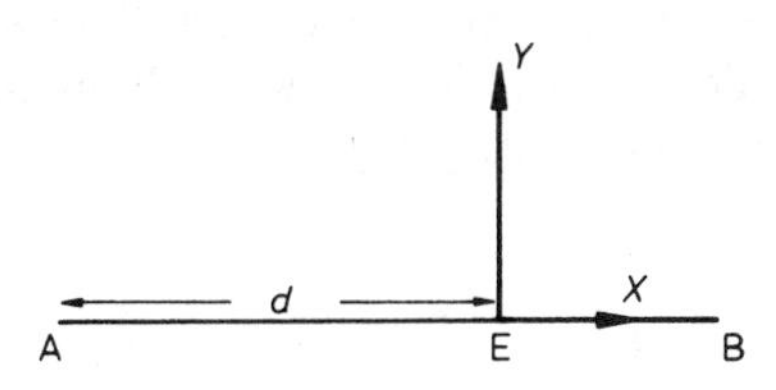

Taking moments about A↺:

$$G = Yd$$

i.e. $\quad 6 = 3d$

$\therefore \quad d = 2.$

Example 5 $ABCDEF$ is a regular hexagon of side 2 m. Forces of magnitude P, 3, 5 and Q newtons act along AB, DC, EF and AE respectively. Show that this system of forces is not in equilibrium.

(a) If the system is equivalent to a couple, find its moment and the values of P and Q.
(b) If the system is equivalent to a single force through E, find P.

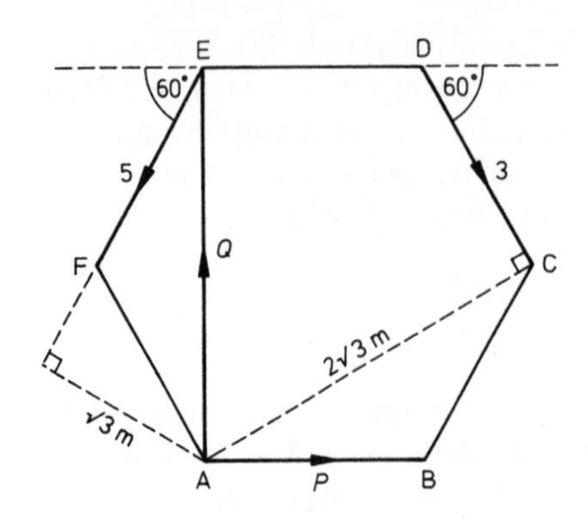

The sum of the moments of the forces about A ↺ (in N m)
$= 3 \times 2\sqrt{3} - 5 \times \sqrt{3} = \sqrt{3}$.
Since this sum is not zero, the forces are not in equilibrium.

(a) If the system is equivalent to a couple, it will have the same moment about any point in the plane,
$\therefore$ the moment of the couple is $\sqrt{3}$ N m.
The sum of the components of the forces in a couple is zero in any direction,
$\therefore$ resolving in the directions of AB and AE:

$$\rightarrow \qquad P + 3\cos 60° - 5\cos 60° = 0$$
$$\uparrow \qquad Q - 3\sin 60° - 5\sin 60° = 0$$

$\therefore \quad P = 2\cos 60° = 1 \quad$ and $\quad Q = 8\sin 60° = 4\sqrt{3}$.

(b) If the system is equivalent to a single force through E, then the sum of the moments about E will be zero.
Taking moments about E ↻: $\quad P \times 2\sqrt{3} - 3 \times \sqrt{3} = 0$

$$\therefore \quad P = \frac{3\sqrt{3}}{2\sqrt{3}} = 1{\cdot}5.$$

Exercise 5.2

In questions 1 to 4 the resultant of the given forces acts through a point T on AB or AB produced. In each case find the magnitude and direction of the resultant and the length of AT. [In question 4 $ABCDEF$ is a regular hexagon.]

1. 2.

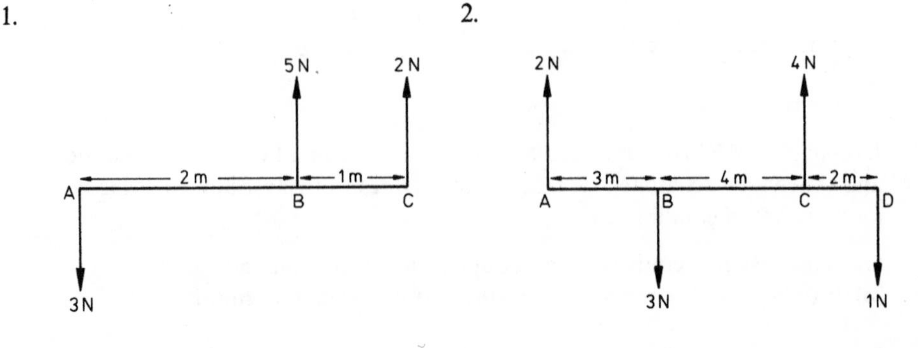

3.

4.

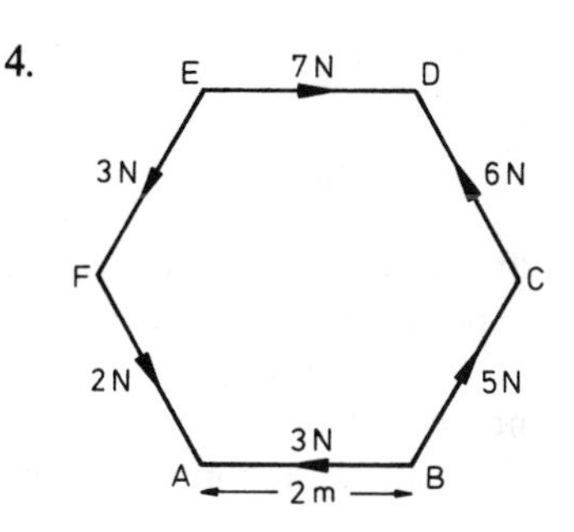

5. Forces of magnitude $8P$, $7P$, $5P$ and $3P$ act along the sides AB, BC, CD and DA respectively of a square $ABCD$ of side $2a$. The resultant of this system of forces cuts AB produced at E. Find
(a) the magnitude and direction of the resultant,
(b) the length of AE.

In questions 6 and 7 find the equation of the line of action of the resultant of the given forces.

6.

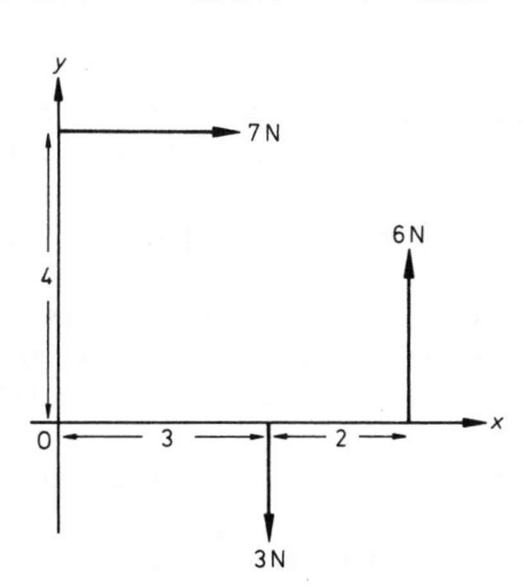

7.

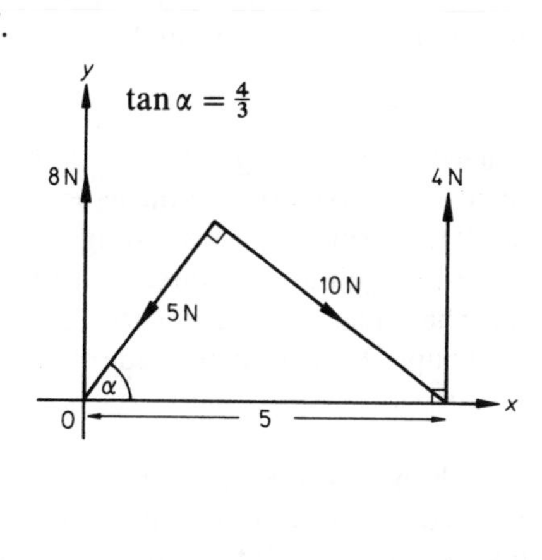

In questions 8 to 19 a system of forces acts in the plane of perpendicular axes Ox and Oy. The sums of the components of the forces in the directions of Ox and Oy are X and Y respectively. The sums of the anti-clockwise moments of the forces about three non-collinear points A, B and C in the plane are G_A, G_B and G_C respectively. In each question use the given facts to decide which of the following statements could be true.

a: the system is in equilibrium,
b: the system is equivalent to a couple,
c: the system is equivalent to a single force.

8. $X = 0$, $Y = 0$. 9. $X = 0$, $Y \neq 0$. 10. $Y = 0$, $G_A = 0$.

11. $X = 0$, $G_B \neq 0$. 12. $Y \neq 0$, $G_C = 0$. 13. $G_A \neq 0$, $G_B = 0$.

14. $G_A = G_B = 0$. 15. $G_A = G_B \neq 0$. 16. $G_A = G_B \neq G_C$.

17. $G_A = G_B = G_C = 0$. 18. $G_A = G_B = G_C \neq 0$ 19. $X = 0$, $G_A = G_B = 0$.

20. In the square $OABC$ of side 2 m, O, A and B are the points (0, 0), (2, 0) and (2, 2) respectively. Forces of magnitude 5, 8, 3 and $5\sqrt{2}$ newtons act along OA, BA, OC and AC respectively. Show that this system of forces is equivalent to a couple and find its moment. If a fifth force of magnitude $6\sqrt{2}$ newtons acts along OB, find the equation of the line of action of the resultant of the enlarged system.

21. In a rectangle $ABCD$, $AB = 12$ m, $BC = 5$ m. Forces of magnitude 6, P, 18, 10 and Q newtons act along AB, CB, CD, AD and DB respectively. Show that this system of forces is not in equilibrium. Find P and Q given that the system reduces to (a) a couple, (b) a single force acting along DC.

22. In a rectangle $ABCD$, $AB = 3$ m, $BC = 2$ m. Forces of magnitude 5, 8, 12, 6 and P newtons act along BA, BC, DC, AD and EA respectively, where E is a point on CD. Given that the forces are in equilibrium, find P and the distance DE. If the force along EA is replaced by an equal force acting through C, show that the system now reduces to a couple and find its moment.

23. $ABCDEF$ is a regular hexagon of side 1 m. Forces of magnitude 4, 6, 10, 8, P and 2 newtons act along AB, BC, CD, ED, FE and AF respectively, directions being indicated by the order of the letters.

(a) If the resultant passes through the centre of the hexagon find the value of P and the magnitude and direction of the resultant.

(b) The force in FE is replaced by another so that the new system reduces to a couple. Find the magnitude of the force and the moment of the couple indicating its sense. (SU)

24. The points A, B and C have coordinates $(9a, 0)$, $(0, -4a)$ and $(6a, 4a)$ respectively referred to the coordinate axes Ox and Oy.

(a) Forces $12P$, $15P$, $5P$ and P act along $\overrightarrow{OA}$, $\overrightarrow{BC}$, $\overrightarrow{CA}$ and $\overrightarrow{OB}$ respectively. Calculate the magnitude of the resultant of these forces and the equation of the line of action of this resultant.

(b) Forces $15P$, S and T act along $\overrightarrow{BC}$, $\overrightarrow{OA}$ and $\overrightarrow{CA}$ respectively. Given that these forces reduce to a couple, calculate (i) the values of S and T, in terms of P, (ii) the magnitude of the couple, in terms of a and P, (iii) the sense of the couple. (AEB 1979)

5.3 Centre of gravity; centre of mass

A solid body may be regarded as a tightly packed collection of small particles. The *weight* of the body is the resultant of the weights of the particles. For relatively small bodies on or near the earth's surface the weights of the constituent particles are assumed to form a system of parallel forces acting vertically downwards.

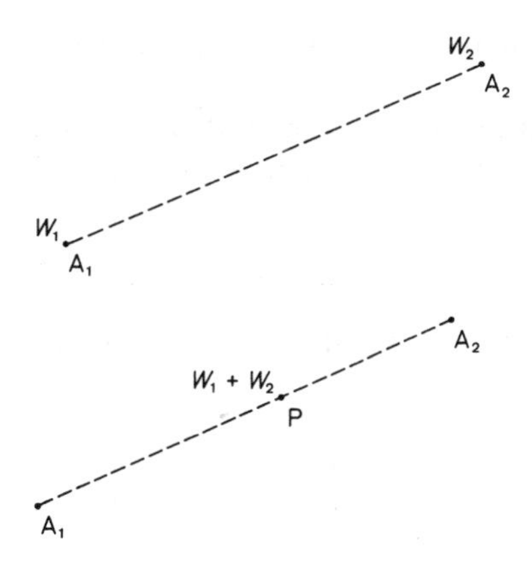

Suppose that at points A_1 and A_2 of a body there are particles of weights W_1 and W_2 respectively. As shown in §5.1, the resultant of the weights of these particles is a force of magnitude $W_1 + W_2$ acting through the point P which divides A_1A_2 in the ratio $W_2 : W_1$.

By extending this argument to include all the particles forming the body, we deduce that the resultant weight force is equal to the sum of the weights of the particles and acts through a fixed point in the body.

The *centre of gravity* of a body is the fixed point through which its weight acts.

The *symmetries* of a body can be used to determine its centre of gravity. If the particles in a body are evenly distributed about an axis or a plane of symmetry, then the centre of gravity of the body must lie on this axis or plane. Thus the centre of gravity of a thin uniform rod is at its mid-point and the centre of gravity of a uniform sphere is at its centre.

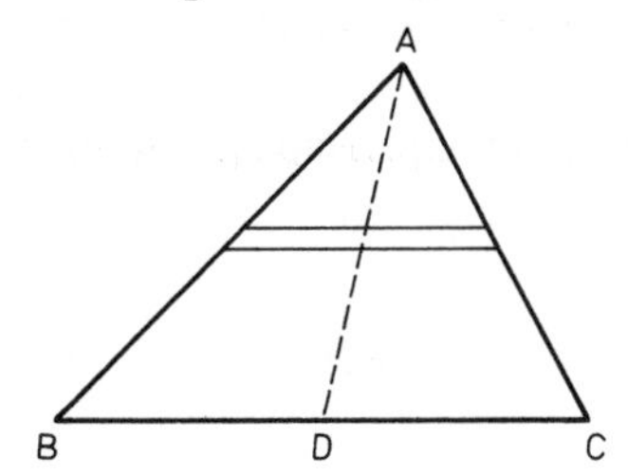

To find the centre of gravity of a *uniform triangular lamina ABC*, suppose that the triangle is divided into n strips parallel to the side BC. If the value of n is increased the strips become thin uniform rods with centres of gravity at their mid-points. Hence the centre of gravity of the whole lamina must lie on the line joining these mid-points i.e. on the median AD. Similarly the centre of gravity must also lie on the medians through B and C.

Thus the centre of gravity of a uniform triangular lamina is at the point of intersection of the medians.

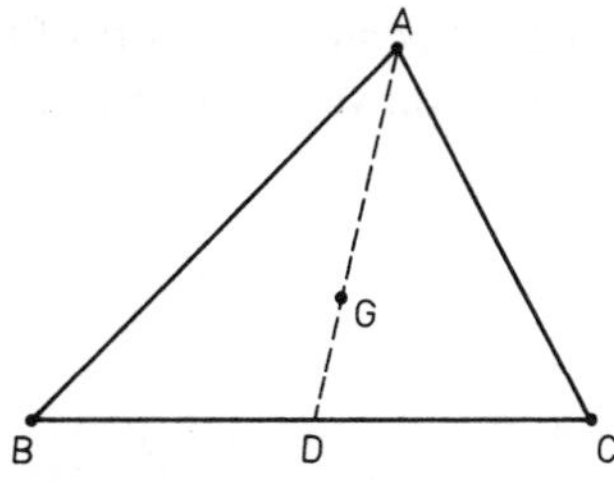

Consider now three particles each of weight W placed at the vertices of a triangle ABC. The resultant of the weights of the particles at B and C is a force of $2W$ acting at D, the mid-point of BC. Hence the resultant of the weights of all three particles is a force of $3W$ acting at the point G on AD which divides

AD in the ratio 2:1. Using similar arguments we find that G is also the point which divides the medians through B and C in the ratio 2:1.

Thus the centre of gravity of the three equal particles at A, B and C is at the point of intersection of the medians of triangle ABC, i.e. at the same point as the centre of gravity of the uniform lamina ABC.

[Note that we have also verified that this point divides each median in the ratio 2:1.]

In general the centre of gravity of a set of particles in a plane is found using the principle of moments.

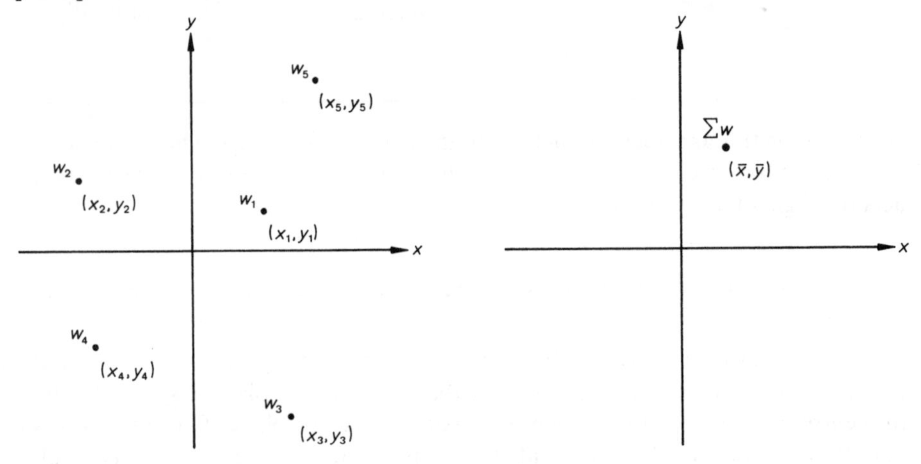

Let particles with weights $w_1, w_2, \ldots, w_n$ have positions $(x_1, y_1), (x_2, y_2), \ldots, (x_n, y_n)$ relative to x- and y-axes in their plane. Suppose that the resultant weight of the set of particles acts through the point $(\bar{x}, \bar{y})$.

Assuming the x, y plane to be horizontal and applying the principle of moments we have,
for moments about the y-axis:

$$w_1x_1 + w_2x_2 + \cdots + w_nx_n = (w_1 + w_2 + \cdots + w_n)\bar{x}$$

and for moments about the x-axis:

$$w_1y_1 + w_2y_2 + \cdots + w_ny_n = (w_1 + w_2 + \cdots + w_n)\bar{y}.$$

Hence the centre of gravity of the particles is at the point $(\bar{x}, \bar{y})$,

where
$$\bar{x} = \frac{\sum wx}{\sum w}, \quad \bar{y} = \frac{\sum wy}{\sum w}.$$

This result can be expressed in vector notation by letting the position vectors of the particles be $\mathbf{r}_1 = x_1\mathbf{i} + y_1\mathbf{j}, \mathbf{r}_2 = x_2\mathbf{i} + y_2\mathbf{j}, \ldots$ and the position vector of the centre of gravity be $\bar{\mathbf{r}} = \bar{x}\mathbf{i} + \bar{y}\mathbf{j}$, so that

$$\bar{\mathbf{r}} = \frac{(w_1x_1 + w_2x_2 + \cdots)\mathbf{i}}{\sum w} + \frac{(w_1y_1 + w_2y_2 + \cdots)\mathbf{j}}{\sum w}$$

$$= \frac{w_1(x_1\mathbf{i} + y_1\mathbf{j}) + w_2(x_2\mathbf{i} + y_2\mathbf{j}) + \cdots}{\sum w}$$

Thus the position vector of the centre of gravity is $\bar{\mathbf{r}} = \dfrac{\sum w\mathbf{r}}{\sum w}$.

Note that in the case of three particles of weight W at the points $A(x_1, y_1)$, $B(x_2, y_2)$, $C(x_3, y_3)$ we have

$$\bar{x} = \frac{Wx_1 + Wx_2 + Wx_3}{3W} = \tfrac{1}{3}(x_1 + x_2 + x_3)$$

and similarly $\qquad \bar{y} = \tfrac{1}{3}(y_1 + y_2 + y_3)$.

Hence the centre of gravity of three equal particles at A, B, C and also of a uniform triangular lamina ABC has coordinates $(\tfrac{1}{3}\{x_1 + x_2 + x_3\}, \tfrac{1}{3}\{y_1 + y_2 + y_3\})$ and position vector $\tfrac{1}{3}\{\mathbf{r}_1 + \mathbf{r}_2 + \mathbf{r}_3\}$.

Example 1 Find the coordinates of the centre of gravity of particles of weights 4 N, 7 N, 3 N and 6 N at the points (1, 2), (−2, −1), (6, −3) and (0, 4) respectively.

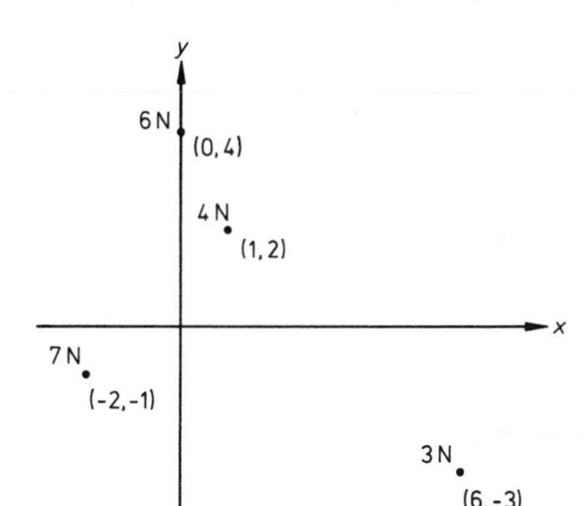

Let the centre of gravity of the particles be at the point $(\bar{x}, \bar{y})$, then the resultant weight of the particles has magnitude 20 N and acts through $(\bar{x}, \bar{y})$.

Taking moments about the y-axis,

$$20\bar{x} = 4 \times 1 + 7 \times (-2) + 3 \times 6 + 6 \times 0$$

$$\therefore \quad \bar{x} = \frac{4 - 14 + 18 + 0}{20} = \frac{8}{20} = \frac{2}{5}$$

Taking moments about the x-axis,

$$20\bar{y} = 4 \times 2 + 7 \times (-1) + 3 \times (-3) + 6 \times 4$$

$$\therefore \quad \bar{y} = \frac{8 - 7 - 9 + 24}{20} = \frac{16}{20} = \frac{4}{5}$$

Hence the centre of gravity has coordinates $\left(\dfrac{2}{5}, \dfrac{4}{5}\right)$.

Example 2 Find the coordinates of the centre of gravity of a uniform lamina in the form of a trapezium with vertices $A(1, 0)$, $B(7, 0)$, $C(3, 4)$ and $D(0, 4)$.

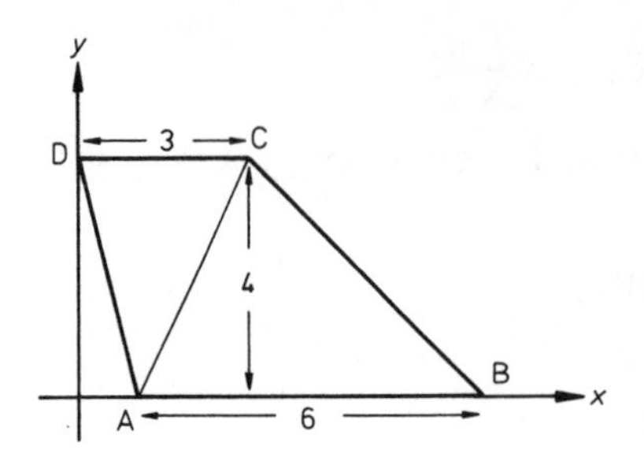

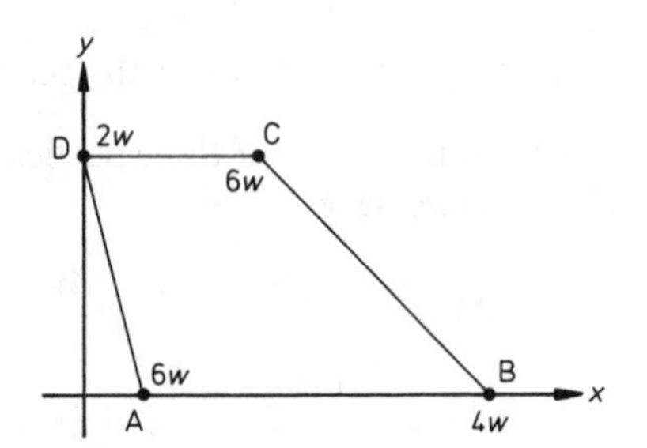

The weight of the lamina $ABCD$ is the resultant of the weights of lamina ABC and lamina ACD. Let w be the weight per unit area of the lamina, then the weight of ABC is $12w$ and is equivalent to three weights of $4w$ at points A, B and C.
Similarly the weight of ACD is $6w$ and is equivalent to three weights of $2w$ at points A, C and D.
Hence the centre of gravity $(\bar{x}, \bar{y})$ of the lamina $ABCD$ is the same as the centre of gravity of particles of weights $6w, 4w, 6w$ and $2w$ at the points A, B, C and D respectively.
Taking moments about the y-axis,

$$18w\bar{x} = 6w \times 1 + 4w \times 7 + 6w \times 3$$

Taking moments about the x-axis,

$$18w\bar{y} = 6w \times 4 + 2w \times 4$$

$$\therefore \bar{x} = \frac{52w}{18w} = \frac{26}{9}, \quad y = \frac{32w}{18w} = \frac{16}{9}.$$

Hence the centre of gravity has coordinates $\left(\frac{26}{9}, \frac{16}{9}\right)$.

The principle of moments is also used to find the centre of gravity of any body which can be divided into smaller bodies, each with known weight and centre of gravity.

Example 3 A thin uniform wire is bent to form a triangle ABC in which $AB = AC = 30$ cm and $BC = 48$ cm. Find its centre of gravity.

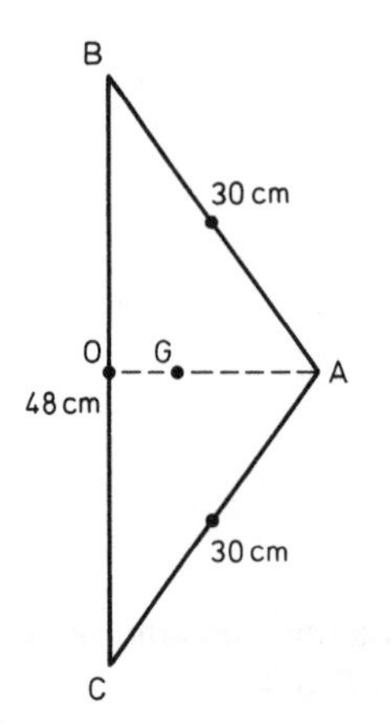

From symmetry the centre of gravity G lies on OA, where O is the mid-point of BC.
Each of the wires AB, AC and BC has its centre of gravity at its mid-point.
By Pythagoras' theorem, $OA = 18$ cm.

Letting w be the weight per cm of the wire, we have:

Body	Weight	Distance of centre of gravity from BC
Wire AB	$30\,w$	9
Wire AC	$30\,w$	9
Wire BC	$48\,w$	0
Whole body	$108\,w$	$\bar{x}$

Taking moments about BC,

$$108w\bar{x} = 30w \times 9 + 30w \times 9 + 48w \times 0$$

$$\therefore \quad \bar{x} = \frac{2 \times 30w \times 9}{108w} = 5$$

$\therefore$ the centre of gravity is 5 cm from BC on the line OA.

Example 4

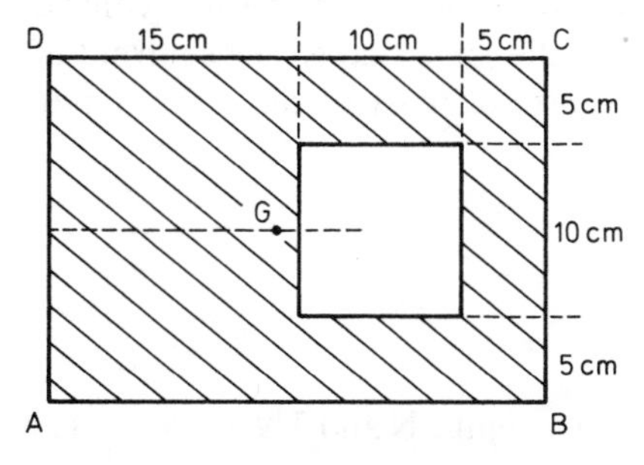

A square hole is cut in a uniform rectangular lamina as shown in the diagram. Find the centre of gravity of the resulting body.

From symmetry the centre of gravity G lies on the line joining the mid-points of AD and BC.

We use the positions of the centres of gravity of the complete rectangle and of the square removed to find the distance $\bar{x}$ of G from AD.
Letting w be the weight per unit area, we have:

Body	Weight	Distance of centre of gravity from AD
Rectangle	$600\,w$	15
Square	$100\,w$	20
Remainder	$500\,w$	$\bar{x}$

Taking moments about AD,

$$500w\bar{x} + 100w \times 20 = 600w \times 15$$

$$\therefore \quad 500w\bar{x} = 600w \times 15 - 100w \times 20$$

$$\therefore \quad \bar{x} = \frac{9000w - 2000w}{500w} = 14$$

$\therefore$ the centre of gravity is 14 cm from AD on the line joining the mid-points of AD and BC.

As shown earlier the centre of gravity of a set of particles in a plane is the point $\left(\frac{\sum wx}{\sum w}, \frac{\sum wy}{\sum w}\right)$. If $m_1, m_2, \ldots$ are the masses of the particles, their *centre of mass* is defined to be the point

$$\left(\frac{\sum mx}{\sum m}, \frac{\sum my}{\sum m}\right).$$

The weight w and the mass m of a typical particle are connected by the relationship $w = mg$, where g is the acceleration due to gravity. Thus, provided that g takes the same value for all the particles, $\frac{\sum wx}{\sum w} = \frac{\sum mgx}{\sum mg} = \frac{\sum mx}{\sum m}$ and similarly $\frac{\sum wy}{\sum w} = \frac{\sum my}{\sum m}$. It follows that, in general, the centre of gravity and the centre of mass of a body are at the same point.

[We note, however, that if a body is so large that the weights of its constituent particles are not parallel forces, then its centre of gravity cannot be found by elementary methods and does not necessarily coincide with the centre of mass.]

A *centroid* is a centre of area or volume. The centroid of a plane area coincides with the centre of mass of a uniform lamina of the same shape. The centroid of a three-dimensional figure coincides with the centre of mass of the corresponding uniform solid.

Exercise 5.3

1. Find the centre of gravity of two particles of weight 2 N and 3 N at points A and B respectively, given that $AB = 30$ cm.

2. Find the centre of gravity of two particles of weight 30 N and 70 N at points P and Q respectively, given that $PQ = 1{\cdot}5$ m.

3. A uniform triangular lamina ABC has $AB = AC = 17$ cm and $BC = 16$ cm. Find the distance of the centre of gravity from BC.

4. A uniform triangular lamina PQR has $PQ = 9$ cm, $QR = 15$ cm and $PR = 12$ cm. Find the distances of the centre of gravity from PR and PQ.

5. Find the coordinates of the centre of gravity of particles of weight 5 N, 7 N, 1 N and 3 N at the points $(1, 0)$, $(3, 1)$, $(6, 3)$ and $(0, 2)$ respectively.

6. Find the coordinates of the centre of mass of particles of mass 9 kg, 4 kg, 6 kg and 5 kg at the points $(4, 3)$, $(6, -6)$, $(-3, 0)$ and $(6, -3)$ respectively.

7. Find the position vector of the centre of mass of a uniform triangular lamina, given that the position vectors of its vertices are $3\mathbf{i} - 5\mathbf{j}$, $7\mathbf{i} + 2\mathbf{j}$ and $-\mathbf{i} + 6\mathbf{j}$.

8. Find the position vector of the centre of gravity of three particles of weight 7 N, 9 N and 4 N with position vectors $4\mathbf{i} + \mathbf{j}$, $3\mathbf{j}$ and $3\mathbf{i} + 4\mathbf{j}$.

9. A uniform lamina in the form of a quadrilateral has vertices $O(0,0)$, $A(4,-6)$, $B(7,0)$ and $C(0,6)$. If the weight per unit area of the lamina is w, show that it has the same centre of gravity as particles of weight $14w$, $7w$, $14w$ and $7w$ at O, A, B and C respectively. Hence find the centre of gravity of the lamina.

10. Use the method of the previous question to find the centre of gravity of a uniform lamina in the form of a pentagon with vertices $O(0,0)$, $A(3,0)$, $B(3,2)$, $C(0,4)$ and $D(-3,1)$.

11. A thin uniform wire is bent to form a triangle ABC in which $AB = 24$ cm, $AC = 10$ cm and $\angle A = 90°$. Find the distances of the centre of gravity from AC and AB.

12. A thin uniform wire is bent into the shape of an isosceles trapezium $ABCD$ in which $AB = 24$ cm, $AD = BC = 25$ cm and $CD = 10$ cm. Find the distance of the centre of gravity from AB.

In questions 13 to 16 find the position of the centre of gravity of the uniform lamina shown in the given diagram.

13.

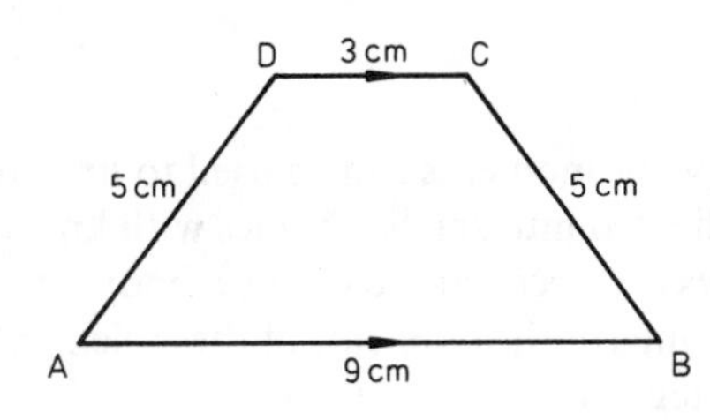

14.

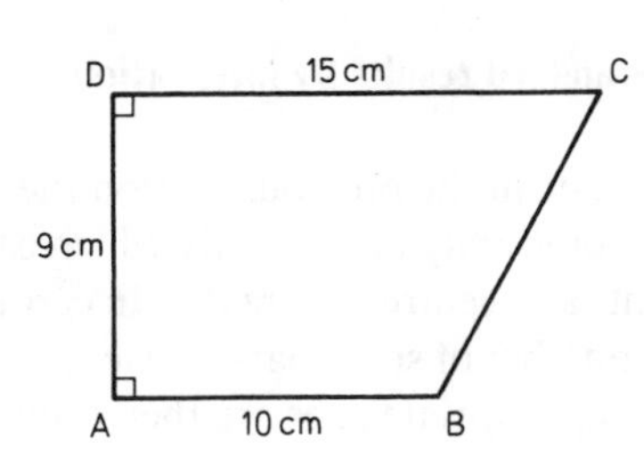

15.

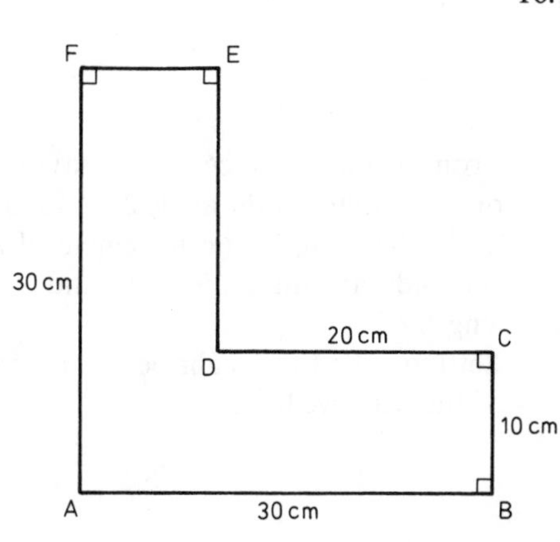

16.

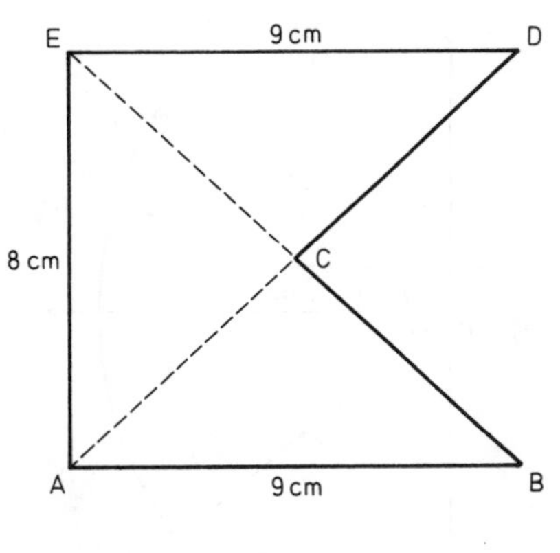

17.

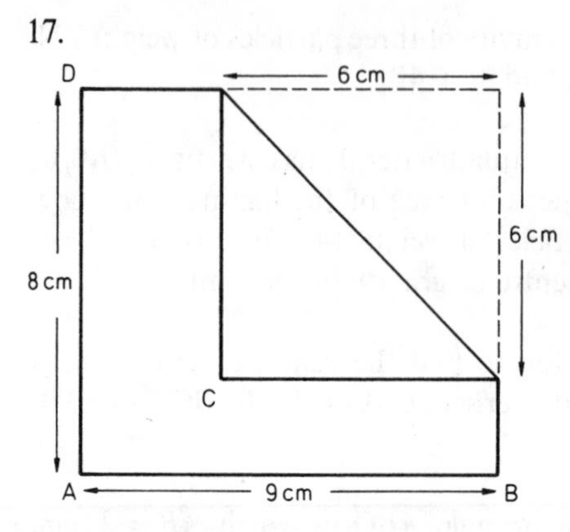

Find the centre of gravity of a uniform rectangular lamina $ABCD$ folded as shown in the diagram.

18.

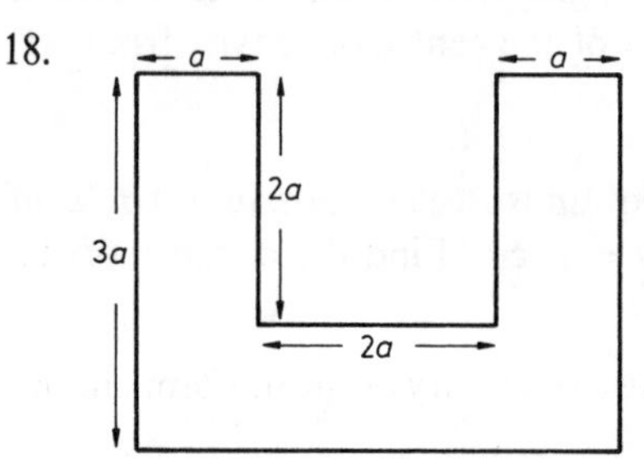

The diagram shows the central cross-section of a uniform cylindrical container weighing $2w$ per unit volume. Find its centre of gravity. If the container is now filled with a liquid weighing w per unit volume, find the new position of the centre of gravity.

5.4 Standard results by integration

As we saw in the previous section the principle of moments can be used to find the centre of gravity of any body which can be divided into smaller bodies with known weights and centres of gravity. In certain cases it is necessary to divide a body into a large number of small parts or elements. The sums of the moments of these elements about appropriate axes are then found by integration.

Example 1 Find the centre of gravity of a uniform wire in the form of an arc of a circle, radius r, subtending an angle 2α at the centre.

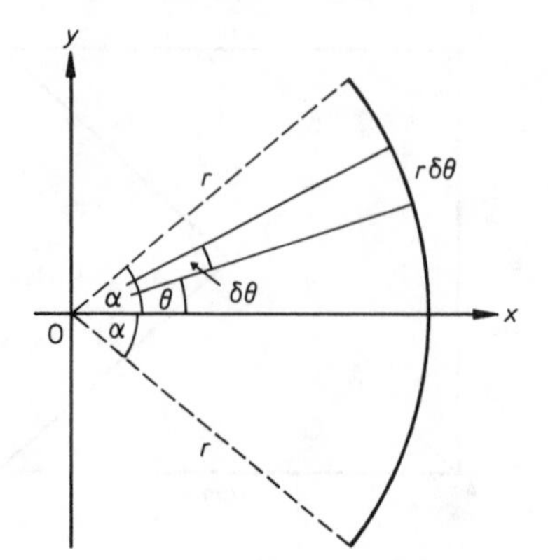

From symmetry the centre of gravity lies on the bisector of the angle 2α. As shown in the diagram, a typical element of arc subtends an angle $\delta\theta$ at O and is of length $r\,\delta\theta$.

Letting w be the weight per unit length of the wire, we have:

Body	Weight	Distance of centre of gravity from the y-axis
Element	$wr\,\delta\theta$	$r\cos\theta$
Whole arc	$wr\,2\alpha$	$\bar{x}$

Taking moments about the y-axis,

$$wr\,2\alpha\bar{x} \approx \sum(wr\,\delta\theta \times r\cos\theta)$$

$$\therefore \quad 2wr\,\alpha\bar{x} = \int_{-\alpha}^{\alpha} wr^2 \cos\theta\, d\theta$$

$$= wr^2\Big[\sin\theta\Big]_{-\alpha}^{\alpha} = 2wr^2 \sin\alpha$$

$$\therefore \quad \bar{x} = \frac{r\sin\alpha}{\alpha}$$

Hence the centre of gravity of the wire is at a distance $(r\sin\alpha)/\alpha$ from the centre.

Example 2 Find the centre of gravity of a uniform lamina in the form of a sector of a circle, radius r, subtending an angle 2α at the centre.

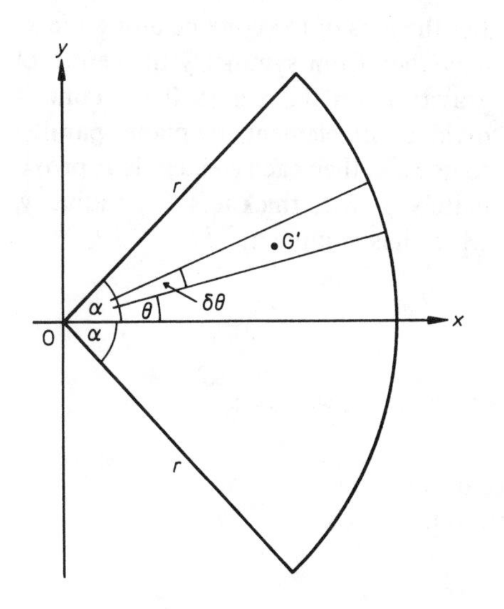

From symmetry the centre of gravity lies on the bisector of the angle 2α. A typical element is a sector subtending an angle $\delta\theta$ at O and having area $\frac{1}{2}r^2\,\delta\theta$. This sector is approximately triangular with its centre of gravity G' at a distance $\frac{2}{3}r$ from O.

Letting w be the weight per unit area of the lamina, we have:

Body	Weight	Distance of centre of gravity from the y-axis
Element	$w\frac{1}{2}r^2\,\delta\theta$	$\frac{2}{3}r\cos\theta$
Whole sector	$w\frac{1}{2}r^2\,2\alpha$	$\bar{x}$

Taking moments about the y-axis,

$$w\tfrac{1}{2}r^2\, 2\alpha\bar{x} \approx \sum(w\tfrac{1}{2}r^2\, \delta\theta \times \tfrac{2}{3}r\cos\theta)$$

$$\therefore \quad wr^2\,\alpha\bar{x} = \int_{-\alpha}^{\alpha} \tfrac{1}{3}wr^3 \cos\theta\, d\theta$$

$$= \tfrac{1}{3}wr^3\Big[\sin\theta\Big]_{-\alpha}^{\alpha} = \tfrac{2}{3}wr^3 \sin\alpha$$

$$\therefore \quad \bar{x} = \frac{2r\sin\alpha}{3\alpha}$$

Hence the centre of gravity of the lamina is at a distance $(\tfrac{2}{3}r\sin\alpha)/\alpha$ from the centre.

[This result can also be produced by showing that a uniform sector of radius r has the same centre of gravity as a uniform arc of radius $\tfrac{2}{3}r$.]

Example 3 Find the centre of gravity of a uniform solid right circular cone of base radius r, height h.

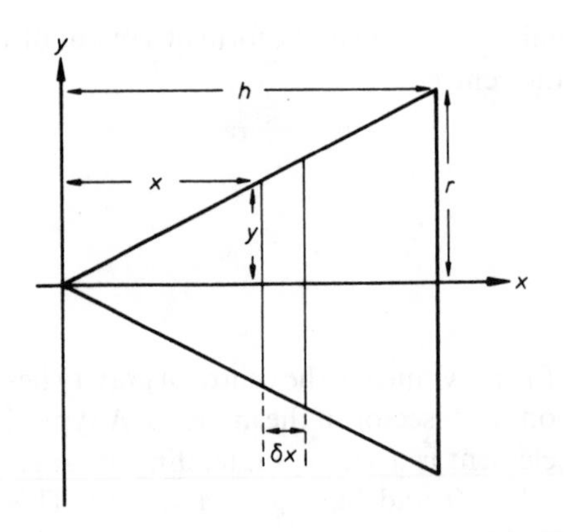

Let the axis of the cone lie along the x-axis, then from symmetry the centre of gravity is on the x-axis. If the cone is divided into elements by planes parallel to its base, then each element is approximately a disc, thickness δx, radius y, which has volume $\pi y^2\,\delta x$.

Letting w be the weight per unit volume of the cone, we have:

Body	Weight	Distance of centre of gravity from the y-axis
Element	$w\pi y^2\,\delta x$	x
Whole cone	$w\tfrac{1}{3}\pi r^2 h$	$\bar{x}$

Taking moments about the y-axis

$$w\tfrac{1}{3}\pi r^2\, h\bar{x} \approx \sum(w\pi y^2\,\delta x \times x)$$

$$\therefore \quad \tfrac{1}{3}\pi r^2\, hw\bar{x} = \int_0^h \pi wxy^2\, dx$$

From similar triangles, $\dfrac{y}{x} = \dfrac{r}{h}$

$$\therefore \quad \tfrac{1}{3}\pi r^2 hw\bar{x} = \int_0^h \frac{\pi r^2 w}{h^2} x^3\,dx$$

$$= \frac{\pi r^2 w}{h^2}\left[\tfrac{1}{4}x^4\right]_0^h = \tfrac{1}{4}\pi r^2 h^2 w$$

$$\therefore \quad \bar{x} = \tfrac{3}{4}h$$

Hence the centre of gravity of the cone is on its axis at a distance $\frac{1}{4}h$ from its base.

Example 4 Find the centre of gravity of a uniform solid tetrahedron.

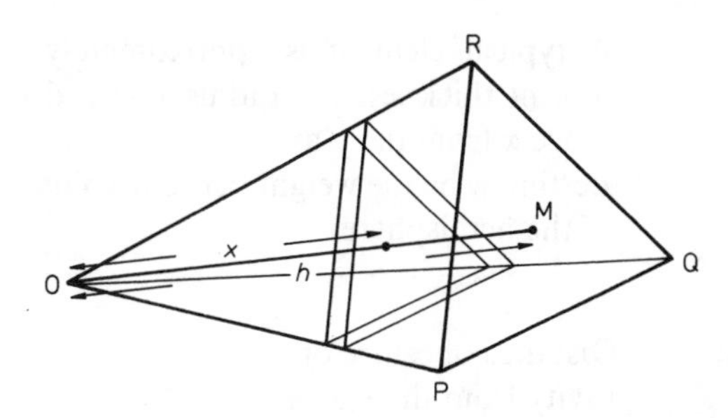

In tetrahedron $OPQR$, let M be the point of intersection of the medians of $\triangle PQR$. Let the length of OM be h and let the area of $\triangle PQR$ be A.
If the tetrahedron is divided into elements by planes parallel to the face PQR, then a typical element is a triangular lamina of thickness δx, which cuts OM at a distance x from O. Since this element is similar to $\triangle PQR$, its medians intersect on OM and its area is Ax^2/h^2.
Letting w be the weight per unit volume, we have:

Body	Weight	Distance of centre of gravity from 0
Element	$wAx^2\,\delta x/h^2$	x
Whole body	$w\frac{1}{3}Ah$	$\bar{x}$

Taking moments about an axis perpendicular to OM

$$w\tfrac{1}{3}Ah\bar{x} \approx \sum\left(\frac{wAx^2\,\delta x}{h^2} \times x\right)$$

$$\therefore \quad \tfrac{1}{3}Ahw\bar{x} = \int_0^h \frac{Aw}{h^2} x^3\,\delta x$$

$$= \frac{Aw}{h^2}\left[\tfrac{1}{4}x^4\right]_0^h = \tfrac{1}{4}Awh^2$$

$$\therefore \quad \bar{x} = \tfrac{3}{4}h$$

Hence the centre of gravity of the tetrahedron lies one quarter of the way up a line joining the centroid of one face to the opposite vertex.

Example 5 Find the centre of gravity of a uniform solid hemisphere of radius r.

If we take x- and y-axes along and perpendicular to the axis of symmetry of the hemisphere, then its centre of gravity will be on the x-axis.

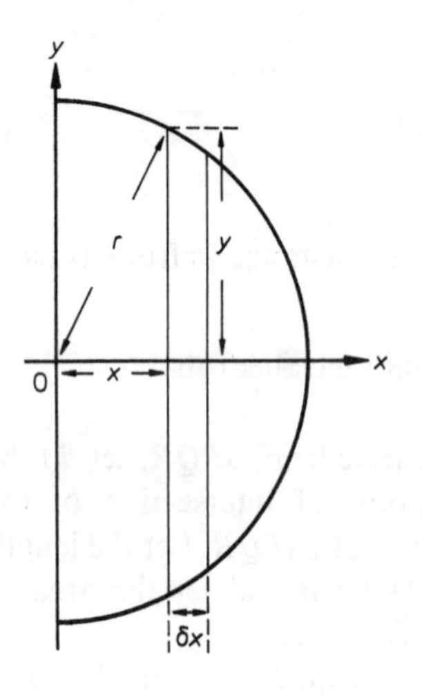

A typical element is approximately a disc of thickness δx, radius y at a distance x from the y-axis.
Letting w be the weight per unit volume of the hemisphere:

Body	Weight	Distance of centre of gravity from the y-axis
Element	$w\pi y^2\,\delta x$	x
Whole hemisphere	$w\frac{2}{3}\pi r^3$	$\bar{x}$

Taking moments about the y-axis,

$$w\tfrac{2}{3}\pi r^3\bar{x} \approx \sum(w\pi y^2\,\delta x \times x)$$

$$\therefore \quad \tfrac{2}{3}\pi r^3 w\bar{x} = \int_0^r \pi w x y^2\,dx$$

By Pythagoras' theorem, $x^2 + y^2 = r^2$,

$$\therefore \quad \tfrac{2}{3}\pi r^3 w\bar{x} = \int_0^r \pi w(r^2x - x^3)\,dx$$

$$= \pi w\left[\tfrac{1}{2}r^2x^2 - \tfrac{1}{4}x^4\right]_0^r = \tfrac{1}{4}\pi r^4 w$$

$$\therefore \quad \bar{x} = 3r/8$$

Hence the centre of gravity of the hemisphere is on its axis of symmetry at a distance $3r/8$ from its plane face.

Example 6 Find the centre of gravity of a uniform hemispherical shell of radius r.

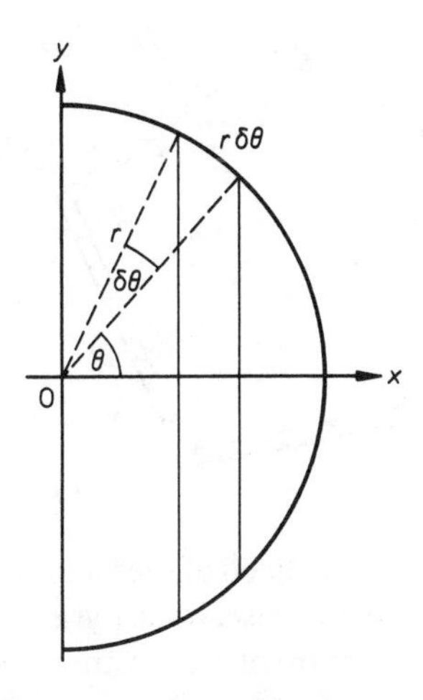

If we take x- and y-axes along and perpendicular to the axis of symmetry of the hemispherical shell, then its centre of gravity will be on the x-axis.

If the shell is divided into narrow bands by planes perpendicular to the x-axis, then a typical elemental band is approximately cylindrical with radius $r \sin \theta$ and width $r\,\delta\theta$. Letting w be the weight per unit area of the shell, we have:

Body	Weight	Distance of centre of gravity from the y-axis
Element	$w2\pi r^2 \sin\theta\,\delta\theta$	$r\cos\theta$
Whole body	$w2\pi r^2$	$\bar{x}$

Taking moments about the y-axis

$$w2\pi r^2\bar{x} \approx \sum(w2\pi r^2 \sin\theta\,\delta\theta \times r\cos\theta)$$

$$\therefore \quad 2\pi w r^2 \bar{x} = \int_0^{\pi/2} 2\pi w r^3 \sin\theta\cos\theta\,d\theta$$

$$= \int_0^{\pi/2} \pi w r^3 \sin 2\theta\,d\theta$$

$$= \pi w r^3[-\tfrac{1}{2}\cos 2\theta]_0^{\pi/2} = \pi w r^3$$

$$\therefore \quad \bar{x} = \frac{r}{2}$$

Hence the centre of gravity of the hemispherical shell is at a distance $\frac{1}{2}r$ from its base.

[For an alternative method see Exercise 5.6, question 11.]

It is not necessary to use integration to find the centres of gravity of hollow pyramids and cones.

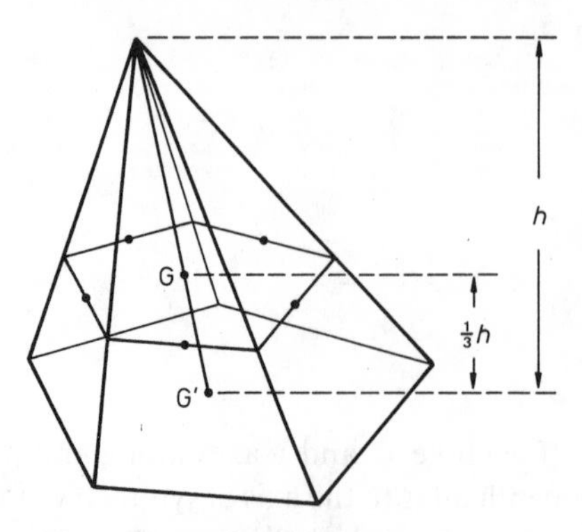

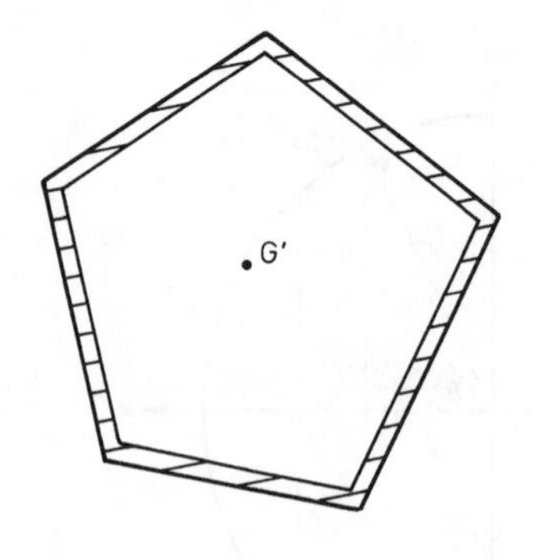

Let us consider a thin hollow pyramid of height h, without base. Each of the faces is a triangular lamina with centre of gravity at a height $\frac{1}{3}h$ above the base. If the pyramid is divided into sections parallel to the base, we see that the centre of gravity must lie on the line joining the centres of gravity of these sections.

Hence the centre of gravity of the *hollow pyramid* lies one third of the way up a line joining the centroid of the base to the vertex.

As the number of sides of the base of the pyramid is increased indefinitely, it tends to become a cone. Hence the centre of gravity of a *hollow right circular cone*, without base, is on the axis of the cone at a distance of one third of the height of the cone from the base.

[Note that the results obtained in this section may be quoted without proof when finding the centres of gravity of composite bodies. A list of standard results will be found in the "Formulae for reference" section near the end of the book.]

Exercise 5.4

1. Prove that the centre of gravity of a uniform wire in the form of a semi-circle of radius r is at a distance $2r/\pi$ from the centre.

2. Prove that the centre of gravity of a uniform semi-circular lamina of radius r is at a distance $4r/3\pi$ from the centre.

In questions 3 to 6 find the distance from AD of the centre of gravity of the given uniform lamina.

3.

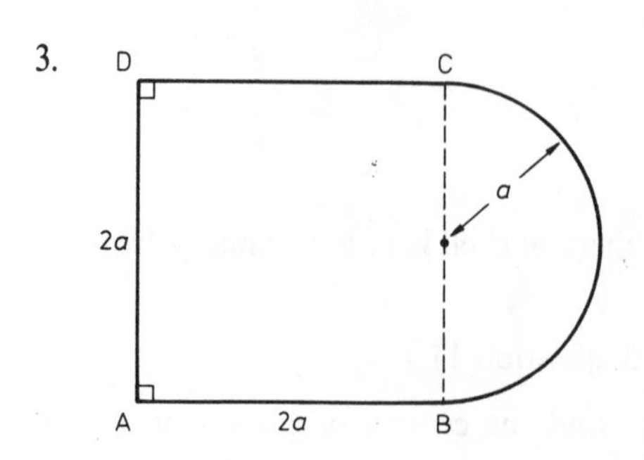

4.

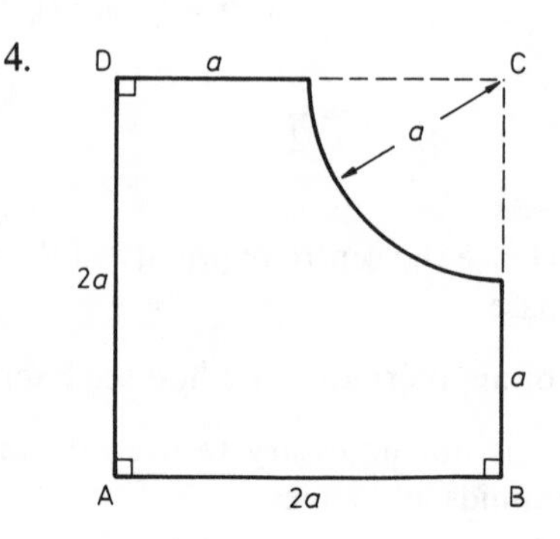

5.

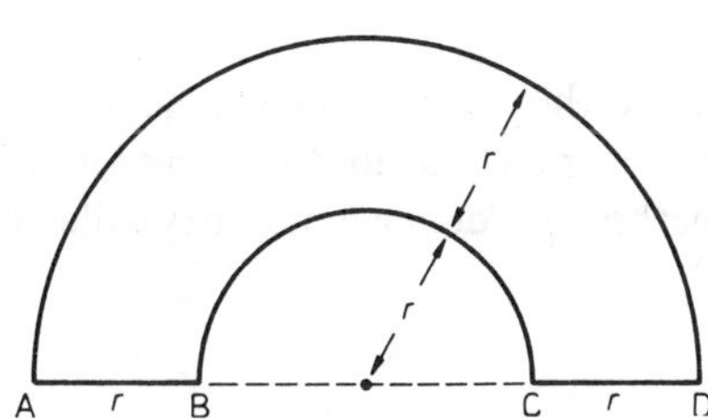

6.

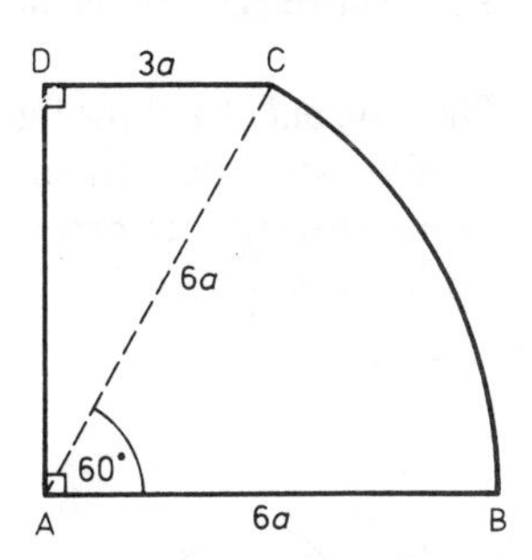

7. A frustum is cut from a solid right circular cone of base radius r and height $2h$ by a plane parallel to the base and at a distance h from it. Find the distance of the centre of gravity of the frustum from the base.

8. A circular lamina of radius $2r$ is divided into two segments by a line at a distance r from the centre. Find the distance from the centre of the centre of gravity of (a) the larger segment, (b) the smaller segment.

9. A solid sphere of radius $3a$ is divided into two parts by a plane at a distance a from the centre. Find, by integration, the volume of the smaller part and the distance of its centre of gravity from the centre of the sphere.

10. Prove that the centre of gravity of a uniform solid pyramid with a square base lies one quarter of the height of the pyramid above the base and on the line joining the vertex to the point of intersection of the diagonals of the base.

11. Prove that the centre of gravity of a uniform solid hemisphere of radius r is a distance $3r/8$ from the centre.

A child's toy is made up from a uniform and solid right circular cone and hemisphere. The radius of the cone is r, and its height $3r$. The radius of the hemisphere is r. The base of the cone and hemisphere are sealed together. The material from which the hemisphere is made is 3 times as heavy per unit volume as the cone material. Find the distance of the centre of gravity of the toy from the vertex of the cone. (SU)

12. Prove that the centre of gravity of a uniform thin hemispherical cup of radius r is at a distance $r/2$ from the centre.

A goblet consists of a uniform thin hemispherical cup of radius r, a circular base of the same material, thickness and radius as the cup, and an intervening stem of length r and whose weight is one-quarter of that of the cup. Show that the height of the centre of gravity above the base is $13r/14$. If the weight of the goblet is W and that of the amount of liquid that fills it is W', show that filling it raises the centre of gravity through a distance $\frac{39}{56}\left(\frac{W'}{W + W'}\right)r$. (W)

5.5 Centroids of areas and volumes

The *centroid* of a plane figure is its centre of area i.e. the point about which the area is evenly distributed. In simple cases we can use symmetry to find the centroid of a figure. However, to arrive at a more general method, we define a quantity called the *first moment of area.*

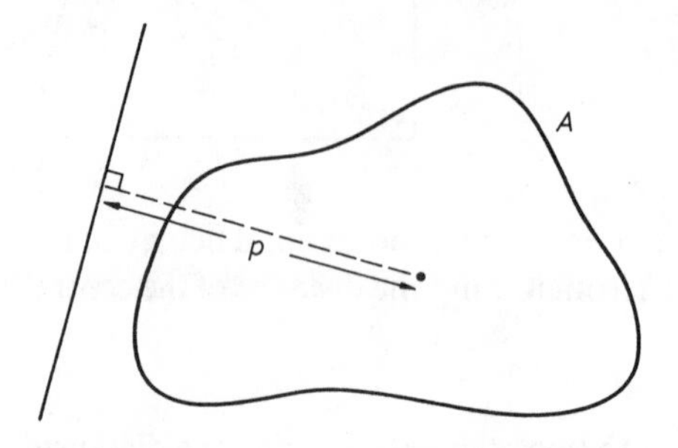

If the centroid of an area A is at a perpendicular distance p from an axis, then the first moment of A about this axis is pA.

If an area is divided into a number of parts, then the sum of the first moments of the parts about any axis is equal to the first moment of the whole area about that axis.

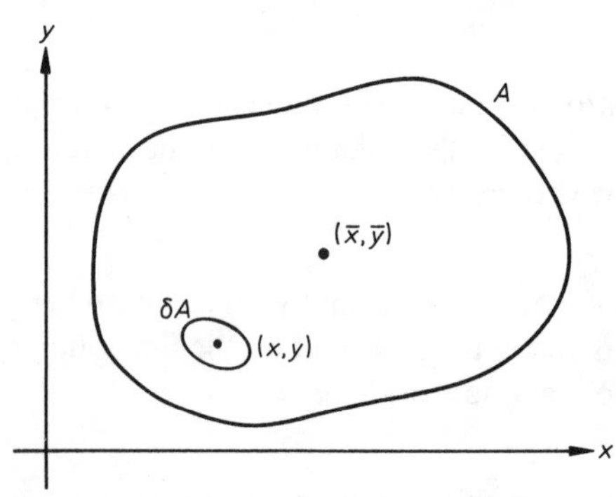

Suppose that a region of the x, y plane has area A and centroid $(\bar{x}, \bar{y})$. If a typical element of area δA has centroid (x, y) then, comparing first moments about the coordinates axes, we have:

$$\bar{x}A = \sum x\,\delta A, \quad \bar{y}A = \sum y\,\delta A.$$

Example 1 Find the centroid of the region bounded by the curve $y^2 = x$ and the line $x = 4$.

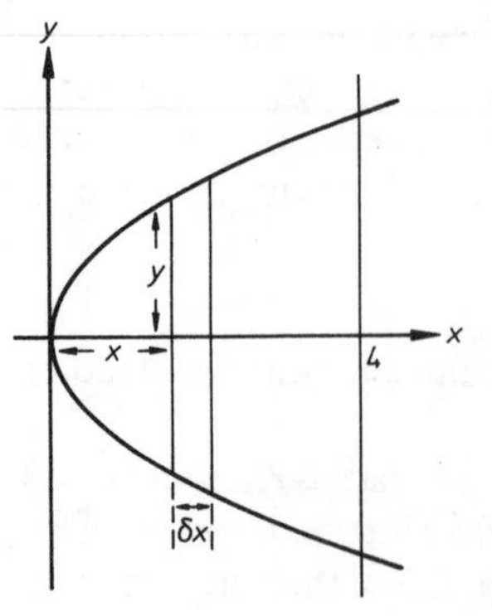

A typical element of area is a strip of width δx, length $2y$ at a distance x from the y-axis. From symmetry the centroids of such strips and of the whole region lie on the x-axis.

Thus, we may write:

	Area	Coordinates of centroid
Element	$2y\,\delta x$	$(x, 0)$
Whole region	A	$(\bar{x}, 0)$

First moment of area about the y-axis

$$= \bar{x}A \approx \sum(x \times 2y\,\delta x)$$

$$\bar{x}A = \int_0^4 2xy\,dx$$

$$= \int_0^4 2x^{3/2}\,dx = 2\left[\frac{2}{5}x^{5/2}\right]_0^4 = \frac{128}{5}$$

But $\quad A = \int_0^4 2y\,dx$

$$= \int_0^4 2x^{1/2}\,dx = 2\left[\frac{2}{3}x^{3/2}\right]_0^4 = \frac{32}{3}$$

$$\therefore \quad \bar{x} = \frac{128}{5}\Big/\frac{32}{3} = \frac{128}{5} \times \frac{3}{32} = \frac{12}{5}$$

Hence the centroid of the region is the point (12/5, 0).

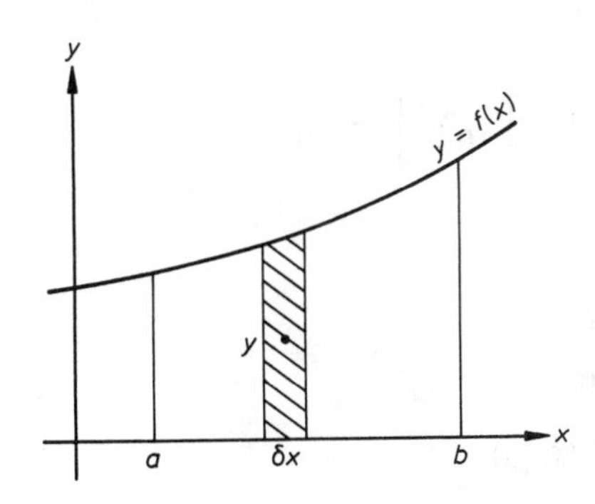

Let us consider now the area bounded by the curve $y = f(x)$, the x-axis and the lines $x = a$, $x = b$. A typical element of area is an approximately rectangular strip of width δx and height y.

	Area	Coordinates of centroid
Element	$y\,\delta x$	$(x, \frac{1}{2}y)$
Whole region	A	$(\bar{x}, \bar{y})$

First moment of area about the y-axis

$$= \bar{x}A \approx \sum(x \times y\,\delta x)$$

$$\therefore \qquad \bar{x}A = \int_a^b xy\,dx$$

First moment of area about the x-axis

$$= \bar{y}A \approx \sum(\tfrac{1}{2}y \times y\,\delta x)$$

$$\therefore \qquad \bar{y}A = \int_a^b \tfrac{1}{2}y^2\,dx$$

But $A = \int_a^b y\,dx$, so the centroid has coordinates

$$\bar{x} = \int_a^b xy\,dx \Big/ \int_a^b y\,dx \qquad \bar{y} = \int_a^b \tfrac{1}{2}y^2\,dx \Big/ \int_a^b y\,dx$$

Example 2 Find the centroid of the area bounded by the curve $y = x^2$, the x-axis and the line $x = 1$.

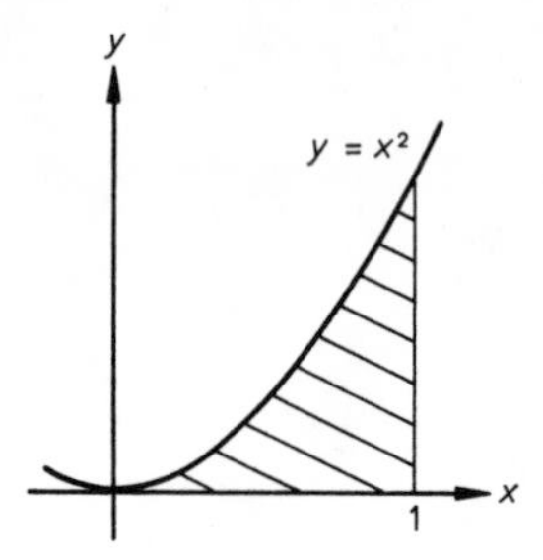

Area of the given region

$$= \int_0^1 y\,dx = \int_0^1 x^2\,dx = \left[\frac{1}{3}x^3\right]_0^1 = \tfrac{1}{3}$$

First moment about the y-axis

$$= \int_0^1 xy\,dx = \int_0^1 x^3\,dx = \left[\frac{1}{4}x^4\right]_0^1 = \tfrac{1}{4}$$

First moment about the x-axis

$$= \int_0^1 \tfrac{1}{2}y^2\,dx = \int_0^1 \tfrac{1}{2}x^4\,dx = \left[\frac{1}{10}x^5\right]_0^1 = \frac{1}{10}$$

$$\therefore \quad \bar{x} = \int_0^1 xy\,dx \Big/ \int_0^1 y\,dx = \frac{1}{4} \div \frac{1}{3} = \frac{3}{4}$$

$$\therefore \quad \bar{y} = \int_0^1 \tfrac{1}{2}y^2\,dx \Big/ \int_0^1 y\,dx = \frac{1}{10} \div \frac{1}{3} = \frac{3}{10}$$

Hence the centroid of the area is the point $(\frac{3}{4}, \frac{3}{10})$.

The centroid of a solid three-dimensional figure is its centre of volume i.e. the point about which the volume is evenly distributed. The methods used to find centroids of volume are similar to those used to find centroids of area.

Example 3 Find the centroid of the volume obtained by rotating completely about the x-axis the area bounded by the curve $y = x^2$, the x-axis and the line $x = 1$.

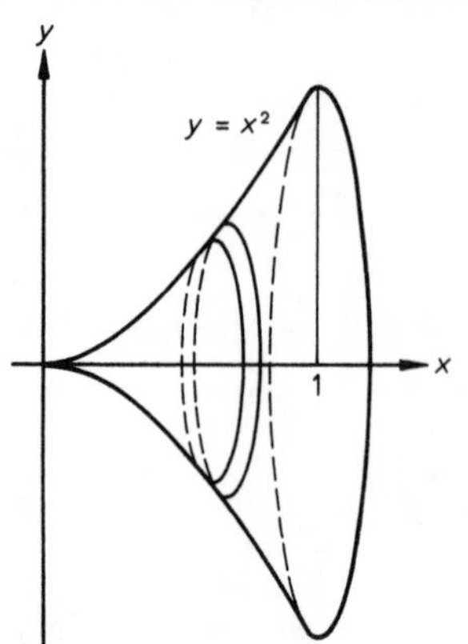

A typical element of volume is approximately a disc of thickness δx, radius y at a distance x from the y-axis. From symmetry the centroids of such discs and of the whole solid lie on the x-axis.

Thus we may write:

	Volume	Coordinates of centroid
Element	$\pi y^2\,\delta x$	$(x, 0)$
Whole solid	V	$(\bar{x}, 0)$

First moment of volume about the y-axis

$$= \bar{x}V \approx \sum(x \times \pi y^2\,\delta x)$$

$$\therefore \quad \bar{x}V = \int_0^1 \pi x y^2\,dx$$

$$= \pi\int_0^1 x^5\,dx = \pi\left[\frac{1}{6}x^6\right]_0^1 = \frac{\pi}{6}$$

$$\text{But} \quad V = \int_0^1 \pi y^2\,dx$$

$$= \pi\int_0^1 x^4\,dx = \pi\left[\frac{1}{5}x^5\right]_0^1 = \frac{\pi}{5}$$

$$\therefore \quad \bar{x} = \frac{\pi}{6}\Big/\frac{\pi}{5} = \frac{\pi}{6} \times \frac{5}{\pi} = \frac{5}{6}$$

Hence the centroid of the solid is at the point $(5/6, 0)$.

Exercise 5.5

Find the first moments about the y-axis and the x-axis of the areas bounded by the following curves and straight lines. In each case use your results to find the coordinates of the centroid of the area.

1. $y = x^2 + 1$, $y = 0$, $x = 0$ and $x = 2$.

2. $y = x(x + 2)$ for $x \geqslant 0$, $y = 0$ and $x = 3$.

3. $y = \dfrac{1}{x^2}$, $y = 0$, $x = 1$ and $x = 2$.

4. $y = x^2$ for $x \geqslant 0$, $x = 0$ and $y = 1$.

Find the centroids of the areas bounded by the following curves and straight lines.

5. $y^2 = x^3$ and $x = 4$.

6. $y^2 = 4x$ and $x = 1$.

7. $y = \cos x$ for $-\frac{1}{2}\pi \leqslant x \leqslant \frac{1}{2}\pi$ and $y = 0$.

8. $x^2y = 4$, $y = 1$ and $y = 4$.

9. $y = e^x$, $y = 0$, $x = 0$ and $x = 1$.

10. $y = \sin x$ for $0 \leqslant x \leqslant \frac{1}{2}\pi$, $y = 0$ and $x = \frac{1}{2}\pi$.

Find the centroids of the solids formed by rotating completely about the x-axis the regions defined by the following inequalities.

11. $y \geqslant x(x - 4)$, $y \leqslant 0$.

12. $y^2 \leqslant 9x$, $y \geqslant 0$, $x \leqslant 1$.

13. $xy \leqslant 4$, $y \geqslant 0$, $1 \leqslant x \leqslant 2$.

14. $0 \leqslant y \leqslant e^{-x}$, $0 \leqslant x \leqslant 1$.

15. Sketch the curve $y^2 = x(3 - x)^2$. Find the centroid of the area enclosed by the loop.

16. Find the centroid of the solid formed by rotating through π radians about the y-axis the region bounded by the curve $y = 4x^2$ and the straight lines $y = 1$ and $y = 4$.

17. Use integration by parts to evaluate

$$\int_1^2 \ln x \, dx, \quad \int_1^2 x \ln x \, dx \quad \text{and} \quad \int_1^2 (\ln x)^2 \, dx.$$

Hence find, correct to 2 decimal places, the coordinates of the centroid of the area bounded by the curve $y = \ln x$, the x-axis and the line $x = 2$.

18. Find the area of the region R bounded by the curve $y = x^2$ and the straight line $y = x$. Show that the first moments of R about the x- and y-axes are given by

$$\int_0^1 \tfrac{1}{2}(x^2 - x^4)\, dx \quad \text{and} \quad \int_0^1 (x^2 - x^3)\, dx \text{ respectively.}$$

Hence find the centroid of the region R.

Exercise 5.6 (miscellaneous)

1. All forces in this question act in the plane of a triangle ABC in which $AB = 4a$, $AC = 3a$ and the angle $A = 90°$. Forces of magnitude $17P$, $15P$, $3P$ act along AB, BC, AC respectively in the directions indicated by the order of the letters. Calculate
(a) the magnitude of the resultant of these three forces and the tangent of the angle made by its line of action with AB,
(b) the distance from A of the point where the line of action of the resultant cuts AB. A couple G is now added to the system and the resultant of this enlarged system acts through the point B. Calculate the magnitude and sense of G. (L)

2. A rigid square lamina $ABCD$ of side a is subject to forces of magnitude $1, 2, 3, 1, 3\sqrt{2}$ and $\lambda\sqrt{2}$ units acting along AB, BC, CD, AD, AC and DB respectively

in the directions indicated by the order of the letters. Given that the direction of the resultant force is parallel to AC, find λ. With this value of λ, find the total moment about A of the forces acting on the lamina. Hence, or otherwise, find AE in terms of a, where E is the intersection of AB with the line of action of the resultant force.

(JMB)

3. (a) State conditions which will ensure that a system of forces in a plane will reduce to a couple.
   (b) State conditions which will ensure that a system of forces in a plane is in equilibrium.
   (c) $ABCD$ is a square of side a. Forces of size 1, 3, 3 and 7 act respectively along AB, BC, CD and DA in the directions indicated by the letters. (i) Find the sizes of the forces which must act along AC and BD so that the six forces will be equivalent to a couple, and find the moment of the couple. (ii) If the forces in DA, AC and BD are to be replaced so that the system is in equilibrium, how will this have to be done? (SU)

4. An equilateral triangle ABC has side $2a$ and D, E are the mid-points of BC, CA respectively. Fixed forces of magnitude $P, 2P, 4P$ act along AB, BC, CA respectively and variable forces of magnitude x, y act along AD, BE respectively, the direction of each force being indicated by the order of the letters. Find (in terms of P) the values of x and y if
(a) the system is equivalent to a force through B parallel to AD, stating the magnitude of this force;
(b) the system is equivalent to a couple, stating the moment of the couple. (O&C)

5. Forces 2, 4, 6, $2p$, $2q$ and 18 newtons act along the sides AB, CD, ED, EF and AF respectively of a regular hexagon $ABCDEF$, the directions of the forces being indicated by the order of the letters. If the system is in equilibrium, find, by resolving parallel and perpendicular to AB, the values of p and q. Check your results by finding the moment of the forces about O, the centre of the hexagon.

The forces along ED, EF and AF are now replaced by a coplanar force through O and a coplanar couple. If the resulting system is in equilibrium and if the length of each side of the hexagon is 2 metres, calculate

(a) the magnitude of this force through O,
(b) the magnitude of the couple. (L)

6. The centre of a regular hexagon $ABCDEF$ of side a is O. Forces of magnitude $P, 2P, 3P, 4P, mP$ and nP act along $\overrightarrow{AB}$, $\overrightarrow{BC}$, $\overrightarrow{CD}$, $\overrightarrow{DE}$, $\overrightarrow{EF}$, and $\overrightarrow{FA}$ respectively. Given that the resultant of these six forces is of magnitude $3P$ acting in a direction parallel to $\overrightarrow{EF}$,
(i) determine the values of m and n,
(ii) show that the sum of the moments of the forces about O is $9Pa\sqrt{3}$.
The mid-point of EF is M.
(iii) Find the equation of the line of action of the resultant referred to OM as x-axis and OA as y-axis.

The forces mP and nP acting along $\overrightarrow{EF}$ and $\overrightarrow{FA}$ are removed from the system. The remaining four forces and an additonal force Q, which acts through O, reduce to a couple. Calculate the magnitude of Q and the moment of the couple. (AEB 1979)

7.

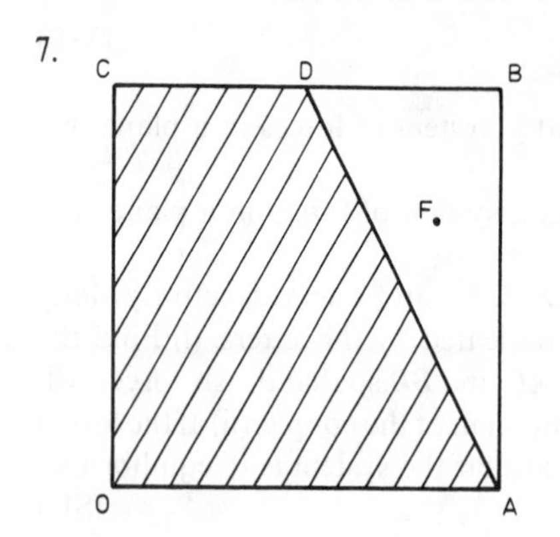

The diagram shows a square $OABC$ of side a. The mid-point of BC is D. Show that, with respect to OA and OC as axes, the coordinates of the centroid F of the triangular region ABD are $(5a/6, 2a/3)$. Find the coordinates of the centre of mass of a uniform lamina in the form of the figure $OADC$. (JMB)

8. A uniform thin sheet of paper has the shape of a rectangle $OABC$, where $OA = 20$ cm and $OC = 12$ cm. The side AB is folded down so as to lie entirely along OA. Find the distances of the centre of gravity of the folded sheet from OA and OC. (O&C)

9. A uniform solid hemisphere of radius a is cut into two parts by a plane parallel to the plane face of the hemisphere and at a distance $\frac{1}{2}a$ from it. Find the volume of each of the two parts and the position of the centre of mass of each of the two parts. (AEB 1980)

10. Prove that the centre of gravity of a uniform circular arc subtending an angle 2θ at the centre of a circle of radius a is $a \sin \theta/\theta$ from the centre. Deduce that the centre of gravity of a uniform sector bounded by that arc and the radii to its extremities is $\frac{2}{3}a \sin \theta/\theta$ from the centre.

Show also that the centre of gravity of a segment of a circular lamina cut off by a chord subtending a right angle at the centre of the circle is $\frac{2}{3}\sqrt{2}a/(\pi - 2)$ from the centre. (W)

11. Find by integration the distance of the centroid of a uniform solid hemisphere of radius a from its plane face.

A hollow sphere of external radius R and internal radius r is cut in half by a plane through its centre. Show that the distance of the centroid of each half from the centre of the sphere is

$$\frac{3(R + r)(R^2 + r^2)}{8(R^2 + Rr + r^2)}.$$

Deduce that the centroid of a hemispherical shell of radius R is at a distance $R/2$ from its centre. (L)

12. Find the area A of the loop of the curve $y^2 = x^2(2 - x)$. Find also the first moment about the y-axis of the volume generated when the area A is rotated through π radians about the x-axis. (AEB 1975)

13. Find the area of the region in the first quadrant enclosed between the curve $y = x^2 + 4$, the line $y = 8$ and the y-axis. The region is rotated through four right-angles about Oy to form a uniform solid. Find the volume and the coordinates of the centre of mass of the solid. (C)

14. Sketch on one diagram the graphs of $x^3y = 1$ and $y = \sqrt{x}$, for positive values of x. Show that the area of the finite region bounded by the straight lines $y = 0$ and $x = 2$ and an arc of each of the above curves is 25/24. Find also the coordinates of the centroid of the region. (C)

15. Find the area of the finite region between the curve $y = x^2 + 1$ and the line $y = 2x + 1$. The region is rotated through four right angles about Ox, to produce a uniform solid of revolution. Find the volume and the coordinates of the centre of gravity of this solid. (C)

16. Sketch the arc of the curve given parametrically by $x = a(\theta - \sin\theta)$, $y = a(1 - \cos\theta)$, for which $0 \leqslant \theta \leqslant 2\pi$. Find the area of the finite region enclosed by this arc and the x-axis. Find also the coordinates of the centroid of the region. (L)

6 Equilibrium of rigid bodies

6.1 Parallel forces in equilibrium

If a set of coplanar forces is in equilibrium then
(i) the sum of the components of the forces in any given direction is zero and
(ii) the sum of the moments of the forces about any given point is zero.
These facts are used to find unknown quantities in many statics problems.

Example 1 A non-uniform rod AB of length 5 m and weight 75 N is supported in a horizontal position by two vertical strings attached to its ends. A weight of 15 N is attached to the rod at the point C, where $AC = 1$ m. If the tensions in the strings at A and B are $2T$ N and T N respectively, find the value of T. Find also the distance, x m, of the centre of gravity of the rod from A.

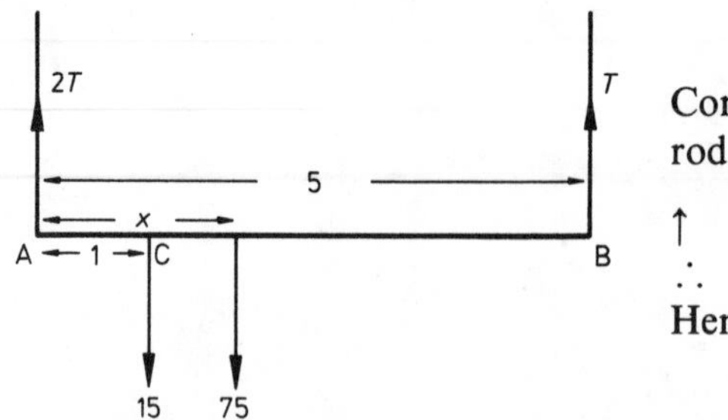

Considering the vertical forces on the rod,

$$\uparrow \qquad 2T - 15 - 75 + T = 0$$
$$\therefore \qquad 3T = 90$$
Hence $\qquad T = 30.$

Taking moments clockwise about A,

$$A\circlearrowright \qquad 15 \times 1 + 75 \times x - T \times 5 = 0$$

$\therefore \qquad 75x = 5T - 15 = 135$

Hence $\qquad x = 1{\cdot}8.$

When a rigid body rests on an object described as *smooth*, (e.g. a smooth support or a smooth table,) then the reaction on the body at any point of contact is assumed to be perpendicular to the surfaces in contact.

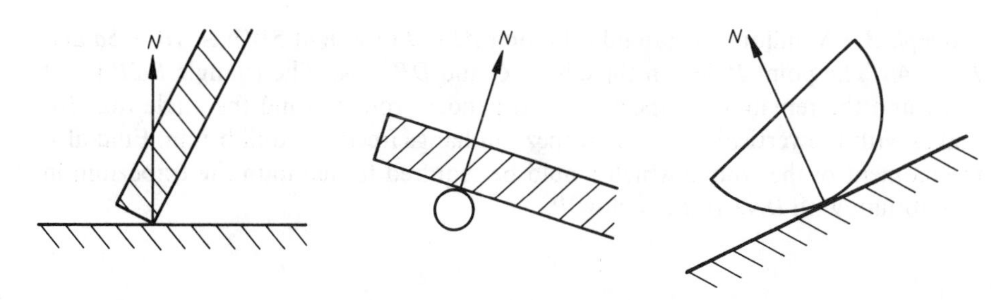

Example 2 A uniform plank AB of length 4 m and weight 300 N rests horizontally on smooth supports at P and Q, where $AP = QB = 1{\cdot}2$ m. Find the maximum weight that can be placed at B without tilting the plank.

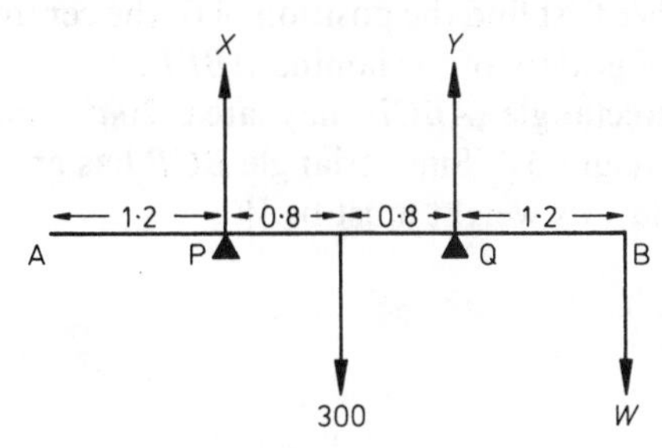

Let us assume that the plank is in equilibrium with reactions of X N and Y N at P and Q respectively and a weight of W N at B.

Taking moments about Q ↻,

$$X \times 1{\cdot}6 + W \times 1{\cdot}2 - 300 \times 0{\cdot}8 = 0$$

$\therefore \quad 1{\cdot}2W = 240 - 1{\cdot}6X$

Since the reaction at P acts vertically upwards, $X \geqslant 0$,

$\therefore \quad 1{\cdot}2W \leqslant 240 \quad$ i.e. $\quad W \leqslant 200.$

Hence the maximum weight that can be placed at B without tilting the plank is 200 N.

[Note that when a weight of 200 N is placed at B, the reaction at P is zero and the plank just fails to rotate about Q.]

When a body is in equilibrium under the action of *two forces*, these forces must be equal in magnitude and act in opposite directions along the same line.

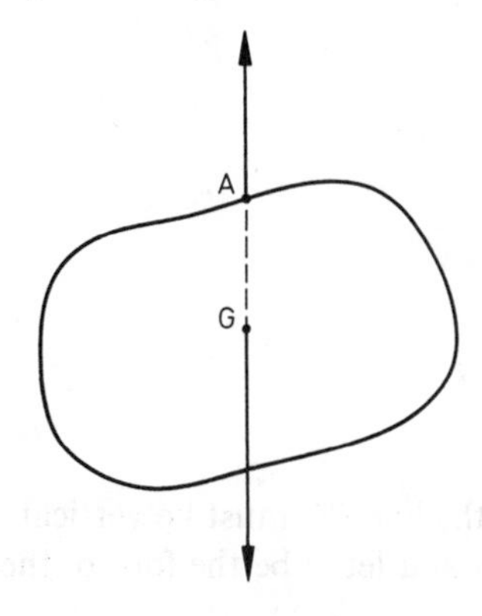

In particular, when a body freely suspended from a point A is in equilibrium, the two forces acting on it, namely its weight and the force at A, must act along the same vertical line. We deduce that the body hangs in equilibrium with its centre of gravity G vertically below A.

Example 3 A uniform rectangular lamina $ABCD$ of weight $5W$ has $AB = 5a$ and $BC = 4a$. The point P lies on the edge DC and $DP = 3a$. The triangle BCP is cut away and the remaining trapezium is suspended from P. Find the angle that DP makes with the vertical when the trapezium hangs freely in equilibrium. Find also the moment of the couple which would be required to maintain the trapezium in equilibrium with D vertically below P.

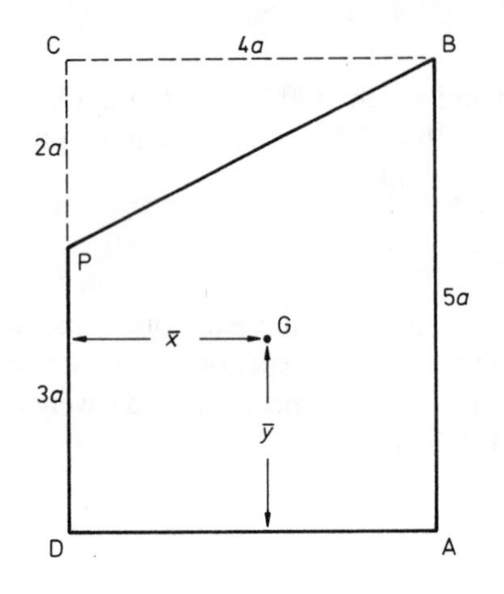

We first find the position of G, the centre of gravity of the lamina $ABPD$. Rectangle $ABCD$ has area $20a^2$ and weight $5W$. Since triangle BCP has area $4a^2$, its weight must be W.

Body	Weight	Distance of centre of gravity from CD	Distance of centre of gravity from DA
Rectangle $ABCD$	$5W$	$2a$	$5a/2$
Triangle BCP	W	$4a/3$	$13a/3$
Remainder	$4W$	$\bar{x}$	$\bar{y}$

Taking moments about CD,

$$4W \times \bar{x} + W \times \frac{4a}{3} = 5W \times 2a$$

$$\therefore \quad \bar{x} = \frac{1}{4}\left(10a - \frac{4a}{3}\right) = \frac{13a}{6}$$

Taking moments about DA,

$$4W \times \bar{y} + W \times \frac{13a}{3} = 5W \times \frac{5a}{2}$$

$$\therefore \quad \bar{y} = \frac{1}{4}\left(\frac{25a}{2} - \frac{13a}{3}\right) = \frac{49a}{24}$$

When the trapezium hangs in equilibrium from P, the line PG must be vertical. Let θ be the angle DP then makes with the vertical and let Q be the foot of the perpendicular from G to DP.

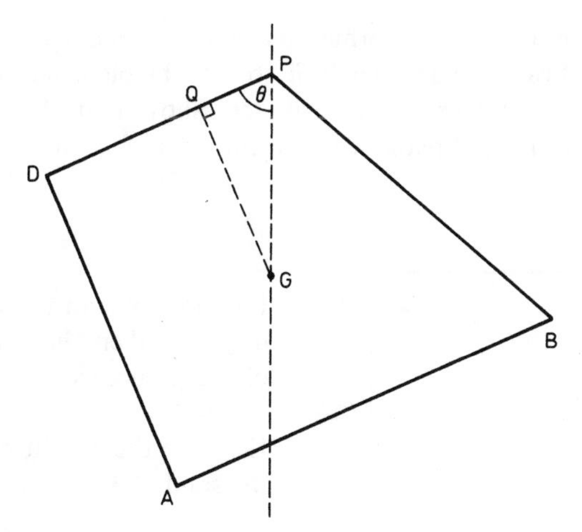

$$GQ = \bar{x} = \frac{13a}{6}, \quad DQ = \bar{y} = \frac{49a}{24}, \qquad \therefore \quad QP = \frac{23a}{24}.$$

Thus $\tan\theta = \dfrac{GQ}{QP} = \dfrac{13}{6} \times \dfrac{24}{23} = \dfrac{52}{23}$ and $\theta \approx 66{\cdot}1°$.

Hence the angle made by DP with the vertical is 66·1°.
Suppose now that the trapezium is in equilibrium with D vertically below P and that a couple of moment M in the sense DAB is required to maintain this position.
Taking moments anti-clockwise about P,

$$M - 4W \times \frac{13a}{6} = 0 \qquad \therefore \quad M = \frac{26Wa}{3}.$$

Hence the moment of the required couple is $26Wa/3$.

When a body rests in equilibrium on a horizontal plane, the weight of the body and the force exerted by the plane on the body must act along the same line, namely the vertical line through the centre of gravity.

In the case of a lamina with an edge AB resting on a smooth horizontal plane, the force N exerted by the plane on the lamina is the resultant of normal reactions at the points of contact along AB. Thus N must act through some point on AB. It follows that, if the vertical line through the centre of gravity G passes through a point on the edge AB, then the force N can act to maintain equilibrium. Otherwise equilibrium is not possible and the lamina will topple.

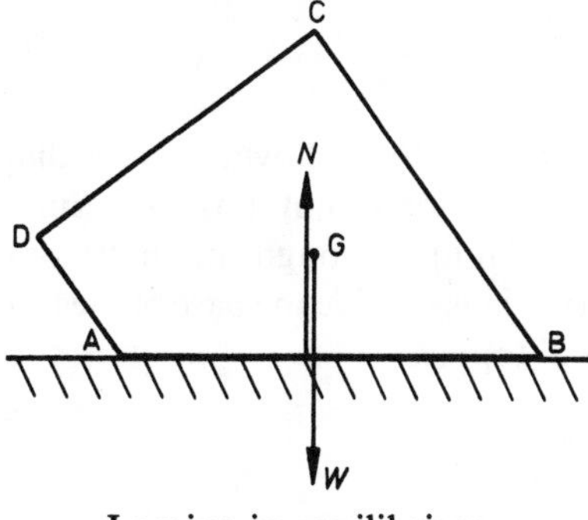

Lamina in equilibrium

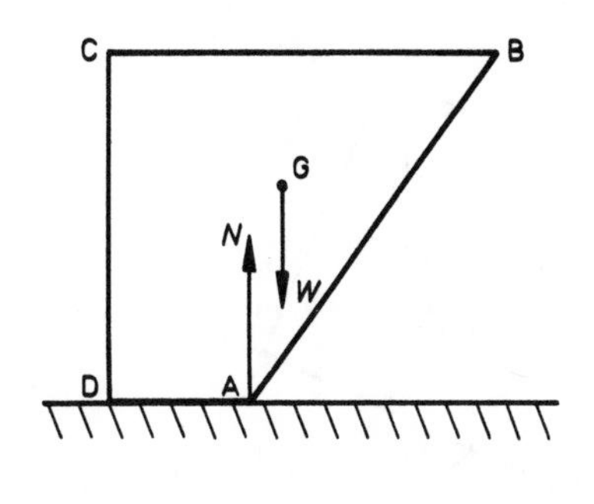

Lamina will topple

Example 4 A uniform solid is formed by joining a hemisphere of radius a to a circular cylinder of radius a and height h, so that the plane face of the hemisphere coincides with a plane face of the cylinder. Show that this solid can rest in equilibrium on a horizontal plane with the curved surface of the cylinder touching the plane if $h^2 \geqslant \frac{1}{2}a^2$.

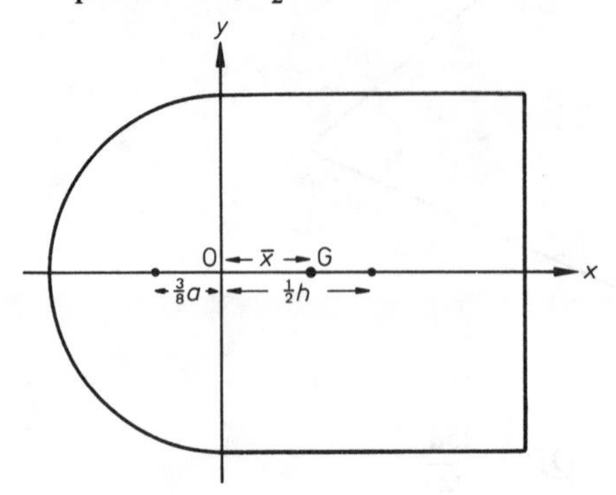

If we take x- and y-axes as shown in the diagram, then the centre of gravity G of the combined solid must lie on the x-axis.
Let w be the weight per unit volume of the solid.

Body	Weight	x-coordinate of centre of gravity
Cylinder	$\pi a^2 h w$	$\frac{1}{2}h$
Hemisphere	$\frac{2}{3}\pi a^3 w$	$-\frac{3}{8}a$
Whole body	$(h + \frac{2}{3}a)\pi a^2 w$	$\bar{x}$

Taking moments about the y-axis,

$$(h + \tfrac{2}{3}a)\pi a^2 w\bar{x} = \pi a^2 h w \times \tfrac{1}{2}h - \tfrac{2}{3}\pi a^3 w \times \tfrac{3}{8}a$$

$$\therefore \quad (h + \tfrac{2}{3}a)\bar{x} = \tfrac{1}{2}h^2 - \tfrac{1}{4}a^2$$

When this solid is placed with the curved surface of the cylinder touching a horizontal plane, it will rest in equilibrium if the vertical line through G passes through a point of contact between the plane and the solid.
This condition will be satisfied if $\bar{x} \geqslant 0$ i.e. if $h^2 \geqslant \frac{1}{2}a^2$.

For a body to rest in equilibrium on a rough inclined plane the force exerted by the plane on the body must act along a vertical line through the centre of gravity of the body.

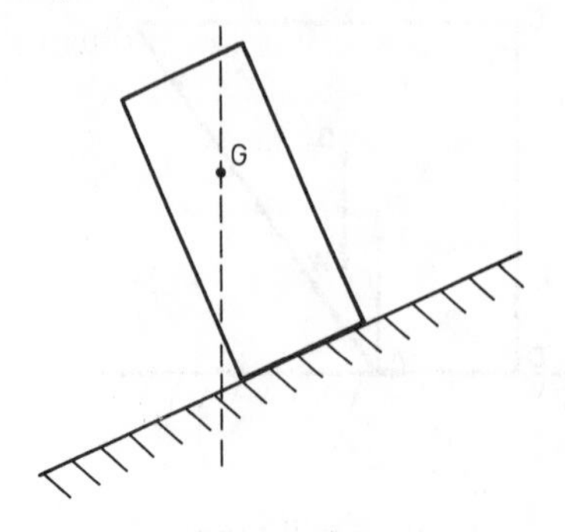

For the block shown in this diagram equilibrium is not possible. Since the vertical line through the centre of gravity G falls outside the base of the block, it will topple.

The reaction R between a body and a rough surface is the resultant of two components, the normal reaction N and a friction force F. According to the laws of friction, if μ is the coefficient of friction between the body and the surface, then $F \leqslant \mu N$. We can also express this condition in terms of λ, the angle of friction, and θ, the angle made by R with the normal to the surface, since $\tan \lambda = \mu$ and $\tan \theta = F/N$.

$$F \leqslant \mu N \;\Leftrightarrow\; \tan \theta \leqslant \mu \;\Leftrightarrow\; \theta \leqslant \lambda.$$

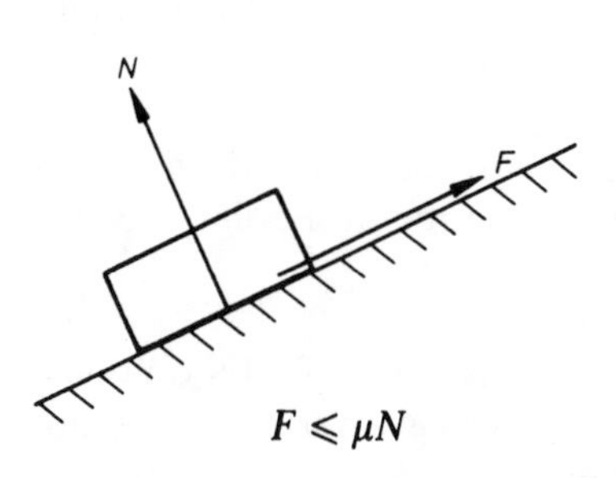

$F \leqslant \mu N$

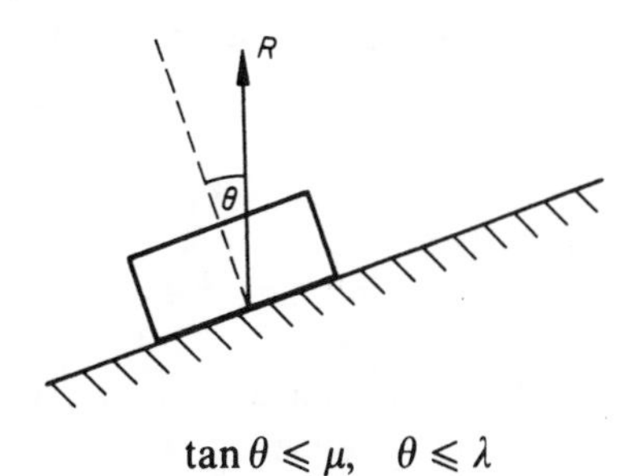

$\tan \theta \leqslant \mu, \quad \theta \leqslant \lambda$

Example 5 A uniform solid cube is in equilibrium with one face resting on a rough inclined plane and one pair of opposite faces vertical. The coefficient of friction between the cube and the plane is μ. The angle of inclination of the plane is gradually increased. Show that equilibrium will be broken by sliding rather than toppling if $\mu < 1$.

(1)

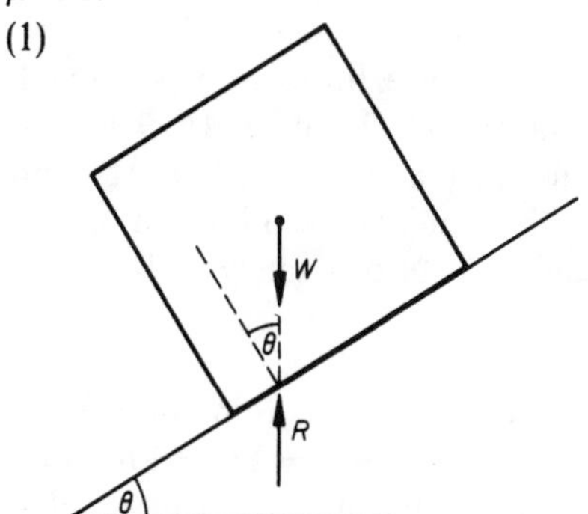

Let θ be the angle of inclination of the plane to the horizontal. When the cube is in equilibrium, as shown in diagram (1), the total reaction R acts vertically. The angle between its line of action and the normal to the plane is θ. When the cube is on the point of sliding, friction is limiting,

$$\therefore \qquad \tan \theta = \mu.$$

(2)

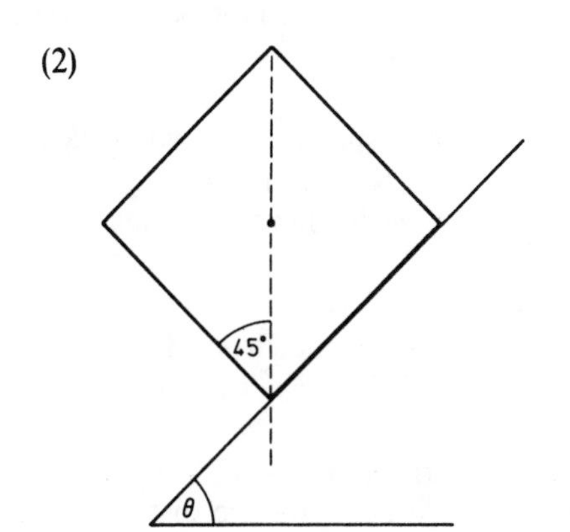

In diagram (2) the cube is on the point of toppling, since the vertical line through its centre of gravity passes through the lower edge of the base. Thus, when the cube is about to topple,

$$\tan \theta = \tan 45° = 1.$$

Hence equilibrium will be broken by sliding if $\tan\theta$ reaches the value μ before it reaches the value 1, i.e. if $\mu < 1$.

Exercise 6.1

1. A beam AB of length 4·5 m is supported in a horizontal position by two vertical cables attached to its ends. The tensions in the cables at A and B are 160 N and 200 N respectively. Find the weight of the beam and the distance of its centre of gravity from A.

2. A uniform rod AB of length 3 m and weight 50 N rests horizontally on smooth supports at A and B. A load of 24 N is attached to the rod at a point C, where $AC = 1$ m. Find the forces exerted on the rod by the supports.

3. A uniform bar of length 2·8 m and weight 80 N has loads of 20 N and 40 N attached to its ends A and B respectively. If the bar balances in a horizontal position when smoothly supported at C, find the distance of C from A.

4. A uniform rod $APQB$ of length 2 m rests horizontally on smooth supports at P and Q, where $PQ = 1{\cdot}2$ m. If the reaction at P is twice the reaction at Q, find the distance AP. Given that when a weight of 5 N is attached to the rod at B the reactions at P and Q are equal, find the weight of the rod.

5. A uniform plank ABC of weight 400 N is held in a horizontal position by a smooth support at B and a vertical rope attached at C, where $AB = 1$ m and $AC = 5$ m. If the rope is likely to break when the tension in it exceeds 750 N, find
(a) the maximum weight that can be placed on the plank at the mid-point of BC,
(b) the maximum distance that a man of weight 800 N can safely walk from B towards C.

6. A non-uniform beam $AXYB$ of length 4 m and weight 400 N is maintained in a horizontal position by smooth supports at X and Y, where $AX = YB = 1$ m. A boy of weight 500 N can just walk to the end A of the beam without overturning it. If the boy walks along the beam towards B, find his distance from B when the beam is just about to tip.

7. A uniform right circular solid cone of weight W is suspended by two vertical strings attached to the ends A and B of a diameter of its base. If the cone hangs in equilibrium with its vertex vertically below A, find the tensions in the strings.

8. $ABCD$ is a uniform rectangular lamina of weight 10 N. A particle of weight W N is attached to the lamina at B. When the system is freely suspended from A it rests in equilibrium with the mid-point of BC vertically below A. Find the value of W.

9. A uniform wire is bent into the form of a trapezium $ABCD$ in which $\angle ABC = \angle BCD = 90°$, $AB = 6a$, $BC = 4a$ and $CD = 3a$. Find the distances of

the centre of gravity from AB and BC. If the trapezium is freely suspended from A, find the angle which AB makes with the vertical.

10. A uniform triangular lamina ABC has $AB = 4a, BC = 5a, CA = 3a$. The lamina is folded through 180° along LN, where L and N are the mid-points of BC and AB respectively, so that the vertex B is at A. Find the coordinates of the centre of mass of the folded lamina relative to AN and AC as x-axis and y-axis respectively. The folded lamina is suspended from C and hangs in equilibrium under gravity. Find the tangent of the angle that CA makes with the vertical. (C)

11. A uniform wire is bent to form a semi-circular arc together with the diameter AB joining the ends of the arc. If the wire hangs in equilibrium from a smooth pivot at A, find the tangent of the angle that AB makes with the vertical.

12. A uniform wire AB of weight $2W$ is bent into a semi-circular arc. A particle of weight W is attached to the end B and the system is freely suspended from A. Find the tangent of the angle that AB makes with the vertical.

13. Three uniform rods AB, BC, CD, each of length $2a$ and weight W are rigidly joined together to form three sides of a square $ABCD$. If the system hangs freely from a smooth pivot at A, find the angle that AB makes with the vertical. Find also, in terms of W and a, the moment of the couple required to maintain the system in equilibrium with BC horizontal.

14. $ABCD$ is a uniform square lamina of side $6a$ and weight W. P is a point on the diagonal AC, such that $AP = 2PC$. The square having diagonal PC is cut away and the remaining body is freely suspended from B. Find the angle which BA makes with the vertical. Find also the moment of the couple required to maintain the body in equilibrium with BA horizontal.

15. $ABCD$ is a uniform rectangular lamina in which $BC = 2a$. P is the mid-point of BC and R is a point on AB such that $AR = a$ and $RB = ka$. The rectangle $BPQR$ is cut away and the remaining body is placed in a vertical plane with AR resting on a horizontal plane. Find the maximum value of k given that the lamina does not topple about R.

16. A uniform lamina of weight $3W$ in the form of a trapezium $ABCD$ has $\angle ABC = \angle BCD = 90°, AB = BC = a$ and $CD = 2a$. The lamina is placed in a vertical plane with AB resting on a horizontal plane. Find, in terms of W, the greatest weight that can be attached to the lamina at D without causing it to topple about A.

17. A solid hemisphere of radius a and a solid right circular cone of height h and base radius a are made from the same uniform material and are joined together with their plane faces completely in contact. O is the centre of the common base. Find the position of the centre of gravity of the whole body.

(a) The body is suspended freely from a point A on the edge of the common base and hangs in equilibrium under gravity. If $h = 3a$, find, correct to the nearest degree, the acute angle made by AO with the vertical.
(b) The body is found to rest in equilibrium when it is placed on a horizontal table with *any* point of its hemispherical surface in contact with the table. Find h in terms of a. (O&C)

18. A uniform thin hemispherical bowl rests in equilibrium with its axis of symmetry horizontal and its curved surface on a rough inclined plane. Find the angle between the plane and the horizontal. Find also the least possible value of the coefficient of friction between the bowl and the plane.

19. A uniform solid hemisphere rests in equilibrium with its curved surface on a rough plane inclined at an angle α to the horizontal, where $\sin\alpha = 0{\cdot}3$. Find the angle between the plane face of the hemisphere and the vertical.

20. A uniform right circular solid cone of radius a and height $3a$ rests in equilibrium with its base in contact with a rough inclined plane. The angle of inclination of the plane is then increased steadily.
(a) Assuming that the cone does not slide down the plane, find the angle of inclination of the plane to the horizontal when the cone is about to topple.
(b) If the coefficient of friction between the cone and the plane is $\frac{1}{2}$, find the angle of inclination of the plane to the horizontal when the cone is about to slide.

21. A uniform solid cylinder of radius a and height h is placed with one plane face resting on a rough inclined plane. The angle of inclination of the plane is then gradually increased. Show that the cylinder will slide before it topples if the coefficient of friction between the cylinder and the plane is less than $\dfrac{2a}{h}$.

6.2 Three force problems

When a body is in equilibrium under the action of *three forces*, the vector sum of these forces must be zero. Hence three non-parallel forces in equilibrium can be represented in magnitude and direction by the sides of a triangle. We also know that the sum of the moments of coplanar forces in equilibrium about any point in their plane is zero.

Thus one approach to three force problems is to
(i) take moments about any suitable point, and
(ii) use a triangle of forces.

Example 1 A uniform rod AB of mass m, which is smoothly hinged at A, is maintained in equilibrium by a horizontal force P acting at B. Given that the rod is inclined at 30° to the horizontal with B below A, find (a) an expression for P, (b) the magnitude and direction of the reaction at the hinge.

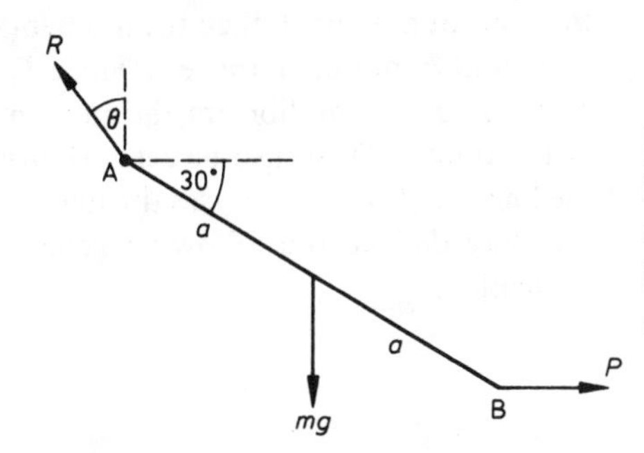

[Note that the reaction at a *smooth hinge* is a single force which may act in any direction. The system of reactions in a hinge which is *not smooth* may reduce to a single force together with a couple.]

Let the length of the rod be $2a$ and let the reaction at the hinge be of magnitude R acting at an angle θ to the vertical.

(a) The weight of the rod is mg and acts through the mid-point of the rod. Thus taking moments anti-clockwise about A, we have

$$P \times 2a \sin 30° - mg \times a \cos 30° = 0$$

$$\therefore \qquad Pa - \frac{\sqrt{3}}{2} mga = 0$$

Hence $P = \dfrac{\sqrt{3}}{2} mg.$

(b)

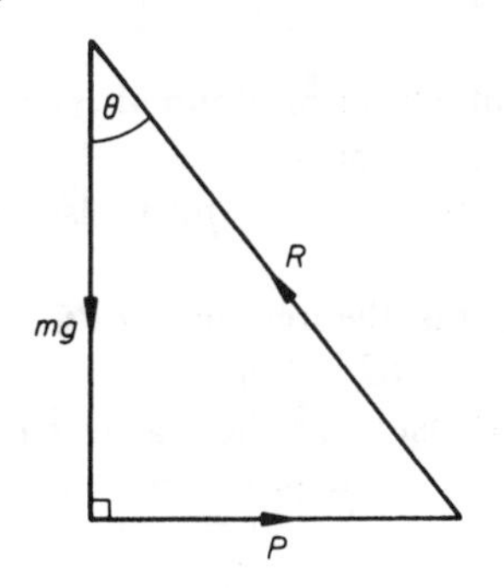

Using the triangle of forces,

$$R = \sqrt{\{(mg)^2 + P^2\}} = mg\sqrt{(1 + \tfrac{3}{4})}$$
$$= \frac{\sqrt{7}}{2} mg$$

$$\tan\theta = \frac{P}{mg} = \frac{\sqrt{3}}{2} \qquad \therefore \quad \theta \approx 40{\cdot}9°$$

Hence the reaction at the hinge has magnitude $\dfrac{\sqrt{7}}{2} mg$ and makes an angle $\tan^{-1}\dfrac{\sqrt{3}}{2} \approx 40{\cdot}9°$ with the vertical.

We now establish a property of three force systems which is a useful aid to problem solving. Suppose that a rigid body is in equilibrium under the action of three coplanar forces F_1, F_2 and F_3. Then either all three forces are parallel or the lines of action of two of them meet.

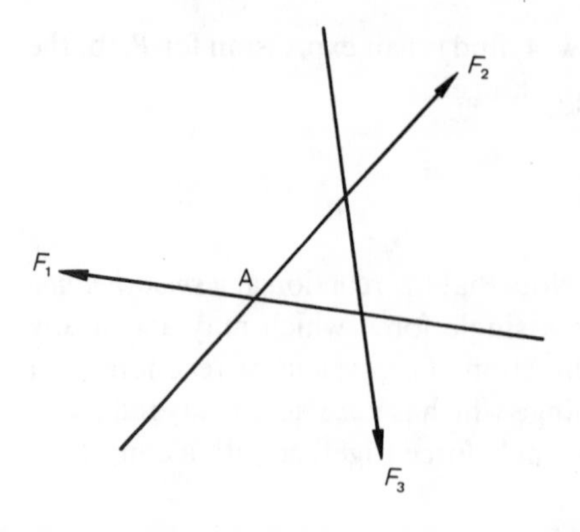

If F_1 and F_2 act along lines which intersect at a point A, then the moments of F_1 and F_2 about A are zero. Since F_1, F_2 and F_3 are in equilibrium, the moment of F_3 about A must also be zero. Hence the line of action of F_3 passes through A. We may deduce the following general principle.

> If a rigid body is in equilibrium under the action of three coplanar forces, then the lines of action of the forces are either parallel or concurrent.

Example 2 A uniform pole AB of length 5 m is smoothly hinged at A and supported at B by a light elastic string BC of natural length 3 m and modulus 60 N. The end C of the string is fixed 3 m vertically above A. Given that the system is in equilibrium with BC horizontal, find in newtons the tension T in the string, the weight W of the pole and magnitude R of the reaction at A.

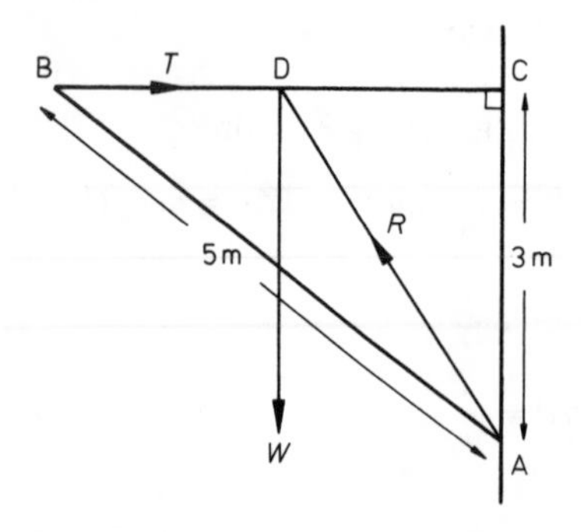

The lines of action of T and W meet at D, the mid-point of BC,

$\therefore$ the third force R must also act through D.

By Pythagoras' theorem in $\triangle ABC$,

$$BC = 4\,\text{m},$$

$\therefore$ the extension in the elastic string

$$= (4 - 3)\,\text{m} = 1\,\text{m}.$$

By Hooke's law, $T = \dfrac{\lambda x}{l}$

where $\lambda = 60$, $l = 3$ and $x = 1$,

$$\therefore \quad T = \frac{60 \times 1}{3} = 20.$$

Hence the tension in the elastic string is 20 N.

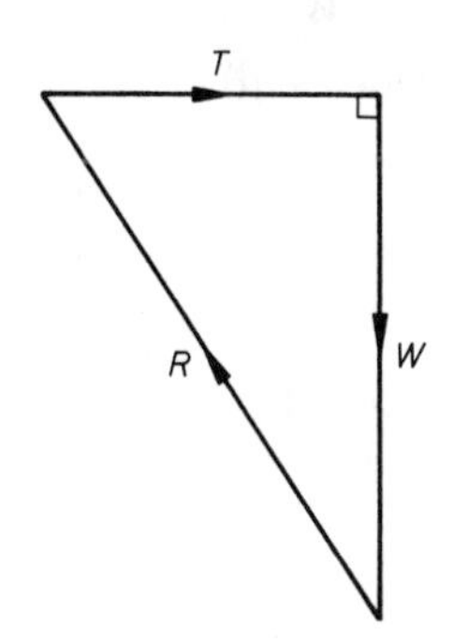

Triangle of forces

Comparing the diagrams we see that the triangle of forces is similar to $\triangle ADC$. By Pythagoras' theorem in $\triangle ADC$,

$$AD = \sqrt{(AC^2 + DC^2)} = \sqrt{(3^2 + 2^2)}\,\text{m} = \sqrt{13}\,\text{m},$$

$$\therefore \quad \frac{T}{2} = \frac{W}{3} = \frac{R}{\sqrt{13}}$$

$$\therefore \quad W = \frac{3}{2}T = 30$$

$$\text{and} \quad R = \frac{\sqrt{13}}{2}T = 10\sqrt{13} \approx 36{\cdot}1.$$

Hence the weight of the pole is 30 N and the magnitude of the reaction at A is 36·1 N.

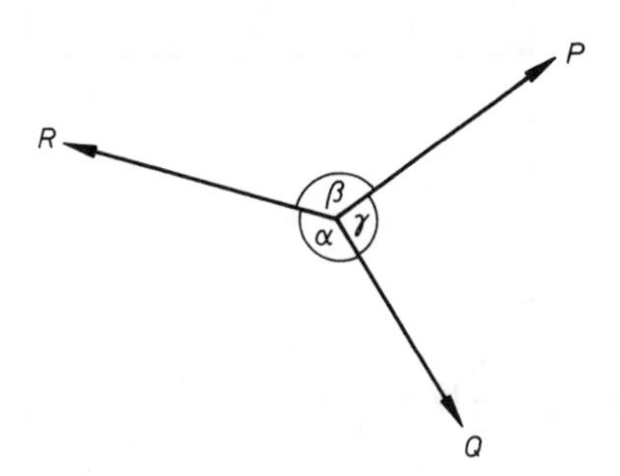

When solving three force problems in which angles rather than lengths are known, *Lami's theorem* may be useful. It states that for forces P, Q and R in equilibrium, as shown in the diagram.

$$\frac{P}{\sin\alpha} = \frac{Q}{\sin\beta} = \frac{R}{\sin\gamma}$$

Example 3 A uniform rod AB of length $5a$ and weight W rests with the end A on a smooth plane inclined at an angle α to the horizontal, where $\sin\alpha = 3/5$. It is maintained in equilibrium by a smooth support at C. If the rod makes an angle β with the horizontal, where $\sin\beta = 7/25$, find

(a) the magnitudes of the reactions at A and C,
(b) the distance AC.

The reaction at A is perpendicular to the inclined plane and the reaction at C is perpendicular to the rod. Let their magnitudes be N_1 and N_2 respectively. Since the forces N_1, N_2 and W are in equilibrium, their lines of action are concurrent.

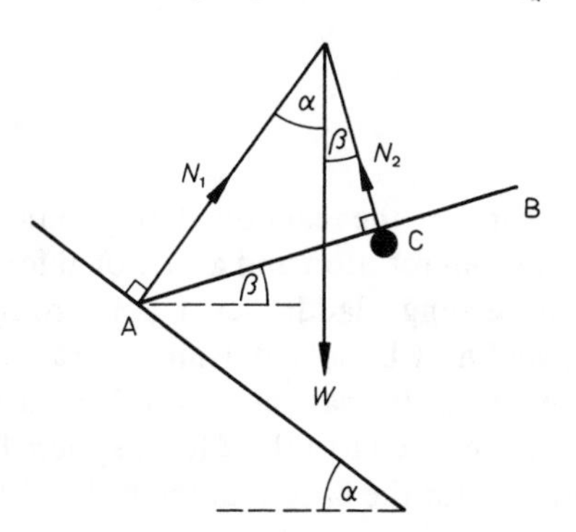

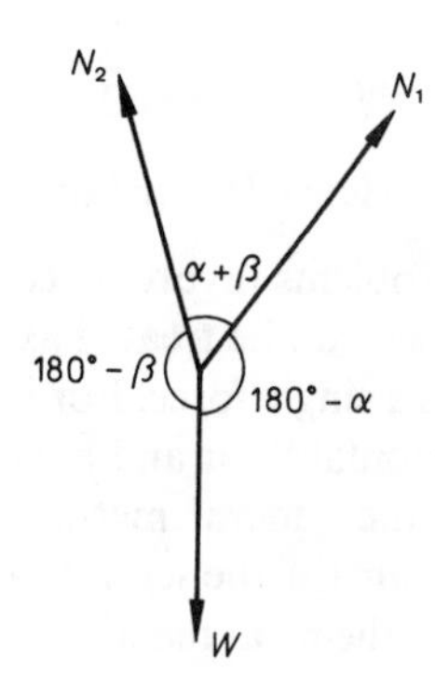

(a) By Lami's theorem, $\dfrac{N_1}{\sin(180° - \beta)} = \dfrac{N_2}{\sin(180° - \alpha)} = \dfrac{W}{\sin(\alpha + \beta)}$

$$\therefore \quad \frac{N_1}{\sin \beta} = \frac{N_2}{\sin \alpha} = \frac{W}{\sin(\alpha + \beta)} \qquad (1)$$

$$\sin \alpha = \frac{3}{5} \Rightarrow \cos \alpha = \frac{4}{5}$$

$$\sin \beta = \frac{7}{25} \Rightarrow \cos \beta = \frac{24}{25}$$

$$\therefore \quad \sin(\alpha + \beta) = \sin \alpha \cos \beta + \cos \alpha \sin \beta$$

$$= \frac{3}{5} \times \frac{24}{25} + \frac{4}{5} \times \frac{7}{25}$$

$$= \frac{72 + 28}{125} = \frac{100}{125} = \frac{4}{5}$$

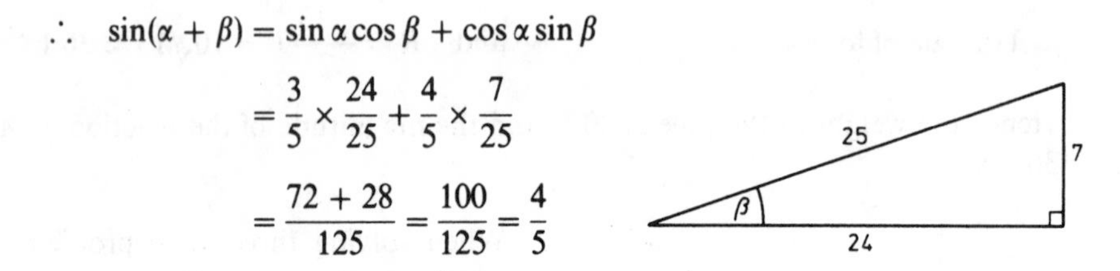

Substituting these values in equation (1),

$$\frac{25}{7}N_1 = \frac{5}{3}N_2 = \frac{5}{4}W$$

$$\therefore \quad N_1 = \frac{7}{25} \times \frac{5}{4}W = \frac{7}{20}W \quad \text{and} \quad N_2 = \frac{3}{5} \times \frac{5}{4}W = \frac{3}{4}W$$

Hence the magnitudes of the reactions at A and C are $7W/20$ and $3W/4$ respectively.

(b) The distance of the centre of gravity of the rod from A is $5a/2$. Let the distance AC be x.

Taking moments anti-clockwise about A,

$$N_1 \times 0 + N_2 \times x - W \times \frac{5}{2}a \cos \beta = 0$$

$$\therefore \quad \frac{3}{4}Wx - \frac{5}{2} \times \frac{24}{25}Wa = 0$$

$$\therefore \quad x = \frac{4}{3} \times \frac{5}{2} \times \frac{24}{25}a = \frac{16}{5}a$$

Hence the distance AC is $16a/5$.

In problems involving contact between rough surfaces the reaction at any point of contact can be treated as the resultant of the normal reaction and a frictional force or as a single force. For instance, a problem concerning a ladder resting on a rough horizontal floor and against a rough vertical wall may be solved using diagram (a) and the general methods to be described in §6.3. Alternatively, as indicated in diagram (b), the set of forces on the ladder can be reduced to a three force system by using the total reactions at the points of contact with the floor and the wall.

(a)

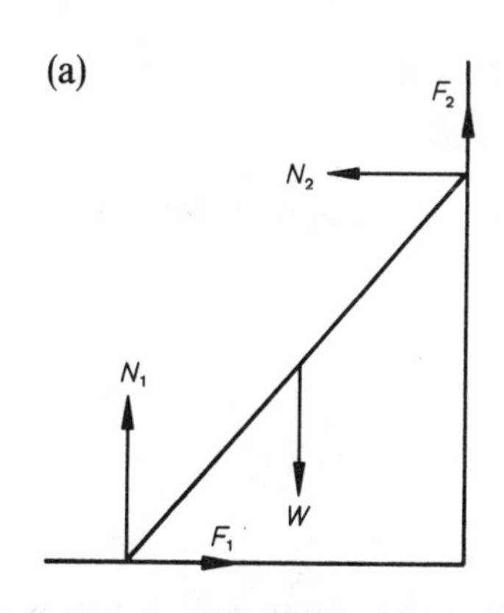

(b)

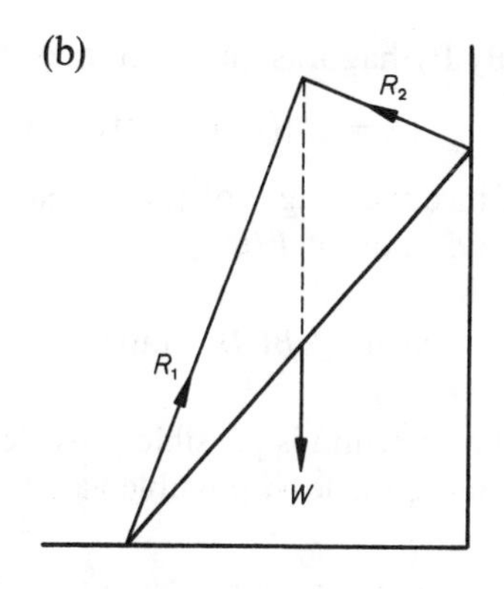

When using this latter approach, it is helpful to remember that if θ is the angle between the total reaction and the normal to the surfaces in contact, then

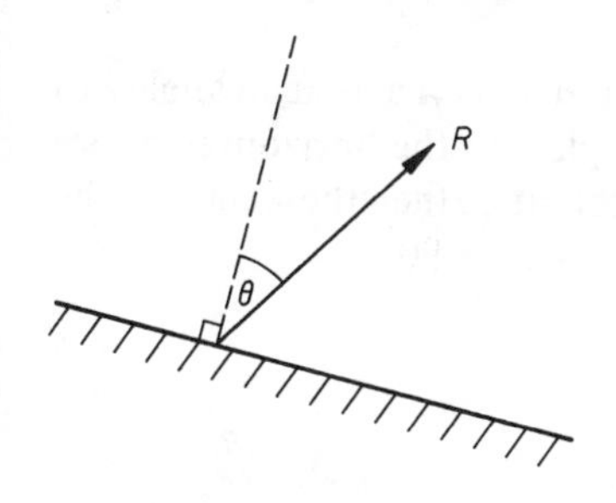

$$\tan\theta \leqslant \mu \quad \text{and} \quad \theta \leqslant \lambda,$$

where μ is the coefficient of friction and λ is the angle of friction.

Example 4 A uniform rod AB of length 1·7 m is placed with the end A against a smooth vertical wall and the end B on rough horizontal ground at a distance of 0·8 m from the wall. If the coefficient of friction between the rod and the ground is μ, find the least value of μ for which equilibrium is possible.

If the rod is in equilibrium under the action of its weight W, the normal reaction N at A and the reaction R at B, then the lines of action of these three forces must be concurrent. Let C be their point of intersection and let θ be the angle the reaction R makes with the vertical.

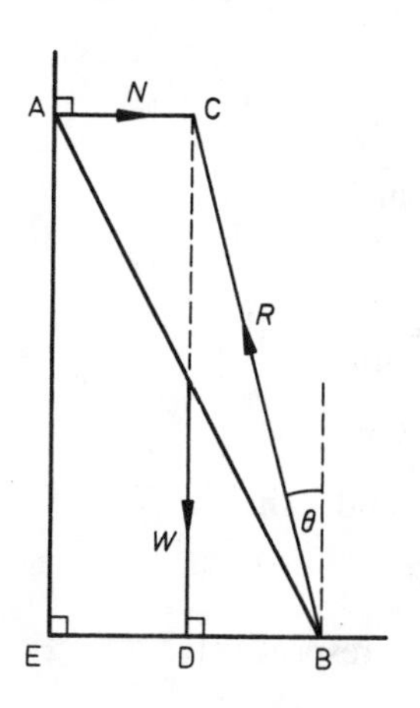

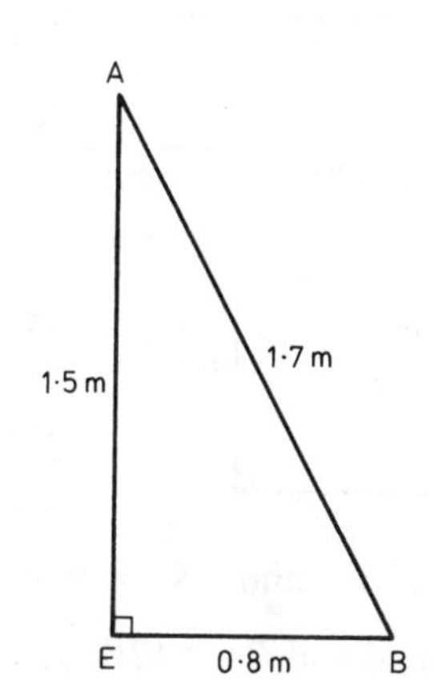

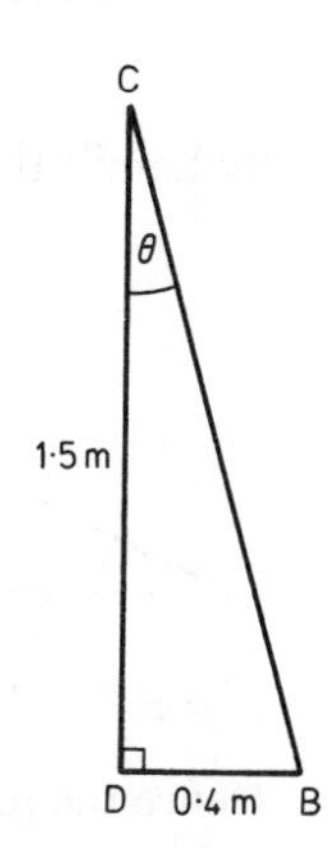

By Pythagoras' theorem in $\triangle ABE$, $AE^2 + EB^2 = AB^2$

$\therefore \quad AE = \sqrt{\{(1{\cdot}7)^2 - (0.8)^2\}}\,\text{m} = 1{\cdot}5\,\text{m}.$

Since the weight of the rod acts vertically through its mid-point, D must be the mid-point of EB,

$$\therefore \quad \text{using } \triangle BCD, \quad \tan\theta = \frac{0{\cdot}4}{1{\cdot}5} = \frac{4}{15}.$$

Equilibrium is possible provided that $\tan\theta \leqslant \mu$.
Hence the least possible value of μ is 4/15.

Example 5 A uniform ladder AB of weight W rests in limiting equilibrium with the end A on rough horizontal ground and the end B against a rough vertical wall. If the coefficient of friction at each end of the ladder is 5/12, find the magnitudes of the reactions at A and B.

Since the ladder is in limiting equilibrium, the reaction R_1 at A acts at an angle λ to the vertical and the reaction R_2 at B acts at an angle λ to the horizontal, where $\tan\lambda = 5/12$. If the lines of action of R_1 and R_2 meets at C, then the weight of the ladder also acts through C and $\angle ACB = \lambda + (90° - \lambda) = 90°$.

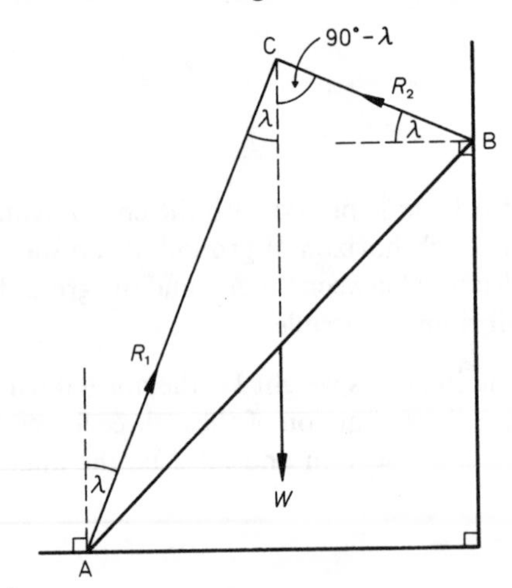

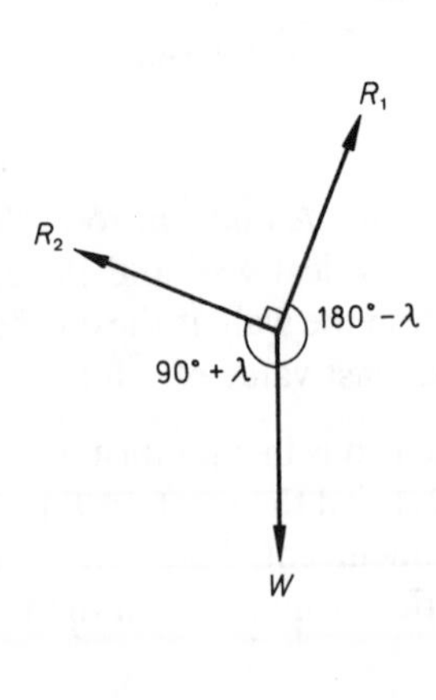

$$\text{By Lami's theorem,} \quad \frac{R_1}{\sin(90° + \lambda)} = \frac{R_2}{\sin(180° - \lambda)} = \frac{W}{\sin 90°}$$

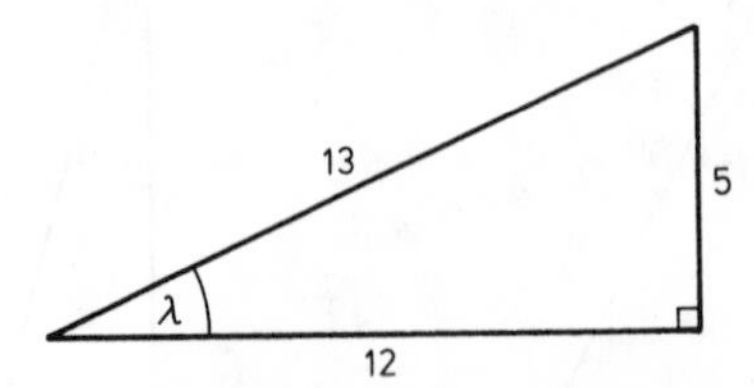

$$\therefore \quad \frac{R_1}{\cos\lambda} = \frac{R_2}{\sin\lambda} = W$$

Since $\tan\lambda = \frac{5}{12}$,

$\cos\lambda = \frac{12}{13}$ and $\sin\lambda = \frac{5}{13}$,

$\therefore \quad R_1 = W\cos\lambda = \frac{12}{13}W$ and $R_2 = W\sin\lambda = \frac{5}{13}W$.

Hence the reactions at A and B are $12W/13$ and $5W/13$ respectively.

Exercise 6.2

[Take g as $10\,\text{m}\,\text{s}^{-2}$.]

1. A uniform rod AB of weight 20 N is smoothly hinged at B and maintained in equilibrium by a horizontal force of P N acting at A. If the rod is inclined at 45° to the horizontal with A above B, find (a) the value of P, (b) the magnitude and direction of the reaction at the hinge.

2. A non-uniform bar AB of length $2a$ and mass m is maintained in a horizontal position by two strings AC and BC attached to a fixed point C above the bar. If $\angle CAB = 60°$ and $\angle ABC = 30°$, find expressions for the tensions in the strings. Find also the distance of the centre of gravity of the bar from A.

3. A uniform rod AB of length 40 cm and weight 8 N rests on a rough peg at C, where $AC = 15$ cm, and against a smooth vertical wall at A. If the rod is in equilibrium at an angle α to the horizontal, where $\sin\alpha = 4/5$, find the magnitudes of the reactions at A and at C.

4. A uniform bar AB of length 2 m and mass 3 kg rests on a smooth peg at X, where $AX = 0{\cdot}5$ m and against a rough vertical wall at A. If the bar is in equilibrium at 60° to the horizontal, find the magnitude and direction of the force exerted by the wall on the bar.

5. A sphere of radius 40 cm and weight 30 N rests in equilibrium on a smooth plane inclined at an angle α to the horizontal, where $\tan\alpha = 8/15$. A horizontal string joins a point on the sphere to a fixed point on the plane. Find the length of the string and the tension in the string.

6. A rod AB of length 40 cm, mass 6 kg and centre of mass G, is suspended by a light inextensible string AXB of length 80 cm, which passes over a smooth peg at X. Find the tension in the string and the angle the rod makes with the vertical when it hangs in equilibrium if (a) $AG = GB$, (b) $AG:GB = 3:5$.

7. A uniform rod of weight W is in equilibrium with one end resting against a smooth vertical plane and the other end against a smooth plane inclined at an angle α to the horizontal, where $\tan\alpha = 5/12$. The line of intersection of the planes is horizontal. If the rod makes an angle θ with the vertical, find $\tan\theta$. Find also the magnitudes of the reactions at the vertical and inclined planes.

8. A non-uniform rod AB of weight $3W$ rests with the end A on a smooth plane inclined at 30° to the horizontal. It is maintained in equilibrium at 30° to the horizontal by a smooth peg at C, where $AC = 3a$. Find (a) the magnitudes of the reactions at A and C, (b) the distance of the centre of gravity of the rod from A.

9. A uniform rod AB of length 1 m is smoothly hinged at A and the end B is attached by means of a light elastic string of natural length 1 m, modulus 20 N to a point C which is 1 m vertically above A. Given that the tension in the string is 12 N,

find (a) the weight of the rod, (b) the magnitude and direction of the reaction at the hinge.

10. A uniform rod AB of weight $4W$ rests in equilibrium with the end A against a smooth vertical wall. A light elastic string XY of natural length 24 cm, modulus $15W$ is attached to a point X vertically above A and a point Y on the rod. Given that $XY = 32$ cm and $\angle XYB = 90°$, find the tension in the string and the length of the rod.

11. A smooth hemispherical bowl of radius $5\sqrt{5}$ cm is fixed with its rim horizontal. A uniform rod AB is in equilibrium with the end A resting inside the bowl and a point C of the rod against the rim, where $AC = 20$ cm. If the rod is inclined at an angle α to the horizontal, find $\tan \alpha$. Find also the length of the rod.

12. A uniform ladder of length 6·5 m rests with one end against a smooth vertical wall and the other end on rough horizontal ground at a distance of 2·5 m from the wall. If the foot of the ladder is about to slip, find the coefficient of friction between the ladder and the ground.

13. A uniform pole of length $2a$ and mass m rests in equilibrium at an angle α to the horizontal, with one end against a smooth vertical wall and the other end on rough horizontal ground. Find expressions for (a) the force exerted on the pole by the wall, (b) the least possible value of the coefficient of friction between the pole and the ground.

14. A uniform ladder of weight W and length $2l$ rests with its upper end against a smooth vertical wall. Its lower end stands on rough horizontal ground, and the coefficient of friction between the ladder and the ground is $\frac{1}{2}$. The ladder is in limiting equilibrium. Find the angle the ladder makes with the horizontal. Find also the reaction at the wall. A man of weight W climbs the ladder. Find how far up it he can go before the ladder will slip. Find also how far up the ladder he can go when a load of weight W is placed on the foot of the ladder. (L)

15. A uniform rod AB of length a and weight W has a light ring attached to the end A. The ring is free to slide along a straight wire inclined at 30° to the horizontal. The rod is maintained in equilibrium by a light inextensible string of length a attached to the end B. The other end of the string is attached to the wire at a fixed point C above A. (a) If the wire is smooth, find the angle θ which the string makes with the vertical. (b) If the string is vertical, find the least value of the coefficient of friction between the ring and the wire.

16. A heavy uniform rod of length $2a$ rests in equilibrium against a small smooth horizontal peg with one end of the rod on a rough horizontal floor. If the height of the peg above the floor is b and friction is limiting when the rod makes an angle $\dfrac{\pi}{6}$ with the floor, show that the coefficient of friction between the floor and the rod is $\dfrac{\sqrt{3}a}{(8b - 3a)}$. (O)

17. A uniform rod AB of length $10a$ rests in limiting equilibrium with the end A on horizontal ground and the end B against a vertical wall. The coefficient of friction at each end of the rod is μ. Given that A is at a distance $6a$ from the wall, find the value of μ.

18. (a)

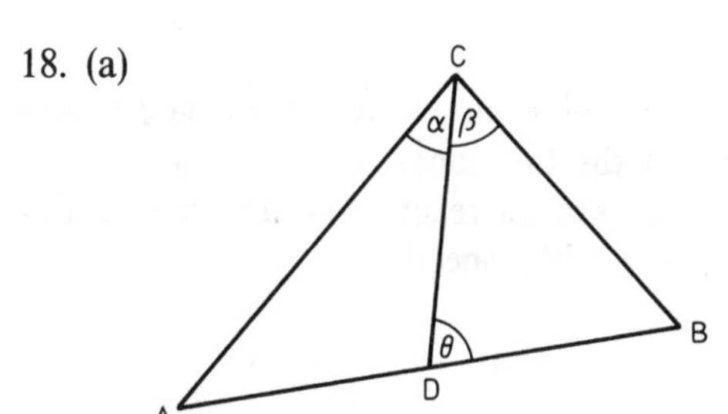

In the given diagram D is the mid-point of AB. Using the sine rule or otherwise prove that

$$2 \cot \theta = \cot \alpha - \cot \beta.$$

(b) A uniform ladder rests at an angle θ to the vertical, with one end on horizontal ground and the other end against a vertical wall. The coefficients of friction between the ladder and the ground and between the ladder and the wall are μ_1 and μ_2 respectively. Given that the ladder is on the point of slipping, use part (a) to prove that $\tan \theta = 2\mu_1/(1 - \mu_1\mu_2)$.

6.3 General conditions for equilibrium

We have already established that if a body is in equilibrium under the action of a set of coplanar forces, then the sum of their components in any direction is zero and the sum of their moments about any point is zero. In this section we consider sets of conditions that are sufficient to ensure equilibrium.

Suppose that a system of coplanar forces is such that
(i) the sum of the components in a direction d_1 is zero,
(ii) the sum of the components in a direction d_2, not parallel to d_1, is zero, and
(iii) the sum of the moments about a point A in the plane is zero.
This system of forces is not equivalent to a single force, because no force can have zero components in two directions. The system does not reduce to a couple, because a couple has the same non-zero moment about every point in its plane. Thus, if a set of forces acting on a rigid body satisfies conditions (i), (ii) and (iii), it follows that the body is in equilibrium.

We can use this set of equilibrium conditions to solve problems involving forces in a *vertical plane* by
(i) resolving horizontally,
(ii) resolving vertically, and
(iii) taking moments about a suitable point.
In this way we obtain up to three equations which we then solve to find any unknown quantities.

Example 1 A uniform rod AB of weight 15 N and length 4 m is smoothly hinged at A and has a particle of weight 30 N attached to it at B. A light inextensible string, attached to the rod at a point C, where $AC = 2{\cdot}5$ m, and to a point D vertically above A, keeps the rod in a horizontal position. If the angle between the rod and the string is 30°, find (a) the tension in the string, (b) the magnitude and direction of the reaction at the hinge.

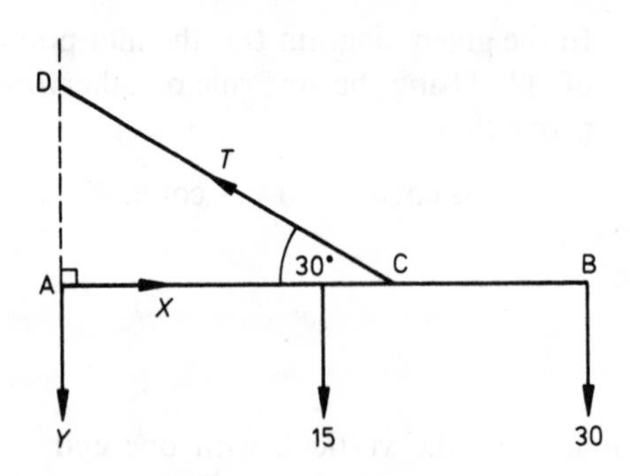

Let the tension in the string be T N and let the horizontal and vertical components of the reaction at the hinge be X N and Y N respectively.

(a) Taking moments anti-clockwise about A,

$$T \times 2{\cdot}5 \sin 30° - 15 \times 2 - 30 \times 40 = 0$$

i.e. $$T \times \tfrac{5}{2} \times \tfrac{1}{2} - 30 - 120 = 0$$

$$\therefore \quad T = 150 \times \tfrac{4}{5} = 120.$$

Hence the tension in the string is 120 N

(b) Resolving horizontally and vertically,

$$\rightarrow \qquad X - T\cos 30° = 0 \qquad (1)$$

$$\downarrow \qquad Y + 15 + 30 - T \sin 30° = 0 \qquad (2)$$

From (1), $X = 120 \times \dfrac{\sqrt{3}}{2} = 60\sqrt{3}$

From (2), $Y = 120 \times \tfrac{1}{2} - 45 = 15$

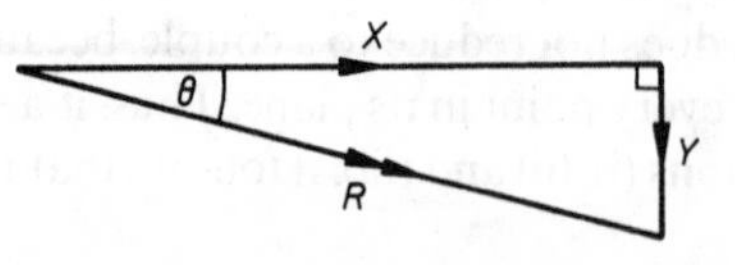

If the resultant reaction R acts at an angle θ to the horizontal, then

$$R^2 = X^2 + Y^2, \tan\theta = \frac{Y}{X}.$$

$$\therefore \quad R = \sqrt{\{(60\sqrt{3})^2 + 15^2\}} = 15\sqrt{\{(4\sqrt{3})^2 + 1^2\}} = 15\sqrt{(48+1)} = 105,$$

$$\tan\theta = \frac{15}{60\sqrt{3}} = \frac{1}{4\sqrt{3}} \quad \text{i.e.} \quad \theta \approx 8{\cdot}2°.$$

Hence the reaction at the hinge has magnitude 105 N and acts at 8·2° to the horizontal.

When a rigid body is kept in equilibrium by frictional forces, then at each point of contact with a rough surface the frictional force F acts along the surface, opposing any tendency to slide. It also satisfies the condition $F \leqslant \mu N$, where N is the normal reaction and μ is the coefficient of friction. If the body is just about to slip, then friction is limiting at *every* point of contact and in each case $F = \mu N$.

Example 2 A uniform solid hemisphere rests in limiting equilibrium with its curved surface in contact with a horizontal floor and a vertical wall. The coefficient of friction at the floor is $\frac{1}{4}$ and at the wall is $\frac{1}{2}$. If θ is the angle of inclination of the plane face of the hemisphere to the horizontal, find $\sin\theta$.

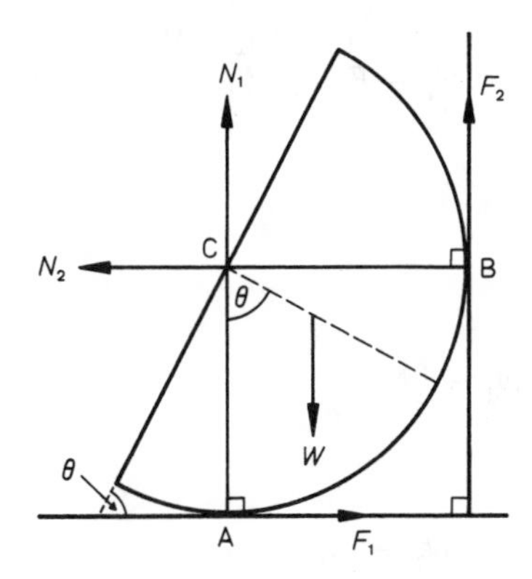

Let the hemisphere have centre C and radius a. Let the forces on it be as shown in the diagram. Since the hemisphere is in limiting equilibrium, friction must be limiting at both A and B,

$$\therefore \quad F_1 = \tfrac{1}{4}N_1, \quad F_2 = \tfrac{1}{2}N_2.$$

Resolving horizontally and vertically,

$$\rightarrow \qquad F_1 - N_2 = 0 \qquad (1)$$
$$\uparrow \qquad N_1 + F_2 - W = 0 \qquad (2)$$

Substituting for F_1 in (1):

$$\tfrac{1}{4}N_1 - N_2 = 0 \qquad \therefore \quad N_1 = 4N_2$$

Substituting for N_1 and F_2 in (2):

$$4N_2 + \tfrac{1}{2}N_2 - W = 0 \qquad \therefore \quad N_2 = \tfrac{2}{9}W$$

Hence $F_1 = N_2 = \frac{2}{9}W$ and $F_2 = \frac{1}{2}N_2 = \frac{1}{9}W$.

The centre of gravity of the hemisphere is on its axis of symmetry at a distance $\frac{3}{8}a$ from C.
Thus, taking moments clockwise about C,

$$W \times \tfrac{3}{8}a\sin\theta - F_1 \times a - F_2 \times a = 0$$
$$\therefore \qquad \tfrac{3}{8}aW\sin\theta - \tfrac{2}{9}aW - \tfrac{1}{9}aW = 0$$

Hence $\frac{3}{8}\sin\theta - \frac{3}{9} = 0$ i.e. $\sin\theta = \frac{8}{9}$.

Example 3 A uniform ladder AB of length $2a$ and weight W rests with the end A on horizontal ground and the end B against a vertical wall. The coefficients of friction at A and B are $\frac{2}{5}$ and $\frac{1}{2}$ respectively. The ladder makes an angle α with the horizontal, where $\tan\alpha = \frac{5}{4}$. If a man of weight W starts to climb the ladder, find his distance from A when the ladder begins to slip.

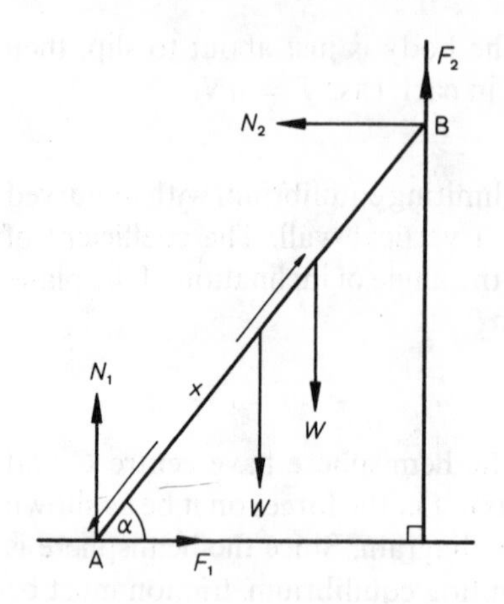

Let the reactions at A and B when the ladder is about to slip be as shown in the diagram. Since friction must be limiting at both A and B,

$$F_1 = \tfrac{2}{5}N_1, \quad F_2 = \tfrac{1}{2}N_2.$$

Resolving horizontally and vertically,

$$\rightarrow \qquad F_1 - N_2 = 0 \qquad (1)$$

$$\uparrow \qquad N_1 + F_2 - 2W = 0 \qquad (2)$$

Substituting for F_1 in (1),

$$\tfrac{2}{5}N_1 - N_2 = 0 \qquad \therefore \quad N_1 = \tfrac{5}{2}N_2$$

Substituting for N_1 and F_2 in (2),

$$\tfrac{5}{2}N_2 + \tfrac{1}{2}N_2 - 2W = 0 \qquad \therefore \quad N_2 = \tfrac{2}{3}W$$

Hence $F_2 = \tfrac{1}{2}N_2 = \tfrac{1}{3}W$.

Let the man be a distance x from A when the ladder begins to slip.
Taking moments clockwise about A,

$$W \times x\cos\alpha + W \times a\cos\alpha - F_2 \times 2a\cos\alpha - N_2 \times 2a\sin\alpha = 0$$

Dividing by $\cos\alpha$,

$$Wx + Wa - 2F_2a - 2N_2\,a\tan\alpha = 0$$

$$\therefore \quad Wx + Wa - \tfrac{2}{3}Wa - \tfrac{4}{3}Wa \times \tfrac{5}{4} = 0$$

$$\therefore \quad x = \tfrac{5}{3}a + \tfrac{2}{3}a - a = \tfrac{4}{3}a.$$

Hence the man is at a distance of $4a/3$ from A when the ladder begins to slip.

The set of conditions discussed earlier, for the equilibrium of a rigid body under the action of coplanar forces, involved the sums of components in two directions and the sum of moments about one point. We now consider whether any other combinations of similar conditions will also ensure equilibrium.

To be certain that a set of forces is not equivalent to a couple, we need to know that the system has zero moment about at least one point. A system with zero moments about two points A and B may have a resultant whose line of action passes through A and B. We can rule out this possibility if we find that the components of the forces in some direction, not perpendicular to AB, have zero sum or that the system has zero moment about a third point not on AB.

We have then three sets of equilibrium conditions. A system of coplanar forces must be in equilibrium if

(a) the sums of the components in two non-parallel directions are each zero, and the sum of the moments about one point is zero;

or (b) the sums of the moments about two points A and B are each zero, and the sum of the components in a direction, not perpendicular to AB, is zero;

or (c) the sums of the moments about three points, not in a straight line, are each zero.

Thus when tackling problems we may

(a) resolve in two directions and take moments about one point;

or (b) take moments about two points A and B then resolve in a direction not perpendicular to AB;

or (c) take moments about three points which are not collinear.

Since each method involves putting three different restrictions on the system of forces concerned, we will obtain three *independent* equations connecting any unknown quantities i.e. three equations each giving information which could not be obtained from the other two. [Note that since these three equations ensure the equilibrium of the system, no new information can be gained from further equations formed by resolving or taking moments.]

The following examples show that, by careful choice of the points about which we take moments and of the directions in which we resolve, we may

(1) avoid equations involving quantities that we do not need to find,

(2) form equations containing as few unknowns as possible.

Example 4 A uniform rod AB of mass 6 kg rests on a smooth peg at a point C. The rod is kept in equilibrium at an angle α to the horizontal, where $\tan\alpha = 5/12$, by a horizontal force of P newtons acting at the upper end B of the rod. Find the value of P.

[Take g as $10\,\text{m}\,\text{s}^{-2}$.]

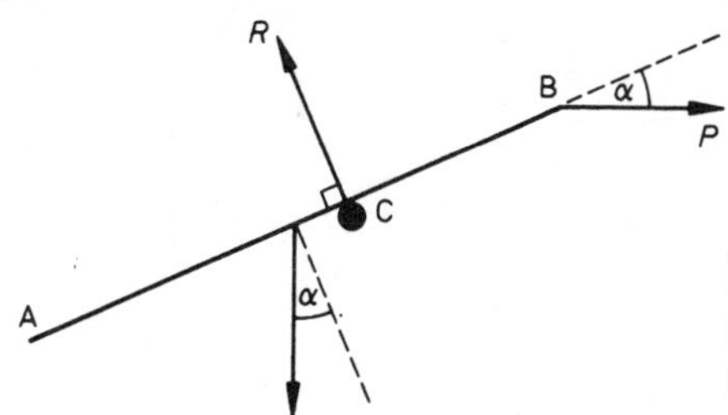

[We try to avoid equations involving distances or the reaction at C, as we do not need to find these quantities.]

Resolving along the rod in the direction AB,

$$P\cos\alpha - 6g\sin\alpha = 0$$

$$\therefore \quad P = 6g\tan\alpha = 6 \times 10 \times \tfrac{5}{12} = 25.$$

Example 5 A uniform bar AB of weight W and length $4a$ rests on a rough peg at a point C, where $AC = a$. The bar is maintained in equilibrium, with A above B, by a force of magnitude $\frac{1}{2}W$ acting at A in a direction perpendicular to the bar. Find the angle α, which the bar makes with the horizontal. Given that the bar is about to slip, find the coefficient of friction μ between the bar and the peg.

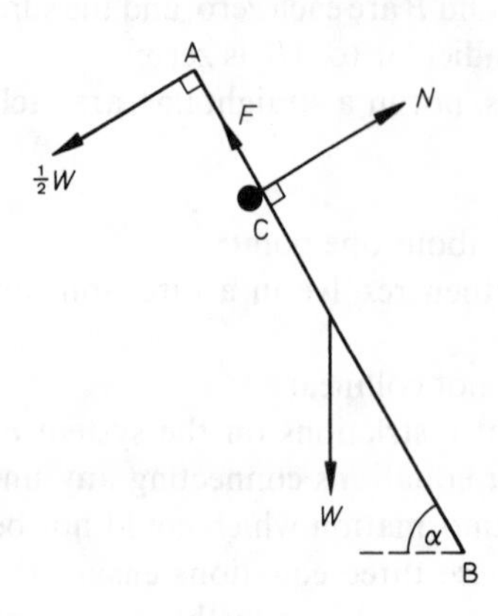

[In this example we try to form equations which simplify the working as much as possible.]

Taking moments clockwise about C,

$$W \times a\cos\alpha - \tfrac{1}{2}W \times a = 0 \qquad \therefore \cos\alpha = \tfrac{1}{2}$$

Hence $\alpha = 60°$.

Let F be the frictional force at C and let N be the normal reaction.
Taking moments clockwise about the mid-point of AB,

$$N \times a - \tfrac{1}{2}W \times 2a = 0 \qquad \therefore \quad N = W$$

Resolving along the bar in the direction BA,

$$F - W\sin\alpha = 0 \qquad \therefore \quad F = \frac{\sqrt{3}}{2}W$$

Since the bar is about to slip, $F = \mu N$.

Hence $\mu = \dfrac{F}{N} = \dfrac{\sqrt{3}}{2}$.

Exercise 6.3

[Take g as $10\,\text{m s}^{-2}$.]

1. A uniform rod AB of weight 20 N and length 2 m is freely hinged at A. The rod is held in equilibrium at 30° to the horizontal, with B higher than A, by a light inextensible string of length 1·25 m. The string joins a point C on the rod to a fixed point D at a distance of 1·25 m vertically above A. If a particle of weight 5 N is suspended from B, find (a) the tension in the string, (b) the horizontal and vertical components of the reaction at A.

2. A uniform rod AB of mass 6 kg and length 3 m is freely hinged at A. A load of mass 2 kg is attached to the rod at B. The rod is kept in equilibrium by a light inextensible string CD of length 1·5 m, which is attached to a point C on the rod and

to a fixed point D on the same horizontal level as A. Given that $AC = 2$ m and $AD = 2{\cdot}5$ m, find (a) the tension in the string, (b) the magnitude and direction of the reaction at the hinge.

3. A uniform square lamina $ABCD$ of mass 6 kg is hinged at A so that it is free to move in a vertical plane. It is maintained in equilibrium, with B vertically below A, by a horizontal force acting at C and a vertical force acting at D each of magnitude P newtons. Find (a) the value of P, (b) the magnitude and direction of the force exerted by the hinge on the lamina.

4. A uniform rod AB of weight 16 N is freely hinged at A. It is kept in equilibrium at an angle of 30° to the horizontal, with B lower than A, by a force acting at B. Find the magnitude and direction of the least force required at B to maintain equilibrium. Find also the corresponding horizontal and vertical components of the reaction at A.

5. A uniform rod AB of length 3 m and weight 40 N rests in equilibrium with the lower end A on a smooth plane inclined at an angle α to the horizontal, where $\tan \alpha = 3/4$, and a point C of the rod against a smooth peg. If the rod makes an angle θ with the horizontal, where $\tan \theta = 7/24$, find (a) the magnitudes of the reactions at A and C, (b) the distance AC.

6. A uniform rod AB of mass 13 kg is in limiting equilibrium with its lower end A on a rough horizontal plane and its upper end B attached to a light inextensible string inclined at an angle θ to the upward vertical, where $\tan \theta = 12/5$. The string passes over a smooth fixed pulley at a point vertically above A and supports at its other end a mass of 6·5 kg. Find (a) the coefficient of friction between the rod and the plane, (b) the angle the rod makes with the horizontal.

7. A smooth hemisphere of radius 60 cm is fixed with its plane face on a rough horizontal floor. A uniform rod AB of length 1 m and weight 10 N rests in equilibrium in a vertical plane with the end A on the floor and a point C of the rod in contact with the hemisphere, where $AC = 80$ cm. Given that the rod is on the point of slipping, find the reaction at C and the coefficient of friction between the rod and the floor.

8. A uniform ladder AB of length 5 m and mass 12 kg rests with the end A on rough horizontal ground and the end B in contact with a smooth vertical wall. The foot of the ladder is at a distance of 1·4 m from the wall. Find the magnitudes of the reactions at A and B. If the ladder is about to slip, find the coefficient of friction between the ladder and the ground.

9. A uniform beam AB of length $2a$ and weight W rests with its end A against a smooth vertical wall and B on a rough horizontal plane. The vertical plane through AB is perpendicular to the wall. When it makes an angle of α with the wall the beam is on the point of slipping. Find the coefficient of friction between the beam and the horizontal plane. Find, also, what horizontal force applied to the mid-point of the

beam (in the vertical plane through AB) will just cause the beam to slip at B towards the wall. (SU)

10. A thin uniform hemispherical bowl, of weight W and radius a, rests in limiting equilibrium with its plane face at an angle θ to the horizontal. Its curved surface is in contact with a rough horizontal floor and a rough vertical wall. If the coefficient of friction at each point of contact is $\frac{1}{3}$, find the magnitudes of the frictional forces at the floor and the wall. Find also the value of $\sin\theta$.

11. A uniform solid hemisphere rests in limiting equilibrium with its plane face at an angle α to the horizontal, where $\sin\alpha = 8/13$. Its curved surface is in contact with rough horizontal ground and a rough vertical wall. If the coefficient of friction at both points of contact is μ, find the value of μ.

12. A uniform rod AB of weight W and length $4a$ rests with its end A on a rough horizontal plane and a point C of the rod, where $AC = 3a$, in contact with a smooth peg. Given that the rod makes an angle θ with the horizontal, where $\sin\theta = 3/5$, find the magnitude of the reaction at C and the horizontal and vertical components of the reaction at A. If μ is the coefficient of friction between the rod and the horizontal plane, show that $\mu \geqslant 24/43$.

13. A uniform ladder of weight W rests at 60° to the horizontal with its foot on rough horizontal ground and the other end against a smooth vertical wall. If the ladder is just about to slip when a man of weight $2W$ stands at the top, find the coefficient of friction between the ladder and the ground. Find the frictional force between the ladder and the ground when the man is standing at the mid-point of the ladder.

14. A uniform ladder of length $2a$ and weight W rests against a smooth vertical wall and has its foot on rough horizontal ground. The coefficient of friction between the ladder and the ground is $\frac{1}{2}$. If the ladder is placed at an angle α to the horizontal, where $\tan\alpha = \frac{3}{2}$, find the distance that a man of weight $3W$ can climb up the ladder before it slips.

15. A uniform pole AB of length 8 m and mass 60 kg rests in limiting equilibrium with the end A on horizontal ground and the end B against a vertical wall. If the coefficient of friction at each end of the pole is $\frac{1}{3}$, find the angle the pole makes with the horizontal. The pole is now placed at an angle α to the horizontal, where $\tan\alpha = 2$, and a body of mass m kg is attached to the pole at B without causing the pole to slip. Find the maximum value of m and the magnitude of the corresponding normal reaction at A.

16. A uniform circular hoop of weight 15 N hangs over a rough horizontal peg A. A horizontal force P N, acting in the plane of the hoop, is applied at the other end B of the diameter through A. If the hoop is on the point of slipping when AB is inclined at

an angle α to the downward vertical, where $\tan \alpha = 4/3$, find the value of P and the coefficient of friction between the hoop and the peg.

17. A uniform rod AB, of weight W and length $6a$, rests with its lower end A on a smooth horizontal plane and a point C of the rod against a smooth peg, where $AC = 4a$. The rod is kept in equilibrium at an angle α to the horizontal by a force of magnitude $\frac{1}{4}W$ acting along the rod. Find the magnitude of the reaction at A and the value of $\sin \alpha$.

18. A uniform ladder, of weight W and length l, rests in a vertical plane with one end against a smooth vertical wall, the wall being perpendicular to the vertical plane through the ladder. The other end of the ladder rests on horizontal ground, the coefficient of friction between the ladder and the ground being 1/4. The ladder is inclined at an angle θ to the horizontal, where $\tan \theta = 24/7$. A man of weight $10W$ climbs up the ladder. Show that the man can reach a height of $6l/7$ above the ground before the ladder begins to slip and calculate the force exerted by the ladder on the wall at this instant. (L)

19. A uniform ladder of length 7 m rests against a vertical wall with which it makes an angle of 45°, the coefficients of friction between the ladder and the wall and the ladder and the (horizontal) ground being $\frac{1}{3}$ and $\frac{1}{2}$ respectively. A boy whose weight is one half that of the ladder slowly ascends the ladder. How far along the ladder will he be when the ladder slips? [Assume the ladder is a uniform rod which is perpendicular to the line of intersection of the wall and ground.] (O&C)

20. A uniform circular disc, of radius a and weight w, stands in a vertical plane upon a rough horizontal floor with the point P of its circumference in contact with an equally rough vertical wall and its plane at right angles to that of the wall. A particle of equal weight w is attached to a point Q of the disc between its centre O and P so that the disc is on the point of slipping. Draw a good diagram indicating clearly the normal reactions and the friction forces acting on the disc. Show that

$$OQ = \frac{2a\mu(1 + \mu)}{1 + \mu^2},$$

where μ is the coefficient of friction. (W)

6.4 Rigid body systems

When two or more rigid bodies are in equilibrium under the action of a set of coplanar forces, then the whole system can be treated as a single body in equilibrium under the action of the external forces. Each separate body in the system is in equilibrium under the action of a set of forces which includes both the external forces on the body and the forces exerted on it by the other bodies in the system.

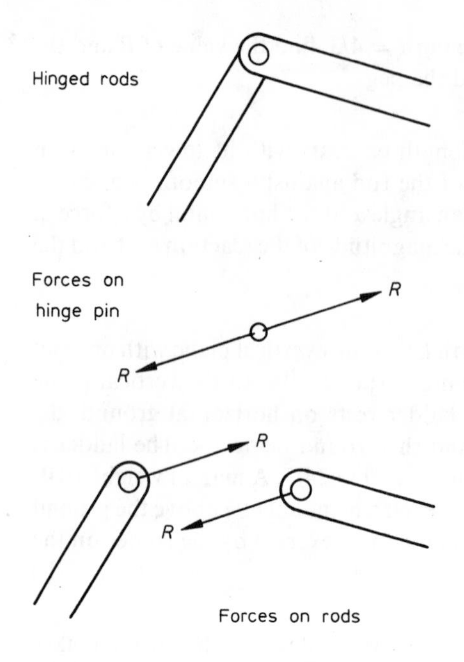

We first consider two rods connected by a smooth light hinge. Each rod exerts a force on the hinge and, according to Newton's third law, the hinge exerts an equal and opposite force on the rod. Thus, provided no external force acts on the hinge, the effect of the hinge is to set up equal and opposite forces on the two rods.

Example 1 Two equal uniform rods AB and BC, each of length $2a$ and weight W, are freely hinged together at B and hang freely from the point A. Equilibrium is maintained by a horizontal force P acting at C. Given that BC makes an angle α with the downward vertical, where $\tan\alpha = 4/3$, find the value of P and the horizontal and vertical components of the force exerted by the rod AB on the rod BC. Find also the angle made by AB with the vertical.

Let the horizontal and vertical components of the reaction at B be X and Y respectively and let θ be the angle AB makes with the vertical.

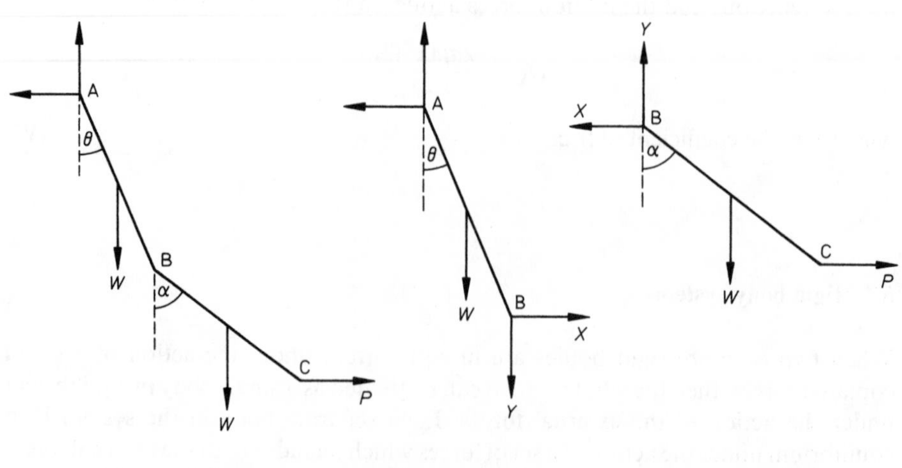

Consider first the forces on the rod BC.
Taking moments about B,

$$P \times 2a\cos\alpha - W \times a\sin\alpha = 0$$
$$\therefore \quad P = \tfrac{1}{2}W\tan\alpha = \tfrac{2}{3}W.$$

Resolving horizontally and vertically,

$$\leftarrow \quad X - P = 0 \quad \therefore \quad X = P = \tfrac{2}{3}W$$
$$\uparrow \quad Y - W = 0 \quad \therefore \quad Y = W$$

Hence the horizontal and vertical components of the force exerted by the rod AB on the rod BC are $\tfrac{2}{3}W$ and W respectively.
Consider now the forces on the rod AB.
Taking moments about A,

$$W \times a\sin\theta + Y \times 2a\sin\theta - X \times 2a\cos\theta = 0$$
$$\therefore \quad Wa\sin\theta + 2Wa\sin\theta - \tfrac{4}{3}Wa\cos\theta = 0$$

$$\therefore \quad 3Wa\sin\theta = \tfrac{4}{3}Wa\cos\theta$$
$$\therefore \quad \tan\theta = \tfrac{4}{9} \quad \text{and} \quad \theta \approx 24°.$$

Hence AB makes an angle of 24° with the vertical.

Example 2 Two uniform rods AB and BC, each of length $2a$, have weights $2W$ and W respectively. The rods are freely jointed at B and rest in a vertical plane with A and C on a rough horizontal plane, the coefficient of friction at both points of contact being $\tfrac{1}{2}$. A particle of weight $2W$ is attached to a point P of the rod AB, where $AP = \tfrac{1}{2}a$. Given that the rods are inclined at an angle θ to the vertical, find the maximum value of θ for which equilibrium is possible.

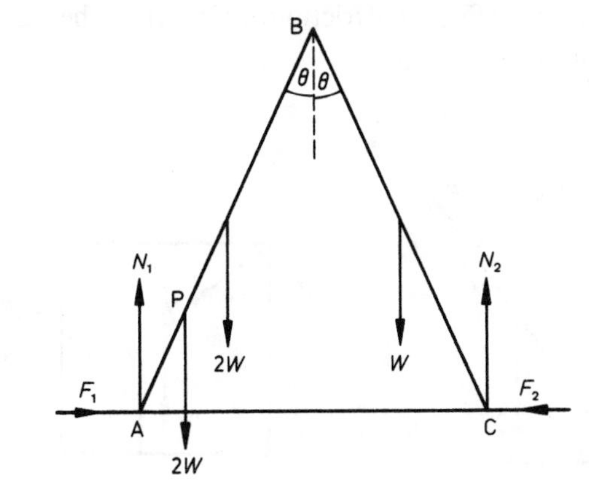

Let the reactions at A and C be as shown in the diagram.

Consider first the forces on the whole system.
Resolving horizontally and vertically,

$$\rightarrow \quad F_1 - F_2 = 0 \qquad (1)$$
$$\uparrow \quad N_1 + N_2 - 5W = 0 \qquad (2)$$

Taking moments about A,

$$N_2 \times 4a\sin\theta - W \times 3a\sin\theta - 2W \times a\sin\theta - 2W \times \tfrac{1}{2}a\sin\theta = 0$$
$$\therefore \quad 4N_2 - 3W - 2W - W = 0 \quad \text{giving} \quad N_2 = \tfrac{3}{2}W.$$

From (2), $\quad N_1 = 5W - N_2 = \tfrac{7}{2}W.$

Using the laws of friction, $F_1 \leqslant \mu N_1$ and $F_2 \leqslant \mu N_2$
Since $\mu = \frac{1}{2}$, $F_1 \leqslant \frac{7}{2}W$ and $F_2 \leqslant \frac{3}{4}W$
However, using (1), we may write $F_1 = F_2 = F$
$\therefore$ in any equilibrium position, $F \leqslant \frac{3}{4}W$ (3)
Taking moments about B for the forces on rod BC,

$$\circlearrowright \quad N_2 \times 2a\sin\theta - W \times a\sin\theta - F_2 \times 2a\cos\theta = 0$$
$$\therefore \quad 3\,Wa\sin\theta - Wa\sin\theta - 2F\,a\cos\theta = 0$$
$$\therefore \quad 2Wa\sin\theta = 2Fa\cos\theta$$
$$\text{Thus} \quad \tan\theta = F/W$$
$$\therefore \text{ using (3),} \quad \tan\theta \leqslant 3/4$$

Hence the maximum value of θ for which equilibrium is possible is $\tan^{-1}(3/4) \approx 36{\cdot}9°$.

[Note that when θ takes this maximum value, slipping is about to occur at C.]

We now consider problems concerning two bodies in contact. By Newton's third law, the forces exerted by each body on the other are equal and opposite. Thus, as in the case of jointed bodies, unknown quantities can be found by resolving and taking moments, considering either the external forces acting on the system as a whole, or both internal and external forces acting on each body separately. In any given problem we try to simplify the working by avoiding equations involving quantities that we do not need to find.

Example 3 A uniform cube of side $12a$ and weight $3W$ rests on a rough horizontal table. A uniform rod AB of length $20a$ and weight $5W$ is freely hinged at A to a fixed point on the table. A point C of the rod rests against the mid-point of one edge of the cube, this edge being perpendicular to the vertical plane containing the rod. When the rod is at an angle α to the horizontal, where $\sin\alpha = 4/5$, the cube is on the point of sliding across the table without toppling. If the coefficient of friction at C is $\frac{1}{2}$, find the coefficient of friction between the cube and the table.

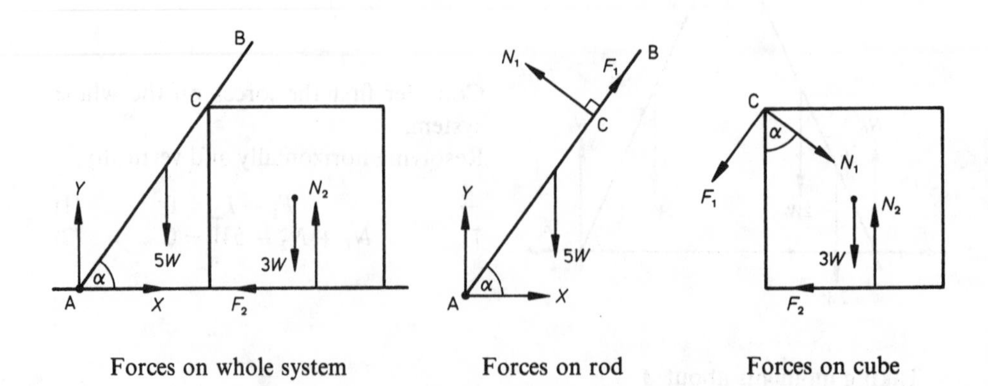

Forces on whole system Forces on rod Forces on cube

Let F_1, N_1 and F_2, N_2 be the frictional forces and normal reactions at C and the base of the cube respectively. Consider first the forces on the rod.

Taking moments about A,

$\circlearrowleft$ $$N_1 \times 12a\operatorname{cosec}\alpha - 5W \times 10a\cos\alpha = 0$$

Using a right-angled triangle,

if $\sin\alpha = \frac{4}{5}$, then $\operatorname{cosec}\alpha = \frac{5}{4}$, $\cos\alpha = \frac{3}{5}$

$\therefore \quad N_1 \times 12a \times \frac{5}{4} - 5W \times 10a \times \frac{3}{5} = 0$

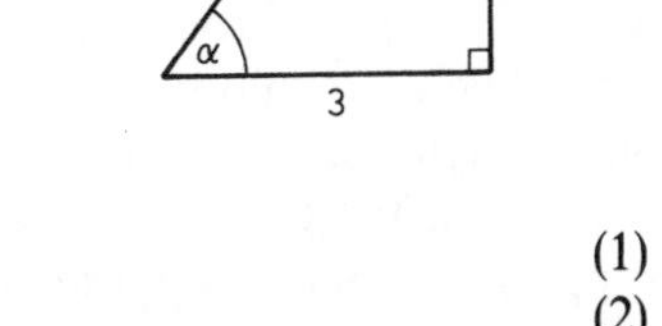

Hence $N_1 = 2W$.
Since the cube is about to slide, $F_1 = \frac{1}{2}N_1 = W$.
Consider now the forces on the cube.
Resolving horizontally and vertically,

$\leftarrow \qquad F_2 + F_1\cos\alpha - N_1\sin\alpha = 0 \qquad (1)$

$\uparrow \qquad N_2 - 3W - F_1\sin\alpha - N_1\cos\alpha = 0 \qquad (2)$

From (1), $\quad F_2 + \frac{3}{5}W - \frac{8}{5}W = 0 \quad \therefore \quad F_2 = W$

From (2), $\quad N_2 - 3W - \frac{4}{5}W - \frac{6}{5}W = 0 \quad \therefore \quad N_2 = 5W$

Since the cube is on the point of sliding, the coefficient of friction between the cube and the table is given by F_2/N_2 and is equal to 1/5.

Exercise 6.4

1. Two uniform rods AB and BC, of weights 5 N and 10 N respectively, are freely hinged at B and hang freely from a fixed hinge at A. The system is kept in equilibrium by a horizontal force of magnitude 5 N acting at C. Find (a) the magnitudes of the reactions at the hinges A and B, (b) the angles the rods AB and BC make with the vertical.

2. Two equal uniform rods, each of weight W, are freely jointed at B and hang freely from a hinge at A. The end C is attached by means of a light inextensible string to a point D on the same horizontal level as A. When the system is in equilibrium the rod AB is inclined at 45° to the vertical and the rod BC is horizontal. Find the tension in the string and the angle it makes with the vertical.

3. Two uniform rods AB and BC, each of length $2a$, have weights W and $3W$ respectively. The rods are smoothly jointed at B. The ends A and C of the rods are smoothly hinged to fixed points at the same horizontal level and a distance $2a$ apart. If the system is in equilibrium with B below AC, find the horizontal and vertical components of the reaction at B.

4. Two equal uniform rods, AB and AC, each of length $2a$ and weight W, are freely jointed at A. The end C is freely hinged to a point on a rough horizontal plane, the end B rests on the plane and is about to slip, and both rods are in a vertical plane. If the coefficient of friction between B and the plane is μ and the angle ABC is θ, (i) prove that $\mu = \frac{1}{2}\cot\theta$, (ii) find, in terms of μ, the magnitude and direction of the reaction at the hinge at C, (iii) find the magnitude of the reaction at the joint A. (SU)

5. Two uniform rods AB and BC, freely jointed together at B, are of equal length, but have weights $4W$ and W respectively. The end A is freely hinged to a fixed point on a rough vertical wall and the end C rests against the wall at a point vertically below A. The system rests in limiting equilibrium with each rod inclined at an angle α to the vertical, where $\tan\alpha = 2$. Find (a) the coefficient of friction between the rod BC and the wall, (b) the horizontal and vertical components of the reaction of AB on BC.

6. Two uniform rods AB and AC, of length $2a$, but of weights $2W$ and W respectively, are smoothly jointed at A. The rods rest in a vertical plane with B and C on a rough horizontal plane, the coefficient of friction between the rods and the plane being 0·4. Given that the system is in limiting equilibrium, determine whether slipping is about to occur at B or at C and find angle BAC.

7. Two equal uniform rods, AB and BC, each of mass $4m$, are smoothly hinged together at B and stand in a vertical plane with A and C on a rough horizontal floor. The angle ABC is 2α where $\tan\alpha = \frac{1}{2}$, and the coefficient of friction at both A and C is μ. A mass m is fixed to the mid-point of AB. Show that equilibrium is possible provided that $\mu \geqslant 9/34$. What happens if $\mu = \frac{1}{4}$?

8. Two smooth uniform spheres of weights W and $2W$ are each suspended from a fixed point O by a light inextensible string and rest against each other. The angle between the strings is 90° and the string supporting the heavier sphere makes an angle α with the vertical, where $\tan\alpha = \frac{3}{4}$. Find (a) the tensions in the strings, (b) the angle of inclination to the horizontal of the line joining the centres of the spheres.

9. A uniform rod AB of length $4a$ and weight W is freely hinged at A. A second uniform rod CD of length $5a$ and weight $2W$ is freely hinged at C, which lies at a distance $4a\sqrt{3}$ from A in the same horizontal plane. The rods rest in equilibrium in a vertical plane with the rod CD in contact with the end B of the rod AB. If the rods are on the point of slipping when $BC = 4a$, find the coefficient of friction between the rods.

10. Two uniform circular cylinders rest in contact on a horizontal table. A third cylinder is supported by the first two. The cylinders are identical and their axes are horizontal and parallel. The coefficient of friction between any pair of cylinders and between the lower cylinders and the table is μ. Find the least value of μ for which equilibrium is possible.

11. A uniform rod OA of weight W and length $2a$ is pivoted smoothly to a fixed point at the end O. A light inextensible string, also of length $2a$, has one end attached to the end A of the rod, its other end being attached to a small heavy ring of weight $\frac{1}{2}W$ which is threaded on to a fixed rough straight horizontal wire passing through O. The system is in equilibrium with the string taut and the rod making an angle α with the downward vertical at O. By taking moments about O for the rod and ring

together, or otherwise, find the normal component R of the reaction exerted by the wire on the ring. Find also the frictional force F on the ring. The equilibrium is limiting when $\alpha = 60°$. Determine the coefficient of friction between the wire and the ring. (O&C)

12. Two uniform rods AB, AC, each of weight W and length $2a$, are smoothly jointed at A. The rods rest on a smooth circular cylinder of radius a with their plane perpendicular to the axis of the cylinder which is horizontal. When the rods are in equilibrium in a vertical plane with A vertically above the axis of the cylinder, angle $BAC = 2\theta$. Show that the force exerted by the cylinder on each rod is of magnitude $W \operatorname{cosec} \theta$. Show also that $\sin^3 \theta = \cos \theta$. Hence show that there is only one such position of equilibrium. (O&C)

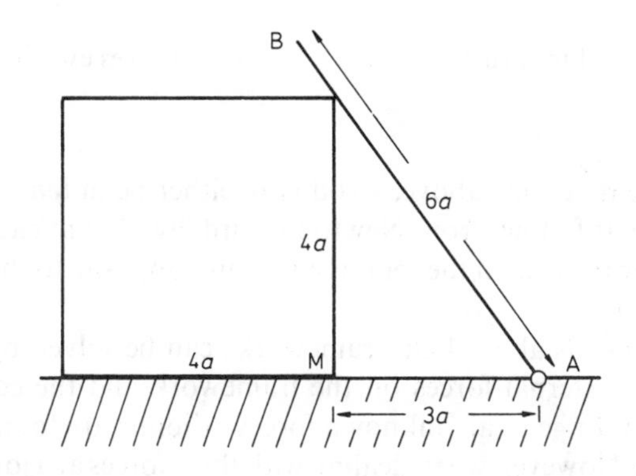

A uniform cube, of edge $4a$ and weight W, rests on a smooth horizontal surface (see diagram). A uniform smooth rod AB, of length $6a$ and weight w, has one end freely hinged to a point A of the surface. The rod rests against the mid-point of a horizontal edge of the top face of the cube. The rod is in a vertical plane through the centre of the cube. A light inelastic string of length $3a$ joins A to the mid-point M of the horizontal edge of the cube nearest to A. The system is in equilibrium. (i) Find the magnitude of the reaction between the rod and the cube. (ii) Find the magnitude of the reaction between the cube and the horizontal surface. (iii) Find the tension in the string AM. (iv) Show that $18w \leqslant 125W$. (C)

6.5 Light frameworks

The expression *light framework* is used to describe a system of rods whose weights are negligible compared with the other forces involved. It is also assumed that the rods are smoothly jointed together at their ends.

Let us consider the two resultant forces acting at the ends of a typical light rod AB in such a framework. If the rod is in equilibrium these forces must be equal and opposite. By taking moments about A and B we also find that both forces act along AB.

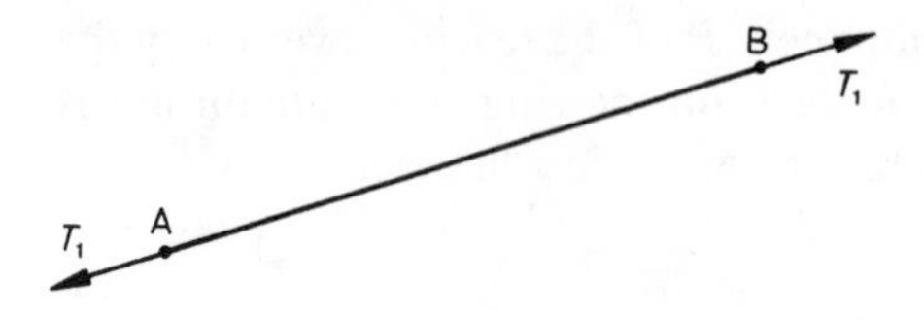

Forces on a rod in tension

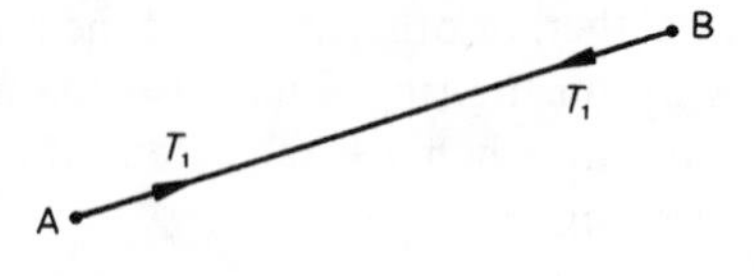

Forces exerted by a rod in tension

Forces on a rod in thrust

Forces exerted by a rod in thrust

As shown in the diagrams above, a rod may either be in *tension* or in *thrust* i.e. under *compression*. It follows from Newton's third law that at each end of the rod, the force exerted by the rod on the joint is equal and opposite to the force exerted by the joint on the rod.

Elementary problems about light frameworks can be solved by considering the equilibrium of the external forces on the framework and the equilibrium of the forces at each joint. In general, unknown forces at joints are found by resolving in suitable directions. However, when dealing with three forces at a joint it is sometimes quicker to use a triangle of forces.

Whatever method is chosen, it is important to draw a large clear diagram showing the forces acting at each joint. Sometimes it is difficult to decide whether a particular rod is in tension or in thrust. In this case we may choose to indicate one type of stress in the diagram, but if calculation leads to a negative value for this stress, we deduce that it is of the other type.

Example 1

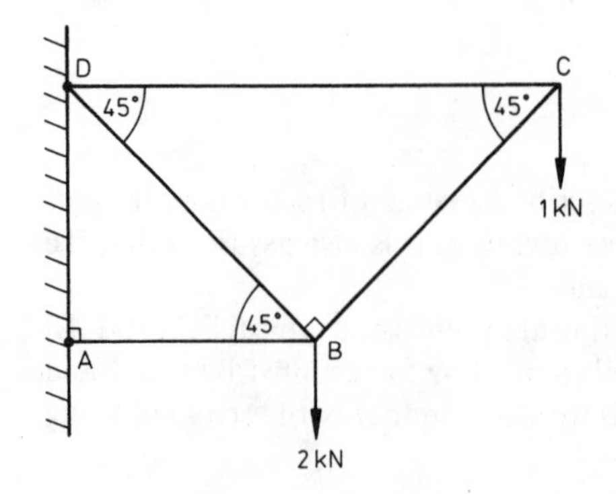

The diagram shows a framework of four smoothly jointed light rods. The framework is freely hinged at A and D to a vertical wall. Loads of 2 kN and 1 kN are hung from B and C respectively. Find the magnitudes of the reactions at A and B. Find also the stresses in the rods, stating whether they are tensions or thrusts.

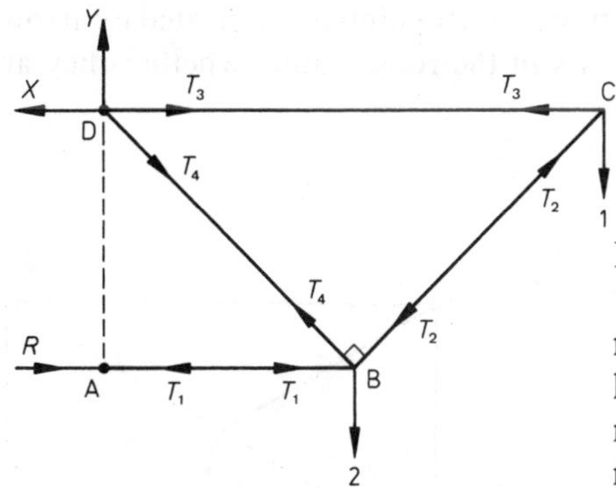

Let the stresses in the rods be thrusts T_1, T_2 and tensions T_3, T_4 as shown. Let the magnitude of the reaction at A be R and let the horizontal and vertical components of the reaction at D be X and Y respectively.

Since there are only two forces acting at A, the reaction R must be a horizontal force such that

$$R - T_1 = 0 \tag{1}$$

Consider now the forces on the whole framework.
Let $AB = AD = a$, then taking moments about D,

$$\circlearrowright \quad R \times a - 2 \times a - 1 \times 2a = 0 \qquad \therefore \quad R = 4$$

Hence the magnitude of the reaction at A is 4 kN.
Resolving horizontally and vertically,

$$\leftarrow \quad X - R = 0 \qquad \therefore \quad X = R = 4$$

$$\uparrow \quad Y - 2 - 1 = 0 \qquad \therefore \quad Y = 3$$

Hence the magnitude of the reaction at D

$$= \sqrt{(X^2 + Y^2)}\,\text{kN} = \sqrt{(4^2 + 3^2)}\,\text{kN} = 5\,\text{kN}$$

From (1), $T_1 = R = 4$.
Using the triangle of forces for the joint C,

$$T_2 = \sqrt{2}, \quad T_3 = 1.$$

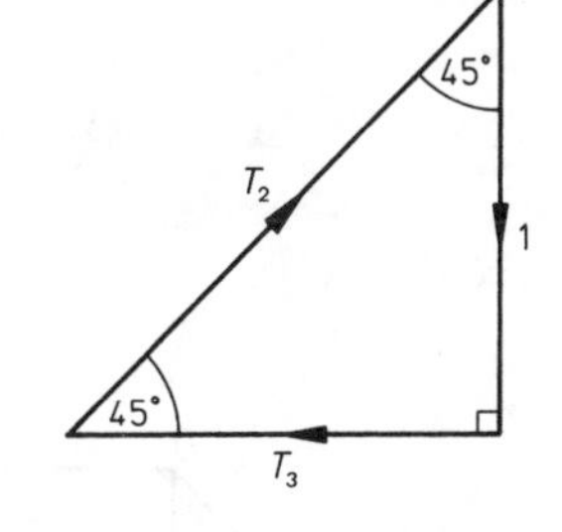

Resolving the forces at B in the direction BD,

$$\nwarrow \quad T_4 - 2\cos 45° - T_1 \cos 45° = 0$$

$$\therefore \quad T_4 = 2 \times \frac{1}{\sqrt{2}} + 4 \times \frac{1}{\sqrt{2}} = 3\sqrt{2}$$

Hence the stresses in AB and BC are thrusts of magnitude 4 kN and $\sqrt{2}$ kN respectively. The stresses in BD and CD are tensions of magnitude $3\sqrt{2}$ kN and 1 kN respectively.

Exercise 6.5

In questions 1 to 8 each diagram shows a light framework consisting of smooth jointed rods. The large dots represent points at which the framework is smoothly hinged to a fixed point. External forces acting at other points are indicated by arrows and given in kilonewtons. Find the stresses in the rods, stating whether they are tensions or thrusts.

1.

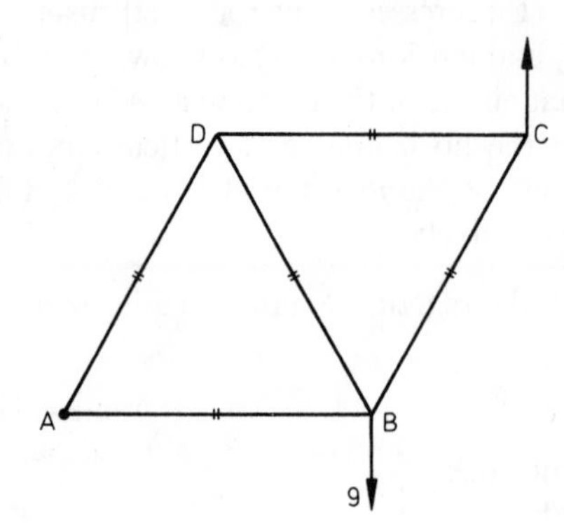

2.

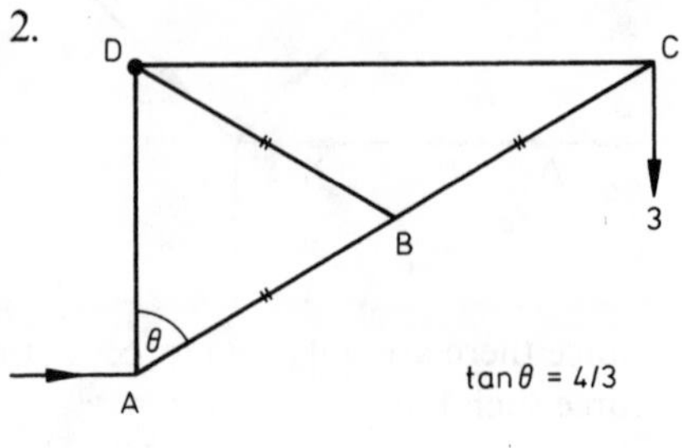

3.

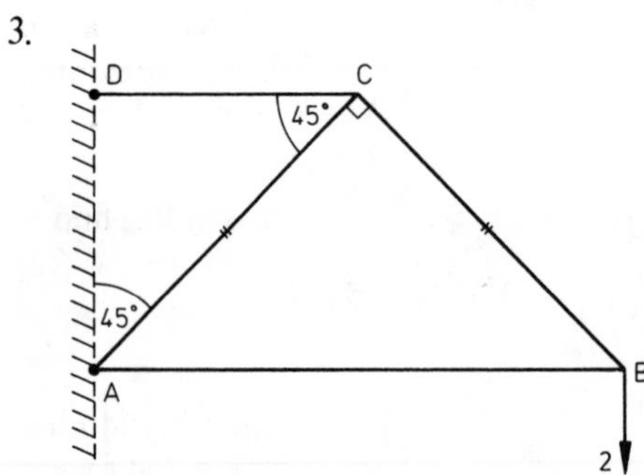

4.

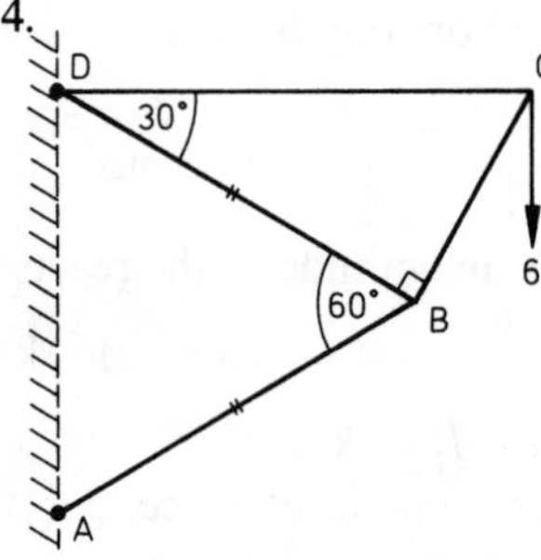

5.

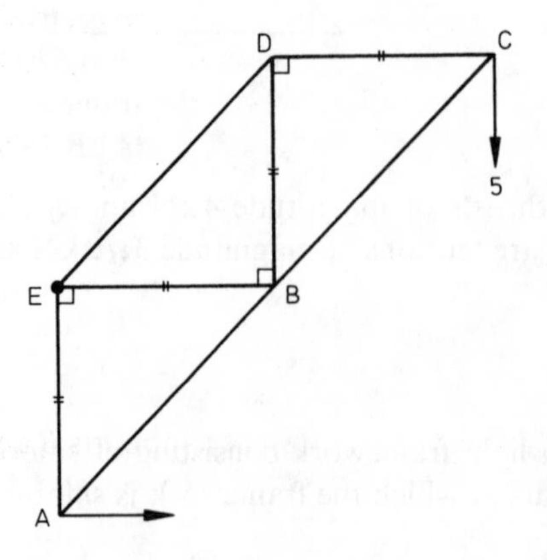

6.

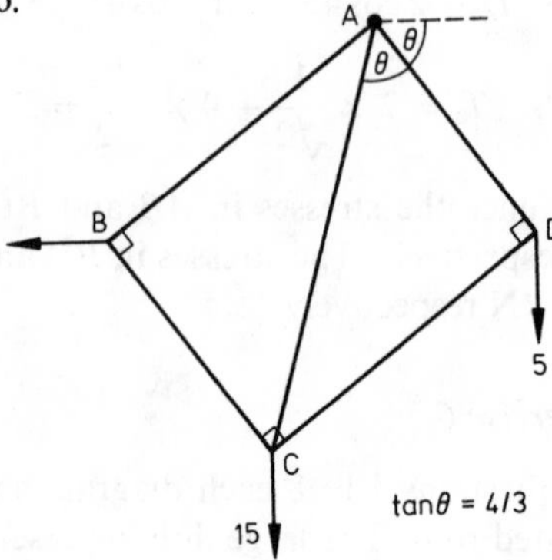

7.

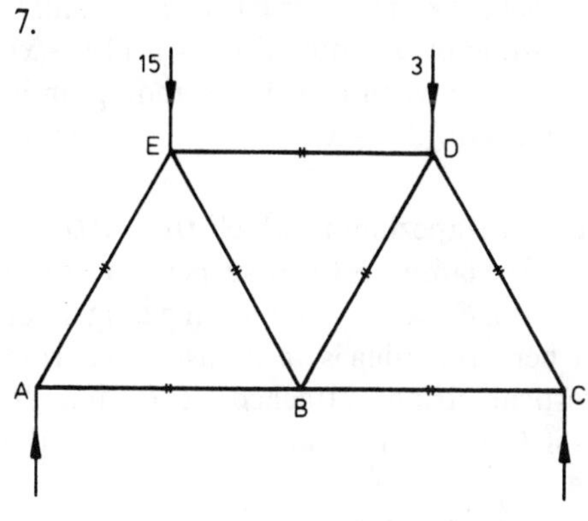

8.

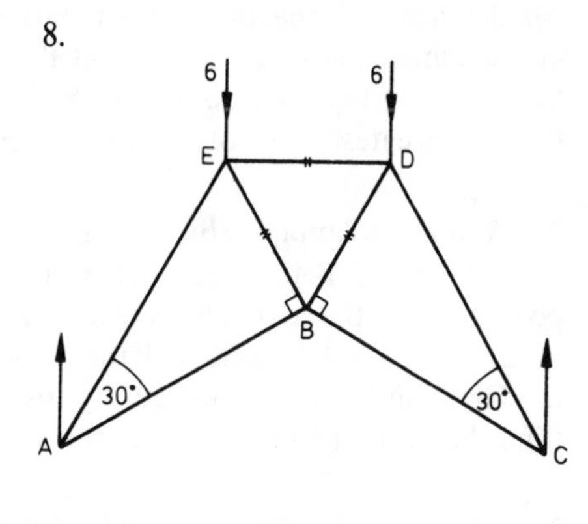

9.

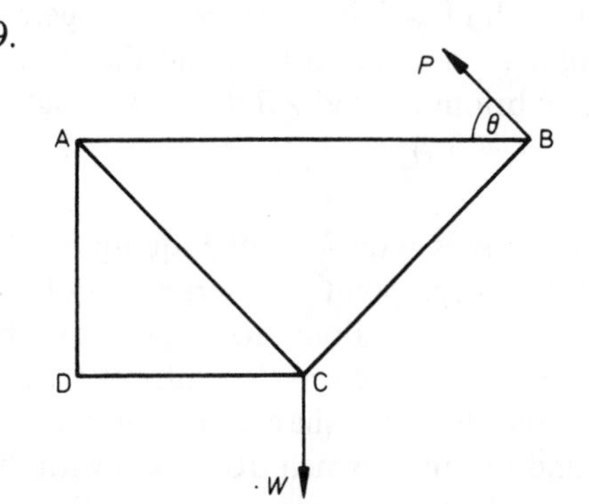

The diagram represents a framework of five smoothly jointed light rods with $AB = \sqrt{2}BC = \sqrt{2}CA = 2CD = 2DA$. The framework is freely hinged to a support at D. A weight W is hung from C and the framework is supported with CD horizontal by a force P at B making an angle θ with the horizontal.

If the magnitude of the stress in CA is twice the magnitude of the stress in BC, find P and the stresses in the rods, specifying whether they are tensions or thrusts. (O)

10.

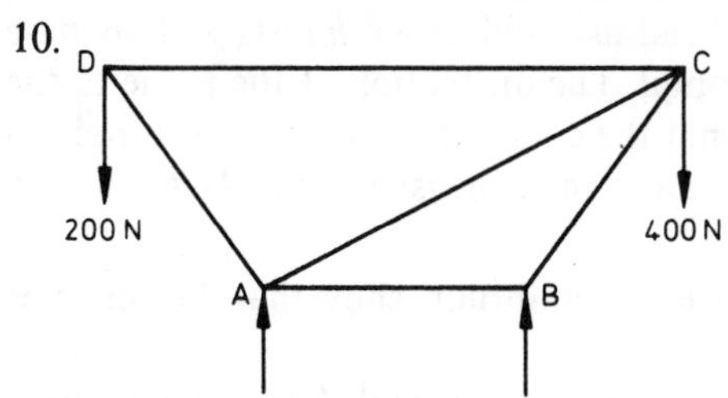

The diagram shows a smoothly jointed framework consisting of five light rigid rods with $AB = BC = AD = 5$ metres and $DC = 11$ metres. The framework is simply supported at A and B and is in equilibrium in a vertical plane with AB and DC horizontal and with loads of 400 N and 200 N at C and D respectively.
(i) Show that the magnitude of the force supporting the framework at B is 520 N.
(ii) Find the magnitude of the force acting in each member of the framework.
(iii) State which members of the framework are in tension. (AEB 1978)

Exercise 6.6 (miscellaneous)

[Take g as $10\,\mathrm{m\,s^{-2}}$.]

1. A uniform plank $ABCD$ of length $6a$ and weight W rests on two supports B and C; $AB = BC = CD = 2a$. A man of weight $4W$ wishes to stand on the plank at A, for

which purpose he places a counter-balancing weight Y at a point P between C and D at a distance x from D. Show that the least value of Y required is $7Wa/(4a - x)$. Show that, if the board is also to be in equilibrium when he is not standing on it, then the greatest value of Y that may be used is $Wa/(2a - x)$. (JMB)

2. A uniform lamina $ABCD$ is in the form of a trapezium in which $DC = AD = a$, $AB = 2a$ and $\angle BAD = \angle CDA = 90°$. When the lamina is freely suspended from a point E in BC the edge AB is vertical. Show that $BE:EC = 4:5$. Find the tangent of the angle which AB makes with the vertical when the lamina is freely suspended from B. The lamina is suspended by two vertical strings attached at C and D. Find the ratio of the tensions in the strings if CD is horizontal. (L)

3. Three equal uniform rods AB, BC, CA, each of weight w, are joined together to form a triangle ABC. A particle of weight W is attached to B and the system is suspended freely from the point A. In equilibrium the rod AB makes an angle θ with the vertical. Prove that $\tan\theta = \sqrt{3}w/(2W + 3w)$. (O)

4. A uniform thin sheet of cardboard of weight W is in the form of a square $ABCD$ of side $6a$. A cut is made along DO, where O is the point of intersection of the diagonals, and the triangular portion AOD is folded over along AO and stuck to the portion AOB with which it now coincides. Find the distances of the centre of gravity of the resulting object from AB and from BC and show that the centre of gravity lies on BD. If this object hangs freely from A, find the angle which BO makes with the horizontal. A particle of weight w is now attached to the object at D so that, in equilibrium, BO is horizontal. Find w. (L)

5. A uniform right circular solid cone of radius r and height h rests with its base on a rough plane which is initially horizontal. The inclination of the plane to the horizontal is gradually increased. Show that the cone will slide before it topples if the coefficient of friction between the plane and the cone is less than $4r/h$. (O)

6. If three non-parallel coplanar forces are in equilibrium, show that they must be concurrent.

A uniform rod AB of length $2a$ and weight W has one end A in contact with a vertical wall and to the other end B is attached one end of a light string. The other end of the string is attached to the wall at C, the point at a height $2a$ vertically above A. The rod makes an angle α ($0° < \alpha < 90°$) with the downward vertical. Show that equilibrium is impossible if the wall is smooth. If $\alpha = 60°$ and the rod is in equilibrium, show that the least possible value of the coefficient of friction between the rod and the wall is $1/\sqrt{3}$. (L)

7. A uniform rectangular lamina $ABCD$ of mass 1 kg has $AB = 0{\cdot}8$ m and $AD = 0{\cdot}6$ m. The lamina is smoothly pivoted at A. It is kept in equilibrium in a vertical plane with the diagonal AC horizontal and with B as the lowest point by a force P acting at right angles to CD. Find the magnitude of P and the magnitude and direction of the force exerted on the lamina by the pivot if (a) the force P acts through

the mid-point of CD, (b) the force P acts at C. Find also the magnitude and the line of action of P when the force exerted by the pivot acts in the direction BA. (L)

8. A uniform rod AB, of length $2l$ and weight W, is smoothly hinged to a vertical wall at A. It is held at rest in a horizontal position by a light inextensible string attached at B and to a point C which is vertically above and at a distance l from A. Calculate the tension in the string and show that the reaction at A is equal in magnitude to the tension. When a load of weight $4W$ is attached to the rod at a point D the horizontal and vertical components of the reaction at A are equal in magnitude. Find AD. (L)

9. A thin uniform rod AB of length 3 m and mass 5 kg is freely pivoted to a fixed point A. A light elastic string BC of modulus 30 N has one end attached to B and its other end C fixed to a point at the same level as A, where $AC = 5$ m. When the system is in equilibrium $BC = 4$ m. Calculate (a) the tension in the string. (b) the natural length of the string. (c) the resultant force, in magnitude and direction, exerted on the rod at A. (L)

10. 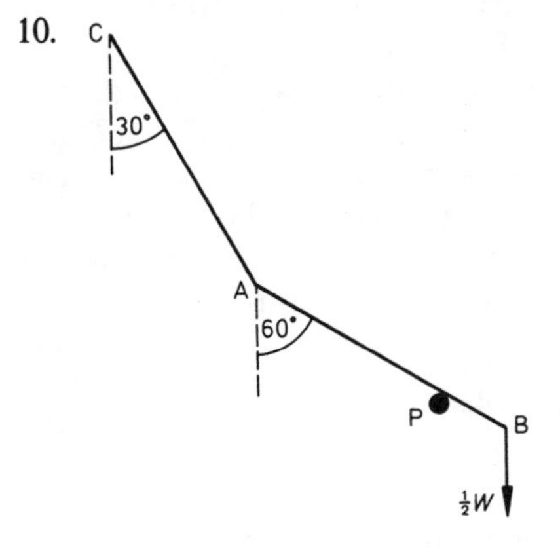

The diagram shows a uniform rod AB, of length $4c$ and weight W, resting on a rough fixed peg P, where $AP = 3c$. A particle of weight $\frac{1}{2}W$ hangs from B. One end of a light string is tied at A and the other end is tied at a fixed point C. The rod and the string are in the same vertical plane and are inclined at angles of 60° and 30° respectively to the vertical. (i) Show that the tension in the string is $\frac{1}{6}W\sqrt{3}$. (ii) Given that equilibrium is limiting, calculate the coefficient of friction between the peg and the rod. (iii) Given that $AC = 4c$ and that the string is elastic with modulus of elasticity $W\sqrt{3}$, calculate the natural length of the string. (AEB 1979)

11. A uniform square lamina $ABCD$ of weight W and side $2a$ rests in a vertical plane with the vertex A in contact with a rough horizontal plane. The lamina is kept in equilibrium by a force P acting at the vertex C in the direction of DC produced. Given that the height of the vertex D above the horizontal plane is $6a/5$, show that $P = W/10$. Find the coefficient of friction between the lamina and the plane if the lamina is in limiting equilibrium. (L)

12. A uniform rod AB, of weight W and length $2l$, rests in equilibrium with the end A on rough horizontal ground and with the end B in contact with a smooth vertical wall, which is perpendicular to the vertical plane containing the rod. If AB makes an angle α with the horizontal, where $\tan \alpha = 4/3$, find the least possible value of μ, the coefficient of friction between the rod and the ground, for equilibrium to be

preserved. If $\mu = \frac{1}{2}$, find the distance from A to the highest point of the rod at which a load of weight W can be attached without equilibrium being disturbed. (L)

13. Prove that the position of the centre of gravity of a uniform solid hemisphere of radius a is distant $\frac{3}{8}a$ from the centre of its plane face.

A uniform solid hemisphere rests with its curved surface on a rough horizontal surface and also against a rough vertical surface. The coefficient of friction at both points of contact is $\frac{1}{4}$. If the hemisphere is just about to slip, show that its top surface is inclined at an angle of just under 52° to the horizontal. (SU)

14. A uniform bar, of length $3a$ and mass m, rests in limiting equilibrium at an angle α to the horizontal with one end A on rough horizontal ground and the other end B against a smooth vertical wall. A horizontal force P is now applied to the bar at a point C, where $AC = a$. Find in terms of m, α and g, the value of P which will just cause the bar to slip towards the wall and the corresponding reaction between the bar and the wall.

15.

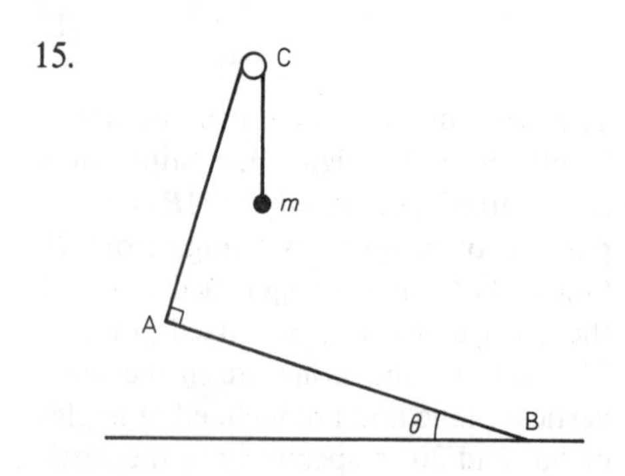

AB is a uniform rod of mass M. The end B rests on a rough horizontal table (coefficient of friction μ), and a light inextensible string is attached to the end A, passes over a smooth pulley at C and carries a mass m at its other end. The system rests in equilibrium with the string at right angles to AB. Find an expression for θ in terms of M and m and show that equilibrium is not possible unless $2m < M$. Show also that $\mu \geqslant \dfrac{m\sqrt{(M^2 - 4m^2)}}{M^2 - 2m^2}$. (SU)

16. A hemisphere of radius a, whose curved surface is smooth, is fixed with its flat base on a rough horizontal table. A uniform rod AB of length $4a$ and weight W rests in equilibrium against the hemisphere, its point of contact with the hemisphere lying between A and B. The end A of the rod is in contact with the table and AB makes an acute angle θ with the horizontal. Prove that the frictional force F at A has magnitude $2W\sin^2\theta$ and find the normal reaction R at A. If the coefficient of friction between the end A and the table is $\frac{1}{2}$, prove that $3\tan^2\theta + 2\tan\theta - 1 \leqslant 0$ and deduce the maximum value of θ consistent with equilibrium. (O&C)

17. $ABCD$ is a rectangular lamina, freely pivoted at a point X on AB. The length of AB is $2a$ and of BC is a. Forces $3W$, $5W$, $2W$ and $7W$ act along the sides AB, BC, DC and AD respectively in the directions indicated by the order of the letters. Find the position of X so that the lamina is in equilibrium and, in this case, determine the reaction of the pivot on the lamina in magnitude and direction. If, instead, the lamina is pivoted at a point Y on CD and is in equilibrium, deduce from your results the position of Y. (O&C)

18. Explain, with the aid of a diagram, the meaning of the expression coefficient of friction.

A straight uniform rod AB of weight W rests in limiting equilibrium with the end A on horizontal ground and the end B against a vertical wall. The vertical plane containing AB is perpendicular to the wall. The coefficient of friction between the rod and the ground is 4/5. The coefficient of friction between the rod and the wall is 3/5. Given that the inclination of AB to the horizontal is α, calculate (i) the normal reactions at A and B in terms of W, (ii) the numerical value of $\tan \alpha$. (AEB 1980)

19. A uniform right circular heavy cylinder, of height h and radius a, rests with one plane face upon a rough horizontal plane. A gradually increasing horizontal force is applied to a point on the edge of the upper face, and the force passes through the axis of the cylinder. Show that the cylinder will slide before it topples if $\mu h < a$, where μ is the coefficient of friction between the cylinder and the horizontal plane. (O)

20. A uniform solid hemisphere of radius a has mass M. A concentric hemisphere of mass $M/8$ is removed. Show that the centre of mass of the remainder is at a distance $45a/112$ from the centre O of the plane base. A particle of mass M is attached to a point X on the outer edge of the plane base of the remainder. Find, relative to OX and the axis of symmetry of the hemisphere as axes, the coordinates of the centre of mass G of the combined solid (i.e. remainder together with attached mass). The combined solid is placed with its curved surface on a horizontal plane and rests in equilibrium. Find the tangent of the angle between the plane base of the solid and the horizontal. (C)

21. Prove by integration that the centre of mass of a uniform solid right circular cone of height h and base radius r is at a distance $\frac{3}{4}h$ from the vertex. Such a cone is joined to a uniform solid right circular cylinder, of the same material, with base radius r and height l, so that the plane base of the cone coincides with a plane face of the cylinder. Find the centre of mass of the solid thus formed. Show that, if $6l^2 \geqslant h^2$, this solid can rest in equilibrium on a horizontal plane, with the curved surface of the cylinder touching the plane. Given that $l = h$, show that the solid can rest in equilibrium with its conical surface touching the plane provided that $r \geqslant \frac{1}{4}h\sqrt{5}$. (JMB)

22. A toy consists of a solid hemisphere of radius a to which is glued a solid circular cylinder of radius a and height $2a$ so that the plane end of the hemisphere is in complete contact with a plane end of the cylinder. The cylinder is made of uniform material of density ρ, and the hemisphere is made of uniform material of density $k\rho$. The toy is designed so that if placed on a horizontal table with the hemisphere downwards and then tilted to one side, it will return to the vertical position. Show that $k > 8$. The toy is placed on a desk of slope α where $\sin \alpha = 1/8$, sufficiently rough to prevent slipping. It rests in equilibrium with the hemisphere in contact with the desk. Find an expression giving the (acute) angle β made by its axis of symmetry with the vertical. Hence deduce that $k \geqslant 13\frac{1}{2}$. (O&C)

23. Using a diagram, explain the meaning of the term angle of friction. A uniform straight pole AB of length 2·5 metres stands in limiting equilibrium with the end A on horizontal ground and the end B against a vertical wall. The vertical plane containing AB is perpendicular to the wall and A is at a distance 0·7 metres from the wall. The angles of friction between the pole and the ground and between the pole and the wall are equal. Show that the line of action of the resultant force exerted on the pole by the ground at A is perpendicular to the line of action of the resultant force exerted on the pole by the wall at B. Hence, or otherwise, show that the coefficient of friction at A and B is $1/7$. Given that the pole weighs 50 N, find the magnitude of the resultant reactions (i) between the pole and the ground, (ii) between the pole and the wall. (AEB 1978)

24. A uniform rod of weight W rests in contact with two parallel, horizontal, smooth pegs, passing over the higher peg and under the lower one. The lower end of the rod rests on a smooth horizontal plane. A force of magnitude P acts on the upper end of the rod in the direction of the rod and tending to move the rod upwards. The length of the rod is l, and its inclination to the horizontal is θ. If the upper and lower pegs are at distances a and b respectively from the lower end of the rod, prove that, so long as the rod remains in equilibrium, the reaction of the upper peg on the rod is $\dfrac{Wl\cos\theta - 2Pb\cot\theta}{2(a-b)}$. Given that $l > 2a$, determine which of the three forces exerted on the rod by the two pegs and the ground will first become zero as P is gradually increased from zero. (C)

25. A uniform rod AB, of length l and weight W, is smoothly hinged at A. The point C is vertically above A and is such that $AC = l$. An elastic string, of unstretched length $\frac{1}{2}l$, obeys Hooke's law and has modulus W. The ends of the string are attached to B and C so that the rod rests in equilibrium in an inclined position. Show that the tension in the string is $W/3$. Find the magnitude of the couple which when applied to the rod will maintain the system in equilibrium with $AB = BC$. (L)

26. A uniform rod AB of weight W and length $2a$ has the end A freely hinged to a fixed point. The other end B is freely hinged to a uniform rod BC of weight W and length $6a$. The ends A and C are joined by a light inelastic string whose length is such that AB is perpendicular to BC, and the system hangs in equilibrium. Show that AB is inclined at 45° to the vertical. Find the tension in AC and the horizontal and vertical components of the reaction on BC at B. (C)

27. Two uniform rods AB and BC of the same thickness and material, and of length 4 metres and 3 metres respectively, are freely hinged together and rest in a vertical plane with the ends A and C on a rough horizontal plane. The system is in limiting equilibrium when the angle ABC is 90°. Determine how equilibrium will be broken when the angle ABC is slightly increased beyond 90° and show that the coefficient of friction between the rods and the ground is $84/163$. (W)

28.

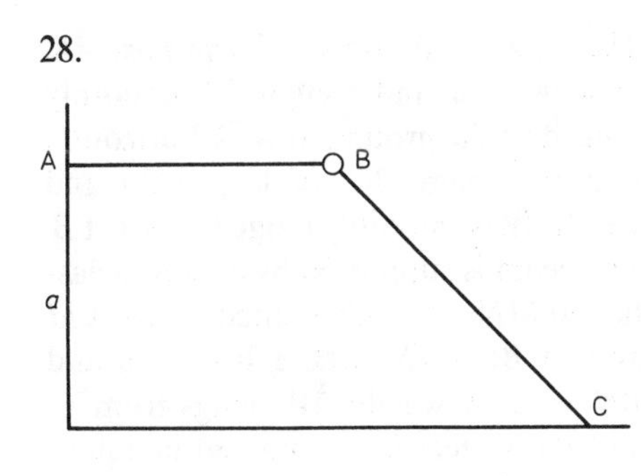

A uniform rod AB of length a and weight W is smoothly jointed at B to a uniform rod BC of length $a\sqrt{2}$ and also of weight W. The system is in equilibrium with C resting on a rough horizontal floor at a distance $2a$ from a rough vertical wall, A resting against the wall, and AB horizontal at a height a above the floor. The plane of the rods is perpendicular to the floor and the wall. (i) Find the frictional force exerted by the wall at A. (ii) Find the vertical and horizontal components of the force exerted on BC at C. (iii) Show that equilibrium is possible only if the coefficient of friction at C is at least $\frac{2}{3}$. (iv) Find the least value of the coefficient of friction at A for equilibrium to be possible. (JMB)

29. Three uniform straight rods, each of weight W, are smoothly jointed together to form an equilateral triangle ABC. The joint B is hinged to a fixed support and the system is free to rotate about B in a vertical plane. The system is held in equilibrium with BC horizontal and A uppermost by a horizontal light string which connects A with a fixed point. Show that the tension in the string is $W\sqrt{3}$ and find the magnitude and direction of the force exerted on the triangle by the support at B. Find the horizontal and vertical components of all the forces which are exerted on the rod AC, and indicate them clearly on your diagram. (JMB)

30. The diagram shows two parallel vertical walls at a distance $3a$ apart, with two uniform smooth cylinders, each of radius a and weight W. The axes of the cylinders are horizontal and parallel to the walls. Each cylinder is in contact with the other and with one wall; the lower cylinder rests on horizontal ground. Find the magnitudes of the reactions between the two cylinders and between each cylinder and the corresponding wall. (C)

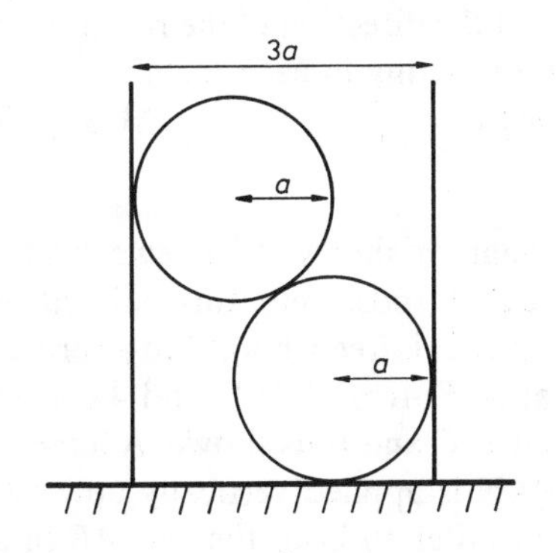

31. A uniform sphere of radius a and weight $W/\sqrt{3}$ rests on a rough horizontal table. A uniform rod AB of weight $2W$ and length $2a$ is freely hinged at A to a fixed point on the table and leans against the sphere so that the centre of the sphere and the rod lie in a vertical plane. The rod makes an angle of 60° with the horizontal. Show that the frictional force between the rod and the sphere is $\frac{1}{3}W$. The coefficient of friction at each point of contact is μ. What is the smallest value of μ which makes equilibrium possible? (O&C)

32.

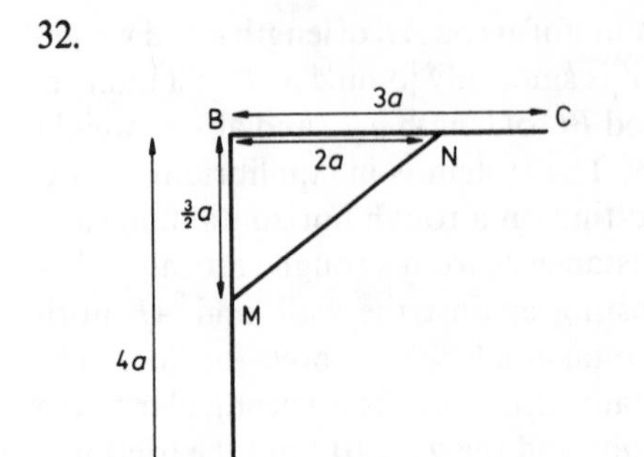

The diagram shows a uniform post AB, of length $4a$ and weight W, smoothly hinged to the ground at A. A horizontal uniform beam BC, of length $3a$ and weight W, is smoothly hinged to AB at B. The beam is supported by a light inelastic rod MN smoothly jointed to AB at M and to BC at N, where $BM = \frac{3}{2}a$ and $BN = 2a$. A weight $3W$ hangs from C, and the system is maintained in equilibrium, with AB vertical, by an external couple of magnitude G applied to AB, as shown.

Find (i) the thrust in the rod MN, (ii) the horizontal and vertical components of the forces on the post AB at A and at B, (iii) the value of G. (C)

33.

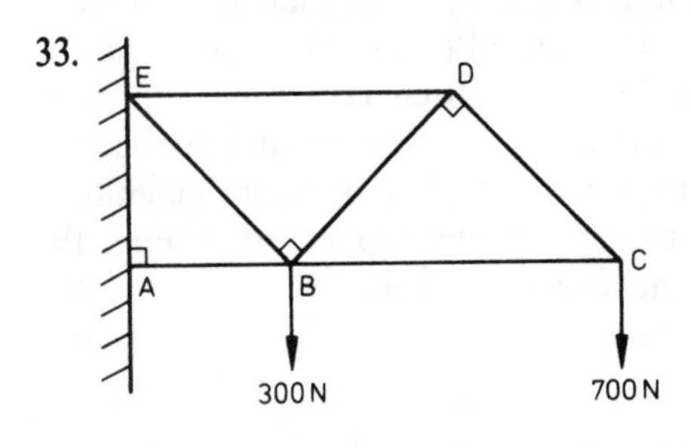

The diagram shows a framework $ABCDE$, consisting of 6 light rigid rods which are smoothly jointed. The framework is smoothly hinged to a vertical wall at A and E and the triangles ABE, BED and BCD are all right-angled and isosceles. When loads of 300 N and 700 N are hung from B and C respectively, the framework is in equilibrium in a vertical plane. (i) Explain why the reaction on the hinge at A is horizontal and find its magnitude. (ii) Find the magnitude and the direction of the reaction on the hinge at E. (iii) Find the magnitude of the force acting in each member of the framework and state which rods are in compression. (AEB 1978)

34.

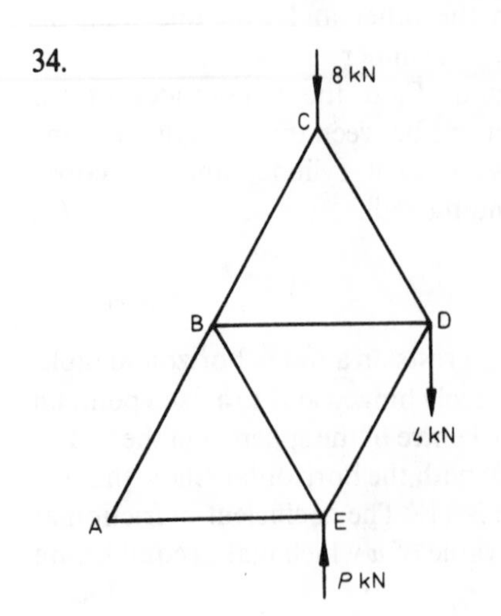

The diagram shows a framework of 7 light equal rods, smoothly jointed at their ends. It is freely hinged to a vertical wall at A. Forces of 8 kN and 4 kN are applied at C and D as shown. A force of size P kN is applied vertically upwards at E in order to keep the rod AE in a horizontal position. Calculate (i) the magnitude of P; (ii) the stresses in the rods, stating if they are in tension or under compression; (iii) the magnitude and direction of the reaction of the support at A on the frame. (SU)

7 Vectors and scalar product

7.1 Properties of vectors

We begin by reviewing some of the basic properties of vectors.

(1) The sum or resultant of two vectors may be found using the triangle law for vector addition

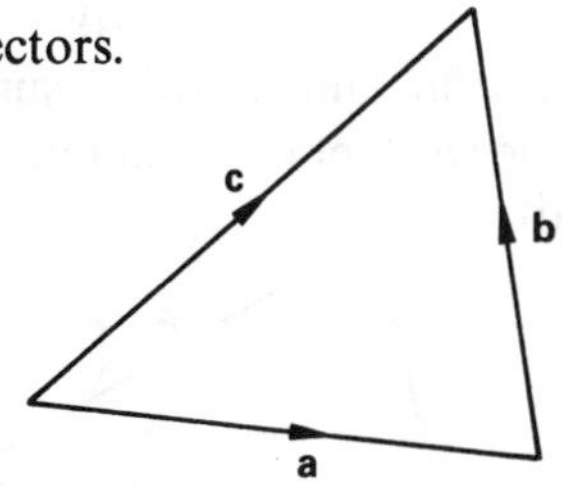

e.g. $\mathbf{a} + \mathbf{b} = \mathbf{c}$.

c

b

a

Vector addition is both commutative and associative

i.e. $\mathbf{a} + \mathbf{b} = \mathbf{b} + \mathbf{a}, \quad (\mathbf{a} + \mathbf{b}) + \mathbf{c} = \mathbf{a} + (\mathbf{b} + \mathbf{c})$.

(2) For any scalar λ, the vector $\lambda\mathbf{a}$ is parallel to $\mathbf{a}$ and has magnitude $|\lambda|$ times the magnitude of $\mathbf{a}$.

It can be shown that

$$\lambda(\mu\mathbf{a}) = (\lambda\mu)\mathbf{a},$$
$$(\lambda + \mu)\mathbf{a} = \lambda\mathbf{a} + \mu\mathbf{a}, \quad \lambda(\mathbf{a} + \mathbf{b}) = \lambda\mathbf{a} + \lambda\mathbf{b}.$$

(3) Given two non-zero vectors $\mathbf{a}$ and $\mathbf{b}$, which are not parallel, any vector $\mathbf{r}$ in the plane of $\mathbf{a}$ and $\mathbf{b}$ can be expressed in the form $\lambda\mathbf{a} + \mu\mathbf{b}$. The vectors $\lambda\mathbf{a}$ and $\mu\mathbf{b}$ are the *component vectors* of $\mathbf{r}$ in the directions of *base vectors* $\mathbf{a}$ and $\mathbf{b}$.

(i) If vectors $\lambda\mathbf{a} + \mu\mathbf{b}$ and $l\mathbf{a} + m\mathbf{b}$ are equal, then $\lambda = l$ and $\mu = m$.

(ii) If vectors $\lambda\mathbf{a} + \mu\mathbf{b}$ and $l\mathbf{a} + m\mathbf{b}$ are parallel, then for some scalar t,

$$\lambda\mathbf{a} + \mu\mathbf{b} = t(l\mathbf{a} + m\mathbf{b}) \text{ and thus } \frac{\lambda}{l} = \frac{\mu}{m} = t.$$

(4) Given three non-zero vectors $\mathbf{a}$, $\mathbf{b}$ and $\mathbf{c}$, which are not coplanar, any vector $\mathbf{r}$ can be expressed in the form $\lambda\mathbf{a} + \mu\mathbf{b} + \nu\mathbf{c}$.

(i) If vectors $\lambda\mathbf{a} + \mu\mathbf{b} + \nu\mathbf{c}$ and $l\mathbf{a} + m\mathbf{b} + n\mathbf{c}$ are equal, then $\lambda = l$, $\mu = m$ and $\nu = n$.

(ii) If vectors $\lambda\mathbf{a} + \mu\mathbf{b} + \nu\mathbf{c}$ and $l\mathbf{a} + m\mathbf{b} + n\mathbf{c}$ are parallel, then for some scalar t,

$$\lambda\mathbf{a} + \mu\mathbf{b} + \nu\mathbf{c} = t(l\mathbf{a} + m\mathbf{b} + n\mathbf{c}) \text{ and thus } \frac{\lambda}{l} = \frac{\mu}{m} = \frac{\nu}{n} = t.$$

(5) If $\mathbf{r} = x\mathbf{i} + y\mathbf{j} + z\mathbf{k}$, where $\mathbf{i}, \mathbf{j}$ and $\mathbf{k}$ are unit vectors in the directions of the x-, y- and z-axes respectively then,
   (i) the magnitude of $\mathbf{r}$ is given by $r = \sqrt{(x^2 + y^2 + z^2)}$,
   (ii) the unit vector in the direction of $\mathbf{r}$ is $\hat{\mathbf{r}} = \frac{x}{r}\mathbf{i} + \frac{y}{r}\mathbf{j} + \frac{z}{r}\mathbf{k}$.

(6)

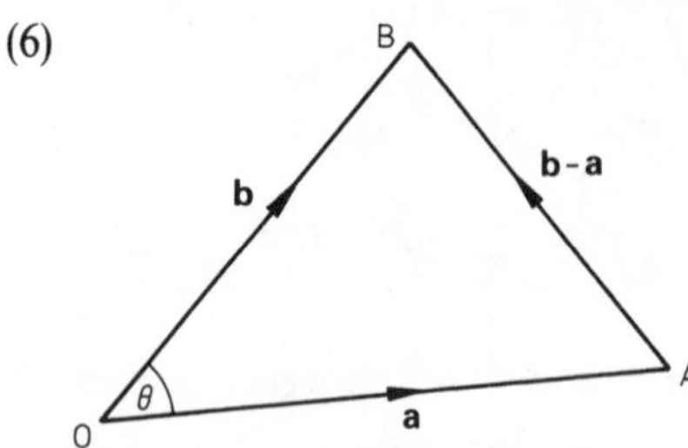

If the position vectors of points A and B with respect to O are $\mathbf{a}$ and $\mathbf{b}$ respectively, then $\mathbf{OA} = \mathbf{a}$, $\mathbf{OB} = \mathbf{b}$ and $\mathbf{AB} = \mathbf{b} - \mathbf{a}$.
   (i) The distance between A and B is given by $|\mathbf{b} - \mathbf{a}|$ (or $|\mathbf{a} - \mathbf{b}|$).
   (ii) If θ is the angle between $\mathbf{a}$ and $\mathbf{b}$, then the cosine rule can be used to express $\cos\theta$ in terms of $|\mathbf{a}|, |\mathbf{b}|$ and $|\mathbf{a} - \mathbf{b}|$.
   (iii) A point C, with position vector $\mathbf{c}$, lies on the straight line passing through A and B if and only if there is a scalar λ such that $\mathbf{AC} = \lambda\mathbf{AB}$ i.e. $\mathbf{c} - \mathbf{a} = \lambda(\mathbf{b} - \mathbf{a})$.

The fundamental ideas summarised in paragraphs (1) to (6) form the foundation for the work of this chapter.

(i)

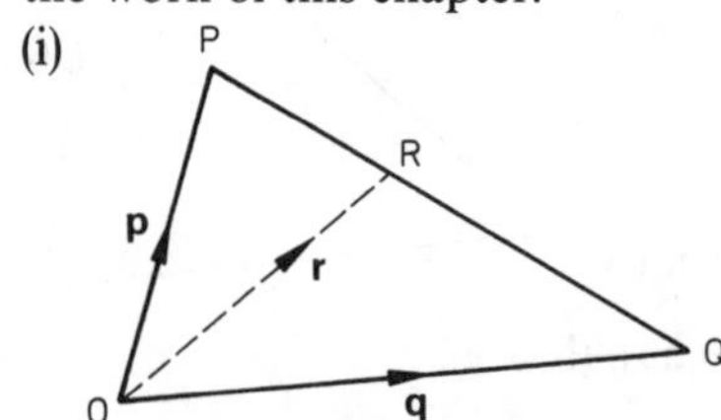

(ii)

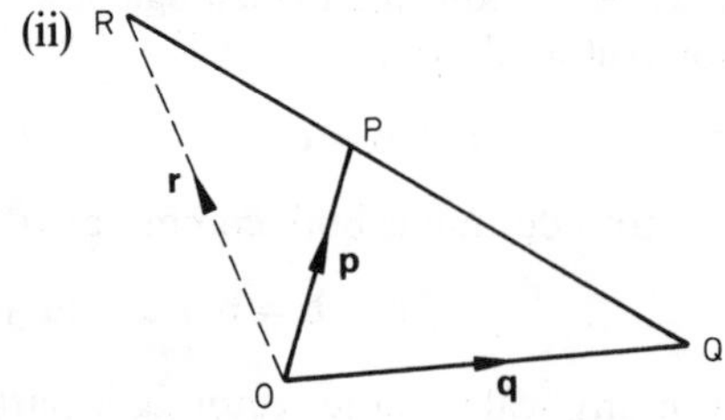

Let us consider three points P, Q and R such that R divides PQ in the ratio $\mu:\lambda$. We will observe the convention that (i) if λ and μ have the same sign then R lies between P and Q, and (ii) if λ and μ have opposite signs then R does not lie between P and Q.

In both cases $\qquad \lambda\mathbf{PR} = \mu\mathbf{RQ}$

$\therefore \qquad \lambda(\mathbf{OR} - \mathbf{OP}) = \mu(\mathbf{OQ} - \mathbf{OR})$

$\therefore \qquad \lambda\mathbf{OR} - \lambda\mathbf{OP} = \mu\mathbf{OQ} - \mu\mathbf{OR}$

Hence $\qquad (\lambda + \mu)\mathbf{OR} = \lambda\mathbf{OP} + \mu\mathbf{OQ}$

This result is called the *ratio theorem* and may also be stated as follows:

> The resultant of the vectors $\lambda\mathbf{OP}$ and $\mu\mathbf{OQ}$ is $(\lambda + \mu)\mathbf{OR}$, where R is the point which divides PQ in the ratio $\mu:\lambda$

If the position vectors of P, Q and R with respect to O are $\mathbf{p}$, $\mathbf{q}$ and $\mathbf{r}$ respectively then

$$(\lambda + \mu)\mathbf{r} = \lambda\mathbf{p} + \mu\mathbf{q}.$$

Thus the point R which divides PQ in the ratio $\mu:\lambda$ has position vector

$$\boxed{\mathbf{r} = \frac{\lambda\mathbf{p} + \mu\mathbf{q}}{\lambda + \mu}}$$

In particular, if R is the *mid-point* of PQ, then $\mathbf{r} = \frac{1}{2}(\mathbf{p} + \mathbf{q})$.

Example 1 The points A and B have position vectors $\mathbf{a} = \mathbf{i} + 3\mathbf{j} - 2\mathbf{k}$ and $\mathbf{b} = 6\mathbf{i} - 2\mathbf{j} + 3\mathbf{k}$ respectively. Find the position vectors of the point P which divides AB internally in the ratio 1:4 and the point Q which divides AB externally in the ratio 3:2.

Observing the usual sign convention, P divides AB in the ratio 1:4 and Q divides AB in the ratio $3:-2$.
Thus the position vector of P

$$= \frac{4\mathbf{a} + \mathbf{b}}{4 + 1} = \frac{4(\mathbf{i} + 3\mathbf{j} - 2\mathbf{k}) + (6\mathbf{i} - 2\mathbf{j} + 3\mathbf{k})}{5} = 2\mathbf{i} + 2\mathbf{j} - \mathbf{k}.$$

Similarly, the position vector of Q

$$= \frac{-2\mathbf{a} + 3\mathbf{b}}{-2 + 3} = -2(\mathbf{i} + 3\mathbf{j} - 2\mathbf{k}) + 3(6\mathbf{i} - 2\mathbf{j} + 3\mathbf{k}) = 16\mathbf{i} - 12\mathbf{j} + 13\mathbf{k}.$$

[Note that the same result will be obtained for the position vector of Q if the ratio $-3:2$ is used instead of the ratio $3:-2$.]

Example 2 Prove that the medians of a triangle meet at a point which divides each median in the ratio 2:1.

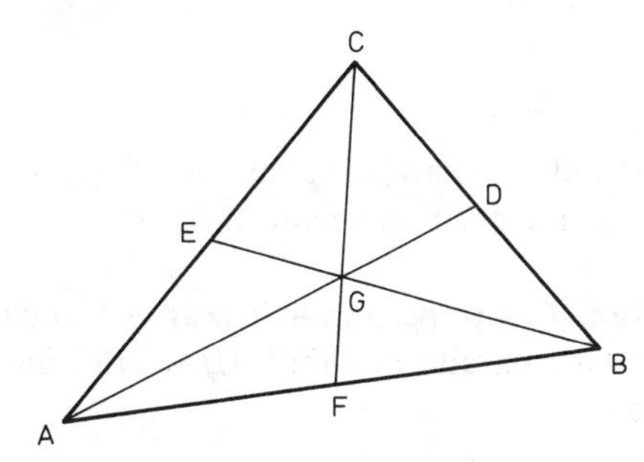

In $\triangle ABC$ let D, E and F be the mid-points of BC, CA and AB respectively. If the position vectors of A, B and C are $\mathbf{a}$, $\mathbf{b}$ and $\mathbf{c}$, then the position vectors of D, E and F are $\frac{1}{2}(\mathbf{b} + \mathbf{c}), \frac{1}{2}(\mathbf{c} + \mathbf{a})$ and $\frac{1}{2}(\mathbf{a} + \mathbf{b})$ respectively.

Let G be the point which divides AD in the ratio 2:1, then its position vector is $\dfrac{1 \times \mathbf{a} + 2 \times \frac{1}{2}(\mathbf{b} + \mathbf{c})}{1 + 2}$ i.e. $\frac{1}{3}(\mathbf{a} + \mathbf{b} + \mathbf{c})$.
Since this expression is symmetrical in $\mathbf{a}, \mathbf{b}$ and $\mathbf{c}$, G must also be the point which divides BE and CF in the ratio 2:1. Hence the medians meet at a point which divides each median in the ratio 2:1.

[As we saw in §5.3 the point G with position vector $\frac{1}{3}(\mathbf{a} + \mathbf{b} + \mathbf{c})$ is the centroid of $\triangle ABC$.]

Let us consider again the points P, Q and R with position vectors $\mathbf{p}, \mathbf{q}$ and $\mathbf{r}$ respectively. It follows from the ratio theorem that if $(\lambda + \mu)\mathbf{r} = \lambda\mathbf{p} + \mu\mathbf{q}$, then R is the point which divides PQ in the ratio $\mu:\lambda$. If $\lambda + \mu = 1$, we obtain the following useful result.

> If $\mathbf{r} = \lambda\mathbf{p} + \mu\mathbf{q}$ for scalars λ and μ such that $\lambda + \mu = 1$, then R is the point which divides PQ in the ratio $\mu:\lambda$.

Example 3 In a parallelogram $OABC$, P is the mid-point of AB and Q divides OP in the ratio 2:1. Show that Q lies on AC and find the ratio $AQ:QC$.

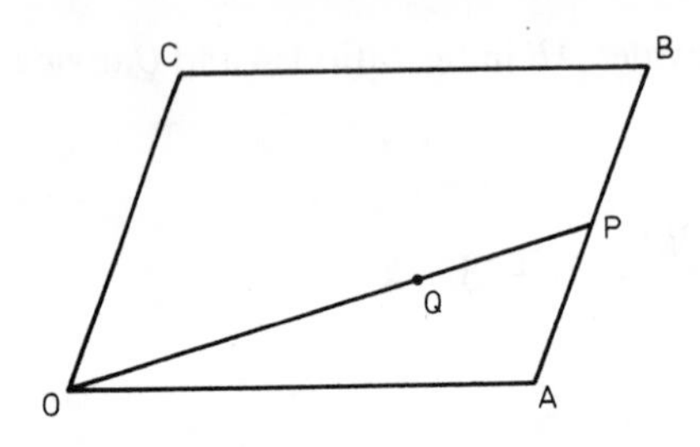

Let the position vectors of A, B, C, P and Q with respect to O be $\mathbf{a}, \mathbf{b}, \mathbf{c}, \mathbf{p}$ and $\mathbf{q}$ respectively.
P is the mid-point of AB

$\therefore \quad \mathbf{p} = \frac{1}{2}(\mathbf{a} + \mathbf{b})$

Q divides OP in the ratio 2:1

$\therefore \quad \mathbf{q} = \frac{2}{3}\mathbf{p} = \frac{2}{3} \times \frac{1}{2}(\mathbf{a} + \mathbf{b}) = \frac{1}{3}\mathbf{a} + \frac{1}{3}\mathbf{b}$

However, since $OABC$ is a parallelogram, $\mathbf{b} = \mathbf{a} + \mathbf{c}$,

$\therefore \quad \mathbf{q} = \frac{1}{3}\mathbf{a} + \frac{1}{3}(\mathbf{a} + \mathbf{c}) = \frac{2}{3}\mathbf{a} + \frac{1}{3}\mathbf{c}$

Hence, by the ratio theorem, Q is the point on AC such that $AQ:QC = 1:2$.

Exercise 7.1

1. The position vectors of the vertices A, B and C of parallelogram $ABCD$ are $\mathbf{a}, \mathbf{b}$ and $\mathbf{c}$ respectively. Find, in terms of $\mathbf{a}, \mathbf{b}$ and $\mathbf{c}$, the position vector of D.

2. The position vectors of the vertices A, B and C of trapezium $ABCD$ are $\mathbf{a}, \mathbf{b}$ and $\mathbf{c}$ respectively. Given that the sides BC and AD are parallel and that $AD = 2BC$, find the position vector of D in terms of $\mathbf{a}, \mathbf{b}$ and $\mathbf{c}$.

3. Find the angle between the vectors $\mathbf{a}$ and $\mathbf{b}$ given that $|\mathbf{a}| = 3$, $|\mathbf{b}| = 8$ and $|\mathbf{a} - \mathbf{b}| = 7$.

4. The angle between the vectors $\mathbf{a}$ and $\mathbf{b}$ is 120°. If $|\mathbf{a}| = 5$ and $|\mathbf{b}| = 3$, find $|\mathbf{a} - \mathbf{b}|$ and $|\mathbf{a} + \mathbf{b}|$.

5. The points A and B have position vectors $\mathbf{a}$ and $\mathbf{b}$ respectively relative to an origin O, where $\mathbf{a} = 2\mathbf{i} + \mathbf{j} - 3\mathbf{k}$ and $\mathbf{b} = -4\mathbf{i} + s\mathbf{j} + t\mathbf{k}$. Find the possible values of s and t if (a) the points O, A and B are collinear, (b) $|\mathbf{AB}| = 7$ and $s = 2t$.

6. Given that $\mathbf{a} = 2\mathbf{i} + 6\mathbf{j}$ and $\mathbf{b} = 3\mathbf{i} + t\mathbf{j}$, find the value of t such that (a) $\mathbf{a}$ and $\mathbf{b}$ are parallel, (b) $\mathbf{a}$ and $\mathbf{b}$ are at right angles.

7. If the points A and B have position vectors $\mathbf{a}$ and $\mathbf{b}$ respectively, find in terms of $\mathbf{a}$ and $\mathbf{b}$ the position vectors of (a) the point P which divides AB internally in the ratio 5:3, (b) the point Q which divides AB externally in the ratio 3:1, (c) the point R such that A is the mid-point of BR.

8. The points A and B have position vectors $\mathbf{a} = 5\mathbf{i} + 4\mathbf{j} + \mathbf{k}$ and $\mathbf{b} = -\mathbf{i} + \mathbf{j} - 2\mathbf{k}$ respectively. Find the position vectors of (a) the point C on AB such that $AC = 2CB$, (b) the point D on AB produced such that $AD = 2BD$.

9. Write down, in terms of $\mathbf{i}$, $\mathbf{j}$ and $\mathbf{k}$, the position vectors of the points $A(2, -5, 3)$ and $B(7, 0, -2)$. Hence find the coordinates of the point C which divides AB internally in the ratio 2:3 and the point D which divides AB externally in the ratio 3:8.

10. The vertices A, B and C of a triangle have position vectors $\mathbf{a}$, $\mathbf{b}$ and $\mathbf{c}$ respectively. Points P, Q and R have position vectors given respectively by $\mathbf{p} = \mathbf{a} - \mathbf{b} + \mathbf{c}$, $\mathbf{q} = \frac{3}{7}\mathbf{a} + \frac{4}{7}\mathbf{c}$, $\mathbf{r} = 3\mathbf{b} - 2\mathbf{c}$. Describe geometrically the positions of P, Q and R relative to A, B and C.

11. The diagonals AC and BD of a quadrilateral $ABCD$ intersect at a point P such that $AP:PC = 2:7$ and $BP:PD = 5:4$. Show that

$$\frac{7\mathbf{a} - 4\mathbf{b}}{7 - 4} = \frac{5\mathbf{d} - 2\mathbf{c}}{5 - 2}$$

where $\mathbf{a}$, $\mathbf{b}$, $\mathbf{c}$ and $\mathbf{d}$ are the position vectors of A, B, C and D respectively. Explain the geometrical significance of this result.

12. The vertices A, B and C of a triangle have position vectors $\mathbf{a}$, $\mathbf{b}$ and $\mathbf{c}$ respectively. The point P divides BC internally in the ratio 3:1, the point Q is the mid-point of CA and the point R divides AB externally in the ratio 1:3. If the position vectors of P, Q and R are $\mathbf{p}$, $\mathbf{q}$ and $\mathbf{r}$ respectively, express $\mathbf{r}$ in terms of $\mathbf{p}$ and $\mathbf{q}$. Hence show that R lies on PQ produced and find the ratio $PR:QR$.

7.2 Equations of lines and planes

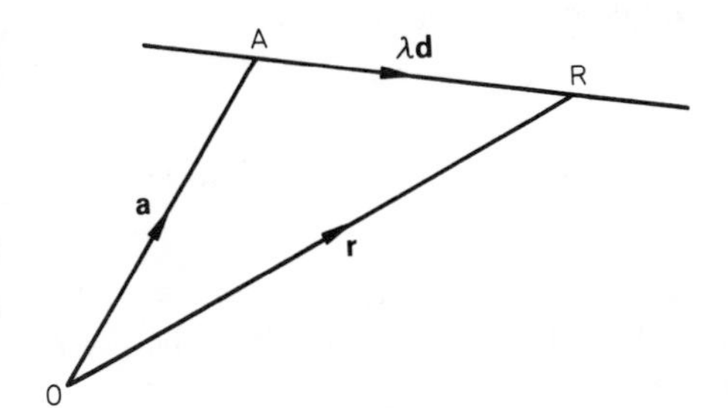

The diagram shows a straight line which passes through a fixed point A with position vector $\mathbf{a}$ and which is parallel to a given vector $\mathbf{d}$. Let $\mathbf{r}$ be the position vector of a point R on the line.

Since **AR** is parallel to **d**, then, for some scalar λ,

$$\mathbf{AR} = \lambda\mathbf{d}$$

$$\therefore \quad \mathbf{r} - \mathbf{a} = \lambda\mathbf{d}$$

$$\text{i.e.} \quad \mathbf{r} = \mathbf{a} + \lambda\mathbf{d}$$

Thus every point on the line has a position vector of the form $\mathbf{a} + \lambda\mathbf{d}$. Moreover, each real value of the parameter λ corresponds to some point on the line.

The equation $\mathbf{r} = \mathbf{a} + \lambda\mathbf{d}$ is called the *vector equation* of the line through A parallel to **d**. The vector **d** is a *direction vector* of the line.

Example 1 Find the vector equation of the line which passes through the point with position vector $2\mathbf{i} - \mathbf{j} + 3\mathbf{k}$ and which is parallel to the vector $2\mathbf{j} - \mathbf{k}$.

The vector equation of the given line is

$$\mathbf{r} = 2\mathbf{i} - \mathbf{j} + 3\mathbf{k} + \lambda(2\mathbf{j} - \mathbf{k})$$

$$\text{i.e.} \quad \mathbf{r} = 2\mathbf{i} + (2\lambda - 1)\mathbf{j} + (3 - \lambda)\mathbf{k}.$$

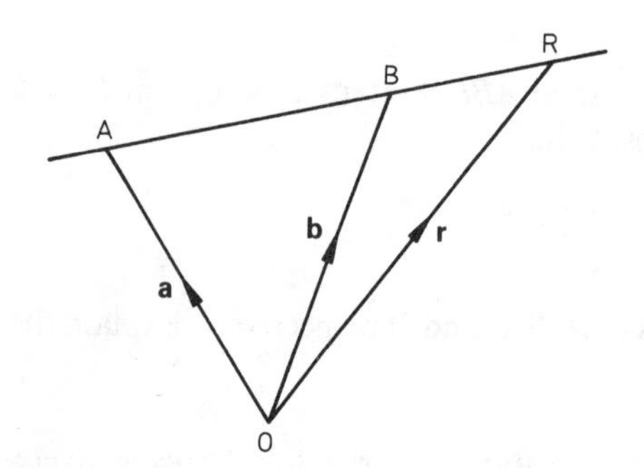

Consider now the line through two points A and B with position vectors **a** and **b** respectively. If a point R, with position vector **r**, lies on this line, then for some scalar λ,

$$\mathbf{AR} = \lambda\mathbf{AB}$$

$$\therefore \quad \mathbf{r} - \mathbf{a} = \lambda(\mathbf{b} - \mathbf{a})$$

$$\therefore \quad \mathbf{r} = \mathbf{a} + \lambda(\mathbf{b} - \mathbf{a}).$$

Hence the vector equation of the line through A and B is $\mathbf{r} = (1 - \lambda)\mathbf{a} + \lambda\mathbf{b}$.

Example 2 Find the vector equation of the straight line which passes through the points $A(1, 0, -2)$ and $B(2, 3, -1)$.

The position vectors of A and B are $\mathbf{i} - 2\mathbf{k}$ and $2\mathbf{i} + 3\mathbf{j} - \mathbf{k}$ respectively.
Hence the vector equation of the line through A and B is

$$\mathbf{r} = (1 - \lambda)(\mathbf{i} - 2\mathbf{k}) + \lambda(2\mathbf{i} + 3\mathbf{j} - \mathbf{k})$$

$$\text{i.e.} \quad \mathbf{r} = (\lambda + 1)\mathbf{i} + 3\lambda\mathbf{j} + (\lambda - 2)\mathbf{k}.$$

Example 3 Three non-collinear points A, B and C have position vectors **a**, **b** and $\frac{4}{3}\mathbf{a} - \frac{2}{3}\mathbf{b}$ respectively, relative to an origin O. Find the position vector of the point D on CA produced, such that BD is parallel to OA.

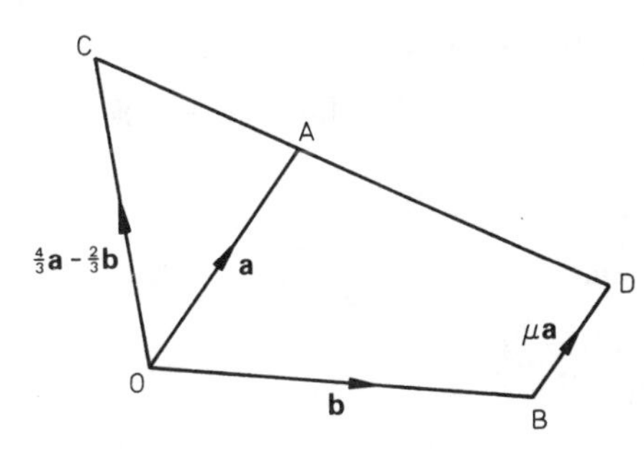

The vector equation of the line through A and C is

$$\mathbf{r} = (1-\lambda)\mathbf{a} + \lambda(\tfrac{4}{3}\mathbf{a} - \tfrac{2}{3}\mathbf{b})$$

i.e. $\mathbf{r} = (1+\tfrac{1}{3}\lambda)\mathbf{a} - \tfrac{2}{3}\lambda\mathbf{b}$.

The vector equation of the line through B parallel to OA is

$$\mathbf{r} = \mathbf{b} + \mu\mathbf{a}$$

At the point D where these lines intersect

$$(1+\tfrac{1}{3}\lambda)\mathbf{a} - \tfrac{2}{3}\lambda\mathbf{b} = \mu\mathbf{a} + \mathbf{b}$$

Since the vectors **a** and **b** are not parallel, it follows that

$$1+\tfrac{1}{3}\lambda = \mu \quad \text{and} \quad -\tfrac{2}{3}\lambda = 1$$

$$\therefore \qquad \lambda = -\tfrac{3}{2}, \quad \mu = \tfrac{1}{2}$$

Hence the position vector of D is $\frac{1}{2}\mathbf{a} + \mathbf{b}$.

In three dimensional space a pair of vector equations such as $\mathbf{r} = \mathbf{a} + \lambda\mathbf{p}$ and $\mathbf{r} = \mathbf{b} + \mu\mathbf{q}$ may represent
(i) the same line, (ii) parallel lines,
(iii) intersecting lines, (iv) skew lines.

Example 4 Show that the lines l_1 and l_2, with vector equations $\mathbf{r} = \mathbf{k} + \lambda(\mathbf{i} - \mathbf{j} - 3\mathbf{k})$ and $\mathbf{r} = 2\mathbf{i} + \mathbf{j} + \mu(3\mathbf{j} + 5\mathbf{k})$ respectively, intersect, and find the position vector of their point of intersection.

The equations of l_1 and l_2 may be written:

$$\mathbf{r} = \lambda\mathbf{i} - \lambda\mathbf{j} + (1-3\lambda)\mathbf{k}$$
$$\mathbf{r} = 2\mathbf{i} + (1+3\mu)\mathbf{j} + 5\mu\mathbf{k}$$

At any point of intersection

$$\lambda\mathbf{i} - \lambda\mathbf{j} + (1-3\lambda)\mathbf{k} = 2\mathbf{i} + (1+3\mu)\mathbf{j} + 5\mu\mathbf{k}$$

Thus

$$\lambda = 2 \qquad (1)$$
$$-\lambda = 1 + 3\mu \qquad (2)$$
$$1 - 3\lambda = 5\mu \qquad (3)$$

From (1) and (2), $\lambda = 2, \mu = -1$.

Since these values also satisfy equation (3), the lines do have a point of intersection. Its position vector, given by substituting $\lambda = 2$ in the equation of l_1, is $2\mathbf{i} - 2\mathbf{j} - 5\mathbf{k}$.

[Note that we can check this result by substituting $\mu = -1$ in the equation of l_2.].

Example 5 Show that the lines l_1 and l_2, with vector equations

$$\mathbf{r} = \lambda\mathbf{i} + (3 - 2\lambda)\mathbf{j} + (2 + \lambda)\mathbf{k} \quad \text{and} \quad \mathbf{r} = (1 - 2\mu)\mathbf{i} + 4\mu\mathbf{j} + (1 - 2\mu)\mathbf{k}$$

respectively, are distinct parallel lines.

The equations of l_1 and l_2 may be written:

$$\mathbf{r} = 3\mathbf{j} + 2\mathbf{k} + \lambda(\mathbf{i} - 2\mathbf{j} + \mathbf{k})$$
$$\mathbf{r} = \mathbf{i} + \mathbf{k} + \mu(-2\mathbf{i} + 4\mathbf{j} - 2\mathbf{k})$$

Thus l_1 passes through a point A with position vector $\mathbf{a} = 3\mathbf{j} + 2\mathbf{k}$ and is parallel to the vector $\mathbf{p} = \mathbf{i} - 2\mathbf{j} + \mathbf{k}$.
Similarly l_2 passes through a point B with position vector $\mathbf{b} = \mathbf{i} + \mathbf{k}$ and is parallel to the vector $\mathbf{q} = -2\mathbf{i} + 4\mathbf{j} - 2\mathbf{k}$.
Since $\mathbf{q} = -2\mathbf{p}$, the vectors $\mathbf{p}$ and $\mathbf{q}$ are parallel.
Hence the lines l_1 and l_2 are parallel or coincident.
Since $\mathbf{AB} = \mathbf{b} - \mathbf{a} = \mathbf{i} - 3\mathbf{j} - \mathbf{k}$, which is not a scalar multiple of $\mathbf{p}$, the line through A and B is not parallel to $\mathbf{p}$.
Hence the lines l_1 and l_2 must be distinct parallel lines.

Example 6 Show that the lines with vector equations $\mathbf{r} = 2\lambda\mathbf{i} - 3\mathbf{j} + (\lambda - 2)\mathbf{k}$ and $\mathbf{r} = (\mu + 1)\mathbf{i} + (2 - \mu)\mathbf{j} + (2\mu - 5)\mathbf{k}$ do not intersect.

At any point of intersection

$$2\lambda\mathbf{i} - 3\mathbf{j} + (\lambda - 2)\mathbf{k} = (\mu + 1)\mathbf{i} + (2 - \mu)\mathbf{j} + (2\mu - 5)\mathbf{k}$$

Thus

$$2\lambda = \mu + 1 \qquad (1)$$
$$-3 = 2 - \mu \qquad (2)$$
$$\lambda - 2 = 2\mu - 5 \qquad (3)$$

From (2), $\mu = 5$
Using (1), $2\lambda = 5 + 1 \quad \therefore \quad \lambda = 3$

Testing these values in (3):

$$\lambda - 2 = 1, \quad 2\mu - 5 = 5$$

$\therefore$ no values of λ and μ can be found which satisfy all three equations.
Hence the given lines do not intersect.

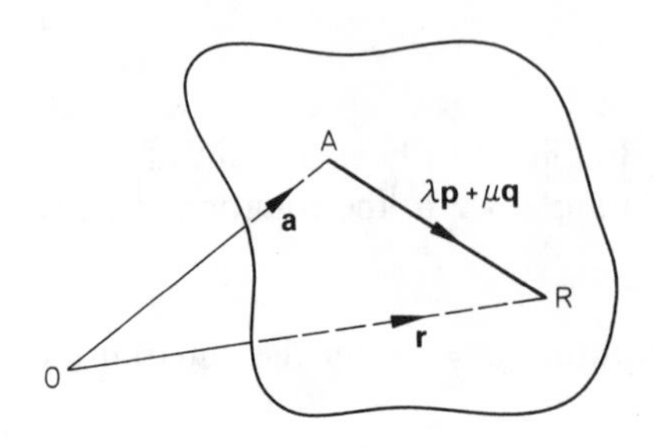

Consider now a plane which passes through a point A with position vector $\mathbf{a}$ and which contains non-parallel vectors $\mathbf{p}$ and $\mathbf{q}$. Let $\mathbf{r}$ be the position vector of a point R in the plane.

Since **AR** lies in the plane, there are scalars λ and μ such that

$$\mathbf{AR} = \lambda\mathbf{p} + \mu\mathbf{q}$$

$\therefore$
$$\mathbf{r} - \mathbf{a} = \lambda\mathbf{p} + \mu\mathbf{q}$$

i.e.
$$\mathbf{r} = \mathbf{a} + \lambda\mathbf{p} + \mu\mathbf{q}.$$

This equation is called the *vector equation* of the plane in *parametric form.*

Example 7 Write down the vector equation of the plane through the point with position vector $\mathbf{i} - 2\mathbf{k}$ which is parallel to the vectors $\mathbf{i} + \mathbf{j}$ and $\mathbf{k}$.

The vector equation of the given plane is

$$\mathbf{r} = \mathbf{i} - 2\mathbf{k} + \lambda(\mathbf{i} + \mathbf{j}) + \mu\mathbf{k}$$

i.e. $\mathbf{r} = (\lambda + 1)\mathbf{i} + \lambda\mathbf{j} + (\mu - 2)\mathbf{k}$.

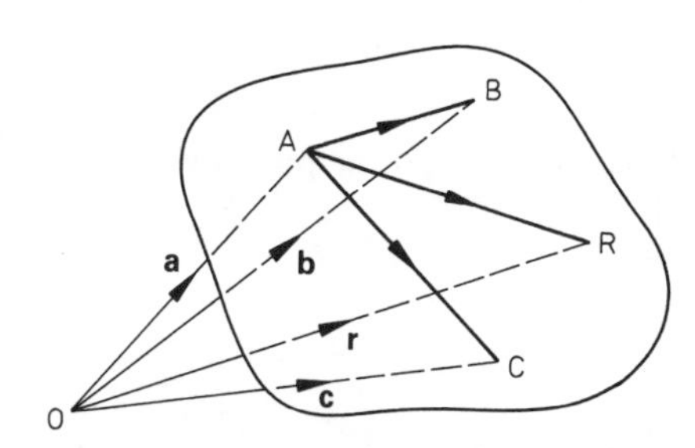

Suppose that $\mathbf{r}$ is the position vector of a point R which lies in the plane containing three non-collinear points A, B and C with position vectors $\mathbf{a}, \mathbf{b}$ and $\mathbf{c}$ respectively.

Since **AR** lies in the plane containing **AB** and **AC**, there are scalars λ and μ such that

$$\mathbf{AR} = \lambda\mathbf{AB} + \mu\mathbf{AC}$$

$\therefore$
$$\mathbf{r} - \mathbf{a} = \lambda(\mathbf{b} - \mathbf{a}) + \mu(\mathbf{c} - \mathbf{a})$$

i.e.
$$\mathbf{r} = (1 - \lambda - \mu)\mathbf{a} + \lambda\mathbf{b} + \mu\mathbf{c}$$

This is the vector equation of the plane through given points A, B and C.

Example 8 Find a vector equation of the plane through the points $A(1, 0, 0)$, $B(2, -6, 1)$ and $C(-3, 0, 4)$. Hence find the coordinates of the point of intersection of the plane ABC and the line with vector equation $\mathbf{r} = 2t\mathbf{i} + (1 - 5t)\mathbf{j} + (t - 2)\mathbf{k}$.

The points A, B and C have position vectors $\mathbf{i}, 2\mathbf{i} - 6\mathbf{j} + \mathbf{k}$ and $-3\mathbf{i} + 4\mathbf{k}$ respectively.

$\therefore$ the vector equation of the plane ABC is

$$\mathbf{r} = (1 - \lambda - \mu)\mathbf{i} + \lambda(2\mathbf{i} - 6\mathbf{j} + \mathbf{k}) + \mu(-3\mathbf{i} + 4\mathbf{k})$$

i.e. $\mathbf{r} = (1 + \lambda - 4\mu)\mathbf{i} - 6\lambda\mathbf{j} + (\lambda + 4\mu)\mathbf{k}$.

The point of intersection of this plane with the line $\mathbf{r} = 2t\mathbf{i} + (1 - 5t)\mathbf{j} + (t - 2)\mathbf{k}$ is given by

$$1 + \lambda - 4\mu = 2t \quad (1)$$
$$-6\lambda \quad = -5t + 1 \quad (2)$$
$$\lambda + 4\mu = t - 2 \quad (3)$$

Adding (1) to (3), $\quad 1 + 2\lambda = 3t - 2$

$\therefore \quad 2\lambda = 3t - 3 \qquad (4)$

Substituting in (2), $\quad -3(3t - 3) = -5t + 1$

i.e. $\quad -9t + 9 = -5t + 1 \qquad \therefore \quad t = 2$

Substituting in the equation of the line, $\mathbf{r} = 4\mathbf{i} - 9\mathbf{j}$.
Hence the coordinates of the point of intersection are $(4, -9, 0)$.

Exercise 7.2

1. Find the vector equation of the line with direction vector $\mathbf{d}$ which passes through the point with position vector $\mathbf{a}$ given that
(a) $\mathbf{a} = \mathbf{i} + 2\mathbf{j} - \mathbf{k}$, $\mathbf{d} = 3\mathbf{i} - \mathbf{k}$,
(b) $\mathbf{a} = 4\mathbf{i} - 3\mathbf{k}$, $\mathbf{d} = \mathbf{i} - 3\mathbf{j} + 3\mathbf{k}$.

2. Find the vector equation of the line which passes through the points with
(a) position vectors $2\mathbf{i} - 3\mathbf{j} + \mathbf{k}$ and $-2\mathbf{i} + \mathbf{j} + \mathbf{k}$,
(b) position vectors $\mathbf{i} + 5\mathbf{j}$ and $3\mathbf{i} - \mathbf{j} + 2\mathbf{k}$,
(c) coordinates $(0, 6, -8)$ and $(5, -7, 2)$,
(d) coordinates $(0, 0, 0)$ and $(5, -2, 3)$.

3. The vertices A, B and C of a triangle have position vectors $\mathbf{a}, \mathbf{b}$ and $\mathbf{c}$ respectively. The point P divides BC internally in the ratio 2:1 and the point Q divides AC externally in the ratio 2:1. Find vector equations for the lines AP and BQ. Hence find the position vector of the point of intersection of the lines AP and BQ.

4. The position vectors of points A and B with respect to an origin O are $\mathbf{a}$ and $\mathbf{b}$ respectively. P is the mid-point of OA and $APBQ$ is a parallelogram. The lines AB and OQ intersect at T. Find the position vectors of P, Q and T.

5. The position vectors relative to an origin O of the points A and B are $\mathbf{a}$ and $\mathbf{b}$ respectively. The point P on AB is such that $3AP = PB$. The point Q on OP produced is such that AQ is parallel to OB. Find the position vector of the point of intersection of the lines OA and BQ.

6. The position vectors of four points A, B, C, D are $\mathbf{a} = 5\mathbf{j}$, $\mathbf{b} = -6\mathbf{i} + \mathbf{j}$, $\mathbf{c} = 6\mathbf{i} - \mathbf{j}$, $\mathbf{d} = -3\mathbf{i} + 5\mathbf{j}$ respectively. The lines AB and CD intersect at a point X. Find the ratio $AX:XB$.

7. Decide whether the following pairs of straight lines are intersecting, parallel or skew. If the lines intersect give the position vector of their point of intersection.
(a) $\mathbf{r} = (1 - \lambda)\mathbf{i} + 3\lambda\mathbf{j} + (2\lambda + 5)\mathbf{k}$,
$\mathbf{r} = 2\mu\mathbf{i} + (\mu - 4)\mathbf{j} + 3\mathbf{k}$.
(b) $\mathbf{r} = -7\mathbf{i} + 4\mathbf{j} + 3\mathbf{k} + \lambda(2\mathbf{i} + \mathbf{k})$,
$\mathbf{r} = 5\mathbf{i} + 8\mathbf{j} + \mathbf{k} + \mu(-4\mathbf{i} + 3\mathbf{k})$.
(c) $\mathbf{r} = 2\lambda\mathbf{i} + 3\mathbf{j} - (2\lambda + 5)\mathbf{k}$,
$\mathbf{r} = (2 - 3\mu)\mathbf{i} - 4\mathbf{j} + 3\mu\mathbf{k}$.
(d) $\mathbf{r} = 2\lambda\mathbf{i} + (4\lambda + 1)\mathbf{j} + (5 - 3\lambda)\mathbf{k}$,
$\mathbf{r} = (\mu + 4)\mathbf{i} + (5\mu + 3)\mathbf{j} + (\mu - 6)\mathbf{k}$.

8. Decide which of the points A, B and C, with position vectors $\mathbf{a} = \mathbf{i} - 2\mathbf{j}$, $\mathbf{b} = 3\mathbf{i} - \mathbf{j} - \mathbf{k}$ and $\mathbf{c} = \mathbf{i} + \mathbf{j} + 2\mathbf{k}$ respectively, lie on the following lines and planes.
(a) $\mathbf{r} = (2\lambda + 5)\mathbf{i} + \lambda\mathbf{j} - (\lambda + 2)\mathbf{k}$,
(b) $\mathbf{r} = 2\mathbf{i} - 3\mathbf{j} + 2\mathbf{k} + \lambda(\mathbf{i} + 2\mathbf{j} - 3\mathbf{k})$,
(c) $\mathbf{r} = (2\lambda - \mu - 2)\mathbf{i} + (1 - \lambda + 2\mu)\mathbf{j} + \lambda\mathbf{k}$,
(d) $\mathbf{r} = -3\mathbf{i} - \mathbf{j} + 4\mathbf{k} + \lambda(2\mathbf{i} + \mathbf{j} - \mathbf{k}) + \mu(3\mathbf{j} + 2\mathbf{k})$.

9. Write down in parametric form the vector equations of the planes through the given points parallel to the given pairs of vectors.
(a) $(1, -2, 0)$; $\mathbf{i} + 3\mathbf{j}$ and $-\mathbf{j} + 2\mathbf{k}$,
(b) the origin; $2\mathbf{i} - \mathbf{j}$ and $-\mathbf{i} + 2\mathbf{j} - 7\mathbf{k}$,
(c) $(3, 1, -1)$; $\mathbf{j}$ and $\mathbf{i} + \mathbf{j} + \mathbf{k}$.

10. Find a vector equation for the plane passing through the points with position vectors $2\mathbf{k}, \mathbf{i} - 3\mathbf{j} + \mathbf{k}$ and $5\mathbf{i} + 2\mathbf{j}$.

11. Write down direction vectors for the lines l_1 and l_2 with equations $\mathbf{r} = (2 - \lambda)\mathbf{i} + 5\mathbf{j} + 3\lambda\mathbf{k}$ and $\mathbf{r} = 4\mathbf{i} + 5\mu\mathbf{j} - 2\mathbf{k}$ respectively. By writing down the position vector of a particular point on l_1, find a vector equation for the plane which contains l_1 and is parallel to l_2.

12. Find a vector equation for the plane through the points $A(1, 0, -2)$ and $B(3, -1, 1)$ which is parallel to the line with vector equation

$$\mathbf{r} = 3\mathbf{i} + (2\lambda - 1)\mathbf{j} + (5 - \lambda)\mathbf{k}.$$

Hence find the coordinates of the point of intersection of this plane and the line $\mathbf{r} = \mu\mathbf{i} + (5 - \mu)\mathbf{j} + (2\mu - 7)\mathbf{k}$.

13. The position vectors of three non-collinear points A, B and C, with respect to an origin O, are $\mathbf{a}, \mathbf{b}$ and $\mathbf{c}$ respectively. Given that O does not lie in the plane ABC, show that
(a) if the point P with position vector $\mathbf{p} = \lambda\mathbf{a} + \mu\mathbf{b}$ lies on the line AB, then $\lambda + \mu = 1$;
(b) if the point Q with position vector $\mathbf{q} = \alpha\mathbf{a} + \beta\mathbf{b} + \gamma\mathbf{c}$ lies in the plane ABC, then $\alpha + \beta + \gamma = 1$.

7.3 Forces and their lines of action

In this section we consider the application of vector methods to systems of forces given in vector form, often in terms of unit vectors $\mathbf{i}, \mathbf{j}$ and $\mathbf{k}$.

[Note that unless otherwise stated we will assume that vector quantities are measured in the appropriate SI units.]

Example 1 A force $\mathbf{P}$ of magnitude 28 N acts at the point A with position vector $2\mathbf{j} + \mathbf{k}$ in the direction of the vector $3\mathbf{i} - 2\mathbf{j} + 6\mathbf{k}$. A second force $\mathbf{Q}$ of magnitude 27 N also acts at A, but in the direction of the vector $\mathbf{i} - 4\mathbf{j} - 8\mathbf{k}$. Find the magnitude of the resultant of the forces $\mathbf{P}$ and $\mathbf{Q}$. Obtain a vector equation for its line of action.

$$|3\mathbf{i} - 2\mathbf{j} + 6\mathbf{k}| = \sqrt{\{3^2 + (-2)^2 + 6^2\}} = \sqrt{\{9 + 4 + 36\}} = 7$$
$$|\mathbf{i} - 4\mathbf{j} - 8\mathbf{k}| = \sqrt{\{1^2 + (-4)^2 + (-8)^2\}} = \sqrt{\{1 + 16 + 64\}} = 9$$

$\therefore$ the unit vectors in the directions of $\mathbf{P}$ and $\mathbf{Q}$ are $\frac{1}{7}(3\mathbf{i} - 2\mathbf{j} + 6\mathbf{k})$ and $\frac{1}{9}(\mathbf{i} - 4\mathbf{j} - 8\mathbf{k})$ respectively.

Thus $\quad \mathbf{P} = 28 \times \frac{1}{7}(3\mathbf{i} - 2\mathbf{j} + 6\mathbf{k}) = 12\mathbf{i} - 8\mathbf{j} + 24\mathbf{k}$
and $\quad \mathbf{Q} = 27 \times \frac{1}{9}(\mathbf{i} - 4\mathbf{j} - 8\mathbf{k}) = 3\mathbf{i} - 12\mathbf{j} - 24\mathbf{k}$
$\therefore \quad \mathbf{P} + \mathbf{Q} = 15\mathbf{i} - 20\mathbf{j} = 5(3\mathbf{i} - 4\mathbf{j})$
$\therefore \quad |\mathbf{P} + \mathbf{Q}| = 5\sqrt{\{3^2 + (-4)^2\}} = 5\sqrt{\{9 + 16\}} = 25$

Hence the resultant of $\mathbf{P}$ and $\mathbf{Q}$ acts in the direction of the vector $3\mathbf{i} - 4\mathbf{j}$ and has magnitude 25 N.
Since $\mathbf{P}$ and $\mathbf{Q}$ act at A, their resultant must also act at A. Thus a vector equation for its line of action is

$$\mathbf{r} = 2\mathbf{j} + \mathbf{k} + \lambda(3\mathbf{i} - 4\mathbf{j})$$
i.e. $\quad \mathbf{r} = 3\lambda\mathbf{i} + (2 - 4\lambda)\mathbf{j} + \mathbf{k}.$

The approach of Example 1 can also be used to find the magnitude and direction of the resultant of a set of coplanar forces with different points of application. However, to find the line of action of the resultant we must use the principle of moments introduced in §5.1.

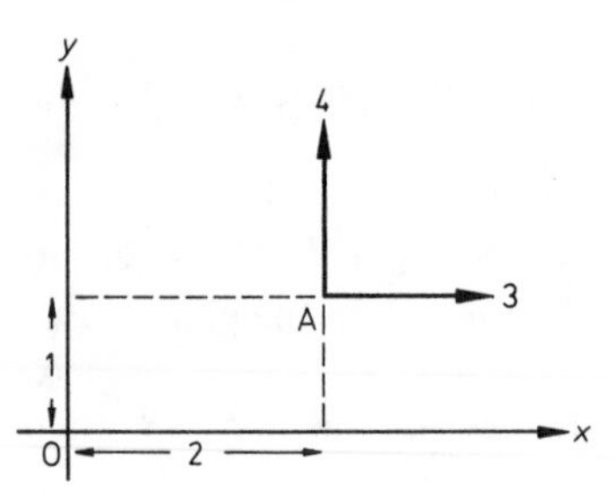

To illustrate how the moments of individual forces are calculated, we suppose that a force $\mathbf{F} = 3\mathbf{i} + 4\mathbf{j}$ acts at the point A with position vector $2\mathbf{i} + \mathbf{j}$.

Taking anti-clockwise moments as positive, the moment of $\mathbf{F}$ about O

$= $ (moment about O of the force $3\mathbf{i}$) $+$ (moment about O of the force $4\mathbf{j}$)
$= -(3 \times 1) + (4 \times 2)$ N m $= 5$ N m.

Example 2 Coplanar forces $\mathbf{F}_1 = 5\mathbf{i} + 2\mathbf{j}$, $\mathbf{F}_2 = 4\mathbf{i} + 4\mathbf{j}$ and $\mathbf{F}_3 = 3\mathbf{i} - \mathbf{j}$ act at points A_1, A_2 and A_3 with position vectors $\mathbf{i} - 3\mathbf{j}, 4\mathbf{i} + 2\mathbf{j}$ and $-2\mathbf{i} + \mathbf{j}$ respectively. Find the magnitude of the resultant of these forces and the Cartesian equation of its line of action.

$$\mathbf{F}_1 + \mathbf{F}_2 + \mathbf{F}_3 = (5 + 4 + 3)\mathbf{i} + (2 + 4 - 1)\mathbf{j} = 12\mathbf{i} + 5\mathbf{j}$$
$$\therefore \quad |\mathbf{F}_1 + \mathbf{F}_2 + \mathbf{F}_3| = \sqrt{(12^2 + 5^2)} = \sqrt{(144 + 25)} = 13.$$

Hence the magnitude of the resultant of the three forces is 13 N.

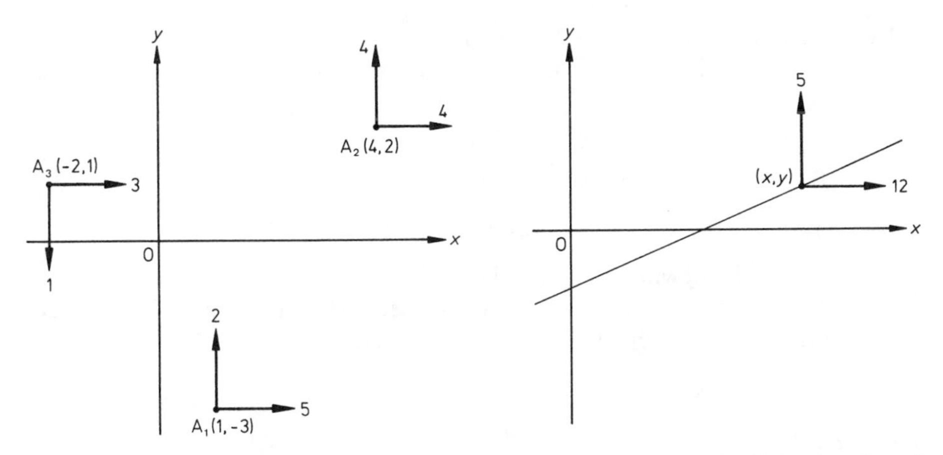

Let the point with position vector $x\mathbf{i} + y\mathbf{j}$ be some point on the line of action of the resultant, then the anti-clockwise moment of the resultant about O is $5x - 12y$.

Taking moments anti-clockwise about O for the original forces,

$$\text{moment of } \mathbf{F}_1 = 5 \times 3 + 2 \times 1 = 17$$
$$\text{moment of } \mathbf{F}_2 = -4 \times 2 + 4 \times 4 = 8$$
$$\text{moment of } \mathbf{F}_3 = -3 \times 1 + 1 \times 2 = -1$$

$\therefore$ using the principle of moments,

$$5x - 12y = 17 + 8 - 1$$

Hence the Cartesian equation of the line of action of the resultant is $5x - 12y = 24$.

The *ratio theorem* derived in §7.1 can be used to find resultant forces.

> If two forces acting at a point O are represented in magnitude and direction by $\lambda\mathbf{OP}$ and $\mu\mathbf{OQ}$, then their resultant is represented by $(\lambda + \mu)\mathbf{OR}$, where R is the point which divides PQ in the ratio $\mu:\lambda$.

Example 3 Forces represented by **AB**, 2**AC** and 4**BC** act along the sides of a triangle ABC. Find their resultant.

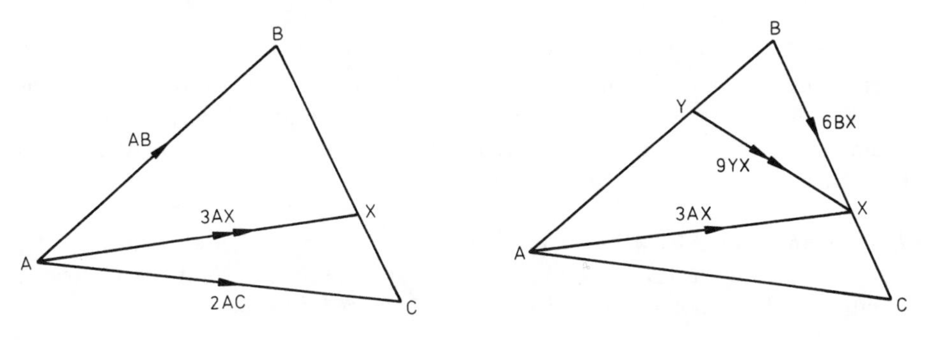

By the ratio theorem, $\mathbf{AB} + 2\mathbf{AC} = 3\mathbf{AX}$,
where X is the point which divides BC in the ratio 2:1.

Since $BX:XC = 2:1$, we have $BC = \frac{3}{2}BX$,

$\therefore \quad \mathbf{AB} + 2\mathbf{AC} + 4\mathbf{BC} = 3\mathbf{AX} + 6\mathbf{BX}$.

By the ratio theorem, $3\mathbf{AX} + 6\mathbf{BX} = 9\mathbf{YX}$,

where Y is the point which divides AB in the ratio 6:3 i.e. 2:1. Hence the resultant of the three forces acts through X and is represented by $9\mathbf{YX}$, where X divides BC in the ratio 2:1 and Y divides AB in the ratio 2:1.

Exercise 7.3

1. Forces of magnitude 10 N and 5 N act in the directions of the vectors $3\mathbf{i} - 4\mathbf{j}$ and $4\mathbf{i} + 3\mathbf{j}$ respectively. Express each force as a vector in terms of $\mathbf{i}$ and $\mathbf{j}$. Hence find the resultant of the forces as a vector.

2. A force $\mathbf{R}$ of magnitude $4\sqrt{2}$ N acting in the direction of the vector $\mathbf{i} - \mathbf{j}$ is the resultant of two forces $\mathbf{P}$ and $\mathbf{Q}$ which act in the directions of the vectors $\mathbf{i} + \mathbf{j}$ and $\mathbf{i} - 7\mathbf{j}$ respectively. Find the magnitudes of $\mathbf{P}$ and $\mathbf{Q}$.

3. In a parallelogram $OABC$ the position vectors of A and B relative to O are $\mathbf{a} = 4\mathbf{i} + 3\mathbf{j}$ and $\mathbf{b} = 3\mathbf{j} + 3\mathbf{k}$ respectively. Forces of magnitude 20 N, $9\sqrt{2}$ N and 5 N act along OA, OB and OC respectively. Find the magnitude of their resultant.

4. In each of the following cases obtain a vector equation for the line of action of the resultant of forces $\mathbf{F}_1$, $\mathbf{F}_2$ and $\mathbf{F}_3$ acting at a point with position vector $\mathbf{a}$.
(a) $\mathbf{F}_1 = 2\mathbf{i} - 3\mathbf{j}$, $\mathbf{F}_2 = \mathbf{i} + 5\mathbf{j}$, $\mathbf{F}_3 = -\mathbf{i} + 2\mathbf{j}$; $\mathbf{a} = \mathbf{i} + \mathbf{j}$.
(b) $\mathbf{F}_1 = -3\mathbf{i} + 2\mathbf{k}$, $\mathbf{F}_2 = \mathbf{i} + \mathbf{j} + \mathbf{k}$, $\mathbf{F}_3 = 4\mathbf{j}$; $\mathbf{a} = 2\mathbf{j} - \mathbf{k}$.
(c) $\mathbf{F}_1 = \mathbf{i} - 5\mathbf{j} + 8\mathbf{k}$, $\mathbf{F}_2 = 2\mathbf{i} - \mathbf{j}$, $\mathbf{F}_3 = -3\mathbf{i} + \mathbf{k}$; $\mathbf{a} = 3\mathbf{i}$.

5. Forces $\mathbf{P} = 3\mathbf{i} + \mathbf{j} - 2\mathbf{k}$ and $\mathbf{Q} = \mathbf{i} - \mathbf{j} + 3\mathbf{k}$ act through points with position vectors $2\mathbf{i} - \mathbf{j}$ and $9\mathbf{i} - \mathbf{k}$ respectively. Write down vector equations for the lines of action of $\mathbf{P}$ and $\mathbf{Q}$. Hence find the position vector of the point through which both forces act. Find also a vector equation for the line of action of the resultant of $\mathbf{P}$ and $\mathbf{Q}$.

6. Forces $\mathbf{F}_1 = \mathbf{i} + 7\mathbf{j} + 4\mathbf{k}$ and $\mathbf{F}_2 = 5\mathbf{i} + 3\mathbf{j} - 4\mathbf{k}$ act through the origin O. A third force $\mathbf{F}_3 = \mathbf{i} - 4\mathbf{j} + 6\mathbf{k}$ acts through the point A with position vector $\mathbf{a} = 3\mathbf{i} + 5\mathbf{j}$. Show that the resultant of $\mathbf{F}_1$ and $\mathbf{F}_2$ also acts through A. Find the magnitude of the resultant of all three forces and obtain a vector equation for its line of action.

7. Coplanar forces $\mathbf{F}_1 = \mathbf{i} - 2\mathbf{j}$, $\mathbf{F}_2 = 3\mathbf{i} + 4\mathbf{j}$, $\mathbf{F}_3 = -5\mathbf{i} + \mathbf{j}$ act at points A_1, A_2, A_3 with position vectors $4\mathbf{i} + 5\mathbf{j}$, $3\mathbf{i} - 2\mathbf{j}$, $\mathbf{i} + 4\mathbf{j}$ respectively. Find the Cartesian equation of the line of action of the resultant of these forces.

8. Coplanar forces $\mathbf{F}_1 = 6\mathbf{i} + 2\mathbf{j}$, $\mathbf{F}_2 = -\mathbf{i} + 7\mathbf{j}$, $\mathbf{F}_3 = 4\mathbf{i} - 3\mathbf{j}$ act at points A_1, A_2, A_3 with position vectors $\mathbf{i} - \mathbf{j}$, $4\mathbf{j}$, $2\mathbf{i} + 3\mathbf{j}$ respectively. Find the magnitude of the resultant and the tangent of the angle it makes with the vector $\mathbf{i}$. Find also the Cartesian equation of its line of action.

9. The following sets of forces act along the edges of a triangle ABC. In each case find the resultant force, using a clear diagram to illustrate your answer.
(a) $2\mathbf{AB}, 3\mathbf{AC}, 6\mathbf{BC}$,
(b) $\mathbf{CB}, 3\mathbf{AB}, 5\mathbf{CA}$,
(c) $2\mathbf{BC}, \mathbf{BA}, 2\mathbf{AC}$,
(d) $6\mathbf{AB}, 4\mathbf{CA}, 3\mathbf{CB}$.

10. Three forces acting at a point P are represented by $\mathbf{PA}$, $\mathbf{PB}$ and $\mathbf{PC}$. Show that their resultant is given by $3\mathbf{PG}$, where G is the centroid of triangle ABC.

11. Forces represented by $3\mathbf{AB}$, $2\mathbf{CB}$, $4\mathbf{CD}$ and $6\mathbf{AD}$ act along the sides of a square $ABCD$. Show that their resultant is $15\mathbf{XY}$, where X divides AC in the ratio 2:3 and Y divides BD in the ratio 2:1.

12. Three forces represented by $\mathbf{PQ}$, $\mathbf{QR}$ and $\mathbf{RP}$ act along the sides of a triangle PQR. Show that these forces are equivalent to a couple whose moment is represented by twice the area of the triangle.

7.4 Use of vectors in kinematics and dynamics

Suppose that a particle P moving with constant velocity $\mathbf{v}$ has position vector $\mathbf{r}_0$ initially and position vector $\mathbf{r}$ at time t.

It follows that $$\mathbf{r} = \mathbf{r}_0 + t\mathbf{v}.$$

This equation gives the position vector of P at any instant and can be taken as the vector equation of the straight line path of the particle.

The next example shows the use of such equations in a problem concerning the *relative motion* of two ships.

Example 1 At noon ships A and B have position vectors $3\mathbf{i} + 2\mathbf{j}$ and $\mathbf{i} - 4\mathbf{j}$ respectively and constant velocities $4\mathbf{i} + 3\mathbf{j}$ and $3\mathbf{i} + 5\mathbf{j}$ respectively, where distances are measured in kilometres and times in hours. Find vectors to represent (a) the velocity of A relative to B, (b) the displacement of A relative to B, t hours after noon. Find also the least distance between the ships.

(a) The velocity of A relative to B

$$= (\text{velocity of } A) - (\text{velocity of } B)$$
$$= (4\mathbf{i} + 3\mathbf{j}) - (3\mathbf{i} + 5\mathbf{j}) = \mathbf{i} - 2\mathbf{j}.$$

(b) The position vector of A after t hours is given by

$$\mathbf{r}_A = 3\mathbf{i} + 2\mathbf{j} + t(4\mathbf{i} + 3\mathbf{j})$$

The position vector of B after t hours is given by

$$\mathbf{r}_B = \mathbf{i} - 4\mathbf{j} + t(3\mathbf{i} + 5\mathbf{j})$$

$\therefore$ the displacement of A relative to B after t hours is

$$\mathbf{r}_A - \mathbf{r}_B = 2\mathbf{i} + 6\mathbf{j} + t(\mathbf{i} - 2\mathbf{j})$$

i.e. $\mathbf{r}_A - \mathbf{r}_B = (2 + t)\mathbf{i} + (6 - 2t)\mathbf{j}$.

The distance between the ships

$$\begin{aligned} = |\mathbf{r}_A - \mathbf{r}_B| &= \sqrt{\{(2 + t)^2 + (6 - 2t)^2\}} \\ &= \sqrt{\{4 + 4t + t^2 + 36 - 24t + 4t^2\}} \\ &= \sqrt{\{5t^2 - 20t + 40\}} \\ &= \sqrt{\{5(t - 2)^2 + 20\}} \end{aligned}$$

Since $(t - 2)^2$ is never negative, this expression takes its least value when $t = 2$. Hence the least distance between the ships is $\sqrt{20}$ km i.e. $5\sqrt{2}$ km.

[Note that the least distance can also be found by writing

$$f(t) = 5t^2 - 20t + 40$$

Differentiating, $\quad f'(t) = 10t - 20, \quad f''(t) = 10$

$\therefore \quad f'(t) = 0 \Leftrightarrow 10t - 20 = 0 \Leftrightarrow t = 2.$

Since $f(2) = 20$ and $f''(2) > 0$, the function $f(t)$ has a minimum value of 20.]

In general, if a moving particle P has position vector $\mathbf{r}$ at time t, then its velocity $\mathbf{v}$ and acceleration $\mathbf{a}$ are given by

$$\mathbf{v} = \frac{d\mathbf{r}}{dt} \quad \text{and} \quad \mathbf{a} = \frac{d\mathbf{v}}{dt} = \frac{d^2\mathbf{r}}{dt^2}$$

or, using a dot to indicate differentiation with respect to time,

$$\mathbf{v} = \dot{\mathbf{r}} \quad \text{and} \quad \mathbf{a} = \ddot{\mathbf{r}}.$$

Example 2 A particle P has position vector $\mathbf{r}$ at time t and moves in a circle with constant angular velocity ω. The centre of the circle is the point C with position vector $\mathbf{c} = 2\mathbf{i} + 3\mathbf{j}$, and when $t = 0$, $\mathbf{CP} = 5\mathbf{i}$. Find an expression for $\mathbf{r}$ and hence for the velocity $\mathbf{v}$ of the particle at time t. Find also the time at which P first moves parallel to the vector $\sqrt{3}\mathbf{i} - \mathbf{j}$.

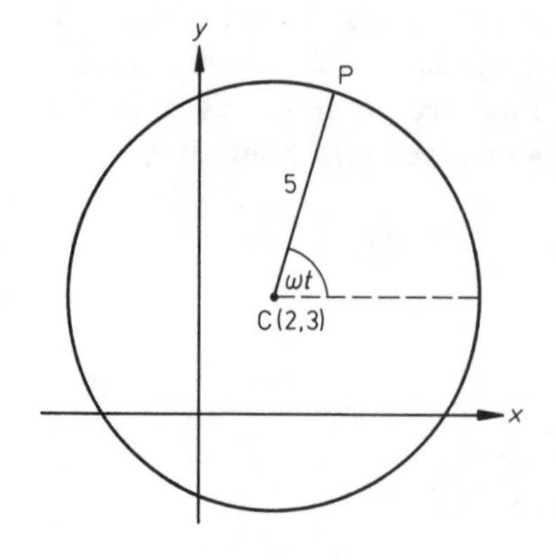

Since P moves on a circle with centre C, the vector $\mathbf{CP}$ has constant magnitude and rotates with constant angular velocity ω. Initially $\mathbf{CP} = 5\mathbf{i}$ and $|\mathbf{CP}| = |5\mathbf{i}| = 5$. Thus, at time t, the vector **CP** has magnitude 5 units and makes an angle ωt with the direction of **i**. It follows that

$$\mathbf{CP} = 5\cos\omega t\mathbf{i} + 5\sin\omega t\mathbf{j}$$
$$\therefore \quad \mathbf{r} - \mathbf{c} = 5\cos\omega t\mathbf{i} + 5\sin\omega t\mathbf{j}$$
$$\therefore \quad \mathbf{r} = 2\mathbf{i} + 3\mathbf{j} + 5\cos\omega t\mathbf{i} + 5\sin\omega t\mathbf{j}$$

Differentiating with respect to t,

$$\mathbf{v} = \dot{\mathbf{r}} = -5\omega\sin\omega t\mathbf{i} + 5\omega\cos\omega t\mathbf{j}.$$

Thus when P is moving parallel to the vector $\sqrt{3}\mathbf{i} - \mathbf{j}$ there is a scalar λ such that

$$-5\omega\sin\omega t\mathbf{i} + 5\omega\cos\omega t\mathbf{j} = \lambda(\sqrt{3}\mathbf{i} - \mathbf{j})$$

i.e. $\quad -5\omega\sin\omega t = \lambda\sqrt{3} \quad$ and $\quad 5\omega\cos\omega t = -\lambda$

$$\therefore \quad \tan\omega t = \frac{\sin\omega t}{\cos\omega t} = \sqrt{3}$$

Hence P first moves parallel to the vector $\sqrt{3}\mathbf{i} - \mathbf{j}$ when $\omega t = \pi/3$ i.e. when $t = \pi/3\omega$.
[Note that at this instant P is not moving in the direction of the vector $\sqrt{3}\mathbf{i} - \mathbf{j}$, but in the opposite direction.]

In problems concerning the forces on moving objects *Newton's second law* is useful. It states that if a particle of mass m moves under the action of a force **F** with acceleration **a**, then

$$\mathbf{F} = m\mathbf{a}.$$

Example 3 A particle of mass 3 kg moves so that its position vector after t seconds is $\mathbf{r} = 8t\mathbf{i} - 5\mathbf{j} + 2t^3\mathbf{k}$. Find (a) the speed of the particle when $t = 1$, (b) the force acting on the particle when $t = 3$.

Let **v** and **a** be the velocity and acceleration of the particle after t seconds, then

$$\mathbf{v} = \dot{\mathbf{r}} = 8\mathbf{i} + 6t^2\mathbf{k}$$
and $\quad \mathbf{a} = \ddot{\mathbf{r}} = 12t\mathbf{k}.$

(a) When $t = 1$, $\mathbf{v} = 8\mathbf{i} + 6\mathbf{k}$

$\therefore \quad v = |8\mathbf{i} + 6\mathbf{k}| = \sqrt{(8^2 + 6^2)} = 10.$

Hence, when $t = 1$, the speed of the particle is $10\,\mathrm{m\,s^{-1}}$.

(b) Let **F** be the force on the particle after t seconds, then by Newton's second law,

$$\mathbf{F} = 3\mathbf{a}.$$

When $t = 3$, $\mathbf{a} = 36\mathbf{k} \qquad \therefore \quad \mathbf{F} = 108\mathbf{k}.$
Hence, when $t = 3$, the force acting on the particle is $108\mathbf{k}$ (in newtons).

Example 4 A particle of unit mass is acted upon at time t by a force **F**, where $\mathbf{F} = 4\mathbf{i} - 6t\mathbf{j}$. At time $t = 0$ the particle is at the point with position vector $\mathbf{i} + \mathbf{j}$ and its velocity is $5\mathbf{i}$. Find the position vector of the particle at time $t = 2$.

Let $\mathbf{r}$, $\mathbf{v}$ and $\mathbf{a}$ be the position vector, velocity and acceleration of the particle at time t.
By Newton's second law, $\mathbf{F} = m\mathbf{a}$.
Since the particle has unit mass, $\mathbf{F} = \mathbf{a}$.

$$\therefore \qquad \mathbf{a} = \frac{d\mathbf{v}}{dt} = 4\mathbf{i} - 6t\mathbf{j}$$

Integrating, $\qquad \mathbf{v} = 4t\mathbf{i} - 3t^2\mathbf{j} + \mathbf{c},$

where $\mathbf{c}$ is a constant vector.
When $t = 0$, $\mathbf{v} = 5\mathbf{i}$, $\quad \therefore \quad \mathbf{c} = 5\mathbf{i}$

$$\therefore \qquad \mathbf{v} = \frac{d\mathbf{r}}{dt} = (4t + 5)\mathbf{i} - 3t^2\mathbf{j}$$

Integrating, $\qquad \mathbf{r} = (2t^2 + 5t)\mathbf{i} - t^3\mathbf{j} + \mathbf{d},$

where $\mathbf{d}$ is a constant vector.
When $t = 0$, $\mathbf{r} = \mathbf{i} + \mathbf{j}$, $\quad \therefore \quad \mathbf{d} = \mathbf{i} + \mathbf{j}$
$\therefore \quad \mathbf{r} = (2t^2 + 5t + 1)\mathbf{i} + (1 - t^3)\mathbf{j}$

Hence, when $t = 2$, the position vector of the particle is

$$\mathbf{r} = (8 + 10 + 1)\mathbf{i} + (1 - 8)\mathbf{j} = 19\mathbf{i} - 7\mathbf{j}.$$

Exercise 7.4

1. Particles A and B are moving with constant velocities $5\mathbf{i} + 2\mathbf{j}$ and $2\mathbf{i} + \mathbf{j}$ respectively. Initially the particle A has position vector $3\mathbf{i} + 4\mathbf{j}$ and the particle B has position vector $5\mathbf{i} + 8\mathbf{j}$. Write down expressions for $\mathbf{r}_A$ and $\mathbf{r}_B$, the position vectors of the particles at time t. Find the vector $\mathbf{AB}$ at time t. Hence find the minimum distance between the particles.

2. Particles P and Q are moving with constant velocities $3\mathbf{i} - 2\mathbf{j}$ and $-\mathbf{i} + 4\mathbf{j}$ respectively. Initially the particle P has position vector $\mathbf{i} + 5\mathbf{j}$ and the particle Q has position vector $3\mathbf{i} + 2\mathbf{j}$. Find the velocity of P relative to Q and the displacement of P relative to Q at time t. Show that the particles collide and state the value of t when the collision occurs.

3. At noon ships A and B have position vectors $10\mathbf{i} + 3\mathbf{j}$ and $12\mathbf{i} - 15\mathbf{j}$ respectively. Both ships move with constant velocity, the velocity of A being $3\mathbf{i} + 8\mathbf{j}$ and the velocity of B relative to A being $-5\mathbf{i} + 4\mathbf{j}$. The unit of distance is the kilometre and the unit of time is the hour. Find an expression for the vector $\mathbf{AB}$, t hours after noon. Hence find the shortest distance between the ships. Find also the position vectors of the ships A and B when they are closest together.

4. A particle P is moving with constant angular velocity ω on a circle of radius 8 units, with centre at the origin. Given that initially P has position vector $8\mathbf{i}$, find its position vector $\mathbf{r}$ at time t. Hence show that $\ddot{\mathbf{r}} = -\omega^2\mathbf{r}$.

5. A particle P moves so that its position vector at time t is given by $\mathbf{r} = \mathbf{i} - 2\mathbf{j} + 3(\mathbf{i}\sin 2t + \mathbf{j}\cos 2t)$. Show that P is moving on a circle, stating the position vector of the centre of the circle and its radius. Find also the speed at which P is moving.

6. Two particles A and B are moving such that, at time t, their position vectors are $\mathbf{r}_A = 3(1 + \cos t)\mathbf{i} + (2\sin t)\mathbf{j}$ and $\mathbf{r}_B = (5 - 6t)\mathbf{i} + (3 + 4t)\mathbf{j}$. Find the value of t when the particles are first moving (a) in the same direction, (b) in opposite directions.

7. The position vector of a particle P at time t is $\mathbf{r} = (1 - t^2)\mathbf{i} + (3t - 5t^2)\mathbf{j}$. Find the time at which P is moving (a) directly towards the origin, (b) directly away from the origin.

8. A particle of mass m moves so that its position vector at time t is $\mathbf{r}$. Find its velocity $\mathbf{v}$ and the force $\mathbf{F}$ acting on it, given that
(a) $\mathbf{r} = 4t^2\mathbf{i} + (3t^3 - 2t)\mathbf{j}$,
(b) $\mathbf{r} = 5\mathbf{i} - 2t^3\mathbf{j} + (t^2 - 1)\mathbf{k}$,
(c) $\mathbf{r} = (2t - t^2)\mathbf{i} + (3\sin 2t)\mathbf{j}$,
(d) $\mathbf{r} = (te^{-t})\mathbf{i} - 2t\mathbf{j}$.

9. A particle of mass 5 kg moves so that its position vector after t seconds is $\mathbf{r} = (\cos 2t)\mathbf{i} + (4\sin 2t + 3)\mathbf{j}$. Find (a) the speed of the particle when $t = \pi/3$, (b) the force acting on the particle when $t = \pi/2$.

10. A particle of mass 2 kg starts from rest at the origin and moves under the action of the force $\mathbf{F}$, where $\mathbf{F} = 8\mathbf{i} + 20\mathbf{j}$. Find the velocity and the position vector of the particle after 3 seconds.

11. A particle of unit mass starts from rest at the point with position vector $2\mathbf{i} - 3\mathbf{j}$. At time t the force acting on the particle is $12t\mathbf{i} - 16\mathbf{j}$. Find the speed of the particle and its position vector at time $t = 2$.

12. A particle of mass 3 kg is acted upon at time t by a force $\mathbf{F}$, where $\mathbf{F} = 6\mathbf{i} - 36t^2\mathbf{j} + 54t\mathbf{k}$. At time $t = 0$ the particle is at the point with position vector $\mathbf{i} - 5\mathbf{j} - \mathbf{k}$ and its velocity is $3(\mathbf{i} + \mathbf{j})$. Find the position vector of the particle at time $t = 1$.

13. A particle of mass 2 kg moves from rest at the origin under the action of two forces. One force $\mathbf{P}$ has magnitude 22 N and acts in the direction of the vector $2\mathbf{i} - 6\mathbf{j} + 9\mathbf{k}$. The other force $\mathbf{Q}$ has magnitude 30 N and acts in the direction of the vector $4\mathbf{j} - 3\mathbf{k}$. Find the acceleration of the particle and the distance it travels in the first 3 seconds of its motion.

14. A particle P is projected vertically upwards from a point O with a speed of $16\,\text{m s}^{-1}$. At the same instant a second particle Q is projected horizontally from a point A, 25 m vertically above O, with speed $12\,\text{m s}^{-1}$. Using $\mathbf{i}$ and $\mathbf{j}$ as unit vectors in the horizontal and upward vertical directions respectively, find expressions for the position vectors of P and Q with respect to O at time t after projection. Hence find the least distance between the particles.

[Take g as $10\,\text{m s}^{-2}$.]

7.5 Scalar product

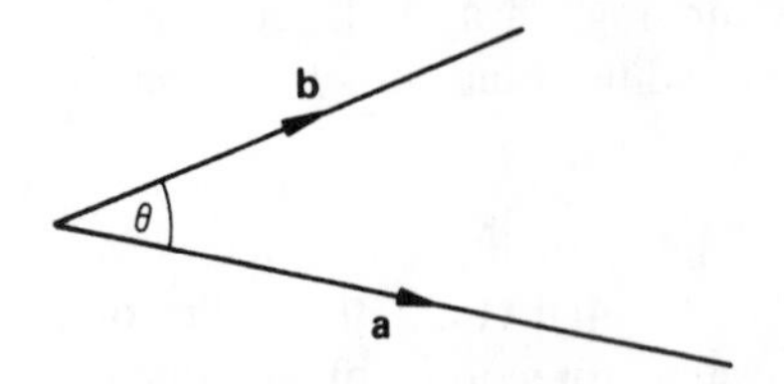

The *scalar product* (or dot product) of two vectors **a** and **b** is defined as the scalar $ab\cos\theta$, where θ is the angle between **a** and **b**.

The scalar product is denoted by $\mathbf{a}.\mathbf{b}$, which is read "**a** dot **b**".

Thus $$\mathbf{a}.\mathbf{b} = ab\cos\theta$$

The scalar product $\mathbf{a}.\mathbf{a}$ is sometimes written as $\mathbf{a}^2$.

[Note that although a product of three real numbers a, b, c can be evaluated by writing $abc = (ab)c$, it is not possible to define a scalar product of three vectors in a similar way. Since $\mathbf{a}.\mathbf{b}$ is a scalar, the expressions $(\mathbf{a}.\mathbf{b}).\mathbf{c}$ and $\mathbf{a}.\mathbf{b}.\mathbf{c}$ are meaningless. Similarly, since $\mathbf{a}.\mathbf{a}$ is a scalar, the expression $\mathbf{a}^3$ cannot be defined as $(\mathbf{a}.\mathbf{a}).\mathbf{a}$.]

The commutative, associative and distributive properties of multiplication of real numbers can be used to establish various properties of the scalar product.

(1) $\mathbf{a}.\mathbf{b} = ab\cos\theta = ba\cos\theta = \mathbf{b}.\mathbf{a}$
Hence the scalar product is *commutative.*

(2) Since we cannot form products such as $(\mathbf{a}.\mathbf{b}).\mathbf{c}$ there is no true associative law for the scalar product. However, for any scalar λ,

$$\lambda(\mathbf{a}.\mathbf{b}) = (\lambda\mathbf{a}).\mathbf{b} = \mathbf{a}.(\lambda\mathbf{b}) = \lambda ab\cos\theta.$$

(3)

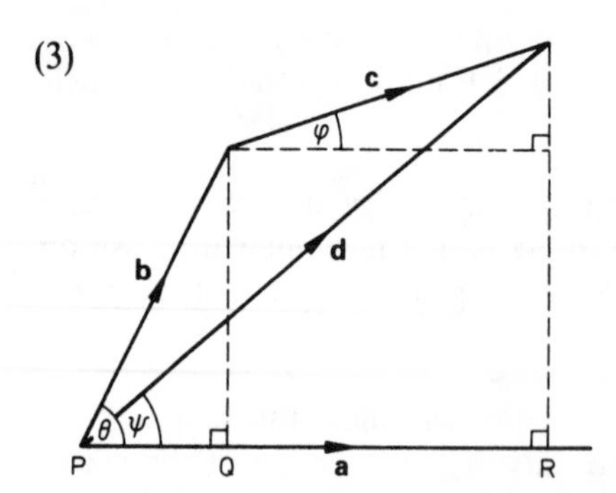

For vectors **a**, **b**, **c** and **d** shown in the diagram,

$$\mathbf{a}.(\mathbf{b}+\mathbf{c}) = \mathbf{a}.\mathbf{d}$$

Since
$$PR = PQ + QR$$
$$d\cos\psi = b\cos\theta + c\cos\phi$$
$$\therefore \quad ad\cos\psi = ab\cos\theta + ac\cos\phi$$
$$\therefore \quad \mathbf{a}.\mathbf{d} = \mathbf{a}.\mathbf{b} + \mathbf{a}.\mathbf{c}$$
$$\therefore \quad \mathbf{a}.(\mathbf{b}+\mathbf{c}) = \mathbf{a}.\mathbf{b} + \mathbf{a}.\mathbf{c}$$

Hence the scalar product is *distributive* over vector addition.

The definition of the scalar product gives rise to further useful properties.

(4) For any vector **a**, $\mathbf{a}.\mathbf{a} = a^2\cos 0° = a^2$.

(5) If non-zero vectors **a** and **b** are perpendicular, then

$$\mathbf{a}.\mathbf{b} = ab\cos 90° = 0.$$

Conversely, if non-zero vectors **a** and **b** are such that $\mathbf{a}.\mathbf{b} = 0$, then **a** is perpendicular to **b**.

These results can be summarised as follows:

(1) $\mathbf{a}.\mathbf{b} = \mathbf{b}.\mathbf{a}$, (2) $\lambda(\mathbf{a}.\mathbf{b}) = (\lambda\mathbf{a}).\mathbf{b} = \mathbf{a}.(\lambda\mathbf{b})$,

(3) $\mathbf{a}.(\mathbf{b}+\mathbf{c}) = \mathbf{a}.\mathbf{b} + \mathbf{a}.\mathbf{c}$, (4) $\mathbf{a}.\mathbf{a} = a^2$.

(5) $\mathbf{a}.\mathbf{b} = 0 \Leftrightarrow \mathbf{a} = \mathbf{0}$ or $\mathbf{b} = \mathbf{0}$ or $\mathbf{a}$ is perpendicular to $\mathbf{b}$.

The scalar product has important applications in geometrical proofs.

Example 1 Prove the cosine rule in the form

$$a^2 = b^2 + c^2 - 2bc\cos A.$$

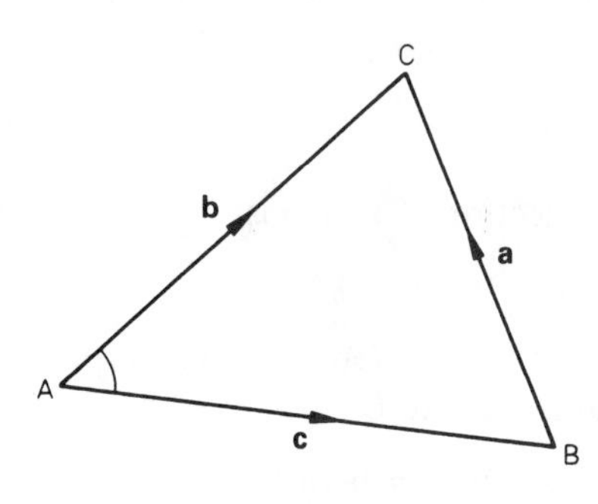

In $\triangle ABC$, let $\mathbf{a} = \overrightarrow{BC}$, $\mathbf{b} = \overrightarrow{AC}$ and $\mathbf{c} = \overrightarrow{AB}$, then

$$\begin{aligned} a^2 &= \mathbf{a}.\mathbf{a} = (\mathbf{b}-\mathbf{c}).(\mathbf{b}-\mathbf{c}) \\ &= \mathbf{b}.(\mathbf{b}-\mathbf{c}) - \mathbf{c}.(\mathbf{b}-\mathbf{c}) \\ &= \mathbf{b}.\mathbf{b} - \mathbf{b}.\mathbf{c} - \mathbf{c}.\mathbf{b} + \mathbf{c}.\mathbf{c} \\ &= b^2 + c^2 - 2\mathbf{b}.\mathbf{c} \end{aligned}$$

$$\therefore \quad a^2 = b^2 + c^2 - 2bc\cos A.$$

Example 2 Show that any angle inscribed in a semi-circle is a right angle.

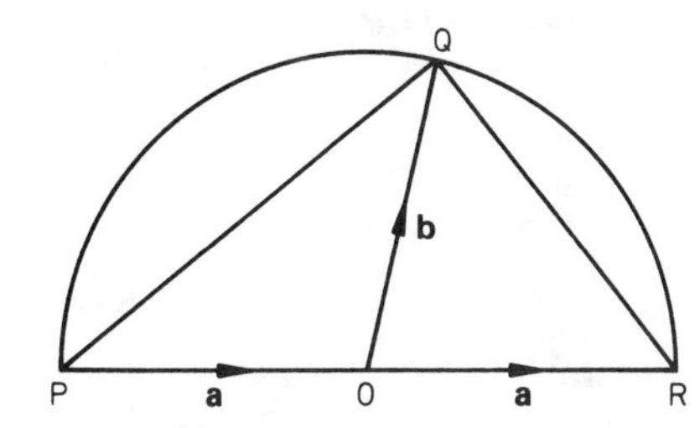

In the diagram, PR is a diameter of a circle with centre O. Let $\overrightarrow{PO} = \overrightarrow{OR} = \mathbf{a}$ and let Q be a point on the circle such that $\overrightarrow{OQ} = \mathbf{b}$, then $\overrightarrow{PQ} = \mathbf{a} + \mathbf{b}$, $\overrightarrow{QR} = \mathbf{a} - \mathbf{b}$.

$$\begin{aligned} \therefore \quad \overrightarrow{PQ}.\overrightarrow{QR} &= (\mathbf{a}+\mathbf{b}).(\mathbf{a}-\mathbf{b}) \\ &= \mathbf{a}.\mathbf{a} - \mathbf{a}.\mathbf{b} + \mathbf{b}.\mathbf{a} - \mathbf{b}.\mathbf{b} = a^2 - b^2 \end{aligned}$$

Since PO and OQ are radii of the circle, $a = b$,

$$\therefore \quad \overrightarrow{PQ}.\overrightarrow{QR} = 0$$

Hence $\overrightarrow{PQ}$ is perpendicular to $\overrightarrow{QR}$ i.e. $\angle PQR = 90°$.

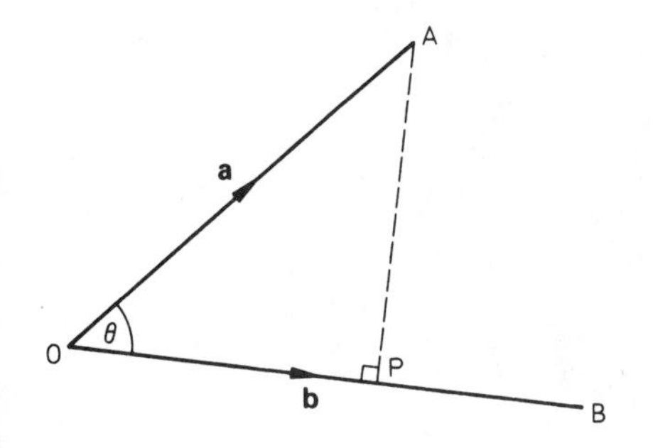

The diagram shows two vectors $\mathbf{a}$ and $\mathbf{b}$ represented by line segments $\overrightarrow{OA}$ and $\overrightarrow{OB}$ respectively. If the vector $\mathbf{a}$ is resolved along and perpendicular to $\mathbf{b}$, then its component in the direction of $\mathbf{b}$ is $a\cos\theta$.

Thus the component of **a** in the direction of **b** is $\frac{\mathbf{a}.\mathbf{b}}{b}$ or $\mathbf{a}.\hat{\mathbf{b}}$, where $\hat{\mathbf{b}}$ is the unit vector in the direction of **b**.

In geometrical work the quantity $\mathbf{a}.\hat{\mathbf{b}}$ is called the *projection* of **a** on **b**. In the diagram the length of this projection is represented by the distance OP.

Example 3 The points A and B have position vectors **a** and **b** with respect to an origin O. Show that the area of triangle OAB is given by $\frac{1}{2}\sqrt{\{a^2b^2 - (\mathbf{a}.\mathbf{b})^2\}}$.

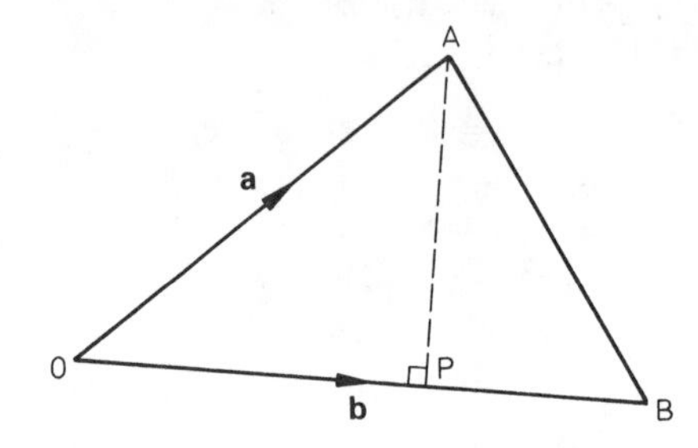

The projection of **a** on **b** is

$$(\mathbf{a}.\mathbf{b})/b,$$

$\therefore$ if P is the foot of the perpendicular from A to OB,

$$OP^2 = (\mathbf{a}.\mathbf{b})^2/b^2$$

Using Pythagoras' theorem in $\triangle OAP$, $OP^2 + AP^2 = OA^2$

$$\therefore \quad AP^2 = OA^2 - OP^2 = a^2 - \frac{(\mathbf{a}.\mathbf{b})^2}{b^2}$$

Hence the area of $\triangle OAB = \frac{1}{2} \times OB \times AP$

$$= \frac{1}{2}\sqrt{\left\{b^2 \times \left(a^2 - \frac{(\mathbf{a}.\mathbf{b})^2}{b^2}\right)\right\}}$$

$$= \frac{1}{2}\sqrt{\{a^2b^2 - (\mathbf{a}.\mathbf{b})^2\}}.$$

When considering the unit vectors **i**, **j** and **k** the definition of scalar product leads to the results

$$\mathbf{i}.\mathbf{i} = \mathbf{j}.\mathbf{j} = \mathbf{k}.\mathbf{k} = 1, \quad \mathbf{i}.\mathbf{j} = \mathbf{j}.\mathbf{k} = \mathbf{k}.\mathbf{i} = 0$$

Thus for vectors $\mathbf{a} = a_1\mathbf{i} + a_2\mathbf{j} + a_3\mathbf{k}$ and $\mathbf{b} = b_1\mathbf{i} + b_2\mathbf{j} + b_3\mathbf{k}$,

$$\begin{aligned}\mathbf{a}.\mathbf{b} &= (a_1\mathbf{i} + a_2\mathbf{j} + a_3\mathbf{k}).(b_1\mathbf{i} + b_2\mathbf{j} + b_3\mathbf{k})\\ &= a_1b_1\mathbf{i}.\mathbf{i} + a_1b_2\mathbf{i}.\mathbf{j} + a_1b_3\mathbf{i}.\mathbf{k}\\ &\quad + a_2b_1\mathbf{j}.\mathbf{i} + a_2b_2\mathbf{j}.\mathbf{j} + a_2b_3\mathbf{j}.\mathbf{k}\\ &\quad + a_3b_1\mathbf{k}.\mathbf{i} + a_3b_2\mathbf{k}.\mathbf{j} + a_3b_3\mathbf{k}.\mathbf{k}\end{aligned}$$

$$\therefore \quad \boxed{\mathbf{a}.\mathbf{b} = a_1b_1 + a_2b_2 + a_3b_3}$$

If the angle between **a** and **b** is θ, then $\mathbf{a}.\mathbf{b} = ab\cos\theta$.

Thus
$$\cos\theta = \frac{\mathbf{a}.\mathbf{b}}{ab} = \frac{a_1b_1 + a_2b_2 + a_3b_3}{ab}$$

It follows that non-zero vectors **a** and **b** are perpendicular if and only if

$$a_1b_1 + a_2b_2 + a_3b_3 = 0.$$

Example 4 Find the angle θ between the vectors $\mathbf{a} = 4\mathbf{i} + 5\mathbf{j} + 3\mathbf{k}$ and $\mathbf{b} = 3\mathbf{i} - 5\mathbf{j} - 4\mathbf{k}$.

$$\mathbf{a}.\mathbf{b} = 4 \times 3 + 5 \times (-5) + 3 \times (-4) = 12 - 25 - 12 = -25$$
$$a = \sqrt{\{4^2 + 5^2 + 3^2\}} = \sqrt{\{16 + 25 + 9\}} = \sqrt{50}$$
$$b = \sqrt{\{3^2 + (-5)^2 + (-4)^2\}} = \sqrt{\{9 + 25 + 16\}} = \sqrt{50}$$
$$\therefore \quad \cos\theta = \frac{\mathbf{a}.\mathbf{b}}{ab} = \frac{-25}{\sqrt{50} \times \sqrt{50}} = -\frac{25}{50} = -\frac{1}{2}$$

Hence the angle between the vectors is 120°.

Example 5 The points A, B and C have position vectors $\mathbf{a} = 5\mathbf{i} + 3\mathbf{j} + 2\mathbf{k}$, $\mathbf{b} = 2\mathbf{i} - \mathbf{j} + 3\mathbf{k}$ and $\mathbf{c} = 7\mathbf{i} - 3\mathbf{j} + 10\mathbf{k}$ respectively. Show that $\angle ABC$ is a right angle.

$$\mathbf{BA} = \mathbf{a} - \mathbf{b} = 3\mathbf{i} + 4\mathbf{j} - \mathbf{k}, \quad \mathbf{BC} = \mathbf{c} - \mathbf{b} = 5\mathbf{i} - 2\mathbf{j} + 7\mathbf{k},$$
$$\therefore \quad \mathbf{BA}.\mathbf{BC} = 3 \times 5 + 4 \times (-2) + (-1) \times 7 = 15 - 8 - 7 = 0$$

Hence BA is perpendicular to BC i.e. $\angle ABC$ is a right angle.

Exercise 7.5

1. Use the properties of the scalar product to simplify
(a) $(\mathbf{p} + \mathbf{q}).(\mathbf{p} + \mathbf{q})$, (b) $(\mathbf{p} + \mathbf{q}).(\mathbf{p} - \mathbf{q})$,
(c) $\mathbf{p}.(\mathbf{q} + \mathbf{r}) - \mathbf{p}.(\mathbf{q} - \mathbf{r})$, (d) $|\mathbf{p} - \mathbf{q}|^2$.

2. Given that **x** is perpendicular to **y**, simplify
(a) $\mathbf{x}.(\mathbf{x} + \mathbf{y})$, (b) $(2\mathbf{x} + \mathbf{y}).(\mathbf{x} - 3\mathbf{y})$.

3. Given that $|\mathbf{a}| = 7, |\mathbf{b}| = 4$ and $\mathbf{a}.\mathbf{b} = 8$, calculate $(\mathbf{a} + \mathbf{b}).(\mathbf{a} + \mathbf{b})$ and $(\mathbf{a} - \mathbf{b}).(\mathbf{a} - \mathbf{b})$. Hence find $|\mathbf{a} + \mathbf{b}|$ and $|\mathbf{a} - \mathbf{b}|$.

4. Given that $|\mathbf{x}| = 5, |\mathbf{y}| = 10$ and $\mathbf{x}.\mathbf{y} = 22$, calculate $|\mathbf{x} + \mathbf{y}|$ and $|\mathbf{x} - \mathbf{y}|$.

5. Show that if **a** and **b** are non-zero parallel vectors, then either $\mathbf{a}.\mathbf{b} = ab$ or $\mathbf{a}.\mathbf{b} = -ab$.

6. Show that if non-zero vectors $\mathbf{a}$ and $\mathbf{b}$ are of the same magnitude, then $(\mathbf{a} + \mathbf{b})$ is perpendicular to $(\mathbf{a} - \mathbf{b})$.

7. Show that if $\mathbf{a}, \mathbf{b}$ and $\mathbf{c}$ are non-zero vectors such that $\mathbf{a}.\mathbf{b} = \mathbf{a}.\mathbf{c}$, then either $\mathbf{b} = \mathbf{c}$ or $\mathbf{a}$ is perpendicular to $(\mathbf{b} - \mathbf{c})$.

8. Vectors $\mathbf{a}, \mathbf{b}$ and $\mathbf{c}$ are such that $\mathbf{a}.\mathbf{c} = 3$ and $\mathbf{b}.\mathbf{c} = 4$. Given that the vector $\mathbf{d} = \mathbf{a} + \lambda\mathbf{b}$ is perpendicular to $\mathbf{c}$, find the value of λ.

9. Coplanar vectors $\mathbf{a}, \mathbf{b}$ and $\mathbf{c}$ are such $|\mathbf{a}| = |\mathbf{b}| = 1$, $\mathbf{a}.\mathbf{b} = \frac{1}{2}$, $\mathbf{b}.\mathbf{c} = -\frac{1}{2}$ and $\mathbf{a}.\mathbf{c} = 2$. Find an expression for $\mathbf{c}$ in terms of $\mathbf{a}$ and $\mathbf{b}$.

10. Use vector methods to prove Pythagoras' theorem.

11. Use vector methods to prove that the diagonals of a rhombus bisect each other at right angles.

12. In triangle ABC the point D is the mid-point of BC. By letting $\overrightarrow{AB} = \mathbf{x}$ and $\overrightarrow{BD} = \mathbf{y}$, prove Apollonius' theorem, which states that

$$AB^2 + AC^2 = 2(AD^2 + BD^2).$$

13. In triangle ABC the altitudes through A and B intersect at a point H. Show that if $\overrightarrow{HA} = \mathbf{a}$, $\overrightarrow{HB} = \mathbf{b}$ and $\overrightarrow{HC} = \mathbf{c}$ then $\mathbf{a}.\mathbf{b} = \mathbf{a}.\mathbf{c}$ and $\mathbf{b}.\mathbf{a} = \mathbf{b}.\mathbf{c}$. Deduce that the altitudes of a triangle are concurrent.

14. Find the angles between the following pairs of vectors
(a) $8\mathbf{i} - \mathbf{j} + 4\mathbf{k}$, $2\mathbf{i} + 2\mathbf{j} - \mathbf{k}$,
(b) $\mathbf{j} + 7\mathbf{k}$, $-5\mathbf{i} + 4\mathbf{j} + 3\mathbf{k}$,
(c) $2\mathbf{i} - 3\mathbf{j} + 8\mathbf{k}$, $6\mathbf{i} - 4\mathbf{j} - 3\mathbf{k}$,
(d) $\mathbf{i} - \mathbf{j} + \mathbf{k}$, $\mathbf{i} + \mathbf{k}$.

15. Find the cosine of the angle BAC, given that
(a) the points A, B and C have position vectors $\mathbf{i} - \mathbf{j}$, $5\mathbf{i} - 2\mathbf{j} - \mathbf{k}$ and $\mathbf{i} - 2\mathbf{j} - \mathbf{k}$ respectively;
(b) the points A, B and C have coordinates $(0, -2, 1)$, $(1, -1, -2)$ and $(-1, 1, 0)$ respectively.

16. Find the component of the vector $\mathbf{i} + 3\mathbf{j} - 2\mathbf{k}$ in the direction of
(a) $\mathbf{j}$,
(b) $-\mathbf{k}$,
(c) $-4\mathbf{i} + 3\mathbf{j}$
(d) $6\mathbf{i} - 3\mathbf{j} + 2\mathbf{k}$.

17. Given the points $A(1, -5, 3)$, $B(0, 2, -4)$, $C(-1, 1, 0)$ and $D(7, -2, -8)$, find the lengths of the projections of AB on AC and BD.

18. If $\mathbf{a} = \lambda\mathbf{i} + \mu\mathbf{j}$ and $\mathbf{b} = \mu\mathbf{i} - \lambda\mathbf{j}$, show that $\mathbf{a}$ is perpendicular to $\mathbf{b}$. Hence write down vectors perpendicular to
(a) $3\mathbf{i} + 4\mathbf{j}$,
(b) $5\mathbf{i} - 2\mathbf{j}$,
(c) $-2\mathbf{i} - 3\mathbf{j}$.

19. Points A and B have position vectors $4\mathbf{i} - 5\mathbf{j} - 2\mathbf{k}$ and $8\mathbf{i} - 5\mathbf{j} + 6\mathbf{k}$ respectively relative to an origin O. Find the angles of triangle AOB and also its area.

20. Given the points $A(7, 3, -1)$, $B(1, 5, 2)$ and $C(3, -1, 1)$, find the cosine of $\angle BAC$ and the area of triangle ABC.

21. Show that the lines with vector equations $\mathbf{r} = (1 + 4\lambda)\mathbf{i} + (1 - \lambda)\mathbf{j} + 2\lambda\mathbf{k}$ and $\mathbf{r} = (5 + 3\mu)\mathbf{i} + 2\mu\mathbf{j} + (2 - 5\mu)\mathbf{k}$ cut at right angles and give the position vector of their point of intersection.

22. The points A, B, C and D have coordinates $(3, -2, 0)$, $(-1, 2, 4)$, $(0, 1, -1)$ and $(5, -4, 4)$ respectively. Find (a) the point of intersection of the lines AB and CD, (b) the acute angle between these lines.

23. $ABCDA'B'C'D'$ is a cube with horizontal faces $ABCD$ and $A'B'C'D'$ and vertical edges AA', BB', CC', DD'. Taking $\mathbf{AB} = \mathbf{i}$, $\mathbf{AD} = \mathbf{j}$, $\mathbf{AA'} = \mathbf{k}$, obtain vectors (in terms of $\mathbf{i}, \mathbf{j}, \mathbf{k}$) perpendicular to the planes $ACC'A'$ and $CDA'B'$, and hence find the angle between these two planes. (O&C)

24. A plane p contains the vectors $\mathbf{a} = \mathbf{i} + 2\mathbf{j} - 3\mathbf{k}$ and $\mathbf{b} = 3\mathbf{i} - \mathbf{j} + 2\mathbf{k}$. If $\mathbf{c} = \mathbf{i} + \lambda\mathbf{j} + \mu\mathbf{k}$, find the values of λ and μ such that $\mathbf{c}$ is perpendicular to the plane p. If $\mathbf{d} = -3\mathbf{i} + \mathbf{j} - 3\mathbf{k}$, find $\mathbf{c}.\mathbf{d}$. Hence find the sine of the angle between the plane p and the vector $\mathbf{d}$.

25. Prove that if the vectors $\mathbf{a}$ and $\mathbf{b}$ are differentiable functions of a scalar variable t, then $\frac{d}{dt}(\mathbf{a}.\mathbf{b}) = \mathbf{a}.\frac{d\mathbf{b}}{dt} + \frac{d\mathbf{a}}{dt}.\mathbf{b}$.

7.6 Further lines and planes

As shown in §7.2 the vector equation of the straight line which passes through the point with position vector $\mathbf{a}$ and has direction vector $\mathbf{d}$ is

$$\mathbf{r} = \mathbf{a} + \lambda\mathbf{d}.$$

Similarly the vector equation of the straight line passing through the points with position vectors $\mathbf{a}$ and $\mathbf{b}$ is

$$\mathbf{r} = (1 - \lambda)\mathbf{a} + \lambda\mathbf{b}.$$

The scalar product can be used to find vectors perpendicular to such lines.

Example 1 The points A, B and C have position vectors $\mathbf{a} = 3\mathbf{i} - \mathbf{j} + 4\mathbf{k}$, $\mathbf{b} = \mathbf{j} - 4\mathbf{k}$ and $\mathbf{c} = 6\mathbf{i} + 4\mathbf{j} + 5\mathbf{k}$ respectively. Find the position vector of the point R on BC such that AR is perpendicular to BC. Hence find the perpendicular distance of A from the line BC.

Since R lies on BC its position vector is of the form

$$\mathbf{r} = (1 - \lambda)(\mathbf{j} - 4\mathbf{k}) + \lambda(6\mathbf{i} + 4\mathbf{j} + 5\mathbf{k})$$

i.e. $$\mathbf{r} = 6\lambda\mathbf{i} + (3\lambda + 1)\mathbf{j} + (9\lambda - 4)\mathbf{k} \qquad (1)$$

Since AR is perpendicular to BC, $\mathbf{AR}.\mathbf{BC} = 0$.

But $\mathbf{BC} = \mathbf{c} - \mathbf{b} = 6\mathbf{i} + 3\mathbf{j} + 9\mathbf{k}$

and $\mathbf{AR} = \mathbf{r} - \mathbf{a} = (6\lambda - 3)\mathbf{i} + (3\lambda + 2)\mathbf{j} + (9\lambda - 8)\mathbf{k}$ (2)

$\therefore \quad 6(6\lambda - 3) + 3(3\lambda + 2) + 9(9\lambda - 8) = 0$

$\therefore \quad 36\lambda - 18 + 9\lambda + 6 + 81\lambda - 72 = 0$

$\therefore \quad 126\lambda - 84 = 0 \quad \therefore \quad \lambda = \frac{2}{3}$

From (1), the position vector of R is $4\mathbf{i} + 3\mathbf{j} + 2\mathbf{k}$

From (2), $\mathbf{AR} = \mathbf{i} + 4\mathbf{j} - 2\mathbf{k}$

Hence the perpendicular distance of A from BC

$= |\mathbf{AR}| = \sqrt{\{1^2 + 4^2 + (-2)^2\}} = \sqrt{\{1 + 16 + 4\}} = \sqrt{21}$.

The parametric form of the vector equation of the plane which passes through a point A with position vector $\mathbf{a}$ and which contains non-parallel vectors $\mathbf{p}$ and $\mathbf{q}$ is

$$\mathbf{r} = \mathbf{a} + \lambda\mathbf{p} + \mu\mathbf{q}.$$

Another form of the equation of a plane is obtained by using the scalar product.

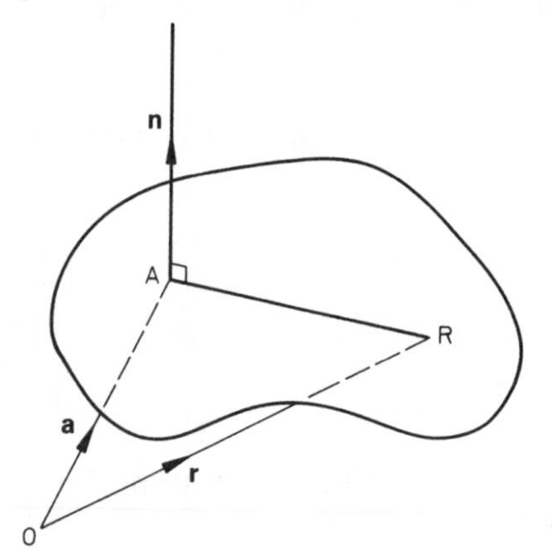

Consider a plane which passes through a point A with position vector $\mathbf{a}$ and which is perpendicular to a vector $\mathbf{n}$. If $\mathbf{r}$ is the position vector of a point R in the plane, then since $\mathbf{AR}$ is perpendicular to $\mathbf{n}$,

$$\mathbf{AR}.\mathbf{n} = 0$$

$$\therefore \quad (\mathbf{r} - \mathbf{a}).\mathbf{n} = 0$$

This is the vector equation of the plane through A perpendicular to $\mathbf{n}$, in scalar product form.

$$(\mathbf{r} - \mathbf{a}).\mathbf{n} = 0 \Leftrightarrow \mathbf{r}.\mathbf{n} - \mathbf{a}.\mathbf{n} = 0$$

$$\Leftrightarrow \mathbf{r}.\mathbf{n} = \mathbf{a}.\mathbf{n}$$

$$\Leftrightarrow \mathbf{r}.\mathbf{n} = p, \quad \text{where} \quad p = \mathbf{a}.\mathbf{n}.$$

Hence the equation of any plane perpendicular to a vector $\mathbf{n}$ may be written in the form $\mathbf{r}.\mathbf{n} = p$.

If the point D with position vector $\mathbf{d}$ is the foot of the perpendicular from the origin to the plane $\mathbf{r}.\mathbf{n} = p$, then the perpendicular distance of the origin from the plane is d.

Since D lies in the plane, $\mathbf{d}.\mathbf{n} = p$.

However, $\mathbf{d}$ is parallel to $\mathbf{n}$,

$\therefore$ either $\mathbf{d}.\mathbf{n} = dn$ or $\mathbf{d}.\mathbf{n} = -dn$.

Hence $dn = |p|$ giving $d = |p|/n$.

Thus the plane $\mathbf{r}.\mathbf{n} = p$ is perpendicular to the vector $\mathbf{n}$ and lies at a distance $\frac{|p|}{n}$ from the origin.

Example 2 Find the vector equation of the plane which passes through the point with position vector $2\mathbf{i} - \mathbf{j} + \mathbf{k}$ and is perpendicular to the vector $3\mathbf{i} + \mathbf{j} - 4\mathbf{k}$.

The vector equation of the given plane is

$$(\mathbf{r} - \{2\mathbf{i} - \mathbf{j} + \mathbf{k}\}).(3\mathbf{i} + \mathbf{j} - 4\mathbf{k}) = 0$$
$$\mathbf{r}.(3\mathbf{i} + \mathbf{j} - 4\mathbf{k}) - (2\mathbf{i} - \mathbf{j} + \mathbf{k}).(3\mathbf{i} + \mathbf{j} - 4\mathbf{k}) = 0$$
$$\mathbf{r}.(3\mathbf{i} + \mathbf{j} - 4\mathbf{k}) - (6 - 1 - 4) = 0$$

i.e
$$\mathbf{r}.(3\mathbf{i} + \mathbf{j} - 4\mathbf{k}) = 1.$$

Example 3 Find in scalar product form the equation of the plane

$$\mathbf{r} = (1 + 3\lambda + 2\mu)\mathbf{i} + (1 + \lambda + 4\mu)\mathbf{j} + (\mu - \lambda)\mathbf{k}.$$

Rearranging the given equation, we have

$$\mathbf{r} = \mathbf{i} + \mathbf{j} + \lambda(3\mathbf{i} + \mathbf{j} - \mathbf{k}) + \mu(2\mathbf{i} + 4\mathbf{j} + \mathbf{k})$$

$\therefore$ the given plane passes through the point with position vectors $\mathbf{i} + \mathbf{j}$ and is parallel to the vectors $3\mathbf{i} + \mathbf{j} - \mathbf{k}$ and $2\mathbf{i} + 4\mathbf{j} + \mathbf{k}$.
Let $\mathbf{n} = n_1\mathbf{i} + n_2\mathbf{j} + n_3\mathbf{k}$ be a vector perpendicular to $3\mathbf{i} + \mathbf{j} - \mathbf{k}$ and $2\mathbf{i} + 4\mathbf{j} + \mathbf{k}$, then

$$\mathbf{n}.(3\mathbf{i} + \mathbf{j} - \mathbf{k}) = 0 \quad \text{i.e.} \quad 3n_1 + n_2 - n_3 = 0 \qquad (1)$$
$$\mathbf{n}.(2\mathbf{i} + 4\mathbf{j} + \mathbf{k}) = 0 \quad \text{i.e.} \quad 2n_1 + 4n_2 + n_3 = 0 \qquad (2)$$

Adding (1) to (2): $5n_1 + 5n_2 = 0 \quad \therefore \quad n_2 = -n_1$
Substituting in (1): $3n_1 - n_1 - n_3 = 0 \quad \therefore \quad n_3 = 2n_1$

$\therefore \quad n_1:n_2:n_3 = n_1:-n_1:2n_1 = 1:-1:2$

Thus one vector perpendicular to the given plane is $\mathbf{i} - \mathbf{j} + 2\mathbf{k}$. Hence the equation of the plane in scalar product form is

$$(\mathbf{r} - \{\mathbf{i} + \mathbf{j}\}).(\mathbf{i} - \mathbf{j} + 2\mathbf{k}) = 0$$
$$\mathbf{r}.(\mathbf{i} - \mathbf{j} + 2\mathbf{k}) - (\mathbf{i} + \mathbf{j}).(\mathbf{i} - \mathbf{j} + 2\mathbf{k}) = 0$$

i.e.
$$\mathbf{r}.(\mathbf{i} - \mathbf{j} + 2\mathbf{k}) = 0$$

Example 4 Find the angle between the planes $\mathbf{r}.(2\mathbf{i} - \mathbf{j} - 3\mathbf{k}) = 10$ and $\mathbf{r}.(\mathbf{i} + 3\mathbf{j} - 2\mathbf{k}) = 16$.

The given planes are perpendicular to the vectors $\mathbf{n}_1 = 2\mathbf{i} - \mathbf{j} - 3\mathbf{k}$ and $\mathbf{n}_2 = \mathbf{i} + 3\mathbf{j} - 2\mathbf{k}$ respectively.
$\therefore$ the angle θ between the planes is equal to the angle between $\mathbf{n}_1$ and $\mathbf{n}_2$.

$$\mathbf{n}_1 . \mathbf{n}_2 = 2 \times 1 + (-1) \times 3 + (-3) \times (-2) = 2 - 3 + 6 = 5$$
$$n_1 = \sqrt{\{2^2 + (-1)^2 + (-3)^2\}} = \sqrt{\{4 + 1 + 9\}} = \sqrt{14}$$
$$n_2 = \sqrt{\{1^2 + 3^2 + (-2)^2\}} = \sqrt{\{1 + 9 + 4\}} = \sqrt{14}$$

$$\therefore \quad \cos\theta = \frac{\mathbf{n}_1 . \mathbf{n}_2}{n_1 n_2} = \frac{5}{\sqrt{14} \times \sqrt{14}} = \frac{5}{14}$$

Hence the angle between the planes is 69·1°.

Example 5 Find the angle between the line $\mathbf{r} = 3\mathbf{k} + \lambda(7\mathbf{i} - \mathbf{j} + 4\mathbf{k})$ and the plane $\mathbf{r}.(2\mathbf{i} - 5\mathbf{j} - 2\mathbf{k}) = 8$.

The given line is parallel to the vector $\mathbf{d} = 7\mathbf{i} - \mathbf{j} + 4\mathbf{k}$ and the given plane is perpendicular to the vector $\mathbf{n} = 2\mathbf{i} - 5\mathbf{j} - 2\mathbf{k}$.

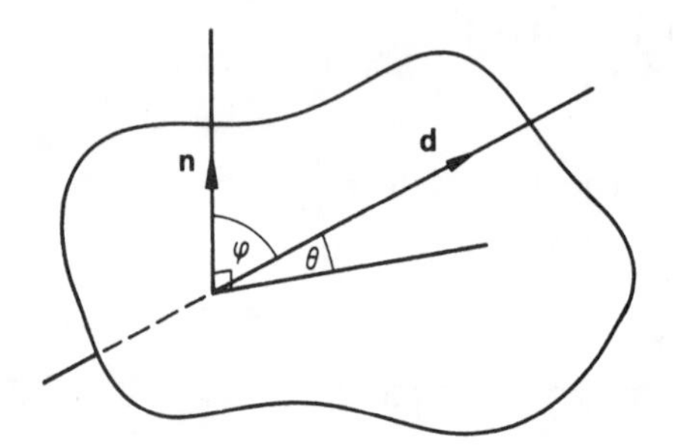

If the angle between the line and the plane is θ and the angle between $\mathbf{d}$ and $\mathbf{n}$ is ϕ then

$$\theta + \phi = 90°,$$

$$\therefore \quad \sin\theta = \cos\phi = \frac{\mathbf{d}.\mathbf{n}}{dn}$$

$$\mathbf{d}.\mathbf{n} = 7 \times 2 + (-1) \times (-5) + 4 \times (-2) = 14 + 5 - 8 = 11$$
$$d = \sqrt{\{7^2 + (-1)^2 + 4^2\}} = \sqrt{\{49 + 1 + 16\}} = \sqrt{66}$$
$$n = \sqrt{\{2^2 + (-5)^2 + (-2)^2\}} = \sqrt{\{4 + 25 + 4\}} = \sqrt{33}$$

$$\therefore \quad \sin\theta = \frac{11}{\sqrt{66} \times \sqrt{33}} = \frac{11}{33\sqrt{2}} = \frac{1}{3\sqrt{2}}$$

Hence the angle between the line and the plane is 13·6°.

Example 6 Find the position vector of the point of intersection of the line $\mathbf{r} = (2\mathbf{i} - \mathbf{k}) + \lambda(\mathbf{i} + 3\mathbf{j})$ and the plane $\mathbf{r}.(5\mathbf{i} - \mathbf{j} + 7\mathbf{k}) = 9$.

At the point of intersection

$$\{(2\mathbf{i} - \mathbf{k}) + \lambda(\mathbf{i} + 3\mathbf{j})\}.(5\mathbf{i} - \mathbf{j} + 7\mathbf{k}) = 9$$
$$\therefore \quad (2\mathbf{i} - \mathbf{k}).(5\mathbf{i} - \mathbf{j} + 7\mathbf{k}) + \lambda(\mathbf{i} + 3\mathbf{j}).(5\mathbf{i} - \mathbf{j} + 7\mathbf{k}) = 9$$
$$\therefore \quad (10 + 0 - 7) + \lambda(5 - 3 + 0) = 9$$
$$\therefore \quad 3 + 2\lambda = 9$$
$$\therefore \quad \lambda = 3$$

Hence the position vector of the point of intersection is

$$2\mathbf{i} - \mathbf{k} + 3(\mathbf{i} + 3\mathbf{j}) \quad \text{i.e.} \quad 5\mathbf{i} + 9\mathbf{j} - \mathbf{k}.$$

By writing $\mathbf{r} = x\mathbf{i} + y\mathbf{j} + z\mathbf{k}$ we can express the equations of lines and planes in Cartesian form.

Consider first the straight line $\mathbf{r} = \mathbf{a} + \lambda\mathbf{d}$, which passes through the point with position vector $\mathbf{a} = a_1\mathbf{i} + a_2\mathbf{j} + a_3\mathbf{k}$ and is parallel to the vector $\mathbf{d} = d_1\mathbf{i} + d_2\mathbf{j} + d_3\mathbf{k}$.

$$\mathbf{r} = \mathbf{a} + \lambda\mathbf{d} \quad \Leftrightarrow \quad \left.\begin{aligned} x &= a_1 + \lambda d_1 \\ y &= a_2 + \lambda d_2 \\ z &= a_3 + \lambda d_3 \end{aligned}\right\} \text{ These are the } \textit{parametric equations} \text{ of the line.}$$

Assuming that d_1, d_2 and d_3 are non-zero, we have

$$\frac{x - a_1}{d_1} = \lambda, \quad \frac{y - a_2}{d_2} = \lambda, \quad \frac{z - a_3}{d_3} = \lambda$$

$$\therefore \qquad \frac{x - a_1}{d_1} = \frac{y - a_2}{d_2} = \frac{z - a_3}{d_3}$$

These are the *Cartesian equations* of the line.
Any line parallel to the vector $d_1\mathbf{i} + d_2\mathbf{j} + d_3\mathbf{k}$ is said to have *direction ratios* $d_1{:}d_2{:}d_3$.

> Thus the equations $\dfrac{x - a_1}{d_1} = \dfrac{y - a_2}{d_2} = \dfrac{z - a_3}{d_3}$ represent the straight line through the point (a_1, a_2, a_3) with direction ratios $d_1{:}d_2{:}d_3$.

By convention the Cartesian equations of a straight line may be given in the above form for all values of d_1, d_2, d_3 including zero. For instance, the equations $\dfrac{x - 2}{3} = \dfrac{y + 1}{0} = \dfrac{z}{-2}$ represent the straight line through the point $(2, -1, 0)$, with direction ratios $3{:}0{:}-2$ i.e. in a direction parallel to the vector $3\mathbf{i} - 2\mathbf{k}$.

Consider now the plane which passes through the point with position vector $\mathbf{a} = a_1\mathbf{i} + a_2\mathbf{j} + a_3\mathbf{k}$ and is perpendicular to the vector $\mathbf{n} = n_1\mathbf{i} + n_2\mathbf{j} + n_3\mathbf{k}$. Writing $\mathbf{r} = x\mathbf{i} + y\mathbf{j} + z\mathbf{k}$, its vector equation $(\mathbf{r} - \mathbf{a}).\mathbf{n} = 0$ becomes

$$\{(x - a_1)\mathbf{i} + (y - a_2)\mathbf{j} + (z - a_3)\mathbf{k}\}.(n_1\mathbf{i} + n_2\mathbf{j} + n_3\mathbf{k}) = 0.$$

Hence the Cartesian equation of this plane is

$$n_1(x - a_1) + n_2(x - a_2) + n_3(x - a_3) = 0.$$

Similarly the vector equation $\mathbf{r}.\mathbf{n} = p$ gives rise to the Cartesian form

$$n_1x + n_2y + n_3z - p = 0.$$

For instance, the Cartesian equation of the plane $\mathbf{r}.(3\mathbf{i} - \mathbf{j} + 4\mathbf{k}) = 5$ is $3x - y + 4z - 5 = 0$.
It follows that the equation $Ax + By + Cz + D = 0$ is the Cartesian equation of a plane perpendicular to the vector $A\mathbf{i} + B\mathbf{j} + C\mathbf{k}$.

> Thus the equation $Ax + By + Cz + D = 0$ represents a plane, the direction ratios of the normal to the plane being $A{:}B{:}C$.

Exercise 7.6

1. The equation of a line l is $\mathbf{r} = \lambda\mathbf{i} - (\lambda + 4)\mathbf{j} + \mathbf{k}$. Find the position vector with respect to the origin O of the point A on l, given that OA is perpendicular to l. Hence find the perpendicular distance of O from the line l.

2. The position vectors of points A and B are $3\mathbf{i} - 8\mathbf{j} + \mathbf{k}$ and $4\mathbf{j} - 2\mathbf{k}$ respectively. Find the position vector of the foot of the perpendicular from the origin O to AB. Hence find the area of the triangle OAB.

3. The points A and B have position vectors $2\mathbf{j} + 7\mathbf{k}$ and $5\mathbf{i} - 3\mathbf{j} + 2\mathbf{k}$ respectively relative to an origin O. Find the position vector of the point R on AB such that OR is perpendicular to AB. Hence find the position vector of the reflection of O in AB.

4. A point A has position vector $2\mathbf{i} + 3\mathbf{j} - 6\mathbf{k}$ and a line l has equation $\mathbf{r} = \mathbf{i} - \mathbf{j} + 7\mathbf{k} + \lambda(4\mathbf{i} - \mathbf{k})$. Find the position vector of a point R on l such that AR is perpendicular to l. Hence find the perpendicular distance of A from the line l.

5. Find, in scalar product form, the equation of the plane through the point with position vector $\mathbf{a}$ and perpendicular to the vector $\mathbf{n}$, given that
(a) $\mathbf{a} = \mathbf{i} - 2\mathbf{j} + \mathbf{k}, \mathbf{n} = 4\mathbf{i} - \mathbf{k}$, (b) $\mathbf{a} = 3\mathbf{j} - 2\mathbf{k}, \mathbf{n} = \mathbf{i} - \mathbf{j} - 2\mathbf{k}$.

6. Find, in the form $\mathbf{r}.\mathbf{n} = p$, the equations of the planes through the given points and perpendicular to the given vectors
(a) $(0, 0, 1), \mathbf{k}$, (b) $(2, -1, 1), 3\mathbf{i} + 4\mathbf{j} - 2\mathbf{k}$.

7. Find the perpendicular distances from the origin of the following planes
(a) $\mathbf{r}.(\mathbf{i} - 2\mathbf{j} + 2\mathbf{k}) = 15$, (b) $\mathbf{r}.\mathbf{i} = 3$,
(c) $\mathbf{r}.(3\mathbf{i} - 4\mathbf{k}) = 20$, (b) $\mathbf{r}.(\mathbf{i} + \mathbf{j} + \mathbf{k}) = 0$.

8. Find a vector which is perpendicular to both $\mathbf{i} + 2\mathbf{k}$ and $3\mathbf{i} + \mathbf{j} + \mathbf{k}$. Hence find the vector equation in scalar product form of the plane parallel to these vectors which passes through the point with position vector $\mathbf{i} - \mathbf{j}$.

9. Find in scalar product form the equation of the plane

$$\mathbf{r} = (3 + 3\lambda + 4\mu)\mathbf{i} + (\lambda + 2\mu)\mathbf{j} + (1 + 3\lambda + 5\mu)\mathbf{k}.$$

10. Find in scalar product form the equation of the plane containing the lines

$$\mathbf{r} = (\lambda + 1)\mathbf{i} - 2\mathbf{j} + 3\mathbf{k} \text{ and } \mathbf{r} = 3\mu\mathbf{i} + (2\mu - 4)\mathbf{j} + (\mu + 2)\mathbf{k}.$$

11. Find the acute angles between the following pairs of planes.
(a) $\mathbf{r}.(\mathbf{i} - \mathbf{j}) = 4, \mathbf{r}.(\mathbf{j} + \mathbf{k}) = 1$,
(b) $\mathbf{r}.(\mathbf{i} - 2\mathbf{j} - 2\mathbf{k}) = 5, \mathbf{r}.\mathbf{j} = 0$,
(c) $\mathbf{r}.(\mathbf{i} + \mathbf{j} + \mathbf{k}) = 1, \mathbf{r}.(\mathbf{i} - \mathbf{j} + \mathbf{k}) = 0$.

12. In each of the following cases find the sine of the angle between the given line and plane.
(a) $\mathbf{r} = \mathbf{i} - 3\mathbf{j} + \lambda(2\mathbf{i} - \mathbf{j} - \mathbf{k})$, $\mathbf{r}.(\mathbf{i} - 2\mathbf{j} - 7\mathbf{k}) = 10$,
(b) $\mathbf{r} = (2 + \lambda)\mathbf{i} - 3\mathbf{j} + (1 - \lambda)\mathbf{k}$, $\mathbf{r}.(4\mathbf{i} + \mathbf{j} - \mathbf{k}) = 6$,
(c) $\mathbf{r} = 3\lambda\mathbf{i} + 2\lambda\mathbf{j} - 6\lambda\mathbf{k}$, $\mathbf{r}.(4\mathbf{i} - 3\mathbf{k}) = 20$.

13. Find the position vectors of points of intersection of the line $\mathbf{r} = \lambda\mathbf{i} - 2\mathbf{j} + (2\lambda - 1)\mathbf{k}$ with the planes $\mathbf{r}.(2\mathbf{i} + \mathbf{j} - 3\mathbf{k}) = 5$ and $\mathbf{r}.(\mathbf{i} + 5\mathbf{j} + 2\mathbf{k}) = -2$.

14. Find the vector equation of the line through the point A with position vector $5\mathbf{i} - 3\mathbf{j} + 4\mathbf{k}$ which is perpendicular to the plane $\mathbf{r}.(3\mathbf{i} - 4\mathbf{j} + \mathbf{k}) = 5$. Hence find the position vector of the foot of the perpendicular from A to this plane.

15. Find the Cartesian equations of the straight line which
(a) has vector equation $\mathbf{r} = \mathbf{i} - 2\mathbf{j} + \mathbf{k} + \lambda(2\mathbf{i} + 3\mathbf{j} + 4\mathbf{k})$,
(b) has vector equation $\mathbf{r} = \lambda\mathbf{i} + (2 - \lambda)\mathbf{j} + (3 + 2\lambda)\mathbf{k}$,
(c) passes through the point $(2, -3, 1)$ and has direction ratios $5:-6:3$,
(d) passes through the point $(5, 0, -4)$ and has direction ratios $1:3:0$.

16. Write down the Cartesian equations of the planes
(a) $\mathbf{r}.(2\mathbf{i} - 3\mathbf{j} + \mathbf{k}) = 5$, (b) $\mathbf{r}.(3\mathbf{i} + \mathbf{j} - 5\mathbf{k}) = 9$.

17. Find the Cartesian equation of the plane which passes through the point with position vector $\mathbf{i} - 2\mathbf{j} + 3\mathbf{k}$ and is perpendicular to the vector $5\mathbf{i} + \mathbf{j} + \mathbf{k}$.

18. Find the Cartesian equation of a plane which passes through the point $(4, 0, -1)$, given that the normal to the plane has direction ratios $2:-1:5$.

19. The points A, B and C have position vectors $3\mathbf{i} + 2\mathbf{j} - \mathbf{k}$, $\mathbf{i} + 2\mathbf{j}$ and $4\mathbf{i} + \mathbf{j} - 3\mathbf{k}$ respectively. Find a vector perpendicular to both $\mathbf{AB}$ and $\mathbf{AC}$. Hence find the equation of the plane containing A, B and C in (a) scalar product form, (b) Cartesian form.

20. Find the Cartesian equation of the plane containing the points $A(2, -1, 1)$, $B(1, -2, 0)$ and $C(-3, 6, -1)$.

21. Find the acute angle between

(a) the lines $\dfrac{x-1}{1} = \dfrac{y+2}{-2} = \dfrac{z}{2}$, $\dfrac{x+3}{3} = \dfrac{y+1}{0} = \dfrac{z-2}{-4}$;

(b) the planes $x - 2y - 3z = 1$, $2x - 4y + z = 3$;

(c) the line $\dfrac{x}{4} = \dfrac{y-1}{-1} = \dfrac{z+3}{-5}$, the plane $x - 2y + 4z + 3 = 0$.

22. The points A and B are fixed points with position vectors $\mathbf{a}$ and $\mathbf{b}$ respectively relative to an origin O. If R is a variable point with position vector $\mathbf{r}$, describe

the locus of R in each of the following cases
(a) $\mathbf{r}.\mathbf{a} = 0$, (b) $(\mathbf{r} - \mathbf{a}).\mathbf{a} = 0$, (c) $(\mathbf{r} - \mathbf{a}).(\mathbf{r} - \mathbf{a}) = 1$,
(d) $\mathbf{r}.\mathbf{a} = \mathbf{r}.\mathbf{b}$, (e) $(\mathbf{r} - \mathbf{a}).\mathbf{r} = 0$, (f) $(\mathbf{r} - \mathbf{a}).(\mathbf{r} - \mathbf{b}) = 0$.

7.7 The scalar product in mechanics

In §7.5 the scalar product was used to find the component of a vector in a given direction and also to calculate the cosine of the angle between two vectors. These methods can be used in mechanics when dealing with vector quantities such as forces and velocities.

Example 1 Find the component of the force $\mathbf{F} = 2\mathbf{i} + 13\mathbf{j} - 8\mathbf{k}$ in the direction of the vector $\mathbf{d} = 2\mathbf{i} - 6\mathbf{j} + 3\mathbf{k}$. Hence resolve $\mathbf{F}$ into component forces parallel and perpendicular to $\mathbf{d}$.

Let $\hat{\mathbf{d}}$ be the unit vector in the direction of $\mathbf{d}$.

$$d = |2\mathbf{i} - 6\mathbf{j} + 3\mathbf{k}| = \sqrt{\{2^2 + (-6)^2 + 3^2\}} = \sqrt{\{4 + 36 + 9\}} = 7$$

$$\therefore \quad \hat{\mathbf{d}} = \frac{\mathbf{d}}{d} = \tfrac{2}{7}\mathbf{i} - \tfrac{6}{7}\mathbf{j} + \tfrac{3}{7}\mathbf{k}$$

Hence the component of $\mathbf{F}$ in the direction of $\mathbf{d}$

$$= \mathbf{F}.\hat{\mathbf{d}} = (2\mathbf{i} + 13\mathbf{j} - 8\mathbf{k}).(\tfrac{2}{7}\mathbf{i} - \tfrac{6}{7}\mathbf{j} + \tfrac{3}{7}\mathbf{k})$$

$$= \tfrac{4}{7} - \tfrac{78}{7} - \tfrac{24}{7} = -\tfrac{98}{7} = -14.$$

Resolving $\mathbf{F}$ parallel and perpendicular to $\mathbf{d}$, the component force parallel to $\mathbf{d}$

$$= (\mathbf{F}.\hat{\mathbf{d}})\hat{\mathbf{d}} = -14(\tfrac{2}{7}\mathbf{i} - \tfrac{6}{7}\mathbf{j} + \tfrac{3}{7}\mathbf{k}) = -4\mathbf{i} + 12\mathbf{j} - 6\mathbf{k}.$$

Thus the component force perpendicular to $\mathbf{d}$

$$= (2\mathbf{i} + 13\mathbf{j} - 8\mathbf{k}) - (-4\mathbf{i} + 12\mathbf{j} - 6\mathbf{k}) = 6\mathbf{i} + \mathbf{j} - 2\mathbf{k}.$$

We next show that the *work done by a force* can be expressed as a scalar product. We recall that if the point of application of a constant force $\mathbf{F}$ moves a distance s in a direction which makes an angle θ with the direction of $\mathbf{F}$, then the work done by the force is $Fs\cos\theta$.

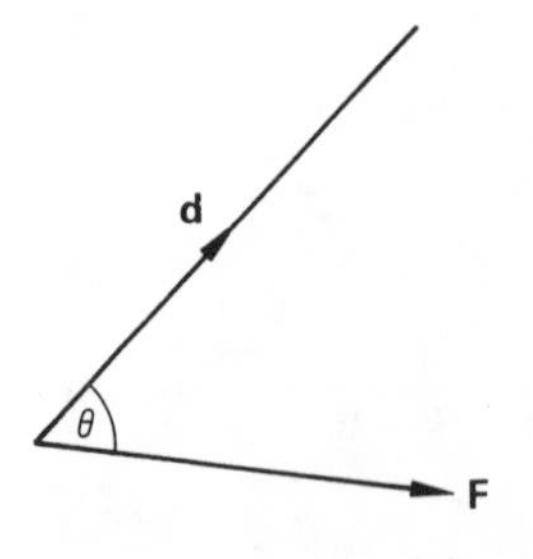

If we now suppose that the point of application of a constant force $\mathbf{F}$ is given a displacement $\mathbf{d}$, where the angle between the vectors $\mathbf{F}$ and $\mathbf{d}$ is θ, then,

$$\text{work done by } \mathbf{F} = Fd\cos\theta = \mathbf{F}.\mathbf{d}.$$

[We will assume that forces, displacements and work are measured in newtons, metres and joules respectively.]

Example 2 A constant force $\mathbf{F} = 3\mathbf{i} - 4\mathbf{j} - 2\mathbf{k}$ acts on a particle. Find the work done by $\mathbf{F}$ when the particle moves from the point with position vector $\mathbf{i} + 3\mathbf{j} - \mathbf{k}$ to the point with position vector $4\mathbf{i} + 2\mathbf{j} + 3\mathbf{k}$.

The displacement of the particle

$$= (4\mathbf{i} + 2\mathbf{j} + 2\mathbf{k}) - (\mathbf{i} + 3\mathbf{j} - \mathbf{k}) = 3\mathbf{i} - \mathbf{j} + 4\mathbf{k}$$

$\therefore$ the work done by $\mathbf{F}$

$$= (3\mathbf{i} - 4\mathbf{j} - 2\mathbf{k}).(3\mathbf{i} - \mathbf{j} + 4\mathbf{k}) = 9 + 4 - 8 = 5.$$

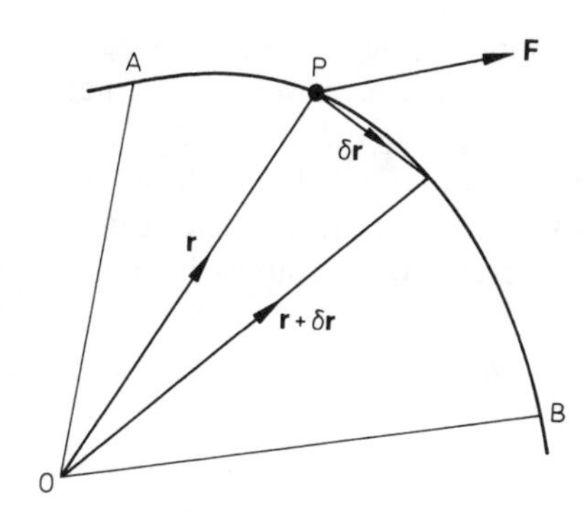

Consider now a variable force $\mathbf{F}$ acting on a particle P as it moves along a curve from a point A to a point B. If the path from A to B is regarded as a series of small displacements, then the work done by $\mathbf{F}$ in a typical displacement $\delta\mathbf{r}$ is approximately equal to $\mathbf{F}.\delta\mathbf{r}$.

$\therefore$ total work done by $\mathbf{F}$

$$\approx \sum_A^B \mathbf{F}.\delta\mathbf{r}$$

Letting $\delta\mathbf{r} \to 0$, we deduce that as the particle moves from A to B,

$$\boxed{\text{the work done by } \mathbf{F} \text{ is } \int_A^B \mathbf{F}.d\mathbf{r}} \qquad (1)$$

If $\mathbf{F}$ and $\mathbf{r}$ are given as functions of the time t, then we write

$$\int_A^B \mathbf{F}.d\mathbf{r} = \int_{t_1}^{t_2} \mathbf{F}.\frac{d\mathbf{r}}{dt}\,dt = \int_{t_1}^{t_2} \mathbf{F}.\mathbf{v}\,dt \qquad (2)$$

where t_1 and t_2 are the times at which the particle passes through A and B respectively.

Example 3 A particle moves along a curve so that its position vector at time t is given by $\mathbf{r} = t^3\mathbf{i} + 2t\mathbf{j}$. If one of the forces acting on the particle is a force $\mathbf{F}$, where $\mathbf{F} = 2\mathbf{i} - t\mathbf{j}$ at time t, find the work done by $\mathbf{F}$ in the interval from $t = 0$ to $t = 3$.

$$\mathbf{r} = t^3\mathbf{i} + 2t\mathbf{j} \Rightarrow \frac{d\mathbf{r}}{dt} = 3t^2\mathbf{i} + 2\mathbf{j}$$

$\therefore$ the work done by **F** from $t = 0$ to $t = 3$

$$= \int_0^3 \mathbf{F} \cdot \frac{d\mathbf{r}}{dt}\,dt$$

$$= \int_0^3 (2\mathbf{i} - t\mathbf{j}) \cdot (3t^2\mathbf{i} + 2\mathbf{j})\,dt$$

$$= \int_0^3 (6t^2 - 2t)\,dt$$

$$= \Big[2t^3 - t^2\Big]_0^3 = 54 - 9 = 45.$$

If a particle of mass m is moving with velocity $\mathbf{v}$ at time t, then its *kinetic energy* at that instant is $\frac{1}{2}mv^2$. We can establish the relationship between work and kinetic energy by considering the motion of this particle along a curved path AB. We will assume that the particle passes through A at time t_1 with velocity $\mathbf{v}_1$ and through B at time t_2 with velocity $\mathbf{v}_2$. If **F** is the resultant force acting on the particle at time t and W is the work done by **F** in the interval $t_1 \leqslant t \leqslant t_2$, then using our earlier results (1) and (2),

$$W = \int_A^B \mathbf{F} \cdot d\mathbf{r} = \int_{t_1}^{t_2} \mathbf{F} \cdot \mathbf{v}\,dt.$$

Using Newton's second law, $\mathbf{F} = m\mathbf{a} = m\dfrac{d\mathbf{v}}{dt}$

$$\therefore \qquad W = \int_{t_1}^{t_2} m\frac{d\mathbf{v}}{dt} \cdot \mathbf{v}\,dt$$

But, using the product rule proved in Exercise 7.5 question 25,

$$\frac{d}{dt}(v^2) = \frac{d}{dt}(\mathbf{v} \cdot \mathbf{v}) = \mathbf{v} \cdot \frac{d\mathbf{v}}{dt} + \frac{d\mathbf{v}}{dt} \cdot \mathbf{v} = 2\frac{d\mathbf{v}}{dt} \cdot \mathbf{v}$$

$$\therefore \qquad \int \frac{d\mathbf{v}}{dt} \cdot \mathbf{v}\,dt = \tfrac{1}{2}v^2 + \text{constant.}$$

Hence
$$W = \Big[\tfrac{1}{2}mv^2\Big]_{t_1}^{t_2} = \tfrac{1}{2}mv_2^2 - \tfrac{1}{2}mv_1^2$$

Thus

work done by **F** = K.E. at B – K.E. at A

Example 4 A particle of mass 10 kg moves under the action of a force **F**, so that its position vector after t seconds is given by $\mathbf{r} = 3t\mathbf{i} - \mathbf{j} + t^2\mathbf{k}$. Find the work done by **F** in the interval $1 \leqslant t \leqslant 5$.

$$\mathbf{r} = 3t\mathbf{i} - \mathbf{j} + t^2\mathbf{k} \;\Rightarrow\; \mathbf{v} = \frac{d\mathbf{r}}{dt} = 3\mathbf{i} + 2t\mathbf{k}$$

$$\Rightarrow\; v^2 = 3^2 + (2t)^2 = 9 + 4t^2$$

$\therefore$ when $t = 1$, $v^2 = 13$ and when $t = 5$, $v^2 = 109$.

Work done by $\mathbf{F}$ = (K.E. when $t = 5$) – (K.E. when $t = 1$)
$= \frac{1}{2} \times 10 \times 109 - \frac{1}{2} \times 10 \times 13$ J
$= 480$ J.

The *power* of a force is the rate at which it does work. Let us suppose that at time t a force $\mathbf{F}$ acts at a point with position vector $\mathbf{r}$. If the point of application of $\mathbf{F}$ is given a displacement $\delta\mathbf{r}$ in a small interval of time δt, then the work done is approximately $\mathbf{F}.\delta\mathbf{r}$ and the average rate of working is $(\mathbf{F}.\delta\mathbf{r})/\delta t$.
$\therefore$ the power of the force $\mathbf{F}$ at time t

$$= \lim_{\delta t \to 0} \frac{\mathbf{F}.\delta\mathbf{r}}{\delta t} = \lim_{\delta t \to 0} \mathbf{F}.\frac{\delta\mathbf{r}}{\delta t} = \mathbf{F}.\frac{d\mathbf{r}}{dt}.$$

Hence if the point of application of a force $\mathbf{F}$ is moving with velocity $\mathbf{v}$, then

the power of $\mathbf{F}$ is $\mathbf{F}.\mathbf{v}$.

Example 5 At time t a particle of unit mass is moving with velocity $\mathbf{v}$ under the action of a force $\mathbf{F} = 4t\mathbf{i} + 3\mathbf{k}$. Given that when $t = 0$, $\mathbf{v} = \mathbf{i} + 2\mathbf{j}$, find an expression for the velocity $\mathbf{v}$ at time t. Hence find an expression for the power of $\mathbf{F}$ at time t.

By Newton's second law, $\mathbf{F} = m\mathbf{a}$.
Since $\mathbf{F} = 4t\mathbf{i} + 3\mathbf{k}$, $m = 1$ and $\mathbf{a} = \frac{d\mathbf{v}}{dt}$,

we have $\frac{d\mathbf{v}}{dt} = 4t\mathbf{i} + 3\mathbf{k}$,

Integrating, $\mathbf{v} = 2t^2\mathbf{i} + 3t\mathbf{k} + \mathbf{c}$,

where $\mathbf{c}$ is a constant vector.

But when $t = 0$, $\mathbf{v} = \mathbf{i} + 2\mathbf{j}$ $\therefore$ $\mathbf{c} = \mathbf{i} + 2\mathbf{j}$.
Hence $\mathbf{v} = (2t^2 + 1)\mathbf{i} + 2\mathbf{j} + 3t\mathbf{k}$.
The power of $\mathbf{F}$ at time $t = \mathbf{F}.\mathbf{v}$
$= (4t\mathbf{i} + 3\mathbf{k}).(\{2t^2 + 1\}\mathbf{i} + 2\mathbf{j} + 3t\mathbf{k})$
$= 4t(2t^2 + 1) + 3 \times 3t$
$= 8t^3 + 4t + 9t = 8t^3 + 13t$.

Exercise 7.7

1. Two particles A and B are moving with constant velocities $8\mathbf{i} + \mathbf{j} - 4\mathbf{k}$ and $4\mathbf{i} - 2\mathbf{j} - 4\mathbf{k}$ respectively. Find the angle between the paths of the particles. Find also the speed of A relative to B.

2. Two forces $\mathbf{F}_1 = 4\mathbf{j} - 5\mathbf{k}$ and $\mathbf{F}_2 = 2\mathbf{i} - 5\mathbf{j} - \mathbf{k}$ act on a particle. Find the magnitude of the resultant force acting on the particle and the angle the resultant makes with the force $\mathbf{F}_1$.

3. Find the component of the force $\mathbf{F}$ in the direction of the vector $\mathbf{d}$ if
(a) $\mathbf{F} = 3\mathbf{i} - 4\mathbf{j} + 4\mathbf{k}$, $\mathbf{d} = 6\mathbf{i} - 2\mathbf{j} - 3\mathbf{k}$,
(b) $\mathbf{F} = \mathbf{i} + 5\mathbf{j} - 3\mathbf{k}$, $\mathbf{d} = 2\mathbf{i} - \mathbf{j} + 2\mathbf{k}$,
(c) $\mathbf{F} = 7\mathbf{i} - \mathbf{j} - 4\mathbf{k}$, $\mathbf{d} = 3\mathbf{i} - 5\mathbf{j} + 4\mathbf{k}$,
(d) $\mathbf{F} = 3\mathbf{i} + 5\mathbf{j} - \mathbf{k}$, $\mathbf{d} = 4\mathbf{i} - 4\mathbf{j} + 7\mathbf{k}$.

4. Resolve the force $\mathbf{F}$ into component forces parallel and perpendicular to the vector $\mathbf{d}$ if
(a) $\mathbf{F} = 11\mathbf{i} + 4\mathbf{j} - 2\mathbf{k}$, $\mathbf{d} = 4\mathbf{i} - 3\mathbf{k}$,
(b) $\mathbf{F} = 5\mathbf{i} - 22\mathbf{j} + 19\mathbf{k}$, $\mathbf{d} = -5\mathbf{j} + 12\mathbf{k}$,
(c) $\mathbf{F} = 18\mathbf{i} + 35\mathbf{j} - 21\mathbf{k}$, $\mathbf{d} = 2\mathbf{i} - 6\mathbf{j} + 9\mathbf{k}$,
(d) $\mathbf{F} = 3\mathbf{i} - 4\mathbf{j} + 10\mathbf{k}$, $\mathbf{d} = 4\mathbf{i} - 8\mathbf{j} + \mathbf{k}$.

5. Find the work done when the point of application of a force $\mathbf{F}$ moves through a displacement $\mathbf{d}$, given that
(a) $\mathbf{F} = 3\mathbf{i} - 7\mathbf{j} + 6\mathbf{k}$, $\mathbf{d} = 5\mathbf{i} + \mathbf{j} - \mathbf{k}$,
(b) $\mathbf{F} = -2\mathbf{i} + 3\mathbf{j} - 8\mathbf{k}$, $\mathbf{d} = 6\mathbf{i} - 4\mathbf{j} - 3\mathbf{k}$,
(c) $\mathbf{F} = 15\mathbf{i} + 7\mathbf{k}$, $\mathbf{d} = 10\mathbf{j} + 3\mathbf{k}$,
(d) $\mathbf{F} = 3\mathbf{i} - 9\mathbf{j} - 4\mathbf{k}$, $\mathbf{d} = \mathbf{i} + 2\mathbf{j} - \mathbf{k}$.

6. At time t a particle of mass m is moving with position vector $\mathbf{r}$ and velocity $\mathbf{v}$ under the action of a force $\mathbf{F}$. In each of the following cases find an expression for $\mathbf{v}$ and hence determine the work done by $\mathbf{F}$ in the given interval of time.
(a) $\mathbf{r} = 3t\mathbf{i} + t^2\mathbf{j}$, $0 \leqslant t \leqslant 2$,
(b) $\mathbf{r} = 5\mathbf{i} - t^2\mathbf{j} + 3t\mathbf{k}$, $2 \leqslant t \leqslant 4$,
(c) $\mathbf{r} = t^3\mathbf{i} - 4t\mathbf{j} + 3\mathbf{k}$, $1 \leqslant t \leqslant 3$,
(d) $\mathbf{r} = t\mathbf{i} - \mathbf{j} + (t^2 - 7t)\mathbf{k}$, $2 \leqslant t \leqslant 5$.

7. At time t a particle of mass m is moving with position vector $\mathbf{r}$ and velocity $\mathbf{v}$ under the action of a force $\mathbf{F}$. Find expressions for $\mathbf{F}$, $\mathbf{v}$ and the power of $\mathbf{F}$ at time t, given that
(a) $\mathbf{r} = 2t^2\mathbf{i} - 5t\mathbf{k}$,
(b) $\mathbf{r} = \mathbf{i} + t^3\mathbf{j} - 3t^2\mathbf{k}$,
(c) $\mathbf{r} = t\mathbf{i} - e^{-t}\mathbf{j}$,
(d) $\mathbf{r} = \sin t\mathbf{j} + 3\cos t\mathbf{k}$.

8. A particle of unit mass is moving under the action of a constant force $\mathbf{F} = \mathbf{i} - 3\mathbf{j} + 4\mathbf{k}$. Find the work done by $\mathbf{F}$ when the particle moves from the point A with position vector $5\mathbf{j} + 2\mathbf{k}$ to the point $\mathbf{B}$ with position vector $3\mathbf{i} + \mathbf{k}$. If the speed of the particle at A is $6\,\mathrm{m\,s^{-1}}$, find the speed of the particle at B.

9. A particle of mass 3 kg moves under the action of a constant force $\mathbf{F}$ with acceleration $\mathbf{a} = 3\mathbf{i} + 2\mathbf{k}$. Find the work done by $\mathbf{F}$ as the particle moves from the point A with position vector $4\mathbf{i} - 9\mathbf{j} + 21\mathbf{k}$ to the point B with position vector $10\mathbf{i} + 15\mathbf{j} - 7\mathbf{k}$. If the particle passes through A with velocity $12\mathbf{j} - 16\mathbf{k}$ and through

B with velocity $\lambda(\mathbf{i} + 2\mathbf{j} - 2\mathbf{k})$, find the value of λ and the power of $\mathbf{F}$ at the points A and B.

10. A particle of unit mass is acted upon at time t by a force $\mathbf{F} = 2t\mathbf{i} - 3\mathbf{j}$. Given that the initial velocity of the particle is $\mathbf{j} + \mathbf{k}$, find an expression for its velocity $\mathbf{v}$ at time t. Hence find the work done by $\mathbf{F}$ in the interval $0 \leqslant t \leqslant 2$ and the power of $\mathbf{F}$ when $t = 2$.

11. A particle of mass 2 kg is acted upon at time t by a force $\mathbf{F} = 8\mathbf{i} - 4\cos t\mathbf{j} + 2t\mathbf{k}$. When $t = 0$ the velocity of the particle is $6\mathbf{i}$. Find expressions for the velocity of the particle and the power of $\mathbf{F}$ at time t.

12. A particle moves along a curve so that its position vector at time t is $\mathbf{r} = (3t - 2)\mathbf{i} + t^4\mathbf{j}$. If one of the forces acting on the particle is $\mathbf{F} = 5t^2\mathbf{i} - \mathbf{j}$, find by integration the work done by $\mathbf{F}$ in the interval $0 \leqslant t \leqslant 2$.

13. A particle moves along a curve with velocity $\mathbf{v} = \mathbf{i} + 10t^2\mathbf{j} - 3t\mathbf{k}$ at time t. If one of the forces acting on the particle is $\mathbf{F} = 2t\mathbf{i} + t^2\mathbf{j} - 2t\mathbf{k}$, find the power of $\mathbf{F}$ in terms of t. Hence find the work done by $\mathbf{F}$ in the interval from $t = 1$ to $t = 3$.

14. A particle moves on a curve with vector equation $\mathbf{r} = (\lambda - 12)\mathbf{i} - \lambda^2\mathbf{j} + \mathbf{k}$, where λ is a parameter. One of the forces acting on the particle is a variable force $\mathbf{F}$. Write down, in the form of a definite integral, an expression for the work done by $\mathbf{F}$ as the particle moves from the point where $\lambda = 1$ to the point where $\lambda = 2$. Evaluate this integral given that $\mathbf{F} = -4\mathbf{r}$.

Exercise 7.8 (miscellaneous)

1. A cube $OPQRABCD$ has OA, PB, QC and RD perpendicular to the base $OPQR$; M is the centre of the face $ABCD$. The position vectors of A, B, C relative to O are $\mathbf{a}, \mathbf{b}, \mathbf{c}$ respectively. Find in terms of $\mathbf{a}, \mathbf{b}$, and $\mathbf{c}$ the vectors $\mathbf{AB}, \mathbf{OM}, \mathbf{OD}$. If $|\mathbf{OP}| = 2$, express $\mathbf{a}, \mathbf{b}, \mathbf{c}, \mathbf{OM}, \mathbf{OD}$ in terms of unit vectors $\mathbf{i}, \mathbf{j}, \mathbf{k}$ along $\mathbf{OP}, \mathbf{OR}, \mathbf{OA}$ respectively. Find the cosine of the angle MOC. (L)

2. Points A and B have position vectors $\mathbf{a}$ and $\mathbf{b}$ respectively with respect to a given origin O. Prove that the position vector of the point P dividing the line segment AB in the ratio λ to μ is $(\mu\mathbf{a} + \lambda\mathbf{b})/(\lambda + \mu)$. The position vectors of A, B, C and D, the vertices of a plane quadrilateral, are $\mathbf{a}, \mathbf{b}, \mathbf{c}$ and $\mathbf{d}$ respectively. Find the position vectors of P, Q, R, S, T and U, the mid-points of AB, BC, CD, DA, AC and BD respectively. Hence show that the point with position vector $\frac{1}{4}(\mathbf{a} + \mathbf{b} + \mathbf{c} + \mathbf{d})$ lies on each of the lines PR, QS and TU and that these lines bisect each other. If $\mathbf{a} = 2\mathbf{i} + 3\mathbf{j}, \mathbf{b} = 6\mathbf{i} + 5\mathbf{j}, \mathbf{c} = 8\mathbf{i} - \mathbf{j}$ and $\mathbf{d} = \mathbf{j}$, where $\mathbf{i}$ and $\mathbf{j}$ are orthogonal unit vectors, show that the acute angle between AC and BD is $\cos^{-1}(5/13)$. Find a unit vector orthogonal to the line TU. (W)

3. The points A, B and C have position vectors $\mathbf{a}, \mathbf{b}$ and $\mathbf{c}$ with respect to an origin O. The point R in the plane ABC has position vector $\mathbf{r}$ where

$\mathbf{r} = \frac{1}{2}(\frac{2}{3}\mathbf{a} + \frac{1}{3}\mathbf{b}) + \frac{1}{2}\mathbf{c}$. Use the ratio theorem to obtain a geometrical description of the position of R with reference to the points A, B and C. Illustrate your answer with a diagram. By writing $\mathbf{r}$ in the form $\lambda\mathbf{a} + \mu(m\mathbf{b} + n\mathbf{c})$ where $\lambda + \mu = 1$, obtain an alternative description of the position of R. Given that the point S has position vector $\frac{2}{3}\mathbf{a} + \frac{1}{12}\mathbf{b} + \frac{1}{4}\mathbf{c}$, show that S bisects the line segment AR. (JMB)

4. 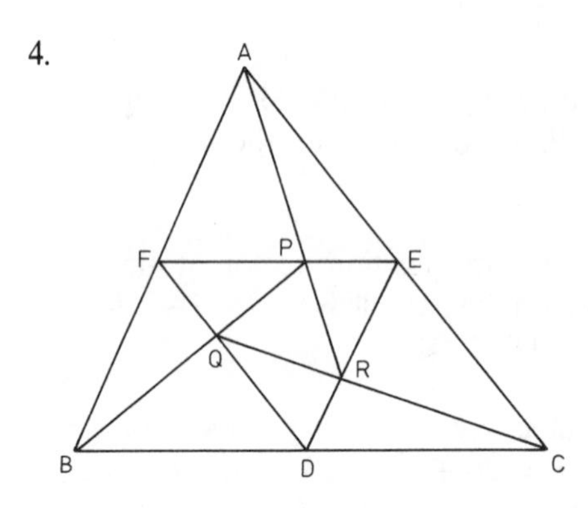

The diagram shows a triangle ABC, with D, E, F as the mid-points of the sides. The points P, Q, R are chosen on EF, FD, DE so that RP, PQ, QR pass through A, B, C. If the position vectors of A, B, C are $\mathbf{a}, \mathbf{b}, \mathbf{c}$ and if

$$\frac{PF}{EP} = \frac{QD}{FQ} = \frac{RE}{DR} = \lambda,$$

find the position vectors of P, Q, R and hence the value of λ. (O)

5. The position vectors of two points A and B relative to an origin O are $\mathbf{a}$ and $\mathbf{b}$ respectively. Show that the vector equation of the line bisecting angle AOB is $\mathbf{r} = \lambda(\hat{\mathbf{a}} + \hat{\mathbf{b}})$, where $\hat{\mathbf{a}}$ and $\hat{\mathbf{b}}$ are unit vectors in the directions of $\mathbf{a}$ and $\mathbf{b}$. Hence prove that the bisector of an interior angle of a triangle divides the opposite side into segments proportional to the adjacent sides.

6. OAD, OBE, OCF are three distinct straight lines. The position vectors with respect to O of the points A, B, C, D, E, F are $\mathbf{a}, \mathbf{b}, \mathbf{c}, p\mathbf{a}, q\mathbf{b}, r\mathbf{c}$ respectively, where p, q, r are scalars, no two of which are equal. Show that any point on BC has position vector given parametrically by $(1 - t)\mathbf{b} + t\mathbf{c}$. The intersections of the pairs of lines BC, EF; CA, FD; AB, DE (produced if necessary) are the points L, M, N respectively. Show that the position vector of L is $[q(1 - r)\mathbf{b} + r(q - 1)\mathbf{c}]/(q - r)$ and find the position vectors of M and N. Taking the values $p = 4, q = 3, r = 2$, show that L, M, N are collinear points. (C)

7. Three lines l_1, l_2 and l_3 have vector equations $\mathbf{r} = 2\mathbf{i} - \mathbf{j} + \mathbf{k} + \lambda(\mathbf{i} + 3\mathbf{j} + 3\mathbf{k})$, $\mathbf{r} = \mathbf{i} - 3\mathbf{j} + \mu(2\mathbf{i} + 5\mathbf{j} + 4\mathbf{k})$ and $\mathbf{r} = 3\mathbf{i} + \nu(\mathbf{j} + 2\mathbf{k})$ respectively. Find the position vector of the point of intersection of l_1 and l_2. Hence show that the three lines are concurrent. Write down, in parametric form, a vector equation of the plane containing l_2 and l_3. Hence show that the three lines are coplanar.

8. Two planes p_1 and p_2 have vector equations $\mathbf{r} = (1 - \lambda)\mathbf{i} + (2\lambda + \mu)\mathbf{j} + (\mu - 1)\mathbf{k}$ and $\mathbf{r} = (s - t)\mathbf{i} + (2s - 3)\mathbf{j} + t\mathbf{k}$ respectively. Show that if a point in p_1 also lies in p_2, then $\mu = 4\lambda + 3$. Hence find a vector equation of the line of intersection of the two planes.

9. Given non-zero vectors $\mathbf{a}$ and $\mathbf{b}$, show that $|\mathbf{a}+\mathbf{b}| = |\mathbf{a}-\mathbf{b}|$ if and only if $\mathbf{a}$ is perpendicular to $\mathbf{b}$. Deduce that a parallelogram is a rectangle if and only if its diagonals are equal in length.

10. Use vector methods to find the acute angle between two diagonals of a cube.

11. A tetrahedron $OABC$ with vertex O at the origin is such that $\overrightarrow{OA} = \mathbf{a}$, $\overrightarrow{OB} = \mathbf{b}$ and $\overrightarrow{OC} = \mathbf{c}$. Show that the line segments joining the mid-points of opposite edges bisect one another. Given that two pairs of opposite edges are perpendicular prove that $\mathbf{a}.\mathbf{b} = \mathbf{b}.\mathbf{c} = \mathbf{c}.\mathbf{a}$ and show that the third pair of opposite edges is also perpendicular. Prove also that, in this case, $OA^2 + BC^2 = OB^2 + AC^2$. (L)

12. ABC is a given triangle, and the centre of the circle through A, B, C is O. Relative to O, the position vectors of A, B, C are $\mathbf{a}, \mathbf{b}, \mathbf{c}$. If H is the point with position vector $\mathbf{a}+\mathbf{b}+\mathbf{c}$, prove that AH, BH, CH are perpendicular to BC, CA, AB respectively. If A', B', C' are the mid-points of BC, CA, AB respectively, prove that AA', BB' and CC' are concurrent in G, and find the position vector of G. Prove that O, G, H lie on a straight line. (O)

13. Let $\mathbf{a} = \mathbf{i} - 2\mathbf{j} + \mathbf{k}, \mathbf{b} = 2\mathbf{i} + \mathbf{j} - \mathbf{k}$. Given that $\mathbf{c} = \lambda\mathbf{a} + \mu\mathbf{b}$ and that $\mathbf{c}$ is perpendicular to $\mathbf{a}$, find the ratio of λ to μ. Let A, B be the points with position vectors $\mathbf{a}, \mathbf{b}$ respectively with respect to an origin O. Write down, in terms of $\mathbf{a}$ and $\mathbf{b}$, a vector equation of the line l through A, in the plane of O, A and B, which is perpendicular to OA. Find the position vector of P, the point of intersection of l and OB. (JMB)

14. The position vectors of A, B, C are $\mathbf{i}, 2\mathbf{j}, 3\mathbf{k}$ respectively. The perpendicular from the origin O to the plane ABC meets the plane at the point D. Find (i) the vector equation of the plane ABC, (ii) the vector equation of the line OD, (iii) the position vector of the point D. (AEB 1980)

15. A straight line passes through the point $(2, 0, -1)$ and is parallel to the vector $\mathbf{i} + 2\mathbf{j} - \mathbf{k}$. The feet of the perpendiculars from the points $A(2, -2, 1)$ and $B(1, 7, 0)$ to this line are P and Q. Find (i) the coordinates of P and Q, (ii) the angles APB and AQB, (iii) the Cartesian equations of the planes PAQ and PBQ, (iv) the angle between the planes PAQ and PBQ.

16. Interpret geometrically the equations $\mathbf{r}.\mathbf{n} = p$ and $\mathbf{r} = \mathbf{a} + t\mathbf{b}$ where $\mathbf{n}, \mathbf{a}$ and $\mathbf{b}$ are constant vectors, p is a constant scalar and t is a parameter which can take all real values. Find the value of t in terms of $\mathbf{a}, \mathbf{b}, \mathbf{n}$ and p if the value of $\mathbf{r}$ given by the second equation also satisfies the first. Explain the geometrical significance of the point given by this value of t, giving special attention to the cases when (i) $\mathbf{b}.\mathbf{n} = 0$ and (ii) $\mathbf{b} = \mathbf{n}$.

The point A has coordinates $(1, 1, 2)$. Determine the coordinates of the point A' which is the reflection of A in the plane $2x + 2y + z = 9$. (AEB 1979)

17. The lines l_1 and l_2 have equations $\mathbf{r}_1 = (2\lambda + 3)\mathbf{i} + (2 - 3\lambda)\mathbf{j} + \lambda\mathbf{k}$ and $\mathbf{r}_2 = (\mu - 4)\mathbf{i} + (2\mu + 1)\mathbf{j} + (4\mu - 3)\mathbf{k}$ respectively. Show that l_1 is perpendicular to l_2. Find the values of λ and μ such that the vector $\mathbf{r}_1 - \mathbf{r}_2$ is perpendicular to both l_1 and l_2. Hence find the position vectors of the ends of the common perpendicular to l_1 and l_2. Find also the length of this common perpendicular.

18. A pyramid has a square base $OABC$ and vertex V. The position vectors of A, B, C, V referred to O as origin are given by $\mathbf{OA} = 2\mathbf{i}, \mathbf{OB} = 2\mathbf{i} + 2\mathbf{j}, \mathbf{OC} = 2\mathbf{j}$, $\mathbf{OV} = \mathbf{i} + \mathbf{j} + 3\mathbf{k}$. (i) Express $\mathbf{AV}$ in terms of $\mathbf{i}, \mathbf{j}$ and $\mathbf{k}$. (ii) Using scalar products, or otherwise, find a vector $\mathbf{x}$ which is perpendicular to both $\mathbf{OV}$ and $\mathbf{AV}$. (iii) Calculate the angle between the vector $\mathbf{x}$, found in (ii), and $\mathbf{VB}$, giving your answer to the nearest degree. (iv) Write down the acute angle between VB and the plane OVA. (C)

19. The forces $\mathbf{F}_1$ and $\mathbf{F}_2$, where $\mathbf{F}_1 = 2\mathbf{i} + 4\mathbf{j} + 5\mathbf{k}$ and $\mathbf{F}_2 = -\mathbf{i} + 3\mathbf{j}$, both act through the point with position vector $\mathbf{j} - \mathbf{k}$. Find the angle between the two forces and the magnitude of their resultant. Obtain, in both vector and Cartesian forms, equations for the line of action of the resultant.

20. Show that the resultant of two forces acting at a point O in directions OA and OB and represented in magnitude by λOA and μOB is represented by $(\lambda + \mu)OC$, where C is the point on AB dividing it internally in the ratio $\mu:\lambda$.

(a) $ABCDEF$ is a regular hexagon of side a. Forces of size AB, AC, AD, AE and AF act along AB, AC, AD, AE and AF respectively. Prove that the resultant of these forces has magnitude $6a$, and give its line of action.

(b) A and B are two fixed points in a plane and P is a variable point in the same plane. Two forces act through P, one of magnitude $2PA$ along PA and the other of magnitude PB along PB. If the resultant of these two forces is always equal to $3PA$ *in magnitude*, find the locus of P. (SU)

21. (i) The force $\mathbf{F}_1$, of magnitude 26 N, acts at the origin in the direction of the vector $(3\mathbf{i} + 4\mathbf{j} + 12\mathbf{k})$. A second force $\mathbf{F}_2$, of magnitude 30 N, also acts at the origin, but in the direction of the vector $(6\mathbf{j} - 8\mathbf{k})$. Calculate the magnitude and the direction of the resultant of the forces $\mathbf{F}_1$ and $\mathbf{F}_2$. Calculate also the cosine of the angle between $\mathbf{F}_1$ and $\mathbf{F}_2$.

(ii) A particle of mass 2 kg starts from rest at the origin and the force $\mathbf{F}_3$, where $\mathbf{F}_3 = (6\mathbf{i} + 26\mathbf{j})$ N, acts on it as it moves. Find the position vector of the particle 2 seconds later. (L)

22. The force acting at time t $(0 \leqslant t < 2)$ on a particle of unit mass is $24t^2\mathbf{i} + 6\mathbf{j}$. At time $t = 0$ the particle is at rest at the point with position vector $-2\mathbf{i} + 3\mathbf{j}$. Find the position vector of the particle at time $t = T$ $(0 \leqslant T < 2)$. For time $t \geqslant 2$ the force acting on the particle is $6\mathbf{j}$. Find the position vector of the particle at time $t = 3$. (L)

23. State the centre and radius of the circle which has vector equation $\mathbf{r} = 6\mathbf{i} + 8\mathbf{j} + 6(\mathbf{i}\cos\theta + \mathbf{j}\sin\theta)$. A particle P of mass 2 kg moves on this circle with

constant angular velocity $\pi/12$ radians per second. Write down the position vector of P at time t given that, at $t = 0$, P is at the point corresponding to $\theta = 0$. Calculate the components parallel to $\mathbf{i}$ and $\mathbf{j}$ of the resultant vector force on the particle when $t = 4$. Find the position vector of the particle when $t = 8$ given that the forces on the particle cease to act when $t = 4$. (L)

24. The position vector, relative to the origin O, of an object of mass m moving in a horizontal plane in which $\mathbf{i}$ and $\mathbf{j}$ are perpendicular unit vectors is given by

$$\mathbf{r} = \frac{\cos\theta}{1+\cos\theta}\mathbf{i} + \frac{\sin\theta}{1+\cos\theta}\mathbf{j}, \quad -\pi < \theta < \pi.$$

Obtain an expression for r, the magnitude of $\mathbf{r}$. Given that $r^2\dfrac{d\theta}{dt} = h$ where h is constant, show that the velocity of the object is $-h\sin\theta\mathbf{i} + h(1+\cos\theta)\mathbf{j}$. Show that the force acting on the object is of magnitude mh^2/r^2, and find the direction of this force. (JMB)

25. A particle is projected from the origin O with initial velocity $u\mathbf{i} + v\mathbf{j}$, where $\mathbf{i}$ and $\mathbf{j}$ are unit vectors in the horizontal and upward vertical directions respectively, and moves freely under gravity. Show that its velocity and position vector with respect to O, at time t after projection, are given by $\dot{\mathbf{r}} = u\mathbf{i} + (v - gt)\mathbf{j}$ and $\mathbf{r} = ut\mathbf{i} + (vt - \frac{1}{2}gt^2)\mathbf{j}$ respectively.

For the remainder of this question components of velocities are measured in m s^{-1} and components of displacements in m. Take g as $10\,\text{m s}^{-2}$.

Two particles, A and B, are projected simultaneously from the origin with velocities $-30\mathbf{i} + 90\mathbf{j}$ and $30\mathbf{i} + 110\mathbf{j}$ respectively. Both particles impinge on a plane which contains the origin, meeting the plane on a line of greatest slope which has the equation $\mathbf{r} = \lambda(3\mathbf{i} + \mathbf{j})$, where λ is a scalar. Show that the times of flight of A and B are equal and that the distance between the points of impact is $400\sqrt{10}\,\text{m}$. Find the cosine of the acute angle between the direction of motion of A at the instant of impact and the line of greatest slope of the plane. (JMB)

26. At noon two ships A and B have the following position and velocity vectors:

	Position vector	*Velocity vector*
Ship A	$10\mathbf{i} + 5\mathbf{j}$	$-2\mathbf{i} + 4\mathbf{j}$
Ship B	$2\mathbf{i} - \mathbf{j}$	$2\mathbf{i} + 7\mathbf{j}$

where $\mathbf{i}$ and $\mathbf{j}$ are unit vectors in the directions East and North respectively, and where the speeds are measured in kilometres per hour and the distances in kilometres. If they continue on their respective courses, (i) find the position vector of A after a time t hours; (ii) show that they will collide, and give the time of the collision; (iii) determine how far ship A will have travelled between noon and the collision. (SU)

27. A ship A is travelling on a course of 060° at a speed of $30\sqrt{3}$ km/h and a ship B is travelling on a course of 030° at 20 km/h. At noon B is 260 km due east of A.

Using unit base vectors **i** and **j** pointing east and north respectively find $\overrightarrow{AB}$ in terms of **i**, **j** and t at time t hours in the afternoon. Hence, or otherwise, calculate the least distance between A and B, to the nearest kilometre, and the time at which they are nearest to one another. (L)

28. An aeroplane A moves in space with a variable velocity **v** and is attacked by a guided missile B which moves with constant speed u, where $u > |v|$. The guidance system of B is such that, if from any instant the velocity of A stayed constant, B would hit A without having to change direction. Show that the velocity of A relative to B is

$$\left[\frac{\mathbf{r}.\mathbf{v} - \sqrt{\{(\mathbf{r}.\mathbf{v})^2 + (u^2 - \mathbf{v}^2)\mathbf{r}^2\}}}{\mathbf{r}^2}\right]\mathbf{r},$$

where **r** is the position vector of A relative to B. (O)

29. At time t a particle is in motion with velocity **v** and is being acted upon by a variable force **F**. Write down expressions for (i) the power at time t, (ii) the work done by **F** during the time interval $0 \leqslant t \leqslant T$.

The particle, of mass m, moves in a plane where **i** and **j** are perpendicular unit vectors so that its position vector at time t is given by $\mathbf{r} = 2a\cos 2t\mathbf{i} + a\sin 2t\mathbf{j}$, where a is a positive constant. Derive expressions for the velocity **v** and the force **F** at time t. Obtain an expression in terms of t for the power at time t and show that the work done by **F** during the interval $0 \leqslant t \leqslant T$ is $3ma^2(1 - \cos 4T)$. If T varies, find the maximum value of the work done by **F** and determine also the smallest value of T for which this maximum value is reached. (JMB)

30. A smooth fixed wire is in the shape of the curve given by

$$\mathbf{r} = a(\theta - \sin\theta)\mathbf{i} + a(1 + \cos\theta)\mathbf{j},\ 0 \leqslant \theta \leqslant 2\pi,$$

where **i**, **j** are unit vectors in the horizontal and upward vertical directions respectively. A particle threaded on the wire is released from rest at the point A $(\theta = 0)$. Write down, in terms of θ and its derivative, an equation expressing the conservation of energy during the motion. Show that the time taken for the particle to reach B $(\theta = 2\pi)$ is $2\pi\sqrt{(a/g)}$. (JMB)

8 Impulse and momentum

8.1 The impulse of a force

When a constant force $\mathbf{F}$ acts for a time t, then the *impulse* of the force is defined as the vector quantity $\mathbf{F}t$.

The SI unit of impulse is the newton second (N s).

Example 1 A force $\mathbf{F} = 4\mathbf{i} - 3\mathbf{j}$ acts on a particle for 3 seconds. Find the magnitude of the impulse of the force.

If the impulse of the force is $\mathbf{I}$, then

$$\mathbf{I} = (4\mathbf{i} - 3\mathbf{j}) \times 3$$
$$\therefore \quad |\mathbf{I}| = 3|4\mathbf{i} - 3\mathbf{j}| = 3\sqrt{(4^2 + 3^2)} = 15.$$

Hence the magnitude of the impulse is 15 N s.

When a variable force $\mathbf{F}$ acts from time t_1 to time t_2, then the impulse of the force is defined as $\int_{t_1}^{t_2} \mathbf{F}\, dt$.

Example 2 A particle is moving under the action of a force $\mathbf{F} = 2\mathbf{i} - 3t^2\mathbf{j} + 4t\mathbf{k}$ at the time t. Find the impulse given to the particle in the interval $1 \leqslant t \leqslant 2$.

$$\text{Impulse} = \int_1^2 (2\mathbf{i} - 3t^2\mathbf{j} + 4t\mathbf{k})\, dt$$

$$= \Big[2t\mathbf{i} - t^3\mathbf{j} + 2t^2\mathbf{k} \Big]_1^2$$
$$= (4\mathbf{i} - 8\mathbf{j} + 8\mathbf{k}) - (2\mathbf{i} - \mathbf{j} + 2\mathbf{k})$$
$$= 2\mathbf{i} - 7\mathbf{j} + 6\mathbf{k}.$$

In some circumstances it is difficult to determine the exact nature of the forces involved. For example, in collisions between moving objects comparatively large unknown forces act for very short intervals of time. We will now show that the impulse of such a force can be found by considering the effect it produces.

Suppose that at time t a particle of mass m is moving with velocity $\mathbf{v}$ under the action of a force $\mathbf{F}$. Let $\mathbf{I}$ be the impulse given to the particle by $\mathbf{F}$ in the interval $t_1 \leqslant t \leqslant t_2$. Whether $\mathbf{F}$ is constant or variable,

$$\mathbf{I} = \int_{t_1}^{t_2} \mathbf{F}\, dt$$

Using Newton's second law in its original form,

$$\mathbf{F} = \frac{d(m\mathbf{v})}{dt},$$

where $m\mathbf{v}$ is the *momentum* of the particle.

$$\therefore \qquad \mathbf{I} = \int_{t_1}^{t_2} \frac{d(m\mathbf{v})}{dt}\, dt$$

$$\mathbf{I} = \Big[m\mathbf{v} \Big]_{t_1}^{t_2}$$

$$\mathbf{I} = (m\mathbf{v})_{at\, t_2} - (m\mathbf{v})_{at\, t_1}$$

i.e. **impulse = change in momentum**

[Because of the relationship between impulse and momentum, both quantities are usually measured in newton seconds. An alternative unit of momentum is the kilogramme metre per second. However, since $1\,\text{N} = 1\,\text{kg}\,\text{m}\,\text{s}^{-2}$, it follows that $1\,\text{N}\,\text{s} = 1\,\text{kg}\,\text{m}\,\text{s}^{-1}$.]

Example 3 A truck of mass 1400 kg moving along a straight horizontal track at $2\,\text{m}\,\text{s}^{-1}$ runs into fixed buffers and rebounds at a speed of $1\,\text{m}\,\text{s}^{-1}$. If the truck and the buffers are in contact for 1·2 s, find the average force exerted by the buffers on the truck.

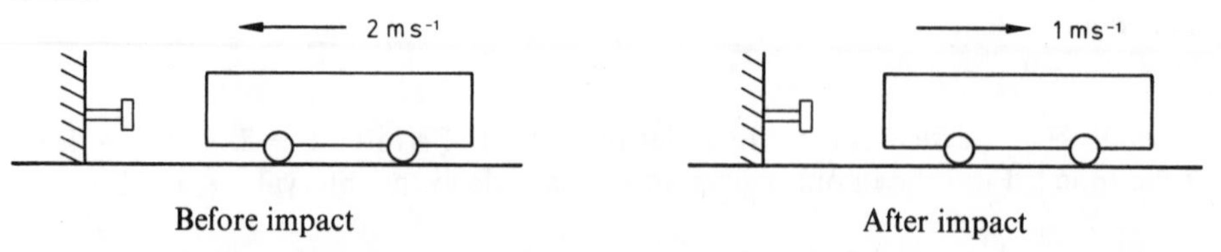

Before impact After impact

Let the direction away from the buffers be positive and let F N be the average force exerted by the buffers.

Impulse = change in momentum

$$\therefore \quad F \times 1{\cdot}2 = 1400 \times 1 - 1400 \times (-2)$$

$$\therefore \quad F = \frac{1400 \times 3}{1{\cdot}2} = 3500$$

Hence the average force exerted by the buffers is 3500 N.

Example 4 A particle of mass 5 kg is moving with velocity $3\mathbf{i} - 7\mathbf{j}$ when it is given an impulse $\mathbf{I} = -7\mathbf{i} + 16\mathbf{j}$. If the velocity of the particle after the impulse is $\mathbf{v}$, find $\mathbf{v}$. Find also the change in kinetic energy.

$$\text{Impulse} = \text{change in momentum}$$
$$\therefore \quad -7\mathbf{i} + 16\mathbf{j} = 5\mathbf{v} - 5(3\mathbf{i} - 7\mathbf{j})$$
$$\therefore \quad 5\mathbf{v} = -7\mathbf{i} + 16\mathbf{j} + 15\mathbf{i} - 35\mathbf{j}$$
$$\text{i.e.} \quad \mathbf{v} = \tfrac{1}{5}(8\mathbf{i} - 19\mathbf{j})$$

Initial speed $= \sqrt{(3^2 + 7^2)} = \sqrt{(9 + 49)} = \sqrt{58}$
New speed $= \frac{1}{5}\sqrt{(8^2 + 19^2)} = \frac{1}{5}\sqrt{(64 + 361)} = \frac{1}{5}\sqrt{425} = \sqrt{17}$.
$\therefore$ there is a loss in kinetic energy of

$$(\tfrac{1}{2} \times 5 \times 58 - \tfrac{1}{2} \times 5 \times 17)\,\text{J} \quad \text{i.e.} \quad 102{\cdot}5\,\text{J}.$$

Example 5 A ball of mass 0·25 kg strikes a wall at a speed of $16\,\text{m}\,\text{s}^{-1}$ and rebounds with a speed of $10\,\text{m}\,\text{s}^{-1}$. If the angle between the velocities of the ball before and after impact is α, where $\tan\alpha = \frac{3}{4}$, find the magnitude of the impulse exerted by the wall on the ball.

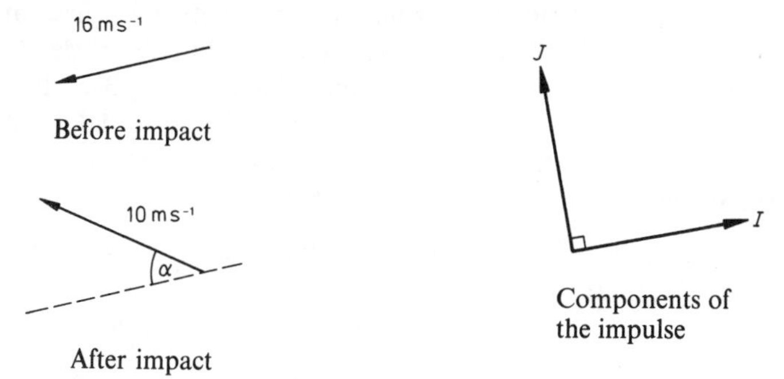

Before impact

After impact

Components of the impulse

Resolving parallel and perpendicular to the initial velocity of the ball,

$$\nearrow \quad I = 0{\cdot}25(-10\cos\alpha) - 0{\cdot}25(-16)$$
$$\nwarrow \quad J = 0{\cdot}25(10\sin\alpha)$$

Using a right-angled triangle,

$$\sin\alpha = \tfrac{3}{5}, \quad \cos\alpha = \tfrac{4}{5}.$$

$$\therefore \quad I = 0{\cdot}25(-8) - 0{\cdot}25(-16) = 2$$
$$J = 0{\cdot}25 \times 6 = 1{\cdot}5$$

Hence the magnitude of the impulse exerted by the wall

$$= \sqrt{(2^2 + 1{\cdot}5^2)}\,\text{N}\,\text{s} = 2{\cdot}5\,\text{N}\,\text{s}.$$

When a jet of water is directed at a wall, the reaction of the wall on the water produces a *continuous change in momentum.* Provided that the water is being delivered at a steady rate and with constant velocity, the force exerted by the wall can be determined by considering the momentum lost by the water each second.

Example 6 A hose delivers water horizontally at a rate of 20 kg per second with a speed of $25\,\text{m s}^{-1}$. The water strikes a vertical wall and does not rebound. Find the force exerted on the wall.

Let F N be the force exerted on the wall, then the force exerted by the wall is also F N,
$\therefore$ impulse exerted in 1 second $= F \times 1\,\text{N s} = F\,\text{N s}$.
Mass of water delivered in 1 second $= 20\,\text{kg}$
$\therefore$ loss of momentum in 1 second $= 20 \times 25\,\text{N s} = 500\,\text{N s}$
However, impulse = change in momentum,
$\therefore$ $F = 500$
Hence the force exerted on the wall is 500 N.

An instantaneous impulse acting along a string is called an *impulsive tension*. It may produce changes in momentum in the direction of the string, but has no effect perpendicular to the string. In particular, if a stationary particle is jerked into motion by the sudden tightening of a string attached to it, then the initial velocity of the particle will be in the direction of the string.

Example 7 One end of a light inextensible string of length 2 m is fixed at a point A on a smooth horizontal plane. The other end is attached to a particle of mass 2 kg. The particle is placed at a point B on the plane, where $AB = \sqrt{3}$ m, then projected horizontally with speed $6\,\text{m s}^{-1}$ in a direction perpendicular to AB. Find the speed of the particle immediately after the string has become taut. Find also the impulsive tension in the string.

If C is the point on the path of the particle at which the string first becomes taut, $\angle CAB = \cos^{-1}(\sqrt{3}/2) = 30°$.

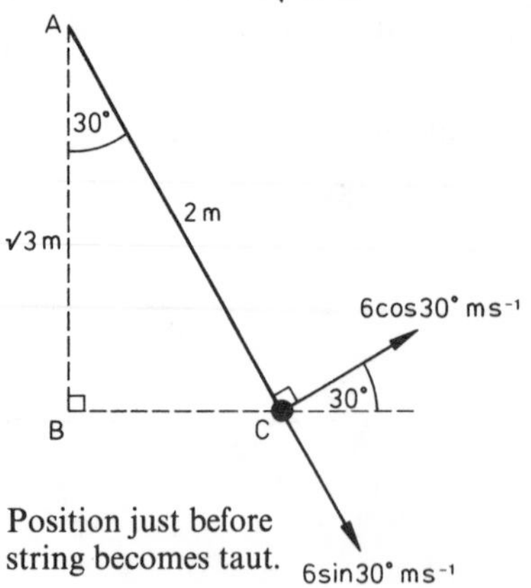

Position just before string becomes taut.

As shown in the diagram, just before the string tightens the velocity of the particle has components $6 \sin 30°\,\text{m s}^{-1}$ along AC and $6 \cos 30°\,\text{m s}^{-1}$ perpendicular to AC. When the string becomes taut the particle loses its momentum in the direction of AC, but its momentum perpendicular to AC is unchanged.

Hence just after the string has become taut, the speed of the particle is $6 \cos 30°\,\text{m s}^{-1}$ i.e. $3\sqrt{3}\,\text{m s}^{-1}$.
The loss of momentum along AC is produced by the impulsive tension in the string,

$\therefore$ impulsive tension $= 2 \times 6 \sin 30°\,\text{N s} = 6\,\text{N s}$.

In problems which involve systems of two or more particles the impulse-momentum relationship can be used both for the system as a whole and for particles taken separately. When considering the impulse given to the whole system, we will see that internal impulses occur in equal and opposite pairs and cancel each other out. Thus we may write:

sum of external impulses = total change in momentum.

The next example concerns two particles connected by a light inextensible string. In this case the impulsive tensions acting on the particles are equal and opposite, so they can be disregarded when dealing with the system as a whole.

Example 8 Particles A and B, of masses 1 kg and 2 kg respectively, are connected by a light inextensible string. The particles are at rest on a smooth horizontal plane with the string taut. A horizontal impulse of magnitude 12 N s is applied to particle B in a direction inclined at 45° to the direction of AB. Find the velocities with which A and B start to move. Find also the impulsive tension in the string.

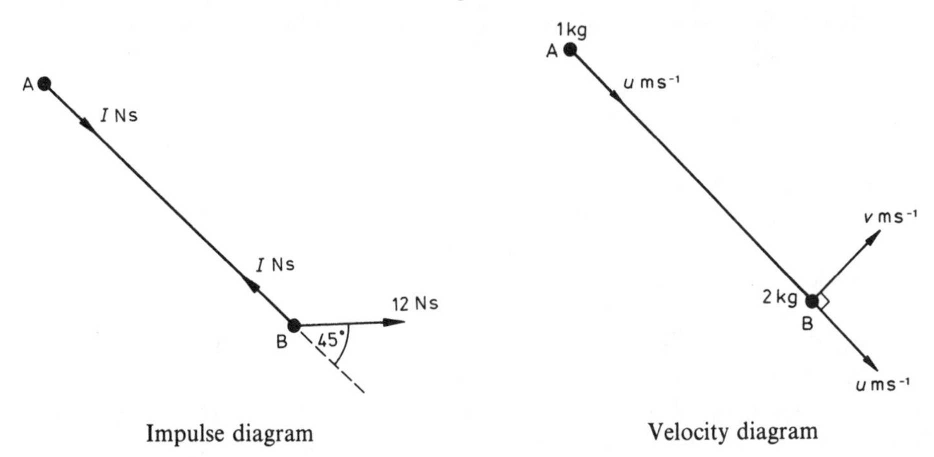

Impulse diagram

Velocity diagram

Let the initial velocity of B have components u and v along and perpendicular to AB respectively. Since the string is taut, A will start to move with speed u along AB.

For A and B together, impulse = change in momentum,

$\therefore$ resolving along and perpendicular to AB,

$$\searrow \qquad 12\cos 45° = 1 \times u + 2 \times u$$
$$\nearrow \qquad 12\sin 45° = 2 \times v$$

$$\therefore \qquad u = 2\sqrt{2} \quad \text{and} \quad v = 3\sqrt{2}$$

Thus the initial velocity of A is $2\sqrt{2}\,\text{m s}^{-1}$ along AB.

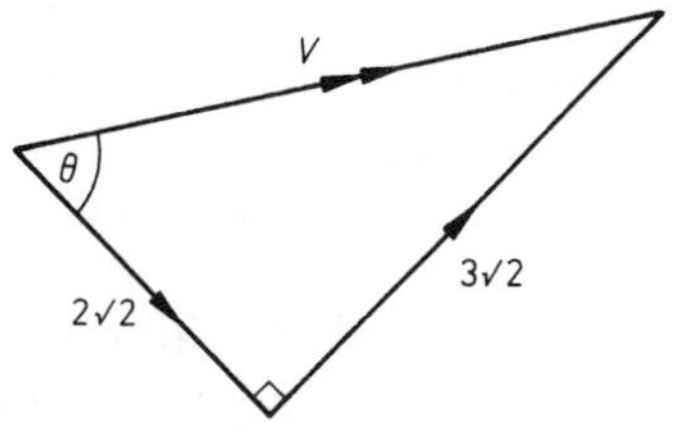

If the initial velocity of B is $V\,\text{m s}^{-1}$ at an angle θ to AB, then

$$V = \sqrt{\{(2\sqrt{2})^2 + (3\sqrt{2})^2\}} = \sqrt{26}$$
$$\theta = \tan^{-1}(3/2) \approx 56{\cdot}3°$$

Thus the initial velocity of B is $\sqrt{26}\,\text{m s}^{-1}$ at an angle of 56·3° to AB.

Let I N s be the impulsive tension in the string, then considering the motion of A,

$$I = 1 \times 2\sqrt{2}.$$

Hence the impulsive tension in the string is $2\sqrt{2}$ N s.

Exercise 8.1

1. When a constant force $\mathbf{F}$ acts for time t, the impulse of the force is $\mathbf{I}$.
(a) If $\mathbf{F} = 2\mathbf{i} - 7\mathbf{j}$ and $t = 4$, find $\mathbf{I}$.
(b) If $\mathbf{I} = 6\mathbf{i} - 2\mathbf{j} + 3\mathbf{k}$ and $t = 3$, find $\mathbf{F}$.
(c) If $\mathbf{F} = -12\mathbf{i} + 5\mathbf{j}$ and $t = 5$, find $|\mathbf{I}|$.
(d) If $\mathbf{I} = 6\mathbf{i} + 12\mathbf{j} - 4\mathbf{k}$ and $t = 3{\cdot}5$, find $|\mathbf{F}|$.

2. Find the impulse of the variable force $\mathbf{F}$ in the stated interval of time.
(a) $\mathbf{F} = 5t\mathbf{i} + 3\mathbf{j}, 0 \leqslant t \leqslant 2$.
(b) $\mathbf{F} = \mathbf{i} - 6t^2\mathbf{j} + 2t\mathbf{k}, 1 \leqslant t \leqslant 3$.
(c) $\mathbf{F} = \cos 2t\mathbf{i} + 3\sin 2t\mathbf{j}, 0 \leqslant t \leqslant \frac{1}{2}\pi$.

3. Find the momentum of
(a) a man of mass 80 kg running at 6 m s^{-1},
(b) a car of mass 1200 kg travelling at 25 m s^{-1},
(c) a bullet of mass 15 g moving at 800 m s^{-1}.

4. A particle of mass m moves so that its position vector at time t is $\mathbf{r}$. Find, as a vector, the momentum of the particle at time T if
(a) $\mathbf{r} = 3\mathbf{i} + t^2\mathbf{j}, m = 2, T = 3$;
(b) $\mathbf{r} = -4t\mathbf{i} + t^3\mathbf{k}, m = 0{\cdot}5, T = 2$;
(c) $\mathbf{r} = 5\cos 3t\mathbf{i} + 2\mathbf{j} + 5\sin 3t\mathbf{k}, m = 4, T = \pi$.

5. A stone of mass 3 kg is falling at a speed of 18 m s^{-1} when it hits the ground. If the stone is brought to rest by the impact, find the impulse exerted by the ground.

6. A ball of mass 0·2 kg strikes a wall when moving horizontally at 12 m s^{-1}. If the ball rebounds horizontally at 8 m s^{-1}, find the impulse exerted by the wall.

7. A bullet of mass 0·03 kg is fired into a fixed block of wood at a speed of 400 m s^{-1}. If the bullet is brought to rest in 0·02 s, find the average resistance exerted by the wood.

8. A particle of mass 5 kg is moving under the action of a constant force of magnitude 8 N. Initially the speed of the particle is 12 m s^{-1}. Find its speed after 3 seconds if the force (a) acts in the direction of motion, (b) directly opposes motion.

9. A particle of mass 2 kg is projected across a rough horizontal plane with a speed of 21 m s^{-1}. Find the time in which the particle comes to rest, given that the coefficient of friction is 2/7.

10. A particle of mass 0·5 kg is moving with velocity $\mathbf{u} = 6\mathbf{i} - 7\mathbf{j}$ when it is given an impulse $\mathbf{I} = 3\mathbf{i} + 11\mathbf{j}$. Find the new velocity of the particle and the average force required to bring it to rest in a further 3 seconds.

11. A particle of mass 0·8 kg is moving with velocity $4\mathbf{i} + 7\mathbf{j}$ when it strikes a fixed obstacle and rebounds with velocity $-6\mathbf{i} + 2\mathbf{j}$. Find the impulse exerted by the particle on the obstacle. Find also the kinetic energy lost by the particle in the collision.

12. A particle of mass 3 kg is moving with speed $6\,\mathrm{m\,s^{-1}}$ when it receives an impulse of magnitude 24 N s. Find the speed of the particle immediately after the impulse and the change in the kinetic energy of the particle if the direction of the impulse and the original direction of motion are (a) the same, (b) opposite, (c) perpendicular.

13. A ball of mass 0·2 kg strikes a wall at a speed of $12\,\mathrm{m\,s^{-1}}$ and rebounds with a speed of $3{\cdot}5\,\mathrm{m\,s^{-1}}$. If its path is diverted through 90° by the impact, find the magnitude of the impulse exerted on the wall.

14. An object of mass 4 kg strikes a barrier at a speed of $10\,\mathrm{m\,s^{-1}}$ and rebounds at a speed of $5\,\mathrm{m\,s^{-1}}$. If the angle between the velocities of the object before and after impact is 60°, find the magnitude of the impulse exerted on the barrier.

15. A hose is discharging water at the rate of 50 litres per second. The jet of water is travelling horizontally with a speed of $12\,\mathrm{m\,s^{-1}}$ when it strikes a vertical wall. Assuming that the water does not rebound, find the force exerted on the wall. [Take the mass of 1 litre of water to be 1 kg.]

16. A stream of water falling vertically from a pipe at the rate of 75 kg per second exerts a force of 900 N on the ground below. Assuming that the ground destroys the momentum of the water, find the speed at which the water hits the ground.

17. In a factory small screws are falling vertically onto a conveyor belt without bouncing at a rate of 48 kg per minute. Just before impact the screws are travelling at $2\,\mathrm{m\,s^{-1}}$. If the conveyor belt moves horizontally at $1{\cdot}5\,\mathrm{m\,s^{-1}}$, find the magnitude and direction of the average force it exerts on the screws.

18. One end of a light inextensible string of length $\sqrt{3}$ m is fixed at a point A on a smooth horizontal plane. The other end is attached to a particle of mass 3 kg. The particle is placed at a point B on the plane, where $AB = 1$ m, then projected horizontally with speed $4\,\mathrm{m\,s^{-1}}$. Find the speed of the particle immediately after the string has become taut and the impulsive tension in the string, given that the angle between the initial velocity of the particle and the direction BA is (a) 60°, (b) 90°.

19. Particles A and B, of masses 3 kg and 2 kg respectively, are connected by a light inextensible string. The particles are at rest on a smooth horizontal plane with the string taut. A horizontal impulse of magnitude 12·5 N s is applied to particle B in a direction inclined at an angle α to the direction of AB, where $\tan\alpha = 4/3$. Find the

velocities with which A and B start to move. Find also the impulsive tension in the string.

20. Particles A and B, both of mass m, are connected by a light inextensible string. A particle C of mass $2m$ is attached to the mid-point of the string. The particles are at rest on a smooth horizontal plane with the string taut and with AC at right angles to BC. A horizontal impulse of magnitude $12m$ is applied to C in a direction inclined at an angle α to AC, such that both parts of the string remain taut. Find the speeds with which the particles begin to move if (a) $\alpha = 45°$, (b) $\alpha = 60°$.

8.2 Conservation of momentum

When two particles collide they exert equal and opposite impulses on each other. Therefore the changes in the momentum of the particles are also equal and opposite. It follows that the total momentum of the two particles together remains unchanged. A similar argument can be used in any situation in which equal and opposite impulses act on two bodies, e.g. when a bullet is fired from a gun or when a light inextensible string connecting two particles suddenly becomes taut.

These examples are special cases of a more general result called the *principle of conservation of linear momentum.*

> If no external force acts on a system of particles in a particular direction, then there can be no change in the total momentum of the system in that direction.

Example 1 Two particles A and B, of masses 3 kg and 2 kg respectively, are travelling in opposite directions along the same straight line when they collide. Just before the impact A is moving with speed $8\,\text{m}\,\text{s}^{-1}$ and B with speed $5\,\text{m}\,\text{s}^{-1}$. After the impact the speeds of A and B are $2\,\text{m}\,\text{s}^{-1}$ and $v\,\text{m}\,\text{s}^{-1}$ respectively. Find the value of v given that after the impact A and B are travelling in (a) the same direction, (b) opposite directions.

(a)

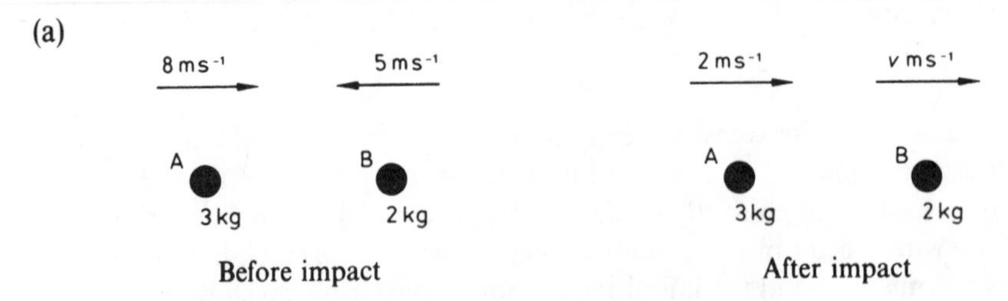

Momentum before impact = momentum after impact
$\therefore$ taking the direction of motion of A as positive,

$$3 \times 8 + 2 \times (-5) = 3 \times 2 + 2 \times v$$

i.e. $$14 = 6 + 2v$$

Thus $$v = 4$$

(b)

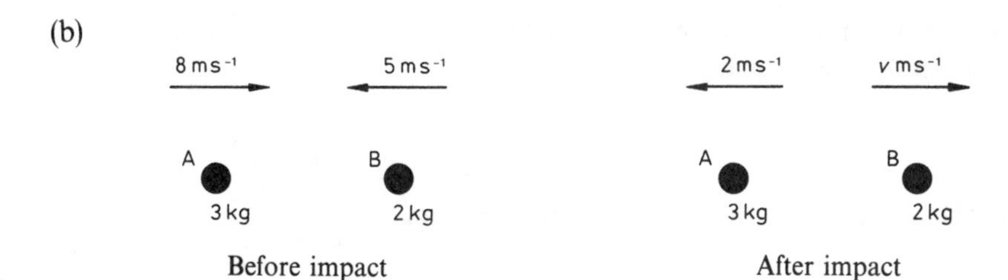

Momentum before impact = momentum after impact

$$\therefore \quad 3 \times 8 + 2 \times (-5) = 3 \times (-2) + 2 \times v$$

i.e. $$14 = -6 + 2v$$

Thus $$v = 10.$$

Example 2 A bullet of mass 0·02 kg is fired from a gun of mass 6 kg with a speed of 450 m s^{-1}. Find the speed of the gun's recoil. Find also the constant force required to bring the gun to rest in 0·5 seconds.

Let v m s^{-1} be the speed of the gun and let F N be the force required to bring the gun to rest in 0·5 seconds.

Momentum after firing = momentum before firing

$$\therefore \quad 6 \times v + 0{\cdot}02 \times (-450) = 0$$

i.e. $$6v - 9 = 0$$

$$v = 1{\cdot}5$$

Hence the speed of the gun's recoil is 1·5 m s^{-1}.
As the gun is brought to rest,

impulse = change in momentum

$$\therefore \quad F \times 0{\cdot}5 = 6 \times 1{\cdot}5$$

i.e. $$F = 18$$

Hence the constant force required is 18 N.

Example 3 A particle A of mass 5 kg slides from rest down a smooth plane inclined at an angle α to the horizontal, where $\sin \alpha = 1/6$. After 3 s it collides with a particle B of mass 1 kg which is sliding down the plane at 2·5 m s^{-1}. If the two particles coalesce, find their common speed immediately after impact.

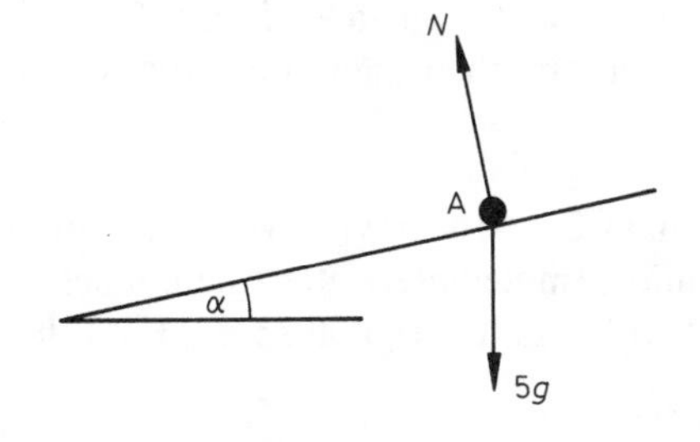

The change in the momentum of A is equal to the impulse received.
$\therefore$ just before the collision the momentum of A down the plane

$$= 5g \sin \alpha \times 3 \text{ N s}$$
$$= 5 \times 9{\cdot}8 \times \tfrac{1}{6} \times 3 \text{ N s}$$
$$= 24{\cdot}5 \text{ N s}$$

Momentum is conserved in the collision between A and B. Thus if $v\,\text{m s}^{-1}$ is their common speed after impact,

$$24{\cdot}5 + 1 \times 2{\cdot}5 = (5 + 1)v$$
$$\therefore \qquad v = 4{\cdot}5$$

Hence the common speed of the particles after impact is $4{\cdot}5\,\text{m s}^{-1}$.

Exercise 8.2

1. A railway truck of mass 12 000 kg is moving along a straight level track at $7\,\text{m s}^{-1}$ when it collides with another truck of mass 8000 kg moving in the same direction at $2\,\text{m s}^{-1}$. After the collision the trucks move on together. Find the common speed of the trucks after impact.

2. A bullet of mass 0·04 kg is fired horizontally into a block of wood of mass 1·4 kg which rests on a smooth horizontal surface. The bullet becomes embedded in the block and after impact they move on together at a speed of $20\,\text{m s}^{-1}$. Find the speed at which the bullet is fired.

3. Two particles, A of mass 5 kg and B of mass 2 kg, are moving horizontally along the same straight line when they collide. Both come to rest instantaneously. If the velocity of A was $3\,\text{m s}^{-1}$ just before the impact, find the velocity of B and the kinetic energy lost in the impact.

4. A gun of mass 4 kg fires a bullet of mass 20 g at a speed of $500\,\text{m s}^{-1}$. Find the speed of the gun's recoil and the energy of the explosion in the gun.

5. Two particles A and B, of mass 3 kg and 5 kg respectively, collide. Just before impact A is moving with speed $8\,\text{m s}^{-1}$ and B is moving with speed $4\,\text{m s}^{-1}$ in the opposite direction. Immediately after the impact A rebounds with speed $7\,\text{m s}^{-1}$. Find the speed of B immediately after impact.

6. Two particles X and Y, of mass 4 kg and 3 kg respectively, collide. Just before impact X is moving with speed $13\,\text{m s}^{-1}$ and Y is moving with speed $5\,\text{m s}^{-1}$ in the opposite direction. Immediately after the impact X moves on, with no change in direction, at $4\,\text{m s}^{-1}$. Find the speed of Y immediately after impact.

7. A particle of mass m and velocity $4\mathbf{i} + \mathbf{j}$ collides with a particle of mass $2m$ and velocity $\mathbf{i} - 8\mathbf{j}$. If the particles coalesce on impact, find their common velocity just after the collision.

8. Two particles, A of mass $3m$ and B of mass $2m$, are moving with velocities $4\mathbf{i} - 6\mathbf{j}$ and $2\mathbf{i} + 7\mathbf{j}$ respectively when they collide. Immediately after the impact A moves on in the same direction as before, but with half its original speed. Find the velocity of B just after the collision.

9. A particle of mass 2 kg moving with speed $10\,\mathrm{m\,s^{-1}}$ collides with a stationary particle of mass 7 kg. Immediately after impact the particles move with the same speed but in opposite directions. Find the kinetic energy lost in the collision.

10. Two particles, of masses $3m$ and $2m$ moving in opposite directions with speeds u and $4u$ respectively, collide. Given that no kinetic energy is lost in the collision, find the speeds of the particles immediately after impact.

11. Three trucks A, B and C, each of mass 2 tonnes, are moving on a straight horizontal track in the same direction with speeds of 10, 5 and $3\,\mathrm{m\,s^{-1}}$ respectively. When any two trucks collide, they become coupled. Thus after two collisions the three trucks move on with a common speed. Find this speed and find also the total loss of kinetic energy in the collisions.

12. A truck A of mass 3000 kg moving at $4\,\mathrm{m\,s^{-1}}$ collides with a truck B of mass 2000 kg moving at $2\,\mathrm{m\,s^{-1}}$ in the same direction. If after impact truck A continues to move in the same direction but at a speed of $3\,\mathrm{m\,s^{-1}}$, find the new speed of truck B. If the trucks are in contact for 1·5 s, find the average force each exerts on the other.

13. A van of mass 1800 kg begins to tow a stationary car of mass 1200 kg. Just before the tow-rope tightens, the van is moving at a speed of $1\,\mathrm{m\,s^{-1}}$. Find the common speed of the vehicles just after the rope tightens. Find also the impulsive tension in the rope.

14. A bullet of mass 0·03 kg is fired from a gun with a horizontal velocity of $600\,\mathrm{m\,s^{-1}}$. If the gun is then brought to rest in 1·2 s by a horizontal force which rises uniformly from zero to P N and then falls uniformly to zero, find the value of P.

15. A particle of mass 2 kg slides from rest down a smooth plane inclined at an angle of 30° to the horizontal. After 5 seconds it collides with a particle of mass 3 kg which is sliding down the plane at a speed of $7\,\mathrm{m\,s^{-1}}$. If the two particles coalesce on impact find their common velocity (a) just after the collision, (b) 2 seconds after the collision.

16. Two small beads A and B, each of mass m, are threaded on a smooth straight wire inclined at an angle α to the horizontal. Bead A is released from rest and begins to slide down the wire. At same instant bead B is projected up the wire towards A with speed u. After a time t the two beads collide and are brought to rest by the impact. Find an expression for t in terms of u, g and α.

8.3 Work, energy and momentum

When faced with a problem in particle dynamics, we will now have to decide whether to consider work-energy relationships, impulse-momentum relationships or a combination of both. The two fundamental equations are:

Work done = change in kinetic energy Impulse = change in momentum

Example 1 A bullet of mass m is fired horizontally with velocity V into a fixed block of wood and penetrates a distance d into it. If the block exerts a constant resistance R, which stops the bullet in a time T, find expressions for R and T in terms of m, V and d.

Work done against resistance = loss of kinetic energy

$$\therefore \quad Rd = \tfrac{1}{2}mV^2$$

Hence $$R = mV^2/2d$$

Impulse = change in momentum

$$\therefore \quad RT = mV$$

Hence $$T = \frac{mV}{R} = mV \times \frac{2d}{mV^2} = \frac{2d}{V}.$$

[Note that as the resistance is constant this problem could also be solved using Newton's second law and the constant acceleration formulae $v^2 = u^2 + 2as$, $v = u + at$.]

Example 2 A pile-driver of mass 3000 kg falls freely from a height of 2·5 m and strikes a pile of mass 1200 kg. After the blow the pile and the driver move on together. If the pile is driven a distance of 0·5 m into the ground, find the speed at which the pile starts to move into the ground and the average resistance of the ground to penetration.

Let $v \text{ m s}^{-1}$ be the downward velocity of the pile-driver just before impact. Using the formula $v^2 = u^2 + 2as$, where $u = 0, a = 9{\cdot}8, s = 2{\cdot}5$, we have

$$v^2 = 0 + 2 \times 9{\cdot}8 \times 2{\cdot}5 = 49$$

$$\therefore \quad v = 7$$

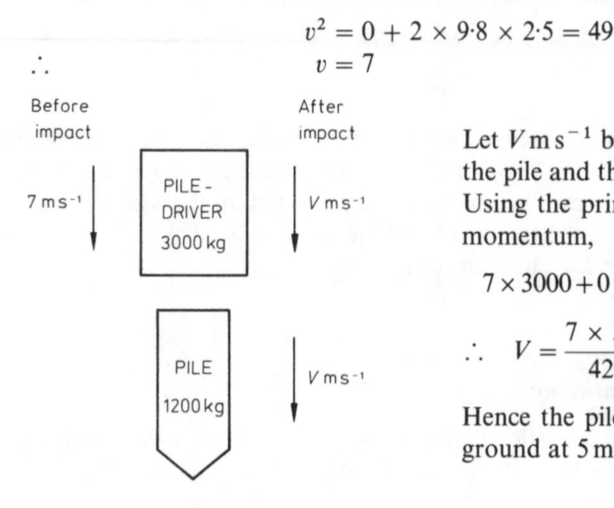

Let $V \text{ m s}^{-1}$ be the common velocity of the pile and the driver after impact. Using the principle of conservation of momentum,

$$7 \times 3000 + 0 \times 1200 = V(3000 + 1200)$$

$$\therefore \quad V = \frac{7 \times 3000}{4200} = 5$$

Hence the pile starts to move into the ground at 5 m s^{-1}.

Let R N be the average resistance of the ground, then the resultant force acting on pile and driver together is $(R - 4200g)$ N.

$$\text{Work done} = \text{change in kinetic energy}$$
$$\therefore \quad (R - 4200g) \times 0{\cdot}5 = \tfrac{1}{2} \times 4200 \times 5^2$$
$$\therefore \quad R - 4200g = 4200 \times 25$$
$$\therefore \quad R = 4200(25 + 9{\cdot}8) = 146\,160$$

Hence the average resistance of the ground is 146 160 N.

When a particle moves in a vertical plane there are changes in its gravitational potential energy. Provided that no force other than gravity is doing work, we may then apply the principle of conservation of mechanical energy in the form:

$$\boxed{\text{K.E.} + \text{P.E.} = \text{constant}}$$

Example 3 A particle A, of mass m, is attached to a fixed point by means of a light inextensible string and hangs in equilibrium. A particle B, of mass $2m$, is travelling horizontally with speed $5u$ when it collides with particle A. Immediately after the impact B continues to move in the same direction but with speed $3u$. If A swings to a height h above its original position before coming instantaneously to rest, find an expression for h in terms of u and g.

Let v be the speed of particle A after impact.

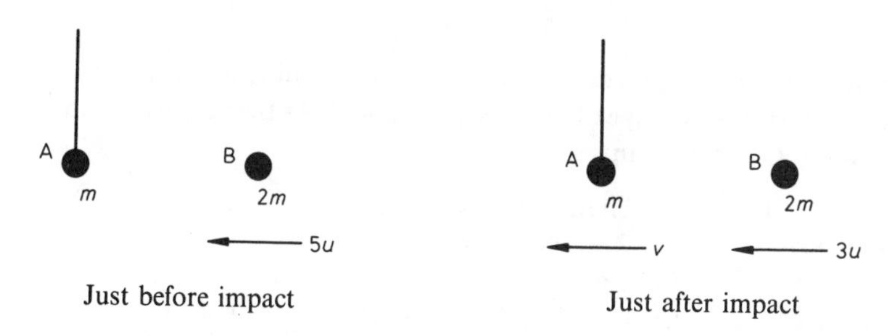

Just before impact Just after impact

Using the principle of conservation of momentum,

$$m \times 0 + 2m \times 5u = m \times v + 2m \times 3u$$
$$10mu = mv + 6mu$$
$$\therefore \quad v = 4u$$

Hence the speed of A immediately after impact is $4u$.
When A is swinging freely its direction of motion is always at right angles to the string. Thus the tension in the string does no work and mechanical energy is conserved.

P.E. after impact = 0, K.E. after impact $= \tfrac{1}{2}m(4u)^2 = 8mu^2$.
P.E. at height $h = mgh$, K.E. at height $h = 0$.

$$\therefore \quad 0 + 8mu^2 = mgh + 0$$
Hence $$h = 8u^2/g.$$

Example 4 Particles A and B, of masses 1 kg and 2 kg respectively, are attached to the ends of a light inextensible string passing over a smooth light pulley. The system is released from rest with both parts of the string taut and vertical. After A has travelled 1·2 m it hits and coalesces with a stationary particle C, of mass 2 kg. Find the new speed of the system and the impulsive tension in the string at the instant when C is picked up.

Taking the initial P.E. of each particle as zero, after the system has travelled 1·2 m the P.E. of A is $1{\cdot}2g$ J and the P.E. of B is $-1{\cdot}2 \times 2g$ J i.e. $-2{\cdot}4g$ J.

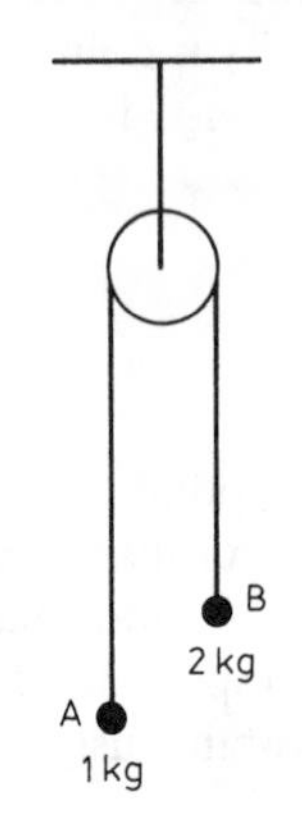

The initial K.E. of the system is zero. If v m s^{-1} is the speed of the particles after they have travelled 1·2 m, the K.E. acquired is $\frac{1}{2}(1 + 2)v^2$ J i.e. $1{\cdot}5v^2$ J.
Since no external force is doing work, the mechanical energy of the system remains constant.

$$\therefore \quad 1{\cdot}5v^2 + (1{\cdot}2g - 2{\cdot}4g) = 0$$
$$\text{i.e.} \quad 1{\cdot}5v^2 = 1{\cdot}2 \times 9{\cdot}8$$
$$\therefore \quad v = 2{\cdot}8$$

Thus, just before C is picked up the system is travelling at 2·8 m s^{-1}.
Let V m s^{-1} be the new speed of the system and J N s the impulsive tension in the string when C is picked up.

Impulse = change in momentum

Thus for B: $J = 2 \times 2{\cdot}8 - 2 \times V$

i.e. $J = 5{\cdot}6 - 2V$ (1)

For A and C: $J = (1 \times V + 2 \times V) - (1 \times 2{\cdot}8 + 2 \times 0)$

i.e. $J = 3V - 2{\cdot}8$ (2)

Subtracting (2) from (1): $0 = 8{\cdot}4 - 5V$

$\therefore$ $V = 1{\cdot}68$

Substituting in (1): $J = 5{\cdot}6 - 2 \times 1{\cdot}68 = 2{\cdot}24$.

Hence the new speed of the system is 1·68 m s^{-1} and the impulsive tension in the string is 2·24 N s.

Example 5 A particle P of mass 4 kg is connected to a particle Q of mass 3 kg by means of a light inextensible string which passes over a smooth light pulley. The system is released from rest with both parts of the string taut and vertical. After 3·5 s particle P hits the ground without rebounding. Assuming that particle Q does not hit the pulley, find the further time which elapses before the string is again taut. Find also the speed with which P leaves the ground.

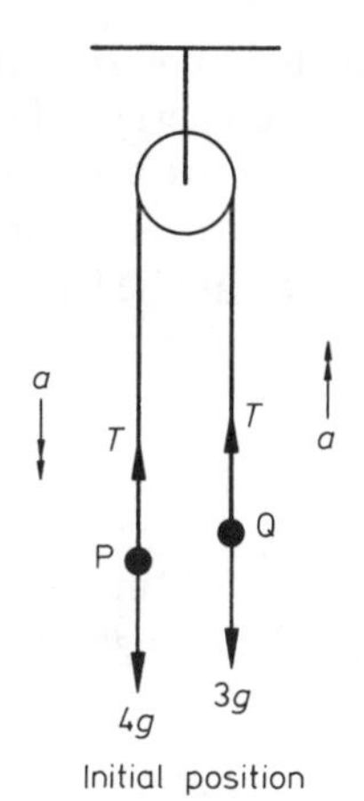

Initial position

Let $a\,\mathrm{m\,s^{-2}}$ be the acceleration of the particles and let $T\,\mathrm{N}$ be the tension in the string.
Using Newton's second law,

$$\begin{aligned} &\text{for } P: & 4g - T &= 4a & (1)\\ &\text{for } Q: & T - 3g &= 3a & (2)\\ &(1) + (2): & g &= 7a \end{aligned}$$

$$\therefore \quad a = \frac{g}{7} = 1{\cdot}4$$

Using the formula $v = u + at$, where $u = 0$, $a = 1{\cdot}4$, $t = 3{\cdot}5$,

$$v = 0 + 1{\cdot}4 \times 3{\cdot}5 = 4{\cdot}9$$

Hence when P hits the ground, Q has an upward velocity of $4{\cdot}9\,\mathrm{m\,s^{-1}}$.

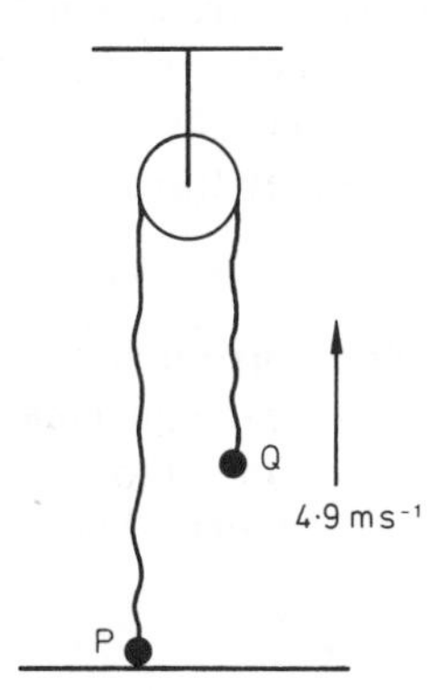

When P hits the ground

While P is on the ground Q moves freely under gravity. Hence the time which elapses before Q returns to the position where the string slackened is given by the formula $s = ut + \frac{1}{2}at^2$, where $s = 0$, $u = 4{\cdot}9$ and $a = -9{\cdot}8$.

$$\therefore \quad 0 = 4{\cdot}9t - 4{\cdot}9t^2$$
$$\therefore \quad t = 0 \quad \text{or} \quad t = 1.$$

Hence the string is again taut after a further time of 1 s.
Since there is no loss of energy in this time, it follows that as the string tightens Q has a downward velocity of $4.9\,\mathrm{m\,s^{-1}}$.

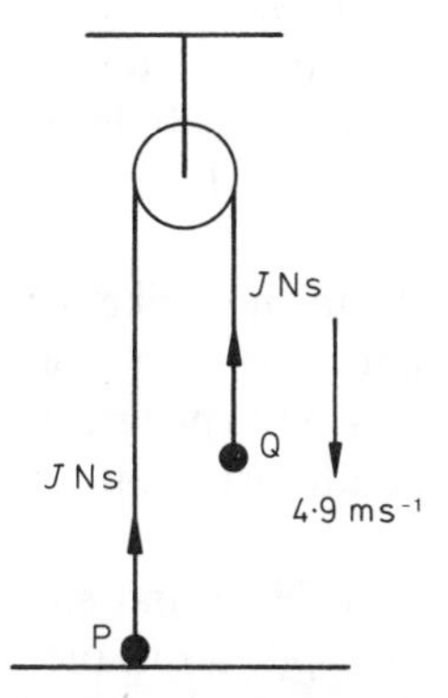

As the string becomes taut again

Let $V\,\mathrm{m\,s^{-1}}$ be the speed with which P is jerked into motion and let J N s be the impulsive tension in the string.

Impulse = change in momentum

$$\begin{aligned} &\text{Thus for } P: & J &= 4V\\ &\text{and for } Q: & J &= 3 \times 4{\cdot}9 - 3V\\ &\text{Subtracting:} & 0 &= 7V - 3 \times 4{\cdot}9\\ &\therefore & V &= 2{\cdot}1 \end{aligned}$$

Hence P leaves the ground at a speed of $2{\cdot}1\,\mathrm{m\,s^{-1}}$.

If we look again at the worked examples in this chapter we will see that a collision or the sudden tightening of a string often results in a loss of kinetic energy. This happens because some of the kinetic energy of the bodies involved is converted into other forms such as heat, sound or light. It is therefore important to stress that the principle of conservation of mechanical energy cannot usually be applied to a system when impulsive forces are acting.

Exercise 8.3

1. A car of mass 1200 kg travelling in a straight line at a speed of $20\,\text{m}\,\text{s}^{-1}$ is brought to rest in a distance of 50 m by a constant braking force. Find the magnitude of this force and the time taken to stop the car.

2. A bullet of mass 25 g is fired horizontally with velocity $600\,\text{m}\,\text{s}^{-1}$ into a fixed block of wood and penetrates a distance of 50 cm into it. Find the constant resistance exerted by the wood and the time taken to bring the bullet to rest.

3. A railway truck of mass 3000 kg moving at $4\,\text{m}\,\text{s}^{-1}$ collides with a truck of mass 2000 kg moving in the opposite direction at $1\,\text{m}\,\text{s}^{-1}$. If the trucks move on together after the collision, find their common speed just after impact. If a constant retarding force of 400 N acts on the trucks after impact, find the distance they travel before coming to rest.

4. A particle is projected at a speed of $7\,\text{m}\,\text{s}^{-1}$ across a rough horizontal plane. After travelling a distance of 2 m it collides with a stationary particle of equal mass. The two particles coalesce and travel a further distance d m before coming to rest. If the coefficient of friction between the particles and the plane is 5/12, find the value of d.

5. A stone of mass 5 kg is dropped from a point 10 m above horizontal ground. Find the impulse exerted by the ground on the stone if (a) it comes to rest without rebounding, (b) it rises to a maximum height of 10 cm after impact.

6. A particle of mass 2 kg is projected vertically upwards from a point A at a speed of $21\,\text{m}\,\text{s}^{-1}$. At the same instant a particle of mass 1 kg is dropped from a point 30 m vertically above A. Find the time which elapses before the two particles collide. If the particles coalesce on impact, find the speed at which they pass through A.

7. A pile-driver of mass 1500 kg falls freely from a height of 1·6 m and strikes a pile of mass 500 kg. After the blow the pile and the driver move on together. If the pile is driven a distance of 0·3 m into the ground, find the speed at which the pile starts to move into the ground and the average resistance of the ground to penetration.

8. A pile-driver of mass 1200 kg falls freely from a height of 3·6 m and strikes without rebounding a pile of mass 800 kg. The blow drives the pile a distance of 36 cm into the ground. Find the resistance of the ground, assumed to be constant. Find also the time for which the pile is in motion.

9. A particle P is released from rest at a point A on a smooth plane inclined at $30°$ to the horizontal. When it has reached a point B on the plane, P collides with a particle Q, of equal mass, moving up the plane at $7\,\text{m s}^{-1}$. After the collision P moves up the plane coming to rest instantaneously at A. Given that Q is brought to rest by the impact, find the distance AB.

10. A bullet of mass 30 g is fired into a stationary block of wood of mass 5 kg and becomes embedded in it. The block is freely suspended from a fixed point and after impact swings to a height of 40 cm above its initial position. Find, correct to 3 significant figures, the velocity of the bullet just before impact.

11. Two beads, A of mass m and B of mass $2m$, are threaded on a smooth circular wire fixed in a vertical plane. Initially A is at rest at the lowest point O of the wire and B is released from rest at a height h above O. Just before B collides with A it is moving with velocity v and immediately after impact A starts to move with velocity v. Find, in terms of h, the height above O at which each bead first comes to rest.

12. A particle P is attached to one end of a light inextensible string of length 1·25 m, the other end being attached to a fixed point A. The string is taut and inclined at an angle of $60°$ to the downward vertical when P is released from rest. As P reaches its lowest point B, it strikes a stationary particle Q, of equal mass, which is free to move horizontally. If the speed of Q just after impact is $2{\cdot}1\,\text{m s}^{-1}$, find the maximum height above B to which P rises after the collision.

13. Particles A and B, of masses 1 kg and 3 kg respectively are attached to the ends of a light inextensible string passing over a smooth light pulley. The system is released from rest with both parts of the string taut and vertical. After A has travelled 1·8 m it hits and coalesces with a stationary particle C, of mass 4 kg. Find the new velocity of the system and the impulsive tension in the string at the instant when C is picked up.

14. Particles A and B, of masses 1 kg and 2 kg respectively, are connected by a light inextensible string which passes over a smooth light pulley. The system is released from rest with the string taut and the hanging parts vertical. After 4·5 s A picks up a stationary particle of mass 3 kg. Find the velocity of the system immediately after impact and the further time which elapses before the system first comes instantaneously to rest.

15. A particle A of mass 5 kg is connected to a particle B of mass 2 kg by means of a light inextensible string which passes over a smooth light pulley. The system is released from rest with both parts of the string taut and vertical. After 3·5 s particle A hits the ground without rebounding. Assuming that particle B does not hit the pulley, find the further time which elapses before the string is again taut. Find also the speed with which A leaves the ground.

16. Particles P and Q, of masses m and $2m$ respectively, are connected by a light inextensible string which passes over a smooth light pulley. The system is released from rest with the string taut and both particles at a height h above the ground.

Assuming that Q hits the ground without rebounding and that P does not hit the pulley, find expressions for (a) the greatest height above the ground reached by P, (b) the impulsive tension in the string when Q leaves the ground again.

17. A particle A, of mass 4 kg, rests on a smooth plane inclined at an angle of 30° to the horizontal. It is connected by a light inextensible string passing over a smooth pulley fixed at the top of the plane to a particle B, of mass 3 kg, which hangs freely. The system is released from rest with both parts of the string taut. When B has fallen a distance of 6·3 m A picks up a particle C, of mass 3·5 kg, which was previously at rest. Find the velocity of the system immediately after C has been picked up. Find also the loss in kinetic energy as C is picked up, giving your answer correct to 3 significant figures.

18. A pile of mass M is being driven into the ground by a pile-driver of mass m. The pile-driver is released from rest at a height h above the top of the pile and falls freely under gravity on to the top of the pile. The pile-driver rebounds vertically from the top of the pile and, after the first impact, it comes first to rest at a point $\frac{3}{4}h$ below the original point of release. (i) Find the kinetic energy lost by the pile-driver in the first impact. (ii) Find the kinetic energy gained by the pile in the first impact, assuming that the momentum of the system is conserved in the impact. (iii) Given that the pile is driven a distance $\frac{1}{16}h$ into the ground by the first impact, show that the frictional resistance, assumed constant, between the pile and the ground is $(36m^2 + M^2)g/M$. (C)

19. Two particles, A and B, of masses $2m$ and $3m$ respectively, are attached to the ends of a light inextensible string of length c and are placed close together on a horizontal table. The particle A is projected vertically upwards with speed $\sqrt{(6gc)}$. (i) Show that, at the instant immediately after the string tightens, B is moving with velocity $\frac{4}{5}\sqrt{(gc)}$. (ii) State the impulse of the tension in the string. (iii) Find the height to which A rises above the table before it comes to instantaneous rest. (iv) Calculate the loss in kinetic energy due to the tightening of the string. (AEB 1979)

20. A cube of mass 100 g and side 14 cm is at rest on a smooth horizontal table. A bullet of mass 20 g is fired with velocity 400 m s^{-1} into the centre of a vertical face of the cube. The bullet travels through the cube in a direction perpendicular to this face and emerges undeflected with velocity 200 m s^{-1}. Find the velocity of the cube as the bullet emerges. Find the kinetic energy gained by the cube and the kinetic energy lost by the bullet. If the cube moves a distance of d cm before the bullet emerges, write down the distance moved by the bullet as it passes through the cube. By considering the work done on the cube and on the bullet, determine the value of d and the magnitude of the constant force exerted by the cube on the bullet.

8.4 Newton's law of restitution

In a collision between two elastic bodies the impulses exerted by each body on the other produce changes in momentum and the bodies bounce away from each other.

If both bodies are moving along the line of action of these impulses at the instant the collision occurs, then the impact is said to be *direct*. Otherwise it is described as *indirect* or *oblique*. Both types of elastic impact were investigated by Sir Isaac Newton and as a result of his experimental work he was able to formulate a law of impact which is known as *Newton's law of restitution* or as *Newton's experimental law*. In its general form the law concerns the indirect impact of two moving bodies. However, we first consider its application to the direct impact of one body with a fixed plane.

Our everyday experience tells us that if several objects made of different materials are dropped from a given height on to horizontal ground, then the various objects will probably rebound to different heights after impact. A lump of plasticine will become flattened and show little tendency to rebound, whereas a "superball" will regain most of its original height. The behaviour of most objects will come somewhere between these two extremes.

Further observation shows that when a body strikes a fixed plane there is a period of time during which the body is compressed and its speed is reduced to zero. This is followed by a period of restitution in which the body regains its original shape and starts to rebound. While the body is being compressed some of its original kinetic energy is stored as elastic potential energy and this is then released as the body rebounds. The remaining kinetic energy is converted into other forms such as heat energy. Thus, the behaviour of a particular body during impact depends on its elasticity.

Experimental evidence suggests that in a series of direct collisions between a given elastic body a fixed plane the ratio of the speed of the body after impact to its speed before impact is approximately constant. Thus, in its simplest form, Newton's law of restitution states that if the speed of the body before impact is u and its speed after impact is v, then

$$v = eu,$$

where e is constant for that particular body. The constant e is known as the *coefficient of restitution* between the body and the plane and its value always lies between 0 and 1. If $e = 0$, collisions between the body and the plane are said to be *inelastic*, but if $e = 1$ they are described as *perfectly elastic*.

Example 1 A ball dropped from a height of 9 m above horizontal ground rebounds to a height of 4 m. Find the coefficient of restitution between the ball and the ground.

Substituting $u = 0, a = g, s = 9$ in the formula $v^2 = u^2 + 2as$,

we have: $v^2 = 0 + 2g \times 9$

$\therefore \quad v = 3\sqrt{(2g)}$

Hence the speed of the ball just before impact is $3\sqrt{(2g)}$ m s^{-1}.

Substituting $v = 0, a = -g, s = 4$ in the formula $v^2 = u^2 + 2as$,

we have: $0 = u^2 - 2g \times 4$

$\therefore \quad u = 2\sqrt{(2g)}$

Hence the speed of the ball just after impact is $2\sqrt{(2g)}$ m s^{-1}.

Thus, if e is the coefficient of restitution,

$$2\sqrt{(2g)} = e \times 3\sqrt{(2g)}$$
$$\therefore \quad e = \tfrac{2}{3}.$$

When considering indirect impacts between an elastic body and a smooth fixed plane, it is found that Newton's law of restitution can be applied to motion perpendicular to the plane.

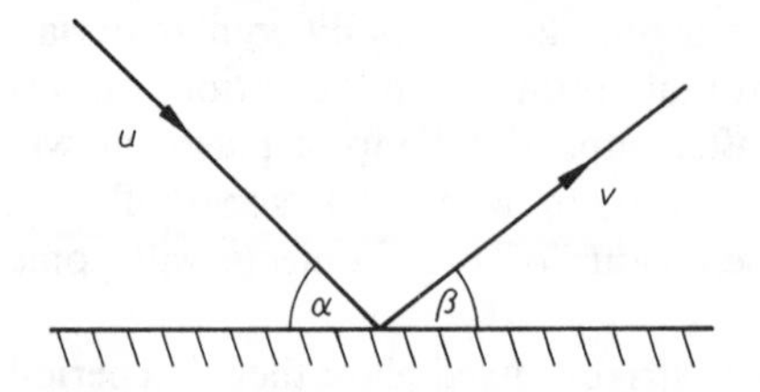

Let us suppose that a body strikes a smooth fixed plane while travelling with speed u in a direction inclined at an angle α to the plane, then rebounds with speed v at an angle β to the plane.

Since the plane is smooth no force acts along the plane and therefore there is no change in the component of velocity along the plane

i.e. $$v \cos \beta = u \cos \alpha.$$

However, the impulse of the normal reaction of the plane changes the component of velocity perpendicular to the plane in accordance with Newton's law,

i.e. $$v \sin \beta = eu \sin \alpha,$$

where e is the coefficient of restitution.

Example 2 A particle of mass 2 kg strikes a fixed smooth plane while moving at a speed of $24\,\text{m}\,\text{s}^{-1}$ in a direction making an angle of 30° with the plane. If the coefficient of restitution between the particle and the plane is $\frac{1}{4}$, find the speed at which the particle rebounds. Find also the impulse exerted by the plane on the particle.

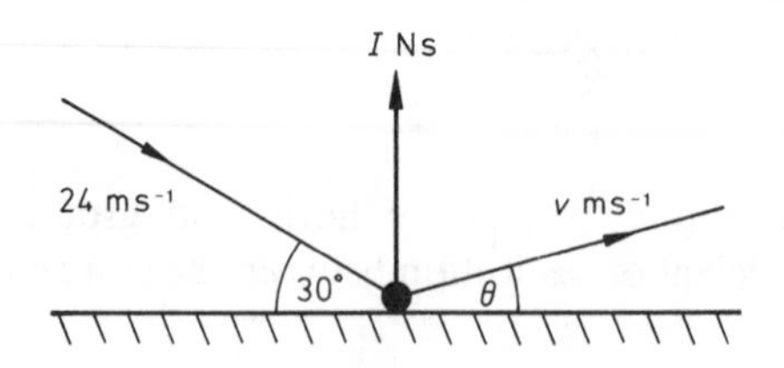

Let the velocity of the particle after impact be $v\,\text{m}\,\text{s}^{-1}$ at an angle θ to the plane and let the impulse exerted by the plane be I N s.

There is no change in velocity parallel to the plane

$$\therefore \quad v \cos \theta = 24 \cos 30° = 12\sqrt{3}.$$

By Newton's law of restitution perpendicular to the plane,

$$v \sin \theta = \tfrac{1}{4} \times 24 \sin 30° = 3$$

Squaring and adding: $v^2 \cos^2 \theta + v^2 \sin^2 \theta = (12\sqrt{3})^2 + 3^2$

$$\therefore \quad v = \sqrt{(432 + 9)} = 21.$$

Hence the speed of the particle after impact is $21\,\text{m}\,\text{s}^{-1}$.

Considering the motion of the particle perpendicular to the plane,

impulse = change in momentum

$$\therefore \qquad I = 2 \times v \sin\theta + 2 \times 24 \sin 30^\circ = 30.$$

Hence the impulse exerted by the plane is 30 N s.

When applied to the direct impact of two moving bodies, Newton's law of restitution states that the ratio of the relative speed of separation after impact to the relative speed of approach before impact is constant.

Before impact ($u_1 > u_2$) After impact ($v_1 < v_2$)

Let us suppose that two spheres of masses m_1 and m_2 collide when moving along the same straight line with velocities u_1 and u_2 respectively. Let the velocities of the spheres after impact be v_1 and v_2 respectively. Provided that all velocities are measured in the same direction, as shown in the diagram, the relative speed of approach is $(u_1 - u_2)$ and the relative speed of separation is $(v_2 - v_1)$.

By Newton's law of restitution

$$v_1 - v_2 = -e(u_1 - u_2)$$

where e is the coefficient of restitution.

Assuming that no external impulse acts during the impact, we can also apply the principle of conservation of momentum to give:

$$m_1u_1 + m_2u_2 = m_1v_1 + m_2v_2$$

Example 3 A sphere A, of mass $2m$ and moving with speed $6u$, collides directly with a sphere B of mass $3m$ moving in the same direction with speed $3u$. After the impact the speed of B is $5u$. Find the coefficient of restitution between the spheres.

Before impact After impact

$6u$ $3u$ v $5u$

A $2m$ $3m$ B A $2m$ $3m$ B

Let the speed of A after impact be v, as shown in the diagram.

Using the principle of conservation of momentum,

$$2m \times 6u + 3m \times 3u = 2m \times v + 3m \times 5u$$
$$\therefore \quad 12mu + 9mu = 2mv + 15mu$$
$$\therefore \quad v = 3u.$$

Using Newton's law of restitution,

$$v - 5u = -e(6u - 3u)$$
$$\therefore \quad 3u - 5u = -3eu$$
$$\therefore \quad e = \tfrac{2}{3}$$

Hence the coefficient of restitution between the spheres is $\frac{2}{3}$.

Example 4 A sphere of mass 5 kg, moving at $8\,\text{m s}^{-1}$, collides directly with a sphere of mass 3 kg moving in the opposite direction at $7\,\text{m s}^{-1}$. If the coefficient of restitution between the spheres is 0·6, find their speeds after impact. Find also the magnitude of the impulse exerted on each sphere during the collision.

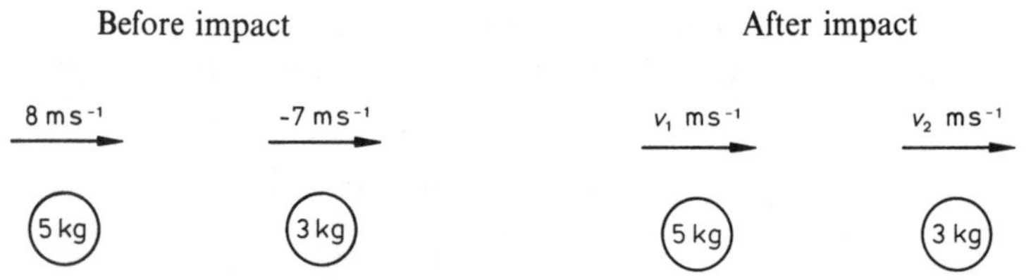

Let the velocities of the spheres after impact be $v_1\,\text{m s}^{-1}$ and $v_2\,\text{m s}^{-1}$, as shown in the diagram.
Using conservation of momentum,

$$5 \times 8 - 3 \times 7 = 5v_1 + 3v_2$$
$$\therefore \quad 5v_1 + 3v_2 = 19 \qquad (1)$$

Using Newton's law of restitution,

$$v_1 - v_2 = -0{\cdot}6\{8 - (-7)\}$$
$$\therefore \quad v_1 - v_2 = -9 \qquad (2)$$

Adding $3 \times (2)$ to (1): $\quad 8v_1 = -8$
Hence $\quad v_1 = -1 \quad \text{and} \quad v_2 = 8.$

Thus after impact the speed of the 5 kg sphere is $1\,\text{m s}^{-1}$ and the speed of the 3 kg sphere is $8\,\text{m s}^{-1}$. The directions of both spheres are reversed in the collision.

Considering the motion of the 3 kg sphere,

$$\text{impulse} = \text{change in momentum}$$
$$= 3 \times 8 - 3 \times (-7)\,\text{N s} = 45\,\text{N s}$$

Hence the magnitude of the impulse exerted on each sphere is 45 N s.

Exercise 8.4

1. A ball of mass 0·2 kg is dropped from a point 10 m above horizontal ground. If the coefficient of restitution between the ball and the ground is $\frac{1}{2}$, find (a) the height above the ground to which the ball rebounds, (b) the loss of kinetic energy in the impact, (c) the impulse exerted by the ground.

2. A ball dropped from a height of 8 m above horizontal ground rebounds to a height of 4·5 m. Find the coefficient of restitution between the ball and the ground. If the ball is then dropped from a height of 12·8 m and allowed to bounce twice, find the maximum height it reaches after the second bounce.

3. A particle is projected vertically upwards with speed $7\,\mathrm{m\,s^{-1}}$ from a point A, which is 2 m above a horizontal plane. After striking this plane the particle first comes instantaneously to rest at A. Find the coefficient of restitution between the particle and the plane.

4. A particle of mass 3 kg strikes a fixed smooth plane while moving at a speed of $20\,\mathrm{m\,s^{-1}}$ in a direction making an angle α with the plane, where $\sin\alpha = 3/5$. If the coefficient of restitution between the particle and the plane is 2/3, find the speed at which the particle rebounds. Find also the impulse exerted by the plane on the particle.

5. A particle strikes a fixed smooth plane while travelling in a direction inclined at 45° to the plane. After impact its direction makes an angle of 30° with the plane. Find the coefficient of restitution between the particle and the plane. If on another occasion the particle is moving in a direction inclined at 60° to the plane, find its direction after impact with the plane.

6. A particle strikes a smooth fixed plane and rebounds in a direction at right angles to its direction just before impact. If the coefficient of restitution is $\frac{1}{3}$, find the acute angle between the initial direction of motion of the particle and the plane.

7. A particle is projected with speed $6{\cdot}5\,\mathrm{m\,s^{-1}}$ at an angle θ to the horizontal, where $\tan\theta = 12/5$, from a point A on a smooth horizontal plane. The particle strikes the plane again at B, then rebounds and strikes the plane for a second time at C. The coefficient of restitution between the particle and the plane is 5/9. Find the speed and direction with which the particle rebounds from the plane at B. Show that the distance AC is 100/21 m.

8. A sphere of mass $4m$ and moving with speed $5u$, collides directly with a sphere of mass $3m$, which is at rest. If the coefficient of restitution between the spheres is 0·4, find their speeds after impact. Find also the loss in kinetic energy due to the impact.

9. A sphere A, of mass 1 kg and moving with speed $12\,\mathrm{m\,s^{-1}}$, collides directly with a sphere B, of mass 4 kg, moving in the same direction with speed $2\,\mathrm{m\,s^{-1}}$. Given that sphere A is brought to rest by the collision, find the coefficient of restitution between the spheres.

10. A sphere moving with speed $7\,\text{m}\,\text{s}^{-1}$ collides directly with an identical sphere moving in the opposite direction with speed $2\,\text{m}\,\text{s}^{-1}$. Find the speeds of the spheres after impact if the coefficient of restitution is (a) $\frac{1}{3}$, (b) $\frac{2}{3}$. In each case state whether the spheres are travelling in the same direction or in opposite directions.

11. A sphere of mass m moving with speed $9u$ collides directly with a sphere of mass M moving in the same direction with speed $5u$. After impact the spheres move in the same direction with speeds $6u$ and $7u$. Find the coefficient of restitution between the spheres and the ratio $m:M$.

12. Three perfectly elastic particles A, B, C, with masses 4 kg, 2 kg, 3 kg respectively, lie at rest in a straight line on a smooth horizontal table. Particle A is projected towards B with speed $15\,\text{m}\,\text{s}^{-1}$ and after A has collided with B, B collides with C. Find the velocities of the particles after the second collision and state whether there will be a third collision.

13. Three identical spheres A, B, C, lie at rest in a straight line on a smooth horizontal table. Particle A is projected towards B with speed u. After the collision between A and B, B moves on to hit C. The coefficient of restitution between any two of the spheres is $\frac{1}{2}$. Find the speeds of the spheres after the second collision and state whether there will be a third collision.

14. A sphere A of mass $2m$, moving on a smooth horizontal floor with speed $5u$, collides directly with a sphere B of mass m, which is at rest. If the coefficient of restitution between the spheres is 4/5, find the speed of B after impact. Sphere B goes on to strike a vertical wall, which is at right angles to the direction of motion of the spheres. A is then brought to rest by a second collision with B. Find the coefficient of restitution between B and the wall.

15. A small smooth sphere moves on a horizontal table and strikes an identical sphere lying at rest on the table at a distance d from a vertical wall, the impact being along the line of centres and perpendicular to the wall. Prove that the next impact between the spheres will take place at a distance $2de^2/(1 + e^2)$ from the wall, where e is the coefficient of restitution for all impacts involved. (L)

16. The coefficient of restitution between two small, smooth spheres A and B is e. The mass of A is m and of B is em. Show that, if A is at rest and B is projected with velocity u to collide directly with A, B is reduced to rest by the collision. If, instead, B is at rest and A is projected with the same speed to hit B directly, find the velocities of A and B after the collision. Show that the loss of kinetic energy is the same in each collision and is greatest when $e = \frac{1}{2}$. (SU)

17. Two small spheres of masses m and $2m$ are connected by a light inextensible string of length $2a$. When the string is taut and horizontal, its mid-point is fixed and the spheres are released from rest. The coefficient of restitution between the spheres is $\frac{1}{2}$. Show that the first impact brings the heavier sphere to rest, and the second impact brings the lighter sphere to rest. Find the velocity of each sphere immediately after the third impact. (L)

18. Three small smooth spheres A, B, C, of equal radius but of mass $m, 2m$ and m respectively, lie at rest and separated from one another on a smooth horizontal table in the order ABC with their centres in a straight line. Sphere A is projected with speed V directly towards sphere B. The coefficient of restitution at each collision is e, where $0 < e < 1$. Find, in terms of V and e, the speeds of the three spheres (a) after the first collision (between A and B), (b) after the second collision (between B and C). Hence show that there will be at least three collisions if $e^2 - 7e + 1 > 0$. (O&C)

Exercise 8.5 (miscellaneous)

1. Two particles of masses $2m$ and $3m$ are moving towards each other with speeds $4u$ and $2u$ respectively. The direction of motion of the heavier particle is reversed by the impact and its speed after impact is u. Find (i) the magnitude of the impulse, (ii) the loss of energy. (C)

2. A particle of mass 4 kg moves under the action of a force $\mathbf{F}$ in the interval of time from $t = 0$ to $t = 2$. When $t = 0$ the velocity of the particle is $2\mathbf{i} - 3\mathbf{j}$. Given that $\mathbf{F} = 6t\mathbf{i} + 2\mathbf{j}$, find (a) the impulse of $\mathbf{F}$, (b) the velocity of the particle when $t = 2$, (c) the gain in kinetic energy.

3. A particle of mass m moves in a horizontal plane Oxy with speed v along the x-axis in the positive direction. It is subjected to a horizontal impulse $\mathbf{I}$ which turns its direction of motion through 30° in an anticlockwise sense and reduces its speed to $v/\sqrt{3}$. Find the vector $\mathbf{I}$. At the same instant an impulse $-\mathbf{I}$ is applied to a particle of mass $3m$ which is at rest. Find the magnitude and direction of the resultant velocity of this particle. (JMB)

4. A pump raises water from an underground reservoir through a height of 10 m and delivers it at a speed of 9 m s^{-1} through a circular pipe of internal diameter 20 cm. Taking 1 litre of water to have a mass of 1 kg and g to have the value 9·8 m s^{-2} find (i) the mass of water raised per second, correct to 3 sig. fig.; (ii) the kinetic energy imparted to the water each second, correct to 3 sig. fig.; (iii) the effective power of the pump correct to 2 sig. fig.; (iv) the actual power of the pump if it is 70% efficient. If the water hits a vertical wall horizontally and does not rebound calculate the magnitude of the force exerted on the wall, correct to 2 sig. fig. (SU)

5. A light inextensible string AB has a particle of mass $2m$ attached at A and a particle of mass m attached at B. The particles are placed on a smooth horizontal table with the string taut. A horizontal impulse is applied to the particle at A of magnitude mu in a direction which makes an angle of 120° with AB. (i) Show that B

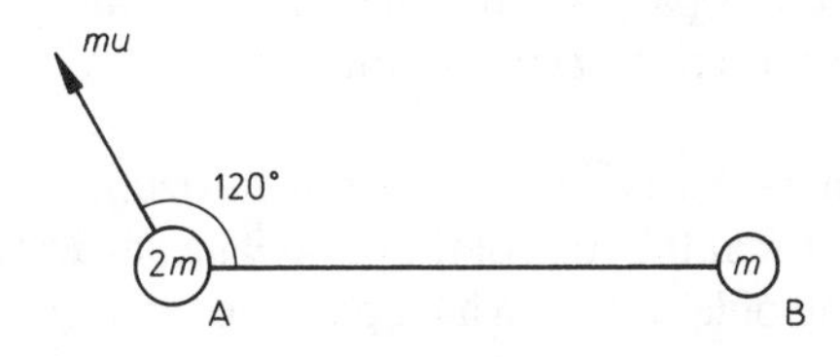

starts to move along BA with speed $u/6$. (ii) Find the components of the initial velocity of the particle at A along BA and perpendicular to BA and state the magnitude of the impulse of the tension in the string. (AEB 1977)

6. A composite particle of mass $m_1 + m_2$ is formed as a result of a head-on collision between two particles of masses m_1, m_2 travelling with speeds u_1, u_2 respectively. Find the speed of the composite particle in terms of m_1, m_2, u_1, u_2. Show that the collision leads to a loss of kinetic energy of amount

$$\frac{1}{2}\left(\frac{m_1 m_2}{m_1 + m_2}\right)(u_1 + u_2)^2. \qquad \text{(W)}$$

7. A gun of mass M fires a shell of mass m horizontally. The recoil of the gun is opposed by a constant force F which brings the gun to rest in t seconds. Show that the energy of the explosion is $\dfrac{F^2t^2(M + m)}{2Mm}$. (O)

8. A particle of mass m is dropped from rest from a point A which is at a height h above a horizontal surface. After hitting the surface the particle rebounds vertically and reaches a maximum height of $\frac{1}{2}h$ above the surface at the point B. Find expressions for the loss of energy and the impulse on the particle at the impact. At the instant of the particle's impact with the surface, a second particle, of mass $4m$, is dropped from rest at A. Show that the two particles collide at B. The two particles coalesce when they collide at B. When the combined particle hits the surface, one-half of its kinetic energy is lost in the impact. Show that, after the impact, the combined particle reaches a greatest height of $41h/100$ above the surface. (C)

9. Two beads, one of mass 3 g and the other of mass 2 g, are threaded on a smooth circular wire which is fixed with its plane vertical. The heavier bead is initially at rest at the lowest point L of the wire. The lighter bead is released from rest and is moving with speed 2 m s^{-1} when it strikes the other bead. There is no loss of energy in the impact. Find the speed of each bead immediately after impact. In the subsequent motion each bead comes to rest instantaneously before reaching the highest point of the wire. Find, to the nearest millimetre, the height above L at which each bead comes first to rest. (C)

10. A pile-driver of mass 2000 kg is raised 1·6 m above a pile of mass 1500 kg and then dropped on to it. After impact the driver and the pile move together and the pile is driven 0·2 m into the ground. Find (i) the velocity of the pile-driver just before it hits the pile, (ii) the common velocity of pile and driver immediately after impact, (iii) the loss of energy on impact, stating the units in which it is measured, (iv) the mean resistance of the ground to penetration. (AEB 1980)

11. A particle A of mass $3m$ lies on a smooth horizontal rectangular table at a distance $5a$ from a point C at the edge of the table. The particle is attached to one end of a light inelastic string of length $6a$ which passes over a small smooth pulley at C.

To the other end of the string is attached a particle B of mass $2m$. The horizontal floor is at a depth $4a$ below C. The particle B is released from rest at C and motion takes place in a vertical plane perpendicular to the edge of the table. Show that the speed of the particles immediately after the string becomes taut is $\frac{2}{5}\sqrt{(2ga)}$. Find (i) the magnitude of the impulse which acts on the particle A, (ii) the loss of kinetic energy of the system due to the string becoming taut, (iii) the speed of the particle B just before it hits the floor. (C)

12. Two masses, one 3 kg and the other 2 kg, are connected by a light inextensible string which passes over a fixed smooth pulley and the system is released from rest. Find the acceleration of the system and show that the tension in the string is 23·52 N. After falling 2 metres the heavier mass strikes a fixed horizontal plane and is brought to rest. If the lighter mass does not reach the pulley, show that after 4/7 seconds the 3 kg mass is again in motion. Find the speed with which it begins to ascend and the impulsive tension in the string. (W)

13. Two particles A and B, each of mass m and connected by a light inextensible string of length $4a$, are resting on a smooth horizontal table with $AB = 2a$. A particle C of mass $3m$ is attached to the midpoint of the string and rests on the table so that $AC = BC$. The particles A and B are then simultaneously projected along the table with equal velocities perpendicular to AB and in the sense away from C. Show that, when the string tightens, the initial velocities of the particles A and B are each inclined at angle $\frac{1}{3}\pi$ with the line AB. (O)

14. A particle P of mass $2m$ is connected to a particle Q of mass $3m$ by means of a light inextensible string which passes over a small smooth pulley. The particles are released from rest with the string taut and its hanging parts vertical. (i) Calculate the acceleration of P and the tension in the string. At the instant when P is moving with velocity v it collides and coalesces with a stationary particle of mass $3m$ to form a particle P_1. (ii) Prove that the velocity of P_1 immediately after the collision is $5v/8$. (iii) Calculate the magnitude of the impulse of the tension in the string. (iv) Calculate the loss in kinetic energy of the whole system due to the collision.

The string breaks and particle Q hits a horizontal floor when moving with speed u and bounces to a height $u^2/(6g)$ above the floor. (v) Calculate the coefficient of restitution between Q and the floor. (AEB 1979)

15. A bullet of mass m travelling horizontally with speed u strikes perpendicularly a wooden block of mass M which is free to move horizontally in the line of motion of the bullet. The resistance to penetration is assumed uniform and the bullet comes to rest after penetrating a distance a into the block. Show that the ultimate common velocity of the system is $mu/(m + M)$ and that the total loss of kinetic energy during the penetration is $mMu^2/2(m + M)$. Deduce the value of the resistance and show that (i) the block moves a distance $am/(m + M)$ before the bullet comes to rest relative to the block, (ii) the penetration by the bullet ceases after a time $2a/u$. (W)

16. A particle of mass m is attached to one end of a light inextensible string of length a, the other end being attached to a fixed point O. Initially the string is taut and

makes an angle α ($\leqslant 90°$) with the downward vertical at O. The particle is released and, when it is at its lowest point A, coalesces with a stationary particle of mass m. The combined body next comes to rest with the string making an angle β with the downward vertical at O. Show that $1 - \cos\beta = \frac{1}{4}(1 - \cos\alpha)$.

The motion continues so that each time the body passes through A it coalesces with a stationary particle of mass m. If $\alpha = 90°$, (a) prove that the speed of the body just before passing through A for the fourth time is $\frac{1}{4}\sqrt{(2ag)}$, (b) find the angle (correct to the nearest degree) made by the string with the downward vertical at O when the body next comes to instantaneous rest after passing through A for the fourth time, (c) find the ratio of the tensions in the string just before and just after the body passes through A for the fourth time. (O&C)

17. Two equal particles are projected at the same instant from points A and B on horizontal ground, the first from A with speed u at an angle of elevation α and the second from B with speed v at an angle of elevation β. They collide directly when they are moving horizontally in opposite directions. Find v in terms of u, α and β, and show that $AB = \dfrac{u^2 \sin\alpha \sin(\alpha + \beta)}{g \sin\beta}$. The particles coalesce on impact. Find, in terms of u, α and β, the speed of the combined particle immediately after the collision. In the case when $\alpha = 30°$ and $\beta = 60°$ find AC, where C is the point where the combined particle returns to the ground. (JMB)

18. A ball is thrown vertically upwards from the floor with velocity V. It rebounds from the ceiling, which is at height h above the floor, and then rebounds from the floor. After this second rebound, it just reaches the ceiling again. If the coefficient of restitution between the ball and ceiling is e, and between the ball and floor is f, prove that

$$V^2 = \frac{2gh(1 - f^2 + e^2 f^2)}{e^2 f^2}.$$

(SU)

19. A particle of mass m strikes a fixed smooth plane at speed u while moving in a direction inclined at an angle α to the plane, then rebounds at an angle β to the plane. Find expressions for (a) the coefficient of restitution between the particle and the plane, (b) the speed of the particle after impact, (c) the impulse exerted on the plane.

20. A particle of mass m, projected from a point A with speed u at an angle of elevation α, strikes a smooth vertical wall at a point B and after rebounding returns to A. Show that the total time taken to move from A to B and then back to A is $(2u \sin\alpha)/g$. Given that e is the coefficient of restitution between the particle and the wall, find expressions for (a) the loss of kinetic energy on impact, (b) the distance of A from the wall.

21. Two elastic spheres of masses m and M collide directly. Their velocities before impact are 12 m/s and 6 m/s respectively in the same direction, and after impact they are 5 m/s and 8 m/s respectively also in the same direction. Find the coefficient of restitution and also the ratio $m{:}M$. Find also the velocities of the spheres after

impact if they collide directly when travelling in opposite directions, the velocity of the heavier sphere being 4 m/s and that of the lighter sphere 14 m/s. (L)

22. Three small spheres A, B and C, of equal radii and masses m, $2m$ and $3m$ respectively, are placed with their centres in a straight line on a smooth horizontal table with B between A and C. The sphere A is given a velocity $5u$ in the direction $\overrightarrow{AB}$ and, as a result of the ensuing collision between A and B, B moves towards C with speed $3u$. Calculate (i) the magnitude and the sense of the velocity of A after the collision, (ii) the coefficient of restitution between A and B, (iii) the loss in kinetic energy due to the collision between A and B.

The sphere B strikes C and as a result C receives an impulse of $4mu$. Calculate (iv) the velocities of B and C after their collision, (v) the coefficient of restitution between B and C. (AEB 1978)

23. A small sphere A, of mass m moving with velocity $2u$ on a smooth horizontal plane, impinges directly on a small sphere B, of equal radius and mass $2m$ moving with velocity u in the same direction on the plane. Given that after impact B moves with velocity $3u/2$, calculate (i) the coefficient of restitution between A and B, (ii) the loss in kinetic energy due to the impact. The sphere B continues to move with velocity $3u/2$ until it hits a vertical wall from which it rebounds and is then brought to rest by a second impact with the approaching sphere A. Calculate (iii) the coefficient of restitution between B and the wall, (iv) the final speed of the sphere. (AEB 1980)

24. Particles A and B of mass m and $2m$ respectively lie at rest on a smooth horizontal plane. They are projected towards each other both with speed u. Show that, if the coefficient of restitution between the particles is $\frac{3}{4}$, their directions of motion are reversed on impact and find their speeds after impact. Find, also, the loss of kinetic energy during impact. If A subsequently hits a smooth vertical plane normally, show that A and B will collide again only if the coefficient of restitution between A and the vertical plane is greater than 1/8. (SU)

25. Two particles A, B of masses M, m respectively are attached to a fixed point O by light inextensible strings of equal length. The particle A is held so that its string is taut and horizontal. It is then released so that it strikes the particle B which was hanging in equilibrium. Prove that if OB just reaches the horizontal then $Me = m$, where e is the coefficient of restitution between the particles. (O)

26. Two particles, A of mass $2m$ and B of mass m, moving on a smooth horizontal table in opposite directions with speeds $5u$ and $3u$ respectively, collide directly. Find their velocities after the collision in terms of u and the coefficient of restitution e. Show that the magnitude of the impulse exerted by B on A is $16mu(1 + e)/3$. Find the value of e for which the speed of B after the collision is $3u$. Moving at this speed B subsequently collides with a stationary particle C of mass km, and thereafter remains attached to C. Find the velocity of the combined particle and find the range of values of k for which a third collision will occur. (JMB)

27. Identical beads P and Q are threaded on a smooth circular hoop fixed in a horizontal plane. Initially Q is at rest at a point A of the hoop when P collides with it. Denoting the coefficient of restitution between P and Q by e, find the values of e for which the next collision takes place at A. (JMB)

9 Series and expansions

9.1 Summation of finite series

There are various standard results that can be used to find the sums of finite series. Two of these are the formula for the sums of arithmetic and geometric progressions obtained in Book 1.

> The sum to n terms of the A.P. with first term a and common difference d is $\frac{1}{2}n\{2a + (n - 1)d\}$.
> The sum to n terms of the G.P. with first term a and common ratio r is $a(1 - r^n)/(1 - r)$.

The series $\sum_{r=1}^{n} r = 1 + 2 + 3 + \ldots + n$ is an arithmetic progression with first term 1 and common difference 1.

$$\therefore \qquad \sum_{r=1}^{n} r = \tfrac{1}{2}n\{2 + (n - 1)\} = \tfrac{1}{2}n(n + 1).$$

The series $\sum_{r=1}^{n} r^2 = 1^2 + 2^2 + 3^2 + \ldots + n^2$ can be summed using the identity

$$(r + 1)^3 - r^3 \equiv 3r^2 + 3r + 1.$$

Substituting $r = 1, 2, 3, \ldots, n$, we have

$$\begin{aligned} 2^3 - 1^3 &= 3.1^2 + 3.1 + 1 \\ 3^3 - 2^3 &= 3.2^2 + 3.2 + 1 \\ 4^3 - 3^3 &= 3.3^2 + 3.3 + 1 \\ &\cdots\cdots\cdots\cdots \\ (n + 1)^3 - n^3 &= 3n^2 + 3n + 1 \end{aligned}$$

Adding together all n equations, we find that most of the terms on the left may be cancelled out to give:

$$(n+1)^3 - 1^3 = 3\sum_{r=1}^{n} r^2 + 3\sum_{r=1}^{n} r + n$$

$$\therefore \quad 3\sum_{r=1}^{n} r^2 = (n+1)^3 - 1^3 - 3\sum_{r=1}^{n} r - n$$

$$= (n+1)^3 - 1 - \tfrac{3}{2}n(n+1) - n$$
$$= n^3 + \tfrac{3}{2}n^2 + \tfrac{1}{2}n = \tfrac{1}{2}n(2n^2 + 3n + 1)$$

Hence $$\sum_{r=1}^{n} r^2 = \tfrac{1}{6}n(n+1)(2n+1).$$

By a similar method it can be shown that

$$\sum_{r=1}^{n} r^3 = \tfrac{1}{4}n^2(n+1)^2 = \left(\sum_{r=1}^{n} r\right)^2$$

To summarise, we now have three further important results:

$$\sum_{r=1}^{n} r = \tfrac{1}{2}n(n+1), \quad \sum_{r=1}^{n} r^2 = \tfrac{1}{6}n(n+1)(2n+1), \quad \sum_{r=1}^{n} r^3 = \tfrac{1}{4}n^2(n+1)^2$$

Example 1 Find the sum of the first n terms of the series

$$1.3.5 + 2.4.6 + 3.5.7 + \dots.$$

The rth term of the series $= r(r+2)(r+4) = r^3 + 6r^2 + 8r$

$\therefore$ the sum of the first n terms

$$= \sum_{r=1}^{n} (r^3 + 6r^2 + 8r)$$

$$= \sum_{r=1}^{n} r^3 + 6\sum_{r=1}^{n} r^2 + 8\sum_{r=1}^{n} r$$

$$= \tfrac{1}{4}n^2(n+1)^2 + 6.\tfrac{1}{6}n(n+1)(2n+1) + 8.\tfrac{1}{2}n(n+1)$$
$$= \tfrac{1}{4}n(n+1)\{n(n+1) + 4(2n+1) + 16\}$$
$$= \tfrac{1}{4}n(n+1)(n^2 + n + 8n + 4 + 16)$$
$$= \tfrac{1}{4}n(n+1)(n^2 + 9n + 20)$$
$$= \tfrac{1}{4}n(n+1)(n+4)(n+5)$$

Example 2 Find the sum of the series $(n+1)^2 + (n+2)^2 + (n+3)^2 + \dots + (2n)^2$.

The sum of the series $= \sum_{r=1}^{2n} r^2 - \sum_{r=1}^{n} r^2$

$$= \tfrac{1}{6}.2n(2n+1)(4n+1) - \tfrac{1}{6}n(n+1)(2n+1)$$
$$= \tfrac{1}{6}n(2n+1)\{2(4n+1) - (n+1)\}$$
$$= \tfrac{1}{6}n(2n+1)(7n+1).$$

Example 3 Find the sum of the series $1^3 + 3^3 + 5^3 + \ldots + (2n-1)^3$.

$$\begin{aligned}
1^3 + 2^3 + 3^3 + \ldots + (2n)^3 &= \sum_{r=1}^{2n} r^3 \\
&= \tfrac{1}{4}(2n)^2(2n+1)^2 \\
&= n^2(2n+1)^2 \\
2^3 + 4^3 + 6^3 + \ldots + (2n)^3 &= 2^3(1^3 + 2^3 + 3^3 + \ldots + n^3) \\
&= 8\sum_{r=1}^{n} r^3 \\
&= 2n^2(n+1)^2
\end{aligned}$$

$$\begin{aligned}
\therefore\quad & 1^3 + 3^3 + 5^3 + \ldots + (2n-1)^3 \\
&= n^2(2n+1)^2 - 2n^2(n+1)^2 \\
&= n^2\{(2n+1)^2 - 2(n+1)^2\} = n^2(2n^2-1)
\end{aligned}$$

In the next example we use the *method of differences* to find the sum of the given series. We express each term of the series as a difference, in such a way that most terms of the series cancel each other out. A similar process was used earlier in the section to show that $\sum_{r=1}^{n} \{(r+1)^3 - r^3\} = (n+1)^3 - 1^3$.

Example 4 Show that $\dfrac{1}{r} - \dfrac{1}{r+1} \equiv \dfrac{1}{r(r+1)}$. Hence evaluate $\sum_{r=1}^{n} \dfrac{1}{r(r+1)}$.

$$\frac{1}{r} - \frac{1}{r+1} \equiv \frac{(r+1)-1}{r(r+1)} \equiv \frac{1}{r(r+1)}$$

$$\begin{aligned}
\therefore\quad \sum_{r=1}^{n} \frac{1}{r(r+1)} &= \sum_{r=1}^{n} \left(\frac{1}{r} - \frac{1}{r+1}\right) \\
&= \left(1 - \frac{1}{2}\right) + \left(\frac{1}{2} - \frac{1}{3}\right) + \left(\frac{1}{3} - \frac{1}{4}\right) + \ldots \\
&\qquad \ldots + \left(\frac{1}{n-1} - \frac{1}{n}\right) + \left(\frac{1}{n} - \frac{1}{n+1}\right) \\
&= 1 - \frac{1}{n+1} = \frac{n}{n+1}
\end{aligned}$$

[Note that as $n \to \infty$, $\dfrac{1}{n+1} \to 0$, $\quad \therefore\ 1 - \dfrac{1}{n+1} \to 1$. Thus the sum to infinity of this series is 1.]

We can extend the method of Example 4 to other series by using partial fractions.

Example 5 Find

$$\sum_{r=1}^{n} \frac{1}{r(r+1)(r+2)} = \frac{1}{1.2.3} + \frac{1}{2.3.4.} + \ldots + \frac{1}{n(n+1)(n+2)}.$$

Let $\dfrac{1}{r(r+1)(r+2)} \equiv \dfrac{A}{r} + \dfrac{B}{r+1} + \dfrac{C}{r+2}$

then $\quad 1 \equiv A(r+1)(r+2) + Br(r+2) + Cr(r+1)$

Putting $r = 0$: $\quad 1 = 2A \quad \therefore \quad A = \frac{1}{2}$

Putting $r = -1$: $\quad 1 = -B \quad \therefore \quad B = -1$

Putting $r = -2$: $\quad 1 = 2C \quad \therefore \quad C = \frac{1}{2}$

$$\text{Hence} \quad \frac{1}{r(r+1)(r+2)} \equiv \frac{1}{2r} - \frac{1}{r+1} + \frac{1}{2(r+2)}$$

$$\begin{aligned}
\therefore \quad \sum_{r=1}^{n} \frac{1}{r(r+1)(r+2)} = \; & \frac{1}{2.1} - \frac{1}{2} + \frac{1}{2.3} \\
+ \; & \frac{1}{2.2} - \frac{1}{3} + \frac{1}{2.4} \\
+ \; & \frac{1}{2.3} - \frac{1}{4} + \frac{1}{2.5} \\
& \cdots\cdots\cdots\cdots\cdots\cdots \\
+ \; & \frac{1}{2(n-1)} - \frac{1}{n} + \frac{1}{2(n+1)} \\
+ \; & \frac{1}{2n} - \frac{1}{n+1} + \frac{1}{2(n+2)}
\end{aligned}$$

$$\begin{aligned}
\therefore \quad \sum_{r=1}^{n} \frac{1}{r(r+1)(r+2)} &= \frac{1}{2.1} - \frac{1}{2} + \frac{1}{2.2} + \frac{1}{2(n+1)} - \frac{1}{n+1} + \frac{1}{2(n+2)} \\
&= \frac{1}{4} - \frac{1}{2(n+1)} + \frac{1}{2(n+2)}
\end{aligned}$$

[Note that the results obtained in this section can also be proved by the method of mathematical induction, as shown in §1.4.]

Exercise 9.1

1. Use standard results to find the sums of the following series.

(a) $\sum_{r=1}^{n} (2r+1)$ (b) $\sum_{r=1}^{n} (r-1)(r+1)$ (c) $\sum_{r=1}^{n} 2^r$

(d) $\sum_{r=1}^{n} r^2(r+1)$ (e) $\sum_{r=1}^{n} 3^{1-r}$ (f) $\sum_{r=1}^{n} (r+4)^3$

2. Use partial fractions to find the sums of the following series.

(a) $\sum_{r=1}^{n} \dfrac{1}{(r+1)(r+2)}$ (b) $\sum_{r=1}^{n} \dfrac{1}{r(r+2)}$ (c) $\sum_{r=1}^{n} \dfrac{1}{(r+1)(r+2)(r+3)}$

3. Find the sums of the following series (i) to $2n$ terms, (ii) to $(2n + 1)$ terms.
(a) $1 - 1 + 1 - 1 + \ldots$ (b) $1 - 2 + 3 - 4 + \ldots$

4. Show that $r(r + 1)(r + 2) - (r - 1)r(r + 1) \equiv 3r(r + 1)$.

Hence find the sum of the series $\sum_{r=1}^{n} r(r + 1)$.

5. Use the method of question 4 to show that

$$\sum_{r=1}^{n} r(r + 1)(r + 2) = \tfrac{1}{4}n(n + 1)(n + 2)(n + 3).$$

6. Show that $(r + 1)^4 - r^4 \equiv 4r^3 + 6r^2 + 4r + 1$. Use this identity and the standard formulae for $\sum_{r=1}^{n} r$ and $\sum_{r=1}^{n} r^2$ to prove that $\sum_{r=1}^{n} r^3 = \tfrac{1}{4}n^2(n + 1)^2$.

7. Use the method of question 6 to find $\sum_{r=1}^{n} r^4$.

8. Use the method of differences to find the sum to n terms of the series $1(1!) + 2(2!) + 3(3!) + \ldots$.

[Hint: express each term in the form $p! - q!$.]

9. Find the sum of (a) the integers from $2n$ to $3n$ inclusive, (b) the squares of the integers from $2n$ to $3n$ inclusive.

10. Find the sum of (a) the first n odd positive integers, (b) the squares of the first n odd positive integers.

11. Find the sums of the following series
(a) $21^3 + 22^3 + 23^3 + \ldots + 35^3$,
(b) $1^3 + 3^3 + 5^3 + \ldots + 25^3$,
(c) $(n + 1)^3 + (n + 2)^3 + (n + 3)^3 + \ldots + (2n)^3$,
(d) $1^3 - 2^3 + 3^3 - 4^3 + \ldots + (2n - 1)^3 - (2n)^3$.

In questions 12 to 19 write down an expression for the rth term of the given series and hence find the sum of the first n terms of the series.

12. $1.4 + 2.5 + 3.6 + \ldots$

13. $2^2 + 5^2 + 8^2 + 11^2 + \ldots$

14. $1.2 + 3.4 + 5.6 + \ldots$

15. $1.2.5 + 2.3.6 + 3.4.7 + \ldots$

16. $\dfrac{1}{1.4} + \dfrac{1}{2.5} + \dfrac{1}{3.6} + \ldots$

17. $\dfrac{1}{1.3} + \dfrac{1}{3.5} + \dfrac{1}{5.7} + \ldots$

18. $\dfrac{1}{1.3.5} + \dfrac{1}{3.5.7} + \dfrac{1}{5.7.9} + \ldots$

19. $\dfrac{1}{2.3.4} + \dfrac{2}{3.4.5} + \dfrac{3}{4.5.6} + \ldots$

20. Find the sum of the series $1 + x + x^2 + \ldots + x^n$, where $x \neq 1$. By differentiating your result, find the sum of the series $1 + 2x + 3x^2 + \ldots + nx^{n-1}$ for $x \neq 1$.

21. Find the sum $S = x + 2x^2 + 3x^3 + \ldots + nx^n$, where $x \neq 1$, by considering the expression $S - xS$.

22. Find the sum $S = 1 - 3a + 5a^2 - \ldots + (2n+1)(-a)^n$, where $a \neq -1$, by considering the expression $(1 + a)S$.

9.2 Maclaurin's theorem

Under certain conditions a function $f(x)$ can be expressed as an infinite series of ascending powers of x. Assuming that such a series exists, we may write

$$f(x) = a_0 + a_1x + a_2x^2 + a_3x^3 + a_4x^4 + \ldots + a_rx^r + \ldots$$

Differentiating with respect to x, we have

$$\begin{aligned}
f'(x) &= a_1 + 2a_2x + 3a_3x^2 + 4a_4x^3 + \ldots + ra_rx^{r-1} + \ldots \\
f''(x) &= 2a_2 + 3.2a_3x + 4.3a_4x^2 + \ldots + r(r-1)a_rx^{r-2} + \ldots \\
f'''(x) &= 3.2a_3 + 4.3.2a_4x + \ldots + r(r-1)(r-2)a_rx^{r-3} + \ldots \\
&\cdots\cdots\cdots\cdots\cdots\cdots\cdots\cdots\cdots\cdots \\
f^{(r)}(x) &= r(r-1)(r-2)\ldots 3.2a_r + \ldots
\end{aligned}$$

Substituting $x = 0$, we find that

$$\begin{aligned}
f(0) &= a_0 & \therefore\quad a_0 &= f(0) \\
f'(0) &= a_1 & \therefore\quad a_1 &= f'(0) \\
f''(0) &= 2a_2 & \therefore\quad a_2 &= \frac{f''(0)}{2!} \\
f'''(0) &= 3.2a_3 & \therefore\quad a_3 &= \frac{f'''(0)}{3!}
\end{aligned}$$

and $f^{(r)}(0) = r(r-1)(r-2)\ldots 3.2a_r \qquad \therefore\quad a_r = \dfrac{f^{(r)}(0)}{r!}$

Thus

$$\boxed{f(x) = f(0) + f'(0)x + \frac{f''(0)}{2!}x^2 + \frac{f'''(0)}{3!}x^3 + \ldots + \frac{f^{(r)}(0)}{r!}x^r + \ldots}$$

This result is called *Maclaurin's theorem*† and the series obtained is known as the *Maclaurin series* for $f(x)$.

It is possible to find a Maclaurin series for any function $f(x)$ whose derivatives $f'(0), f''(0), f'''(0), \ldots$ can be determined. The series obtained may converge to the

†*Maclaurin, Colin* (1698–1746) Scottish mathematician. He became professor of Mathematics at Aberdeen at the age of 19. His most important work was his *Treatise on fluxions* (1742).

sum $f(x)$ for all values of x. However, for many functions, Maclaurin's theorem holds only within a restricted range of values of x.

We now use Maclaurin's theorem to obtain various important expansions.

Let $\quad f(x) = (1+x)^n$, where n is a real number,

then
$$f'(x) = n(1+x)^{n-1},$$
$$f''(x) = n(n-1)(1+x)^{n-2},$$
$$f'''(x) = n(n-1)(n-2)(1+x)^{n-3},$$
$$\dots\dots\dots$$
$$f^{(r)}(x) = n(n-1)(n-2)\dots(n-r+1)(1+x)^{n-r},$$

$$\therefore \quad f(0) = 1, \quad f'(0) = n, \quad f''(0) = n(n-1), \quad f'''(0) = n(n-1)(n-2)$$

and
$$f^{(r)}(0) = n(n-1)(n-2)\dots(n-r+1).$$

Thus
$$(1+x)^n = 1 + nx + \frac{n(n-1)}{2!}x^2 + \frac{n(n-1)(n-2)}{3!}x^3 + \dots$$
$$\dots + \frac{n(n-1)(n-2)\dots(n-r+1)}{r!}x^r + \dots$$

This result was introduced in Book 1 as an extension of the binomial theorem. When n is a positive integer the expansion is valid for all values of x. It can be shown that, for other values of n, the result holds only when $|x| < 1$, i.e. when $-1 < x < 1$.

If
$$f(x) = e^x, \quad \text{then} \quad f(0) = 1$$
$$f'(x) = e^x, \qquad f'(0) = 1$$
$$f''(x) = e^x, \qquad f''(0) = 1$$
$$\dots\dots \qquad \dots\dots$$
$$f^{(r)}(x) = e^x, \qquad f^{(r)}(0) = 1$$

Thus
$$e^x = 1 + x + \frac{x^2}{2!} + \dots + \frac{x^r}{r!} + \dots$$

This expansion is valid for all values of x.

The function $\ln x$ is not defined for $x = 0$ and therefore has no Maclaurin series. However, an expansion of $\ln(1+x)$ can be obtained.

If
$$f(x) = \ln(1+x), \quad \text{then} \quad f(0) = \ln 1 = 0$$
$$f'(x) = (1+x)^{-1}, \qquad f'(0) = 1$$
$$f''(x) = -(1+x)^{-2}, \qquad f''(0) = -1$$
$$f'''(x) = 2(1+x)^{-3}, \qquad f'''(0) = 2$$
$$f^{(4)}(x) = -3.2(1+x)^{-4}, \qquad f^{(4)}(0) = -3!$$
$$\dots\dots \qquad \dots\dots$$
$$f^{(r)}(x) = (-1)^{r+1}(r-1)!(1+x)^{-r}, \qquad f^{(r)}(0) = (-1)^{r+1}(r-1)!$$

Thus
$$\ln(1+x) = 0 + x - \frac{x^2}{2!} + \frac{2x^3}{3!} - \frac{3!x^4}{4!} + \dots + \frac{(-1)^{r+1}(r-1)!x^r}{r!} + \dots$$

i.e.

$$\ln(1+x) = x - \frac{x^2}{2} + \frac{x^3}{3} - \frac{x^4}{4} + \ldots + (-1)^{r+1}\frac{x^r}{r} + \ldots$$

This result is valid for $-1 < x \leqslant 1$.

If

$$\begin{aligned}
f(x) &= \sin x, & \text{then}\quad f(0) &= 0\\
f'(x) &= \cos x, & f'(0) &= 1\\
f''(x) &= -\sin x, & f''(0) &= 0\\
f'''(x) &= -\cos x, & f'''(0) &= -1\\
f^{(4)}(x) &= \sin x, & f^{(4)}(0) &= 0\\
f^{(5)}(x) &= \cos x, & f^{(5)}(0) &= 1\\
&\ldots\ldots & &\ldots\ldots
\end{aligned}$$

Since we can now see a pattern emerging, we may write

$$\sin x = x - \frac{x^3}{3!} + \frac{x^5}{5!} - \frac{x^7}{7!} + \ldots$$

A similar result can be obtained for $\cos x$:

$$\cos x = 1 - \frac{x^2}{2!} + \frac{x^4}{4!} - \frac{x^6}{6!} + \ldots$$

The expansions of $\sin x$ and $\cos x$ are valid for all values of x (in radians).

Exercise 9.2

In questions 1 to 9 use Maclaurin's theorem to expand the given function in ascending powers of x as far as the term in x^4.

1. $\cos x$
2. $\tan x$
3. $\sec x$
4. $\ln(\cos x)$
5. $\sin^2 x$
6. $\ln(1 + \sin x)$
7. $\tan^{-1} x$
8. $e^{\sin x}$
9. $\sin^{-1} x$
10. If $y = e^x \sin x$, show that $\dfrac{d^2y}{dx^2} = 2\left(\dfrac{dy}{dx} - y\right)$. By further differentiation of this result, find the Maclaurin expansion of the function $e^x \sin x$ as far as the term in x^6.

9.3 The binomial series

The *binomial series* is the expansion of $(1 + x)^n$ in ascending powers of x, as follows:

$$(1 + x)^n = 1 + nx + \frac{n(n-1)}{2!}x^2 + \frac{n(n-1)(n-2)}{3!}x^3 + \ldots$$
$$\ldots + \frac{n(n-1)(n-2)\ldots(n-r+1)}{r!}x^r + \ldots$$

where n is a real number and $|x| < 1$.

Simple applications of this expansion were discussed in Book 1, §16.7. We now consider further uses of the series.

First we recall that the expression $(a + x)^n$ cannot be expanded directly using the binomial series. Provided that $|x| < |a|$, we write

$$(a + x)^n = a^n\left(1 + \frac{x}{a}\right)^n = a^n\left\{1 + n\left(\frac{x}{a}\right) + \frac{n(n-1)}{2!}\left(\frac{x}{a}\right)^2 + \ldots\right\}$$

Example 1 Expand $(8 + 3x)^{1/3}$ in ascending powers of x as far as the term in x^3, stating the values of x for which the expansion is valid. Hence obtain an approximate value for $\sqrt[3]{8{\cdot}72}$.

$$\begin{aligned}(8 + 3x)^{1/3} &= 8^{1/3}\left(1 + \frac{3}{8}x\right)^{1/3} \\ &= 2\left\{1 + \frac{1}{3}\left(\frac{3}{8}x\right) + \frac{1}{2!} \times \frac{1}{3}\left(\frac{1}{3} - 1\right)\left(\frac{3}{8}x\right)^2\right. \\ &\qquad \left. + \frac{1}{3!} \times \frac{1}{3}\left(\frac{1}{3} - 1\right)\left(\frac{1}{3} - 2\right)\left(\frac{3}{8}x\right)^3 - \ldots\right\} \\ &= 2\left\{1 + \frac{1}{8}x - \frac{1}{2}\cdot\frac{1}{3}\cdot\frac{2}{3}\cdot\frac{3^2}{8^2}x^2 + \frac{1}{6}\cdot\frac{1}{3}\cdot\frac{2}{3}\cdot\frac{5}{3}\cdot\frac{3^3}{8^3}x^3 - \ldots\right\} \\ &= 2\left\{1 + \frac{1}{8}x - \frac{1}{8^2}x^2 + \frac{5}{3.8^3}x^3 - \ldots\right\} \\ &= 2 + \frac{1}{4}x - \frac{1}{32}x^2 + \frac{5}{768}x^3 - \ldots\end{aligned}$$

The expansion is valid for $|\frac{3}{8}x| < 1$ i.e. for $|x| < \frac{8}{3}$.
Substituting $x = 0{\cdot}24$, we have

$$\begin{aligned}(8{\cdot}72)^{1/3} &= 2 + \tfrac{1}{4}(0{\cdot}24) - \tfrac{1}{32}(0{\cdot}24)^2 + \tfrac{5}{768}(0{\cdot}24)^3 - \ldots \\ &= 2 + 0{\cdot}06 - 0{\cdot}0018 + 0{\cdot}00009 - \ldots \\ &\approx 2{\cdot}05829.\end{aligned}$$

Hence the approximate value of $\sqrt[3]{8{\cdot}72}$ is $2{\cdot}05829$.

[To decide whether it is reasonable to give this result to 5 decimal places, we may consider the sum, S_n, of the first n terms in the expansion, for $n = 1, 2, 3, \ldots$.

$$S_1 = 2$$
$$S_2 = 2{\cdot}06$$
$$S_3 = 2{\cdot}0582$$
$$S_4 = 2{\cdot}05829$$
$$S_5 = 2{\cdot}0582846$$

Each term in the expansion of $(8{\cdot}72)^{1/3}$ is smaller than the previous term but opposite in sign. It follows that $\sqrt[3]{8{\cdot}72}$ lies between S_4 and S_5. Thus the difference between 2·05829 and the true value of $\sqrt[3]{8{\cdot}72}$ must be less than 1×10^{-5}.]

Example 2 Given that the first three terms in the expansion in ascending powers of x of $(1 - 8x)^{1/4}$ are the same as the first three terms in the expansion of $(1 + ax)/(1 + bx)$, find the values of a and b. Hence find an approximation to $(0{\cdot}6)^{1/4}$ in the form p/q, where p and q are integers.

$$(1 - 8x)^{1/4} = 1 + \frac{1}{4}(-8x) + \frac{1}{2!} \times \frac{1}{4}\left(\frac{1}{4} - 1\right)(-8x)^2 + \ldots$$

$$= 1 - 2x - \frac{1}{2}.\frac{1}{4}.\frac{3}{4}.8^2x^2 - \ldots$$

$$= 1 - 2x - 6x^2 - \ldots$$

$$\frac{1 + ax}{1 + bx} = (1 + ax)(1 + bx)^{-1}$$

$$= (1 + ax)\left\{1 + (-1)bx + \frac{1}{2!}(-1)(-2)b^2x^2 + \ldots\right\}$$

$$= (1 + ax)(1 - bx + b^2x^2 + \ldots)$$
$$= 1 + ax - bx - abx^2 + b^2x^2 + \ldots$$
$$= 1 + (a - b)x + (b^2 - ab)x^2 + \ldots$$

Since the first three terms of the expansions are the same.

$$a - b = -2, \quad b^2 - ab = -6$$

i.e. $b - a = 2, \quad b(b - a) = -6$

Hence $b = -3$ and $a = -5$.

Thus $$(1 - 8x)^{1/4} \approx \frac{1 - 5x}{1 - 3x}.$$

Substituting $x = 0{\cdot}05$, we have

$$(1 - 0{\cdot}4)^{1/4} \approx \frac{1 - 0{\cdot}25}{1 - 0{\cdot}15} = \frac{0{\cdot}75}{0{\cdot}85}.$$

Hence $(0{\cdot}6)^{1/4}$ is approximately equal to 15/17.

Example 3 Assuming that x is so small that terms in x^3 and higher powers may be neglected, find a quadratic approximation to $\sqrt{\left(\frac{1 - x}{1 + 2x}\right)}$.

$$\sqrt{\left(\frac{1-x}{1+2x}\right)} = (1-x)^{1/2}(1+2x)^{-1/2}$$

$$(1-x)^{1/2} = 1 + \frac{1}{2}(-x) + \frac{1}{2!} \times \frac{1}{2}\left(-\frac{1}{2}\right)(-x)^2 + \ldots$$

$$= 1 - \frac{1}{2}x - \frac{1}{8}x^2 - \ldots$$

$$(1+2x)^{-1/2} = 1 + \left(-\frac{1}{2}\right)(2x) + \frac{1}{2!} \times \left(-\frac{1}{2}\right)\left(-\frac{3}{2}\right)(2x)^2 + \ldots$$

$$= 1 - x + \frac{3}{2}x^2 - \ldots$$

$\therefore$ neglecting terms in x^3 and higher powers,

$$\sqrt{\left(\frac{1-x}{1+2x}\right)} \approx \left(1 - \frac{1}{2}x - \frac{1}{8}x^2\right)\left(1 - x + \frac{3}{2}x^2\right)$$

$$\approx 1 - x + \frac{3}{2}x^2 - \frac{1}{2}x + \frac{1}{2}x^2 - \frac{1}{8}x^2$$

Hence $\sqrt{\left(\frac{1-x}{1+2x}\right)} \approx 1 - \frac{3}{2}x + \frac{15}{8}x^2.$

Example 4 Expand $\frac{1}{1+x+2x^2}$ in ascending powers of x up to and including the term in x^3.

$$\frac{1}{1+y} = (1+y)^{-1}$$

$$= 1 + (-1)y + \frac{(-1)(-2)}{2!}y^2 + \frac{(-1)(-2)(-3)}{3!}y^3 + \ldots$$

$$= 1 - y + y^2 - y^3 + \ldots$$

Substituting $y = x + 2x^2$, we obtain

$$\frac{1}{1+x+2x^2} = 1 - (x+2x^2) + (x+2x^2)^2 - (x+2x^2)^3 + \ldots$$

$$= 1 - x - 2x^2 + x^2 + 4x^3 - x^3 + \ldots$$
$$= 1 - x - x^2 + 3x^3 + \ldots$$

Suppose that we now wish to expand $\frac{1}{(1-x)(1+2x)}$ in ascending powers of x. We could use the method of Example 4:

$$\frac{1}{(1-x)(1+2x)} = (1+x-2x^2)^{-1} = 1 - (x-2x^2) + (x-2x^2)^2 - \ldots$$

or we could write:

$$\frac{1}{(1-x)(1+2x)} = (1-x)^{-1}(1+2x)^{-1}$$
$$= (1+x+x^2+\dots)(1-2x+4x^2-\dots)$$

We may quickly obtain the first few terms of the expansion using either of these methods. However, if a general formula for the term in x^r is required, we must adopt a different approach. This further method involves the use of partial fractions. (See §2.6)

Example 5 Expand $\frac{1}{(1-x)(1+2x)}$ in ascending powers of x, giving the first three terms and the term in x^r. State the values of x for which the expansion is valid.

$$(1-x)^{-1} = 1 + (-1)(-x) + \frac{(-1)(-2)}{2!}(-x)^2 + \dots + \frac{(-1)(-2)\dots(-r)}{r!}(-x)^r + \dots$$

Thus $(1-x)^{-1} = 1 + x + x^2 + \dots + x^r + \dots$
Similarly $(1+2x)^{-1} = 1 - 2x + 4x^2 + \dots + (-2)^r x^r + \dots$

Let $\frac{1}{(1-x)(1+2x)} \equiv \frac{A}{1-x} + \frac{B}{1+2x}$

then $1 \equiv A(1+2x) + B(1-x)$
Putting $x = 1$: $1 = 3A \quad \therefore \quad A = \frac{1}{3}$
Putting $x = -\frac{1}{2}$: $1 = \frac{3}{2}B \quad \therefore \quad B = \frac{2}{3}$

$$\therefore \quad \frac{1}{(1-x)(1+2x)} = \tfrac{1}{3}(1-x)^{-1} + \tfrac{2}{3}(1+2x)^{-1}$$
$$= \tfrac{1}{3}(1 + x + x^2 + \dots + x^r + \dots) + \tfrac{2}{3}(1 - 2x + 4x^2 + \dots + (-2)^r x^r + \dots)$$
$$= 1 - x + 3x^2 + \dots + \{\tfrac{1}{3} + \tfrac{2}{3}(-2)^r\}x^r + \dots$$
$$= 1 - x + 3x^2 + \dots + \tfrac{1}{3}\{1 - (-2)^{r+1}\}x^r + \dots$$

The expansion of $(1-x)^{-1}$ is valid for $|x| < 1$ and the expansion of $(1+2x)^{-1}$ is valid for $|2x| < 1$, i.e. $|x| < \frac{1}{2}$.

Thus the expansion of $\frac{1}{(1-x)(1+2x)}$ is valid when $|x| < \frac{1}{2}$.

Exercise 9.3

In questions 1 to 9 expand the given function in ascending powers of x up to and including the term in x^3.

1. $(1+x)^{-3}$
2. $(1+2x)^{1/2}$
3. $(4+x)^{-3/2}$

4. $\dfrac{1+2x}{1-2x}$ 5. $\dfrac{1-x}{(2+x)^2}$ 6. $\dfrac{1-2x}{\sqrt{(1-4x)}}$

7. $\dfrac{1}{1+x+x^2}$ 8. $\dfrac{1}{\sqrt{(1+x-2x^2)}}$ 9. $\sqrt[3]{(1+x+3x^2)}$

In questions 10 to 15 assume that x is so small that terms in x^3 and higher powers of x may be neglected. Hence find a quadratic approximation to the given function, stating the values of x for which your answer is valid.

10. $\sqrt{\left(\dfrac{1-x}{1+x}\right)}$ 11. $\sqrt[3]{\left(\dfrac{8+x}{1-3x}\right)}$ 12. $\dfrac{(1+4x)^{1/4}}{(1+5x)^{1/5}}$

13. $\dfrac{1}{(1+x)(3-x)}$ 14. $\dfrac{1}{(1-x)(1+2x)^2}$ 15. $\dfrac{1}{(1-2x)\sqrt{(1-x)}}$

In questions 16 to 21 use partial fractions to find the first five non-zero terms in the expansion of the given function in ascending powers of x. State the values of x for which the expansion is valid.

16. $\dfrac{3-5x}{(1-3x)(1+x)}$ 17. $\dfrac{4x}{(1-x)(3+x)}$ 18. $\dfrac{1+x}{(1+x^2)(1-x)}$

19. $\dfrac{4-x}{(1-x)^2(1+2x)}$ 20. $\dfrac{2}{(1+x)(1+x^2)}$ 21. $\dfrac{8(2x-1)}{(x-2)^2(x^2+2)}$

In questions 22 to 24 expand the given function in ascending powers of x, giving the first three non-zero terms and the term in x^r.

22. $\dfrac{3}{(1+x)(1-2x)}$ 23. $\dfrac{4x}{(1-x)(1+3x)}$ 24. $\dfrac{4+3x}{(2-x)(1+2x)}$

25. Expand $(1+2x)^{1/4}$ in ascending powers of x as far as the term in x^3, stating the values of x for which the expansion is valid. Hence obtain approximate values for (a) $\sqrt[4]{1{\cdot}4}$, (b) $\sqrt[4]{1{\cdot}08}$. Obtain the values of $\sqrt[4]{1{\cdot}4}$ and $\sqrt[4]{1{\cdot}08}$ by calculator, (using the function $x^{1/y}$ if available). Explain why the error in answer (a) is greater than the error in answer (b).

26. Expand $(a+2x)\sqrt{(1-4x)}+\dfrac{b}{(1-x)^2}$ in ascending powers of x as far as the term in x^3. Given that the coefficients of x and x^2 are zero, find the first term in the expansion and the coefficient of x^3.

27. Given that the first three terms in the expansion in ascending powers of x of $(1+x+x^2)^n$ are the same as the first three terms in the expansion of $\left(\dfrac{1+ax}{1-3ax}\right)^3$,

find the non-zero values of n and a. Show that the coefficients of x^3 in the two expansions differ by 7·5.

28. If x is so small that x^3 and higher powers of x may be neglected, find the values of a and b such that

$$\sqrt{(1+4x)} \approx \frac{1+ax}{1+bx}.$$

By letting $x = 0{\cdot}04$ find an approximation to $\sqrt{29}$ in the form p/q, where p and q are integers.

9.4 Exponential and logarithmic series

As shown in §9.2 the expansion in ascending powers of x of the exponential function e^x is as follows:

$$e^x = 1 + x + \frac{x^2}{2!} + \frac{x^3}{3!} + \ldots + \frac{x^r}{r!} + \ldots \text{ for all real values of } x.$$

Example 1 Expand e^{2x} in ascending powers of x giving the first three terms and the term in x^r.

$$e^{2x} = 1 + (2x) + \frac{(2x)^2}{2!} + \ldots + \frac{(2x)^r}{r!} + \ldots$$

$$= 1 + 2x + 2x^2 + \ldots + \frac{2^r}{r!}x^r + \ldots$$

Example 2 Find the first three terms in the expansion of e^{x+x^2} in ascending powers of x.

$$e^{x+x^2} = 1 + (x + x^2) + \frac{(x+x^2)^2}{2!} + \ldots$$

$$= 1 + x + x^2 + \frac{1}{2}x^2 + \text{higher powers of } x$$

$$= 1 + x + \frac{3}{2}x^2 + \ldots$$

Example 3 Expand e^{2-x} in ascending powers of x as far as the term in x^3.

$$e^{2-x} = e^2 \times e^{-x} = e^2\left\{1 + (-x) + \frac{(-x)^2}{2!} + \frac{(-x)^3}{3!} + \ldots\right\}$$

$$= e^2\left\{1 - x + \frac{1}{2}x^2 - \frac{1}{6}x^3 + \ldots\right\}$$

Example 4 Find the coefficient of x^r in the expansion of $(x^2 + 2x + 3)e^x$ in ascending powers of x.

Considering the terms $\ldots + \dfrac{x^{r-2}}{(r-2)!} + \dfrac{x^{r-1}}{(r-1)!} + \dfrac{x^r}{r!} + \ldots$ in the expansion of e^x, we find that the terms in x^r in the expansion of $(x^2 + 2x + 3)e^x$ must be

$$\ldots + x^2 \times \frac{x^{r-2}}{(r-2)!} + 2x \times \frac{x^{r-1}}{(r-1)!} + 3 \times \frac{x^r}{r!} + \ldots$$

Hence the coefficient of x^r in the expansion

$$= \frac{1}{(r-2)!} + \frac{2}{(r-1)!} + \frac{3}{r!}$$

$$= \frac{r(r-1)}{r!} + \frac{2r}{r!} + \frac{3}{r!}$$

$$= \frac{r^2 - r + 2r + 3}{r!}$$

$$= \frac{r^2 + r + 3}{r!}$$

In §9.2 we also obtained the logarithmic series:

$$\ln(1 + x) = x - \frac{x^2}{2} + \frac{x^3}{3} - \ldots + (-1)^{r+1}\frac{x^r}{r} + \ldots \text{ where } -1 < x \leqslant 1.$$

Replacing x by $-x$ in this expansion gives:

$$\ln(1 - x) = -x - \frac{x^2}{2} - \frac{x^3}{3} - \ldots - \frac{x^r}{r} - \ldots \text{ where } -1 \leqslant x < 1.$$

When these series are used to find numerical values of logarithms, large numbers of terms are needed to produce accurate results. A much better expansion for this purpose is given by:

$$\ln\left(\frac{1+x}{1-x}\right) = \ln(1 + x) - \ln(1 - x)$$

$$= \left(x - \frac{x^2}{2} + \frac{x^2}{3} - \ldots\right) - \left(-x - \frac{x^2}{2} - \frac{x^3}{3} - \ldots\right)$$

i.e.

$$\ln\left(\frac{1+x}{1-x}\right) = 2\left(x + \frac{x^3}{3} + \frac{x^5}{5} + \ldots\right)$$

We demonstrate the advantages of using this series rather than the expansion of $\ln(1+x)$ by calculating a sequence of approximate values for ln 1·5.

	$\ln(1+x)$, $x=\frac{1}{2}$	$\ln\left(\frac{1+x}{1-x}\right)$, $x=\frac{1}{5}$
1 term	0·5	0·4
2 terms	0·375	0·40533333
3 terms	0·41666667	0·40546133
4 terms	0·40104167	0·40546499
5 terms	0·40729167	0·40546510
6 terms	0·40468750	0·40546511

Using the figures on the left we can give ln 1·5 to 1 decimal place, but not to 2. However, the figures on the right give a result correct to 8 decimal places.

We now use the laws of logarithms to obtain some further expansions.

Example 5 Expand $\ln(2+3x)$ in ascending powers of x as far as the term in x^3. State the values of x for which the expansion is valid.

$$\ln(2+3x) = \ln 2\left(1+\frac{3}{2}x\right)$$

$$= \ln 2 + \ln\left(1+\frac{3}{2}x\right)$$

$$= \ln 2 + \left(\frac{3}{2}x\right) - \frac{1}{2}\left(\frac{3}{2}x\right)^2 + \frac{1}{3}\left(\frac{3}{2}x\right)^3 - \ldots$$

$$= \ln 2 + \frac{3}{2}x - \frac{9}{8}x^2 + \frac{9}{8}x^3 - \ldots$$

The expansion is valid for $-1 < \frac{3}{2}x \leqslant 1$ i.e. $-\frac{2}{3} < x \leqslant \frac{2}{3}$.

Example 6 Expand $\ln\dfrac{1+x}{\sqrt{(1-2x)}}$ in ascending powers of x as far as the term in x^3, stating the values of x for which the expansion is valid.

$$\ln\frac{1+x}{\sqrt{(1-2x)}} = \ln(1+x) - \frac{1}{2}\ln(1-2x)$$

$$= \left\{x - \frac{1}{2}x^2 + \frac{1}{3}x^3 - \ldots\right\}$$

$$-\frac{1}{2}\left\{(-2x) - \frac{1}{2}(-2x)^2 + \frac{1}{3}(-2x)^3 - \ldots\right\}$$

$$= \left\{x - \frac{1}{2}x^2 + \frac{1}{3}x^3 - \ldots\right\} + \left\{x + x^2 + \frac{4}{3}x^3 + \ldots\right\}$$

$$= 2x + \frac{1}{2}x^2 + \frac{5}{3}x^3 + \ldots$$

The expansion of $\ln(1+x)$ is valid for $-1 < x \leqslant 1$ and the expansion of $\ln(1-2x)$ is valid for $-1 < -2x \leqslant 1$, i.e. $-\frac{1}{2} \leqslant x < \frac{1}{2}$.

Hence the expansion of $\ln\dfrac{1+x}{\sqrt{(1-2x)}}$ is valid for $-\frac{1}{2} \leqslant x < \frac{1}{2}$.

Example 7 Assuming that x is so small that x^3 and higher powers may be neglected, obtain a quadratic approximation to the function

$$\frac{1+e^x}{1+\ln(1+x)}.$$

$$1+e^x = 1+\left(1+x+\frac{x^2}{2!}+\dots\right) \approx 2+x+\frac{1}{2}x^2$$

$$1+\ln(1+x) = 1+\left(x-\frac{x^2}{2}+\dots\right) \approx 1+x-\frac{1}{2}x^2$$

Using the binomial expansion,

$$(1+y)^{-1} = 1+(-1)y+\frac{(-1)(-2)}{2!}y^2+\dots = 1-y+y^2-\dots$$

Substituting $y = x-\frac{1}{2}x^2$,

$$\begin{aligned}\frac{1}{1+\ln(1+x)} &\approx \left\{1+\left(x-\frac{1}{2}x^2\right)\right\}^{-1}\\ &= 1-\left(x-\frac{1}{2}x^2\right)+\left(x-\frac{1}{2}x^2\right)^2-\dots\\ &= 1-x+\frac{1}{2}x^2+x^2+\dots\\ &\approx 1-x+\frac{3}{2}x^2\end{aligned}$$

$$\begin{aligned}\therefore \quad \frac{1+e^x}{1+\ln(1+x)} &\approx \left(2+x+\frac{1}{2}x^2\right)\left(1-x+\frac{3}{2}x^2\right)\\ &\approx 2-2x+3x^2+x-x^2+\frac{1}{2}x^2\end{aligned}$$

i.e. $$\frac{1+e^x}{1+\ln(1+x)} \approx 2-x+\frac{5}{2}x^2.$$

Exercise 9.4

1. Find the first three non-zero terms and the general terms in the expansions in ascending powers of x of:

(a) e^{-x} (b) e^{3x} (c) e^{x^2} (d) e^{1+x}

2. Find the first four terms in the expansions of:
(a) $(1 + 2x)e^{2x}$ (b) $(1 - x)^2 e^{-x}$ (c) $e^{(x^2 - x)}$

3. Use the exponential series to calculate the values of e and $1/e$ correct to 4 decimal places.

4. Find the first three non-zero terms and the general terms in the expansions of the following functions, stating the values of x for which the expansions are valid:
(a) $\ln(1 + 3x)$ (b) $\ln(1 - \frac{1}{2}x)$ (c) $\ln(3 + x)$ (d) $\ln(e - x)$

5. Find the first four terms in the expansions of:

(a) $\ln\left(\dfrac{1 + x}{1 + x^2}\right)$ (b) $\ln(1 + x - 2x^2)$ (c) $\ln\dfrac{(1 - x)^2}{\sqrt{(1 + 4x)}}$

6. Use the expansion of $\ln\left(\dfrac{1 + x}{1 - x}\right)$ to calculate the value of $\ln 3$ correct to 4 decimal places.

7. Find the coefficient of x^r in the expansions of each of the following functions in ascending powers of x:
(a) $(1 - x)e^x$ (b) $(1 + x - 3x^2)e^{2x}$ (c) $(1 + x)^2 e^{x/2}$

8. Given that x is so small that terms in x^3 and higher powers of x may be neglected, find quadratic approximations to the following functions:
(a) $e^x \ln(1 + x)$ (b) $(1 + e^{-x})^2$ (c) $(1 + x)^8 e^{-x/2}$

(d) $\dfrac{e^{3x} - e^x}{e^{2x}}$ (e) $\dfrac{1 - \ln(1 - x)}{1 - x}$ (f) $\dfrac{\ln(1 - 2x)}{1 - e^x}$

9. Show that, if $n > 1$, $\ln\left(\dfrac{n + 1}{n - 1}\right) = 2\left(\dfrac{1}{n} + \dfrac{1}{3n^3} + \dfrac{1}{5n^5} + \dots\right)$. Hence calculate the value of $\ln 2$ correct to 4 decimal places.

10. The expansion of $(1 - x)^n$ in ascending powers of x is the same as the expansion of $e^{ax + x^2}$ as far as the term in x^2. Find the values of a and n.

11. The expansion of $(3x + ax^2)/(3 + bx)$ in ascending powers of x is the same as the expansion of $\ln(1 + x)$ as far as the term in x^3. Find the values of a and b. Hence find a rational approximation to $\ln 1{\cdot}2$.

12. (a) Use the expansion of e^x to show that if n is a positive integer, then $x^n e^{-x} \to 0$ as $x \to \infty$.
(b) By letting $x = e^y$, show that $\dfrac{1}{x}\ln x \to 0$ as $x \to \infty$.

13. Show that, if $n > 1, \ln\left(1 + \frac{1}{n}\right)^n = 1 - \frac{1}{2n} + \frac{1}{3n^2} - \ldots$

Use the expansion of e^x to deduce that

$$\left(1 + \frac{1}{n}\right)^n = e\left(1 - \frac{1}{2n} + \frac{11}{24n^2} - \ldots\right).$$

Deduce that, as $n \to \infty$, $\left(1 + \frac{1}{n}\right)^n \to e$.

9.5 The expansions of sin *x* and cos *x*

$$\sin x = x - \frac{x^3}{3!} + \frac{x^5}{5!} - \ldots, \quad \cos x = 1 - \frac{x^2}{2!} + \frac{x^4}{4!} - \ldots$$

for all real values of x.

The expansions of $\sin x$ and $\cos x$ are used in the same ways as the various series discussed in previous sections.

Example 1 Expand $e^{\sin x}$ in ascending powers of x, neglecting terms in x^4 and higher powers of x.

$$\sin x = x - \frac{x^3}{3!} + \ldots = x - \frac{1}{6}x^3 + \ldots$$

$$e^y = 1 + y + \frac{y^2}{2!} + \frac{y^3}{3!} + \ldots = 1 + y + \frac{1}{2}y^2 + \frac{1}{6}x^3 + \ldots$$

Substituting $y = x - \frac{1}{6}x^3$,

$$e^{\sin x} = 1 + \left(x - \frac{1}{6}x^3\right) + \frac{1}{2}\left(x - \frac{1}{6}x^3\right)^2 + \frac{1}{6}\left(x - \frac{1}{6}x^3\right)^3 + \ldots$$

$$= 1 + x - \frac{1}{6}x^3 + \frac{1}{2}x^2 + \frac{1}{6}x^3 + \ldots$$

$$= 1 + x + \frac{1}{2}x^2 + \ldots$$

Example 2 Expand $\sec x$ in ascending powers of x, as far as the term in x^4.

$$\sec x = (\cos x)^{-1} = \left(1 - \frac{x^2}{2!} + \frac{x^4}{4!} - \ldots\right)^{-1} = \left(1 - \frac{x^2}{2} + \frac{x^4}{24} - \ldots\right)^{-1}$$

Using the binomial expansion, $(1 + y)^{-1} = 1 - y + y^2 - \ldots$

Thus, substituting $y = -\frac{x^2}{2} + \frac{x^4}{24}$,

$$\sec x = 1 - \left(-\frac{x^2}{2} + \frac{x^4}{24}\right) + \left(-\frac{x^2}{2} + \frac{x^4}{24}\right)^2 - \ldots$$

$$= 1 + \frac{x^2}{2} - \frac{x^4}{24} + \frac{x^4}{4} + \ldots$$

$$= 1 + \frac{x^2}{2} + \frac{5x^4}{24} + \ldots$$

Exercise 9.5

1. Evaluate, correct to 4 decimal places,
(a) $\sin 0{\cdot}3$ (b) $\cos 0{\cdot}4$ (c) $\cos 2$ (d) $\sin 1$

2. Find the first three non-zero terms in the expansions of the following functions in ascending powers of x:
(a) $\sin 3x$ (b) $\cos 2x$ (c) $\ln(1 + \sin x)$
(d) $e^{\cos x}$ (e) $x \operatorname{cosec} x$ (f) $\tan x$

3. Find approximations to the following functions, given that powers of x higher than the fourth are negligible:
(a) $\sin x(x + \cos x)$ (b) $\sin\{\ln(1 + 2x)\}$
(c) $\cos(\sin 2x) - e^{-2x^2}$ (d) $e^{2x}\cos^2 x$

Exercise 9.6 (miscellaneous)

1. Find the sums of the following series
(a) $\sum_{r=2}^{n} r(r-1)$ (b) $\sum_{r=0}^{n} (2r+1)(2r+3)$

(c) $\sum_{r=n}^{2n} \frac{1}{r(r+1)}$ (d) $\sum_{r=2}^{n} \frac{1}{r^2-1}$

2. Find the sum to n terms of each of the following series
(a) $1.2^2 + 2.3^2 + 3.4^2 + \ldots$ (b) $1.2.3 + 3.4.5 + 5.6.7 + \ldots$

3. Find the sum of all possible products of the form rs^2, where r and s are distinct positive integers such that $1 \leqslant r, s \leqslant n$.

4. The sum of the first n terms of the series $1 + 2x^2 + 3x^4 + \ldots$, where $x^2 \neq 1$, is S. Find an expression for S by considering $S - x^2S$.

5. Using the remainder theorem, or otherwise, factorise $x^3 + 6x^2 + 11x + 6$. Express $\frac{4x+6}{x^3+6x^2+11x+6}$ in partial fractions. Hence show that

$$\sum_{n=0}^{18} \frac{4n+6}{n^3+6n^2+11n+6} = 2\frac{43}{140}.$$ (L)

6. (a) Find the sum of the first n terms of the series whose rth term is $2^r + 2r - 1$.
   (b) If x is so small that terms in x^n, $n \geqslant 3$, can be neglected and $\dfrac{3 + ax}{3 + bx} = (1 - x)^{1/3}$, find the values of a and b. Hence, without the use of tables, find an approximation in the form p/q, where p and q are integers, for $\sqrt[3]{0{\cdot}96}$. (AEB 1979)

7. Find the expansions of $\left(1 + \dfrac{1}{x^2}\right)^{1/2}$ and $\left(1 - \dfrac{1}{x^2}\right)^{1/2}$ in ascending powers of $1/x^2$ up to and including the term in $1/x^4$. Hence, or otherwise, show that, if the positive number x is very large compared with 1, then $(x^2 + 1)^{1/2} - (x^2 + 1)^{1/2} \approx \dfrac{1}{x}$. (C)

8. Expand $\sqrt{(4 - x)}$ as a series in ascending powers of x up to and including the terms in x^2. If terms in $x^n, n \geqslant 3$, can be neglected, find the quadratic approximation to $\sqrt{\left(\dfrac{4 - x}{1 - 2x}\right)}$. State the range of values of x for which this approximation is valid. (AEB 1980)

9. Expand the function $(2 - x)\sqrt{(1 + 2x + 2x^2)}$ in ascending powers of x as far as the term in x^3.

10. Use partial fractions to expand the following functions in ascending powers of x, giving the first three non-zero terms and the term in x^r. State the values of x for which the expansions are valid.

(a) $\dfrac{3x}{(1 + 2x)(1 - x)}$ (b) $\dfrac{x}{(x - 3)(x - 2)}$ (c) $\dfrac{1 - x - x^2}{(1 - 3x)(1 - 2x)(1 - x)}$

11. Expand the following functions in ascending powers of x, as far as the term in x^4. State the values of x for which the expansions are valid.

(a) $\dfrac{1 - 2x}{\sqrt{(1 + 2x^2)}}$ (b) $\dfrac{1 + 2x^2}{(1 - 2x)^2}$ (c) $\dfrac{3}{(1 + 2x^2)(1 - 2x)}$

12. (a) Expand $e^{x/2}\log_e(1 + x)$ in ascending powers of x as far as the term in x^4 and hence show that, for certain values of x to be stated, $e^{x/2}\log_e(1 + x) + e^{-x/2}\log_e(1 - x) = ax^4 + \ldots$ and give the value of a.
    (b) Expand $(1 + x)^{-1/2}$ in ascending powers of x, giving the first four terms and the general term. Hence, without using tables or calculator, obtain the value of $1/\sqrt{101}$, correct to 6 decimal places, showing all working. (SU)

13. (a) Draw rough sketches of the following graphs, showing quite clearly their general shape, and where they cross any axis:
    (i) $y = e^x$ (ii) $y = e^{-x}$ (iii) $y = \log_e x$ (iv) $y = \log_e(1 + x)$

(b) Expand $\log_e \dfrac{(1+2x)^2}{1-3x}$ in ascending powers of x, as far as the term in x^4, giving the range of values of x for which the expansion is valid. What is the nth term?

(c) Prove that the coefficient of x^n in the expansion of $e^{2x}(x^2+3x-7)$ is $\dfrac{2^{n-2}(n^2+5n-28)}{n!}$ (SU)

14. Obtain the series expansion of $\dfrac{e^{2x}}{1+x}$ in ascending powers of x up to and including the term in x^3. Using these terms find an approximation to the value of $\displaystyle\int_0^{1/4} \frac{e^{2x}\,dx}{1+x}$, giving your answer correct to three places of decimals. (O)

15. State the series expansion in ascending powers of x, including the general term, of $\log_e(1+x)$ for $|x| < 1$. Show that

(i) $\log_e\left(\dfrac{1+x}{1-x}\right) = 2\left[x + \dfrac{1}{3}x^3 + \dfrac{1}{5}x^5 + \ldots\right]$,

(ii) $\log_e m = 2\left[\left(\dfrac{m-1}{m+1}\right) + \dfrac{1}{3}\left(\dfrac{m-1}{m+1}\right)^3 + \dfrac{1}{5}\left(\dfrac{m-1}{m+1}\right)^5 + \ldots\right]$.

Hence evaluate $\log_e 2$ to four decimal places, showing all your working.

If $\log_e\left(\dfrac{1+x}{1-x}\right) - \left(\dfrac{e^{2x}-1}{e^x}\right) = ax^3 + bx^5$, when terms in x^n, $n \geqslant 6$, are neglected, find the values of a and b. (AEB 1978)

16. Given that $1 - px + x^2 \equiv (1-\alpha x)(1-\beta x)$, obtain expressions for $\alpha^2 + \beta^2$ and $\alpha^3 + \beta^3$ in terms of p. Hence, or otherwise, find the first three terms in the expansion of $\log_e(1 - px + x^2)$ in ascending powers of x. (JMB)

17. Find the values of positive constants a, b and c given that the expansions of $e^{-ax}\sin 2x$ and $\cos bx \ln(1+cx)$ in ascending powers of x are the same as far as the term in x^3.

10 Some numerical methods

10.1 Numerical integration

When a function $f(x)$ is difficult to integrate, it may be possible to estimate the value of the definite integral $\int_a^b f(x)\,dx$ by numerical methods. Such methods are also useful when, instead of a formula for $f(x)$, we have a set of values obtained experimentally.

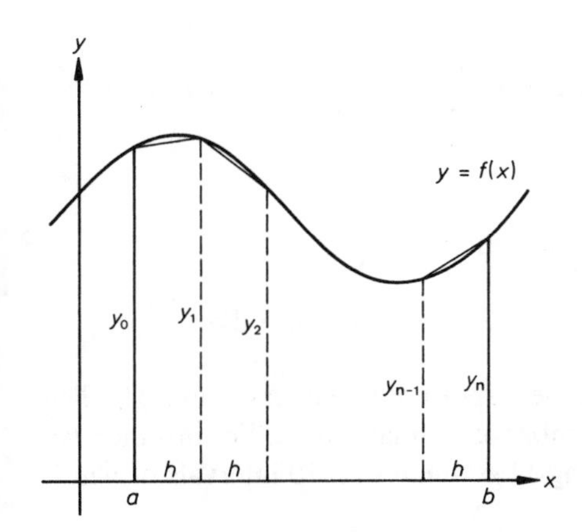

The integral $\int_a^b f(x)\,dx$ is represented by the area under the curve $y = f(x)$ between $x = a$ and $x = b$. Suppose that the area is divided into n strips of width h, as shown in the diagram. The ordinates $y_0, y_1, y_2, \ldots, y_n$ represent the values of $f(x)$ at $x_0 = a, x_1 = a + h, x_2 = a + 2h, \ldots, x_n = b$.

An approximate value for the area is obtained by regarding each strip as a trapezium, so that

$$\int_a^b f(x)\,dx \approx \tfrac{1}{2}h(y_0 + y_1) + \tfrac{1}{2}h(y_1 + y_2) + \ldots + \tfrac{1}{2}h(y_{n-1} + y_n)$$

i.e.
$$\int_a^b f(x)\,dx \approx \tfrac{1}{2}h(y_0 + 2y_1 + 2y_2 + \ldots + 2y_{n-1} + y_n)$$

Thus we have the *trapezium rule* for n strips ($n + 1$ ordinates)

$$\int_a^b f(x)\,dx \approx \tfrac{1}{2}h\{(y_0 + y_n) + 2(y_1 + y_2 + \ldots + y_{n-1})\},$$

$$\text{where } h = (b - a)/n.$$

Example 1 Use the trapezium rule with 5 ordinates to find an approximate value for $\int_0^2 \frac{1}{1 + x^2}\,dx$.

Using 5 ordinates in the interval from $x = 0$ to $x = 2$ gives 4 strips of width $h = 0{\cdot}5$.

n	x_n	y_n (rounded to 5 d.p.)	
0	0	1	
1	0·5		0·8
2	1		0·5
3	1·5		0·307 69
4	2	0·2	
		1·2	1·607 69
		3·215 38	× 2
		4·415 38	3·215 38

By the trapezium rule,

$$\int_0^2 \frac{1}{1 + x^2}\,dx \approx \tfrac{1}{2}h\{(y_0 + y_4) + 2(y_1 + y_2 + y_3)\}$$
$$\approx \tfrac{1}{2} \times 0{\cdot}5 \times 4{\cdot}415\,38$$
$$\approx 1{\cdot}104$$

This integral can also be evaluated directly as follows:

$$\int_0^2 \frac{1}{1 + x^2}\,dx = \left[\tan^{-1} x\right]_0^2 = \tan^{-1} 2 = 1{\cdot}107\,15 \text{ (to 5 d.p.)}$$

Thus the error in the value obtained using the trapezium rule is less than 0·5%. The error may be reduced by increasing the number of ordinates used. For instance, we can obtain a better approximation by using 11 ordinates i.e. 10 intervals of width $h = 0{\cdot}2$.

n	x_n	y_n (rounded to 5 d.p.)	
0	0	1	
1	0·2		0·961 54
2	0·4		0·862 07
3	0·6		0·735 29
4	0·8		0·609 76
5	1		0·5
6	1·2		0·409 84
7	1·4		0·337 84

8	1·6		0·280 90
9	1·8		0·235 85
10	2	0·2	
		1·2	4·933 09
		9·866 18	× 2
		11·066 18	9·866 18

By the trapezium rule, $\displaystyle\int_0^2 \frac{1}{1+x^2}\,dx \approx \tfrac{1}{2}h\{(y_0 + y_{10}) + 2(y_1 + y_2 + \ldots + y_9)\}$

$$\approx \tfrac{1}{2} \times 0{\cdot}2 \times 11{\cdot}066\,18$$
$$\approx 1{\cdot}1066$$

Using 21 ordinates the value 1·1070 is obtained. However, rounding errors in the values of y_n will limit further improvements in accuracy.

Evaluating the integral $\displaystyle\int_a^b f(x)\,dx$ by the trapezium rule is equivalent to replacing the curve $y = f(x)$ by a series of straight lines joining the tops of the ordinates $y_0, y_1, y_2, \ldots$ A better estimate of the area under the curve is usually obtained using *Simpson's rule.*† In this case the curve $y = f(x)$ is regarded as a series of parabolic arcs joining the tops of the ordinates y_0, y_1 and y_2; y_2, y_3 and y_4;....

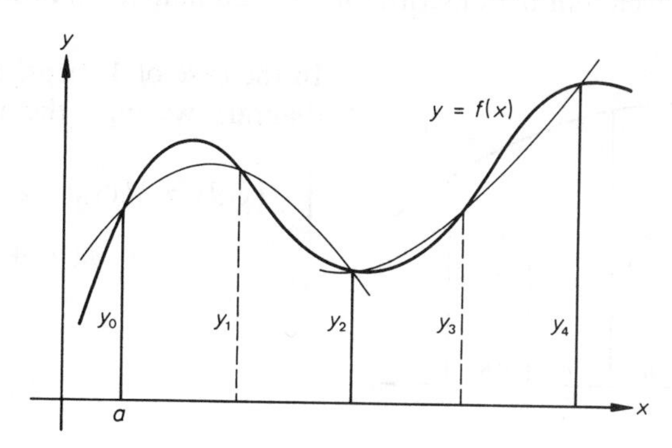

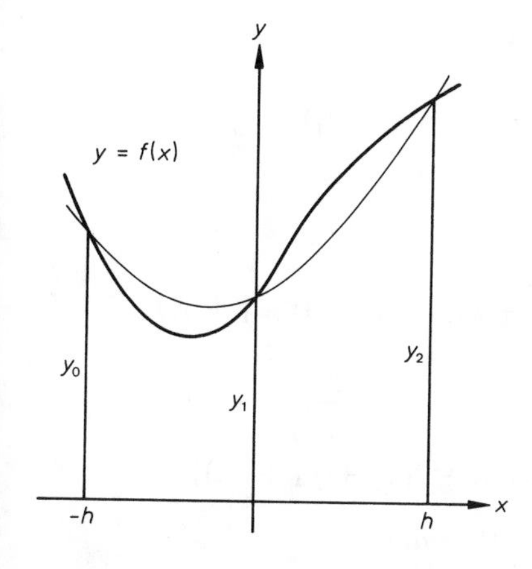

We consider first the integral $\displaystyle\int_{-h}^{h} f(x)\,dx$, letting y_0, y_1, y_2 be the ordinates at $x_0 = -h, x_1 = 0$ and $x_2 = h$. If $y = px^2 + qx + r$ is the parabola which joins the tops of the ordinates y_0, y_1, y_2, then

$$y_0 = ph^2 - qh + r,$$
$$y_1 = r,$$
$$y_2 = ph^2 + qh + r.$$

†*Simpson, Thomas* (1710–1761) English mathematician known as 'the oracle of Nuneaton'. The rule which bears his name was used by earlier mathematicians, but Simpson was credited with its discovery when he published it in 1743.

The area under this parabola between $x = -h$ and $x = h$

$$= \int_{-h}^{h} (px^2 + qx + r)\,dx = \left[\tfrac{1}{3}px^3 + \tfrac{1}{2}qx^2 + rx\right]_{-h}^{h}$$

$$= (\tfrac{1}{3}ph^3 + \tfrac{1}{2}qh^2 + rh) - (-\tfrac{1}{3}ph^3 + \tfrac{1}{2}qh^2 - rh)$$

$$= \tfrac{2}{3}ph^3 + 2rh$$

$$= \tfrac{1}{3}h(2ph^2 + 6r)$$

$$= \tfrac{1}{3}h(y_0 + 4y_1 + y_2)$$

Thus $$\int_{-h}^{h} f(x)\,dx \approx \tfrac{1}{3}h(y_0 + 4y_1 + y_2).$$

Since this result depends only on the values of the ordinates y_0, y_1, y_2 and the interval h between them, we may also write:

$$\int_a^b f(x)\,dx \approx \tfrac{1}{3}h(y_0 + 4y_1 + y_2)$$

where $h = \tfrac{1}{2}(b - a)$ and $y_0 = f(a)$, $y_1 = f(a + h)$, $y_2 = f(b)$.
This is Simpson's rule in its simplest form. The rule for 2 strips (3 ordinates) can be extended to any even number of strips (or any odd number of ordinates.)

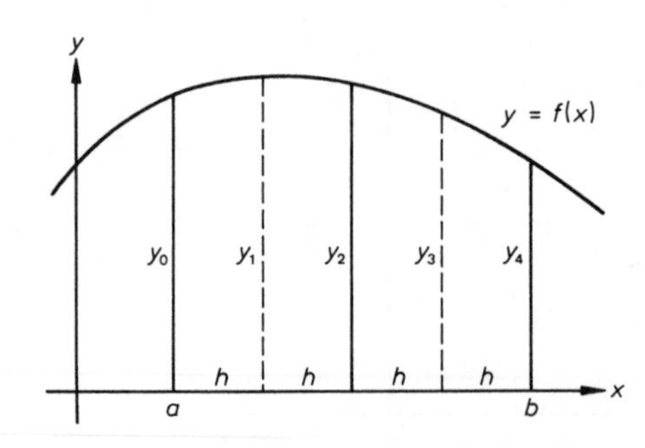

In the case of 4 strips, as shown in the diagram, we apply the basic rule twice:

$$\int_a^b f(x)\,dx \approx \tfrac{1}{3}h(y_0 + 4y_1 + y_2) + \tfrac{1}{3}h(y_2 + 4y_3 + y_4)$$

Thus Simpson's rule for 4 strips (5 ordinates) is

$$\int_a^b f(x)\,dx \approx \tfrac{1}{3}h(y_0 + 4y_1 + 2y_2 + 4y_3 + y_4)$$

More generally, Simpson's rule states that:

$$\int_a^b f(x)\,dx \approx \tfrac{1}{3}h(y_0 + 4y_1 + 2y_2 + 4y_3 + 2y_4 + \ldots + 4y_{n-1} + y_n)$$

i.e. $$\int_a^b f(x)\,dx \approx \tfrac{1}{3}h\{(y_0 + y_n) + 4(y_1 + y_3 + \ldots) + 2(y_2 + y_4 + \ldots)\}$$

where n is even and $h = (b - a)/n$.

Example 2 Use Simpson's rule with 5 ordinates to find an approximate value for $\int_0^2 \frac{1}{1+x^2}\,dx$.

Using 5 ordinates in the interval from $x = 0$ to $x = 2$ gives 4 strips of width $h = 0{\cdot}5$.

n	x_n	y_n (rounded to 5 d.p.)		
0	0	1		
1	0·5		0·8	
2	1			0·5
3	1·5		0·307 69	
4	2	0·2		
		1·2	1·107 69	0·5
		4·430 76	× 4	× 2
		1·0	4·430 76	1·0
		6·630 76		

By Simpson's rule,

$$\int_0^2 \frac{1}{1+x^2}\,dx \approx \tfrac{1}{3}h\{(y_0 + y_4) + 4(y_1 + y_3) + 2y_2\}$$

$$\approx \tfrac{1}{3} \times 0{\cdot}5 \times 6{\cdot}630\,76$$

$$\approx 1{\cdot}105$$

This is closer to the true value of the integral than the result obtained in Example 1 using the trapezium rule. Accuracy can again be improved by increasing the number of ordinates to 11.

n	x_n	y_n (rounded to 5 d.p.)		
0	0	1		
1	0·2		0·961 54	
2	0·4			0·862 07
3	0·6		0.735 29	
4	0·8			0·609 76
5	1		0·5	
6	1·2			0·409 84
7	1·4		0·337 84	
8	1·6			0·280 90
9	1·8		0·235 85	
10	2	0·2		
		1·2	2·770 52	2·162 57
		11·082 08	× 4	× 2
		4·325 14	11·082 08	4·325 14
		16·607 22		

By Simpson's rule,

$$\int_0^2 \frac{1}{1+x^2}\,dx \approx \tfrac{1}{3}h\{(y_0 + y_{10}) + 4(y_1 + y_3 + \ldots) + 2(y_2 + y_4 + \ldots)\}$$
$$\approx \tfrac{1}{3} \times 0{\cdot}2 \times 16{\cdot}607\,22$$
$$\approx 1{\cdot}1071$$

[Unless more accurate values of y_n can be obtained, there is little point in making further increases in the number of ordinates.]

Under certain conditions an approximate value for $\int_a^b f(x)\,dx$ can be obtained by expanding $f(x)$ in ascending powers of x, then integrating term by term. Clearly the method cannot be used unless the expansion of $f(x)$ is valid throughout the interval $a \leqslant x \leqslant b$.

Since the binomial series for $(1 + x^2)^{-1}$ is valid only when $|x| < 1$, this expansion will not produce a good approximation to the value of $\int_0^2 \frac{1}{1+x^2}\,dx$. We consider instead $\int_0^{0.4} \frac{1}{1+x^2}\,dx$. Using the binomial expansion,

$$(1 + x^2)^{-1} = 1 - x^2 + x^4 - x^6 + \cdots$$

$$\therefore \quad \int_0^{0{\cdot}4} \frac{1}{1+x^2}\,dx \approx \int_0^{0{\cdot}4} (1 - x^2 + x^4 - x^6 + \ldots)\,dx$$
$$\approx \left[x - \frac{x^3}{3} + \frac{x^5}{5} - \frac{x^7}{7} + \ldots\right]_0^{0{\cdot}4}$$
$$\approx 0{\cdot}4 - \frac{(0{\cdot}4)^3}{3} + \frac{(0{\cdot}4)^5}{5} - \frac{(0{\cdot}4)^7}{7}$$
$$\approx 0{\cdot}3805.$$

Exercise 10.1

Use the trapezium rule to obtain approximate values of the following integrals, giving your answers to 3 decimal places.

1. $\int_1^3 \frac{1}{1+x}\,dx$, 5 ordinates.
2. $\int_0^3 \sqrt{(1 + x^2)}\,dx$, 4 ordinates.
3. $\int_0^2 10^x\,dx$, 5 strips.
4. $\int_1^8 \lg x\,dx$, 7 strips.
5. $\int_0^{\pi/6} \tan x\,dx$, 7 ordinates.
6. $\int_0^4 xe^{-x}\,dx$, 5 ordinates.

Questions 7 to 12. Use sketches to decide whether the results obtained in questions 1 to 6 are greater than or less than the true values of the given integrals.

Use Simpson's rule to obtain approximate values of the following integrals, giving your answers to 4 decimal places.

13. $\int_0^1 \frac{1}{1+x^3}$, 5 ordinates.

14. $\int_0^{\frac{1}{2}} \sqrt{(1-x^2)}\,dx$, 4 strips.

15. $\int_0^{\pi/3} \sqrt{(\sin x)}\,dx$, 6 strips.

16. $\int_0^4 xe^{-x}\,dx$, 5 ordinates.

17. $\int_1^2 x \lg x\,dx$, 11 ordinates.

18. $\int_0^2 \sqrt{(1+x^3)}\,dx$, 11 ordinates.

19. Expand $(1+x^2)^{12}$ in ascending powers of x as far as the term in x^8. Hence find an approximate value for the integral $\int_0^{0 \cdot 5} (1+x^2)^{12}\,dx$. Using the trapezium rule with 6 ordinates, obtain a second estimate for the value of this integral. Give your answers to 4 decimal places.

20. Use Simpson's rule with 5 ordinates to estimate the value of $\int_0^{0.4} \ln(1+x)dx$. By expanding $\ln(1+x)$ in ascending powers of x as far as the term in x^3 and integrating term by term, obtain a second estimate for the value of this integral. Give your answers to 4 decimal places.

21. Values of a continuous function $f(x)$ were obtained experimentally as follows:

x	1	1·5	2	2·5	3
$f(x)$	8·01	6·02	4·69	3·80	3·27

Estimate the value of $\int_1^3 f(x)\,dx$ using (a) the trapezium rule, (b) Simpson's rule.

22. A car of mass 1000 kg starts from rest and accelerates continuously along a straight road for 20 seconds. After t seconds the resultant force acting on the car is P newtons. The table below gives values of P at intervals of 5 seconds.

t	0	5	10	15	20
P	800	740	590	460	360

Use the trapezium rule to estimate (a) the speed of the car after 10 seconds, (b) the speed of the car after 20 seconds. Use the trapezium rule and your answers to (a) and (b) to estimate the distance travelled in the first 20 seconds.

23. A particle moves in a straight line under the action of a variable force of magnitude F newtons. The value of F after the particle has travelled a distance of s metres is given in the table below.

s	0	1	2	3	4	5	6
F	54	61	67	72	76	79	81

Estimate the work done when the particle has travelled a distance of 6 metres using (a) the trapezium rule, (b) Simpson's rule.

24. Construct a flow diagram to evaluate $\int_0^1 e^{-x^2}\,dx$ using the trapezium rule with 6 ordinates. Test your procedure using a pocket calculator.

25. Construct a flow diagram to evaluate $\int_2^3 \sqrt{(x^2-4)}\,dx$ using Simpson's rule with 11 ordinates. Test your procedure using a pocket calculator.

26.

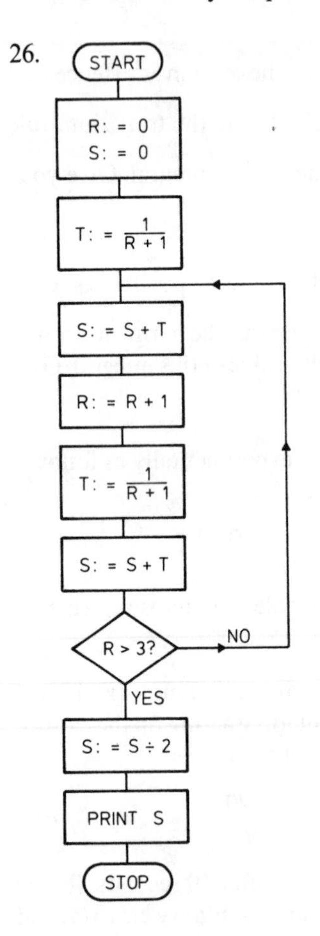

The following flow diagram is designed for the calculation of an approximate value of $\int_0^4 \frac{1}{1+x}\,dx$ using the trapezium rule. Carry out the procedure indicated, tabulating the successive value of R, S and T in simplified fractional form as you proceed. State the printed value of S giving two decimal places in your answers.
By referring to a sketch graph of
$y = \frac{1}{x+1}$ for $0 \leqslant x \leqslant 4$, give a reason why the printed value of S will be an over-estimate of the value of the integral.
(JMB)

27. Show that the formula $\int_0^{2h} f(x)\,dx \approx \frac{1}{3}h\{f(0) + 4f(h) + f(2h)\}$ is exact when $f(x)$ is a cubic polynomial in x. Use the formula given above to obtain an approximate value of I where $I = \int_0^{\frac{1}{2}} \frac{1}{\sqrt{(1-x^2)}}\,dx$, giving your result to three decimal places. Show, by direct integration, that $I = \pi/6$. (L)

10.2 Approximate solutions to equations

The roots of the equation $f(x) = 0$ are the values of x at which the curve $y = f(x)$ cuts the x-axis. Thus approximate values for the roots can be obtained by drawing the graph of $y = f(x)$. However, it is sometimes better to estimate the roots by rearranging the equation and finding the points of intersection of two graphs.

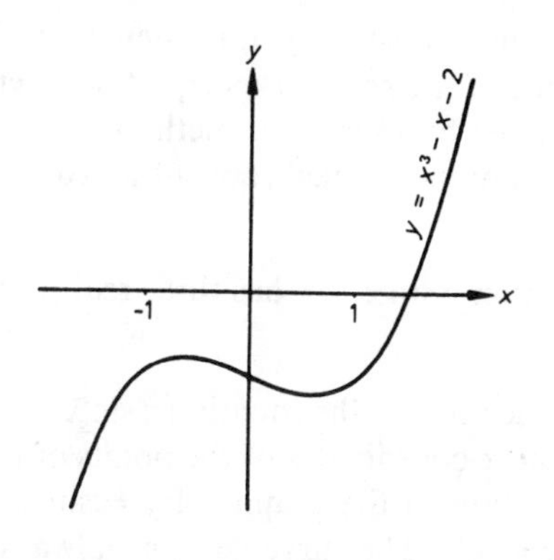

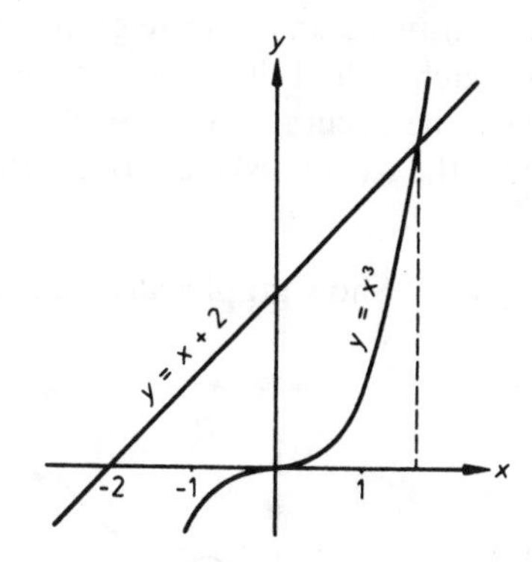

Consider, for instance, the equation $x^3 - x - 2 = 0$. An accurate graph of $y = x^3 - x - 2$ shows that the equation has one real root approximately equal to $1 \cdot 5$. Rearranging the equation as $x^3 = x + 2$, we see that we can also find this root by drawing the graphs of $y = x^3$ and $y = x + 2$.

Example 1 By drawing the graphs of $y = \ln x$ and $y = 2 - x$ in the interval $1 \leqslant x \leqslant 2$, find the solution of the equation $\ln x + x - 2 = 0$, correct to 1 decimal place.

x	1·0	1·2	1·4	1·6	1·8	2·0
$\ln x$	0	0·182	0·336	0·470	0·588	0·693

x	1	1·5	2
$2 - x$	1	0·5	0

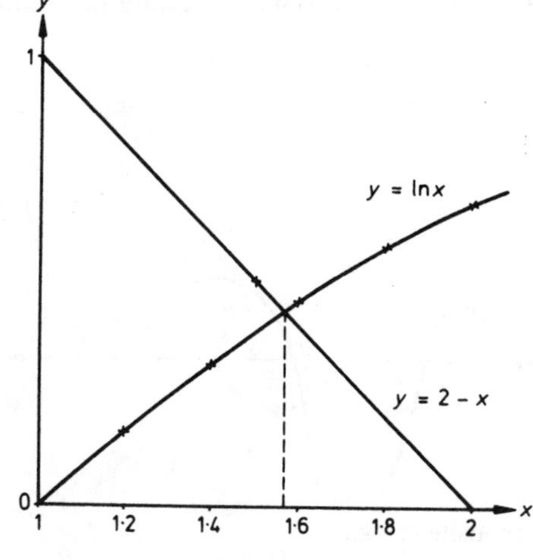

At the point of intersection of the graphs of $y = \ln x$ and $y = 2 - x$ the approximate value of x is $1 \cdot 6$.

Hence if $\ln x + x - 2 = 0$, then $x \approx 1{\cdot}6$.

We can check that this result is accurate to one decimal place as follows:

If $x = 1{\cdot}55, \ln x \approx 0{\cdot}438$ and $2 - x = 0{\cdot}45$ $\therefore$ $\ln x < 2 - x$.
If $x = 1{\cdot}65, \ln x \approx 0{\cdot}501$ and $2 - x = 0{\cdot}35$ $\therefore$ $\ln x > 2 - x$.

Hence $\ln x = 2 - x$ for some value of x between $1{\cdot}55$ and $1{\cdot}65$ i.e. when $x = 1{\cdot}6$ correct to 1 decimal place.

Since drawing an accurate graph is rather time-consuming, it is often quicker to use a sketch to find the approximate locations of the roots of an equation, then to improve the accuracy of these rough estimates by numerical methods. First we consider the use of a sketch to determine the number of real roots of an equation.

Example 2 Show graphically that the equation $\pi \sin x = x$ has three real roots.

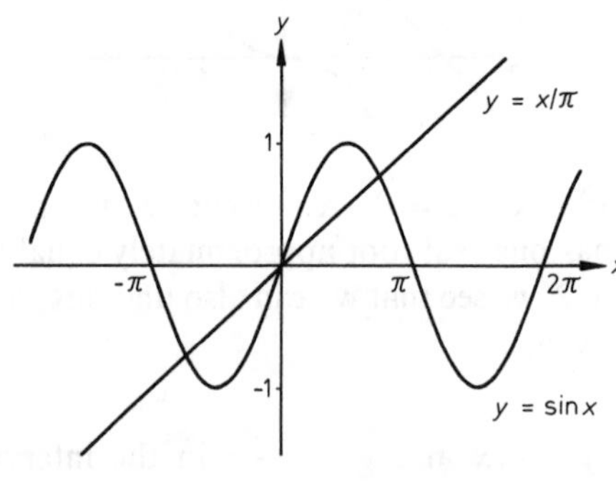

The roots of the equation $\pi \sin x = x$ are the x-coordinates of the points of intersection of the graphs of $y = \sin x$ and $y = x/\pi$. The curve $y = \sin x$ always lies between the lines $y = -1$ and $y = 1$. The straight line $y = x/\pi$ passes through the origin and the points $(-\pi, -1)$, $(\pi, 1)$. Thus, as shown in the sketch, the graphs intersect at three points. Hence the equation $\pi \sin x = x$ has three real roots.

Roots of the equation $f(x) = 0$ can be located by considering the sign changes of the function $f(x)$. Each of the following diagrams shows a function $f(x)$ which is continuous between $x = a$ and $x = b$. We see that if $f(a)$ and $f(b)$ are *opposite* in sign then, in the interval $a < x < b$, the equation $f(x) = 0$ must have at least one root, but may have any *odd* number of roots. If $f(a)$ and $f(b)$ have the *same* sign then, in the interval $a < x < b$, the equation $f(x) = 0$ may have no roots, but could have any *even* number of roots.

[Some of the roots may be coincident.]

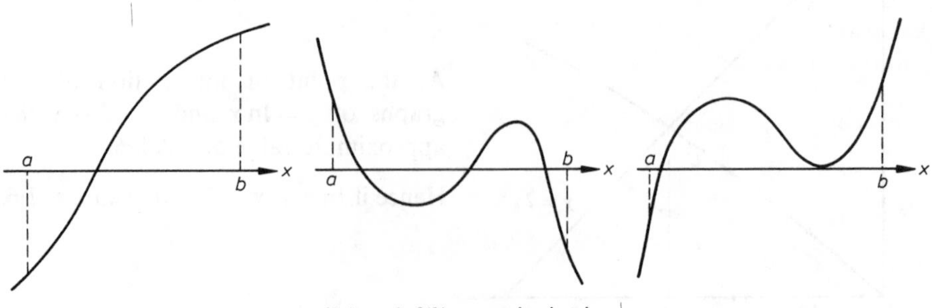

$f(a)$ and $f(b)$ opposite in sign

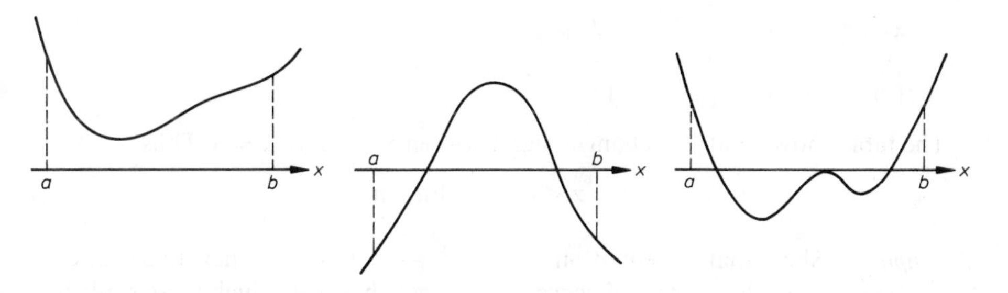

$f(a)$ and $f(b)$ have the same sign

Example 3 Show that the equation $x^3 - 3x^2 + 5 = 0$ has only one real root and that this root lies between -2 and -1.

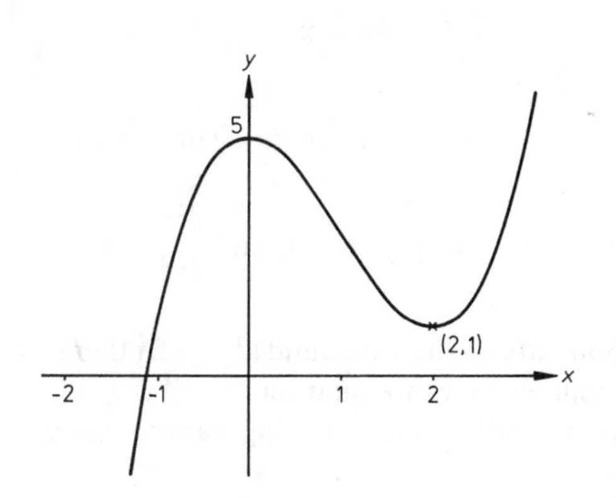

Let $\quad y = x^3 - 3x^2 + 5,$

then $\quad \dfrac{dy}{dx} = 3x^2 - 6x = 3x(x - 2)$

and $\quad \dfrac{d^2y}{dx^2} = 6x - 6 = 6(x - 1)$

$\therefore \quad \dfrac{dy}{dx} = 0$ when $x = 0$ or $x = 2$.

When $x = 0$, $y = 5$ and $\dfrac{d^2y}{dx^2} < 0$.

When $x = 2, y = 1$ and $\dfrac{d^2y}{dx^2} > 0$.

Hence on the curve $y = x^3 - 3x^2 + 5$ there is a maximum point at $(0, 5)$ and a minimum point at $(2, 1)$.
Since both turning points are above the x-axis, if follows that the equation $x^3 - 3x^2 + 5 = 0$ has only one real root.
When $x = -2$, $y = -15$ and when $x = -1$, $y = 1$
$\therefore$ y changes sign in the interval $-2 < x < -1$.
Hence the root of the given equation lies between -2 and -1.

Example 4 Show that the equation $x^3 + x - 16 = 0$ has only one real root α and find the integer n such that $n \leqslant \alpha < n + 1$.

Let $f(x) = x^3 + x - 16$, then $f'(x) = 3x^2 + 1$.
Since $f'(x)$ is always positive, the curve $y = f(x)$ has no turning points and cannot cut the x-axis more than once. For large negative values of $x, f(x)$ is negative and for large positive values of x, $f(x)$ is positive,
$\therefore$ the curve $y = f(x)$ cuts the x-axis exactly once.
Hence the equation $x^3 + x - 16 = 0$ has only one real root α.
When $x \leqslant 0, f(x) < 0 \qquad \therefore \quad \alpha > 0$.

x	1	2	3
$f(x)$	-14	-6	14

The table shows that $f(x)$ changes sign between $x = 2$ and $x = 3$. Thus

$$n \leqslant \alpha < n+1 \quad \text{for} \quad n = 2.$$

Example 5 Show that the equation $x^3 - 4x^2 + 4 = 0$ has one negative root and two positive roots. Find pairs of successive integers between which these roots lie.

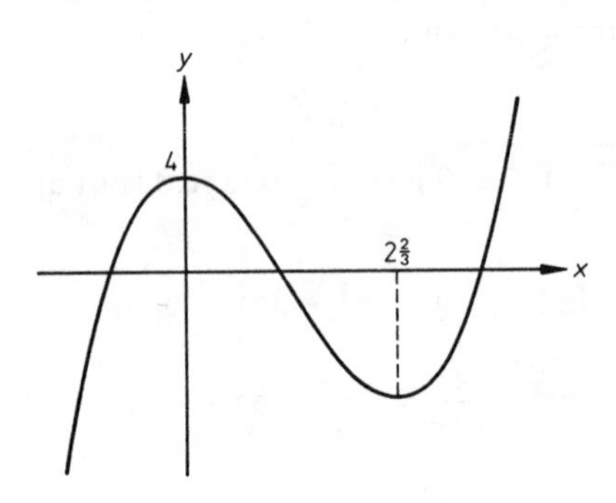

Let $f(x) = x^3 - 4x^2 + 4$ and consider the curve $y = f(x)$.

$$\frac{dy}{dx} = 3x^2 - 8x = x(3x - 8)$$

$$\frac{d^2y}{dx^2} = 6x - 8$$

$$\therefore \quad \frac{dy}{dx} = 0 \text{ when } x = 0 \text{ or } x = 2\tfrac{2}{3}$$

When $x = 0$, $y = 4$ and $\dfrac{d^2y}{dx^2} < 0$. When $x = 2\frac{2}{3}$, $y \approx -5\frac{1}{2}$ and $\dfrac{d^2y}{dx^2} > 0$.

Hence at $x = 0$ there is a maximum point above the x-axis and at $x = 2\frac{2}{3}$ there is a minimum point below the x-axis. It follows that the equation $x^3 - 4x^2 + 4 = 0$ must have one negative root and two positive roots. Testing various integral values of x, we have:

x	-1	0	1	2	3	4
$f(x)$	-1	4	1	-4	-5	4

Using the sign changes shown in the table, we deduce that the roots of the given equation lie between -1 and 0, 1 and 2, 3 and 4.

When it has been shown that an equation $f(x) = 0$ has a root α in the interval $a < x < b$, then *linear interpolation* can be used to estimate the value of α.

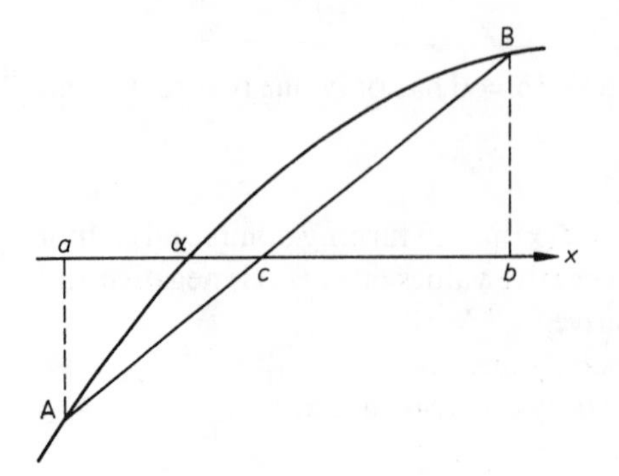

Let us suppose that A and B are the points on the curve $y = f(x)$ at which $x = a$ and $x = b$ respectively. If the straight line joining A to B cuts the x-axis at $(c, 0)$, then c is taken as an approximate value of the root α.

We illustrate the method by letting $f(x) = x^3 + x - 16$. As shown in Example 4, the equation $x^3 + x - 16 = 0$ has a root α in the interval $2 < x < 3$ and the

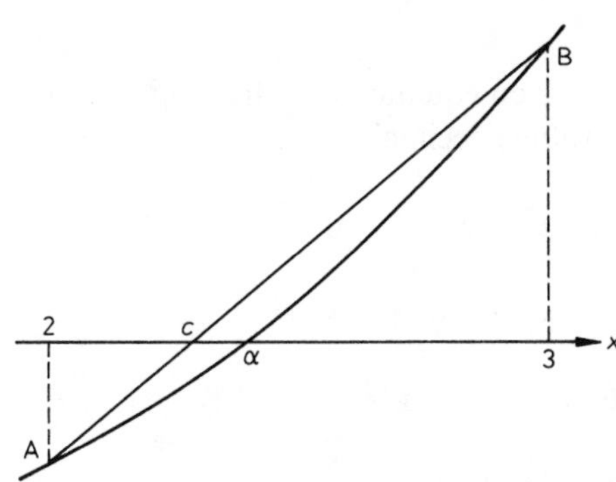

curve $y = x^3 + x - 16$ passes through the points $A(2, -6)$ and $B(3, 14)$. If the straight line joining A to B cuts the x-axis at $(c, 0)$, then using similar triangles, we find that c divides the interval from $x = 2$ to $x = 3$ in the ratio 6:14

Hence, by linear interpolation,

$$\alpha \approx 2 + \frac{6}{6 + 14} = 2{\cdot}3.$$

[Note that since $f(2{\cdot}3) \approx -1{\cdot}5$, the true value of α lies between 2·3 and 3.]

If the equation $f(x) = 0$ is known to have a root so small that, for values of x close to the root, x^3 and higher powers of x are negligible, then the root can be estimated using a *quadratic approximation* to $f(x)$.

Example 6 Given that the equation $(1 - x)e^x = 0{\cdot}7(2x + 1)$ has a small root, use a quadratic approximation to estimate its value.

$$(1 - x)e^x = (1 - x)\left(1 + x + \frac{x^2}{2!} + \dots\right)$$

$$= 1 + x + \tfrac{1}{2}x^2 - x - x^2 + \dots$$

$$\therefore \quad (1 - x)e^x \approx 1 - \tfrac{1}{2}x^2.$$

Hence the small root of the equation $(1 - x)e^x = 0{\cdot}7(2x + 1)$ is approximately equal to the small root of the equation $1 - \frac{1}{2}x^2 = 0{\cdot}7(2x + 1)$.
Multiplying both sides of this equation by 10, we have

$$10 - 5x^2 = 7(2x + 1)$$

i.e. $5x^2 + 14x - 3 = 0$

$$(5x - 1)(x + 3) = 0$$

$\therefore \quad x = 0{\cdot}2 \quad \text{or} \quad x = -3.$

Hence the approximate value of the small root of the given equation is 0·2.

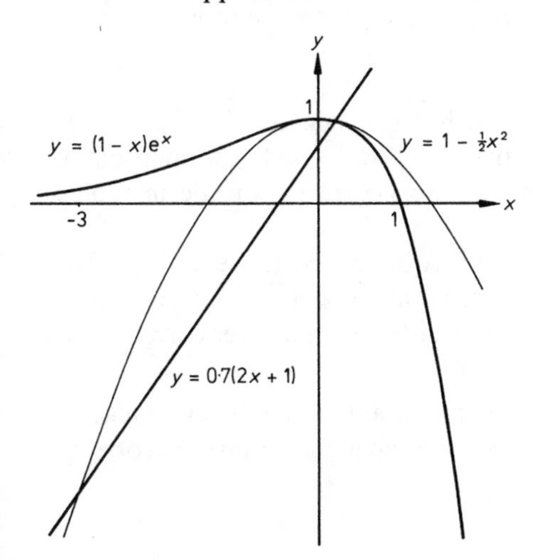

[We see from the sketch that the approximation $(1 - x)e^x \approx 1 - \frac{1}{2}x^2$ is valid for only a small range of values of x. Thus the second root, $x = -3$, obtained using this approximation has no real significance.]

Exercise 10.2

In questions 1 to 10 obtain the roots of the given equations by drawing suitable accurate graphs. Give your answers correct to one decimal place.

1. $x^2 - 2x - 2 = 0$
2. $x^3 + x - 1 = 0$
3. $x^3 - 3x + 3 = 0$
4. $x^3 - 2x^2 + 1 = 0$
5. $x^3 + x^2 - 2x - 1 = 0$
6. $e^x = 3 - x$
7. $2 \sin x = x$
8. $4 \cos x = x$
9. $\dfrac{4}{x^2} = 2x + 1$
10. $(x - 2)\ln x = x - 1$

In questions 11 to 14 use a sketch to find the number of real roots of the given equation.

11. $e^{-x} = x - 1$
12. $2e^x = 5x + 1$
13. $\cos 2x = 1 - x$
14. $\sin \pi x = \frac{1}{2}(x - 1)$

In questions 15 to 22 determine the number of real roots of the given equation and find pairs of successive integers between which the roots lie.

15. $x^3 - 6x^2 + 36 = 0$
16. $x^3 + 5x - 10 = 0$
17. $x^3 - 12x + 12 = 0$
18. $x^3 + 2x^2 - 4x - 4 = 0$
19. $2x^3 + 4x - 15 = 0$
20. $x^3 - x - 3 = 0$
21. $e^x = x + 3$
22. $\sin x = \ln x$

23. Show that the equation $x^3 - 7x - 12 = 0$ has one real root and show that this root lies between 3 and 4. Use linear interpolation to estimate the root, giving your answer to one decimal place.

24. Show that the equation $x^4 - 4x - 4 = 0$ has one negative root α and one positive root β. Show that $-1 < \alpha < 0$ and that $1 < \beta < 2$. Use linear interpolation to estimate the values of α and β, giving your answers correct to one decimal place.

25. Show that the equation $x^4 + 3x^3 + x^2 - 9 = 0$ has one negative root α and one positive root β. Find α and the integer k such that $k < \beta < k + 1$. Use linear interpolation to find an approximate value for β, giving your answer correct to one decimal place.

In questions 26 to 31 use linear interpolation between suitable successive integers to obtain estimates for the roots of the given equations, giving your answers correct to one decimal place.

26. $2x^3 - 25 = 0$

27. $x^3 + 3x^2 - 14x - 10 = 0$

28. $4x^3 - 9x^2 + 3 = 0$

29. $x^3 + 3x^2 - 4x + 3 = 0$

30. $\ln x = x^2 - 6x + 8$

31. $6 \cos x + x = 0$

32. Draw an accurate graph of $y = 2x^3 - 5x + 2$ for values of x between -2 and $+2$. Estimate the roots of the equation $2x^3 - 5x + 2 = 0$ (a) from your graph, (b) by linear interpolation between suitable successive integers. Why does linear interpolation give poor approximations to the roots in this case?

33. Each of the following equations has a root so small that, for values of x close to the root, x^3 and higher powers of x are negligible. Use quadratic approximations to estimate these roots.
(a) $2(1 - x)^5 = 18x + 5$
(b) $8 \ln(1 + x) = 3x + 1$
(c) $10(1 - \cos x) = 2 - 3x$
(d) $5\sqrt[3]{(1 + 3x)} = 14x + 3$

10.3 The Newton-Raphson method

If $x = a$ is an approximate value of a root of an equation $f(x) = 0$, then the tangent to the curve $y = f(x)$ at $x = a$ can be used to find a second approximation to the root. As shown in the diagrams, the x-coordinate of the point of intersection of this tangent with the x-axis is usually closer to the true root than the first approximation $x = a$.

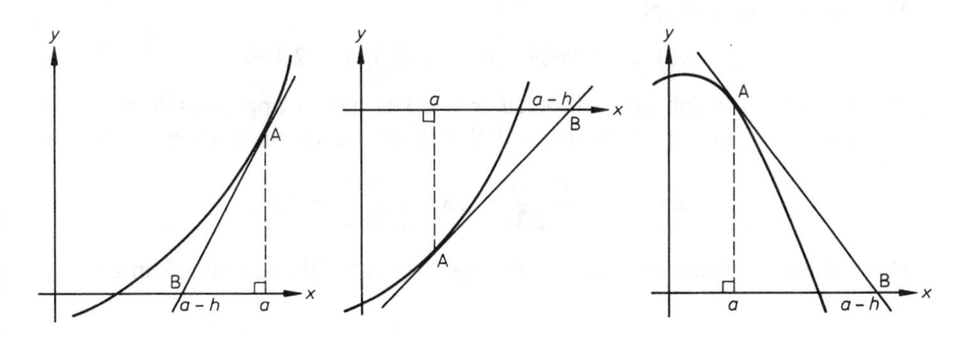

Let A be the point $(a, f(a))$ on the curve $y = f(x)$ and let the point of intersection of the tangent at A with the x-axis be the point B with coordinates $(a - h, 0)$, then:

$$\text{gradient of } AB = \frac{f(a) - 0}{a - (a - h)} = \frac{f(a)}{h}.$$

However, since AB is the tangent to the curve $y = f(x)$ at $x = a$, its gradient is also equal to $f'(a)$,

$$\therefore \qquad f'(a) = \frac{f(a)}{h} \quad \text{which gives} \quad h = \frac{f(a)}{f'(a)}.$$

Thus the tangent at A cuts the x-axis where $x = a - \dfrac{f(a)}{f'(a)}$. This result forms the basis of the procedure known as the *Newton-Raphson*† *method* (or simply Newton's method).

> If $x = a$ is a first approximation to a root of the equation $f(x) = 0$, then a second approximation to the root is given by $x = a - \dfrac{f(a)}{f'(a)}$.

The following diagrams show that the method may fail to produce a better approximation to the root when the value of $f'(a)$ is small or when the first approximation to the root is poor.

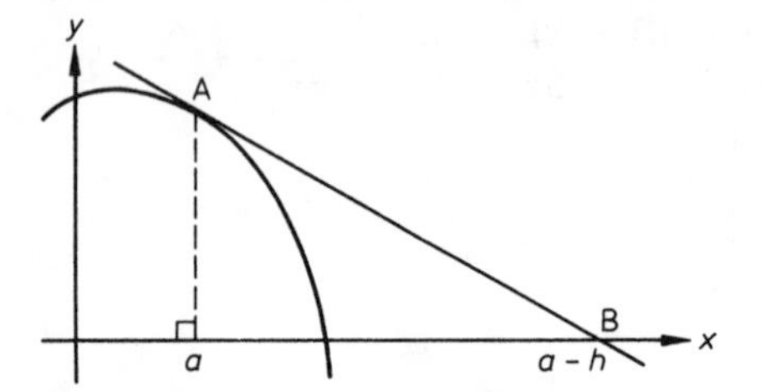

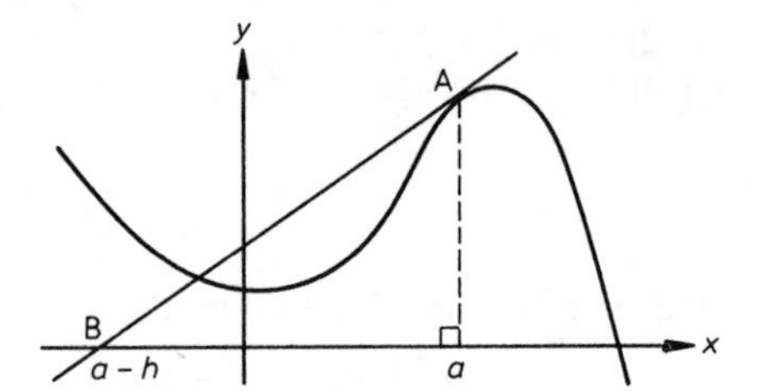

Example 1 Taking $x = 2{\cdot}3$ as a first approximation to the positive root of the equation $3 \sin x = x$, use the Newton-Raphson method to find a second approximation to this root, giving your answer to 2 decimal places.

Let $f(x) = 3 \sin x - x$, then $f'(x) = 3 \cos x - 1$.
Working to 4 decimal places,

$$f(2{\cdot}3) \approx -0{\cdot}0629 \quad \text{and} \quad f'(2{\cdot}3) \approx -2{\cdot}9988.$$

By the Newton-Raphson method, if $x = 2{\cdot}3$ is a first approximation to the positive root of the equation $f(x) = 0$, then a second approximation is given by

$$x = 2{\cdot}3 - \frac{f(2{\cdot}3)}{f'(2{\cdot}3)} \approx 2{\cdot}3 - \frac{0{\cdot}0629}{2{\cdot}9988} \approx 2{\cdot}28.$$

Hence the second approximation to the positive root of the equation $3 \sin x = x$ is $x = 2{\cdot}28$.

If x_0 is a first approximation to a root α of the equation $f(x) = 0$, then the Newton-Raphson method can be used to produce a sequence of approximations $x_1, x_2, x_3, \ldots$ by writing:

$$x_{n+1} = x_n - \frac{f(x_n)}{f'(x_n)}$$

The following diagrams show that, in general, the sequence $x_0, x_1, x_2, x_3, \ldots$ converges quickly to the required root.

†*Raphson, Joseph* (1648–1715) English mathematician. His account of the Newton-Raphson method was published in 1690 and was based on earlier work by Newton. Raphson produced many other mathematical works including a *History of Fluxions* (1715).

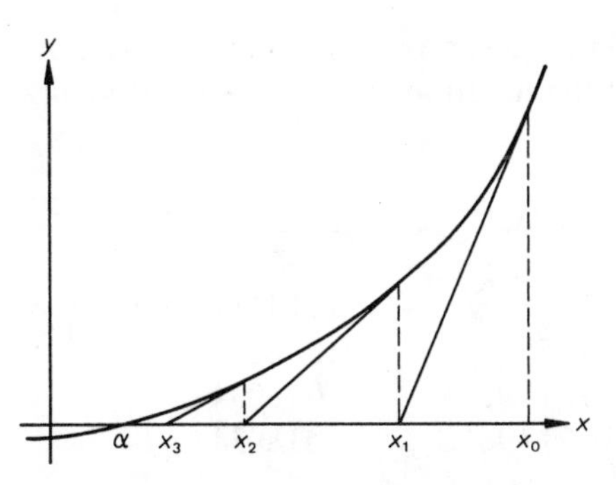

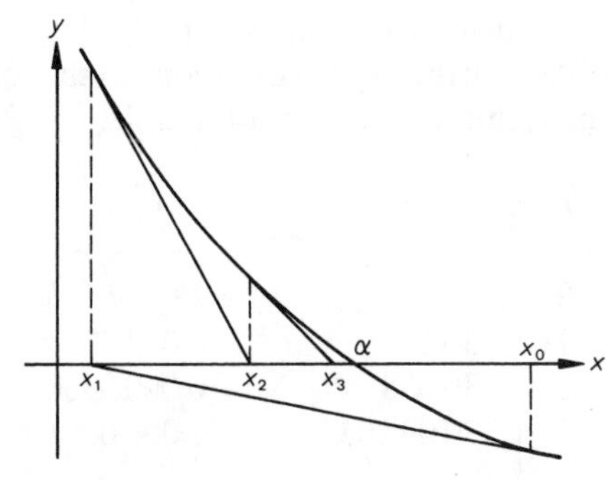

The next two diagrams show that in certain circumstances the method fails. In (1) the sequence x_0, x_1, x_2, x_3, ... converges to a different root β and in (2) the sequence diverges.

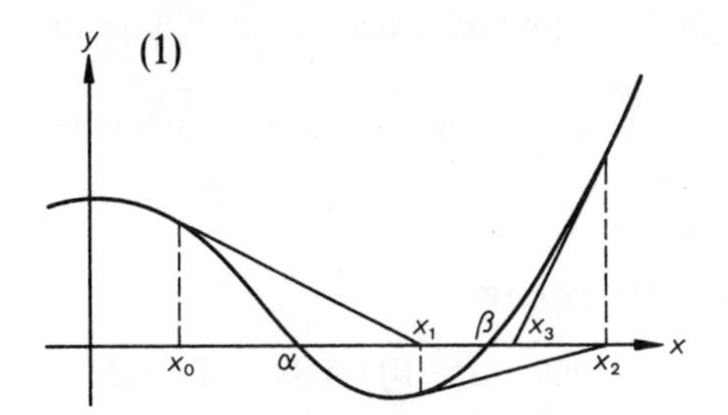

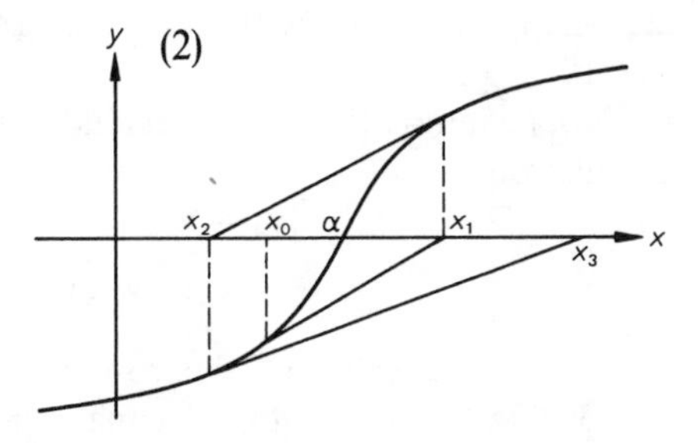

Example 2 The equation $x^3 - 3x^2 - 1 = 0$ has one real root. Starting with $x = 3$, use two iterations of Newton's method to find an approximate value for this root, correct to 3 decimal places.

Let $f(x) = x^3 - 3x^2 - 1$, then $f'(x) = 3x^2 - 6x$.
Using Newton's method and working to 5 decimal places,

let $\quad x_0 = 3, \quad$ then $\quad f(x_0) = -1,\ f'(x_0) = 9$

$$\therefore \quad x_1 = x_0 - \frac{f(x_0)}{f'(x_0)} = 3 - \frac{(-1)}{9} \approx 3{\cdot}11111$$

Let $\ x_1 = 3{\cdot}11, \quad$ then $\quad f(x_1) \approx 0{\cdot}063\,93,\ f'(x_1) = 10{\cdot}3563$

$$\therefore \quad x_2 = x_1 - \frac{f(x_1)}{f'(x_1)} = 3{\cdot}11 - \frac{0{\cdot}063\,93}{10{\cdot}3563} \approx 3{\cdot}103\,83$$

Hence an approximate value for the root is 3·104.

Checking the accuracy of this result we find that

$$f(3{\cdot}1035) \approx -0{\cdot}003 \quad \text{and} \quad f(3{\cdot}1045) \approx 0{\cdot}007,$$

$\therefore$ the root is 3·104 correct to 3 decimal places.

[Note that when evaluating a function $f(x)$ by calculator it is usually best to store the value of x in the memory. If $f(x)$ is a polynomial more accurate results will be obtained by using nested multiplication as described in Book 1, §4.4.]

To demonstrate how rapidly the Newton-Raphson process converges we now obtain further approximations to the root of the equation $x^3 - 3x^2 - 1 = 0$, giving the results in the form of a table.

n	x_n	$f(x_n)$	$f'(x_n)$	x_{n+1}
0	3	-1	9	3·111 1111
1	3·11	0·063 931	10·3563	3·103 8269
2	3·1038	-0·000 0350	10·277 9233	3·103 8034
3	3·103 8034	-0·000 000 03	10·277 9662	3·103 8034

When using the Newton-Raphson process it is usually reasonable to assume that if $|x_{n+1} - x_n| < \varepsilon$, then the error in the approximation x_{n+1} is less than ε.

In the given table the values of x_3 and x_4 agree to 7 decimal places. It follows that $|x_4 - x_3| < 1 \times 10^{-7}$ and thus the error in the approximation x_4 is less than 1×10^{-7}.

Closer examination of the results shows that, allowing for rounding errors in the calculations,

$$|x_4 - x_3| = |f(x_3)/f'(x_3)| < 1 \times 10^{-8}$$

$$\therefore \qquad 3{\cdot}103\,8034 < x_4 < 3{\cdot}103\,803\,41$$

Thus the true value of the root is unlikely to exceed 3·103 803 42. Therefore $x = 3{\cdot}103\,8034$ correct to 7 decimal places.

Exercise 10.3

In questions 1 to 8 use the Newton-Raphson method to find a second approximation to a root of the given equation taking the stated value of a as a first approximation. Give your answers to 2 decimal places.

1. $x^3 - 4x^2 + 4 = 0$, $a = 1$.
2. $x^3 - 3x^2 + 5 = 0$, $a = -1$.
3. $x^3 - x + 2 = 0$, $a = -1{\cdot}5$.
4. $x^3 + x - 16 = 0$, $a = 2{\cdot}3$.
5. $\ln x + x - 2 = 0$, $a = 1{\cdot}6$.
6. $e^x + x - 3 = 0$, $a = 0{\cdot}8$.
7. $2 \sin x = x$, $a = -2$.
8. $3 \cos x = x$, $a = 1{\cdot}2$.

In questions 9 to 14 use two iterations of the Newton-Raphson method to find an approximate value for a root of the given equation, taking the stated value of x_0 as a first approximation. Give your answers to 3 decimal places.

9. $x^3 - 5x - 8 = 0$, $x_0 = 3$.
10. $x^3 - 9x^2 + 20 = 0$, $x_0 = 2$.
11. $4x^3 + 6x^2 - 5x - 6 = 0$, $x_0 = -1$.
12. $x^3 + x + 8 = 0$, $x_0 = -2$.
13. $e^{-x} = x - 1$, $x_0 = 1{\cdot}2$.
14. $\sin x = \ln x$, $x_0 = 2{\cdot}3$.

15. Show that the equation $x^3 + 2x - 1 = 0$ has only one real root and that this root lies between 0 and 1. Find the approximate value of this root correct to 3 decimal places.

16. Show that the equation $x^3 - 12x - 6 = 0$ has one positive and two negative roots. Use linear interpolation between suitable successive integers to find a first approximation to the positive root. Hence find the approximate value of this root correct to 4 decimal places.

17. Give, on the same diagram, a sketch of the graph of $y = 3e^{x/2}$ and of the graph of $y = 4x + 6$. State the number of roots of the equation $3e^{x/2} = 4x + 6$. Taking 4 as a first approximation to one root α of this equation, find a second approximation to α, giving three significant figures in your answer and showing clearly how this answer has been obtained. Using a suitable integer as first approximation to another root β of the equation, find a second approximation to β, again giving three significant figures in your answer and showing clearly how this answer has been determined. (C)

18. Show that the equation $e^x = 2(1 + x)$ has only one positive root. Find the nearest integer to this root. Rewrite the flow chart shown below and complete it to specify a process which will find the root of the above equation, with an error of less than 0·01, starting from the integer nearest to the root. Evaluate the root to this degree of accuracy.

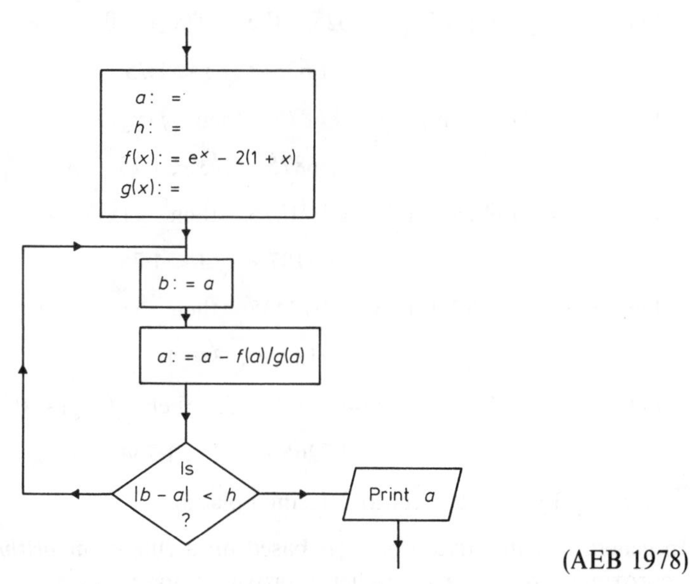

(AEB 1978)

10.4 Other iterative processes

An *iterative process* is a repetitive procedure designed to produce a sequence of approximations to some numerical quantity. The Newton-Raphson method described in the previous section is one example of such a process.

Interval bisection is another iterative process used to obtain approximations to a root α of an equation $f(x) = 0$. To use this method it is necessary to locate an interval containing the root, i.e. to find values x_0 and x_1 such that $x_0 < \alpha < x_1$. The length of this interval is halved by taking $x_2 = \frac{1}{2}(x_0 + x_1)$ and then determining whether α lies between x_0 and x_2 or between x_2 and x_1. The new interval is then bisected to produce a further approximation x_3. The bisection process is repeated until the interval containing α is small enough to ensure a result to the required degree of accuracy. We will see in the following example that this method converges much more slowly than the Newton-Raphson method.

Example 1 Use interval bisection to obtain a sequence of approximations to the positive root of the equation $x^2 - 3 = 0$. Hence evaluate $\sqrt{3}$ correct to 2 decimal places.

Let $f(x) = x^2 - 3$, then the positive root of the equation $f(x) = 0$ is $\sqrt{3}$.
Since $f(1) < 0$ and $f(2) > 0$, we have $1 < \sqrt{3} < 2$.
Taking $x_0 = 1$ and $x_1 = 2$ as first approximations to $\sqrt{3}$,

let $x_2 = \frac{1}{2}(x_0 + x_1) = \frac{1}{2}(1 + 2) = 1{\cdot}5$, then $f(x_2) < 0$

$$\therefore \qquad 1{\cdot}5 < \sqrt{3} < 2$$

Let $x_3 = \frac{1}{2}(1{\cdot}5 + 2) = 1{\cdot}75$, then $f(x_3) > 0$

$$\therefore \qquad 1{\cdot}5 < \sqrt{3} < 1{\cdot}75$$

Let $x_4 = \frac{1}{2}(1{\cdot}5 + 1{\cdot}75) = 1{\cdot}625$, then $f(x_4) < 0$

$$\therefore \qquad 1{\cdot}625 < \sqrt{3} < 1{\cdot}75$$

Let $x_5 = \frac{1}{2}(1{\cdot}625 + 1{\cdot}75) = 1{\cdot}6875$, then $f(x_5) < 0$

$$\therefore \qquad 1{\cdot}6875 < \sqrt{3} < 1{\cdot}75$$

Let $x_6 = \frac{1}{2}(1{\cdot}6875 + 1{\cdot}75) = 1{\cdot}71875$, then $f(x_6) < 0$

$$\therefore \qquad 1{\cdot}7187 < \sqrt{3} < 1{\cdot}75$$

Let $x_7 = \frac{1}{2}(1{\cdot}7187 + 1{\cdot}75) = 1{\cdot}73435$, then $f(x_7) < 0$

$$\therefore \qquad 1{\cdot}7187 < \sqrt{3} < 1{\cdot}7344$$

Let $x_8 = \frac{1}{2}(1{\cdot}7187 + 1{\cdot}7344) = 1{\cdot}72655$, then $f(x_8) < 0$

$$\therefore \qquad 1{\cdot}7265 < \sqrt{3} < 1{\cdot}7344$$

Hence $\sqrt{3} = 1{\cdot}73$ correct to 2 decimal places.

In general, an iterative process is based on a rule or *algorithm* for obtaining an approximation x_{n+1} from earlier approximations $x_0, x_1, x_2, \ldots, x_n$. In many cases this rule can be expressed as an iterative formula giving x_{n+1} in terms of x_n. In section §10.3 we showed that the Newton-Raphson method can be used to produce such a formula. We now discuss an alternative method.

Let us consider the general iterative formula

$$x_{n+1} = f(x_n)$$

If the sequence $x_0, x_1, x_2, x_3, \ldots$ converges to a finite limit α, then as $n \to \infty$, $x_{n+1} \to \alpha$ and $f(x_n) \to f(\alpha)$

giving $$\alpha = f(\alpha).$$

Thus, if the iteration $x_{n+1} = f(x_n)$ is convergent, it will produce a sequence of approximations to a root of the equation $x = f(x)$.

When solving an equation we may be able to obtain a suitable iterative formula by writing the equation in the form $x = f(x)$. For any particular equation several different rearrangements will be possible, but not all the corresponding iterative processes will converge to the required root. It is difficult to predict which rearrangements will be most successful. However, it can be shown that, in general, the iteration $x_{n+1} = f(x_n)$ converges to a root α, provided that (i) x_0 is a fairly good approximation to α, and that (ii) $-1 < f'(\alpha) < 1$.

To illustrate the various possibilities we consider the equation $x^2 - 4x - 8 = 0$. This has one root α between -2 and -1 and another root β between 5 and 6. We will attempt to calculate these roots using iterative formulae based on the rearrangements:

(a) $x = \frac{1}{4}x^2 - 2$, (b) $x = \dfrac{8}{x-4}$, (c) $x = 2\sqrt{(x+2)}$.

(a) The following table shows the sequences obtained using the formula $x_{n+1} = \frac{1}{4}x_n^2 - 2$, taking x_0 as -2, -1, 5 and 6, then working to 3 decimal places.

x_0	-2	-1	5	6
x_1	-1	$-1{\cdot}75$	4·25	7
x_2	$-1{\cdot}75$	$-1{\cdot}234$	2·516	10·25
x_3	$-1{\cdot}234$	$-1{\cdot}619$	$-0{\cdot}417$	24·266
x_4	$-1{\cdot}619$	$-1{\cdot}345$	$-1{\cdot}957$	145·210
x_5	$-1{\cdot}345$	$-1{\cdot}548$	$-1{\cdot}043$	⋮
x_6	$-1{\cdot}548$	$-1{\cdot}401$	$-1{\cdot}728$	⋮

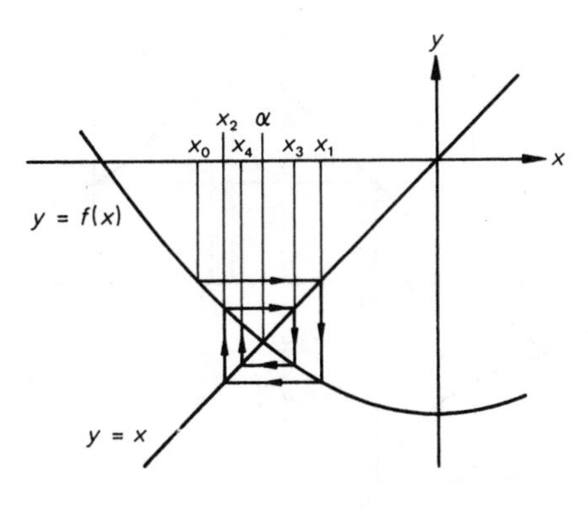

$x_0 = -2$, $f(x) = \frac{1}{4}x^2 - 2$

The sketch graphs show that three of the sequences converge slowly to the root α and that the other sequence diverges. Writing $f(x) = \frac{1}{4}x^2 - 2$, we find that

$$f'(\alpha) \approx f'(-1{\cdot}5) = -0{\cdot}75$$

and $$f'(\beta) \approx f'(5{\cdot}5) = 2{\cdot}75.$$

These results tend to confirm that the rate of convergence depends on the value of $f'(x)$.

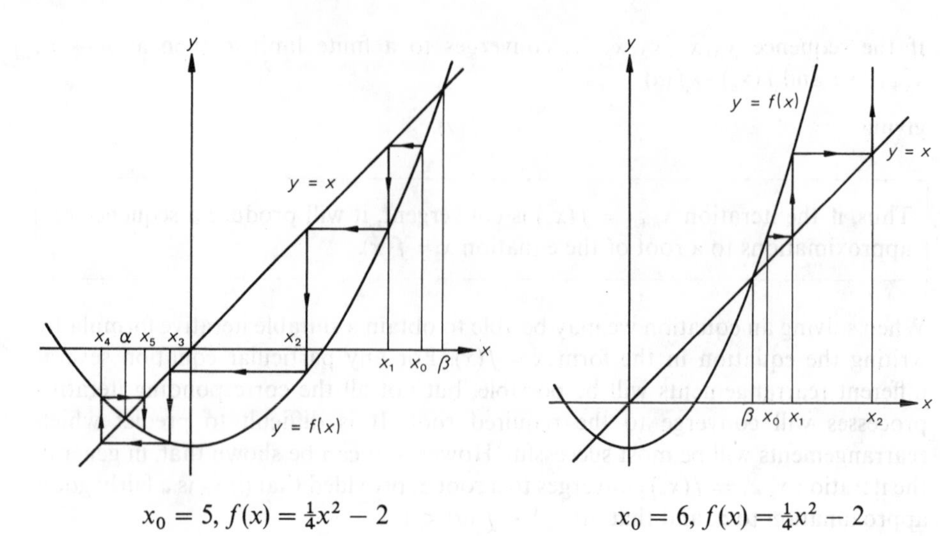

$x_0 = 5, f(x) = \frac{1}{4}x^2 - 2$ $\qquad$ $x_0 = 6, f(x) = \frac{1}{4}x^2 - 2$

(b) Next we work with the formula $x_{n+1} = \dfrac{8}{x_n - 4}$

x_0	-2	-1	5	6
x_1	$-1{\cdot}333$	$-1{\cdot}6$	8	4
x_2	$-1{\cdot}500$	$-1{\cdot}429$	2	—
x_3	$-1{\cdot}455$	$-1{\cdot}474$	-4	
x_4	$-1{\cdot}467$	$-1{\cdot}461$	-1	
x_5	$-1{\cdot}463$	$-1{\cdot}465$	$-1{\cdot}6$	
x_6	$-1{\cdot}464$	$-1{\cdot}464$	$-1{\cdot}429$	

In this case three of the sequences converge fairly quickly to the root α. In the remaining sequence the value of x_2 is undefined. Writing $g(x) = 8/(x-4)$, we have $g'(\alpha) \approx -0{\cdot}25$ and $g'(\beta) \approx -4$.

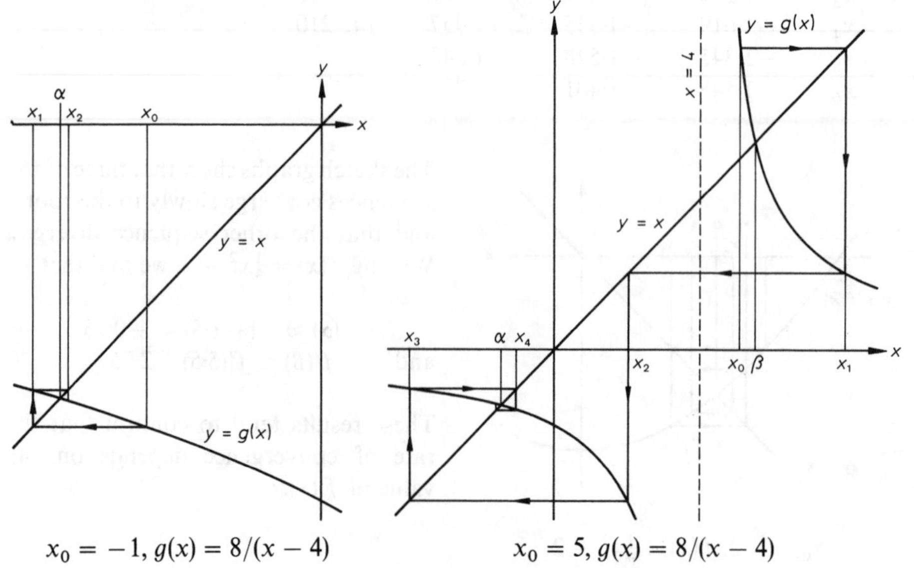

$x_0 = -1, g(x) = 8/(x-4)$ $\qquad$ $x_0 = 5, g(x) = 8/(x-4)$

(c) Since the rearrangement $x = 2\sqrt{(x+2)}$ is not valid for the negative root α of the original equation, we hope to use the formula $x_{n+1} = 2\sqrt{(x_n + 2)}$ to obtain the positive root β.

x_0	5	6
x_1	5·292	5·657
x_2	5·401	5·534
x_3	5·441	5·490
x_4	5·456	5·474
x_5	5·461	5·468
x_6	5·463	5·466

Both sequences converge to the root β, one from below, the other from above. Letting $h(x) = 2\sqrt{(x+2)}$, we find that $h'(\beta) \approx 0{\cdot}37$.

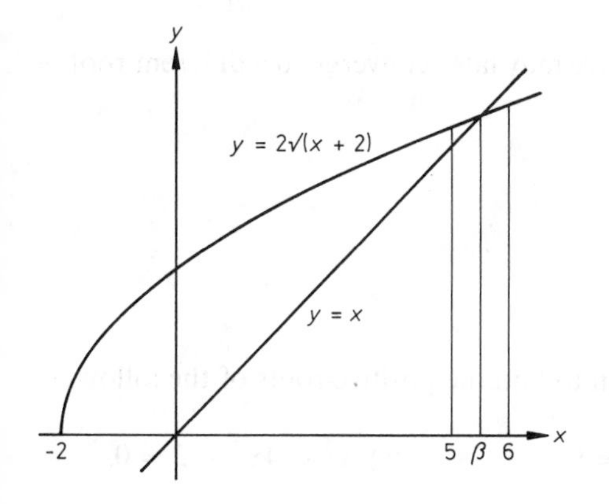

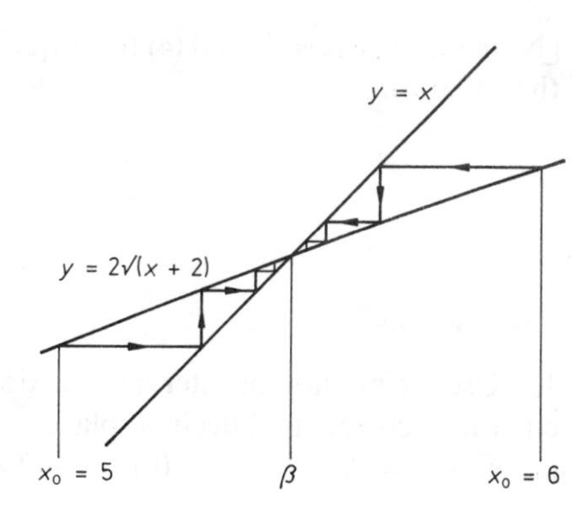

Returning to the general formula $x_{n+1} = f(x_n)$, we conclude that there are four ways in which the sequence of approximations $x_0, x_1, x_2, \ldots$ may behave close to a root α of the equation $x = f(x)$.

(1)

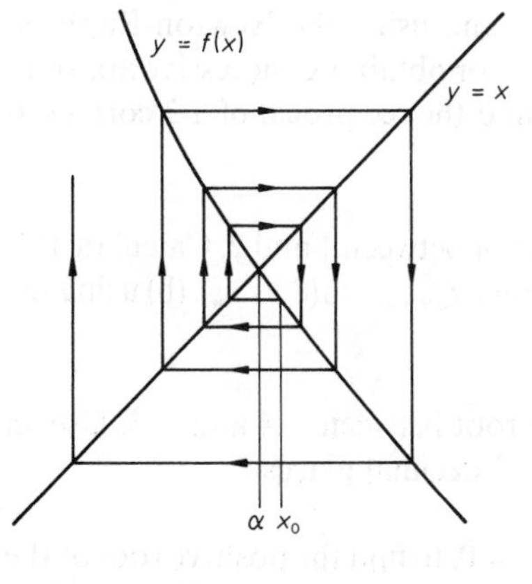

$f'(\alpha) < -1$

(2)

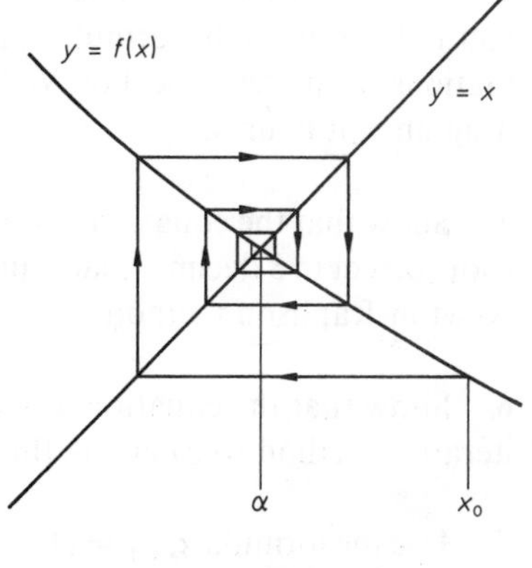

$-1 < f'(\alpha) < 0$

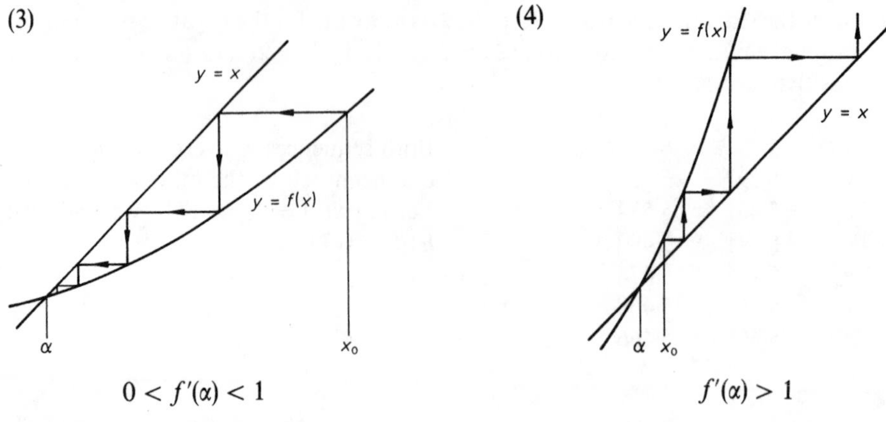

$0 < f'(\alpha) < 1$ $\qquad\qquad$ $f'(\alpha) > 1$

[Note that in cases (1) and (4) the sequence may later converge to a different root of the equation $x = f(x)$.]

Exercise 10.4

1. Use the method of interval bisection to find the positive roots of the following equations correct to 2 decimal places.
(a) $x^2 - 5 = 0$, (b) $x^3 - 2 = 0$, (c) $x^3 - 4x^2 + 2 = 0$.

2. The iterative formula $x_{n+1} = \frac{1}{2}(x_n + a/x_n)$ gives successive approximations to the square root of a. Use this formula and the given values of x_0 to calculate correct to 3 decimal places.
(a) $\sqrt{2}, x_0 = 1$, (b) $\sqrt{10}, x_0 = 3$, (c) $\sqrt{13}, x_0 = 4$.

3. The iterative formula $x_{r+1} = \frac{1}{2}(x_r + N/x_r^2)$ can be used to find an approximate cube root of a number N. Letting $x_0 = 2$, find $\sqrt[3]{5}$ correct to 3 significant figures.

4. By considering the function $f(x) = N - 1/x$ and using the Newton-Raphson method, establish the formula $x_{r+1} = x_r(2 - Nx_r)$ for obtaining successive approximations to the reciprocal of N. Taking $x_0 = 1$, find the reciprocal of 1·3 correct to 3 significant figures.

5. Show that the equation $x = \ln(8 - x)$ has a root between 1 and 2. Calculate this root correct to 3 decimal places (a) using the formula $x_{n+1} = \ln(8 - x_n)$, (b) using the Newton-Raphson method.

6. Show that the equation $x = \cos x - 3$ has a root between -4 and -3. Use an iterative method to calculate this root correct to 3 decimal places.

7. Use the formula $x_{r+1} = (1 - x_r^2)/5$, where $x_0 = 0$, to find the positive root of the equation $x^2 + 5x - 1 = 0$, correct to 4 decimal places. Discuss with the aid of

sketch graphs the behaviour of the sequence $x_0, x_1, x_2, \ldots$ when $x_0 = -6$ and when $x_0 = -5$.

8. Show graphically, or otherwise, that the equation $x^3 - x - 1 = 0$ has only one real root and find the integer n such that the root α satisfies $n < \alpha < n + 1$. An iterative process for finding this root is defined by

$$x_1 = 1, \quad x_{m+1} = (x_m + 1)^{1/3} \quad \text{for all} \quad m \in \mathbb{N}.$$

Obtain, to 3 places of decimals, the values of x_2 and x_3. Show, on a sketch graph, the line $y = x$ and the curve $y = (x + 1)^{1/3}$, indicating on this graph the relation between x_1, x_2, x_3 and the root α. (L)

9. Verify by sketching graphs of $y = x$ and $y = 2(1 + 1/x)$ that, in general, the sequence defined by $x_{n+1} = 2(1 + 1/x_n)$ converges to the root $1 + \sqrt{3}$ of the equation $x^2 - 2x - 2 = 0$. Hence find $\sqrt{3}$ to 4 significant figures, taking a suitable integer as first approximation.

10. Demonstrate graphically that neither

$$x_{n+1} = \tfrac{2}{5}(4 - x_n^3) \quad \text{nor} \quad x_{n+1} = (8 - 5x_n)/2x_n^2$$

can be used to find the real root of the equation $2x^3 + 5x - 8 = 0$. Derive a suitable iteration formula and hence obtain the root to 3 decimal places.

11. The flow diagram shown represents the steps involved in finding an estimate of a solution of the equation $x^2 - 3x + 1 = 0$, using an iterative method. Follow the steps to their conclusion and obtain the estimate of the root. Indicate clearly everything that is to be printed and work to three decimal places. Indicate the purpose of the decision box.

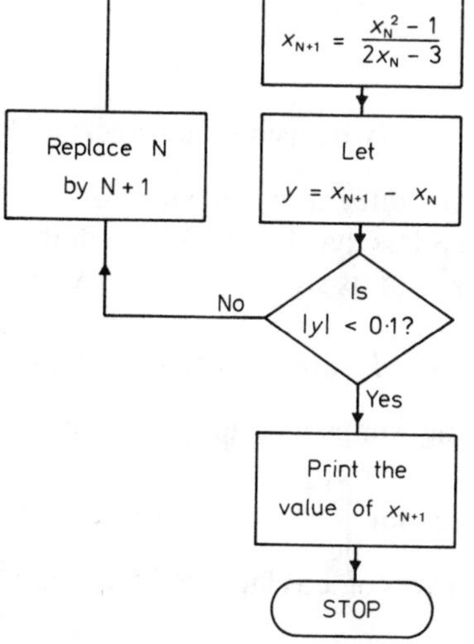

Let N = 1

Let $x_N = 2{\cdot}5$

Let $x_{N+1} = \dfrac{x_N^2 - 1}{2x_N - 3}$

Let $y = x_{N+1} - x_N$

Is $|y| < 0{\cdot}1$?

No: Replace N by N + 1

Yes: Print the value of x_{N+1}

STOP

(L)

12. The equation $f(x) = 0$ is known to have precisely one root in the interval $a \leqslant x \leqslant b$. Construct a flow diagram giving an algorithm to estimate this root, using the method of interval bisection. The following are given: the numbers a and b (with $a < b$), the function f, and the positive number δ which is the maximum permissible magnitude of the error in the root. Demonstrate the working of your flow diagram by applying it to the equation $\cos x - x = 0$ with $a = 0$, $b = 1$ and $\delta = 0{\cdot}1$, giving your answer correct to one decimal place. (C)

Exercise 10.5 (miscellaneous)

1. Values of a continuous function f were found experimentally as given below.

t	0	0·3	0·6	0·9	1·2	1·5	1·8
$f(t)$	2·72	3·00	3·32		4·06	4·48	4·95

Use linear interpolation to estimate $f(0{\cdot}9)$. Then use Simpson's rule with seven ordinates to estimate $\int_0^{1\cdot 8} f(t)\,dt$, tabulating your working and giving your answer to two places of decimals. (JMB)

2. The following pairs of values of x and y satisfy approximately a relation of the form $y = ax^n$, where a and n are integers. By plotting the graph of $\lg y$ against $\lg x$ find the values of the integers a and n.

x	0·7	0·9	1·1	1·3	1·5
y	1·37	2·92	5·32	8·80	13·50

Estimate the value of the integral $\int_{0\cdot 7}^{1\cdot 5} y\,dx$ (a) by Simpson's rule, using five ordinates and clearly indicating your method, (b) by using the relation $y = ax^n$ with the values found for a and n. (L)

3. Find an approximate value of $\int_0^{1/2} \sqrt{(1 - x^2)}\,dx$ by using the trapezoidal rule with intervals of 0·1. Show, by working out the integral by substitution, that the magnitude of the error in your approximation is less than 0·001. Draw a diagram to illustrate the reason for this error, having regard to its sign. (AEB 1979)

4. If $f(x) = a + bx + cx^2 + kx^3$, where a, b, c, k are constants, show that $\int_{-h}^{h} f(x)\,dx = \frac{1}{3}h\{f(-h) + 4f(0) + f(h)\}$. Using Simpson's rule with 5 ordinates, obtain an approximate value, to 3 decimal places, for $\int_0^{0\cdot 4} \sqrt{(1 + x^2)}\,dx$. Check your result by expanding $\sqrt{(1 + x^2)}$ in ascending powers of x as far as the term in x^6 and integrating term by term. (O&C)

5. Plot the graph of $y = x^3$ for x from -3 to $+3$. Use the graph to obtain solutions to the following equations (i) $x^3 - 2x - 5 = 0$, (ii) $2x^3 - 4x + 1 = 0$. By writing the equation $5x^3 + 2x^2 - 1 = 0$ in terms of $1/x$, obtain its solution, correct to one decimal place. (AEB 1979)

6. Sketch the graph of $y = \tan x$ for values of x between 0 and 2π. Use your sketch graph, together with additional straight line graphs, to solve the following problems:
(i) If $a > 0$ and $b > 0$, find the number of roots, between 0 and 2π, of the equation $\tan x = ax + b$.
(ii) If m can take all real values, find the set of values of m such that the equation $\tan x = mx + \pi$ has exactly three roots lying in the interval $0 < x < 2\pi$. (C)

7. Sketch the graph of $y = x^2 e^{-x}$ giving the coordinates of the turning points. Explain, from your graph, why, if $0 < k < 4/e^2$, the equation $ke^x = x^2$ has three real roots. (O&C)

8. By considering the graph of $y = x^3 - 3px - q$, where p and q are real constants, show that the condition for the equation $x^3 - 3px - q = 0$ to have three distinct real roots is $4p^3 > q^2$. Find the additional condition necessary for two of the roots to be positive. (JMB)

9. The least positive root of the equation $x^3 - 10x^2 + 12x + 16 = 0$ is denoted by α. Use linear interpolation between two suitable successive integers to obtain an estimate for α, giving your answer to one decimal place. Will your answer be an underestimate or overestimate of the exact root? Explain your reasoning. (O&C)

10. Sketch the graph of $y = \cos x$ for $0 \leqslant x \leqslant 4\pi$. Use your sketch to find the number of positive roots of the equation $\cos x = x/10$. Prove that the equation of the tangent to the curve $y = \cos x$ at $(\frac{1}{2}\pi, 0)$ is $y = \frac{1}{2}\pi - x$. By using $\frac{1}{2}\pi - x$ as an approximation for $\cos x$ (valid for x near $\frac{1}{2}\pi$), calculate an estimate of the smallest positive root of the equation $\cos x = x/10$, giving your answer to 2 decimal places. Show on a new sketch the part of the curve $y = \cos x$ near $x = \frac{1}{2}\pi$ and the tangent at $x = \frac{1}{2}\pi$. Hence determine whether the estimate of the root calculated above is larger or smaller than the correct value. (C)

11. The given flow diagram is designed to search for a root of the equation $f(x) = 0$ in the range $0 < x < 10$. (i) If x reaches 10 explain why the message printed has to be "No root found" and not "There is no root between 0 and 10". (ii) Work through the flow diagram in the case where $f(x) = e^x - 3x - 4$. (iii) Extend the flow diagram so that, when a root is located between two consecutive integers, a further search is carried out which will narrow the range in which the root lies to 0·1.

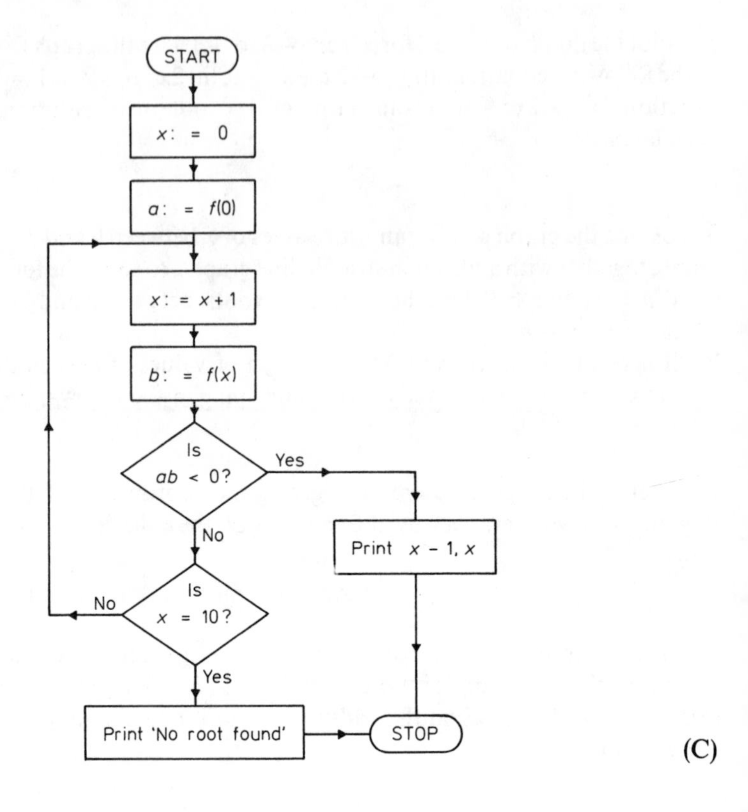

(C)

12. Using the same axes draw accurate graphs of $y = \ln x$ and $y = 3 - x$ in the interval $1 \leqslant x \leqslant 4$. Deduce that the equation $x + \ln x - 3 = 0$ has a root near $x = 2{\cdot}2$. Clearly showing your method, obtain alternative approximations to the root of the equation (a) by linear interpolation between $x = 2{\cdot}2$ and $x = 2{\cdot}3$, (b) by one application of the Newton-Raphson procedure using $x = 2{\cdot}2$ as the initial value. (L)

13. Given that $P(x) = x^3 - 6x^2 + 9x + a$ for all real x, find the values of x for which $P'(x) = 0$. Determine the values of the constant a such that the equation $P(x) = 0$ has a repeated root. By sketching graphs of $y = P(x)$ for these values of a, or otherwise, find the set of values of a for which $P(x) = 0$ has only one real root. Given now that $a = -5$, and that 4 is an approximation to the root of the equation $P(x) = 0$, use Newton's method once to obtain another approximation to the root, giving your answer correct to one decimal place. (JMB)

14. Use linear interpolation to find an approximation x_1 to the root of the equation $f(x) = 0$ lying between 3 and 4, where $f(x) \equiv x^4 - 4x^3 + x + 12$. Use one application of the Newton-Raphson process applied to $f(x)$ starting with $x = 4$ to find another approximation x_2 to the same root. Use one application of the same process starting with $x = 3$ to find another approximation x_3 to the same root. Correct to 2 decimal places the root is $x = 3{\cdot}69$. Indicate on a diagram how each of the approximations is related to the graph of $f(x)$ for $3 \leqslant x \leqslant 4$. (C)

15. Explain, with the use of diagrams, the use of Newton's method for obtaining the numerical solutions of an equation. Draw a diagram to show how the method can break down.

Obtain the real factors, with integer coefficients, of $x^4 - 4x^3 + 3x^2 - 4x + 12$. Hence or otherwise obtain the greater real root of the equation

$$x^4 - 4x^3 + 3x^2 - 4x + 11 = 0,$$

correct to 3 decimal places. (AEB 1979)

16. Show that the equation $x^3 + 3x - 3 = 0$ has only one real root and that it lies between $x = 0{\cdot}8$ and $x = 1$. Obtain approximations to the root (a) by performing one application of the Newton-Raphson procedure using $x = 0{\cdot}8$ as the first approximation, (b) by performing two iterations, using the procedure defined by $x_{n+1} = (3 - x_n^3)/3$, and starting with $x = 0{\cdot}8$. (L)

17. Using the same axes and scales, draw graphs of $y = \cos x$ and $y = x$ for the interval $0 \leqslant x \leqslant 1{\cdot}4$. (A scale of 10 cm to a unit on each axis is recommended.) On your graph mark as A, B, C and D the first four points whose coordinates are printed when the following section of a flow diagram is executed.

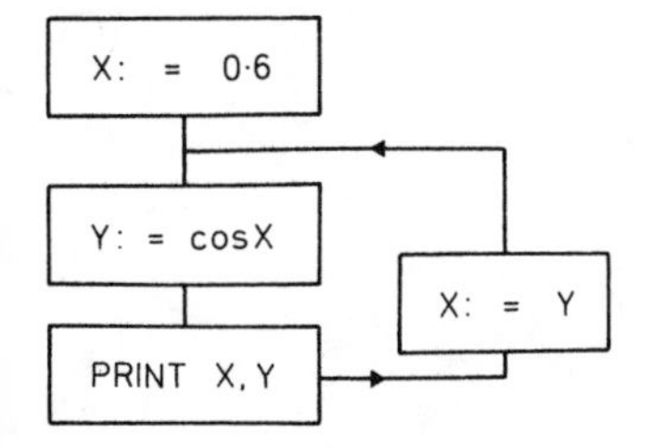

The procedure indicated in the flow diagram is an iterative method of solving a particular equation. Write down this equation and obtain from your graph an approximation solution to it. (JMB)

18. Given that $f(x) = x^3 + x^2 + 2x - 2$, show that the equation $f(x) = 0$ has only one root and that this root lies between 0 and 1. Use the iterative procedure

$$x_{n+1} = \frac{2}{x_n^2 + x_n + 2}, \quad x_1 = 1,$$

to obtain x_2 and x_3. By means of a sketch, illustrate how the successive values of x_n converge to the desired root. (L)

19. A solution of the equation $x = \phi(x)$ is to be attempted by using the iteration $x_{n+1} = \phi(x_n)$, starting with an initial estimate x_1. Draw sketch graphs showing $y = x$ and $y = \phi(x)$ to illustrate the following possibilities regarding convergence towards, or divergence from, the root $x = \alpha$. (i) $x_1 > \alpha$ and successive iterates steadily decrease, with the value α as a limit. (ii) $x_1 > \alpha$ and successive iterates are alternately less than α and greater than α, but approach α as a limit. (iii) $x_1 > \alpha$ and successive iterates get steadily larger. Use an iterative method to find a non-zero root of the equation $x = \tan^{-1}(2x)$ correct to 2 significant figures. (C)

20. 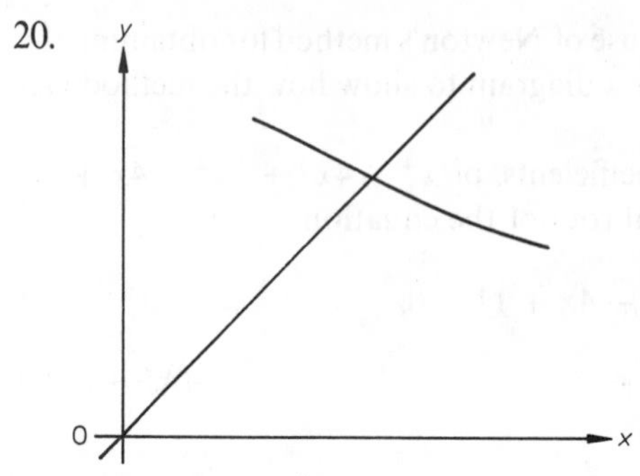

The equation $\sin x = \cos^{-1} x$ has a root in the neighbourhood of $x = \frac{1}{4}\pi$, and the iteration $x_{n+1} = \cos(\sin x_n)$ is to be employed to calculate this root. The diagram shows the appropriate parts of the graphs of $y = x$ and $y = \cos(\sin x)$. Explain with reference to this diagram why successive iterates are alternately greater than and less than the root. Use the given iteration to determine the root correct to 2 decimal places, showing clearly that you have achieved this degree of accuracy. Construct a flow diagram to show how the iteration may be continued until successive iterates differ by less than a given quantity δ. (C)

21. Show that the equation $x^3 - 3x + 1 = 0$ has one negative root and two positive roots. Rearrange the equation in the form $x = F(x)$ in such a way that the iteration $x_{n+1} = F(x_n)$ will converge to the larger positive root, showing clearly that the condition for convergence is satisfied by your function F. Show by means of a sketch how the iterates converge to the root if the initial estimate x_1 is taken to be 1. Use your iteration to calculate the root correct to two decimal places, taking 1·5 as your initial approximation. (C)

11 Differential equations

11.1 Differential equations and their solution curves

An equation connecting a function and one or more of its derivatives is called a *differential equation.* We shall see in §11.4 that many physical laws, such as the laws of natural growth or decay, can be expressed as differential equations.

The *order* of a differential equation is the order of the highest derivative which appears in it. For instance, the equations $\dfrac{dx}{dt} = -4x$ and $\left(\dfrac{dy}{dx}\right)^2 = x + y$ are first order equations and the equation $\dfrac{d^2y}{dx^2} - 3\dfrac{dy}{dx} = e^{2x}$ is a second order equation.

Let us consider a simple differential equation of the first order:

$$\frac{dy}{dx} = x - 1.$$

Any equation connecting x and y which satisfies this differential equation is a solution. Thus $y = \frac{1}{2}x^2 - x$, $y = \frac{1}{2}x^2 - x + 5$ and $y = \frac{1}{2}x^2 - x - 28$ are all solutions. The *general solution* of the differential equation is $y = \frac{1}{2}x^2 - x + c$, where c is an arbitrary constant. Since c can take any real value, this general solution can be represented graphically by an infinite set of curves called the *solution curves* or *integral curves* of the equation. Each member of this "family" of curves represents a particular solution, or *particular integral*, of the differential equation.

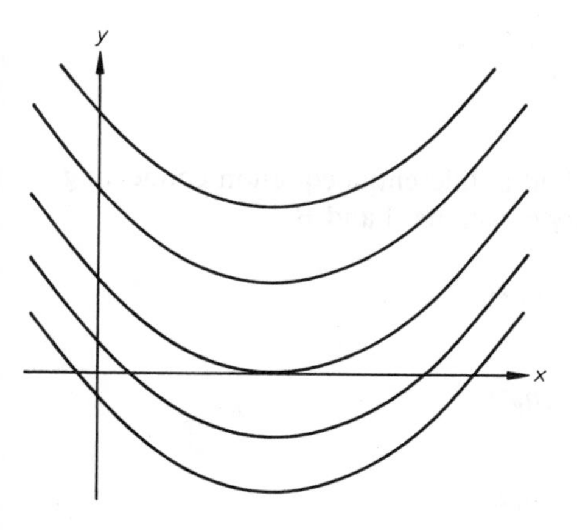

The solution curves of $\dfrac{dy}{dx} = x - 1$.

Given the equation of a family of curves, we can obtain the corresponding differential equation by differentiating and then eliminating any arbitrary constants.

Example 1 Find the differential equation representing the family of curves, $y=Ax^3$, where A is an arbitrary constant.

If $y = Ax^3$, then $\dfrac{dy}{dx} = 3Ax^2$

$$\therefore \quad x\frac{dy}{dx} = 3Ax^3 = 3y$$

Hence the required differential equation is $x\dfrac{dy}{dx} = 3y$.

Example 2 Sketch some of the curves given by the equation $x^2 + y^2 - 2kx = 0$. Obtain the differential equation which represents this family of curves.

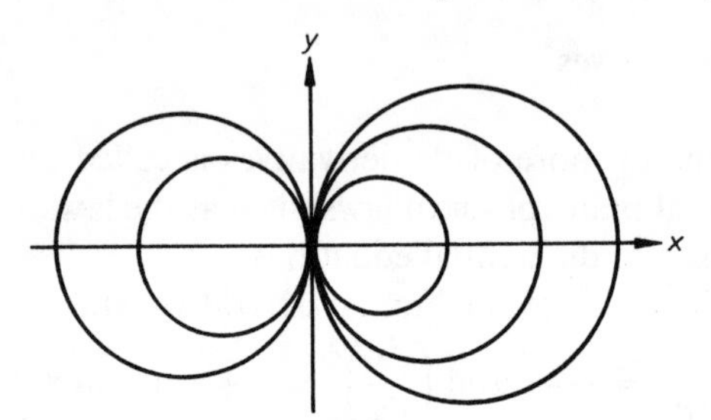

The equation $x^2 + y^2 - 2kx = 0$ may be written in the form

$$(x-k)^2 + y^2 = k^2.$$

Hence each curve in the family is a circle, centre $(k, 0)$ and radius $|k|$.

Differentiating the equation $\quad x^2 + y^2 - 2kx = 0$

we obtain $\quad 2x + 2y\dfrac{dy}{dx} - 2k = 0$

$$\therefore \quad 2x^2 + 2xy\frac{dy}{dx} = x^2 + y^2 \quad (= 2kx)$$

Hence this family of circles is represented by the differential equation

$$2xy\frac{dy}{dx} = y^2 - x^2.$$

Example 3 Given that $y = Ae^x + Be^{2x}$, find a differential equation connecting x and y which is independent of the arbitrary constants A and B.

If $y = Ae^x + Be^{2x}$, then $\quad \dfrac{dy}{dx} = Ae^x + 2Be^{2x}$

and $\quad \dfrac{d^2y}{dx^2} = Ae^x + 4Be^{2x}$

$$\therefore \quad \frac{dy}{dx} - y = Be^{2x} \quad \text{and} \quad \frac{d^2y}{dx^2} - y = 3Be^{2x}$$

$$\therefore \quad \frac{d^2y}{dx^2} - y = 3\left(\frac{dy}{dx} - y\right)$$

Hence the required differential equation is

$$\frac{d^2y}{dx^2} - 3\frac{dy}{dx} + 2y = 0.$$

In Examples 1 and 2 equations involving one arbitrary constant gave rise to differential equations of the first order. However, in Example 3 it was necessary to introduce two derivatives to eliminate the two arbitrary constants. This suggests that in general an equation containing n arbitrary constants should lead to an nth order differential equation. Conversely, it seems reasonable to assume that the general solution of a differential equation of order n should contain n arbitrary constants.

We now consider ways of sketching the solution curves of a differential equation when the general solution of the equation is not available. We take as an example the differential equation

$$\frac{dy}{dx} = x - y.$$

A general impression of the solution curves can be gained by calculating values of dy/dx at various points of the x, y plane, then plotting the directions on a sketch graph.

y \ x	-3	-2	-1	0	1	2	3
3	-6	-5	-4	-3	-2	-1	0
2	-5	-4	-3	-2	-1	0	1
1	-4	-3	-2	-1	0	1	2
0	-3	-2	-1	0	1	2	3
-1	-2	-1	0	1	2	3	4
-2	-1	0	1	2	3	4	5
-3	0	1	2	3	4	5	6

Table showing values of $\frac{dy}{dx}$

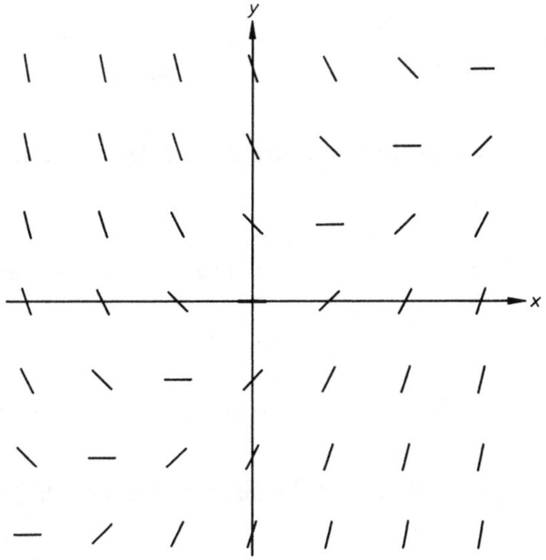

Sketch graph showing directions of the solution curves

We see that $\frac{dy}{dx} = 0$ on the line $y = x$. Above this line $\frac{dy}{dx} < 0$ and below the line $\frac{dy}{dx} > 0$. Hence the points on the line $y = x$ are minimum points on the solution curves.

The diagram also suggests that the line $y = x - 1$ could be one of the solution curves. Substituting $y = x - 1$ in the differential equation confirms that this is so.

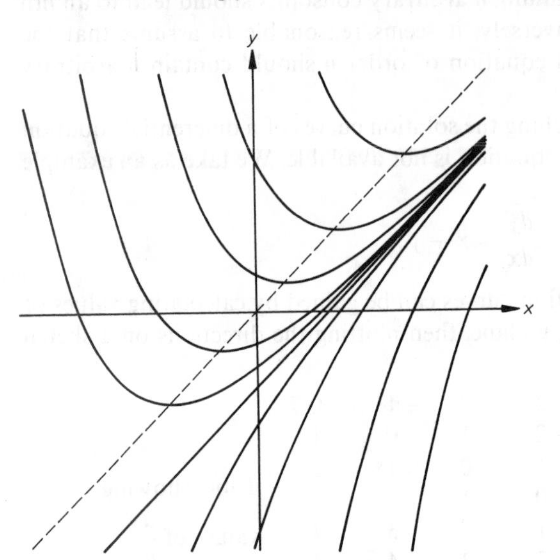

It is now possible to make a rough sketch of the solution curves.

Exercise 11.1

1. Verify that the following are solutions of the differential equation

$$x\frac{dy}{dx} = x^2 + y,$$

(a) $y = x^2$, (b) $y = x^2 - x$, (c) $y = x^2 + 5x$.

2. Verify that the following are solutions of the differential equation $\frac{d^2y}{dx^2} = 1 - y$,

(a) $y = 1 + 2\cos x$, (b) $y = 1$, (c) $y = 1 - 3\sin x$.

3. Decide which of the following are solutions of the differential equation

$$\frac{d^2y}{dx^2} + \frac{dy}{dx} = 6y.$$

(a) $y = 3e^{2x}$, (b) $y = 6e^{x}$, (c) $y = 2e^{3x}$, (d) $y = e^{2x} - e^{-3x}$.

4. Given that $y = ax + b$ is a solution of the differential equation $\frac{dy}{dx} = 4x - 2y$, find a and b.

5. Given that $y = ax^2 + bx + c$ is a solution of the differential equation

$$\frac{d^2y}{dx^2} = y - x^2,$$

find a, b and c.

In questions 6 to 11 sketch some of the curves represented by the given equation and obtain a differential equation for y which does not contain the arbitrary constant A.

6. $y = x^3 + A$
7. $y = x^2 - Ax$
8. $y = A/x$
9. $y = Ae^{-x}$
10. $x^2 + y^2 = A^2$
11. $y^2 = 4Ax$

In questions 12 to 19 form differential equations by eliminating the arbitrary constant A, B and C.

12. $y = x + \frac{A}{x}$
13. $y = x^2 \ln x + Ax^2$
14. $y^2 = A \cos x$
15. $y = x^A$
16. $y = A \cos x + B \sin x$
17. $y = Ae^{-x} + Be^{3x}$
18. $y^2 = Ax^2 + Bx + C$
19. $y = (Ax + B)e^x + C$

In questions 20 to 25 sketch the solution curves of the given differential equation.

20. $\frac{dy}{dx} = \frac{x}{y}$
21. $\frac{dy}{dx} = \frac{y}{x}$
22. $\frac{dy}{dx} = -\frac{x}{2y}$
23. $\frac{dy}{dx} = x + y$
24. $x\frac{dy}{dx} = x - y$
25. $2y\frac{dy}{dx} = x + y$

11.2 Elementary methods of solution

Differential equations of the form $\frac{dy}{dx} = f(x)$ may be solved by direct integration.

Example 1 Solve the differential equation $\frac{dy}{dx} = 6x^2 - 5$.

By integration the general solution is $y = 2x^3 - 5x + c$, where c is an arbitrary constant.

Example 2 Solve the differential equation $\cos^2 x \dfrac{dy}{dx} = \sin x$.

Rearranging, we have: $\dfrac{dy}{dx} = \dfrac{\sin x}{\cos^2 x} = \sec x \tan x$

Integrating, we obtain the general solution $y = \sec x + c$, where c is an arbitrary constant.

Sometimes we require a solution which satisfies certain given conditions. We then use these conditions to determine the values of any arbitrary constants in the general solution.

Example 3 Solve the differential equation $e^x \dfrac{dy}{dx} + x = 0$, given that $y = 2$ when $x = 0$.

Rearranging we obtain: $\dfrac{dy}{dx} = -xe^{-x}$

Integrating by parts using the formula

$$\int u\frac{dv}{dx}\,dx = uv - \int v\frac{du}{dx}\,dx,$$

where $\quad u = x \quad$ and $\quad \dfrac{dv}{dx} = -e^{-x}$

$$\frac{du}{dx} = 1 \qquad\qquad v = e^{-x}$$

$$\int(-xe^{-x})\,dx = xe^{-x} - \int e^{-x}\,dx$$

$$= xe^{-x} + e^{-x} + c$$

$\therefore$ the general solution of the differential equation is

$$y = xe^{-x} + e^{-x} + c.$$

Since $y = 2$ when $x = 0$, $\quad 2 = 0 \times 1 + 1 + c \qquad \therefore \quad c = 1$
Hence the required solution is $y = xe^{-x} + e^{-x} + 1$.

When the integration of a differential equation leads to a general solution involving logarithmic functions, any restrictions on the values of x and y should be clearly stated.

Example 4 Solve the differential equation $\dfrac{dy}{dx} = \dfrac{x}{x-1}$, given that $y = 2$ when $x = 2$.

Rearranging: $\dfrac{dy}{dx} = \dfrac{x}{x-1} = 1 + \dfrac{1}{x-1}$

[Since we require a solution valid for $x = 2$, we will assume that x is greater than 1 when integrating.]

For $x > 1$, the general solution of this differential equation is

$$y = x + \ln(x - 1) + c$$

Since $y = 2$ when $x = 2$, $\quad 2 = 2 + \ln 1 + c \qquad \therefore \quad c = 0$
Hence the required solution is $y = x + \ln(x - 1)$.

Second order differential equations of the form $\dfrac{d^2y}{dx^2} = f(x)$ can also be solved by direct integration.

Example 5 Solve the differential equation $\dfrac{d^2y}{dx^2} = 6x^3 + \dfrac{1}{x^3}$

Integrating we have: $\dfrac{dy}{dx} = \dfrac{3}{2}x^4 - \dfrac{1}{2x^2} + A$

Integrating again we obtain the general solution

$$y = \frac{3}{10}x^5 + \frac{1}{2x} + Ax + B,$$

where A and B are arbitrary constants.

Example 6 Solve the differential equation $\dfrac{d^2x}{dt^2} = -\sin t + \cos t$, subject to the conditions that $x = 0$, $\dfrac{dx}{dt} = 4$ when $t = 0$.

Integrating: $\dfrac{dx}{dt} = \cos t + \sin t + A$

$$x = \sin t - \cos t + At + B$$

Since $\dfrac{dx}{dt} = 4$ when $t = 0$, $4 = 1 + 0 + A \qquad \therefore \quad A = 3$
Since $x = 0$ when $t = 0$, $\quad 0 = 0 - 1 + 0 + B \qquad \therefore \quad B = 1$
Hence the required solution is $x = \sin t - \cos t + 3t + 1$.

To solve differential equations of the form $\dfrac{dy}{dx} = f(y)$, we use the fact that $\dfrac{dy}{dx} = 1\bigg/\dfrac{dx}{dy}$.

Example 7 Solve the differential equation $\dfrac{dy}{dx} = y^2$.

The equation may be written as: $\dfrac{dx}{dy} = \dfrac{1}{y^2}$

Integrating with respect to y: $x = -\dfrac{1}{y} + c$

$\therefore \quad \dfrac{1}{y} = c - x$

Hence the general solution is $y = \dfrac{1}{c - x}$.

A substitution is sometimes used to reduce a differential equation to a simpler form.

Example 8 Use the substitution $y = vx$, where v is a function of x, to solve the differential equation $x\dfrac{dy}{dx} = x + y$, given that $y = -1$ when $x = 1$.

If $y = vx$, then using the product rule, $\dfrac{dy}{dx} = v + x\dfrac{dv}{dx}$

Substituting in the given differential equation we have,

$$x\left(v + x\frac{dv}{dx}\right) = x + vx$$

i.e. $$vx + x^2\frac{dv}{dx} = x + vx$$

$$\therefore \quad \frac{dv}{dx} = \frac{1}{x}$$

For $x > 0$, the general solution of this differential equation is

$$v = \ln x + c$$
$$\therefore \quad y = vx = x(\ln x + c)$$

Since $y = -1$ when $x = 1$, $-1 = 1(0 + c) \quad \therefore \quad c = -1$
Hence the required solution is $y = x(\ln x - 1)$.

In Example 8 we again dealt with an integration involving logarithms by considering a restricted set of values of x. However, in some problems it is better to avoid placing restrictions on the values of x and y by using modulus signs, as described in §3.4.

For instance, in the case of the differential equation

$$\frac{dy}{dx} = \frac{1}{x - 5}$$

the general solution may be given as

$$y = \ln|x - 5| + c \qquad (1)$$

Suppose now that we require the particular solution which satisfies the condition $y = -2$ when $x = 6$.
Substituting these values in the general solution (1),

$$-2 = \ln 1 + c \qquad \therefore \quad c = -2$$

Hence $$y = \ln|x - 5| - 2.$$

But, using the method of Examples 4 and 8,

for $x > 5$, $$y = \ln(x - 5) + c \tag{2}$$

As before, since $y = -2$ when $x = 6$, we have $c = -2$

Hence $$y = \ln(x - 5) - 2.$$

Thus, although equation (1) provides a neat general solution, equation (2) leads to a simpler particular integral.

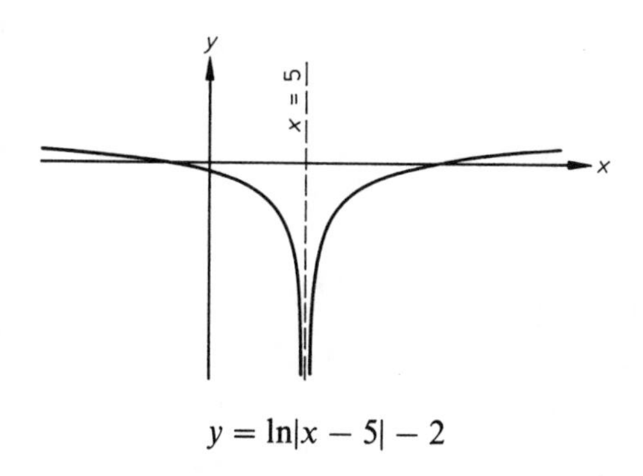

$y = \ln|x - 5| - 2$

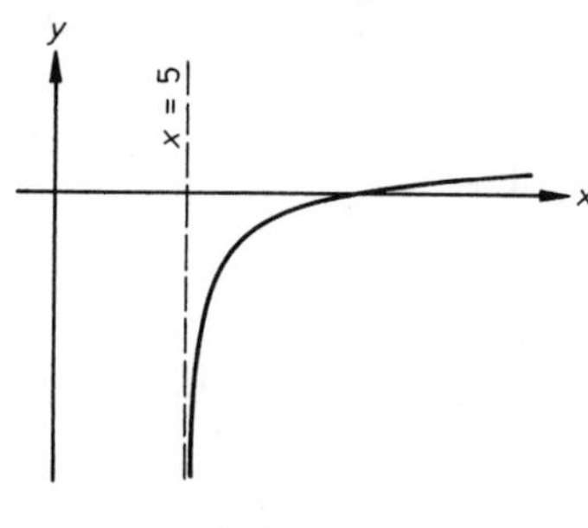

$y = \ln(x - 5) - 2$

The sketch graphs show that solution (1) represents a family of curves with two branches, whereas solution (2) produces single curves lying to the right of the line $x = 5$. A question requiring a particular integral is probably best answered by a solution of type (2).

Exercise 11.2

Solve the following differential equations giving y in terms of x.

1. $\dfrac{dy}{dx} = 3x^2 + 1$

2. $\dfrac{dy}{dx} = \cos\tfrac{1}{2}x$

3. $x\dfrac{dy}{dx} = x^2 + 1$

4. $(x - 1)\dfrac{dy}{dx} = x + 1$

5. $\dfrac{d^2y}{dx^2} + \cos x = 2x$

6. $e^x\dfrac{d^2y}{dx^2} = 1 - e^{2x}$

7. $\dfrac{dy}{dx} + 2y^2 = 0$

8. $\dfrac{dy}{dx} = \cos^2 y$

9. $\dfrac{dy}{dx} = e^{-y}$

10. $3\dfrac{dy}{dx} = y^4$

Solve the following differential equations, subject to the given conditions.

11. $\dfrac{dy}{dx} = (x + 1)^2$; $y = 0$ when $x = 2$

12. $\dfrac{dy}{dx} = \dfrac{1}{\sqrt{(2x + 3)}}$; $y = 4$ when $x = 3$

13. $\sec x\dfrac{dy}{dx} = x$; $y = 0$ when $x = \pi$

14. $(1 + x^2)\dfrac{dy}{dx} = 2x$; $y = 0$ when $x = -1$

15. $\dfrac{d^2y}{dx^2} = \sin 2x$; $y = 0$, $\dfrac{dy}{dx} = -\dfrac{1}{4}$ when $x = 0$

16. $x^2\dfrac{d^2y}{dx^2} = 2(x^2 - 1)$; $y = \dfrac{dy}{dx} = 0$ when $x = 1$

17. $\dfrac{dy}{dx} + 2y = 0$; $y = 3$ when $x = 0$

18. $\dfrac{dy}{dx} = y - 1$; $y = 0$ when $x = 0$

19. $\dfrac{dy}{dx} = 1 + y^2$; $y = 1$ when $x = 0$

20. $2\dfrac{dy}{dx} = y^2 - 1$; $y = 2$ when $x = 0$

21. Use the substitution $y = 1/z$ to solve the differential equation $x^2\dfrac{dy}{dx} = y^2$.

22. Use the substitution $y = vx$, where v is a function of x, to solve the differential equation $x\dfrac{dy}{dx} - y = x^2 \cos x$.

23. Use the substitution $y = vx^2$, where v is a function of x, to solve the differential equation $x\dfrac{dy}{dx} - 2y = x$.

24. Use the substitution $dy/dx = p$, where p is a function of x, to solve the differential equation $\dfrac{d^2y}{dx^2} + \dfrac{dy}{dx} = 1$, given that $y = \dfrac{dy}{dx} = 2$ when $x = 1$.

25. If $z = xe^y$, where y is a function of x, find an expression for dz/dx. Hence find y in terms of x given that $xe^y\dfrac{dy}{dx} + e^y = 2x$ and that $y = 0$ when $x = 2$.

11.3 Equations with separable variables

If a first order differential equation can be expressed in the form

$$f(y)\frac{dy}{dx} = g(x)$$

then the variables x and y are said to be *separable*.
Integrating both sides of the equation with respect to x:

$$\int f(y)\frac{dy}{dx}\,dx = \int g(x)\,dx$$

As shown in §4.3, $\dfrac{dy}{dx}dx$ may be replaced by dy.

Thus $$\int f(y)\,dy = \int g(x)\,dx$$

Example 1 Find the general solution of the equation $2y\dfrac{dy}{dx} = \cos x$.

Integrating both sides of the equation: $\displaystyle\int 2y\,dy = \int \cos x\,dx$

Hence the general solution of the equation is

$$y^2 = \sin x + c.$$

Example 2 Find the general solution of the equation

$$\frac{dy}{dx} = 2xy^2 - y^2.$$

Rearranging: $$\frac{dy}{dx} = (2x - 1)y^2$$

$$\frac{1}{y^2}\frac{dy}{dx} = 2x - 1$$

$$\therefore \quad \int \frac{1}{y^2}\,dy = \int (2x - 1)\,dx \qquad \therefore \quad -\frac{1}{y} = x^2 - x + c$$

Hence the general solution of the equation is

$$y = -1/(x^2 - x + c).$$

Example 3 Find y in terms of x given that $2\frac{dy}{dx} = e^{x-2y}$ and that $y = 1$ when $x = 2$.

Rearranging: $$2\frac{dy}{dx} = e^x \times e^{-2y}$$

$$2e^{2y}\frac{dy}{dx} = e^x$$

$\therefore$ $$\int 2e^{2y}\,dy = \int e^x\,dx$$

$\therefore$ $$e^{2y} = e^x + c$$

Since $y = 1$ when $x = 2$, $\quad e^2 = e^2 + c \quad \therefore \quad c = 0$

Thus $$e^{2y} = e^x$$

$\therefore$ $$2y = x$$

Hence $$y = \tfrac{1}{2}x.$$

Example 4 Express y in terms of x given that $\frac{dy}{dx} = xy - x$ and that $y = 4$ when $x = 0$.

Rearranging: $$\frac{dy}{dx} = x(y-1)$$

$$\frac{1}{(y-1)}\frac{dy}{dx} = x$$

$\therefore$ $$\int \frac{1}{y-1}\,dy = \int x\,dx$$

$\therefore$ $$\ln|y-1| = \tfrac{1}{2}x^2 + c$$

$\therefore$ $$|y-1| = e^{x^2/2+c}$$

Thus $$y - 1 = Ae^{x^2/2}, \quad \text{where} \quad |A| = e^c$$

Since $y = 4$ when $x = 0$, $3 = Ae^0 = A$

Hence $\quad y - 1 = 3e^{x^2/2}$

i.e. $$y = 1 + 3e^{x^2/2}.$$

Example 5 Find the general solution of the equation

$$(x^2-1)\frac{dy}{dx} = xy.$$

Rearranging: $$\frac{1}{y}\frac{dy}{dx} = \frac{x}{x^2-1}$$

$\therefore$ $$\int \frac{1}{y}\,dy = \int \frac{x}{x^2-1}\,dx$$

$$\therefore \qquad \ln|y| = \tfrac{1}{2}\ln|x^2 - 1| + c$$

$$\therefore \quad 2\ln|y| - \ln|x^2 - 1| = 2c$$

Using the laws of logarithms:

$$\ln\left|\frac{y^2}{x^2 - 1}\right| = 2c$$

$$\therefore \qquad \frac{y^2}{x^2 - 1} = A, \quad \text{where} \quad \ln|A| = 2c$$

Hence the general solution of the equation is $y^2 = A(x^2 - 1)$.

Example 6 Find the general solution of the equation

$$2x\frac{dy}{dx} - (2y + 1)(x + 1) = 0$$

Rearranging:
$$2x\frac{dy}{dx} = (2y + 1)(x + 1)$$

$$\frac{2}{(2y + 1)}\frac{dy}{dx} = 1 + \frac{1}{x}$$

$$\therefore \qquad \int \frac{2}{2y + 1}\,dy = \int\left(1 + \frac{1}{x}\right)dx$$

$$\therefore \qquad \ln|2y + 1| = x + \ln|x| + c$$

$$\therefore \quad \ln|2y + 1| - \ln|x| = x + c$$

i.e.
$$\ln\left|\frac{2y + 1}{x}\right| = x + c$$

$$\therefore \qquad \left|\frac{2y + 1}{x}\right| = e^{x + c}$$

Thus
$$\frac{2y + 1}{x} = Ae^x, \quad \text{where} \quad |A| = e^c$$

$$\therefore \qquad 2y + 1 = Axe^x$$

Hence the required general solution is $y = \tfrac{1}{2}(Axe^x - 1)$.

Example 7 Find y in terms of x given that $x(x - 1)\dfrac{dy}{dx} = y$ and that $y = 1$ when $x = 2$.

Rearranging:
$$\frac{1}{y}\frac{dy}{dx} = \frac{1}{x(x - 1)}$$

$$\therefore \qquad \int \frac{1}{y}\,dy = \int \frac{1}{x(x-1)}\,dx$$

$$\therefore \qquad \int \frac{1}{y}\,dy = \int \left\{\frac{1}{x-1} - \frac{1}{x}\right\} dx$$

$$\therefore \qquad \ln|y| = \ln|x-1| - \ln|x| + c$$

i.e. $\ln|y| - \ln|x-1| + \ln|x| = c$

Thus $\qquad \ln\left|\frac{xy}{x-1}\right| = c$

$$\therefore \qquad \frac{xy}{x-1} = A, \quad \text{where} \quad \ln|A| = c$$

Since $y = 1$ when $x = 2$, $\qquad 2 = A$

Hence $\dfrac{xy}{x-1} = 2$

i.e. $\qquad y = 2\left(1 - \dfrac{1}{x}\right).$

Exercise 11.3

In questions 1 to 10 find the general solutions of the given differential equations, expressing y in terms of x in each case.

1. $3y^2\dfrac{dy}{dx} = 2x - 1$
2. $\dfrac{dy}{dx} = 6xy^2$
3. $\dfrac{dy}{dx} = e^y \sin x$
4. $\dfrac{dy}{dx} = e^{x-y}$
5. $\dfrac{dy}{dx} = x \sec y$
6. $\dfrac{dy}{dx} = 3\cos^2 y$
7. $x\dfrac{dy}{dx} = y$
8. $(1-x)\dfrac{dy}{dx} = y$
9. $\dfrac{dy}{dx} = \dfrac{4xy}{x^2+1}$
10. $\dfrac{dy}{dx} = \dfrac{2y}{x^2-1}$

In questions 11 to 20 find expressions for y in terms of x.

11. $x^2\dfrac{dy}{dx} - y^2 = 0$; $y = -1$ when $x = 1$.

12. $\dfrac{dy}{dx} + 2xy = 0$; $y = 5$ when $x = 0$.

13. $\cot x \dfrac{dy}{dx} = y$; $y = 2$ when $x = 0$.

14. $\dfrac{dy}{dx} = xe^{-2y}$; $y = 0$ when $x = 0$.

15. $\dfrac{dy}{dx} - 2xy = 2x$; $y = 1$ when $x = 0$.

16. $x\dfrac{dy}{dx} = xy + y$; $y = 1$ when $x = 1$.

17. $(1 + x^3)\dfrac{dy}{dx} = 3x^2 \tan y$; $y = \frac{1}{2}\pi$ when $x = 0$.

18. $x \cos y \dfrac{dy}{dx} = 1 + \sin y$; $y = 0$ when $x = 1$.

19. $x\dfrac{dy}{dx} = 2y(y - 1)$; $y = 2$ when $x = \frac{1}{2}$.

20. $2x\dfrac{dy}{dx} = 1 - y^2$; $y = 0$ when $x = 1$.

In questions 21 to 24 find the general solutions of the given differential equations, expressing y in terms of x.

21. $(1 - x)\dfrac{dy}{dx} = xy$

22. $(x^2 - 1)\dfrac{dy}{dx} = (x^2 + 1)y$

23. $\dfrac{dy}{dx} = e^x(1 + y^2)$

24. $e^y\dfrac{dy}{dx} + 2x = 2xe^y$

25. Find y in terms of x given that $e^{2x}y\dfrac{dy}{dx} + 2x = 0$ and that $y = 1$ when $x = 0$.

26. Find y in terms of x given that $xy\dfrac{dy}{dx} = \sqrt{(y^2 - 9)}$ and that $y = 5$ when $x = e^4$.

27. Use the substitution $x + y = z$ to obtain the general solution of the differential equation $(x + y - 1)\dfrac{dy}{dx} = x - y + 1$.

28. Use the substitution $y = vx$, where v is a function of x, to obtain the general solution of the differential equation $xy\dfrac{dy}{dx} = 2x^2 - y^2$.

11.4 Formulation of differential equations

In our first example information about the tangent to a curve is expressed in the form of a differential equation. This equation is then solved to find the equation of the curve.

Example 1 The tangent at any point P on a curve in the first quadrant cuts the x-axis at Q. Given that $OP = PQ$, where O is the origin, and that the point $(1, 4)$ lies on the curve, find the equation of the curve.

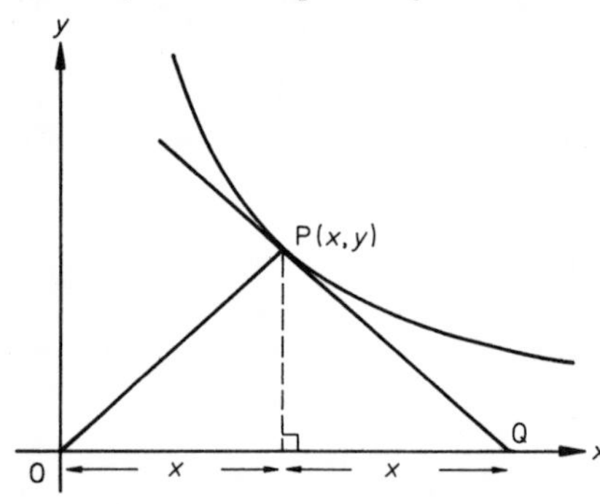

Since $OP = PQ$ the triangle OPQ is isosceles. Thus if P is the point (x, y), the coordinate of Q must be $(2x, 0)$.

$\therefore$ the gradient of PQ

$$= \frac{y - 0}{x - 2x} = -\frac{y}{x}$$

However, since PQ is the tangent at P its gradient is also given by dy/dx.

$$\therefore \quad \frac{dy}{dx} = -\frac{y}{x}$$

$$\text{i.e.} \quad \frac{1}{y}\frac{dy}{dx} = -\frac{1}{x}$$

$$\therefore \quad \int \frac{1}{y}\,dy = -\int \frac{1}{x}\,dx$$

Since the curve lies in the first quadrant, $x > 0$ and $y > 0$

$$\therefore \quad \ln y = -\ln x + c$$

$$\text{i.e.} \quad \ln y + \ln x = c$$

$$\therefore \quad \ln xy = c$$

$$\therefore \quad xy = k, \quad \text{where} \quad \ln k = c$$

Since the curve passes through the point $(1, 4)$, $1 \times 4 = k$
Hence the equation of the curve is $xy = 4$.

Many scientific laws can be expressed as differential equations. For instance, as we saw in §3.1, Newton's law of cooling and certain laws of growth and decay lead to differential equations of the form $dx/dt = \pm kx$, where k is a positive constant. The solutions of such equations are used to make predictions about the behaviour of the variables involved.

Example 2 A liquid is being heated in an oven maintained at a constant temperature of 180°C. It is assumed that the rate of increase in the temperature of the liquid is proportional to $(180 - \theta)$, where θ°C is the temperature of the liquid at time t minutes. If the temperature of the liquid rises from 0°C to 120°C in 5 minutes, find the temperature of the liquid after a further 5 minutes.

The rate of increase in the temperature of the liquid is $d\theta/dt$,

$$\therefore \quad \frac{d\theta}{dt} = k(180 - \theta), \text{ where } k \text{ is constant.}$$

Rearranging, $$\frac{1}{(180 - \theta)}\frac{d\theta}{dt} = k$$

$$\therefore \quad \int \frac{1}{180 - \theta}\,d\theta = \int k\,dt$$

$$\therefore \quad \int \frac{-1}{180 - \theta}\,d\theta = -\int k\,dt$$

Thus for $\theta < 180$, $\ln(180 - \theta) = -kt + c$

$\therefore$ $180 - \theta = e^{-kt+c}$

Writing $A = e^c$, $180 - \theta = Ae^{-kt}$

Hence $\theta = 180 - Ae^{-kt}$

Letting $t = 0$ when $\theta = 0$, $0 = 180 - A$

$\therefore$ $\theta = 180(1 - e^{-kt})$

Since $\theta = 120$ when $t = 5$, $120 = 180(1 - e^{-5k})$

$\therefore$ $\frac{2}{3} = 1 - e^{-5k}$

i.e. $e^{-5k} = \frac{1}{3}$

Hence when $t = 10$, $\theta = 180(1 - e^{-10k})$

$= 180(1 - \{e^{-5k}\}^2)$

$= 180(1 - \{\frac{1}{3}\}^2) = 160$

Thus the temperature of the liquid after a further 5 minutes is 160°C.

Exercise 11.4

1. The tangent at any point P on a curve cuts the x-axis at the point Q. Given that $\angle OPQ = 90°$, where O is the origin, and that the point $(1, 2)$ lies on the curve, find the equation of the curve.

2. The tangent at any point P on a curve cuts the x-axis at A and the y-axis at B. Given that $2\overrightarrow{AP} = \overrightarrow{PB}$ and that the curve passes through the point $(1, 1)$, find the equation of the curve.

3. At time t, the rate of increase in the concentration C of micro-organisms in a controlled environment is equal to k times the concentration, where k is a positive constant. When $t = 0$, $C = C_0$. Write down a differential equation involving C, t and k. Hence find C in terms of C_0, t and k. Find also, in terms of k, the time at which the concentration has increased by 50% from its value at $t = 0$. (L)

4. A body is placed in a room which is kept at a constant temperature. The temperature of the body falls at a rate $k\theta$ C per minute, where k is a constant and θ is the difference between the temperature of the body and that of the room at time t. Express this information in the form of a differential equation and hence show that $\theta = \theta_0 e^{-kt}$, where θ_0 is the temperature difference at time $t = 0$. The temperature of the body falls 5°C in the first minute and 4°C in the second minute. Show that the fall of temperature in the third minute is 3·2°C. (L)

5. A rectangular tank with its base horizontal is filled with water to a depth h at time $t = 0$. Water leaks out of the tank from a small hole in the base at a rate proportional to the square root of the depth of the water. If the depth of water is $\frac{1}{2}h$ at time t, find the further time it will take before the tank is empty. (O)

6. A radioactive substance decays so that the rate of decrease of mass at any time is proportional to the mass present at that time. Denoting by x the mass remaining at time t, write down a differential equation satisfied by x, and show that $x = x_0 e^{-kt}$, where x_0 is the initial mass and k is a constant. The mass is reduced to 4/5 of its initial value in 30 days. Calculate, correct to the nearest day, the time required for the mass to be reduced to half its initial value. A mass of 625 milligrammes of the substance is prepared. Determine the mass which is present 90 days after the preparation. (JMB)

7. According to Newton's law, the rate of cooling of a body in air is proportional to the difference between the temperature T of the body and the temperature T_0 of the air. If the air temperature is kept constant at 20°C and the body cools from 100°C to 60°C in 20 minutes, in what further time will the body cool to 30°C? (O)

8. Two liquids, X and Y, are flowing into a trough at the constant rate of 10 and 20 litres per minute respectively. The liquid in the trough is stirred continuously and pumped out at the rate of 30 litres per minute. Initially the trough contains 200 litres of X and 100 litres of Y. After t minutes the tank contains x litres of X. By considering the change in x in a small interval of time δt, show that

$$\frac{dx}{dt} = 10 - \frac{x}{10}.$$

Hence find an expression for x in terms of t. Find, correct to the nearest litre, the quantity of liquid X in the trough after 10 minutes. After how long, to the nearest second, will there be less of liquid X in the mixture than of liquid Y?

9. A tank being cleaned initially contains 3900 litres of water and 100 kg of dissolved dye. The mixture is kept uniform by stirring. Pure water is run in at the rate

of 130 litres per minute and the mixture is removed at the same rate. If the mass of dissolved dye remaining after t minutes is x kg, show that

$$\frac{dx}{dt} = -\frac{x}{30}.$$

How long will it be before half the dye has been removed? Determine how much dye remains after 30 minutes. (L)

10. In a certain chemical reaction, in which a compound X is formed from a compound Y, the masses of X, Y present at time t are x, y respectively. The sum of the masses of X and Y is a, where a is constant, and at any time the rate at which x is increasing is proportional to the product of the two masses at that time. Show that $\frac{dx}{dt} = kx(a - x)$, where k is constant. If $x = a/5$ at $t = 0$ and $x = a/2$ at $t = \ln 2$, show that $k = 2/a$. Hence find t, correct to three significant figures, when $x = 99a/100$. (O & C)

11. A plant grows in a pot which contains a volume V of soil. At time t the mass of the plant is m and the volume of soil utilised by the roots is αm, where α is constant. The rate of increase of the mass of the plant is proportional to the mass of the plant times the volume of soil not yet utilised by the roots. Obtain a differential equation for m, and verify that it can be written in the form $V\beta\frac{dt}{dm} = \frac{1}{m} + \frac{\alpha}{V - \alpha m}$, where β is a constant. The mass of the plant is initially $V/4\alpha$. Find, in terms of V and β, the time taken for the plant to double its mass. Find also the mass of the plant at time t. (JMB)

12. At any instant, a spherical meteorite is gaining mass because of two effects: (i) mass is condensing onto it at a rate which is proportional to the surface area of the meteorite at that instant, (ii) the gravitational field of the meteorite attracts mass onto itself, the rate being proportional to the meteorite mass at that instant. Assuming that the two effects can be added together and that the meteorite remains spherical and of constant density, show that its radius r at time t satisfies the differential equation $\frac{dr}{dt} = A + Br$, where A and B are constants. If $r = r_0$ at $t = 0$, show that $r = r_0e^{Bt} + \frac{A}{B}(e^{Bt} - 1)$. (L)

11.5 The equations $d^2y/dx^2 \pm n^2y = constant$

We now consider some second order differential equations which have important applications especially in mechanics.

Type 1: $\dfrac{d^2y}{dx^2} + n^2y = 0$, where n is constant.

Multiplying both sides by $2\dfrac{dy}{dx}$ we obtain:

$$2\frac{dy}{dx}\frac{d^2y}{dx^2} + n^2\left(2y\frac{dy}{dx}\right) = 0.$$

But $$2\frac{dy}{dx}\frac{d^2y}{dx^2} = \frac{d}{dx}\left\{\left(\frac{dy}{dx}\right)^2\right\} \quad \text{and} \quad 2y\frac{dy}{dx} = \frac{d}{dx}(y^2),$$

so integrating both sides with respect to x:

$$\left(\frac{dy}{dx}\right)^2 + n^2y^2 = \text{constant}$$

Letting $y = a$ when $\dfrac{dy}{dx} = 0$, this becomes

$$\left(\frac{dy}{dx}\right)^2 + n^2y^2 = n^2a^2$$

$$\therefore \quad \frac{dy}{dx} = \pm n\sqrt{(a^2 - y^2)}$$

Hence $$\int \frac{1}{\sqrt{(a^2 - y^2)}}\,dy = \pm \int n\,dx$$

$$\therefore \quad \sin^{-1}\frac{y}{a} = \pm(nx + b)$$

$$\therefore \quad \frac{y}{a} = \sin\{\pm(nx + b)\}$$

$$\therefore \quad y = \pm a\sin(nx + b)$$

Since a is an arbitrary constant which may take any real value, positive or negative, there is no loss of generality in writing:

$$y = a\sin(nx + b)$$

Letting $b = c + \frac{1}{2}\pi$, we obtain:

$$y = a\sin(nx + c + \tfrac{1}{2}\pi)$$

i.e. $$y = a\cos(nx + c)$$

But $$a\cos(nx + c) = a\cos nx\cos c - a\sin nx\sin c,$$

thus the solution can also be expressed in the form

$$y = A\cos nx + B\sin nx$$

Thus the general solution of the differential equation

$$\frac{d^2y}{dx^2} + n^2y = 0$$

may be expressed in three possible forms:

$$y = a\sin(nx + b), \quad y = a\cos(nx + c)$$

or

$$y = A\cos nx + B\sin nx,$$

where a, b, c, A and B are arbitrary constants.

Example 1 Solve the differential equation $\frac{d^2y}{dx^2} + 9y = 0$ given that $y = 1$ when $x = 0$ and that $y = 2$ when $x = \frac{1}{2}\pi$.

The general solution of the given equation is

$$y = A\cos 3x + B\sin 3x$$

Since $y = 1$ when $x = 0$, $\quad 1 = A + 0$ i.e. $A = 1$
Since $y = 2$ when $x = \frac{1}{2}\pi$, $\quad 2 = 0 - B$ i.e. $B = -2$

Hence the required solution is $y = \cos 3x - 2\sin 3x$.

Example 2 Solve the differential equation $9\frac{d^2y}{dx^2}+4y=0$ given that $y=\frac{dy}{dx}=2$ when $x = 0$.

Rearranging the equation, $\frac{d^2y}{dx^2} + \frac{4}{9}y = 0$

Hence the general solution is

$$y = A\cos\tfrac{2}{3}x + B\sin\tfrac{2}{3}x$$

$$\therefore \quad \frac{dy}{dx} = -\tfrac{2}{3}A\sin\tfrac{2}{3}x + \tfrac{2}{3}B\cos\tfrac{2}{3}x$$

Since $y = 2$ when $x = 0$, $\qquad 2 = A$
Since $\frac{dy}{dx} = 2$ when $x = 0$, $\qquad 2 = \frac{2}{3}B$
$\therefore \quad A = 2$ and $B = 3$
Hence the required solution is $y = 2\cos\frac{2}{3}x + 3\sin\frac{2}{3}x$.

Type 2: $\frac{d^2y}{dx^2} + n^2y = k$, where n and k are constant.

Rearranging: $\qquad \frac{d^2y}{dx^2} + n^2\left(y - \frac{k}{n^2}\right) = 0$

Letting $z = y - \frac{k}{n^2}$, we find that $\frac{d^2z}{dx^2} = \frac{d^2y}{dx^2}$

so the equation becomes: $$\frac{d^2z}{dx^2} + n^2z = 0$$

$\therefore$ the general solution is $$z = A\cos nx + B\sin nx$$

i.e. $$y - \frac{k}{n^2} = A\cos nx + B\sin nx$$

Thus the general solution of the differential equation

$$\frac{d^2y}{dx^2} + n^2y = k$$

is $$y = A\cos nx + B\sin nx + k/n^2,$$

where A and B are arbitrary constants.

Type 3: $\dfrac{d^2y}{dx^2} - n^2y = 0$, where n is constant.

Since $y = e^{nx}$ is a solution of this equation, we try the substitution $y = e^{nx}v$, where v is a function of x.

Differentiating: $$\frac{dy}{dx} = e^{nx}\frac{dv}{dx} + ne^{nx}v$$

$\therefore$ $$\frac{d^2y}{dx^2} = e^{nx}\frac{d^2v}{dx^2} + 2ne^{nx}\frac{dv}{dx} + n^2e^{nx}v$$

Hence the differential equation becomes

$$e^{nx}\frac{d^2v}{dx^2} + 2ne^{nx}\frac{dv}{dx} + n^2e^{nx}v - n^2e^{nx}v = 0$$

i.e. $$\frac{d^2v}{dx^2} + 2n\frac{dv}{dx} = 0$$

Integrating with respect to x,

$$\frac{dv}{dx} + 2nv = a$$

Rearranging: $$\frac{dv}{dx} = a - 2nv$$

$\therefore$ $$\frac{1}{(a-2nv)}\frac{dv}{dx} = 1$$

$\therefore$ $$\int\frac{-2n}{(a-2nv)}\,dv = \int(-2n)\,dx$$

$\therefore$ $$\ln|a - 2nv| = -2nx + c$$

i.e. $$|a - 2nv| = e^{-2nx+c}$$

Thus $$a - 2nv = ke^{-2nx}, \quad \text{where } |k| = e^c.$$

$\therefore$ $$v = \frac{a}{2n} - \frac{k}{2n}e^{-2nx}$$

$\therefore$ $$y = e^{nx}v = \frac{a}{2n}e^{nx} - \frac{k}{2n}e^{-nx}$$

> Thus the general solution of the differential equation
> $$\frac{d^2y}{dx^2} - n^2y = 0$$
> is of the form $$y = Ae^{nx} + Be^{-nx},$$
> where A and B are arbitrary constants.

Example 3 Solve the differential equation $\dfrac{d^2y}{dx^2} - 100y = 0$, given that $y = 6$, $\dfrac{dy}{dx} = 40$ when $x = 0$.

The general solution of the given equation is

$$y = Ae^{10x} + Be^{-10x}$$

$$\therefore \quad \frac{dy}{dx} = 10Ae^{10x} - 10Be^{-10x}$$

Since $y = 6$ when $x = 0$, $\quad 6 = A + B$

Since $\dfrac{dy}{dx} = 40$ when $x = 0$, $\quad 40 = 10A - 10B$

$\therefore \quad A = 5 \quad \text{and} \quad B = 1$

Hence the required solution is $y = 5e^{10x} + e^{-10x}$.

Type 4: $\dfrac{d^2y}{dx^2} - n^2y = k$, where n and k are constant.

By the method applied to equations of type 2 it can be shown that

$$y + \frac{k}{n^2} = Ae^{nx} + Be^{-nx}$$

> Thus the general solution of the differential equation
> $$\frac{d^2y}{dx^2} - n^2y = k$$
> is $$y = Ae^{nx} + Be^{-nx} - \frac{k}{n^2},$$
> where A and B are arbitrary constants.

Exercise 11.5

[In this exercise the general solutions of the given differential equations may be quoted without proof.]

Write down the general solutions of the following differential equations:

1. $\dfrac{d^2y}{dx^2} = -16y$

2. $4\dfrac{d^2y}{dx^2} + 7y = 0$

3. $\dfrac{d^2y}{dx^2} + 9y = 18$

4. $\dfrac{d^2y}{dx^2} = 10 - 10y$

5. $\dfrac{d^2y}{dx^2} = 36y$

6. $\dfrac{d^2y}{dx^2} - 8y = 0$

7. $9\dfrac{d^2y}{dx^2} = 1 + 27y$

8. $\dfrac{d^2y}{dx^2} - 4y + 1 = 0$

Solve the following differential equations subject to the given conditions.

9. $\dfrac{d^2y}{dx^2} + 4y = 0$; $y = 4$ when $x = 0$ and $y = 3$ when $x = 3\pi/4$.

10. $\dfrac{d^2y}{dx^2} - 16y = 0$; $y = 10$ and $\dfrac{dy}{dx} = -16$ when $x = 0$.

11. $36\dfrac{d^2y}{dx^2} + y = 0$; $y = 6$ and $\dfrac{dy}{dx} = 0$ when $x = 0$.

12. $9\dfrac{d^2y}{dx^2} + 16y = 0$; $y = -5$ when $x = 0$ and $y = 4$ when $x = 3\pi/8$.

13. $4\dfrac{d^2y}{dx^2} - 25y = 0$; $y = -9$ and $\dfrac{dy}{dx} = 12\frac{1}{2}$ when $x = 0$.

14. $\dfrac{d^2y}{dx^2} - 12y = 0$; $y = 4/e^2$ when $x = 1/\sqrt{3}$ and $y = 4$ when $x = 0$.

15. $9\dfrac{d^2y}{dx^2} - 49y = 28$; $y = 8\frac{3}{7}$ and $\dfrac{dy}{dx} = -7$ when $x = 0$.

16. $4\dfrac{d^2y}{dx^2} + 25y = 40$; $y = 5{\cdot}6$ and $\dfrac{dy}{dx} = 22{\cdot}5$ when $x = 0$.

17. $\dfrac{d^2y}{dx^2} + 3y = 6$; $y = 2$ when $x = 0$ and $y = 3$ when $x = \pi/2\sqrt{3}$.

18. $16\dfrac{d^2y}{dx^2} - 81y = 243$; $y = -1$ and $\dfrac{dy}{dx} = 0$ when $x = 0$.

19. $\dfrac{d^2y}{dx^2} - 18y = 54$; $y = -3 + \sqrt{2}$ and $\dfrac{dy}{dx} = 3$ when $x = 0$.

20. $\dfrac{d^2y}{dx^2} + 12y = 6$; $\dfrac{dy}{dx} = 24$ when $x = \pi/4\sqrt{3}$ and $\dfrac{dy}{dx} = -18$ when $x = 0$.

Exercise 11.6 (miscellaneous)

1. Sketch some of the straight lines represented by the equation $y = k(2x - k)$, where k is any constant. Obtain the differential equation for this family of lines.

2. Find a second order differential equation of which the general solution is $y = Axe^x + Be^{-x}$, where A and B are arbitrary constants.

3. The differential equation of a family of curves is $\dfrac{dy}{dx} = y - x^2$. Find the equations of the loci of points where $\dfrac{dy}{dx} = 0$ and where $\dfrac{d^2y}{dx^2} = 0$. Hence sketch the integral curves of the differential equation.

4. (a) Find, in any form, the solution of the differential equation $\dfrac{du}{dt} - 7u + 3 = 0$, given that $u = 0$ when $t = 5$.
   (b) Solve the differential equation $\dfrac{dy}{dx} = x\sqrt{(x^2 + 9)}$ given that $y = -9$ when $x = 0$. (C)

5. (a) Solve the differential equation $\dfrac{dx}{dt} = \dfrac{1}{x + 2} - \dfrac{3}{3x + 5}$, given that $x = 1$ when $t = 2$.
   (b) Solve the differential equation $\dfrac{dy}{dx} = \dfrac{2y^2 + 3}{4y}$, given that $y = 1$ when $x = -1$. Given your answer in a form expressing y in terms of x. (C)

6. The temperature y degrees of a body, t minutes after being placed in a certain room, satisfies the differential equation $6\dfrac{d^2y}{dt^2} + \dfrac{dy}{dt} = 0$. By using the substitution $z = dy/dt$, or otherwise, find y in terms of t, given that $y = 63$ when $t = 0$ and $y = 36$ when $t = 6 \ln 4$. Find after how many minutes the rate of cooling of the body will have fallen below one degree per minute, giving your answer correct to the nearest minute. How cool does the body get? (O&C)

7. By substituting $z = (1 + x)y$, solve the differential equation

$$\frac{dy}{dx} + \frac{y}{1+x} = \cos x,$$

given that $y = 0$ when $x = 0$. In your answer give y in terms of x.

8. (a) Solve the differential equation $xy\dfrac{dy}{dx} = 1 + y^2$, given that, when $x = 2$, $y = 0$. Find the equation of the tangent to the curve at the point in the first quadrant where $x = 4$.
(b) If $Ax^2 - By^2 = C$, where A, B and C are constants, prove that

$$xy\frac{d^2y}{dx^2} + x\left(\frac{dy}{dx}\right)^2 - y\frac{dy}{dx} = 0. \qquad \text{(SU)}$$

9. (a) Solve the differential equation $dy/dx = \sin x \cos^2 y$, given that, when $x = \pi/2$, $y = \pi/4$.
(b) A curve passes through the origin and is such that $dy/dx = e^{2x-y}$. Obtain the equation of the curve in the form $y = f(x)$ and give a rough sketch of its graph. (SU)

10. (i) Find y in terms of x given that $(1 + x)\dfrac{dy}{dx} = (1 - x)y$ and that $y = 1$ when $x = 0$.
(ii) A curve passes through $(2, 2)$ and has gradient at (x, y) given by the differential equation $ye^{y^2}\dfrac{dy}{dx} = e^{2x}$. Find the equation of the curve. Show that the curve also passes through the point $(1, \sqrt{2})$ and sketch the curve.
(L)

11. If y is a solution of the differential equation $\dfrac{dy}{dx} = \dfrac{1 + y^2}{1 + x^2}$ deduce that

$$y = x + k(1 + xy)$$

where k is a constant.

(You may assume that $\tan^{-1} a - \tan^{-1} b = \tan^{-1}\left(\dfrac{a - b}{1 + ab}\right) + n\pi$, n an integer.)
(W)

12. Obtain the general solution of the differential equation $2x^2\dfrac{dy}{dx} + x^2y^2 = y^2$.
Find the three particular solutions such that (a) when $x = 1$, $y = 1$, (b) when $x = 1$, $y = \frac{2}{3}$, (c) when $x = 1$, $y = \frac{1}{2}$. Show that each of the integral curves given by these particular solutions has a maximum point on the line $x = 1$.
Sketch the three curves on the same axes. (L)

13. In a chemical reaction there are present, at time, t, x kg of substance X and y kg of substance Y, and initially there is 1 kg of X and 2 kg of Y. The variables x and y satisfy the equations $\dfrac{dx}{dt} = -x^2y, \dfrac{dy}{dt} = -xy^2$. Find $\dfrac{dy}{dx}$ in terms of x and y, and express y in terms of x. Hence obtain a differential equation in x and t only, and so find an expression for x in terms of t. (JMB)

14. (a) The equation of the family of all parabolas whose axes lie along the x-axis has the form $y^2 = 4a(x - h)$ where each parabola is specified by values of a and h. Form a differential equation (*not* involving a or h) satisfied by any point (x, y) lying on any one of these parabolas.

(b) The tangent to a curve at the point (x, y) cuts the x-axis at Q. If $OQ = kx$ (where O is the origin of the coordinate system) show that points on the curve satisfy the differential equation

$$\frac{dy}{dx} = \frac{y}{x - kx}.$$

By solving this differential equation show that the equation of the curve has the form $y = Ax^{1/(1-k)}$ where A is an arbitrary constant. (AEB 1975)

15. (i) Obtain the general solution of the differential equation $x \ln x \dfrac{dy}{dx} = 1$

(a) where $0 < x < 1$ and (b) where $x > 1$.

(ii) Sketch the family of integral curves for the differential equation

$$\frac{dy}{dx} = x(y + 1).$$

(iii) Find the equations of the two curves which pass through the point $(1, 1)$ and which satisfy the differential equation

$$\tan(y - 1)\frac{dy}{dx} = \tan(x - 1). \qquad \text{(L)}$$

16. According to Newton's law of cooling, the rate at which a hot object cools is proportional to the difference between the temperature of the object and the temperature of the surrounding air (assumed to be constant). If an object cools from 100° to 80° in 10 minutes, and from 80° to 65° in another 10 minutes, find the temperature of the surrounding air and the temperature of the object after a further 10 minutes. (O)

17. (a) Given that $\dfrac{dy}{dx} = 9y^2 - 4$ and that $y = 1$ when $x = 2$, find an equation expressing x in terms of y.

(b) In a certain country, the price p of a particular commodity increases with the time t at a rate equal to kp where k is a positive constant. Write down a differential equation expressing this information. Show that if $p = 1$ when $t = 0$ and $p = \alpha$ when $t = 1$, then, at time t, $p = \alpha^t$. (C)

18. (a) Solve the differential equation $\dfrac{dy}{dx} = 3y + 2$ given that $y = 1$ when $x = 2$, expressing your answer in a form giving y as a function of x.

(b) The rate of decomposition of a radioactive substance is proportional to the mass of the substance remaining. Write down a differential equation which expresses this law, stating the meaning of each symbol used. If one third of the mass is left after 12 years, how much was left after 3 years? (C)

19. Social scientists studying the growth of population in an urban region use a model in which the rate of growth of population is proportional to $P(N - P)$, where P is the current size of the population (regarded as a continuous variable) and N is a constant. Set up a differential equation for this model and find its general solution expressing P as a function of time t. In a particular region, the population at a certain time was 100 000. Fifty years later the population was 200 000 and fifty years after that it was 300 000. Using the above model determine the value of N in this case. What is the significance of N? (C)

20. A water reservoir is in the shape of the surface obtained by rotating a parabola $y = kx^2$ about its vertical axis, $x = 0$. Show that when the central depth of the water in the reservoir is h, the surface area A of the water is proportional to h and the volume V is proportional to h^2. In a dry period the rate of loss of water, due entirely to evaporation, is λA per day. Obtain a differential equation for h after t days and deduce that the depth decreases at a constant rate. If, in addition to the evaporation a constant volume C of water is now used per day, obtain the new differential equation and show that the time to empty the reservoir from a depth H will be $\displaystyle\int_0^H \frac{\pi h\, dh}{Ck + \pi\lambda h}$ days. Verify that when $H = 20$, $\lambda = \frac{1}{2}$, $C = 500$ and $k = 0{\cdot}0025$ this amounts to just under 35 days. (AEB 1979)

21. (i) An electrically charged body loses its charge, Q coulombs, at a rate kQ coulombs per second, where k is a constant. Write down a differential equation involving Q and t, where t seconds is the time since the discharge started. Solve this equation for Q, given that the initial charge was Q_0 coulombs. If $Q_0 = 0{\cdot}001$, and $Q = 0{\cdot}0005$ when $t = 10$, find the value of Q when $t = 20$.

(ii) Solve the differential equation $\dfrac{d^2y}{dx^2} - 25y = 0$, given that $y = 4$, $\dfrac{dy}{dx} = 10$ when $x = 0$. (L)

22. Find the solution of the differential equation $\dfrac{d^2y}{dx^2} + 9y = 18$, for which y has

a maximum at $(\frac{1}{2}\pi, 6)$. Find the minimum value of y, and the values of x for which $y = 0$.

(O)

23. (i) Solve the differential equation $\dfrac{dy}{dx} = \left(\dfrac{x}{y}\right)e^{x+y}$ in the form $f(y) = g(x)$, given that $y = 0$ when $x = 0$.

(ii) Solve the differential equation $\dfrac{d^2y}{dx^2} + 4y = 16$, given that $y = 4$ and $\dfrac{dy}{dx} = -4$ when $x = 0$. For which values of x does the graph of y have points of inflexion?

(O)

24. (i) Find the general solution of the differential equation

$$(1 + x)\frac{dy}{dx} = xy.$$

(ii) By using the substitution $z = dy/dx$, or otherwise, solve the differential equation $\dfrac{d^3y}{dx^3} + 4\dfrac{dy}{dx} = 0$, given that y takes the values $0, 0, 1$ for $x = 0, \frac{1}{4}\pi, \frac{1}{2}\pi$ respectively.

(O&C)

25. Prove that, if $y = \cos^2(x^2)$, then

$$x\frac{d^2y}{dx^2} - \frac{dy}{dx} + 16x^3y = 8x^3.$$

Prove that changing the independent variable in this differential equation from x to t, where $t = x^2$, gives the equation

$$\frac{d^2y}{dt^2} + ky = C,$$

where k, C are constants and hence, or otherwise, find the general solution of the first equation in terms of x. What is the general solution of the differential equation

$$x\frac{d^2y}{dx^2} - \frac{dy}{dx} - 16x^3y = 8x^3?$$

(O&C)

12 Further particle dynamics

12.1 Acceleration as a function of displacement

Let us consider a particle moving along a straight line so that at time t its displacement from a fixed point on the line is s. The velocity v and the acceleration a of the particle at time t are given by:

$$v = \frac{ds}{dt} \quad \text{and} \quad a = \frac{dv}{dt} = \frac{d^2s}{dt^2}$$

Suppose now that the acceleration a is regarded as a function of the displacement s, then, using the chain rule,

$$a = \frac{dv}{dt} = \frac{dv}{ds} \times \frac{ds}{dt} = \frac{dv}{ds} \times v = v\frac{dv}{ds}$$

$\therefore$ if $a = f(s)$, then $\qquad v\dfrac{dv}{ds} = f(s),$

giving $\qquad \displaystyle\int v\,dv = \int f(s)\,ds$

$\therefore \qquad \displaystyle\tfrac{1}{2}v^2 = \int f(s)\,ds$

i.e. $\qquad \displaystyle v^2 = 2\int f(s)\,ds$

> Thus if the acceleration a is a function of the displacement s, the relationship $a = v\dfrac{dv}{ds}$ may be used to obtain an expression for v^2 in terms of s.

Example 1 A particle of mass m moves in a straight line under the action of a force directed towards a fixed point O of the line. The magnitude of the force is $3ms^2$, where

s is the displacement of the particle from O at time t. Given that the particle starts from rest at a distance $2a$ from O, find the speed of the particle as it reaches O.

If v is the velocity of the particle at time t, then its acceleration is $v\dfrac{dv}{ds}$ in the direction of the displacement s. Since the force of magnitude $3ms^2$ acts in the opposite direction,

by Newton's second law, $$mv\frac{dv}{ds} = -3ms^2$$

$$\therefore \quad v\frac{dv}{ds} = -3s^2$$

$$\therefore \quad \int v\,dv = -\int 3s^2\,ds$$

$$\tfrac{1}{2}v^2 = -s^3 + c$$

But $v = 0$ when $s = 2a$,

$\therefore$ $0 = -8a^3 + c$ i.e. $c = 8a^3$

Thus $v^2 = 2(8a^3 - s^3)$,

and when $s = 0$, $v^2 = 16a^3$.

Hence when the particle reaches O its speed is $4a^{3/2}$.

Example 2 Assuming that the earth is a sphere of radius R, the acceleration due to gravity of an object at a distance x from the centre of the earth is inversely proportional to x^2 when above the earth's surface and equal to g on the surface. Given that an object projected vertically upwards from the earth's surface with speed u rises to a maximum height R above the surface, find u in terms of g and R.

The acceleration of the object is inversely proportional to the square of the displacement x and in the opposite direction,

$$\therefore \quad v\frac{dv}{dx} = -\frac{k}{x^2},$$

where v is the velocity and k is constant.
Since the acceleration is $-g$ when $x = R$,

$$-g = -\frac{k}{R^2} \quad \text{giving} \quad k = gR^2.$$

$$\therefore \quad v\frac{dv}{dx} = -\frac{gR^2}{x^2}$$

$$\therefore \quad \int v\,dv = -\int \frac{gR^2}{x^2}\,dx$$

$$\tfrac{1}{2}v^2 = \frac{gR^2}{x} + c$$

Given that $v = 0$ when $x = 2R$,

$$0 = \tfrac{1}{2}gR + c \quad \text{i.e.} \quad c = -\tfrac{1}{2}gR$$

$$\therefore \quad v^2 = \frac{2gR^2}{x} - gR$$

Since $v = u$ when $x = R$, $\quad u^2 = \dfrac{2gR^2}{R} - gR = gR$

Hence $\quad u = \sqrt{(gR)}$.

Exercise 12.1

In questions 1 to 4 a particle moves in a straight line so that after t seconds its displacement from a fixed point on the line is s metres and its velocity is $v\,\text{m s}^{-1}$.

1. Given that the acceleration of the particle is $1/(1+t)\,\text{m s}^{-2}$ and that the particle starts from rest, find its speed after 2 seconds.

2. Given that the particle is subject to a retardation of $5e^{-t}\,\text{m s}^{-2}$ and that its initial speed is $4\,\text{m s}^{-1}$, find the time at which the particle first comes to rest.

3. Given that the acceleration of the particle is $2s\,\text{m s}^{-2}$ and that $v = 2$ when $s = 0$, find the speed of the particle when $s = 4$.

4. Given that the particle is subject to a retardation of $2\sin^2 s\,\text{m s}^{-2}$ and that $v = 1$ when $s = \pi/4$, find, correct to 3 significant figures, the speed of the particle when $s = \pi/3$.

5. A particle of mass m moving along a straight line is subject to a force $4mx$ acting towards a fixed point O on the line, where x is the displacement of the particle from O. Given that the particle starts from rest at a distance a from O, find the speed of the particle as it passes through O.

6. The acceleration due to gravity of an object at a distance x from the centre of the earth is inversely proportional to x^2 when above the earth's surface and equal to g on the surface. The earth is assumed to be a sphere of radius R. If an object falls from rest at a height $2R$ above the earth's surface, find its speed when it reaches a height R above the surface.

7. A particle of mass m is moving in a straight line under the action of a force $4m/x^3$ directed towards a fixed point O of the line, x being the displacement of the particle from O. Given that initially the particle is at a distance h from O, moving away from O at a speed of $2/h$, find expressions for (a) dx/dt in terms of x, and (b) x in terms of t and h.

8. A particle of mass m is moving in a straight line under the action of a force of magnitude $\tfrac{1}{4}e^{2x}$, where x is the displacement of the particle from a fixed

point O on the line. Initially the particle is at a point A travelling towards O at speed u. Given that the force on the particle acts in the direction of $\overrightarrow{OA}$ and that $OA = \ln(2u)$, find an expression for dx/dt in terms of x. Hence show that $x = \ln\left(\frac{2u}{1 + ut}\right)$.

12.2 Acceleration as a function of velocity

We consider again a particle moving in a straight line so that at time t its displacement from a fixed point on the line is s and its velocity is v. If the acceleration of the particle is a function of the velocity, $f(v)$ say, then we may write:

(1) $\frac{dv}{dt} = f(v)$, or (2) $v\frac{dv}{ds} = f(v)$.

Equation (1) is used when a relationship between velocity and time is required. However, equation (2) may be used to obtain a relationship between velocity and displacement.

Example 1 A particle of mass m is projected vertically upwards from a point O with speed u and after time t its velocity is v. Given that the particle moves under gravity in a medium which exerts a resistance kv per unit mass, where k is a positive constant, find the time taken by the particle to reach its greatest height.

As the particle rises to its highest point the forces acting on it are its weight mg and the resistance mkv, both directed vertically downwards.

Thus, by Newton's second law, $-mg - mkv = m\frac{dv}{dt}$

$$\therefore \quad -(g + kv) = \frac{dv}{dt}$$

$$1 = -\frac{1}{(g + kv)}\frac{dv}{dt}$$

$$\therefore \quad \int dt = -\int \frac{1}{g + kv}\,dv$$

$$t = -\frac{1}{k}\ln(g + kv) + c$$

Since $v = u$ when $t = 0$, $c = \frac{1}{k}\ln(g + ku)$

$$\therefore \quad t = -\frac{1}{k}\ln(g + kv) + \frac{1}{k}\ln(g + ku) = \frac{1}{k}\ln\left(\frac{g + ku}{g + kv}\right)$$

$\therefore$ when $v = 0$, $t = \frac{1}{k}\ln\left(\frac{g + ku}{g}\right)$

Hence the time taken by the particle to reach its greatest height is $\frac{1}{k}\ln\left(1 + \frac{ku}{g}\right)$.

Example 2 A particle of mass m moves in a straight line under the action of a retarding force $mv^4/2k^4$, where k is a positive constant and v is the velocity of the particle. Initially the particle is at a distance k^2 from a fixed point A, travelling away from A at speed k. Find the speed of the particle when its displacement from A is x. Find also the time which elapses before the particle reaches the point B, where $AB = 4k^2$.

Writing the acceleration of the particle in the form $v\,dv/dx$ and using Newton's second law we have:

$$-\frac{mv^4}{2k^4} = mv\frac{dv}{dx}$$

$$\therefore \qquad \frac{1}{k^4} = -\frac{2}{v^3}\frac{dv}{dx}$$

$$\therefore \qquad \int \frac{1}{k^4}\,dx = \int\left(-\frac{2}{v^3}\right)dv$$

Thus
$$\frac{x}{k^4} = \frac{1}{v^2} + c$$

Since $v = k$ when $x = k^2$, $\dfrac{1}{k^2} = \dfrac{1}{k^2} + c$ i.e. $c = 0$

$$\therefore \qquad \frac{x}{k^4} = \frac{1}{v^2}$$

$$v^2 = \frac{k^4}{x}$$

Hence when the displacement of the particle from A is x, its speed is $k^2/\sqrt{x}$.

$\therefore$ if x is the displacement from A at time t,

$$\frac{k^2}{\sqrt{x}} = \frac{dx}{dt}$$

$$\therefore \qquad \int k^2\,dt = \int x^{1/2}\,dx$$

$$k^2t = \tfrac{2}{3}x^{3/2} + C$$

Since $x = k^2$ when $t = 0$, $C = -\tfrac{2}{3}k^3$

$$\therefore \qquad k^2t = \tfrac{2}{3}(x^{3/2} - k^3)$$

Thus when $x = 4k^2$, $\quad k^2t = \tfrac{2}{3}(8k^3 - k^3) = 14k^3/3$

Hence the time which elapses before the particle reaches B is $14k/3$.

Example 3 A car of mass m moving on a straight horizontal road is subject to a constant resistance R. The engine is working at a constant rate kR. Given that in time t the car accelerates to a speed v from rest, where $v < k$, find an expression for t in terms of m, R, k and v. Find also an expression for the distance x travelled by the car in this time.

If the tractive force exerted by the engine is F when the car is travelling at speed v, then its rate of working $= kR = Fv$,

$$\therefore \quad F = kR/v$$

Using Newton's second law, the equation of motion is

$$\frac{kR}{v} - R = m\frac{dv}{dt}$$

$$\therefore \quad R\left(\frac{k-v}{v}\right) = m\frac{dv}{dt} \qquad (1)$$

$$\therefore \quad \int R\,dt = \int m\left(\frac{v}{k-v}\right)dv$$

$$= \int m\left(\frac{k}{k-v} - 1\right)dv$$

$$\therefore \quad Rt = m\{-k\ln(k-v) - v\} + c$$

Since $v = 0$ when $t = 0$, $c = mk\ln k$

$$\therefore \quad Rt = m\{-k\ln(k-v) - v + k\ln k\}$$

Hence
$$t = \frac{m}{R}\left\{k\ln\left(\frac{k}{k-v}\right) - v\right\}$$

Substituting $\frac{dv}{dt} = v\frac{dv}{dx}$ in equation (1),

$$R\left(\frac{k-v}{v}\right) = mv\frac{dv}{dx}$$

$$\therefore \quad \int R\,dx = \int m\left(\frac{v^2}{k-v}\right)dv$$

$$\begin{array}{r} -v-k \\ -v+k\overline{)v^2\qquad} \\ \underline{v^2 - kv} \\ kv \\ \underline{kv - k^2} \\ k^2 \end{array}$$

By long division

$$\frac{v^2}{k-v} = -v - k + \frac{k^2}{k-v}$$

$$\therefore \quad \int R\,dx = \int m\left\{\frac{k^2}{k-v} - v - k\right\}dv$$

$$\therefore \quad Rx = m\{-k^2\ln(k-v) - \tfrac{1}{2}v^2 - kv\} + C$$

Since $v = 0$ when $x = 0$, $C = mk^2\ln k$

$$\therefore \quad Rx = m\{-k^2\ln(k-v) - \tfrac{1}{2}v^2 - kv + k^2\ln k\}$$

Hence
$$x = \frac{m}{R}\left\{k^2\ln\left(\frac{k}{k-v}\right) - \tfrac{1}{2}v^2 - kv\right\}.$$

In certain types of problem it is quicker to use definite integrals rather then indefinite integrals. For instance, if in Example 3 we had been asked to find the time which elapses as the car accelerates from speed $\frac{1}{3}k$ to speed $\frac{2}{3}k$, we could have deduced from equation (1) that:

$$\int_{t_1}^{t_2} R\,dt = \int_{k/3}^{2k/3} m\left(\frac{v}{k-v}\right) dv$$

where $t = t_1$ when $v = \frac{1}{3}k$ and $t = t_2$ when $v = \frac{2}{3}k$

$$\therefore \quad R\int_{t_1}^{t_2} dt = m\int_{k/3}^{2k/3}\left(\frac{k}{k-v} - 1\right) dv$$

$$\therefore \quad R\Big[\,t\,\Big]_{t_1}^{t_2} = m\Big[-k\ln(k-v) - v\Big]_{k/3}^{2k/3}$$

$$\therefore \quad R(t_2 - t_1) = m(-k\ln\tfrac{1}{3}k - \tfrac{2}{3}k) - m(-k\ln\tfrac{2}{3}k - \tfrac{1}{3}k)$$

$$= mk(\ln\tfrac{2}{3}k - \ln\tfrac{1}{3}k + \tfrac{1}{3} - \tfrac{2}{3})$$

$$= mk(\ln 2 - \tfrac{1}{3})$$

Hence the time taken is $\dfrac{mk}{R}(\ln 2 - \frac{1}{3})$.

Exercise 12.2

1. A particle of unit mass, moving in a straight line with speed v at time t, is subject to a retarding force of magnitude $(3 + 2v)$. Given that initially $v = 66$, find the value of t when $v = 6$.

2. A particle travelling in a straight line is subject to a retardation of $(1 + v^2)$, where v is its speed. If the initial speed of the particle is u, find an expression for the distance it travels before coming to rest.

3. A particle is moving in a straight line so that when the displacement of the particle from a fixed point on the line is x, its velocity is v. Given that the acceleration of the particle is $k(1 - v)$, where k is constant and that $x = 0$ when $v = 0$, find an expression for x in terms of v.

4. A particle is projected horizontally from a point O on a horizontal plane with an initial velocity $2a/b$ where a and b are positive constants. The particle is subject to a retardation $a + bv$ where v is the velocity at time t from the commencement of the motion. Show that $bv = a(3e^{-bt} - 1)$ and, using this result or otherwise, find the distance gone by the particle when it comes to instantaneous rest. (O&C)

5. The retardation of a particle moving in a straight line with speed v is proportional to v^3. The initial speed of the particle is u and s is the displacement of the particle from its initial position at time t. Given that $v = \frac{1}{2}u$ when $s = 2/u$, find an expression for v in terms of u and s. Hence find an expression for t in terms of u and s.

6. Show that the acceleration of an object moving along a straight line may be written as $v\,dv/ds$.

A vehicle of mass 2500 kg moving on a straight course is subject to a single resisting force in the line of motion of magnitude kv newtons, where v metres per second is the velocity and k is constant. At 100 km/h this force is 2000 N. The vehicle is slowed down from 100 km/h to 50 km/h. Find (i) the distance travelled, (ii) the time taken. (AEB 1980)

7. A particle falls from rest and, at any moment, it experiences a resistance of size kmv, where m is the mass of the particle, v is its velocity and k is a constant.

(i) Show that, from the equation of motion we can deduce

$$\int \frac{dv}{g - kv} = \int dt.$$

(ii) Using this equation, show that the velocity at any time is

$$\frac{g}{k}(1 - e^{-kt}).$$

(iii) Show that, no matter for how long the particle falls, its velocity never exceeds a certain value. If this value is V, show that k is g/V. (SU)

8. A particle of mass m moving under gravity is subject to a resistance of magnitude mkv^2, where v is its speed and k is constant. Initially the particle is travelling vertically downwards at speed u, where $u^2 < g/k$. Find the speed of the particle when it has fallen a distance s.

9. A train, when braking, is subject to a retardation of $\left(1 + \frac{v}{100}\right)\text{m s}^{-2}$, where v is its speed at any time. The brakes are applied when it is travelling at $20\,\text{m s}^{-1}$.

(i) Show that it takes just over 18 seconds to come to rest.

(ii) Show that it will travel approximately 177 m after the brakes are applied.

$\left[\text{It may prove useful to know that } \frac{v}{v + 100} \text{ can be written as } 1 - \frac{100}{v + 100}\right].$

(iii) If it then accelerates with acceleration $\left(1 + \frac{v^2}{100}\right)\text{m s}^{-2}$, find, correct to 3 significant figures, how long it takes to attain a speed of $20\,\text{m s}^{-1}$. (SU)

10. A particle of mass m moves, in the direction of x increasing, along the positive x-axis under the action of two forces. These are (i) a force toward the origin O of magnitude kv, and (ii) a force away from O of magnitude kU^2/v, where k and U are positive constants and v is the speed of the particle at time t. Write down the equation of motion of the particle in the form of a differential equation in v and t, and show that the time taken for the particle to accelerate from speed $\frac{1}{4}U$ to speed $\frac{1}{2}U$ is $\frac{m}{2k}\ln\frac{5}{4}$.

Write the equation of motion in the form of a differential equation in v and x, and hence show that the distance travelled by the particle while it accelerates from speed $\frac{1}{4}U$ to speed $\frac{1}{2}U$ is given by

$$\frac{m}{k}\int_{U/4}^{U/2}\left(\frac{U^2}{U^2-v^2}-1\right)dv.$$

Show that this distance is $\dfrac{mU}{4k}\left(2\ln\dfrac{9}{5}-1\right)$. (C)

11. A particle of mass m moves in a straight line under the action of a retarding force of magnitude $k(4+v^2)$, where v is the speed of the particle at time t and k is constant. If the initial speed of the particle is u, find an expression of the time taken by the particle to come to rest. Find also an expression, in terms of m, k and u, for the distance travelled in this time.

12. An engine of mass m kilograms, driven by a constant tractive force of P newtons, experiences a resistance which is k times its speed in metres/second. Show that, when moving horizontally from rest, its speed t seconds after the start is $\dfrac{P}{k}(1-e^{-kt/m})$ metres/second. Hence deduce that it will have travelled mP/ek^2 metres in m/k seconds. What power is being developed by the engine when the time is m/k seconds after the start? (W)

13. A train is pulled along a level track by an engine which exerts a constant pull P newtons at all speeds and the total resistance to motion varies as the square of the speed $v\,\mathrm{m\,s^{-1}}$. The mass of the engine and train combined is 161 700 kg, the maximum speed attained is $120\,\mathrm{km\,h^{-1}}$ and the power then developed is 1078 kW. Show that (a) $P = 32\,340$, (*b*) $5\times 10^4 v\dfrac{dv}{ds} = 10^4 - 9v^2$, where s metres is the distance travelled. Hence show that the distance travelled from rest while attaining a speed of $20\,\mathrm{m\,s^{-1}}$ is about 1240 metres. (C)

14. A train of mass M moves on a straight horizontal track. At speeds less than V the resultant force on the train is constant and equal to P; at speeds not less than V the rate of working of the force is constant and equal to PV. Show that a speed $v(>V)$ is attained from rest in a time $M(V^2+v^2)/2PV$ and that the distance travelled in this time is $M(V^3+2v^3)/6PV$. (W)

15. A car of mass m moving on a level road is subject to a constant resistance R. The engine is working at a constant rate H. Find the time taken for the speed of the car to increase from u to $2u$, where $2Ru < H$. Find also, in terms of m, R, H and u, the distance travelled in this time.

16. A car of mass m moving on a level road is subject to a resisting force kv, where v is the speed of the car at time t and k is constant. When the engine of the car is working at full power S, the car can travel at a steady speed U. Find the

time taken by the car to accelerate under full power from rest to speed V, where $V < U$. Find also the distance travelled in this time. Give both answers in terms of m, S, U and V.

12.3 Simple harmonic motion

Any motion which satisfies a differential equation of the form

$$\ddot{x} = -\omega^2 x$$

where ω is constant, is described as *simple harmonic motion* or S.H.M.

For a particle moving in a straight line, the variable x is the displacement of the particle from a fixed point on the line. However, x may also be a displacement measured along a curved path. In the case of circular motion or the rotation of a rigid body x could represent an angle.

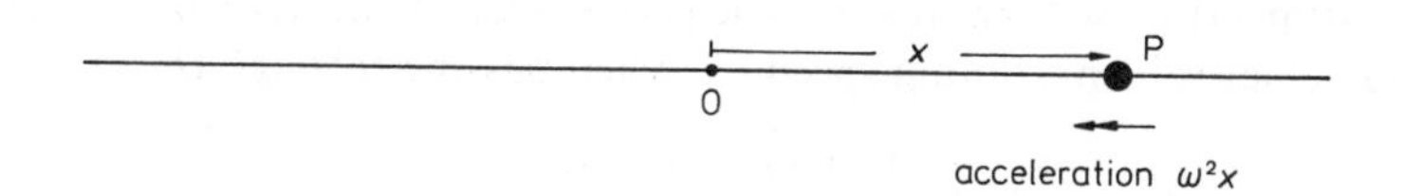

The diagram shows a particle P executing simple harmonic motion along a straight line. If x is the displacement of P from a fixed point O and $\ddot{x} = -\omega^2 x$, then the magnitude of the acceleration of P is a constant multiple of the distance OP. Since when $x > 0, \ddot{x} < 0$ and when $x < 0, \ddot{x} > 0$, the acceleration is always directed towards O. This leads to an alternative definition of simple harmonic motion.

When a particle moves in a straight line with acceleration always directed towards a fixed point of the line and proportional to its distance from that point, the particle is said to be executing simple harmonic motion.

The differential equation $\ddot{x} = -\omega^2 x$, i.e. $\dfrac{d^2x}{dt^2} = -\omega^2 x$, is of a type discussed in §11.5. In this case the method of solution can be simplified by expressing the acceleration of the particle in the form $v\,dv/dx$, where v is the velocity at time t, to give:

$$v\frac{dv}{dx} = -\omega^2 x$$

$$\therefore \quad \int 2v\,dv = -\int 2\omega^2 x\,dx$$

$$v^2 = -\omega^2 x^2 + c$$

If $v = 0$ when $x = a$, we find that

$$v^2 = \omega^2(a^2 - x^2) \tag{1}$$

As shown in §11.5, this leads to the general solution:

$$x = a\sin(\omega t + \varepsilon) \tag{2}$$

or the equivalent form:

$$x = A\cos\omega t + B\sin\omega t \tag{3}$$

From equations (1) and (2) we deduce that the particle is oscillating between points A and A', which lie on opposite sides of O at a distance a from O. The point O is called the *centre* or *central point* of the oscillation and a is the *amplitude.*

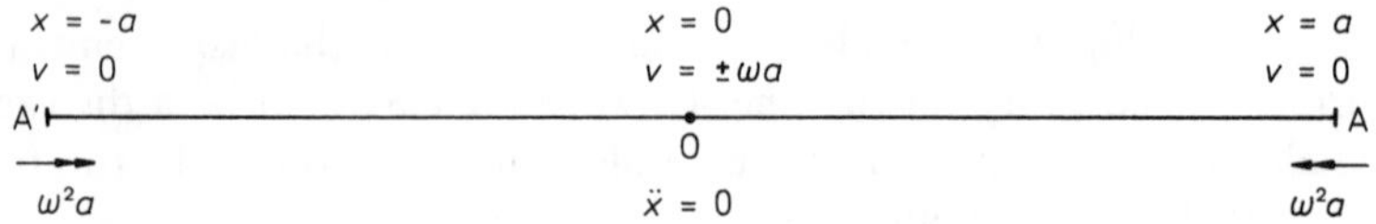

$x = -a$ $v = 0$ A' ; $\omega^2 a$ — $x = 0$ $v = \pm\omega a$ O $\ddot{x} = 0$ — $x = a$ $v = 0$ A ; $\omega^2 a$

From equation (1) we see that the particle is momentarily at rest when it reaches A or A', and achieves its maximum speed ωa when passing through O.

Since $\quad x = A\cos\omega t + B\sin\omega t$
and $\quad v = -A\omega\sin\omega t + B\omega\cos\omega t,$

when ωt is increased to $\omega t + 2\pi$ the same values of x and v are obtained. This means that at times t and $t + 2\pi/\omega$ the particle passes through the same point with the same velocity. Hence $2\pi/\omega$ is the time taken to perform a complete oscillation and is called the *period* of the motion. In particular, $2\pi/\omega$ is the time taken by the particle to move from A to A' and back again to A.

The *frequency*, i.e. the number of complete oscillations per unit time, is $\omega/2\pi$. Summarising the properties of S.H.M. we have:

$\ddot{x} = -\omega^2 x,$
$v^2 = \omega^2(a^2 - x^2)$, where a is the amplitude,
$x = a\sin(\omega t + \varepsilon)$, or
$x = A\cos\omega t + B\sin\omega t,$
$T = 2\pi/\omega$, where T is the period of oscillations.

When solving problems involving S.H.M. standard formulae may be quoted without proof. It is usually best to begin be determining a and ω.

Example 1 A particle is performing simple harmonic oscillations between points A and A' which are 6 m apart. When the particle is at a distance $\sqrt{5}$ m from the mid-point O of AA' its speed is 10 m s^{-1}. Find (a) the period of the motion, (b) the maximum speed of the particle, (c) its maximum acceleration.

Let the equation of motion of the particle be

$$\ddot{x} = -\omega^2 x,$$

then $$v^2 = \omega^2(a^2 - x^2),$$

where x m is the displacement of the particle from O, $v\,\text{m s}^{-1}$ is its speed and a m is the amplitude of the motion.
But $AA' = 6\,\text{m} = 2a\,\text{m}$, which gives $a = 3$

$\therefore$ $$v^2 = \omega^2(9 - x^2)$$

Since $v = 10$ when $x = \sqrt{5}$, $100 = \omega^2(9 - 5)$ $\quad\therefore\quad \omega = 5$

(a) The period of the motion $= \dfrac{2\pi}{\omega}$ seconds $= \dfrac{2\pi}{5}$ seconds.

(b) The speed of the particle is greatest when $x = 0$
$\therefore$ the maximum speed $= \omega a\,\text{m s}^{-1} = 15\,\text{m s}^{-1}$.

(c) The acceleration of the particle is greatest when the magnitude of $\omega^2 x$ is greatest,
$\therefore$ the maximum acceleration $= \omega^2 a\,\text{m s}^{-2} = 75\,\text{m s}^{-2}$.

In S.H.M. problems involving time we need to find suitable expressions for x and v in terms of t. The forms of these expressions depend on the starting point chosen for the motion. One way of approaching such problems is to write:

$$x = A\cos\omega t + B\sin\omega t, \qquad (3)$$
$$v = -A\omega\sin\omega t + B\omega\cos\omega t, \qquad (4)$$

then to use any given conditions to determine A and B. We first apply this method to two important special cases.

Case I: $x = a$ when $t = 0$.

Substituting in (3): $a = A$
Since $v = 0$ when $x = a$, from (4): $B = 0$

$\therefore$ $$\boxed{x = a \quad \text{when} \quad t = 0 \;\Rightarrow\; x = a\cos\omega t.}$$

Case II: $x = 0$, $v > 0$ when $t = 0$.

Substituting in (3): $0 = A$
$\therefore$ $x = B\sin\omega t$
Since the amplitude of the motion is a and $v > 0$ when $t = 0$, we must have $B = a$.

$\therefore$ $$\boxed{x = 0, v > 0 \quad \text{when} \quad t = 0 \;\Rightarrow\; x = a\sin\omega t.}$$

Example 2 A particle moving in a straight line performs simple harmonic oscillations about a point O with amplitude 2 m and maximum speed $1\,\text{m s}^{-1}$. If P

and Q are the two points on the line which lie at a distance $\sqrt{2}$ m from O, find the time taken by the particle to move directly from P to Q.

Let the equation of motion be $\ddot{x} = -\omega^2 x$, then, assuming that $x = a$ when $t = 0$, $x = a \cos \omega t$.
Since the amplitude of the motion is 2 m, $a = 2$.
The maximum speed of the particle $= \omega a \text{ m s}^{-1} = 1 \text{ m s}^{-1}$

$\therefore \quad \omega = \frac{1}{2}$
Thus $\quad x = 2\cos\frac{1}{2}t$.

Let P and Q be the points at which $x = \sqrt{2}$ and $x = -\sqrt{2}$ respectively.

For $x = \sqrt{2}$: $\quad \sqrt{2} = 2\cos\frac{1}{2}t$
i.e. $\quad \cos\frac{1}{2}t = 1/\sqrt{2}$

$\therefore$ the particle first passes through P when $\frac{1}{2}t = \frac{1}{4}\pi$ i.e. when $t = \frac{1}{2}\pi$.

For $x = -\sqrt{2}$: $\quad -\sqrt{2} = 2\cos\frac{1}{2}t$
i.e. $\quad \cos\frac{1}{2}t = -1/\sqrt{2}$

$\therefore$ the particle first passes through Q when $\frac{1}{2}t = \frac{3}{4}\pi$ i.e. when $t = \frac{3}{2}\pi$.
Hence the particle moves directly from P to Q in a time of π seconds.

Example 3 A particle is performing simple harmonic motion about a point O with amplitude 5 m and period $\frac{1}{2}\pi$ seconds. If P is the point at which the speed of the particle is 10 m s^{-1}, find the time taken by the particle to move directly from O to P.

Let the equation of motion of the particle be $\ddot{x} = -\omega^2 x$, and let $x = a \sin \omega t$, so that the particle is at O when $t = 0$.
Since the amplitude of the motion is 5 m, $a = 5$.

The period of oscillation $= \dfrac{2\pi}{\omega}\text{ s} = \dfrac{\pi}{2}\text{ s}, \qquad \therefore \quad \omega = 4.$

Thus $x = 5 \sin 4t$ and $v = 20 \cos 4t$.

Substituting $v = 10$: $\quad 10 = 20 \cos 4t$
$\therefore \quad \cos 4t = \frac{1}{2}$

$\therefore$ the particle first passes through P when $4t = \pi/3$. Hence the particle moves directly from O to P in a time of $\pi/12$ seconds.

Example 4 A particle moves on a straight line through a fixed point O so that at time t seconds its displacement from O is x metres and its equation of motion is $\ddot{x} = -9x$. Given that $x = 5$ and $\dot{x} = -6$ when $t = \pi/6$, find the position and speed of the particle when $t = 2\pi/3$.

The general solution of the equation of motion is

$$x = A\cos 3t + B \sin 3t$$
$$\therefore \quad \dot{x} = -3A \sin 3t + 3B \cos 3t$$

Since $x = 5$ and $\dot{x} = -6$ when $t = \pi/6$,

$$5 = A\cos\tfrac{1}{2}\pi + B\sin\tfrac{1}{2}\pi$$

i.e. $\quad 5 = B$

and $\quad -6 = -3A\sin\tfrac{1}{2}\pi + 3B\cos\tfrac{1}{2}\pi$

i.e. $\quad -6 = -3A$

$\therefore \quad A = 2 \quad \text{and} \quad B = 5$

Thus $\quad x = 2\cos 3t + 5\sin 3t,$

$$\dot{x} = -6\sin 3t + 15\cos 3t.$$

$\therefore$ when $t = 2\pi/3$, $x = 2\cos 2\pi + 5\sin 2\pi = 2$

and $\quad \dot{x} = -6\sin 2\pi + 15\cos 2\pi = 15$

i.e. the particle is 2 m from O moving at $15\,\text{m}\,\text{s}^{-1}$.

Exercise 12.3

[Standard formulae may be quoted without proof.]

1. A particle moves in a straight line with simple harmonic motion of amplitude 2·5 m. If the period of oscillation is π seconds, find the maximum speed of the particle and its maximum acceleration.

2. A particle moves in a straight line with simple harmonic motion about the point O as centre. The maximum speed of the particle is $6\,\text{m}\,\text{s}^{-1}$ and its maximum acceleration is $18\,\text{m}\,\text{s}^{-2}$. Find (a) the amplitude of the motion, (b) the period of the motion, (c) the speed of the particle when it is 1 m from O.

3. A particle of mass 3 kg moves in a straight line with simple harmonic motion between two points A and A' which are 15 m apart. Given that when the particle is at a distance of 3 m from A its speed is $2\,\text{m}\,\text{s}^{-1}$, find the period of the motion. Find also the greatest force exerted on the particle during the motion.

4. A mass of 10 kg moves with simple harmonic motion. When it is 2 m from the centre of the oscillation, the velocity and acceleration of the body are $12\,\text{m}\,\text{s}^{-1}$ and $162\,\text{m}\,\text{s}^{-2}$ respectively. Calculate (i) the number of oscillations per minute; (ii) the amplitude of the motion; (iii) the force being applied to the body when it is at the extremities of its motion. (SU)

5. A particle P moves in a straight line so that its acceleration is always directed towards a point O in its path and is of magnitude proportional to the distance OP. When P is at the point A, where $OA = 1$ m, its speed is $3\sqrt{3}$ m/s and when P is at the point B, where $OB = \sqrt{3}$ m, its speed is 3 m/s. Calculate the maximum speed attained by P and the maximum value of OP. Show that P takes $\pi/18$ seconds to move directly from A to B. Find, in m/s correct to 2 significant figures, the speed of P one second after it passes O. (L)

6. A particle, A, is performing simple harmonic oscillations about a point O with amplitude 2 m and period 12π s. Find the least time from the instant when A passes through O until the instant when (i) its displacement is 1 m, (ii) its velocity is half that at O, (iii) its kinetic energy is half that at O. (AEB 1978)

7. A particle P is describing simple harmonic motion in the horizontal line $ADCB$, where $AD = DC = \frac{1}{2}CB$. The speed of P as it passes through C is 5 m/s and P is instantaneously at rest at A and B. Given that P performs 3 complete oscillations per second, calculate (i) the distance AB, (ii) the speed of P as it passes through D, (iii) the distance of P from C at an instant when the acceleration of P is 18π m/s^2, (iv) the time taken by P to go directly from D to A. (AEB 1978)

8. A particle P moves in a straight line with simple harmonic motion of period 2 seconds and maximum speed 4 m/s. Find the speed of P when it is at the point A which is $2/\pi$ metres from O, the centre of the path. Find also the time taken by P to move directly from O to A. When P is passing through O it strikes and adheres to a stationary particle which is free to move. If each particle is of mass 2 kg, find the kinetic energy lost in the collision. (L)

9. A particle moves in a straight line so that at time t its displacement from a fixed point on the line is x and its equation of motion is $\ddot{x} = -4x$. Given that $x = 3$ and $\dot{x} = -6$ when $t = \pi/4$, find (a) x in terms of t, (b) the values of x and $\dot{x}$ when $t = 3\pi/4$, (c) the least positive value of t for which $x = 0$.

10. A particle moves in a straight line so that at time t its displacement from a fixed point on the line is x and its equation of motion is $\ddot{x} = -16x$. Given that $x = 3$ and $\dot{x} = 4\sqrt{3}$ when $t = 0$, find (a) x in terms of t, (b) the least positive value of t for which $\dot{x} = 0$, (c) the amplitude of the oscillation.

11. A particle P moves round a circle, diameter AB, with constant angular speed ω. A second particle Q moves along AB so that PQ is always perpendicular to AB, i.e. Q is the projection of P on AB. Show that the motion of Q is simple harmonic.

12.4 Forces producing simple harmonic motion

To show that a given set of forces acting on a particle produce simple harmonic motion we must prove that the equation of motion of the particle can be expressed in the form $\ddot{x} = -\omega^2 x$.

Since the tension in an elastic string is proportional to the extension in the string, it may be possible for a particle attached to an elastic string or a spring to perform simple harmonic motion.

Let us suppose that a light elastic string of natural length l and modulus of elasticity λ has one end fastened to a fixed point O. A particle of mass m is

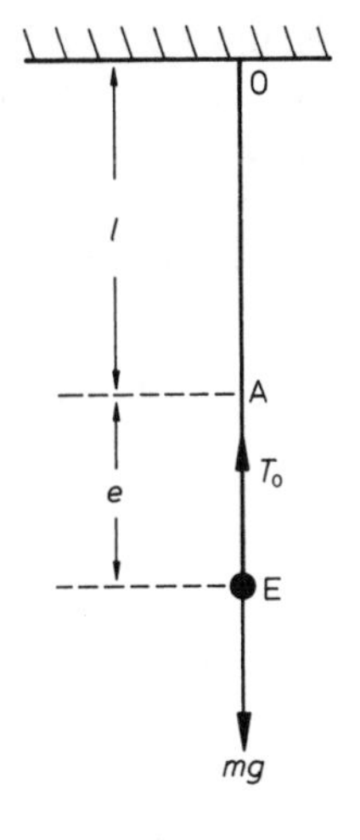

attached to the other end. Let e be the extension in the string and T_0 the tension when the particle hangs in equilibrium at the point E.

By Hooke's law, $T_0 = \frac{\lambda e}{l}$

Since the particle is in equilibrium,

$$T_0 - mg = 0$$

$$\therefore \quad \frac{\lambda e}{l} = mg$$

Consider now the forces on the particle when its displacement from E is x vertically downwards.

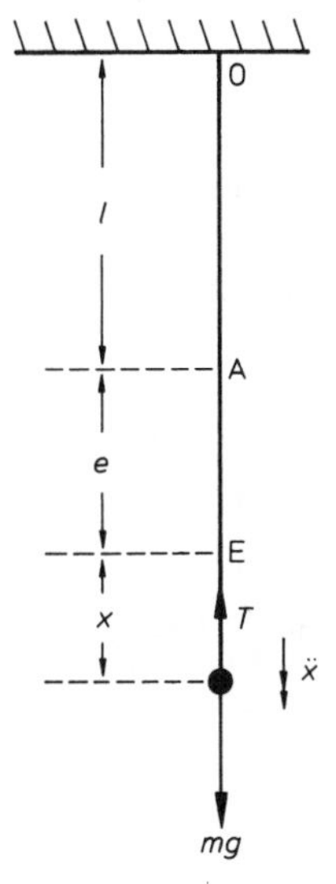

By Hooke's law, $T = \frac{\lambda(x+e)}{l}$

where T is the tension in the string. Applying Newton's second law vertically downwards:

$$mg - T = m\ddot{x}$$

$$\therefore \quad \frac{\lambda e}{l} - \frac{\lambda(x+e)}{l} = m\ddot{x}$$

$$\therefore \quad -\frac{\lambda x}{l} = m\ddot{x}$$

Hence the equation of motion of the particle is $\ddot{x} = -\frac{\lambda}{lm}x$. Since this equation is of the form $\ddot{x} = -\omega^2 x$, where $\omega = \sqrt{(\lambda/lm)}$, the particle can perform simple harmonic motion with centre E and period $2\pi\sqrt{(lm/\lambda)}$. The amplitude a of the motion is the maximum distance below E reached by the particle.

Provided that $a < e$ the string will remain taut throughout the motion and complete simple harmonic oscillations will be possible. However, if $a > e$ the particle will perform simple harmonic motion when it is below A, but move freely under gravity when the string becomes slack.

[Note that in the case of a particle suspended from a spring of natural length l, complete S.H.M. involving both extension and compression of the spring is theoretically possible when $a > e$.]

Example 1 A light elastic string of natural length l and modulus mg has one end fastened to a fixed point O. The other end of the string is attached to a particle of mass m which hangs in equilibrium at a point E. Find the distance OE. The particle is now pulled down to a point A at a distance $5l/2$ vertically below O. If at time $t = 0$ the particle is released from rest at A, show that the subsequent motion is simple harmonic and find the speed of the particle as it passes through E. Find also the time at which the particle first passes through the point B which lies at a distance $7l/4$ vertically below O.

When the particle is at E let the tension in the string be T_0 and the extension e.

By Hooke's law: $T_0 = \dfrac{mg \times e}{l}$

But since the particle is in equilibrium, $T_0 = mg$

$$\therefore \quad \frac{mg \times e}{l} = mg$$

$$e = l$$

Hence $OE = l + e = 2l$.

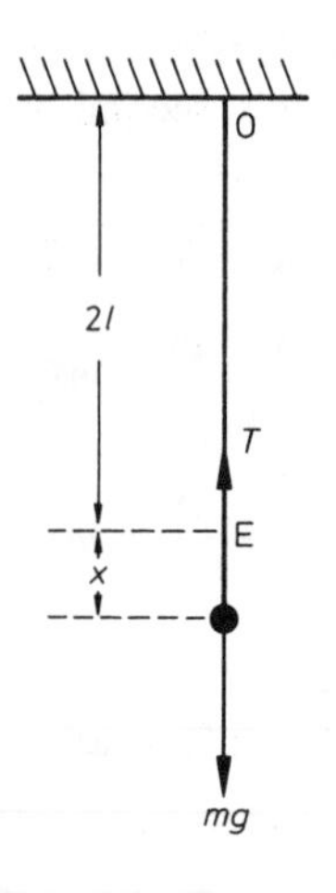

Let T be the tension in the string when the displacement of the particle from E is x vertically downwards.

By Hooke's law: $T = \dfrac{mg(x+l)}{l}$

Applying Newton's second law vertically downwards:

$$mg - T = m\ddot{x}$$

$$\therefore \quad mg - \frac{mg(x+l)}{l} = m\ddot{x}$$

$$\therefore \quad \ddot{x} = -\frac{g}{l}x$$

Since this equation is of the form $\ddot{x} = -\omega^2 x$, where $\omega = \sqrt{(g/l)}$, the particle performs simple harmonic motion with centre E.
The speed v of the particle is given by the formula

$$v^2 = \omega^2(a^2 - x^2)$$

When the particle is at A, $x = \frac{1}{2}l$ and $v = 0$ $\quad \therefore \quad a = \frac{1}{2}l$

Hence $\qquad v^2 = \dfrac{g}{l}(\frac{1}{4}l^2 - x^2)$.

$\therefore$ when $x = 0$, $v^2 = \dfrac{g}{l} \times \dfrac{l^2}{4} = \dfrac{gl}{4}$

Hence as the particle passes through E its speed is $\frac{1}{2}\sqrt{(gl)}$.

Since $x = \frac{1}{2}l$ and $v = 0$ when $t = 0$,

$$x = \tfrac{1}{2}l\cos\omega t, \quad \text{where} \quad \omega = \sqrt{(g/l)}.$$

When $x = -\frac{1}{4}l$, $\quad -\frac{1}{4}l = \frac{1}{2}l\cos\omega t$

i.e. $\quad \cos\omega t = -\frac{1}{2}$

$\therefore$ the particle first passes through B when $\omega t = 2\pi/3$

i.e. when $t = \dfrac{2\pi}{3}\sqrt{\left(\dfrac{l}{g}\right)}$.

We include the following example as a reminder that many problems involving particles attached to elastic strings can be solved using energy considerations rather than the theory of simple harmonic motion.

Example 2 An elastic string of natural length l and modulus $4mg$ has one end fastened to a fixed point O. The other end of the string is attached to a particle of mass m. If the particle is released from rest at O, find the distance it falls before coming instantaneously to rest at a point A.

Let x be the extension in the string when the particle reaches A. The kinetic energy of the particle is zero when it is released from O and when it comes to rest at A. Since $OA = l + x$, the loss in gravitational potential energy is $mg(l + x)$.
The energy stored in the elastic string as it is stretched to the length $(l + x)$ is $4mgx^2/2l$ i.e. $2mgx^2/l$.
Thus, using the principle of conservation of mechanical energy,

$$\frac{2mgx^2}{l} = mg(l + x)$$

$$\therefore \quad 2x^2 = l(l + x)$$

$$2x^2 - lx - l^2 = 0$$

$$(x - l)(2x + l) = 0$$

Since $x > 0$, we must have $x = l$.
Hence the particle falls a distance $2l$ before coming to rest.

In the next example we consider oscillations in a horizontal plane.

Example 3 A particle P of mass m lies on a smooth horizontal table and is attached to two fixed points A, B on the table by two elastic strings each of natural length l. The strings AP, PB have moduli $2mg$ and mg respectively and $AB = 5l$. If E is the equilibrium position of the particle, find AE. Given that the particle is released from rest at the point C on AB such that $AC = l$, show that it performs simple harmonic motion, stating the period and the amplitude of the oscillations.

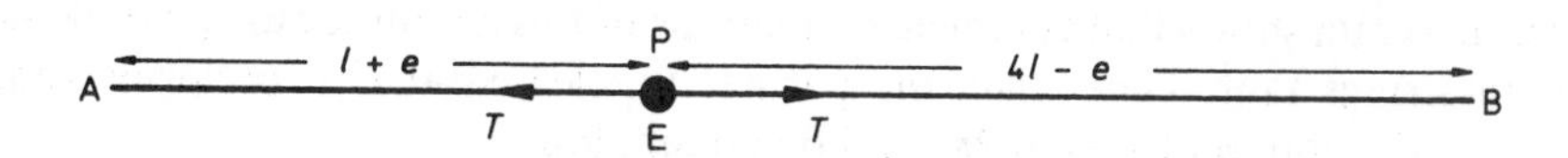

Let T be the tension in both strings when the particle is in equilibrium at E.
If $AE = l + e$, then the extensions in the strings AP and PB are e and $3l - e$ respectively.

By Hooke's law for AP: $T = \dfrac{2mge}{l}$

By Hooke's law for BP: $T = \dfrac{mg(3l - e)}{l}$

$$\therefore \quad \frac{2mge}{l} = \frac{mg(3l - e)}{l}$$

$$2e = 3l - e$$

$$\therefore \quad e = l$$

Hence $AE = l + e = 2l$.

Suppose now that the displacement of the particle P from E is x, as shown in the diagram.
Assuming that both strings are taut, i.e. that P lies between the points C and D, the extension in AP is $(l + x)$ and the extension in PB is $(2l - x)$.
Let T_A and T_B be the tensions in the strings AP and PB respectively, then applying Hooke's law,

$$T_A = \frac{2mg(l + x)}{l}, \quad T_B = \frac{mg(2l - x)}{l}$$

Using Newton's second law in the direction AB,

$$T_B - T_A = m\ddot{x}$$

$$\therefore \quad \frac{mg(2l - x)}{l} - \frac{2mg(l + x)}{l} = m\ddot{x}$$

$$-\frac{3mgx}{l} = m\ddot{x}$$

$$\therefore \quad \ddot{x} = -\frac{3g}{l}x$$

Thus the equation of motion of the particle is of the form $\ddot{x} = -\omega^2 x$, where $\omega = \sqrt{(3g/l)}$. Hence when both strings are taut the particle performs simple harmonic motion with centre E and period $2\pi\sqrt{(l/3g)}$.
Since the particle starts from rest at C, where $CE = l$, the amplitude of the simple harmonic motion is l.
The diagram shows that complete oscillations of this amplitude are possible with both strings taut. Thus when the particle is released at C it performs simple harmonic motion of period $2\pi\sqrt{(l/3g)}$ and amplitude l.

These examples illustrate the fact that in simple harmonic motion the centre of oscillation is an equilibrium position. If a displacement x from this equilibrium position is considered, then the equation of motion is obtained in the standard form

$$\ddot{x} = -\omega^2 x.$$

However, if a displacement from a point which is not an equilibrium position is used, then the simple harmonic motion equation takes the more general form

$$\ddot{x} + \omega^2 x = \text{constant}.$$

Solutions to this type of differential equation were discussed in §11.5.

Finally, we show that the motion of a *simple pendulum* is approximately simple harmonic. A simple pendulum consists of a particle, known as the "bob", suspended from a fixed point by a light inextensible string. When the particle swings in a vertical plane with the string taut, the bob moves along a circular arc.

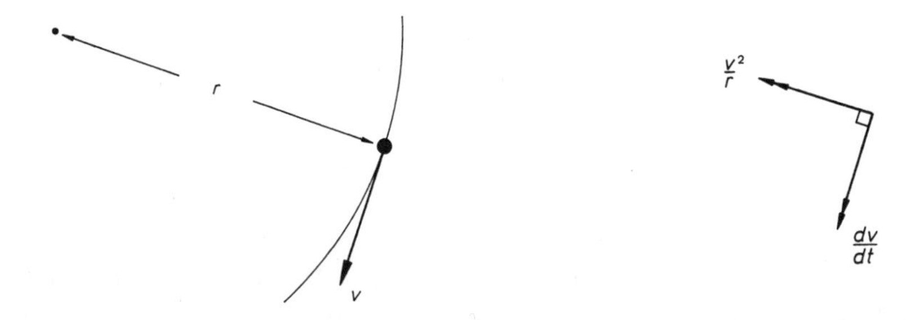

It was shown in Book 1 that when a particle moves in a circle with variable speed v, then the acceleration of the particle has components v^2/r towards the centre of the circle and dv/dt in the direction of motion. These results can now be used to find the equation of motion of a simple pendulum.

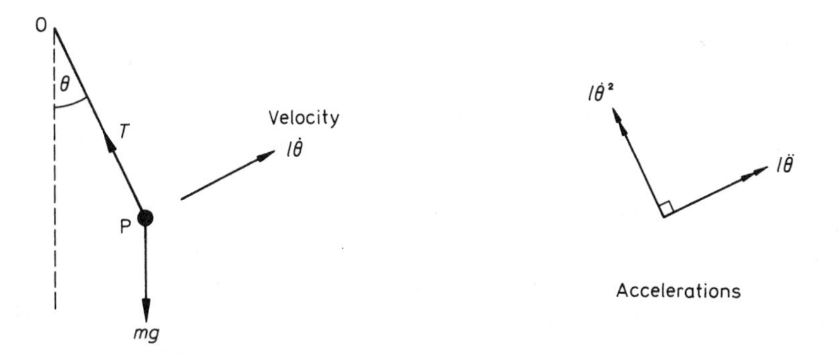

Let the bob of the pendulum be a particle P of mass m and let the length of the string be l. At the instant when the string makes an angle θ with the downward vertical, P is moving perpendicular to the string with velocity $l\dot{\theta}$. Hence the acceleration of P has components $l\dot{\theta}^2$ and $l\ddot{\theta}$ along and perpendicular to the string.

Applying Newton's second law perpendicular to the string,

$$mg \sin \theta = -ml\ddot{\theta}$$

Hence $$\ddot{\theta} = -\frac{g}{l} \sin \theta$$

However, if θ is small, $\sin \theta \approx \theta$

$\therefore$ $$\ddot{\theta} \approx -\frac{g}{l} \theta$$

Thus small oscillations of a simple pendulum are approximately simple harmonic.

The period of complete oscillations is $2\pi \sqrt{\frac{l}{g}}$.

A pendulum which beats seconds is called a *seconds pendulum.* Such a pendulum swings through its equilibrium position once every second and therefore each complete oscillation takes 2 seconds. Hence the length of a seconds pendulum is found by writing:

$$2\pi \sqrt{\frac{l}{g}} = 2.$$

Exercise 12.4

1. A particle of mass m is attached to one end of a light elastic string of length a and modulus $\frac{3}{2}mg$. The other end of the string is attached to a fixed point O and the particle hangs in equilibrium under gravity at the point E. Find the distance OE. If the particle is a *further* distance x below E show that the resultant force acting on the particle is proportional to x. The particle is pulled down to the point at a distance a below E and released from rest. Show that, in the subsequent motion and while the string is taut, the particle executes Simple Harmonic Motion and that its distance below E at time t after being released is

$$a \cos \left\{ \left(\frac{3g}{2a} \right)^{1/2} t \right\}. \qquad \text{(O\&C)}$$

2. A light elastic string of natural length l has one end fastened to a fixed point O. The other end of the string is attached to a particle of mass m. When the particle hangs in equilibrium the length of the string is $\frac{7}{4}l$. The particle is displaced from equilibrium so that it moves vertically with the string taut. Show that the motion is simple harmonic with period $\pi\sqrt{(3l/g)}$.

At time $t = 0$ the particle is released from rest at a point A at a distance $\frac{3}{2}l$ vertically below O. Find (i) the depth below O of the lowest point L of the motion, (ii) the time taken to move from A to L, (iii) the depth below O of the particle at time $t = \frac{1}{3}\pi \sqrt{\frac{3l}{g}}$. (C)

3. A particle of mass m is attached to one end of a light elastic string of natural length l and modulus $2mg$. The other end of the string is fixed at a point A. The

particle rests on a support B vertically below A, with $AB = 5l/4$. Find the tension in the string and the reaction exerted on the particle by the support B. The support B is suddenly removed. Show that the particle will execute simple harmonic motion and find (i) the depth below A of the centre of oscillation, (ii) the period of the motion. (JMB)

4. A light elastic spring AB of natural length l has the end A attached to a fixed point and hangs vertically under gravity with a particle of mass m attached at B. The system is set in motion and performs small vertical oscillations. Show that the motion is simple harmonic. The particle of mass m at B is replaced by a particle of mass αm, and it is found that the period of oscillation is doubled. Find the value of α. (C)

5. One end of an elastic string of modulus mg and natural length a is attached to a fixed point O. To the other end A are attached two particles P and Q, P having mass $2m$ and Q having mass m. The particles hang down in equilibrium under gravity. If Q falls off, show that P subsequently performs simple harmonic motion and state the period and amplitude of this motion. If on the other hand P falls off, find the distance from O of the highest point reached by Q. (O&C)

6. An elastic string of natural length $2a$ and modulus λ has its ends attached to two points A, B on a smooth horizontal table. The distance AB is $4a$ and C is the mid-point of AB. A particle of mass m is attached to the mid-point of the string. The particle is released from rest at D, the mid-point of CB. Denoting by x the displacement of the particle from C, show that the equation of motion of the particle is

$$\frac{d^2x}{dt^2} + \frac{2\lambda}{ma}x = 0.$$

Find the maximum speed of the particle and show that the time taken for the particle to move from D directly to the mid-point of CD is $\dfrac{\pi}{3}\left(\dfrac{ma}{2\lambda}\right)^{1/2}$. (JMB)

7. A particle P of mass m is attached by two light elastic strings to two fixed points A and B on a smooth horizontal table. Each string is of natural length l and $AB = 4l$. If the strings AP and PB have moduli of elasticity λ and 3λ respectively, find the length of AP when the particle is in equilibrium. If the particle is given a small displacement from the equilibrium position along the line AB, show that while both strings are taut the particle executes simple harmonic motion. Find the period of the oscillations. Given that the string PB just becomes slack in the ensuing motion, find the maximum speed of the particle.

8. Two fixed points A and B on a smooth horizontal table are at a distance $5l$ apart. A particle of mass m lies between A and B. It is attached to A by means of a light elastic string of modulus mg and natural length l and to B by means of a light elastic string of modulus $2mg$ and natural length $2l$. If O is the point at which the particle would rest in equilibrium, find the distance OA. The particle is projected towards O

from the mid-point of AB with speed u. Show that while both strings remain taut the particle performs simple harmonic motion. Find the period and the amplitude of this motion. Show that the particle will perform complete simple harmonic oscillations if $2u^2 \leqslant 3gl$.

9. A particle of mass m moving on a smooth horizontal plane is attached to a point O of the plane by an elastic string of natural length l and modulus λ. Initially the particle is at the point A which lies at a distance l from O and is moving away from O at speed u. If the particle first comes to rest at the point B, use energy considerations to find the distance AB. Show that when the particle is moving in the line AB its equation of motion is of the form $\ddot{x} = -\omega^2 x$, where x is its displacement from A. Find the time taken by the particle to move directly from A to B. Find also the further time which elapses before the particle next comes to rest.

10. A particle of mass m is attached to two elastic strings each of natural length a. The first string has modulus $3mg$ and its other end is attached to a fixed point A. The second string has modulus $6mg$ and its other end is attached to a fixed point B which is at a distance $3a$ vertically below A. Show that the particle can rest in equilibrium at a point O between A and B with both strings taut. Find the distance y of O from A. The particle is in motion in the line AB with both strings taut. At time t, the particle is at P where $OP = x$. Prove that $\dfrac{d^2x}{dt^2} = -\dfrac{9g}{a}x$.

The particle is projected vertically upwards from the point at a distance $2a$ below A with speed u. Find the maximum value of u for which the upper string does not become slack in the subsequent motion. For this maximum value of u find the time taken for the particle to travel from O to its highest point. (O&C)

11. Find, to the nearest millimetre, the length of a seconds pendulum, (a) at Greenwich, where $g \approx 9{\cdot}81$, (b) at the equator, where $g \approx 9{\cdot}78$, (c) at the North Pole, where $g \approx 9{\cdot}83$, (d) on the surface of the moon, where $g \approx 1{\cdot}62$.

12. A musician wishes to use a simple pendulum as a makeshift metronome. Regarding a complete oscillation as two "beats", find, to the nearest centimetre, the length of pendulum required to produce (a) 50 beats per minute, (b) 80 beats per minute, (c) 100 beats per minute, (d) 160 beats per minute. [Take g as $9{\cdot}8\,\text{m}\,\text{s}^{-2}$.]

12.5 Motion in a vertical circle

It was established in Book 1 that the acceleration of a particle moving in a circle, radius r, with variable speed v has components v^2/r towards the centre of the circle and dv/dt in the direction of motion.

When dealing with motion in a vertical circle we can obtain equations of motion by applying Newton's second law in these two directions. However, it may sometimes be more convenient to use an energy equation.

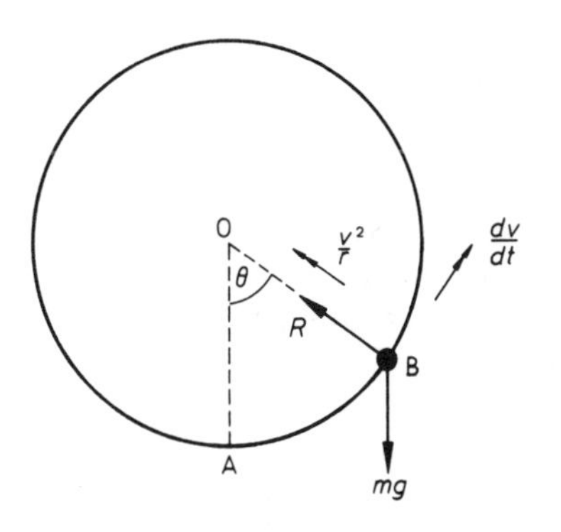

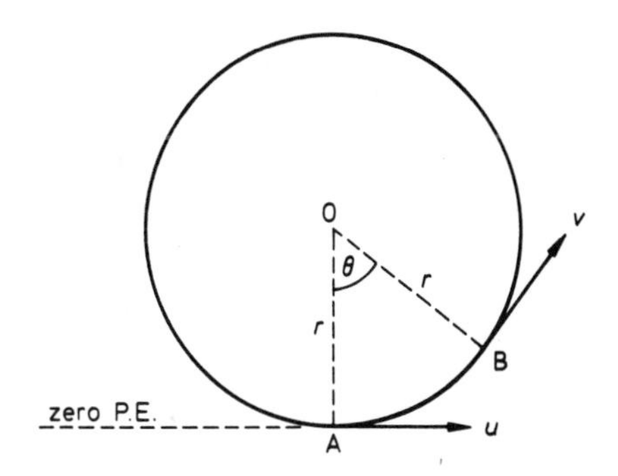

We consider first a small bead of mass m sliding on a smooth circular wire, of centre O and radius r, fixed in a vertical plane. Let u be the speed of the bead as it passes through the lowest point A of the wire and let v be the speed of the bead when it is at a point B such that $\angle AOB = \theta$. Since the wire is smooth the reaction R of the wire on the bead is perpendicular to the wire. Using Newton's second law radially and tangentially,

$$\nwarrow \qquad R - mg\cos\theta = m\frac{v^2}{r}$$

$$\nearrow \qquad mg\sin\theta = -m\frac{dv}{dt}$$

The reaction R is perpendicular to the direction of motion and therefore does no work. This means that we can apply the principle of conservation of mechanical energy:

$$(\text{K.E.} + \text{P.E.}) \text{ at } A = (\text{K.E.} + \text{P.E.}) \text{ at } B$$

$$\therefore \qquad \tfrac{1}{2}mu^2 = \tfrac{1}{2}mv^2 + mgr(1 - \cos\theta) \qquad (1)$$

As θ increases the speed of the bead decreases. If the bead comes to rest before it reaches the highest point of the wire it will slide back down the wire and perform oscillations about A. However, if the speed of the bead is greater than zero for all values of θ, it will execute complete circles.

From (1): $\qquad v^2 = u^2 - 2gr(1 - \cos\theta)$

$\therefore$ if $\qquad v^2 > 0$ for all θ

then $\qquad u^2 > 2gr(1 - \cos\theta)$ for all θ

Since the greatest value taken by $(1 - \cos\theta)$ is 2, the condition for performing complete circles is

$$u^2 > 4gr.$$

Example 1 A small ring of mass 0·02 kg is threaded on a smooth circular wire of centre O fixed in a vertical plane. The ring is slightly displaced from rest at the highest point A of the wire. Find the magnitude and direction of the reaction of the wire on the ring when the radius to the ring makes an angle of 45° with (a) the upward vertical, (b) the downward vertical.

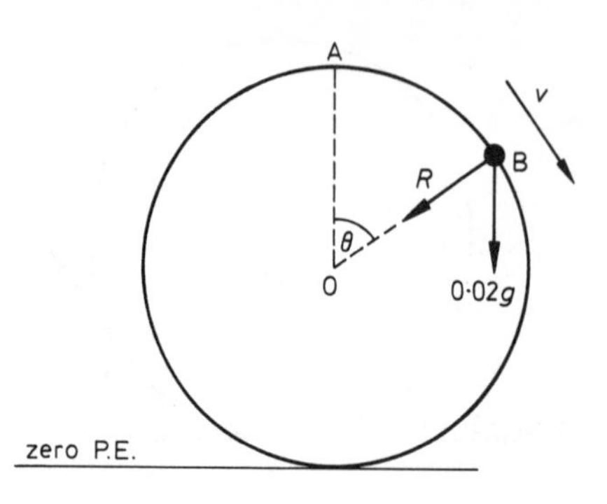

Let v m s^{-1} be the speed of the ring when it is at a point B such that $\angle AOB = \theta$. Let the reaction of the wire on the ring be R N towards O and let the radius of the wire be r m.

Using the principle of conservation of mechanical energy,

$$(\text{K.E.} + \text{P.E.}) \text{ at } A = (\text{K.E.} + \text{P.E.}) \text{ at } B$$

$$\therefore \quad 0 + 0{\cdot}02g \times 2r = \tfrac{1}{2} \times 0{\cdot}02 \times v^2 + 0{\cdot}02g \times r(1 + \cos\theta)$$

$$\therefore \quad 4gr = v^2 + 2gr(1 + \cos\theta)$$

Thus $$v^2 = 2gr(1 - \cos\theta).$$

Applying Newton's second law towards O,

$$R + 0{\cdot}02g\cos\theta = 0{\cdot}02 \times \frac{v^2}{r}$$

$$\therefore \quad R + 0{\cdot}02g\cos\theta = 0{\cdot}02 \times 2g(1 - \cos\theta)$$

Thus $$R = 0{\cdot}02g(2 - 3\cos\theta).$$

(a) When $\theta = 45°$, $R = 0{\cdot}02 \times 9{\cdot}8 \times (2 - 3\cos 45°) \approx -0{\cdot}024$.
Hence when the radius to the ring makes an angle of 45° with the upward vertical the reaction of the wire is 0·024 N *away from O*.

(b) When $\theta = 135°$, $R = 0{\cdot}02 \times 9{\cdot}8 \times (2 - 3\cos 135°) \approx 0{\cdot}808$.
Hence when the radius to the ring makes an angle of 45° with the downward vertical the reaction of the wire is 0·808 N *towards O*.

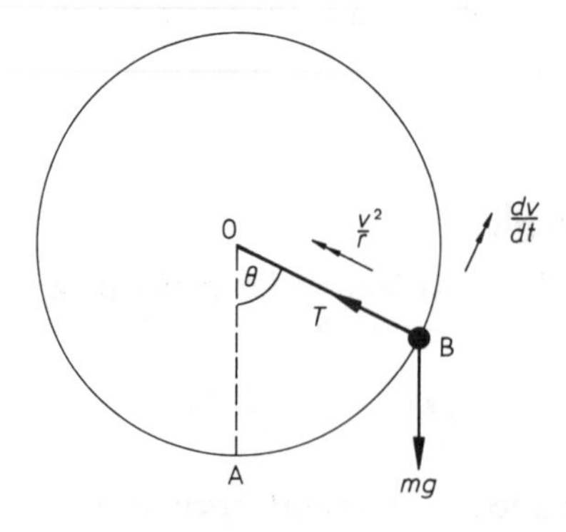

Suppose now that a particle of mass m, suspended from a fixed point O by a light inextensible string of length r, moves in a vertical circle. Let u be the speed of the particle as it passes through its lowest point A and let v be the speed of the particle when it is at a point B such that $\angle AOB = \theta$.

Applying Newton's second law towards O,

$$T - mg\cos\theta = m\frac{v^2}{r} \qquad (1)$$

As in the case of a bead sliding on a circular wire, it can also be shown that

$$v^2 = u^2 - 2gr(1 - \cos\theta) \tag{2}$$

$$\therefore \quad T = \frac{m}{r}\{u^2 - 2gr(1 - \cos\theta)\} + mg\cos\theta$$

i.e.

$$T = \frac{mu^2}{r} - 2mg + 3mg\cos\theta \tag{3}$$

Thus when the particle passes through A its speed v and the tension T in the string take their maximum values. As θ increases, both v and T decrease, but the exact nature of the subsequent motion depends on the magnitude of u. We now consider in more detail the following possibilities:

(i) v reaches zero before T,
(ii) T reaches zero before v,
(iii) both v and T are positive for all values of θ.

Case (i): The particle comes to rest with the string taut.

Substituting $v = 0$ in (1) and (2), then rearranging:

$$T = mg\cos\theta \tag{4}$$
$$2gr - u^2 = 2gr\cos\theta \tag{5}$$

Using (4):

$$T > 0 \quad \text{when} \quad v = 0$$
$$\Leftrightarrow \quad \cos\theta > 0 \quad \text{when} \quad v = 0$$

Hence the particle comes to rest at a point in the lower half of the circle.

Using (5):

$$\cos\theta > 0 \quad \text{when} \quad v = 0$$
$$\Leftrightarrow \quad 2gr - u^2 > 0$$
$$\Leftrightarrow \quad 2gr > u^2$$

Hence the particle performs oscillations below the level of O with the string taut if and only if $u^2 < 2gr$.

Case (ii): The string becomes slack while the particle is moving.

Substituting $T = 0$ in (1) and (3), then rearranging:

$$v^2 = -gr\cos\theta \tag{6}$$
$$2gr - u^2 = 3gr\cos\theta \tag{7}$$

Using (6):

$$v^2 > 0 \quad \text{when} \quad T = 0$$
$$\Leftrightarrow \quad -1 < \cos\theta < 0 \quad \text{when} \quad T = 0$$

Hence the string can become slack only at a point in the upper half of the circle.

Using (7):

$$-1 < \cos\theta < 0 \quad \text{when} \quad T = 0$$
$$\Leftrightarrow \quad -3gr < 3gr\cos\theta < 0 \quad \text{when} \quad T = 0$$
$$\Leftrightarrow \quad -3gr < 2gr - u^2 < 0$$
$$\Leftrightarrow \quad 0 < u^2 - 2gr < 3gr$$
$$\Leftrightarrow \quad 2gr < u^2 < 5gr$$

Hence the string becomes slack at a point above the level of O if and only if $2gr < u^2 < 5gr$. The particle will then leave its circular path and move freely under gravity until the string is again taut.

Case (iii): The particle performs complete circles

Using (2): $\quad v > 0$ for all θ

$\Leftrightarrow \quad u^2 > gr(2 - 2\cos\theta)$ for all θ

Using (3): $\quad T > 0$ for all θ

$\Leftrightarrow \quad u^2 > gr(2 - 3\cos\theta)$ for all θ

Since $\cos\theta$ takes values from -1 to $+1$, both conditions are satisfied if and only if $u^2 > 5gr$. Hence the particle performs complete circles if and only if $u^2 > 5gr$.

Lastly we illustrate, by means of a worked example, a further way in which motion in a vertical circle may arise.

Example 2 A particle of mass m is released from rest at a point A on the outer surface of a smooth fixed sphere of centre O and radius r. Given that OA makes an angle α with the upward vertical, find an expression for the speed at which the particle is travelling when it leaves the surface of the sphere.

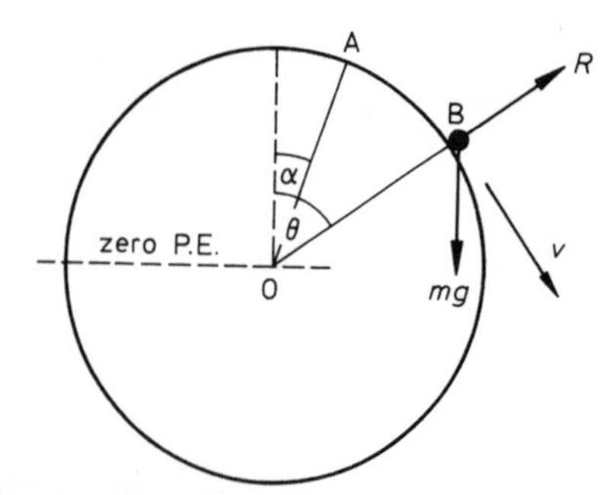

Let v be the speed of the particle when it reaches a point B such that OB makes an angle θ with the upward vertical. Since the sphere is smooth the reaction R of the sphere on the particle acts in the direction OB.

Since the reaction does no work mechanical energy is conserved during the motion,

$\therefore \quad$ (K.E. + P.E.) at A = (K.E. + P.E.) at B

$\therefore \quad 0 + mg \times r\cos\alpha = \frac{1}{2}mv^2 + mg \times r\cos\theta.$

i.e. $$2gr\cos\alpha = v^2 + 2gr\cos\theta \qquad (1)$$

Applying Newton's second law towards O,

$$mg\cos\theta - R = m\frac{v^2}{r} \qquad (2)$$

The particle will leave the surface of the sphere when R vanishes.
Putting $R = 0$ in equation (2) gives $gr\cos\theta = v^2$

$\therefore \quad$ from (1): $2gr\cos\alpha = v^2 + 2v^2$

i.e. $\quad v^2 = \frac{2}{3}gr\cos\alpha$

Hence the speed at which the particle leaves the surface is $\sqrt{(\frac{2}{3}gr\cos\alpha)}$.

Exercise 12.5

1. A particle P of mass 2 kg is suspended from a fixed point O by a light inextensible string of length 2·5 m and hangs in equilibrium at a point A. The particle is projected from A with horizontal speed $5\ \text{m s}^{-1}$. Given that when $\angle AOP = \theta$, the speed of the particle is $v\ \text{m s}^{-1}$ and the tension in the string is T N, find expressions for v^2 and T in terms of θ. Hence find the values of v and T when $\theta = 40°$. Find also to the nearest degree, the greatest value attained by θ in the subsequent motion.

2. A small bead of mass 0·5 kg moves on a smooth circular wire of radius 2 m which is fixed in a vertical plane. When the radius to the bead makes an angle of 60° with the downward vertical the reaction between the bead and the wire is of magnitude 12 N. Find the speed of the bead at this instant and the magnitude of the reaction when the bead passes through the lowest point of the wire.

3. A particle of mass m is suspended from a fixed point O by a light inextensible string of length a. When the particle is in its lowest position it is given a horizontal speed of $\sqrt{(3ag)}$. If v is the speed of the particle when the string makes an angle θ with the upward vertical, find $\cos\theta$ and an expression for v^2 at the instant when the string becomes slack.

4. A particle of mass m is placed inside a smooth hollow sphere of radius a. The particle is set in motion at the lowest point of the sphere with speed $\sqrt{(\lambda ag)}$. Show that the particle will perform complete circles if $\lambda > 5$.

5. A small ring P of mass m is threaded on a smooth circular wire, of centre O and radius r, fixed in a vertical plane. Initially the ring is slightly displaced from the highest point of the wire. Find the magnitude of the force exerted by the ring on the wire when OP makes an angle θ with the upward vertical. Find the range of values of $\cos\theta$ for which this force acts towards O.

6. A particle of mass m slides down the smooth outside surface of a fixed sphere of radius a. At the top of the sphere its velocity is horizontal and of magnitude u. If θ is the angle that the radius to the particle makes with the upward vertical, show that the reaction between the particle and the sphere is

$$mg(3\cos\theta - 2) - mu^2/a.$$

Show that the particle, when just displaced from rest at the top, leaves the surface when $\theta = \cos^{-1}\frac{2}{3}$, and find its speed at that instant. (W)

7. A light rod OP of length a is free to rotate in a vertical plane about a fixed point O, and a mass m is attached to it at P. Denote by θ the angle between OP and the downward vertical. When $\theta = 0$, P is moving horizontally with speed u. Find the tension in the rod as a function of θ, and show that the value of the function is positive throughout the motion if $u^2 > 5ag$. Find the least value of the tension when $u^2 = 6ag$. Show that, if $u^2 < 2ag$, P remains below the level of O and deduce that, in this case also, the value obtained for the tension is positive throughout the motion. (JMB)

8. A particle P is projected horizontally with speed u from the lowest point A of the smooth inside surface of a fixed hollow sphere of internal radius a.
(i) In the case when $u^2 = ga$ show that P does not leave the surface of the sphere. Show also that, when P has moved halfway along its path from A towards the point at which it first comes to rest, its speed is $\sqrt{\{ga(\sqrt{3} - 1)\}}$.
(ii) Find u^2 in terms of ga in the case when P leaves the surface at a height $3a/2$ above A, and find, in terms of a and g, the speed of P as it leaves the surface. (JMB)

9. A particle of mass m, attached to a fixed point O by an inelastic string of length l, is allowed to fall from a point in the horizontal through O at a distance $l \cos \alpha$ from O. Find the speed of the particle when it first begins to move in a circle. In the subsequent motion, show that, when the string first makes an angle θ $(< \frac{1}{2}\pi - \alpha)$ with the downward vertical, the speed v of the particle is $\sqrt{\{2gl(\cos\theta - \sin^3\alpha)\}}$ and the tension in the string is $3mv^2/2l + mg\sin^3\alpha$. Will the tension vanish during the subsequent motion? (W)

10. One end of a light inextensible string of length a is attached to a particle of mass m. The other end is attached to a fixed point O which is at a height $5a/2$ above the horizontal ground. Initially the string is taut and horizontal. The particle is then projected vertically downward with velocity $(2ag)^{\frac{1}{2}}$. When the string has turned through an angle θ $(< \pi)$ find the velocity v of the particle and show that the tension in the string is $mg(2 + 3\sin\theta)$.
(a) If the string can withstand a tension of at least $5mg$, prove that the string will not break.
(b) If the string can withstand a tension of at most $7mg/2$, find the values of θ and v when the string breaks. In this case find the time to reach the vertical through O. Hence show that the particle strikes the ground at the point vertically below O. (O&C)

11. Two small smooth pegs O and C are fixed at the same level and $OC = a$. A light inextensible string of length $3a$ has one end attached to O and a particle P of mass m hanging at its other end vertically below O. The particle is projected with speed u parallel to OC. Find the speed of P when OP first makes an acute angle θ with the downward vertical and show that the tension in the string is then $mg(3\cos\theta - 2) + mu^2/3a$. Show that the string will reach a horizontal position if $u^2 \geqslant 6ga$. Given that $u^2 > 6ga$, find the tension when, after the string has struck the peg C, the moving portion CP makes an angle ϕ above the horizontal, and show that P will complete a semicircle with centre C if $u^2 \geqslant 12ga$. (JMB)

12.

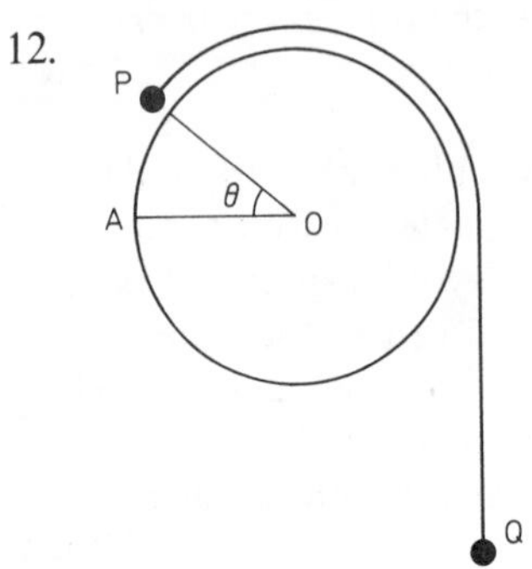

A particle P of mass m is at rest at a point A on the smooth outer surface of a fixed sphere of centre O and radius a, OA being horizontal. The particle is attached to one end of a light inextensible string which is taut and passes over the sphere, its other end carrying a particle Q of mass $2m$ which hangs freely. The string lies in the vertical plane containing OA. The system is released from rest, and after time t the angle POA is θ, as shown in the diagram. Show that, while P remains in contact with the sphere,

$$3a\left(\frac{d\theta}{dt}\right)^2 = 2g(2\theta - \sin\theta).$$

Find the tension in the string and the reaction of the sphere on P in terms of m, g and θ. (JMB)

Exercise 12.6 (*miscellaneous*)

1. A particle is in motion along a straight line. When at a distance x from a fixed point of the line its velocity is v. Show that its acceleration is $v\,dv/dx$.

The gravitational force per unit mass at a distance $x(> a)$ from the earth's centre is ga^2/x^2, where a is the radius of the earth, and g is constant. A particle is projected vertically upwards from the earth's surface at A with speed u. Show that if $u^2 < 2ag$, the particle rises to a maximum height $au^2/(2ag - u^2)$ above A. What happens if $u^2 > 2ag$? (O&C)

2. A man pushes a loaded cart of total mass m along a straight horizontal path against a constant resisting force λm. The cart starts from rest, and the man exerts a force, in the direction of the motion, of magnitude $Ke^{-\alpha x}$, where x is the distance moved, α is a positive constant and K is a constant greater than λm. Write down a differential equation, involving x and the speed v, for the motion of the cart (while it continues to move), and hence obtain v in terms of x and the given constants. (JMB)

3. A particle moves in a straight line in such a manner that when the particle is at the point P, at a distance x from some fixed point O on the line, the only force acting on the particle is in the direction **OP** and has magnitude $m\mu\left\{\frac{a}{x^3} - \frac{1}{x^2}\right\}$, where m is the mass of the particle and μ a positive constant.

If the particle is projected from the point $x = a$ with speed V in the direction **OP** find the set of values to which V must belong if the particle reaches the point $x = 2a$ but does not reach the point $x = 4a$. (C)

4. A particle moves along the x-axis so that its displacement x from the origin O at time t satisfies the differential equation $\frac{d^2x}{dt^2} + 9x - 18 = 0$. Its speed when $t = \pi/12$ is $21/\sqrt{2}$ in the direction of the positive x-axis and its speed when $t = \pi/6$ is 15 in the *opposite* direction. Find x at any time t. What is the greatest distance from O reached by the particle and what is its greatest speed? If the motion starts at $t = 0$, find the time at which the particle first passes through O. [The general solution of the differential equation may be quoted.] (O&C)

5. A particle is projected vertically upwards under (constant) gravity g from a point A on the earth's surface with an initial speed k. At time t after projection, the

particle is at height x above A and its speed is v; the air resistance is gv/k per unit mass. Show that, before the particle reaches its highest point, $\frac{dv}{dt} = -\frac{g}{k}(v + k)$, and $v = k(2e^{-gt/k} - 1)$. Hence find the corresponding value of x as a function of t.
(O&C)

6. A bullet of mass 40 g is fired horizontally with velocity 800 m/s into a fixed block of material which exerts a constant resistance, R newtons, which stops it in 0·5 m. Show that $R = 2{\cdot}56 \times 10^4$ and find the time taken. An identical bullet is fired at the same velocity into a second fixed block which offers a resistance which varies as the square root of the speed of the bullet. The initial value of this resistance is $2{\cdot}56 \times 10^4$ N. Write down the equation of motion for the bullet and find the time taken to stop it. Explain why the bullet could never be stopped by a material which offered a resistance proportional to the velocity. (AEB 1979)

7. With the usual notation, prove that $\frac{dv}{dt} = v\frac{dv}{ds}$. A particle P of mass m moves in a straight line and starts from a point O with velocity u. When $OP = x$, where $x \geqslant 0$, the velocity v of P is given by $v = u + \frac{x}{T}$, where T is a positive constant. Show that, at any instance, the force acting on P is proportional to v. Given that the velocity of P at the point A is $3u$, calculate (i) the distance OA in terms of u and T, (ii) the time, in terms of T, taken by P to move from O to A, (iii) the work done, in terms of m and u, in moving P from O to A. (AEB 1980)

8. A point P is travelling in the positive direction on the x-axis with acceleration proportional to the square of its speed v. At time $t = 0$ it passes through the origin with speed gT and with acceleration g. Show that $dv/dx = v/(gT^2)$, and hence obtain an expression for v in terms of x, g and T. Prove that, at time t,

$$x = gT^2 \ln\left(\frac{T}{T-t}\right).$$

Sketch the graph of x against t for $0 \leqslant t < T$. (L)

9. A particle is projected vertically upwards with speed u from a point O. The air resistance is proportional to the fourth power of the velocity and is initially equal to the weight of the particle. If v is the speed of the particle when it is at a height z above O, show that $\frac{d}{dz}(\tfrac{1}{2}v^2) + g\left(1 + \frac{v^4}{u^4}\right) = 0.$
Solve this equation by writing w for v^2 and so prove that the particle is instantaneously at rest when its height above O is $\pi u^2/8g$. (W)

10. A body of mass m falls from rest under gravity through a resisting medium. The resistance is kv^2 per unit mass where k is constant and v is the velocity when the body has fallen a distance x.

(i) Given that u is the limiting value of the velocity, i.e. the velocity for which there would be zero acceleration, establish the differential equation $v\dfrac{dv}{dx} = k(u^2 - v^2)$.

(ii) By solving this differential equation obtain an expression for v in terms of x. (JMB)

11. A locomotive of mass M works at a constant rate H. Its motion is resisted by a force Mkv, where v is the speed at time t and k is constant. If $c^2 = H/(Mk)$, show that the equation of motion of the locomotive can be written $v\dfrac{dv}{dt} = k(c^2 - v^2)$.
Prove that the locomotive will increase its speed from $\frac{1}{3}c$ to $\frac{1}{2}c$ in time $\{1/(2k)\}\ln(32/27)$, and in this interval of time will travel a distance $\dfrac{c}{2k}\ln(1{\cdot}5) - \dfrac{c}{6k}$. (O&C)

12. A car of mass M is driven by an engine working at constant power and the resistance to motion is proportional to the square of the speed. If R is the resistance at full speed V on a horizontal road, prove that the distance covered as the speed increases from v_1 to v_2 is $\dfrac{MV^2}{3R}\log_e\left(\dfrac{V^3 - v_1^3}{V^3 - v_2^3}\right)$. (O)

13. (a) A particle travelling in a straight line is subject to a retardation of $(v^2 + 25)\,\text{m s}^{-2}$, where $v\,\text{m s}^{-1}$ is its speed at t seconds. If its initial speed is $12\,\text{m s}^{-1}$, show that it travels a distance of $\log_e(13/5)$ metres before coming to rest.

(b) A particle moving with SHM oscillates between points A and B which are 10 m apart. When it is 1 m from A, its speed is $16\,\text{m s}^{-1}$. Find (i) the periodic time; (ii) the maximum speed it attains; (iii) the maximum acceleration; (iv) the shortest time it takes to travel from A to C, where AC is $\frac{1}{4}AB$. (SU)

14. A particle moves along a straight line with simple harmonic motion of amplitude 0·5 m and of period 4 seconds. Find the maximum speed in m/s and the maximum acceleration in m/s^2 of the particle. Find also in m/s the speed of the particle when it is 0·25 m from its central position. (Answers may be left in terms of π.) Find also the least time which elapses between two instants when the speed of the particle is half its maximum speed. (L)

15. A particle of mass m moves in simple harmonic motion about O in a straight line, under the action of a restoring force of magnitude proportional to the distance from O. At time $t = 0$ the speed of the particle is zero. After 1 second the speed of the particle is $2\,\text{m s}^{-1}$, after a further second the speed is $2\sqrt{3}\,\text{m s}^{-1}$ and subsequently the particle passes through O for the first time. Show that the speed of particle when it passes through O is $4\,\text{m s}^{-1}$ and find (i) the period and amplitude of the motion, (ii) the time at which the speed of the particle is equal to $2\,\text{m s}^{-1}$ after it first passes through O.

A stationary particle of mass m is placed at O and the moving particle collides and coalesces with this particle. The combined particle moves under the action of the same restoring force as before. Find the period of the subsequent motion and the speed when the displacement is $12/\pi$ m from O. (C)

16. At time t the displacement from the origin of a particle moving along the x-axis under the action of a single force is given by $x = a\cos(\omega t + \alpha)$ where a, ω and α are constants. Find the acceleration of the particle in terms of its displacement from the origin and describe the force acting on the particle. What is the periodic time of this motion?

Two particles, each moving in simple harmonic motion, pass through their centres of oscillation at time $t = 0$. They are next at their greatest distances from their centres at times $t = 2$ s and $t = 3$ s respectively, having been at the same distance from their respective centres at time $t = 1$ s. Show that the amplitudes of their motions are in the ratio $1:\sqrt{2}$. When will the particles next be at their centres, simultaneously? When will the initial conditions next be repeated? (W)

17. A particle moves on a straight line through a fixed point O so that at time t its displacement from O is x and its equation of motion is $\ddot{x} = -\omega^2 x$. If when $t = 0$, $x = d$ and $\dot{x} = u$, show that $x = d\cos\omega t + (u/\omega)\sin\omega t$. Given that the first three times at which the particle passes through the point $x = d$ are $t = 0, t = 2$ and $t = 8$, find d in terms of u.

18. A rough horizontal platform moves horizontally in simple harmonic motion. A particle P of mass 2 kg rests on the platform without slipping and is at O, the centre of oscillations when the motion starts; the initial speed of P is 4 m s^{-1}. The period of the S.H.M. is 8 s. Find (i) the amplitude of the motion correct to 3 sig. fig., (ii) the distance OP, 1 s after the motion starts, correct to 3 sig. fig., (iii) an expression for the time taken for the particle to be first displaced a distance 1 m from O. Do not evaluate the expression.

Draw a diagram showing the forces acting on P with its acceleration and obtain the maximum frictional force acting on P. Hence give the range of values of the coefficient of friction in order that P should not slip. (SU)

19. The end A of a light elastic string AB, obeying Hooke's law and of natural length 0·5 m, is fixed. When a particle of mass 2 kg is attached to the string at B and hangs freely under gravity, the extension of the string in the equilibrium position is 0·075 m. Calculate, in newtons, the modulus of elasticity of the string. The particle is now pulled down vertically a further 0·1 m and released. Show that, until the string becomes slack, the motion of the particle is simple harmonic. Show that the time that elapses before the particle first passes through the equilibrium position is $(\pi\sqrt{3})/40$ s and find the speed of the particle when it is in this position. [Take g as 10 m/s^2]. (L)

20. One end of a light elastic string of natural length l and modulus λ is attached to a fixed point A, and a small mass m, attached to the other end, hangs in equilibrium at B. A small mass $2m$ is made to adhere to the under side of the first mass, and the combined mass is released from rest at B. Show that the combined mass oscillates

about a point C which is at a depth $l + \frac{3mgl}{\lambda}$ below A, and that the equation of motion is

$$\frac{d^2x}{dt^2} = -\frac{\lambda}{3ml}x,$$

where x is the depth below C of the combined mass at time t. Show that the speed of the combined mass when $x = 0$ is $2g\sqrt{(ml/3\lambda)}$. Find the force of adhesion between the masses in terms of m, g, λ and x. (JMB)

21. A particle P of mass m lies on a smooth horizontal table and is attached to two fixed points A, B on the table by two light elastic strings, each obeying Hooke's law and of natural lengths $2l, 3l$ and of moduli $2mg, mg$, respectively. If $AB = 7l$, show that, when P is in equilibrium, $AP = 5l/2$. The particle P is held at rest at the point C in the line AB, where $AC = 3l$ and C lies between A and B, and is then released. Show that the motion of P is simple harmonic of period $\pi\sqrt{(3l/g)}$. Find the maximum speed of P during the motion. (O&C)

22. One end of a light elastic string, of natural length $2a$ and modulus $4mg$, is attached to a fixed point A of a smooth horizontal table. A particle of mass m is attached to the other end of the string. The particle is released from rest from a point C of the table, where $AC = 3a$. Show that the particle reaches C again after time $(4 + \pi)\sqrt{(2a/g)}$. (L)

23. A scale pan of mass m is suspended from a fixed point by an elastic spring, of modulus $2mg$ and natural length l. When the scale pan is hanging at rest, a particle of mass $\frac{1}{2}m$ falls vertically on to the scale pan, striking it with speed $\sqrt{(lg)}$. Find the amplitude of the oscillations which the scale pan starts to make if the particle rebounds, the coefficient of restitution between the particle and the scale pan being $\frac{1}{4}$. (L)

24. A particle rests on the *outside* of a fixed smooth vertical hoop, of radius a, at its highest point A, and a second particle rests on the *inside* of the hoop at its lowest point B. The particle at A is just disturbed from rest and the particle at B is given a horizontal velocity v such that both particles leave the hoop at the same point P. Show that (i) the radius OP is inclined to the vertical at an angle $\cos^{-1}\frac{2}{3}$, (ii) $v = 2\sqrt{(ag)}$. (W)

25. A smooth wire in the form of a circle of centre O and radius a is fixed in a vertical plane. Two beads, P of mass $2m$ and Q of mass m, are threaded on to the wire. Initially Q is at rest at A, the lowest point of the wire, and P is held at a point B on the wire, where the angle AOB is $\frac{1}{3}\pi$.

The particle P is then released. Find the speed of P when the angle AOP is θ and P has not yet reached A, and find the reaction then exerted on P by the wire. Show that, when P reaches A, its speed is $\sqrt{(ga)}$. Show that, when the particles collide, the impulse of the reaction between them is of magnitude $\frac{2}{3}m(1 + e)\sqrt{(ga)}$, where e is the coefficient of restitution. In the case when $e = 1/5$ find the height above A which Q subsequently reaches before coming instantaneously to rest. (JMB)

26. A particle is at rest at the lowest point A inside a smooth hollow sphere of radius a. It is projected horizontally with velocity $\sqrt{(7ga/2)}$. Show that it loses contact with the sphere at a height $3a/2$ above A and will next strike the sphere at A. (C)

27.

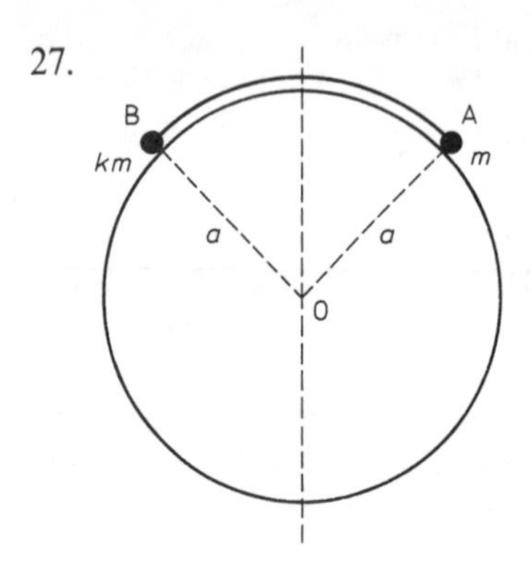

Two particles A and B of masses m and km respectively, where $k < 1$, are joined by a light inextensible string of length $\frac{1}{2}\pi a$. The string is taut and the particles are held, at the same horizontal level, in contact with a fixed smooth cylinder of radius a. The axis of the cylinder is horizontal and the string is in a vertical plane perpendicular to the axis of the cylinder. The system is released from rest. If θ is the angle made with the vertical by the radius OA of the cylinder when the particle A is about to lose contact with the cylinder, show that $(3 + k)\cos\theta + 2k\sin\theta = (1 + k)\sqrt{2}$. (C)

28. Two small equal beads, threaded on a vertical smooth circular wire of radius a, are connected by a light inextensible string of length a. One bead is held at the highest point of the wire with the string taut and is then released. Find the speed of the beads at the instant the string first becomes slack. (O)

29. A particle of mass m is projected horizontally with velocity $\mathbf{u}$ from a point O in a medium producing a resisting force $-mk\dot{\mathbf{r}}$ where k is a positive constant and $\mathbf{r}$ is the position vector of the particle relative to O. The acceleration due to gravity is $\mathbf{g}$. Write down the equation of motion (involving $\ddot{\mathbf{r}}$) and show that at time t

(i) $\dfrac{d}{dt}(\mathbf{r}e^{kt}) = (\mathbf{u} + \mathbf{g}t)e^{kt}$,

(ii) $\mathbf{r} = \dfrac{1}{k}(1 - e^{-kt})\mathbf{u} + \dfrac{1}{k^2}(e^{-kt} + kt - 1)\mathbf{g}$.

If k is so small that k^2 may be neglected compared with k, show that approximately $\mathbf{r} = \mathbf{u}t(1 - \frac{1}{2}kt) + \mathbf{g}t^2(\frac{1}{2} - \frac{1}{6}kt)$. (JMB)

13 Coordinate geometry

13.1 Angles, distances and areas

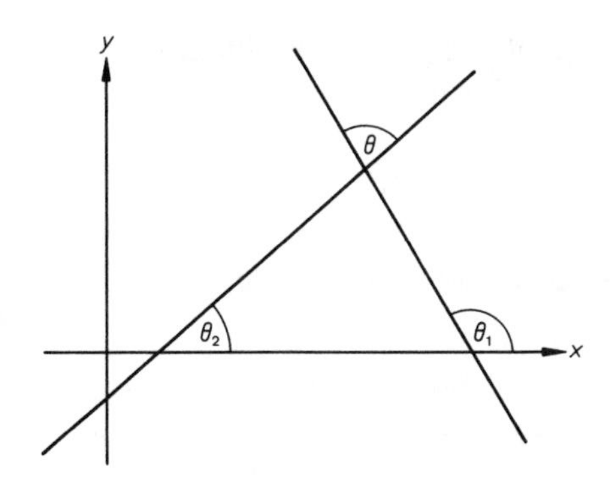

The diagram shows two lines which make angles θ_1 and θ_2 with the positive direction of the x-axis. If the gradients of these lines are m_1 and m_2 respectively, then

$$m_1 = \tan\theta_1, \qquad m_2 = \tan\theta_2.$$

One angle between the lines is θ, where $\theta = \theta_1 - \theta_2$,

$$\therefore \qquad \tan\theta = \tan(\theta_1 - \theta_2) = \frac{\tan\theta_1 - \tan\theta_2}{1 + \tan\theta_1\tan\theta_2}.$$

Hence the angle θ between two lines with gradients m_1 and m_2 is given by

$$\boxed{\tan\theta = \frac{m_1 - m_2}{1 + m_1 m_2}}$$

Example 1 Find the acute angle between the straight lines $3x - y + 2 = 0$ and $x - 2y - 1 = 0$.

The equations of the lines may be re-written as:

$y = 3x + 2$ and $y = \frac{1}{2}x - \frac{1}{2}$.

Hence the gradients of the lines are 3 and $\frac{1}{2}$.

If θ is one angle between the lines, then

$$\tan\theta = \frac{3-\frac{1}{2}}{1+3\times\frac{1}{2}} = \frac{2\frac{1}{2}}{2\frac{1}{2}} = 1$$

$\therefore$ the acute angle between the lines is $45°$.

[Note that the angle between two curves at any point of intersection is defined as the angle between the tangents to the curves at the point.]

Consider now the line joining the points $A(x_1, y_1)$ and $B(x_2, y_2)$. Let P be the point which divides AB in the ratio $\mu:\lambda$. Then if C, D and Q are the feet of the perpendiculars from A, B and P to the x-axis, Q must divide CD in the ratio $\mu:\lambda$. It follows that the x-coordinate of P

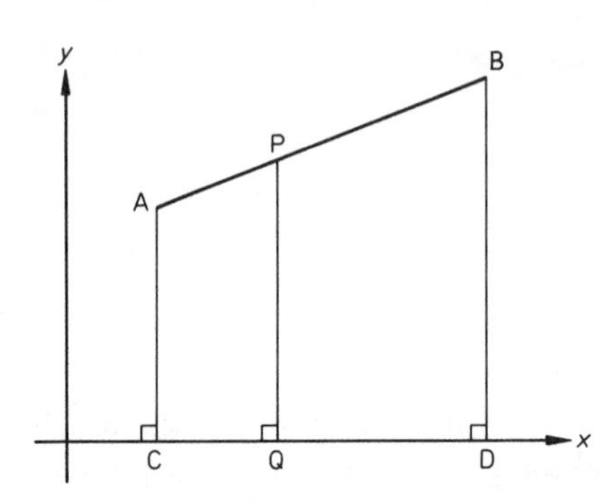

$$= x_1 + \frac{\mu}{\lambda+\mu}(x_2 - x_1) = \frac{\lambda x_1 + \mu x_2}{\lambda + \mu}$$

A similar expression is obtained for the y-coordinate of P.

i.e. The point dividing the line AB in the ratio $\mu:\lambda$ has coordinates

$$\left(\frac{\lambda x_1 + \mu x_2}{\lambda+\mu}, \frac{\lambda y_1 + \mu y_2}{\lambda+\mu}\right).$$

This formula can be used when P divides AB externally in a given ratio if μ and λ are taken to have opposite signs.
[We recall that the corresponding formula for the position vector of a point which divides a line in the ratio $\mu:\lambda$ was obtained in §7.1.]

Example 2 If A and B are the points $(2, 1)$ and $(-4, 4)$ respectively, find the coordinates of the point C which divides AB internally in the ratio $2:1$ and of the point D which divides AB externally in the ratio $3:2$.

The coordinates of C are

$$\left(\frac{1\times 2 + 2(-4)}{1+2}, \frac{1\times 1 + 2\times 4}{1+2}\right) \quad \text{i.e.} \quad (-2, 3)$$

Regarding D as the point which divides AB in the ratio $3:-2$, the coordinates of D are

$$\left(\frac{(-2)2 + 3(-4)}{-2+3}, \frac{(-2)1 + 3\times 4}{-2+3}\right) \quad \text{i.e.} \quad (-16, 10).$$

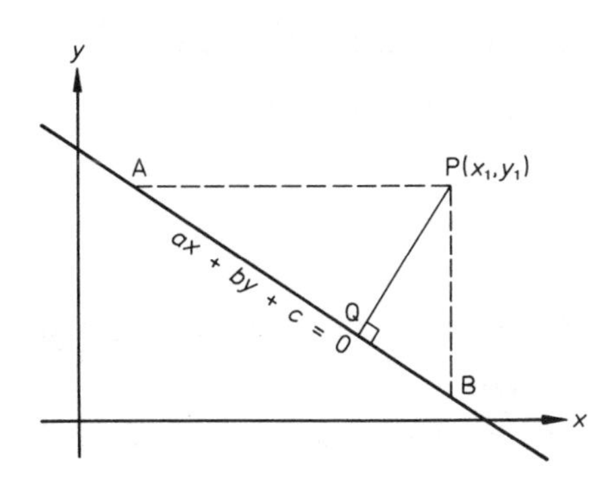

Next we derive a formula for the perpendicular distance from a point $P(x_1, y_1)$ to a straight line $ax + by + c = 0$. In the diagram A and B are points on the line $ax + by + c = 0$, such that AP is parallel to the x-axis and BP is parallel to the y-axis. The point Q is the foot of the perpendicular from P to the line and so PQ is the required perpendicular distance.

Substituting $y = y_1$ in the equation $ax + by + c = 0$, we find that the x-coordinate of A is $-(by_1 + c)/a$.

$$\therefore \qquad AP = \left|x_1 - \left\{-\frac{(by_1 + c)}{a}\right\}\right| = \left|\frac{ax_1 + by_1 + c}{a}\right|$$

Similarly $BP = \left|\dfrac{ax_1 + by_1 + c}{b}\right|$

$$\begin{aligned} \therefore \qquad AB &= \sqrt{\{AP^2 + BP^2\}} \\ &= \sqrt{\left\{\frac{(ax_1 + by_1 + c)^2}{a^2} + \frac{(ax_1 + by_1 + c)^2}{b^2}\right\}} \\ &= \sqrt{\left\{\frac{(ax_1 + by_1 + c)^2(a^2 + b^2)}{a^2b^2}\right\}} \\ &= \left|\frac{(ax_1 + by_1 + c)\sqrt{(a^2 + b^2)}}{ab}\right| \end{aligned}$$

However, the area of $\triangle APB = \frac{1}{2} \times PQ \times AB = \frac{1}{2} \times AP \times BP$

$$\begin{aligned} \therefore \qquad PQ &= \frac{AP \times BP}{AB} \\ &= \left|\frac{(ax_1 + by_1 + c)^2}{ab} \times \frac{ab}{(ax_1 + by_1 + c)\sqrt{(a^2 + b^2)}}\right| \\ &= \left|\frac{ax_1 + by_1 + c}{\sqrt{(a^2 + b^2)}}\right| \end{aligned}$$

i.e.

> The perpendicular distance from the point (x_1, y_1) to the line
>
> $ax + by + c = 0$ is $\left|\dfrac{ax_1 + by_1 + c}{\sqrt{(a^2 + b^2)}}\right|$

Example 3 Find the distance of the point $(3, -5)$ from the line $2x - y = 1$.

The distance of the point $(3, -5)$ from the line $2x - y - 1 = 0$ is

$$\left|\frac{2 \times 3 - (-5) - 1}{\sqrt{\{2^2 + (-1)^2\}}}\right| = \frac{10}{\sqrt{5}} = 2\sqrt{5}.$$

Example 4 Find the locus of points equidistant from the lines $y = 2x$ and $2x + 4y - 3 = 0$.

Let $P(x_1, y_1)$ be a point equidistant from the lines $2x - y = 0$ and $2x + 4y - 3 = 0$, then

$$\left|\frac{2x_1 - y_1}{\sqrt{\{2^2 + (-1)^2\}}}\right| = \left|\frac{2x_1 + 4y_1 - 3}{\sqrt{\{2^2 + 4^2)}}\right|$$

$$\therefore \quad \frac{1}{\sqrt{5}}\left|2x_1 - y_1\right| = \frac{1}{\sqrt{20}}\left|2x_1 + 4y_1 - 3\right|$$

$$\therefore \quad 2|2x_1 - y_1| = |2x_1 + 4y_1 - 3|$$

Thus either

$$2(2x_1 - y_1) = 2x_1 + 4y_1 - 3$$
$$4x_1 - 2y_1 = 2x_1 + 4y_1 - 3$$
$$2x_1 - 6y_1 + 3 = 0$$

or

$$2(2x_1 - y_1) = -(2x_1 + 4y_1 - 3)$$
$$4x_1 - 2y_1 = -2x_1 - 4y_1 + 3$$
$$6x_1 + 2y_1 - 3 = 0$$

Hence the locus of all points P equidistant from the given lines is the pair of straight lines

$$2x - 6y + 3 = 0, \qquad 6x + 2y - 3 = 0.$$

It follows that these lines are the bisectors of the angles between the given lines.

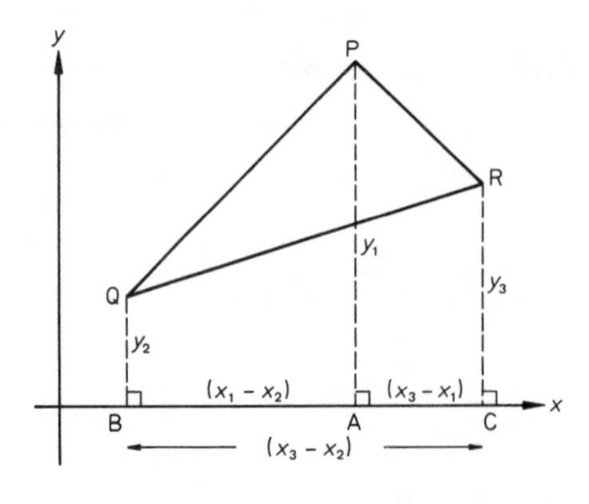

Finally we consider the area of the triangle whose vertices are $P(x_1, y_1)$, $Q(x_2, y_2)$ and $R(x_3, y_3)$.

As shown in the diagram, the area of ΔPQR = area $PQBA$ + area $PACR$ − area $QBCR$.

Since each of these areas is a trapezium,

$$\text{area of } \Delta PQR = \tfrac{1}{2}(x_1 - x_2)(y_1 + y_2) + \tfrac{1}{2}(x_3 - x_1)(y_1 + y_3) - \tfrac{1}{2}(x_3 - x_2)(y_2 + y_3)$$
$$= \tfrac{1}{2}\{x_1(y_2 - y_3) + x_2(y_3 - y_1) + x_3(y_1 - y_2)\}$$

It is found that, although this result holds for any points P, Q and R, it sometimes gives a negative value for the area. We must therefore include modulus signs in the general formula.

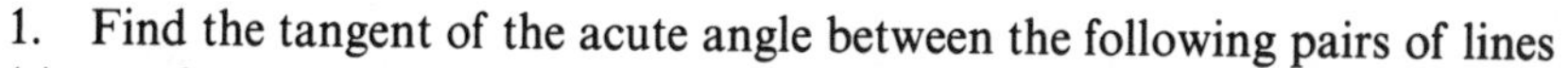

$$\text{Area of triangle} = \tfrac{1}{2}|x_1(y_2 - y_3) + x_2(y_3 - y_1) + x_3(y_1 - y_2)|$$

[Note that this formula is rarely needed in elementary work. Problems on area usually involve right-angled triangles, isosceles triangles or other triangles whose heights and bases are easily calculated.]

Exercise 13.1

1. Find the tangent of the acute angle between the following pairs of lines
(a) $y = 2x - 3$, $y = 3x + 1$
(b) $3x - 2y + 1 = 0$, $2x - y - 7 = 0$
(c) $x + 3y - 2 = 0$, $5x - 6y + 2 = 0$
(d) $7x + 4y + 5 = 0$, $y = 0$
(e) $y = 4x$, $2x - 8y - 5 = 0$
(f) $4x + 3y + 2 = 0$, $x = 0$.

2. Find the coordinates of the points which divide the lines joining the given pairs of points internally in the stated ratio.
(a) $(7, 3)$, $(1, -6)$; $1:2$
(b) $(2, -4)$, $(-8, 1)$; $2:3$
(c) $(0, 6)$, $(2\tfrac{1}{2}, -4)$; $4:1$
(d) $(-3, 5)$, $(1, -1)$; $5:3$

3. Find the coordinates of the points which divide the lines joining the given pairs of points externally in the stated ratio.
(a) $(5, -1)$, $(-2, 3)$; $1:2$
(b) $(2, -3)$, $(0, 5)$; $2:3$
(c) $(4, 3)$, $(-2, 0)$; $4:1$
(d) $(-3, 4)$, $(1, -1)$; $5:3$

4. Find the distances between the given points and the given lines.
(a) $(1, 2)$, $4x - 3y - 3 = 0$
(b) $(-3, -1)$, $5x + 12y + 1 = 0$
(c) $(2, 5)$, $x - y + 5 = 0$
(d) $(0, 0)$, $2x + 3y = 13$
(e) $(-6, 10)$, $x = 3$
(f) $(7, -4)$, $3y - x - 1 = 0$

5. Find the areas of the triangles with the following vertices:
(a) $(2, 3)$, $(5, -1)$, $(2, -1)$
(b) $(-3, 4)$, $(5, 1)$, $(-3, -2)$
(c) $(-1, 3)$, $(2, -3)$, $(4, 5)$
(d) $(3, -1)$, $(-2, 0)$, $(1, 4)$
(e) $(5, 7)$, $(-4, -11)$, $(0, -3)$
(f) $(-2, 6)$, $(3, -7)$, $(4, 5)$

6. Find the equations of the bisectors of the angles between the given pairs of lines
(a) $3x - y - 3 = 0$
$x + 3y + 1 = 0$
(b) $2x + 3y = 2$
$3x + 2y = 3$
(c) $3x + 6y - 1 = 0$
$x - 2y + 1 = 0$
(d) $y = 7x - 1$
$y = x + 5$

7. Find the angles of the triangle PQR with vertices $P(1, 4)$, $Q(3, -2)$ and $R(5, 2)$.

8. Find the angles of the triangle ABC with vertices $A(1,-3)$, $B(4,6)$ and $C(-1,1)$.

9. A triangle has vertices $A(-4,1)$, $B(3,0)$ and $C(1,2)$. If the internal bisector of $\angle B$ meets AC at P, use the fact that $AP:PC = AB:BC$ to find the coordinates of P.

10. A triangle has vertices $P(1,-2)$, $Q(5,1)$ and $R(6,10)$. Find the coordinates of the point on QR which is equidistant from the sides PQ and PR.

11. Find the tangent of the acute angle between the parabola $y^2 = 4x$ and the circle $x^2 + y^2 = 5$ at their points of intersection.

12. The curves $y = x^2$ and $y^2 = 8x$ intersect at the origin and at a point A. Find the angle between the curves at A.

13. Find the incentre of the triangle with vertices $(-1,4)$, $(2,-2)$ and $(7,8)$.

14. Find the incentre of the triangle whose sides have equations $x - y + 1 = 0$, $x + 7y + 9 = 0$ and $7x - y - 2 = 0$.

15. The points $A(-8,9)$ and $C(1,2)$ are opposite vertices of a parallelogram $ABCD$. The sides BC, CD of the parallelogram lie along the lines $x + 7y - 15 = 0$, $x - y + 1 = 0$, respectively. Calculate (i) the coordinates of D, (ii) the tangent of the acute angle between the diagonals of the parallelogram, (iii) the length of the perpendicular from A to the side CD, (iv) the area of the parallelogram. (JMB)

13.2 Further work on circles

When a problem involves two circles, of radii a and b, the relative positions of the circles can be determined by finding the distance c between their centres. Assuming that $a > b$, there are various different possibilities.

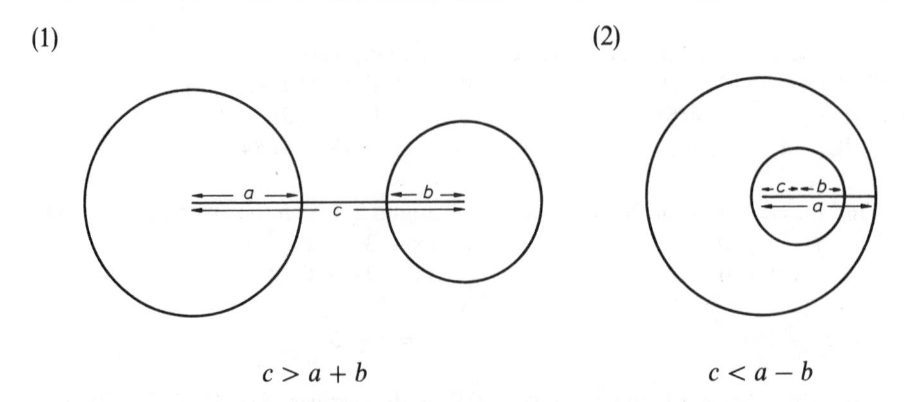

In diagrams (1) and (2) the circles do not intersect.

(3) (4)

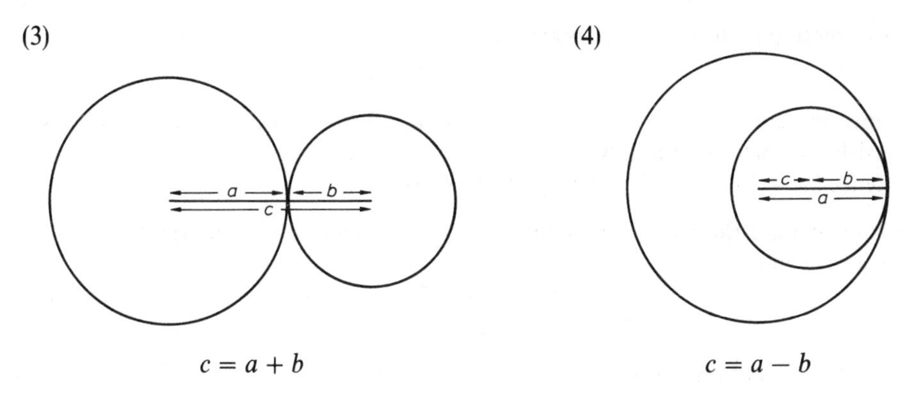

$c = a + b$ $c = a - b$

In diagrams (3) and (4) the circles touch either externally or internally.

(5)

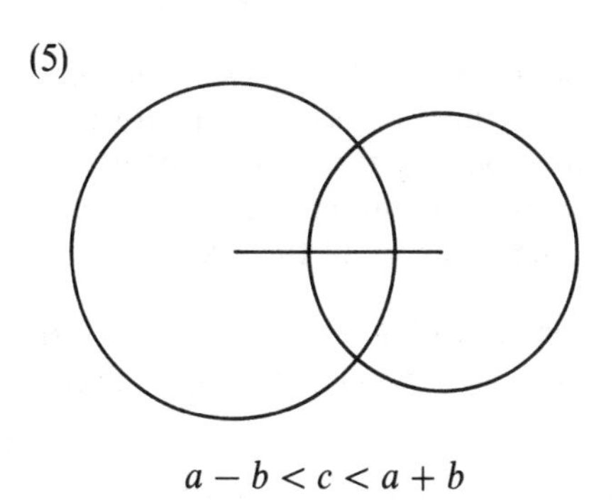

$a - b < c < a + b$

In diagram (5) the circles have two distinct points of intersection.

(6)

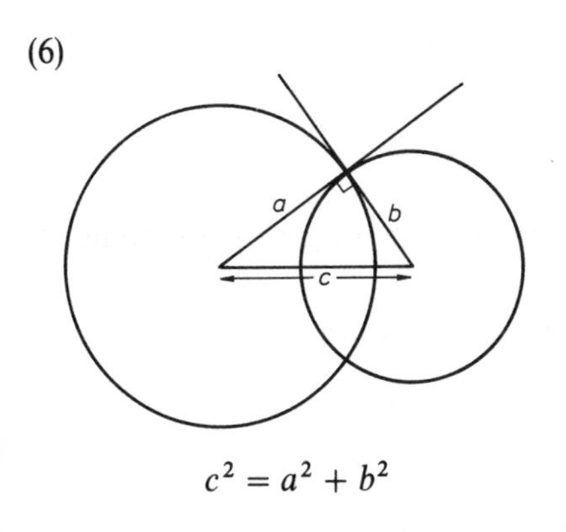

$c^2 = a^2 + b^2$

Two circles which cut at right-angles are called *orthogonal* circles. Diagram (6) shows two such circles.

The *common chord* of two circles is the line joining their points of intersection. The equation of this chord can be obtained without finding the coordinates of the points of intersection.

Example 1 The circles $x^2 + y^2 + 4x - 3y + 1 = 0$ and $x^2 + y^2 + x - y - 2 = 0$ intersect at the points A and B. Find the equation of the common chord AB.

$$x^2 + y^2 + 4x - 3y + 1 = 0 \quad (1)$$
$$x^2 + y^2 + x - y - 2 = 0 \quad (2)$$

Subtracting (2) from (1) we obtain:

$$3x - 2y + 3 = 0$$

Since the coordinates of A and B satisfy equations (1) and (2), they must also satisfy this new equation, which represents a straight line.
Hence $3x - 2y + 3 = 0$ is the equation of AB.

We now extend the method of Example 1 by considering two intersecting circles with equations:

$$x^2 + y^2 + 2gx + 2fy + c = 0 \qquad \text{(i)}$$
$$x^2 + y^2 + 2Gx + 2Fy + C = 0 \qquad \text{(ii)}$$

From these equations we can form a new equation:

$$x^2 + y^2 + 2gx + 2fy + c + k(x^2 + y^2 + 2Gx + 2Fy + C) = 0 \qquad \text{(iii)}$$

i.e. $$(1 + k)x^2 + (1 + k)y^2 + 2(g + kG)x + 2(f + kF)y + c + kC = 0$$

This is the equation of a circle for all values of k, except $k = -1$. Moreover, any pair of values of x and y which satisfy equations (i) and (ii) must also satisfy equation (iii). Hence, in general, equation (iii) represents a circle passing through the points of intersection of circles (i) and (ii). When $k = -1$ equation (iii) represents a straight line through the points of intersection i.e. the common chord of the circles.

Example 2 The circles $x^2 + y^2 + 3x - y - 5 = 0$ and $x^2 + y^2 - 2x + y - 1 = 0$ intersect at points A and B. Find the equation of the circle which passes through the origin and the points A and B.

Any circle which passes through A and B has an equation of the form:

$$x^2 + y^2 + 3x - y - 5 + k(x^2 + y^2 - 2x + y - 1) = 0$$

If this circle also passes through the origin,

$$-5 + k(-1) = 0 \quad \text{i.e.} \quad k = -5$$

Hence the equation of the circle which passes through the origin and the points A and B is

$$x^2 + y^2 + 3x - y - 5 - 5(x^2 + y^2 - 2x + y - 1) = 0$$

i.e. $$4x^2 + 4y^2 - 13x + 6y = 0.$$

Exercise 13.2

1. Two circles have equations $x^2 + y^2 - 2x - 6y - 54 = 0$ and $x^2 + y^2 - 8x + 2y + 13 = 0$.

Show that one circle lies entirely inside the other.

2. Prove that the circles whose equations are $x^2 + y^2 - 2x - 2y - 2 = 0$, $x^2 + y^2 - 6x - 10y + 33 = 0$ lie entirely outside one another.

3. A circle C_1 has equation $x^2 + y^2 - 32x - 24y + 300 = 0$ and a second circle C_2 has as diameter the line joining the points $(8, 0)$ and $(0, 6)$. Show that the circles C_1 and C_2 touch externally.

4. A circle C has equation $x^2 + y^2 - 4x + 2y = 40$. Find the equations of the circles with centre $(3, 1)$ which touch circle C. Find also the coordinates of the points of contact.

5. Show that the circles $x^2 + y^2 - 2x - 2y - 2 = 0$ and

$$x^2 - y^2 - 8x - 10y + 32 = 0$$

touch externally and find the coordinates of the point of contact.

6. Find the equations of the two circles with centres $A(1, 1)$ and $B(9, 7)$ which have equal radii and touch each other externally. Find also the equations of the common tangents to these circles.

7. Two circles with centres $(3, -5)$ and $(-6, 7)$ pass through the point $(4, 2)$. Find the equations of these circles. Find also the equation and the length of their common chord.

8. Prove that the circles whose equations are $x^2 + y^2 - 4y - 5 = 0$, $x^2 + y^2 - 8x + 2y + 1 = 0$ cut orthogonally and find the equation of the common chord.

9. Find the equation of the circle with centre $(3, -2)$ and radius 5 units. Find also the equation of the circle with centre $(-7, 3)$ which intersects the original circle at right angles.

10. Show that any circle which passes through the points $(1, 0)$ and $(-1, 0)$ has an equation of the form $x^2 + y^2 - 2\lambda y - 1 = 0$. Prove also that if the circles given by $\lambda = \lambda_1$ and $\lambda = \lambda_2$ cut orthogonally then $\lambda_1\lambda_2 = -1$.

11. Show that if the circles $x^2 + y^2 + 2gx + 2fy + c = 0$ and

$$x^2 + y^2 + 2Gx + 2Fy + C = 0$$

cut orthogonally, then $2gG + 2fF = c + C$.

12. The circles $x^2 + y^2 - 2x = 0$ and $x^2 + y^2 + 4x - 6y - 3 = 0$ intersect at the points A and B. Find (a) the equation of AB, (b) the equation of the circle which passes through A, B and the point $(1, 2)$.

13. Given that the circles $x^2 + y^2 - 3y = 0$ and $x^2 + y^2 + 5x - 8y + 5 = 0$ intersect at P and Q, find the equation of the circle which passes through P, Q and (a) the point $(1, 1)$, (b) the origin, (c) touches the x-axis.

14. The circle $x^2 + y^2 + 3x - 5y - 4 = 0$ and the straight line $y = 2x + 5$ intersect at the points A and B. Find the equation of the circle which passes through A, B and (a) the point $(3, 1)$, (b) the origin, (c) has its centre on the y-axis.

15. Show that the circles $x^2 + y^2 + 8x + 2y + 8 = 0$, $x^2 + y^2 - 2x + 2y - 2 = 0$ have three common tangents, and find their equations. (C)

13.3 Loci using parametric forms

Many problems in coordinate geometry are best solved using the parametric equations of curves. The basic techniques involved were introduced in Book 1 and used to investigate the properties of various curves including the parabola $y^2 = 4ax$. We now consider further applications of these methods. As clear diagrams are often useful in this type of work, we begin by giving sketches of several important curves together with the most commonly used parametric forms of their equations.

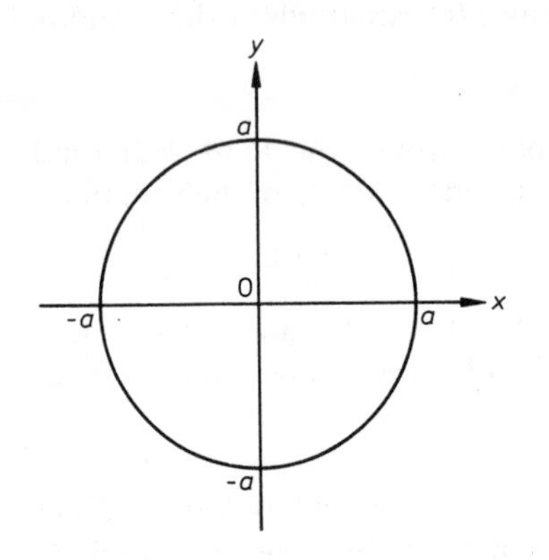

The circle $x^2 + y^2 = a^2$.
Parametric equations:

$$x = a\cos\theta,$$
$$y = a\sin\theta.$$

This is the circle with centre O and radius a.

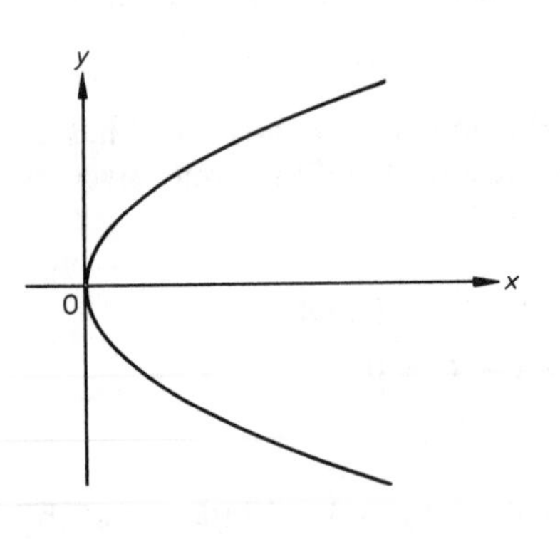

The parabola $y^2 = 4ax$.
Parametric equations:

$$x = at^2,$$
$$y = 2at.$$

This parabola has vertex O and is symmetrical about the x-axis.

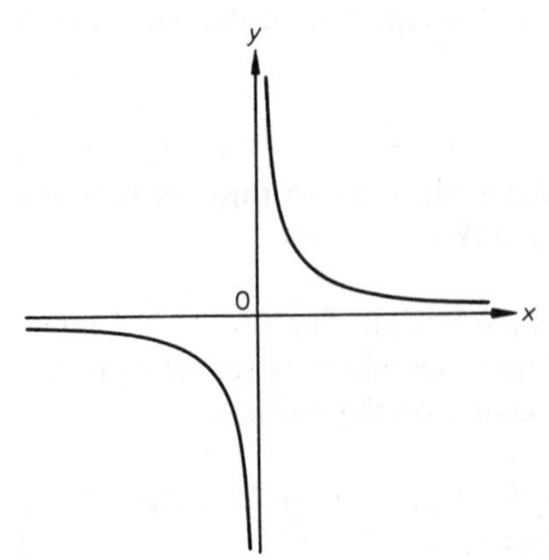

The rectangular hyperbola $xy = c^2$.
Parametric equations:

$$x = ct,$$
$$y = c/t.$$

The x- and y-axes are asymptotes to the curve and the curve is symmetrical about the lines $y = \pm x$.

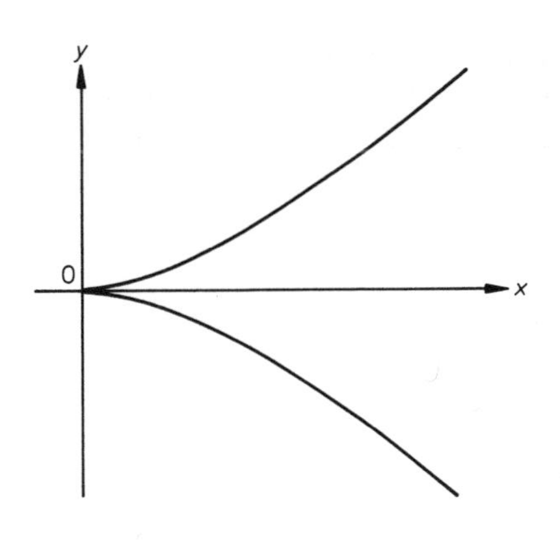

The semi-cubical parabola $ay^2 = x^3$.
Parametric equations:

$$x = at^2,$$
$$y = at^3.$$

The curve has a *cusp* at O and is symmetrical about the x-axis.

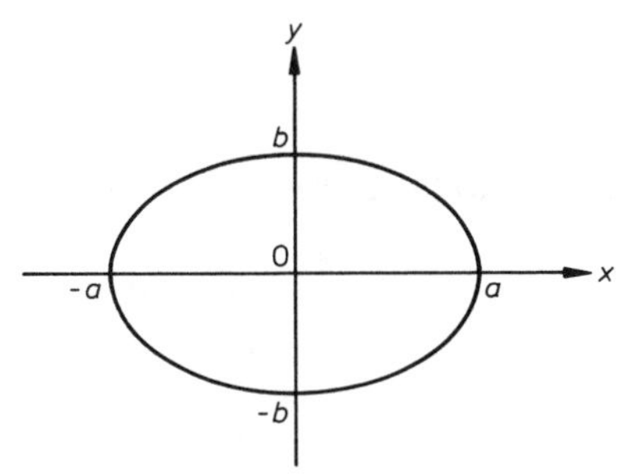

The ellipse $\dfrac{x^2}{a^2} + \dfrac{y^2}{b^2} = 1$.
Parametric equations:

$$x = a\cos\theta,$$
$$y = b\sin\theta.$$

The curve is symmetrical about the x- and y-axes.

Example 1 Find the equation of the chord of the ellipse $\dfrac{x^2}{a^2} + \dfrac{y^2}{b^2} = 1$ joining the points $P(a\cos\theta, b\sin\theta)$ and $Q(a\cos\phi, b\sin\phi)$.

$$\text{Gradient of } PQ = \frac{b\sin\theta - b\sin\phi}{a\cos\theta - a\cos\phi}$$

$$= \frac{2b\cos\frac{1}{2}(\theta+\phi)\sin\frac{1}{2}(\theta-\phi)}{-2a\sin\frac{1}{2}(\theta+\phi)\sin\frac{1}{2}(\theta-\phi)} = -\frac{b\cos\frac{1}{2}(\theta+\phi)}{a\sin\frac{1}{2}(\theta+\phi)}$$

Thus the equation of the chord PQ is

$$y - b\sin\theta = -\frac{b\cos\frac{1}{2}(\theta+\phi)}{a\sin\frac{1}{2}(\theta+\phi)}(x - a\cos\theta)$$

i.e. $ay\sin\frac{1}{2}(\theta+\phi) - ab\sin\theta\sin\frac{1}{2}(\theta+\phi)$

$$= -bx\cos\tfrac{1}{2}(\theta+\phi) + ab\cos\theta\cos\tfrac{1}{2}(\theta+\phi)$$

$\therefore \quad bx\cos\frac{1}{2}(\theta+\phi) + ay\sin\frac{1}{2}(\theta+\phi)$

$$= ab\{\cos\theta\cos\tfrac{1}{2}(\theta+\phi) + \sin\theta\sin\tfrac{1}{2}(\theta+\phi)\}$$

$$= ab\cos\{\theta - \tfrac{1}{2}(\theta+\phi)\} = ab\cos\tfrac{1}{2}(\theta-\phi).$$

Hence the equation of the chord PQ may be written:

$$\frac{x\cos\frac{1}{2}(\theta+\phi)}{a} + \frac{y\sin\frac{1}{2}(\theta+\phi)}{b} = \cos\tfrac{1}{2}(\theta-\phi).$$

Example 2 Find the equations of the tangent and the normal to the curve $xy = c^2$ at the point $P(ct, c/t)$. Given that the normal at P meets the curve again at Q, find the coordinates of Q. If the tangent at P meets the y-axis at R, find the equation of the locus of the mid-point M of PR.

The parametric equations of the curve are

$$x = ct, \qquad y = \frac{c}{t}$$

Differentiating with respect to t we have

$$\frac{dx}{dt} = c, \qquad \frac{dy}{dt} = -\frac{c}{t^2}$$

$$\therefore \quad \frac{dy}{dx} = \frac{dy}{dt}\bigg/\frac{dx}{dt} = \left(-\frac{c}{t^2}\right)\bigg/c = -\frac{1}{t^2}$$

Hence the gradients of the tangent and the normal to the curve at P are $-1/t^2$ and t^2 respectively.

$\therefore$ the equation of the tangent at P is

$$y - \frac{c}{t} = -\frac{1}{t^2}(x - ct)$$

i.e. $$t^2y - ct = -x + ct$$

i.e. $$x + t^2y = 2ct \qquad (1)$$

The equation of the normal at P is

$$y - \frac{c}{t} = t^2(x - ct)$$

i.e. $$ty - c = t^3x - ct^4$$

i.e. $$t^3x - ty = c(t^4 - 1) \qquad (2)$$

Let $(cu, c/u)$ be a point which lies on the curve $xy = c^2$ and on the normal at P, then from equation (2)

$$t^3 \times cu - t \times \frac{c}{u} = c(t^4 - 1)$$

$$\therefore \quad t^3u^2 + (1 - t^4)u - t = 0$$

$$\therefore \quad (u - t)(t^3u + 1) = 0$$

$\therefore$ either $u = t$ or $u = -1/t^3$

Since the solution $u = t$ gives the point P, the remaining solution $u = -1/t^3$ must give the point Q.

Hence the coordinates of Q are $(-c/t^3, -ct^3)$.

Substituting $x = 0$ in equation (1) we have,

$$t^2y = 2ct \quad \text{i.e.} \quad y = 2c/t$$

$\therefore$ the coordinates of the point R are $(0, 2c/t)$.

Thus the coordinates of the mid-point M of PR are

$$\left(\frac{1}{2}\{ct+0\}, \frac{1}{2}\left\{\frac{c}{t}+\frac{2c}{t}\right\}\right) \quad \text{i.e.} \quad \left(\frac{1}{2}ct, \frac{3c}{2t}\right)$$

Hence as t varies the parametric equations of the locus of M are

$$x=\frac{1}{2}ct, \qquad y=\frac{3c}{2t}$$

Eliminating t from these equations we obtain the Cartesian equation of the locus, namely $xy = \frac{3}{4}c^2$.

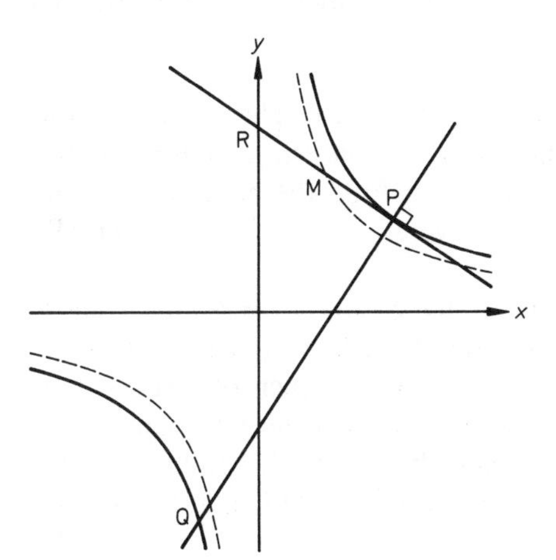

As shown in the diagram the locus of M is another rectangular hyperbola.

Exercise 13.3

1. Write down parametric coordinates for a point on each of the following curves and sketch the curves.
(a) $y^2 = 12x$, (b) $xy = 9$, (c) $y^2 = x^3$,
(d) $4xy = 25$, (e) $4y^2 = x^3$, (f) $y^2 = 10x$.

2. Write down parametric coordinates for a point on each of the following curves and sketch the curves.
(a) $\dfrac{x^2}{9}+\dfrac{y^2}{4}=1$, (b) $x^2+y^2=9$, (c) $\dfrac{x^2}{5}+y^2=1$,
(d) $4x^2+9y^2=4$, (e) $9x^2+y^2=9$, (f) $x^2+(y-1)^2=25$.

3. Find the equations of the tangent and normal to the curve $xy = 16$ at the point $(4t, 4/t)$.

4. Find the equations of the tangent and normal to the curve $x = t$, $y = 1/t$ at the point where $t = 2$.

5. Find the equations of the tangent and normal to the curve $x = 2\cos\theta$, $y = \sin\theta$ at the point where $\theta = \frac{1}{3}\pi$.

6. P is the point $(5\cos\theta, 4\sin\theta)$ on the curve $x^2/25 + y^2/16 = 1$. Given that S and S' are the points with coordinates $(-3, 0)$ and $(3, 0)$, show that the value of $PS + PS'$ is independent of θ.

7. Find the equations of the tangents to the curve $xy = c^2$ which have gradient -4 and write down the coordinates of their points of contact.

8. The points $P(cp, c/p)$, $Q(cq, c/q)$ and $R(cr, c/r)$ lie on the curve $xy = c^2$. Show that if $\angle RPQ$ is a right angle, then $p^2qr = -1$. Prove further that QR is perpendicular to the tangent at P to the curve.

9. The parametric equations of a curve are $x = 2t$, $y = 2/t$. The tangent to the curve at $P(2t, 2/t)$ meets the x-axis at A; the normal to the curve at P meets the y-axis at B. The point Q divides AB in the ratio 3:1. Find the parametric equations of the locus of Q as P moves on the given curve. (JMB)

10. Let P and Q be points on the rectangular hyperbola $xy = a^2$, with coordinates $(ap, a/p)$ and $(aq, a/q)$. Obtain the equation of the line PQ, and deduce, or obtain otherwise, the equation of the tangent at P to the hyperbola. Show that, if the line PQ meets the coordinate axes at A and B, then the mid-point M of AB is the mid-point of PQ also. Obtain the coordinates of the point T where the tangents at P and Q meet, and show that the line MT passes through the origin. (W)

11. Prove that the normal to the hyperbola $xy = c^2$ at the point $P(ct, c/t)$ has equation $y = t^2x + \dfrac{c}{t} - ct^3$. If the normal at P meets the line $y = x$ at N, and O is the origin, show that $OP = PN$ provided that $t^2 \neq 1$. The tangent to the hyperbola at P meets the line $y = x$ at T. Prove that $OT \, . \, ON = 4c^2$. (O&C)

12. The normal at the point $P(ct, c/t)$ on the rectangular hyperbola $xy = c^2$ meets the curve again at Q. If the normal meets the x-axis at A and the y-axis at B, show that the mid-point of AB is also the mid-point of PQ.

13. Show that the tangent at $P(cp, c/p)$ to the rectangular hyperbola $xy = c^2$ has the equation $p^2y + x = 2cp$. The perpendicular from the origin to this tangent meets it at N, and meets the hyperbola again at Q and R. Prove that (i) the angle QPR is a right angle, (ii) as p varies, the point N lies on the curve whose equation is $(x^2 + y^2)^2 = 4c^2xy$. (C)

14. The tangents at the points $P(cp, c/p)$ and $Q(cq, c/q)$ on the rectangular hyperbolar $xy = c^2$ intersect at the point R. Given that R lies on the rectangular hyperbola $xy = \frac{1}{2}c^2$, find the equation of the locus of the mid-point M of PQ as p and q vary.

15. Find the equation of the tangent to the curve $ay^2 = x^3$ at the point (at^2, at^3) and prove that, apart from one exceptional case, the tangent meets the curve again. Find the coordinates of the point of intersection. What is the exceptional case? (O&C)

16. The tangents to the curve $9y^2 = x^3$ at the points $P(9p^2, 9p^3)$ and $Q(9q^2, 9q^3)$ intersect at the point R. Find the coordinates of R. If p and q vary in such a way that $\angle PRQ$ is always a right angle, find the equation of the locus of R.

17. Show that the equation of the tangent to the curve $x = a\cos t$, $y = b\sin t$ at the point $P(a\cos p, b\sin p)$ is $\frac{x}{a}\cos p + \frac{y}{b}\sin p = 1$. This tangent meets the curve $x = 2a\cos\theta$, $y = 2b\sin\theta$ at the points Q and R, which are given by $\theta = q$ and $\theta = r$ respectively. Show that p differs from each of q and r by $\pi/3$. (JMB)

18. Show that if the tangents to the ellipse $x^2/a + y^2/b = 1$ at the points $P(a\cos\theta, b\sin\theta)$ and $Q(a\cos\phi, b\sin\phi)$ intersect at the point R, then the coordinates of R are $\left(\frac{a\cos\frac{1}{2}(\theta+\phi)}{\cos\frac{1}{2}(\theta-\phi)}, \frac{b\sin\frac{1}{2}(\theta+\phi)}{\cos\frac{1}{2}(\theta-\phi)}\right)$. If P and Q move on the ellipse in such a way that $\phi = \theta + \frac{1}{2}\pi$, find the equation of the locus of R.

19. Find the equation of the normal to the curve $x^2 - y^2 = 1$ at the point $P(\sec\theta, \tan\theta)$, where $-\frac{1}{2}\pi < \theta < \frac{1}{2}\pi$. If this normal cuts the x-axis at A and the y-axis at B, show that P is the centre of the circle which passes through A, B and the origin O. If the line OP cuts this circle again at the point Q, find the equation of the locus of Q as θ varies.

20. (a) The points $A(d, 0)$, $B(-d, 0)$ and $P(d\cos\theta,\ d\sin\theta)$ lie on the circle $x^2 + y^2 = d^2$. Show that the equation of the tangent to the circle at P is $x\cos\theta + y\sin\theta = d$. Show that the sum of the perpendicular distances from A and B to this tangent is independent of θ, and calculate the product of these distances.
(b) The point $P(p, p^3)$ lies on the curve $y = x^3$. Show that the equation of the tangent to the curve at P is $y - 3p^2x + 2p^3 = 0$. Show that when $p = 3$ this tangent passes through the point $A(7/3, 9)$, and find the other two values of p for which the tangent passes through A. (W)

21. A curve has the parametric equations $x = a(\theta - \sin\theta)$, $y = a(1 - \cos\theta)$, where $0 \leqslant \theta \leqslant \pi$. A point P of the curve has parameter ϕ, where $\phi \neq 0$.

(i) Show that, at P, $\frac{dy}{dx} = \frac{\sin\phi}{1 - \cos\phi}$.

(ii) The normal to the curve at P meets the x-axis at G, and O is the origin. Show that $OG = a\phi$.

(iii) The tangent to the curve at P meets at K the line through G parallel to the y-axis. Show that $GK = 2a$. (C)

13.4 Conic sections

Some of the curves we have been studying arise naturally as cross-sections of a circular cone. The following diagrams show the sections produced when a double cone of semi-vertical angle θ is cut by a plane which does not pass through the vertex. The nature of the curve obtained in each case is determined by the value of the angle α between the plane and the axis of the cone, where $0 \leqslant \alpha \leqslant \frac{1}{2}\pi$.

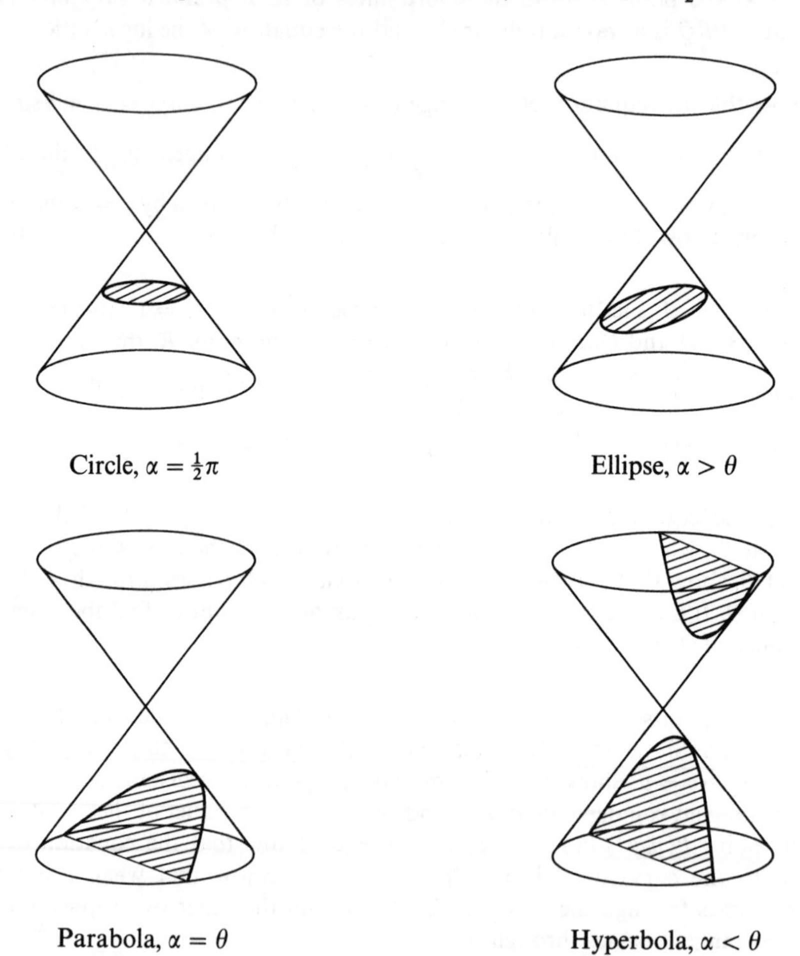

Circle, $\alpha = \frac{1}{2}\pi$ Ellipse, $\alpha > \theta$

Parabola, $\alpha = \theta$ Hyperbola, $\alpha < \theta$

The complete set of conics includes the sections produced by planes through the vertex of the double cone, namely:

for $\alpha < \theta$, a pair of straight lines;
for $\alpha = \theta$, a single straight line;
for $\alpha > \theta$, a single point.

It can be shown that the definition of a parabola as a conic section is consistent with the locus definition used in Book 1. We now consider the corresponding locus definitions of the ellipse and hyperbola.

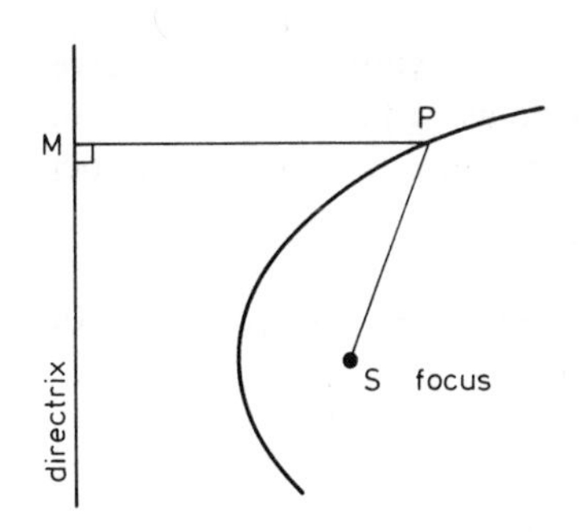

Suppose that P is a point which moves so that its distance from a fixed point S, the *focus*, is a constant multiple of its distance from a fixed line called the *directrix*. If M is the foot of the perpendicular from P to the directrix then

$$PS = ePM$$

where e is a positive constant called the *eccentricity*. The nature of the locus of P depends on the value taken by e, as follows:

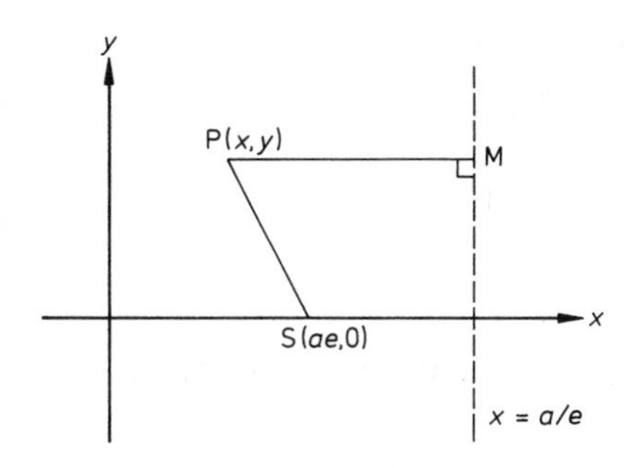

for an ellipse, $0 < e < 1$;
for a parabola, $e = 1$;
for a hyperbola, $e > 1$.

The standard form of the equation of the ellipse is obtained by taking the point $(ae, 0)$ as focus and the line $x = a/e$ as directrix.

$$PS^2 = e^2PM^2$$

$\therefore$

$$(x - ae)^2 + y^2 = e^2\left(\frac{a}{e} - x\right)^2$$

$$x^2 - 2aex + a^2e^2 + y^2 = a^2 - 2aex + e^2x^2$$

$$x^2(1 - e^2) + y^2 = a^2(1 - e^2)$$

i.e.

$$\frac{x^2}{a^2} + \frac{y^2}{a^2(1 - e^2)} = 1$$

Hence the equation of the ellipse is:

$$\frac{x^2}{a^2} + \frac{y^2}{b^2} = 1, \quad \text{where} \quad b^2 = a^2(1 - e^2).$$

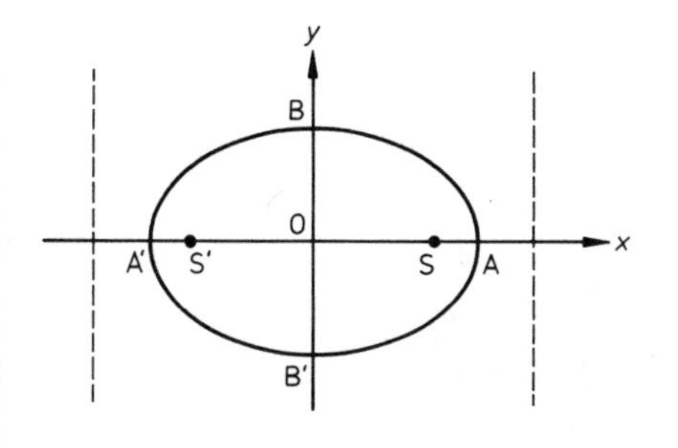

From this equation we see that the ellipse is symmetrical about both x- and y-axes. It cuts the x-axis at the points $A(a, 0)$ and $A'(-a, 0)$. It cuts the y-axis at the points $B(b, 0)$ and $B'(-b, 0)$. AA' is called the *major axis* and BB' the *minor axis*. The origin O is the *centre*

of the ellipse and any chord through O is a *diameter*. From symmetry of the curve it is clear that the same locus would have been produced using the point $S'(-ae, 0)$ as focus and the line $x = -a/e$ as directrix. Thus the ellipse is said to have foci $(\pm ae, 0)$ and directrices $x = \pm ae$.

[Note that in terms of a and b the coordinates of the foci are $(\pm\sqrt{\{a^2 - b^2\}}, 0)$ and the equations of the directrices are $x = \pm a^2/\sqrt{(a^2 - b^2)}$.]

Example 1 Find the eccentricity, the foci and the directrices of the ellipse $\dfrac{x^2}{9} + y^2 = 1$.

For the ellipse $\dfrac{x^2}{a^2} + \dfrac{y^2}{b^2} = 1$, we have $b^2 = a^2(1 - e^2)$.
Substituting $a = 3$, $b = 1$ we find that for the given ellipse,

$$1 = 9(1 - e^2)$$
$$\therefore \quad 1 = 9 - 9e^2$$
$$\therefore \quad e^2 = 8/9$$

Hence the eccentricity of the ellipse is $\frac{2}{3}\sqrt{2}$. Thus the foci are the points $(\pm 2\sqrt{2}, 0)$ and the directrices are the lines $x = \pm\frac{9}{4}\sqrt{2}$.

The standard form of the equation of the hyperbola is

$$\frac{x^2}{a^2} - \frac{y^2}{b^2} = 1, \quad \text{where} \quad b^2 = a^2(e^2 - 1).$$

Using the methods applied to the ellipse it can be shown that this hyperbola has foci $(\pm ae, 0)$ and directrices $x = \pm a/e$.

From the equation we see that the hyperbola is also symmetrical about both x- and y-axes. It cuts the x-axis at the points $A(a, 0)$ and $A'(-a, 0)$, but does not cut the y-axis.
Rearranging the equation we have:

$$\frac{y^2}{b^2} = \frac{x^2}{a^2} - 1$$

$$\therefore \quad \frac{y^2}{x^2} = \frac{b^2}{a^2} - \frac{b^2}{x^2}$$

$$\therefore \quad \text{as } |x| \to \infty, \quad \frac{y^2}{x^2} \to \frac{b^2}{a^2} \quad \text{and} \quad \frac{y}{x} \to \pm\frac{b}{a}$$

i.e. as $|x|$ increases the curve approaches the lines $y = \pm\dfrac{b}{a}x$.

Hence the lines $y = \pm\dfrac{b}{a}x$ are asymptotes to the hyperbola.

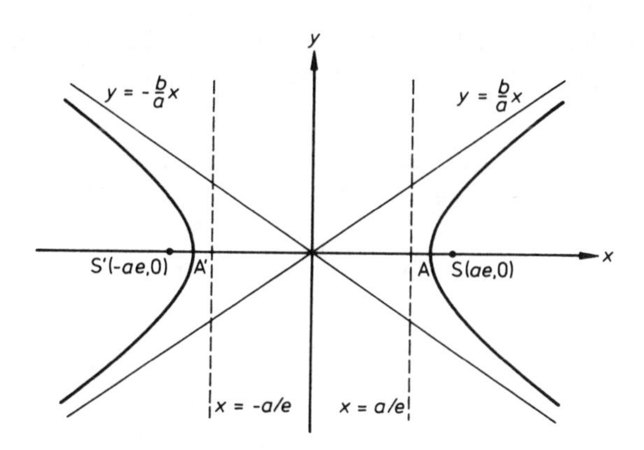

A hyperbola with perpendicular asymptotes is called a *rectangular hyperbola.* Since the lines $y = \pm\frac{b}{a}x$ are perpendicular when $a = b$, the equation of a rectangular hyperbola is of the form

$$x^2 - y^2 = a^2.$$

When the asymptotes are used as x- and y-axes the equation of the rectangular hyperbola takes the more familiar form $xy = c^2$.

Exercise 13.4

1. Find the eccentricities, the foci and the directrices of the following ellipses.

(a) $\dfrac{x^2}{25} + \dfrac{y^2}{9} = 1$, (b) $16x^2 + 25y^2 = 100$, (c) $\dfrac{x^2}{4} + y^2 = 1$,

(d) $\dfrac{x^2}{9} + \dfrac{y^2}{5} = 1$, (e) $\dfrac{x^2}{4} + \dfrac{y^2}{9} = 1$, (f) $4x^2 + y^2 = 1$.

2. Find the eccentricities, the foci, the directrices and the asymptotes of the following hyperbolas.

(a) $\dfrac{x^2}{16} - \dfrac{y^2}{9} = 1$, (b) $x^2 - y^2 = 4$, (c) $x^2 - 4y^2 = 4$

3. Find the eccentricities of the following ellipses

(a) $x = 2\cos\theta$, $y = \sin\theta$, (b) $x = 5\cos\theta$, $y = 3\sin\theta$,
(c) $x = 2\cos\theta$, $y = 3\sin\theta$, (d) $x = 5\sin\theta$, $y = 3\cos\theta$.

4. Find in the form $x^2/a^2 + y^2/b^2 = 1$ the equations of the ellipses with (a) eccentricity 1/2, foci $(\pm 2, 0)$, (b) eccentricity 3/5, foci $(\pm 9, 0)$.

5. Use the locus definition of the ellipse to find the equation of the ellipse with eccentricity $\frac{2}{3}$, focus (2, 1) and directrix $x = -\frac{1}{2}$.

6. Find the equations of the tangent and the normal to each of the following ellipses at the given point:

(a) $\dfrac{x^2}{9} + \dfrac{y^2}{4} = 1$; (3, 0),

(b) $\dfrac{x^2}{8} + \dfrac{y^2}{2} = 1$; $(-2, 1)$,

(c) $9x^2 + 16y^2 = 40$; $(2, \frac{1}{2})$,

(d) $4x^2 + 5y^2 = 120$; $(-5, -2)$.

7. Find the equations of the tangents at the point (x_1, y_1) to the following curves:

(a) $4x^2 + 9y^2 = 36$,

(b) $\dfrac{x^2}{a^2} + \dfrac{y^2}{b^2} = 1$,

(c) $\dfrac{x^2}{3} - \dfrac{y^2}{2} = 1$.

8. Use the locus definition of the hyperbola to find the equation of the hyperbola with eccentricity $\sqrt{2}$, focus $(2k, 2k)$ and directrix $x + y = k$.

9. The line $y = mx + c$ is a tangent to the ellipse $\dfrac{x^2}{a^2} + \dfrac{y^2}{b^2} - 1 = 0$, $(a > b > 0)$. Show that $c^2 = a^2m^2 + b^2$. The perpendicular distances from the points $(\sqrt{(a^2 - b^2)}, 0)$ and $(-\sqrt{(a^2 - b^2)}, 0)$ to any tangent to the ellipse are p_1, p_2. Show that $p_1p_2 = b^2$. (C)

10. A line with gradient m is drawn through the fixed point $C(h, 0)$, where $0 < h < a$, to meet the ellipse $x^2/a^2 + y^2/b^2 = 1$ at the points P and Q. Prove that the mid-point R of PQ has the coordinates $\left(\dfrac{a^2hm^2}{a^2m^2 + b^2}, \dfrac{-b^2hm}{a^2m^2 + b^2}\right)$. Show that, as m varies, R always lies on the curve whose equation is $\dfrac{x^2}{a^2} + \dfrac{y^2}{b^2} = \dfrac{hx}{a^2}$. (C)

11. The ellipse $x^2/a^2 + y^2/b^2 = 1$ intersects the positive x-axis at A and the positive y-axis at B. Determine the equation of the perpendicular bisector of AB. (i) Given that this line intersects the x-axis at P and that M is the mid-point of AB, prove that the area of triangle PMA is $b(a^2 + b^2)/8a$. (ii) If $a^2 = 3b^2$, find, in terms of b, the coordinates of the points where the perpendicular bisector of AB intersects the ellipse. (JMB)

12. Prove that the hyperbola H_1, with equation $x^2 - y^2 = a^2$, cuts the hyperbola H_2, with equation $xy = c^2$, at right angles at two points P and Q. If the distance between the tangents to H_1 at P and Q is equal to the distance between the tangents to H_2 at P and Q, find the relation between a and c. (O)

Exercise 13.5 (miscellaneous)

1. A straight line parallel to the line $2x + y = 0$ intersects the x-axis at A and the y-axis at B. The perpendicular bisector of AB cuts the y-axis at C. Prove that the gradient of the line AC is $-\frac{3}{4}$. Find also the tangent of the acute angle between the line AC and the bisector of the angle AOB, where O is the origin. (JMB)

2. The triangle ABC has vertices $A(0, 12)$, $B(-9, 0)$, $C(16, 0)$. Find the equations of the internal bisectors of the angles ABC and ACB. Hence, or otherwise, find the equation of the inscribed circle of the triangle ABC. Find also the equation of the circle passing through A, B and C. (C)

3. Write down the perpendicular distance from the point (a, a) to the line $4x - 3y + 4 = 0$. The circle, with centre (a, a) and radius a, touches the line $4x - 3y + 4 = 0$ at the point P. Find a, and the equation of the normal to the circle at P. Show that P is the point $(1/5, 8/5)$. Show that the equation of the circle which has centre P and which passes through the origin is $5(x^2 + y^2) - 2x - 16y = 0$. (L)

4. Find the centre and the radius of the circle C which passes through the points $(4, 2)$, $(2, 4)$ and $(2, 6)$. If the line $y = mx$ is a tangent to C, obtain the quadratic equation satisfied by m. Hence or otherwise find the equations of the tangents to C which pass through the origin O. Find also (i) the angle between the two tangents, (ii) the equation of the circle which is the reflection of C in the line $y = 3x$. (AEB 1977)

5. Show that the equation of the tangent to the curve $y^2 = 4ax$ at the point $(at^2, 2at)$ is $ty = x + at^2$ and the equation of the tangent to the curve $x^2 = 4by$ at the point $(2bp, bp^2)$ is $y = px - bp^2$. The curves $y^2 = 32x$ and $x^2 = 4y$ intersect at the origin and at A. Find the equation of the common tangent to these curves and the coordinates of the points of contact B and C between the tangent and the curves. Calculate the area of the triangle ABC. (AEB 1976)

6. On the same diagram sketch the circles $x^2 + y^2 = 4$ and $x^2 + y^2 - 10x = 0$. The line $ax + by + 1 = 0$ is a tangent to both these circles. State the distances of the centres of the circles from this tangent. Hence, or otherwise, find the possible values of a and b and show that, if 2ϕ is the angle between the common tangents, then $\tan \phi = \frac{3}{4}$. (L)

7. Find the centre and radius of each of the circles C_1 and C_2 whose equations are $x^2 + y^2 - 16y + 32 = 0$ and $x^2 + y^2 - 18x + 2y + 32 = 0$ respectively and show that the circles touch externally. Find the coordinates of their point of contact and show that the common tangent at that point passes through the origin. The other tangents from the origin, one to each circle, are drawn. Find, correct to the nearest degree, the angle between these tangents. (SU)

8. Two circles, C_1 and C_2, have equations $x^2 + y^2 - 4x - 8y - 5 = 0$ and $x^2 + y^2 - 6x - 10y + 9 = 0$, respectively. Find the x-coordinates of the points P and Q at which the line $y = 0$ cuts C_1, and show that this line touches C_2. Find the tangent of the acute angle made by the line $y = 0$ with the tangents to C_1 at P and Q. Show that, for all values of the constant λ, the circle C_3 whose equation is $\lambda(x^2 + y^2 - 4x - 8y - 5) + x^2 + y^2 - 6x - 10y + 9 = 0$ passes through the points of intersection of C_1 and C_2. Find the two possible values of λ for which the line $y = 0$ is a tangent to C_3. (JMB)

9. The circles whose equations are $x^2 + y^2 - x + 6y + 7 = 0$

and $x^2 + y^2 + 2x + 2y - 2 = 0$

intersect at the points A and B. Find (i) the equation of the line AB, (ii) the coordinates of A and B. Show that the two given circles intersect at right angles and obtain the equation of the circle which passes through A and B and which also passes through the centres of the two circles. (AEB 1978)

10. A curve C_1 has equation $2y^2 = x$; a curve C_2 is given by the parametric equations $x = 4t$, $y = 4/t$. Sketch C_1 and C_2 on the same diagram and calculate the coordinates of their point of intersection, P. The tangent to C_1 at P crosses the x-axis at T and meets C_2 again at Q. (a) Show that T is a point of trisection of PQ. (b) Find the area of the finite region bounded by the x-axis, the line PT and the curve C_1. (L)

11. Show that the tangent at the point P, with parameter t, on the curve $x = ct$, $y = c/t$ has equation $x + t^2y = 2ct$. This tangent meets the x-axis in a point Q and the line through P parallel to the x-axis cuts the y-axis in a point R. Show that, for any position of P on the curve, QR is a tangent to the curve with parametric equations $x = ct$, $y = c/(2t)$. (L)

12. Prove that the equation of the normal to the rectangular hyperbola $xy = c^2$ at the point $P(ct, c/t)$ is $ty - t^3x = c(1 - t^4)$. The normal at P and the normal at the point $Q(c/t, ct)$, where $t > 1$, intersect at the point N. Show that $OPNQ$ is a rhombus, where O is the origin. Hence, or otherwise, find the coordinates of N. If the tangents to the hyperbola at P and Q intersect at T, prove that the product of the lengths of OT and ON is independent of t. (JMB)

13. The line of gradient $m(\neq 0)$ through the point $A(a, O)$ is a tangent to the rectangular hyperbola $xy = c^2$ at the point P. Find m in terms of a and c, and show that the coordinates of P are $(\frac{1}{2}a, 2c^2/a)$. The line through A parallel to the y-axis meets the hyperbola at Q, and the line joining Q to the origin O intersects AP at R. Given that OQ and AP are perpendicular to each other, find the numerical value of c^2/a^2 and the numerical value of the ratio $AR:RP$. (JMB)

14. The point P in the first quadrant lies on the curve with parametric equations $x = t^2$, $y = t^3$. The tangent to the curve at P meets the curve again at Q and is normal to the curve at Q. Find the coordinates of P and of Q. (L)

15. Sketch the parabola whose parametric equations are $x = t^2$, $y = 2t$ and on the same diagram sketch the curve with parametric equations $x = 10(1 + \cos\theta)$, $y = 10\sin\theta$. These curves touch at the origin and meet again at two other points A and B. The normals at A and B to the parabola meet at P and the tangents to the other curve at A and B meet at Q. Calculate the length of PQ. (L)

16. Obtain an equation of the tangent, at the point with parameter t, to the curve $\mathscr{C}$ whose parametric equations are given by

$$x = 2\sin^3 t, \quad y = 2\cos^3 t, \quad 0 \leqslant t \leqslant \pi/2.$$

Show that, if the tangent meets the coordinate axes in points R and S, then RS is of constant length. Sketch the curve $\mathscr{C}$. Find the area of the finite region enclosed by the curve $\mathscr{C}$ and the coordinate axes. (L)

17. Sketch the curve whose parametric equations are $x = a\cos\phi$, $y = b\sin\phi$, $0 \leqslant \phi \leqslant 2\pi$, where a and b are positive constants. The point P is given by $\phi = \pi/4$. Find (a) the equation of the tangent to the curve at P, (b) the equation of the normal to the curve at P. By evaluating a suitable integral, calculate the area of the region in the first quadrant between the curve and the coordinate axes. Hence deduce the area of the region enclosed by the curve. (L)

18. Prove that the equation of the tangent at the point (x_1, y_1) on the ellipse $\dfrac{x^2}{a^2} + \dfrac{y^2}{b^2} = 1$ is $\dfrac{xx_1}{a^2} + \dfrac{yy_1}{b^2} = 1$. The tangent at the point $(2\cos\theta, \sqrt{3}\sin\theta)$ on the ellipse $\dfrac{x^2}{4} + \dfrac{y^2}{3} = 1$ passes through the point $P(2, 1)$. Show that $\sqrt{3}\cos\theta + \sin\theta = \sqrt{3}$. Without using tables, calculator or slide rule, find all the solutions of this equation which are in the range $0° \leqslant \theta < 360°$. Hence obtain the coordinates of the points of contact, Q and R, of the tangents to the ellipse from P. Verify that the line through the origin and the point P passes through the mid-point of the line QR. (JMB)

19. Prove that the equation of the normal at $(\alpha\cos\phi, \beta\sin\phi)$ to the ellipse $\dfrac{x^2}{\alpha^2} + \dfrac{y^2}{\beta^2} = 1$ is $\alpha x\sec\phi - \beta y\operatorname{cosec}\phi = \alpha^2 - \beta^2$. P is the point $(a\cos\theta, b\sin\theta)$ on the ellipse $\dfrac{x^2}{a^2} + \dfrac{y^2}{b^2} = 1$. M and N are the feet of the perpendiculars from P to the axes. Find the equation of MN. Prove that, for variable θ, MN is always normal to a fixed concentric ellipse and find the equation of this ellipse. (O&C)

20. Prove that the equation of the tangent to the curve $xy^2 = c^3$ $(c > 0)$ at the point $P(ct^2, c/t)$ is $2t^3y + x = 3ct^2$. Prove that the parameters of the points of contact of the tangents which pass through the point $Q(h, k)$ $(k \neq 0)$ satisfy a cubic equation and, by considering the turning values of this cubic, or otherwise, prove that there are three distinct tangents to the curve which pass through Q if $hk^2 < c^3$

and $h > 0$. State a necessary and sufficient condition on the parameters t_1, t_2, t_3 for the tangents at the corresponding points to be concurrent.

When Q lies on the curve, the tangent at Q cuts the curve again at R; if the parameter of Q is t, determine the parameter of R. Prove that, if the tangents Q_1, Q_2, Q_3 are concurrent and cut the curve again at R_1, R_2, R_3 then the tangents at R_1, R_2, R_3 are concurrent. (O&C)

14 Complex numbers

14.1 Algebraic form

We are already familiar with various types of real number, such as integers, rationals and irrationals. However, since the square of any non-zero real number is positive, there are no solutions within the real number system of the equations such as $x^2 + 2 = 0$ and $x^2 + 2x + 2 = 0$. To deal with these equations we must extend the number system to include the square roots of negative numbers, e.g. $\sqrt{(-1)}$, $-\sqrt{(-5)}$, $\sqrt{(-7/3)}$. Numbers of the form $\pm\sqrt{(-k^2)}$, where k is real, are called *imaginary numbers*. If the number $\sqrt{(-1)}$ is denoted by i, so that $i^2 = -1$, then any imaginary number can be expressed as a multiple of i. For instance:

$$\sqrt{(-4)} = \sqrt{\{4 \times (-1)\}} = \sqrt{4} \times \sqrt{(-1)} = 2i,$$
$$\sqrt{(-7)} = \sqrt{\{7 \times (-1)\}} = \sqrt{7} \times \sqrt{(-1)} = i\sqrt{7}.$$

The set of all numbers of the form $a + ib$, where a and b are real, is called the set of *complex numbers* and denoted by $\mathbb{C}$. The number a is called the *real part* of $a + ib$ and b is the *imaginary part* of $a + ib$. Since $a + ib$ is real when $b = 0$, the set of real numbers $\mathbb{R}$ is a subset of $\mathbb{C}$.

Let us now consider the roots of the quadratic equation $ax^2 + bx + c = 0$, given by $x = \dfrac{-b \pm \sqrt{(b^2 - 4ac)}}{2a}$. As we know, when $b^2 - 4ac \geqslant 0$, this equation has real roots, but when $b^2 - 4ac < 0$ the formula for x involves the square root of a negative number. We can write this square root as a multiple of i, thus obtaining the roots of the quadratic equations in the form of complex numbers.

Example 1 Find the roots of the equations (a) $x^2 + 9 = 0$, (b) $x^2 + 2x + 2 = 0$, (c) $2x^2 - 2x + 5 = 0$.

(a) $x^2 + 9 = 0 \;\Rightarrow\; x^2 = -9 \;\Rightarrow\; x = \pm\sqrt{(-9)} = \pm 3i.$

(b) $x^2 + 2x + 2 = 0 \Rightarrow x = \dfrac{-2 \pm \sqrt{(4-8)}}{2}$

$$= \frac{-2 \pm \sqrt{(-4)}}{2}$$

$$= \frac{-2 \pm 2i}{2} = -1 \pm i.$$

(c) $2x^2 - 2x + 5 = 0 \Rightarrow x = \dfrac{2 \pm \sqrt{(4-40)}}{4}$

$$= \frac{2 \pm \sqrt{(-36)}}{4}$$

$$= \frac{2 \pm 6i}{4} = \frac{1}{2} \pm \frac{3}{2}i.$$

Each of the equations in Example 1 has a pair of roots of the form $p \pm iq$. The numbers $p + iq, p - iq$ are called *conjugate complex numbers* and $p - iq$ is described as the *conjugate* of $p + iq$. Further consideration of the general formula for the roots of a quadratic equation with real coefficients shows that complex roots must always occur in conjugate pairs. Thus if $p + iq$ is one root of a given quadratic equation, then the other root is $p - iq$.

It is assumed that complex numbers can be combined according to the usual laws of algebra, together with the property $i^2 = -1$.

Example 2 Express, in the form $a + ib$, (a) $(2 - 3i) + (-1 + 4i)$, (b) $(2 - 3i) - (-1 + 4i)$, (c) $(2 - 3i)(-1 + 4i)$, (d) $(2 - 3i)(2 + 3i)$.

(a) $(2 - 3i) + (-1 + 4i) = (2 - 1) + (-3 + 4)i = 1 + i,$
(b) $(2 - 3i) - (-1 + 4i) = (2 + 1) + (-3 - 4)i = 3 - 7i,$
(c) $(2 - 3i)(-1 + 4i) \quad = -2 + 8i + 3i - 12i^2$
$\qquad = -2 + 8i + 3i + 12 = 10 + 11i,$
(d) $(2 - 3i)(2 + 3i) \quad = 4 + 6i - 6i - 9i^2 = 4 + 9 = 13.$

This last result illustrates the fact that the product of a pair of conjugate complex numbers is real. In general,

$$(a + ib)(a - ib) = a^2 + b^2.$$

This property is used to simplify a quotient of complex numbers.

Example 3 Express $\dfrac{-1 + 2i}{1 + 3i}$ in the form $a + ib$.

Multiplying numerator and denominator by the conjugate of $1 + 3i$:

$$\frac{(-1 + 2i)(1 - 3i)}{(1 + 3i)(1 - 3i)} = \frac{-1 + 3i + 2i - 6i^2}{1 - 3i + 3i - 9i^2} = \frac{-1 + 5i + 6}{1 + 9} = \frac{5 + 5i}{10}$$

$$\therefore \quad \frac{-1 + 2i}{1 + 3i} = \frac{1}{2} + \frac{1}{2}i$$

We next consider two *equal* complex numbers $a + ib$ and $c + id$, where a, b, c, d are real.

$$\begin{aligned} a + ib = c + id \quad &\Rightarrow \quad a - c = i(d - b) \\ &\Rightarrow \quad (a - c)^2 = i^2(d - b)^2 \\ &\Rightarrow \quad (a - c)^2 = -(d - b)^2 \\ &\Rightarrow \quad a - c = d - b = 0 \\ &\Rightarrow \quad a = c \quad \text{and} \quad b = d \end{aligned}$$

Hence $a + ib$ and $c + id$ are equal if and only if $a = c$ and $b = d$. This means that if two complex numbers are known to be equal, we may equate real and imaginary parts.

Example 4 Find the square roots of the complex number $5 - 12i$, giving your answers in the form $a + ib$.

Let $\quad (a + ib)^2 = 5 - 12i$
then $\quad a^2 - b^2 + 2abi = 5 - 12i$
Equating real and imaginary parts:

$$a^2 - b^2 = 5, \quad 2ab = -12$$
i.e. $\quad a^2 - b^2 = 5, \quad ab = -6$

By inspection, either $a = 3, b = -2$ or $a = -3, b = 2$.
$\therefore$ the square roots of $5 - 12i$ are $3 - 2i$ and $-3 + 2i$.

[Note that the equations $a^2 - b^2 = 5$, $ab = -6$ could also be solved by eliminating b to obtain a quadratic equation in a^2.]

It is often convenient to use a single letter, such as z, to represent a complex number. The complex conjugate of z is then denoted by z^* (or sometimes $\bar{z}$). The real and imaginary parts of z are denoted by $\text{Re}(z)$ and $\text{Im}(z)$ respectively.

Thus, if $z = x + iy$, $z^* = x - iy$, $\text{Re}(z) = x$ and $\text{Im}(z) = y$.

Exercise 14.1

1. Express the following in terms of i:
(a) $\sqrt{(-25)}$, (b) $\sqrt{(-5)}$, (c) $\sqrt{9} - \sqrt{(-16)}$, (d) $2 - \sqrt{(-18)}$

2. Write down the roots of the following equations:
(a) $x^2 - 4 = 0$, (b) $x^2 + 4 = 0$, (c) $x^2 + 2 = 0$,
(d) $4x^2 + 9 = 0$, (e) $x^3 + 3x = 0$, (f) $x^4 - 1 = 0$.

3. Find the roots of the following equations:
(a) $x^2 - 2x + 2 = 0$, (b) $x^2 - 6x + 10 = 0$, (c) $x^2 - 4x + 13 = 0$,
(d) $4x^2 - 4x + 5 = 0$, (e) $2x^2 + 3x + 2 = 0$, (f) $x^2 - x + 1 = 0$.

4. Write down the conjugates of the following complex numbers:
(a) $3 + 5i$, (b) $4 - 7i$, (c) $3i$, (d) -4, (e) $-1 + i$.

5. Find $z_1 + z_2$ and $z_1 - z_2$ in each of the following cases:
(a) $z_1 = 5 + 8i, z_2 = 2 + i$,
(b) $z_1 = 2 + 3i, z_2 = 1 - 4i$,
(c) $z_1 = 3 - i, z_2 = 5 - 3i$,
(d) $z_1 = -1 + 2i, z_2 = 3 + 5i$.

6. Express in the form $a + ib$:
(a) $(3 + i)(2 + 3i)$,
(b) $(1 - 2i)(5 + i)$,
(c) $(2 - i)(3 - 2i)$,
(d) $(3 - 4i)(3 + 4i)$,
(e) $(2 - 5i)^2$,
(f) $(1 - i)^3$.

7. Simplify (a) i^3, (b) i^5, (c) i^8, (d) $\frac{1}{i^2}$, (e) $\frac{1}{i}$, (f) $\frac{1}{i^3}$.

8. Express in the form $a + ib$:

(a) $\frac{3 + 4i}{1 - 2i}$,
(b) $\frac{2i}{3 + i}$,
(c) $\frac{3 - 2i}{1 - i}$,

(d) $\frac{\sqrt{3} - i}{\sqrt{3} + i}$,
(e) $\frac{1}{(1 + i)^2}$,
(f) $\frac{1}{(2 - i)(1 + 2i)}$.

9. Express in the form $a + ib$:

(a) $\frac{1}{2 + 3i} + \frac{1}{2 - 3i}$,
(b) $\frac{1}{3 + i} - \frac{1}{1 + 7i}$,
(c) $3 + 4i + \frac{25}{3 + 4i}$.

10. Find the values of a and b given that
(a) $(a + ib)(2 - i) = a + 3i$,
(b) $(a + i)(1 + ib) = 3b + ia$.

11. Find the square roots of the following complex numbers
(a) $3 - 4i$,
(b) $2i$,
(c) $8 + 6i$,
(d) $21 - 20i$.

12. Find the values of real numbers p and q given that one root of the equation $x^2 + px + q = 0$ is
(a) $2 + i$,
(b) $-1 + 3i$,
(c) $4i$,
(d) $3 - 5i$.

13. Find the complex factors of the following quadratic expressions.
(a) $x^2 + 1$,
(b) $x^2 + 25$,
(c) $9x^2 + 4$,
(d) $x^2 + 2x + 5$,
(e) $x^2 + 6x + 13$,
(f) $4x^2 - 12x + 25$.

14. Given that the complex number $a + ib$ is conjugate to its own square, find all possible pairs of values of a and b.

15. Given that $\frac{5}{x + iy} + \frac{2}{1 + 3i} = 1$, where x and y are real, find x and y.

16. If $(1 + i)z - iw + i = iz + (1 - i)w - 3i = 6$, find the complex numbers z, w, expressing each in the form $a + bi$ where a, b are real. (O&C)

17. Express $(6+5i)(7+2i)$ in the form $a+ib$. Write down $(6-5i)(7-2i)$ in a similar form. Hence find the prime factors of 32^2+47^2. (JMB)

18. The roots of the quadratic equation $z^2+pz+q=0$ are $1+i$ and $4+3i$. Find the complex numbers p and q. It is given that $1+i$ is also a root of the equation $z^2+(a+2i)z+5+ib=0$, where a and b are real. Determine the values of a and b. (JMB)

19. Given that $z=(p+i)^4$, where p is real, find the values of p for which (a) z is real, (b) z is a real multiple of i.

20. Find the two square roots of $5+12i$. Hence find the roots of the equation $z^2-(1+6i)z-10=0$.

21. Prove that for any complex number z,

$$\mathrm{Re}(z)=\tfrac{1}{2}(z+z^*) \quad \text{and} \quad \mathrm{Im}(z)=\frac{1}{2i}(z-z^*).$$

14.2 Roots of polynomial equations

In the previous section we saw that a quadratic equation with real coefficients has either real roots or a pair of conjugate complex roots. This result can be extended to polynomial equations of higher degree by considering further the properties of conjugate complex numbers.

For two complex numbers $z_1=x_1+iy_1$, $z_2=x_2+iy_2$, we have

$$\begin{aligned}(z_1+z_2)^* &= \{(x_1+iy_1)+(x_2+iy_2)\}^*\\ &= \{(x_1+x_2)+i(y_1+y_2)\}^*\\ &= (x_1+x_2)-i(y_1+y_2)\\ &= (x_1-iy_1)+(x_2-iy_2)=z_1^*+z_2^*\\ (z_1z_2)^* &= \{(x_1+iy_1)(x_2+iy_2)\}^*\\ &= \{(x_1x_2-y_1y_2)+i(x_1y_2+x_2y_1)\}^*\\ &= (x_1x_2-y_1y_2)-i(x_1y_2+x_2y_1)\\ &= (x_1-iy_1)(x_2-iy_2)=z_1^*z_2^*\end{aligned}$$

i.e.

$$\boxed{(z_1+z_2)^*=z_1^*+z_2^*, \qquad (z_1z_2)^*=z_1^*z_2^*}$$

It follows that $(z^2)^*=(z^*)^2$ and $(z^3)^*=(z^*)^3$.
Thus, if $f(z)=az^3+bz^2+cz+d$, where a, b, c, d are real, then

$$\begin{aligned}\{f(z)\}^* &= \{az^3+bz^2+cz+d\}^*\\ &= (az^3)^*+(bz^2)^*+(cz)^*+d^*\\ &= a(z^3)^*+b(z^2)^*+cz^*+d\\ &= a(z^*)^3+b(z^*)^2+cz^*+d=f(z^*)\end{aligned}$$

More generally, it can be shown that $(z^n)^*=(z^*)^n$, where n is a positive integer.

Hence if $P(z)$ is a polynomial of degree n with real coefficients, then

$$\{P(z)\}^* = P(z^*)$$

Suppose now that α is a complex root of the polynomial equation $P(z) = 0$.

$$P(\alpha) = 0 \Leftrightarrow \{P(\alpha)\}^* = \{0\}^* \Leftrightarrow P(\alpha^*) = 0$$

$\therefore$ if α is a root of the equation $P(z) = 0$, its conjugate α^* is also a root.

Thus complex roots of polynomial equations with real coefficients occur in conjugate pairs.

Much of the general theory of polynomial equations is too difficult to discuss here. However, because of its importance, we will state one basic theorem.

Every polynomial equation of degree n has exactly n roots of the form $a + ib$, where a and b are real. Some of these roots may be repeated.

If the polynomial equation $P(z) = 0$ has a real root a, then $P(z)$ has a linear factor $z - a$. Thus, if the equation $P(z) = 0$ has r real roots, then $P(z)$ has r real linear factors.

Similarly, if the equation has a complex root $a + ib$, then since $a - ib$ is also a root, the polynomial $P(z)$ has factors $z - a - ib$ and $z - a + ib$. The product of these factors is $z^2 - 2az + a^2 + b^2$, which is a quadratic factor with real coefficients. Thus, if the equation $P(z) = 0$ has k pairs of conjugate complex roots, then $P(z)$ has k real quadratic factors which cannot be split into real linear factors.

Hence any polynomial with real coefficients is a product of real linear and quadratic factors.

When solving problems it is useful to remember that the formulae for sums and products of roots of polynomial equations remain valid for equations with complex roots.

Example 1 Find the real root of the equation $3z^3 - 10z^2 + 7z + 10 = 0$ given that one root is $2 - i$.

Since $2 - i$ is a root, its complex conjugate $2 + i$ is also a root.

The equation has three roots with sum $-\dfrac{(-10)}{3}$ i.e. $\dfrac{10}{3}$

$\therefore$ the real root of the equation $= \frac{10}{3} - (2 - i + 2 + i) = -\frac{2}{3}$

One cubic equation of special interest is the equation $z^3 = 1$, whose roots are the *cube roots of unity*.

$$z^3 - 1 = 0$$
$$\Rightarrow \quad (z - 1)(z^2 + z + 1) = 0$$
$$\Rightarrow \quad \text{either } z - 1 = 0 \quad \text{or} \quad z^2 + z + 1 = 0$$

Hence the complex cube roots of unity are the roots of the equation $z^2 + z + 1 = 0$. Let ω be one of these complex roots, then

$$\omega^3 = 1 \quad \text{and} \quad \omega^2 + \omega + 1 = 0.$$

But $(\omega^2)^3 = \omega^6 = (\omega^3)^2 = 1^2 = 1$,
$\therefore \quad \omega^2$ must also be a complex cube root of unity.

> Thus the cube roots of 1 are 1, ω, ω^2, where $\omega^3 = 1$ and $\omega^2 + \omega + 1 = 0$.

By solving the equation $z^2 + z + 1 = 0$ we find that the complex cube roots of 1 are $\frac{1}{2}(-1 \pm i\sqrt{3})$. However, when solving problems involving these roots it is often easier to use ω and its properties, rather than the numerical values of the roots.

Example 2 If ω is a complex cube root of unity, find the value of $\omega^3 + \omega^4 + \omega^5$.

Since $\omega^3 = 1$ and $\omega^2 + \omega + 1 = 0$,
$\omega^3 + \omega^4 + \omega^5 = \omega^3(1 + \omega + \omega^2) = 0$.

Example 3 Given that ω is a complex cube root of unity find the equation whose roots are i, $i\omega$, $i\omega^2$.

The roots of the required equation are given by $z = i\alpha$, where α is a cube root of unity.
$\therefore \quad z^3 = (i\alpha)^3 = i^3\alpha^3 = -i$
Hence the equation with roots i, $i\omega$, $i\omega^2$ is $z^3 + i = 0$.

Exercise 14.2

1. Prove that a number z is real if and only if $z = z^*$.

2. Prove that if z is a complex number and k is real, $(kz)^* = kz^*$.

3. If z_1 and z_2 are two complex numbers, prove that

(a) $(z_1 - z_2)^* = z_1^* - z_2^*$, (b) if $z_2 \neq 0$, then $\left(\dfrac{z_1}{z_2}\right)^* = \dfrac{z_1^*}{z_2^*}$.

4. Prove by induction that if z is a complex number then $(z^n)^* = (z^*)^n$, where n is a positive integer.

5. Find the real root of the equation $z^3 + z + 10 = 0$ given that one root is $1 - 2i$.

6. Given that $3 + i$ is a root of the equation $z^3 - 3z^2 - 8z + 30 = 0$, find the remaining roots.

7. Given that $1 + i$ is a root of the equation $z^3 - 2z + k = 0$, find the other two roots and the value of the real constant k.

8. Given that $2 - 3i$ is a root of the equation $z^3 + pz^2 + qz + 13 = 0$, find the other two roots and the value of the real constants p and q.

9. Show that $z = i$ is a root of the equation $z^4 + z^3 + z - 1 = 0$. Find the three other roots.

10. Show that $z = -1 + i$ is a root of the equation $z^4 - 2z^3 - z^2 + 2z + 10 = 0$. Find the remaining roots.

11. Given that $1, \omega_1, \omega_2$ are the roots of the equation $z^3 = 1$, express ω_1 and ω_2 in the form $a + ib$. Hence show that
(a) $(\omega_1)^2 = \omega_2$, (b) $(\omega_2)^2 = \omega_1$, (c) $1/\omega_1 = \omega_2$.

12. Express the roots of the equation $z^3 - \alpha^3 = 0$ in terms of α and ω, where ω is a complex cube root of unity. Use your answer to find the roots of the following equations in the form $a + ib$:
(a) $z^3 - 27 = 0$, (b) $z^3 + 8 = 0$, (c) $z^3 - i = 0$.

13. Find the roots of the following equations in the form $a + ib$:
(a) $z^3 - 8 = 0$, (b) $z^3 + 1 = 0$, (c) $(z + 1)^3 = 1$.

14. Evaluate the following expressions given that ω is a complex cube root of unity.

(a) $\omega + \omega^3 + \omega^5$, (b) $\dfrac{1}{\omega^2 + \omega^4}$, (c) $\dfrac{\omega^2}{\omega + \omega^3}$.

15. Given that ω is a complex cube root of unity, show that $\omega^* = \omega^2$. Hence write down in terms of ω the conjugates of
(a) $1 + \omega$, (b) $1 - \omega^2$, (c) $3 + 4\omega + 5\omega^2$.

16. Given that ω is a complex cube root of unity, simplify the expression

$$(a + b + c)(a + \omega b + \omega^2 c)(a + \omega^2 b + \omega c).$$

17. Given that $1, \omega, \omega^2$ are the cube roots of unity, find the equation whose roots are $\dfrac{1}{3}, \dfrac{1}{2 + \omega}, \dfrac{1}{2 + \omega^2}$.

18. Given that ω denotes either one of the non-real roots of the equation $z^3 = 1$, show that (i) $1 + \omega + \omega^2 = 0$, and (ii) the other non-real root is ω^2.

Show that the non-real roots of the equation $\left(\frac{1-u}{u}\right)^3 = 1$ can be expressed in the form $A\omega$ and $B\omega^2$, where A and B are real numbers, and find A and B.

(JMB)

14.3 Geometrical representation

The complex number $z = x + iy$ is completely specified by the ordered pair of real numbers (x, y). Thus to any complex number z there corresponds a unique point $P(x, y)$ in the x, y plane. This suggests that z may be represented by

(1) the point P with coordinates (x, y),
or (2) the vector $\overrightarrow{OP}$, where O is the origin,
or (3) any vector equal to $\overrightarrow{OP}$ in magnitude and direction.

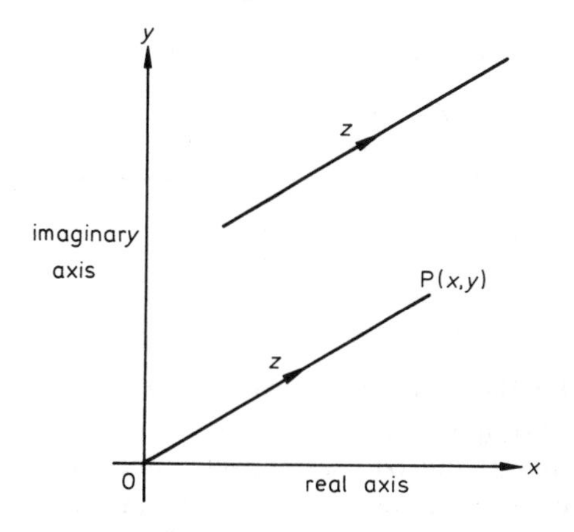

A diagram showing complex numbers represented by points or vectors is called an *Argand*† *diagram*.
Since real numbers are represented along the x-axis and imaginary numbers along the y-axis, in an Argand diagram these axes are often referred to as the *real axis* and the *imaginary axis* respectively.

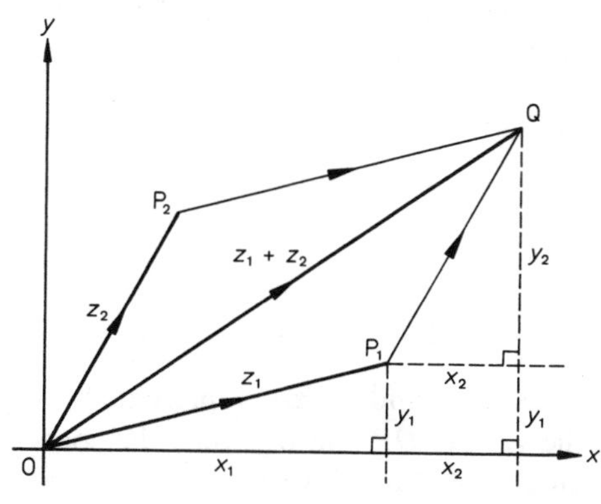

If $z_1 = x_1 + iy_1$ and $z_2 = x_2 + iy_2$, then $z_1 + z_2 = (x_1 + x_2) + i(y_1 + y_2)$. Let z_1, z_2 and $z_1 + z_2$ be represented by the vectors $\overrightarrow{OP_1}$, $\overrightarrow{OP_2}$ and $\overrightarrow{OQ}$ respectively. The diagram shows that $\overrightarrow{P_1Q}$ is equal to $\overrightarrow{OP_2}$ in magnitude and direction

$$\therefore \quad \overrightarrow{OQ} = \overrightarrow{OP_1} + \overrightarrow{P_1Q} = \overrightarrow{OP_1} + \overrightarrow{OP_2}.$$

Thus the sum of two complex numbers z_1 and z_2 is represented in an Argand diagram by the sum of the corresponding vectors $\overrightarrow{OP_1}$ and $\overrightarrow{OP_2}$.

† *Argand, Jean Robert* (1768–1822) Swiss mathematician. He published the geometrical representation of complex numbers in an "Essai" in 1806.

There are two ways of obtaining a geometrical representation of $z_1 - z_2$. If the complex number $(-z_2)$ is represented by the vector $\overrightarrow{OP_3}$, then $z_1 - z_2$ is represented by the sum of the vectors $\overrightarrow{OP_1}$ and $\overrightarrow{OP_3}$, as shown in diagram (i). However, since $\overrightarrow{OP_1} - \overrightarrow{OP_2} = \overrightarrow{P_2P_1}$, $z_1 - z_2$ can also be represented by $\overrightarrow{P_2P_1}$, as shown in diagram (ii).

(i)

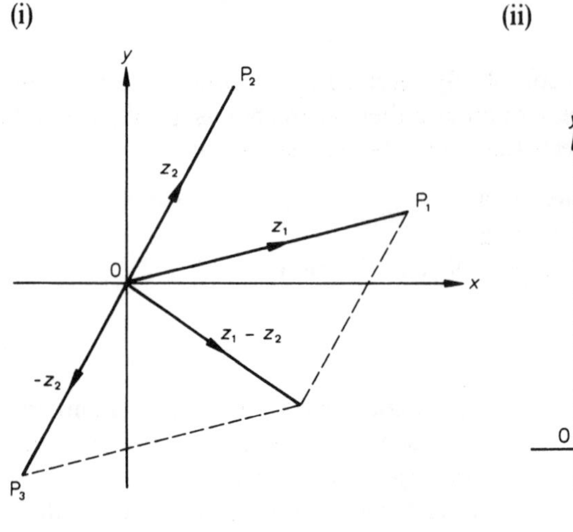

(ii)

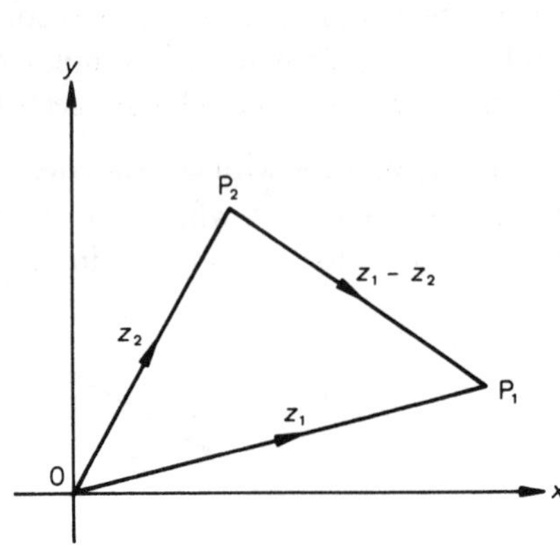

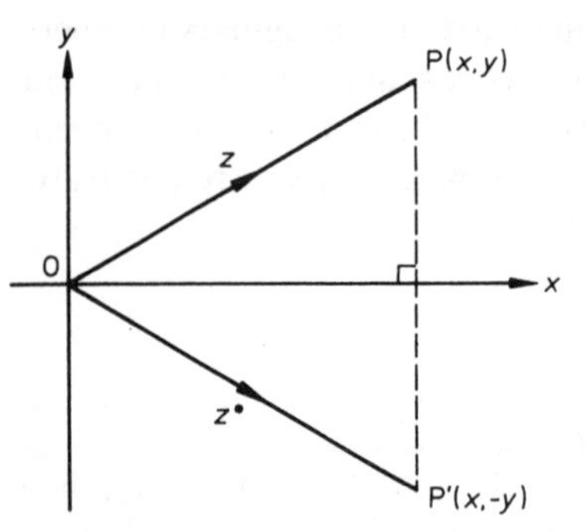

Let the complex number $z = x + iy$ and its conjugate $z^* = x - iy$ be represented by $\overrightarrow{OP}$ and $\overrightarrow{OP'}$ respectively. Since the coordinates of P and P' are (x, y) and $(x, -y)$ respectively, P' must be the reflection of P in the real axis.

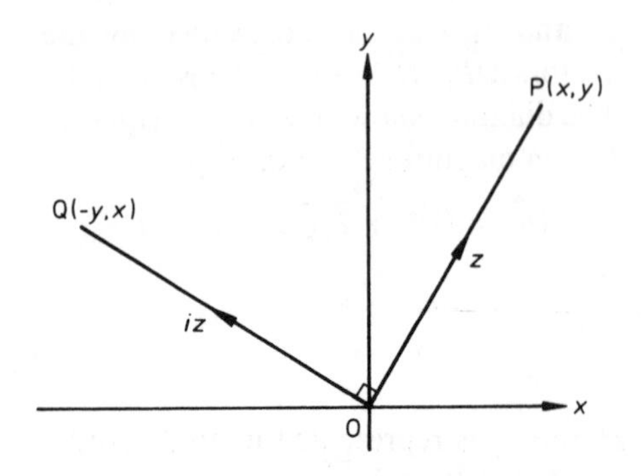

If $z = x + iy$, then

$$iz = i(x + iy) = -y + ix.$$

Thus if z and iz are represented by the vectors $\overrightarrow{OP}$ and $\overrightarrow{OQ}$, then the coordinates of P and Q are (x, y) and $(-y, x)$ respectively.

Since the gradient of OP is y/x and the gradient of OQ is $-x/y$, OP must be perpendicular to OQ. Hence multiplication of a complex number z by i results in an anti-clockwise rotation of the vector $\overrightarrow{OP}$ through $\frac{1}{2}\pi$ radians.

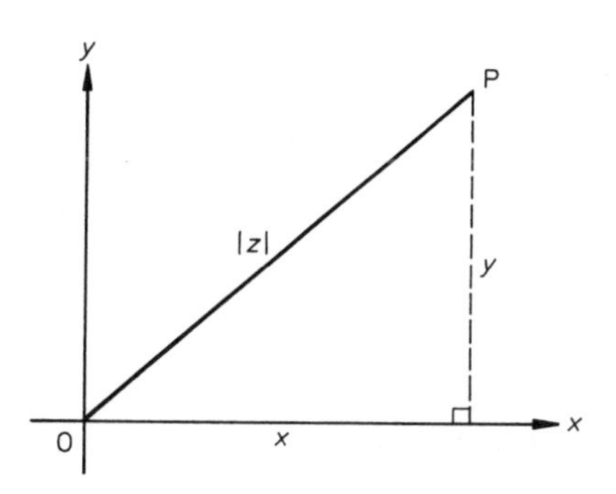

The *modulus* of a complex number z is defined as the length of the corresponding vector $\overrightarrow{OP}$ and is denoted by $|z|$.

Thus if $z = x + iy$, $|z| = \sqrt{(x^2 + y^2)}$.

For instance, $\qquad |3 + 4i| = \sqrt{\{3^2 + 4^2\}} = 5$

and $\qquad |1 - 2i| = \sqrt{\{1^2 + (-2)^2\}} = \sqrt{5}.$

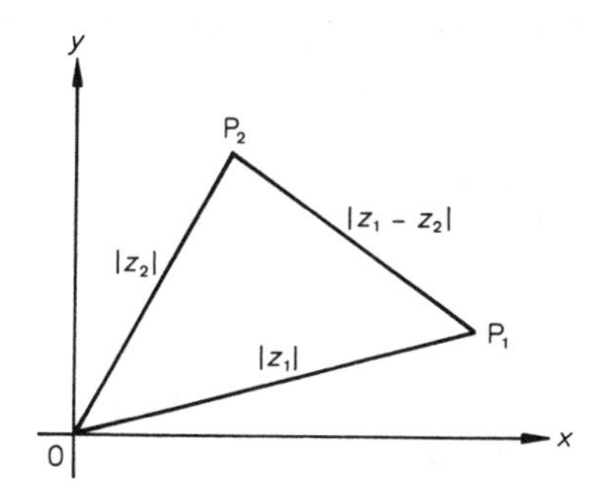

If complex numbers z_1 and z_2 are represented by the vectors $\overrightarrow{OP_1}$ and $\overrightarrow{OP_2}$ respectively, then $z_1 - z_2$ is represented by $\overrightarrow{P_2P_1}$.

$$\therefore \qquad |z_1 - z_2| = |\overrightarrow{P_2P_1}|$$

Thus $|z_1 - z_2|$ is the distance between the points P_1 and P_2.

Example 1 Given that the complex numbers $2 + 5i$ and $3 - 2i$ are represented in an Argand diagram by the points A and B, find the length of AB.

$$AB = |(2 + 5i) - (3 - 2i)|$$
$$= |-1 + 7i| = \sqrt{\{(-1)^2 + 7^2\}} = \sqrt{50} = 5\sqrt{2}.$$

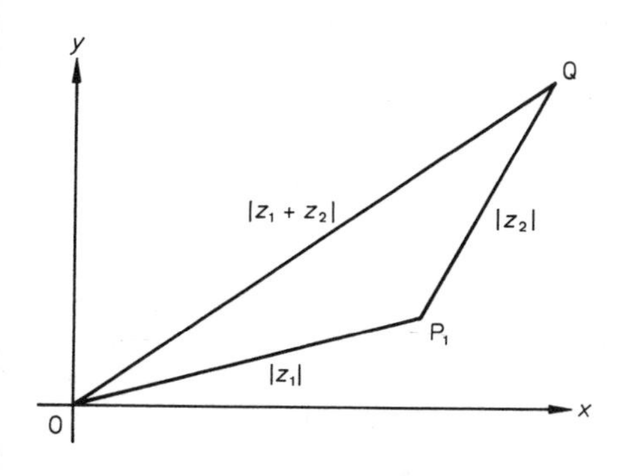

If complex numbers z_1, z_2 and $z_1 + z_2$ are represented by the vectors $\overrightarrow{OP_1}$, $\overrightarrow{OP_2}$ and $\overrightarrow{OQ}$, then as shown earlier, z_2 is also represented by $\overrightarrow{P_1Q}$. We now use triangle OP_1Q to produce a set of results called the *triangle inequalities*.

Since $\quad OQ \leqslant OP_1 + P_1Q, \qquad |z_1 + z_2| \leqslant |z_1| + |z_2|$

Since $\quad OP_1 \leqslant P_1Q + OQ, \qquad |z_1| \leqslant |z_2| + |z_1 + z_2|$

$\therefore \qquad |z_1| - |z_2| \leqslant |z_1 + z_2|$

Since $P_1Q \leqslant OP_1 + OQ$, $\quad |z_2| \leqslant |z_1| + |z_1 + z_2|$

$\therefore \quad |z_2| - |z_1| \leqslant |z_1 + z_2|$

Combining these three results, we have

$$\big||z_1| - |z_2|\big| \leqslant |z_1 + z_2| \leqslant |z_1| + |z_2|$$

[For an alternative method of proof see Exercise 14.4 question 14.]

Exercise 14.3

1. Write down the complex numbers corresponding to the following points in an Argand diagram.
(a) $(3, 2)$, (b) $(-2, 4)$, (c) $(0, 5)$, (d) $(1, -2)$, (e) $(-2, 0)$.

2. Represent the following complex numbers by vectors in an Argand diagram and find the modulus of the complex number in each case.
(a) $4 + 3i$, (b) $2 - 5i$, (c) $3i$, (d) $-1 - i$, (e) -4.

3. Represent the complex number $z = 2 + i$ by a vector in an Argand diagram. Find and represent in the same diagram $-z, z^*, -z^*, iz$ and $-iz$.

4. Represent the complex number $z = 1 + i$ by a vector in an Argand diagram. Find and represent in the same diagram z^2, z^3 and $1/z$.

5. If $z_1 = 1 + 3i$ and $z_2 = 3 - i$, find $z_1 z_2$ and z_1/z_2 in the form $a + ib$. Show on an Argand diagram the vectors representing $z_1, z_2, z_1 z_2$ and z_1/z_2.

6. Find and represent in an Argand diagram the square roots of
(a) -1, (b) $-2i$, (c) $3 + 4i$, (d) $15 + 8i$.

7. In each of the following cases represent z_1, z_2 and $z_1 + z_2$ in an Argand diagram and verify that

$$\big||z_1| - |z_2|\big| \leqslant |z_1 + z_2| \leqslant |z_1| + |z_2|.$$

(a) $z_1 = 12 + 5i, z_2 = -3 + 4i$, (b) $z_1 = -1 - i, z_2 = 7 - i$.

8. If $z_1 = 1 + 2i$, find the set of values of z_2 for which
(a) $|z_1 + z_2| = |z_1| + |z_2|$, (b) $|z_1 + z_2| = |z_1| - |z_2|$,
(c) $|z_1 + z_2| = |z_2| - |z_1|$.

9. If z_1 and z_2 are complex numbers, show geometrically that

$$\big||z_1| - |z_2|\big| \leqslant |z_1 - z_2| \leqslant |z_1| + |z_2|.$$

10. In each of the following cases find the distance between points in the Argand diagram representing the given pair of complex numbers.
(a) $2 + i, 5 - 3i$, (b) $1 - 2i, 3 + 7i$, (c) $2 - 5i, -1 - 3i$.

11. The points A, B, C and D in the Argand diagram correspond to the complex numbers $8 - i$, $3 + 11i$, $-9 + 6i$, $-4 - 6i$. Prove that $ABCD$ is a square.

12. Given that $z = 1 + i\sqrt{2}$, express in the form $a + ib$ each of the complex numbers $p = z + \frac{1}{z}$, $q = z - \frac{1}{z}$. In an Argand diagram, P and Q are the points which represent p and q respectively, O is the origin, M is the midpoint of PQ and G is the point on OM such that $OG = \frac{2}{3}OM$. Prove that the angle PGQ is a right angle. (JMB)

14.4 Modulus-argument form

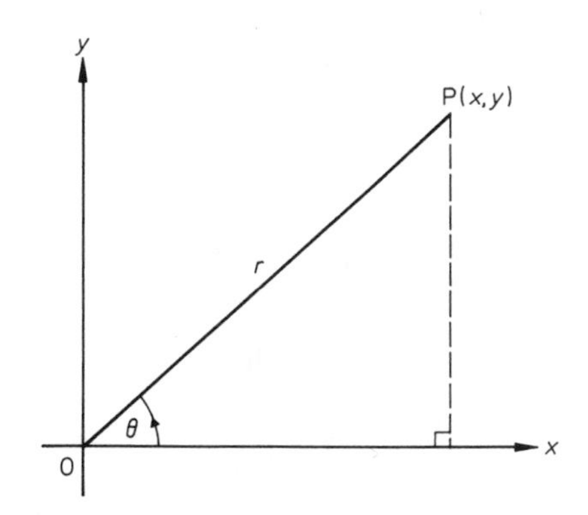

Let us suppose that a complex number $z = x + iy$ is represented in an Argand diagram by the vector $\overrightarrow{OP}$, where r is the length of $\overrightarrow{OP}$ and θ is the angle $\overrightarrow{OP}$ makes with the positive direction of the real axis.

From the diagram we see that

$$x = r\cos\theta, \qquad y = r\sin\theta,$$
$$r^2 = x^2 + y^2, \quad \tan\theta = y/x.$$

As we already know, the length r is called the modulus of z and written $|z|$,

$$\therefore \qquad \boxed{|z| = r = \sqrt{(x^2 + y^2)}}$$

The angle θ is called the *argument* of z (or less commonly, the amplitude), positive values of θ being measured anti-clockwise from the real axis and negative values clockwise. The problem with this definition is that for any given z there is an infinite set of possible values for θ. To avoid confusion we usually work with the value of θ for which $-\pi < \theta \leqslant \pi$. This is called the *principal argument* of z and denoted by arg z.

$$\therefore \qquad \boxed{\arg z = \theta, \quad \text{where} \quad \sin\theta = \frac{x}{r}, \qquad \cos\theta = \frac{y}{r} \quad \text{and} \quad -\pi < \theta \leqslant \pi.}$$

In practice the formula $\tan\theta = y/x$ is often used to find the principal argument of a complex number z, despite the fact that it leads to two possible values for θ in the permitted range. The correct value of arg z is then chosen with the aid of a sketch showing the approximate position of the corresponding vector $\overrightarrow{OP}$.

Example 1 Find the moduli and principal arguments of the complex numbers (a) $1 + i$, (b) $-1 - i\sqrt{3}$, (c) -5.

(a)

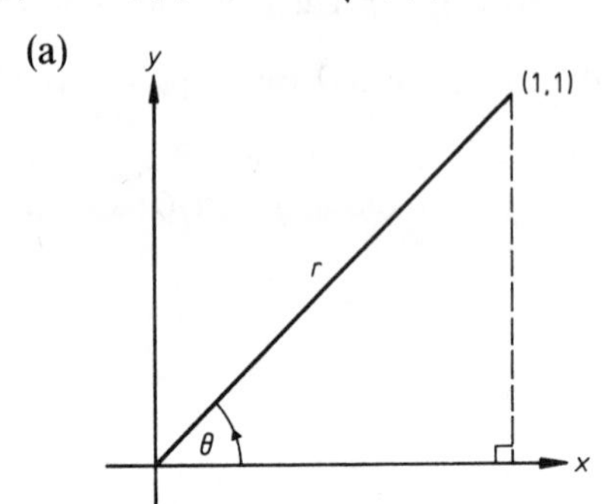

$|1 + i| = \sqrt{(1^2 + 1^2)} = \sqrt{2}$
If $\arg(1 + i) = \theta$,
then $\tan\theta = \frac{1}{1} = 1 = \tan\frac{1}{4}\pi$
Since the point representing $1 + i$ lies in the first quadrant, $0 < \theta < \frac{1}{2}\pi$.

$\therefore \quad \theta = \frac{1}{4}\pi$

Hence $|1 + i| = \sqrt{2}$ and $\arg(1 + i) = \frac{1}{4}\pi$.

(b)

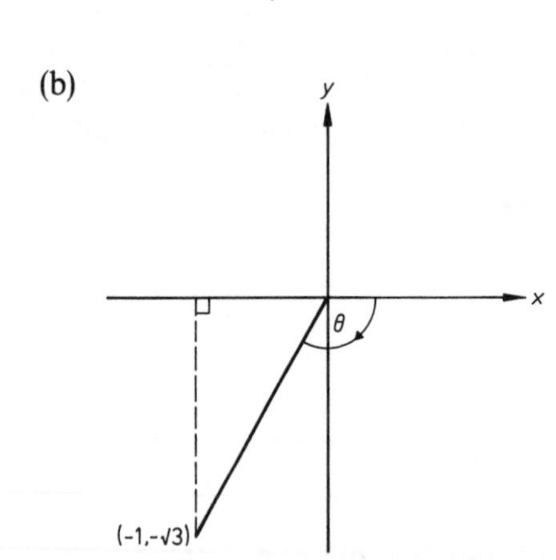

$$|-1 - i\sqrt{3}| = \sqrt{\{(-1)^2 + (-\sqrt{3})^2\}} = \sqrt{\{1 + 3\}} = 2$$

If $\arg(-1 - i\sqrt{3}) = \theta$, then

$$\tan\theta = \frac{-\sqrt{3}}{-1} = \sqrt{3} = \tan\tfrac{1}{3}\pi$$

Since the point representing $-1 - i\sqrt{3}$ lies in the third quadrant,

$$-\pi < \theta < -\tfrac{1}{2}\pi.$$

$$\therefore \quad \theta = \tfrac{1}{3}\pi - \pi = -\tfrac{2}{3}\pi.$$

Hence $|-1 - i\sqrt{3}| = 2$ and $\arg(-1 - i\sqrt{3}) = -\frac{2}{3}\pi$.

(c)

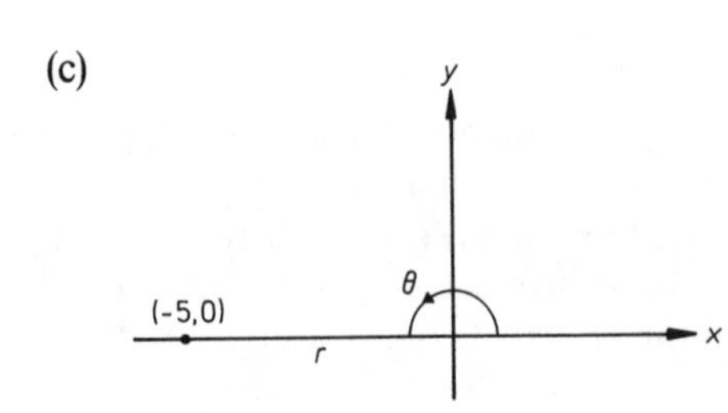

Since the point representing -5 lies on the negative part of the real axis, at a distance of 5 units from the origin, $|-5| = 5$ and $\arg(-5) = \pi$.

When obtaining the argument of a complex number by calculator, care must be taken to use the inverse tangent function properly. Since this function always takes values between $-\frac{1}{2}\pi$ and $+\frac{1}{2}\pi$, it gives the correct value of arg z only when z is represented by a point in the first or fourth quadrants. When the point representing z lies in the second or third quadrants, it will be necessary to add $\pm\pi$ to the value obtained for $\tan^{-1}(y/x)$.

y

arg $z = \tan^{-1}(y/x) + \pi$ | arg $z = \tan^{-1}(y/x)$

x

arg $z = \tan^{-1}(y/x) - \pi$ | arg $z = \tan^{-1}(y/x)$

Example 2 Find in radians the principal arguments of the complex numbers $1 - 2i$ and $-1 + 2i$.

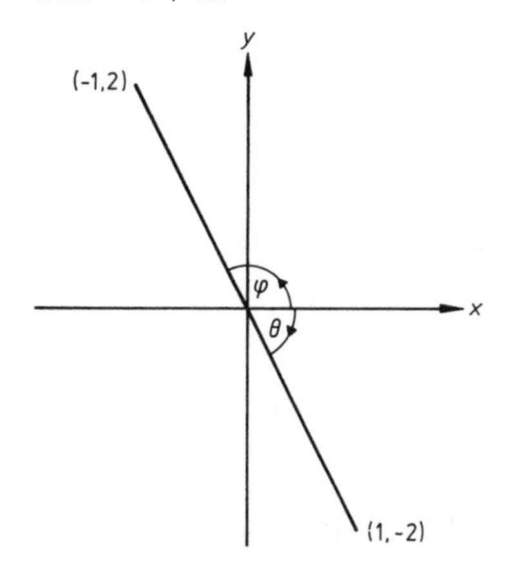

If $\arg(1-2i)=\theta$, then $\tan\theta=\dfrac{-2}{1}=-2$

Since the point $(1, -2)$ lies in the 4th quadrant,

$\arg(1 - 2i) = \tan^{-1}(-2) \approx -1{\cdot}107$

If $\arg(-1 + 2i) = \phi$, then

$$\tan\phi = \frac{2}{-1} = -2$$

Since the point $(-1, 2)$ lies in the 2nd quadrant,
$\arg(-1 + 2i) = \tan^{-1}(-2) + \pi \approx 2{\cdot}034.$

If a complex number $z = x + iy$ has modulus r and argument θ, then $x = r\cos\theta$ and $y = r\sin\theta$

$$\therefore \qquad z = r(\cos\theta + i\sin\theta)$$

This is called the *modulus-argument* form of the number.
Using the results of Example 1 the numbers $1 + i$, $-1 - i\sqrt{3}$ and -5 have expressions in modulus-argument form as follows:

$$1 + i = \sqrt{2}(\cos\tfrac{1}{4}\pi + i\sin\tfrac{1}{4}\pi),$$
$$-1 - i\sqrt{3} = 2\{\cos(-\tfrac{2}{3}\pi) + i\sin(-\tfrac{2}{3}\pi)\},$$
$$-5 = 5(\cos\pi + i\sin\pi).$$

Example 3 Write down the moduli and arguments of (a) $3(\cos 2\alpha + i\sin 2\alpha)$, (b) $\cos\frac{2}{5}\pi + i\sin\frac{2}{5}\pi$, (c) $a(\cos\phi - i\sin\phi)$.

(a) Since $r(\cos\theta + i\sin\theta)$ has modulus r and argument θ, $3(\cos 2\alpha + i\sin 2\alpha)$ has modulus 3 and argument 2α.
(b) Since $\cos\frac{2}{5}\pi + i\sin\frac{2}{5}\pi = 1(\cos\frac{2}{5}\pi + i\sin\frac{2}{5}\pi)$, this number has modulus 1 and argument $\frac{2}{5}\pi$.
(c) Since $a(\cos\phi - i\sin\phi) = a\{\cos(-\phi) + i\sin(-\phi)\}$, its modulus is a and its argument is $(-\phi)$.

We now investigate the general properties of complex numbers in modulus-argument form.
If $z = r(\cos\theta + i\sin\theta)$, then $z^* = r(\cos\theta - i\sin\theta)$.
$\therefore\quad zz^* = r^2(\cos\theta + i\sin\theta)(\cos\theta - i\sin\theta) = r^2(\cos^2\theta + \sin^2\theta) = r^2$.
Since z^* may also be written as $r\{\cos(-\theta) + i\sin(-\theta)\}$, the modulus of z^* is r and its argument is $(-\theta)$. It follows that:

$$zz^* = |z|^2, |z^*| = |z| \text{ and, for } -\pi < \arg z < \pi, \arg z^* = -\arg z.$$

If $z = r(\cos\theta + i\sin\theta)$, then $\dfrac{1}{z} = \dfrac{1}{r(\cos\theta + i\sin\theta)}$

$$\therefore\quad \frac{1}{z} = \frac{\cos\theta - i\sin\theta}{r(\cos\theta + i\sin\theta)(\cos\theta - i\sin\theta)}$$

$$= \frac{\cos\theta - i\sin\theta}{r(\cos^2\theta + \sin^2\theta)} = \frac{1}{r}\{\cos(-\theta) + i\sin(-\theta)\}$$

$$\therefore\quad \left|\frac{1}{z}\right| = \frac{1}{|z|} \text{ and, for } -\pi < \arg z < \pi, \arg\left(\frac{1}{z}\right) = -\arg z.$$

If $z_1 = r_1(\cos\theta_1 + i\sin\theta_1)$ and $z_2 = r_2(\cos\theta_2 + i\sin\theta_2)$, then

$$\begin{aligned} z_1 z_2 &= r_1 r_2(\cos\theta_1 + i\sin\theta_1)(\cos\theta_2 + i\sin\theta_2) \\ &= r_1 r_2\{(\cos\theta_1\cos\theta_2 - \sin\theta_1\sin\theta_2) + i(\sin\theta_1\cos\theta_2 + \cos\theta_1\sin\theta_2)\} \\ &= r_1 r_2\{\cos(\theta_1 + \theta_2) + i\sin(\theta_1 + \theta_2)\} \end{aligned}$$

Hence $z_1 z_2$ has modulus $r_1 r_2$ and argument $\theta_1 + \theta_2$.

$$\begin{aligned} \frac{z_1}{z_2} &= \frac{r_1(\cos\theta_1 + i\sin\theta_1)(\cos\theta_2 - i\sin\theta_2)}{r_2(\cos\theta_2 + i\sin\theta_2)(\cos\theta_2 - i\sin\theta_2)} \\ &= \frac{r_1\{(\cos\theta_1\cos\theta_2 + \sin\theta_1\sin\theta_2) + i(\sin\theta_1\cos\theta_2 - \cos\theta_1\sin\theta_2)\}}{r_2(\cos^2\theta_2 + \sin^2\theta_2)} \\ &= \frac{r_1}{r_2}\{\cos(\theta_1 - \theta_2) + i\sin(\theta_1 - \theta_2)\} \end{aligned}$$

Hence z_1/z_2 has modulus r_1/r_2 and argument $\theta_1 - \theta_2$.

$\therefore$
$$|z_1 z_2| = |z_1||z_2| \quad \text{and} \quad \left|\frac{z_1}{z_2}\right| = \frac{|z_1|}{|z_2|}$$

If $\theta_1 = \arg z_1$ and $\theta_2 = \arg z_2$, it follows that $z_1 z_2$ has argument

$$\theta_1 + \theta_2 = \arg z_1 + \arg z_2.$$

However, since $-\pi < \theta_1 \leqslant \pi$ and $-\pi < \theta_2 \leqslant \pi$, $\theta_1 + \theta_2$ may itself lie outside the interval from $-\pi$ to π. In such cases $\arg z_1 + \arg z_2$ does not give the principal argument of $z_1 z_2$ and it is necessary to add or subtract 2π to obtain $\arg(z_1 z_2)$. Similar adjustments may be needed when finding $\arg(z_1/z_2)$.

$\therefore$
$$\arg(z_2 z_2) = \arg z_1 + \arg z_2 \ (\pm 2\pi \text{ if necessary}),$$
$$\arg(z_1/z_2) = \arg z_1 - \arg z_2 \ (\pm 2\pi \text{ if necessary}).$$

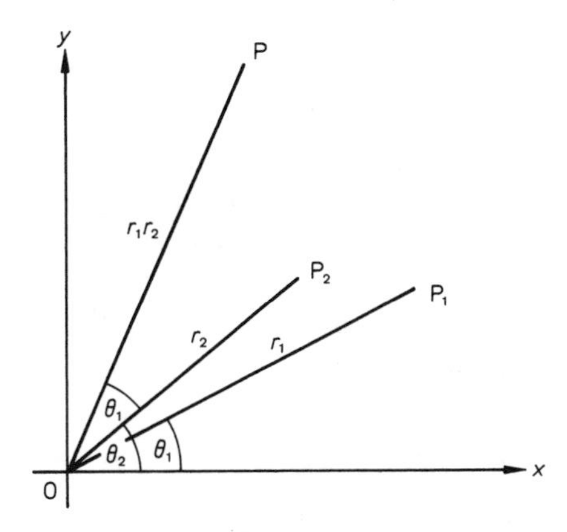

Let z_1 and z_2 be represented in an Argand diagram by the vectors $\overrightarrow{OP_1}$ and $\overrightarrow{OP_2}$ respectively. When z_2 is multiplied by z_1 the vector $\overrightarrow{OP_2}$ is rotated through an angle θ_1 and its length is multiplied by r_1 to produce the vector $\overrightarrow{OP}$ representing $z_1 z_2$.

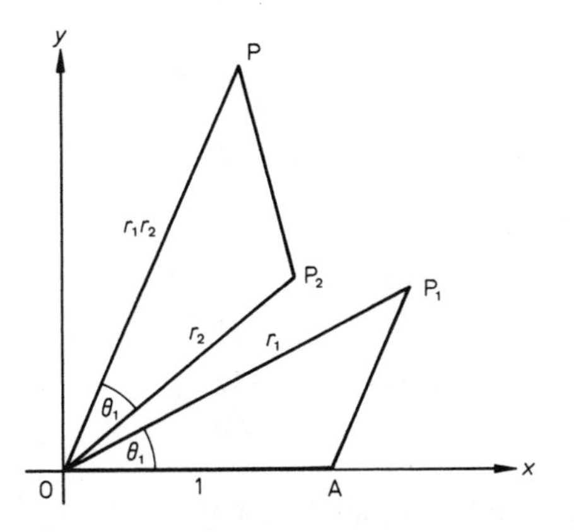

In this diagram we see that if A is the point $(1, 0)$, then triangles OAP_1 and OP_2P are similar.

[There are similar geometrical representations of the quotient z_1/z_2.]

Example 4 If $z_1 = 6(\cos\frac{5}{12}\pi + i\sin\frac{5}{12}\pi)$ and $z_2 = 3(\cos\frac{1}{4}\pi + i\sin\frac{1}{4}\pi)$, find z_1z_2 and z_1/z_2 in the form $a + ib$.

$$z_1z_2 = 6 \times 3\{\cos(\tfrac{5}{12}\pi + \tfrac{1}{4}\pi) + i\sin(\tfrac{5}{12}\pi + \tfrac{1}{4}\pi)\}$$

$$= 18\{\cos\tfrac{2}{3}\pi + i\sin\tfrac{2}{3}\pi\} = 18\left\{-\frac{1}{2} + \frac{\sqrt{3}}{2}i\right\} = -9 + 9\sqrt{3}i$$

$$\frac{z_1}{z_2} = \frac{6}{3}\{\cos(\tfrac{5}{12}\pi - \tfrac{1}{4}\pi) + i\sin(\tfrac{5}{12}\pi - \tfrac{1}{4}\pi)\}$$

$$= 2\{\cos\tfrac{1}{6}\pi + i\sin\tfrac{1}{6}\pi\} = 2\left\{\frac{\sqrt{3}}{2} + \frac{1}{2}i\right\} = \sqrt{3} + i.$$

Exercise 14.4

In questions 1 to 4 find the modulus and principal argument of each of the given complex numbers. Give the argument in radians, either as a multiple of π or correct to 3 decimal places.

1. (a) $\sqrt{3} + i$ (b) $1 - i$ (c) $4i$ (d) -3

2. (a) $3 - 4i$ (b) $-2 + i$ (c) $-1 - 3i$ (d) $5 - 3i$

3. (a) $\dfrac{10}{\sqrt{3} - i}$ (b) $\dfrac{1}{1 + i\sqrt{3}}$ (c) $\dfrac{2 + 3i}{5 + i}$ (d) $\dfrac{7 - i}{-4 - 3i}$

4. (a) $5(\cos\frac{1}{3}\pi + i\sin\frac{1}{3}\pi)$ (b) $\cos\frac{3}{2}\pi + i\sin\frac{3}{2}\pi$
(c) $\cos\frac{2}{3}\pi - i\sin\frac{2}{3}\pi$ (d) $-2(\cos\frac{1}{4}\pi + i\sin\frac{1}{4}\pi)$

5. Express the following in the form $r(\cos\theta + i\sin\theta)$, where $r > 0$ and $-\pi < \theta \leqslant \pi$,
(a) $\sqrt{3} - 3i$ (b) 8 (c) $-2 - 2i$ (d) $-i$

6. If $z = r(\cos\theta + i\sin\theta)$, where $r > 0$ and $0 < \theta < \frac{1}{2}\pi$, find in terms of r and θ the modulus and principal argument of
(a) $-z$ (b) iz (c) z^2 (d) z/z^*

7. Simplify the following expressions:
(a) $(\cos\frac{1}{4}\pi + i\sin\frac{1}{4}\pi)(\cos\frac{3}{4}\pi + i\sin\frac{3}{4}\pi)$ (b) $(\cos\frac{5}{12}\pi + i\sin\frac{5}{12}\pi)^2$
(c) $(\cos\frac{2}{3}\pi + i\sin\frac{2}{3}\pi)(\cos\frac{7}{12}\pi + i\sin\frac{7}{12}\pi)$ (d) $(\cos\frac{2}{9}\pi + i\sin\frac{2}{9}\pi)^3$
(e) $\dfrac{(\cos\frac{1}{3}\pi + i\sin\frac{1}{3}\pi)}{(\cos\frac{5}{6}\pi + i\sin\frac{5}{6}\pi)}$ (f) $\dfrac{(\cos\frac{1}{4}\pi + i\sin\frac{1}{4}\pi)^2}{(\cos\frac{1}{6}\pi + i\sin\frac{1}{6}\pi)}$

8. Find the moduli and principal arguments of w, z, wz and w/z, given that:
(a) $w = 10i$, $z = 1 + i\sqrt{3}$, (b) $w = -2\sqrt{3} + 2i$, $z = 1 - i$.

9. Given that $z = 2/(1 + \cos 2\theta - i \sin 2\theta)$, where $-\frac{1}{2}\pi < \theta < \frac{1}{2}\pi$, find expressions for $|z|$ and $\arg z$ in terms of θ.

10. Given that $z = \cos\theta + i \sin\theta$, where $0 < \theta < \pi$, find the moduli and principal arguments of $1 - z$, $z - 1$, $1/(1 - z)$, $z/(z - 1)$.

11. If $z_1 = 1 + i\sqrt{3}$ and $z_2 = 2i$, find $|z_1|$ and $\arg z_1$, $|z_2|$ and $\arg z_2$. Using an Argand diagram, deduce that $\arg(z_1 + z_2) = 5\pi/12$. Hence show that $\tan(5\pi/12) = 2 + \sqrt{3}$.

12. Two non-zero complex numbers z_1 and z_2 are such that $|z_1 + z_2| = |z_1 - z_2|$. Represent $z_1, z_2, z_1 + z_2$ and $z_1 - z_2$ by vectors on an Argand diagram. Hence, or otherwise, find the possible values of $\arg(z_1/z_2)$. (JMB)

13. (a) Express the complex number $z = \dfrac{1 + i}{3 - 2i}$ in the form $a + bi$ $(a, b \in \mathbb{R})$, and find the argument of z, giving your answer in radians correct to 3 significiant figures. Write down the argument of z^2, and find the exact value of $|z^2|$.
(b) The conjugate of the non-zero complex number z is denoted by z^*. In an Argand diagram with origin O, the point P represents z and Q represents $1/z^*$. Prove that O, P and Q are collinear, and find the ratio $OP:OQ$ in terms of $|z|$. (C)

14. Show that, for any complex number z, (a) $zz^* = |z|^2$, (b) $z + z^* = 2\,\mathrm{Re}(z)$, (c) $\mathrm{Re}(z) \leqslant |z|$. Hence show that
(i) $|z_1 + z_2|^2 = |z_1|^2 + |z_2|^2 + 2\,\mathrm{Re}(z_1 z_2^*)$,
(ii) $|z_1 + z_2| \leqslant |z_1| + |z_2|$.

14.5 Loci in the Argand diagram

Let us consider again a complex number $z = x + iy$ represented by the vector $\overrightarrow{OP}$, where P is the point (x, y). If z varies subject to some given condition, then the corresponding set of points in the Argand diagram is called the *locus* of P. Since many loci can be defined using either distances or angles, equations of loci in the Argand diagram often involve moduli or arguments of complex variables.

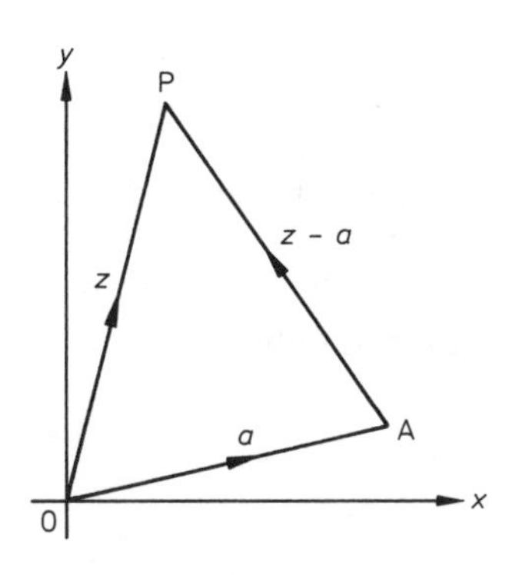

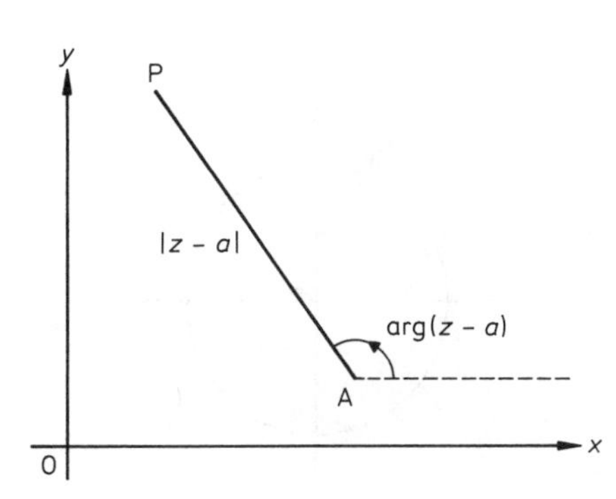

If a is constant complex number represented by a fixed vector $\overrightarrow{OA}$, then $z - a$ is represented by $\overrightarrow{AP}$. Hence $|z - a|$ is represented by the distance AP and $\arg(z - a)$ is the angle $\overrightarrow{AP}$ makes the positive direction of the real axis.

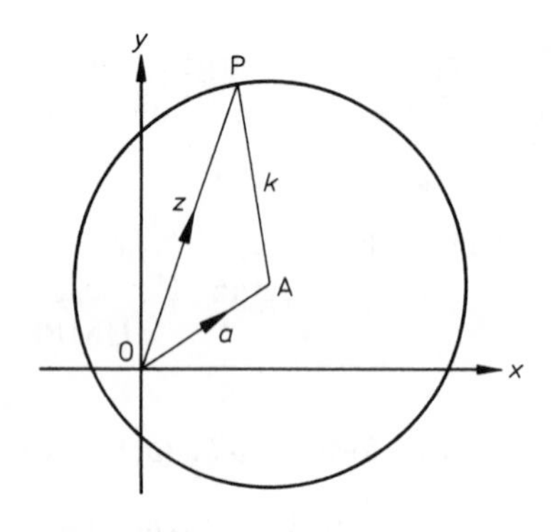

It follows that if $|z - a| = k$, where k is a positive constant, then the distance of P from A is constant. Thus

> $|z - a| = k$ is the equation of the circle with centre at the point A and radius k.

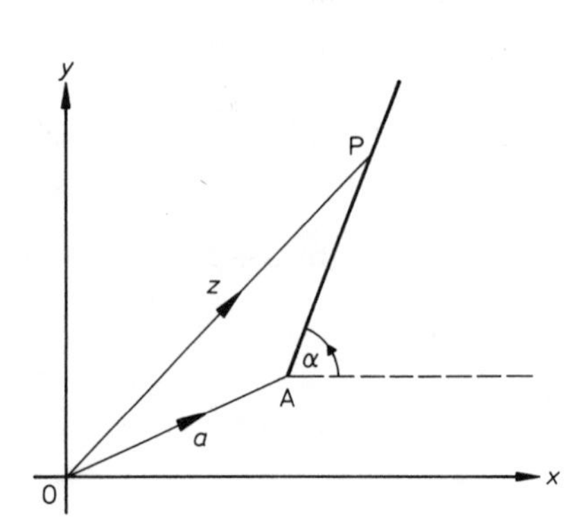

However, if $\arg(z - a) = \alpha$, where α is constant and $-\pi < \alpha \leqslant \pi$, then the direction of $\overrightarrow{AP}$ is constant. Thus

> $\arg(z - a) = \alpha$ is the equation of a *half-line* with end-point A, inclined at an angle α to the real axis.

Example 1 Sketch the loci defined by the equations (a) $|z| = 2$, (b) $|z - 3 + 2i| = 5$.

Let z be represented in the Argand diagram by the point P.

(a)

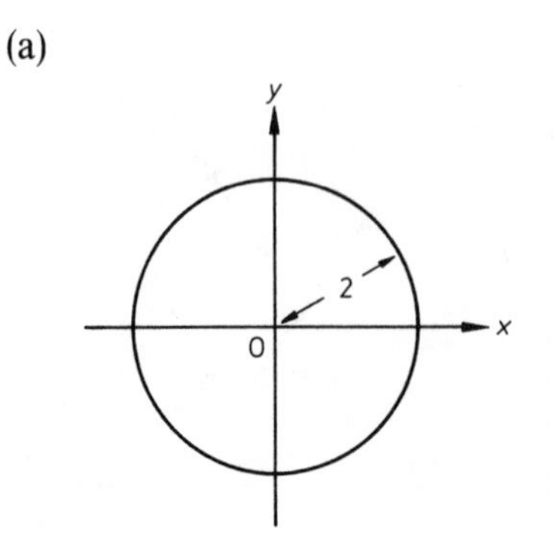

If $|z| = 2$, then the distance of P from the origin is 2 units.
Hence the locus of P is a circle with centre the origin and radius 2.

(b)

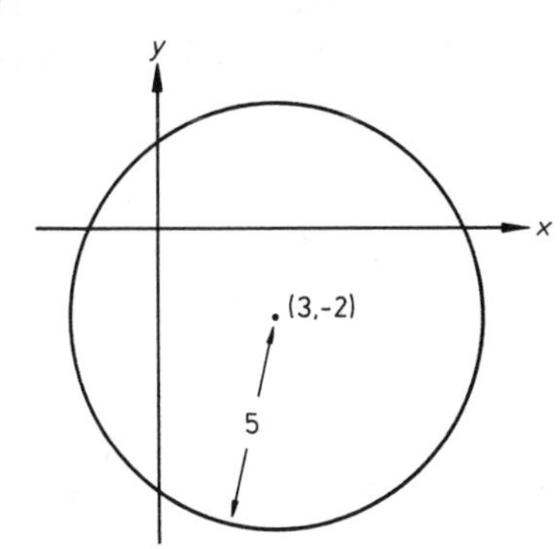

If $|z - 3 + 2i| = 5$,
then $|z - (3 - 2i)| = 5$,
i.e. the distance of P from the point representing $3 - 2i$ is 5 units.
Hence the locus of P is a circle with centre $(3, -2)$ and radius 5.

Example 2 Sketch the loci defined by the equations (a) $\arg(z - 1 - 2i) = \frac{1}{4}\pi$, (b) $\arg(z + 2) = -\frac{2}{3}\pi$.

Let z be represented in the Argand diagram by the point P.

(a)

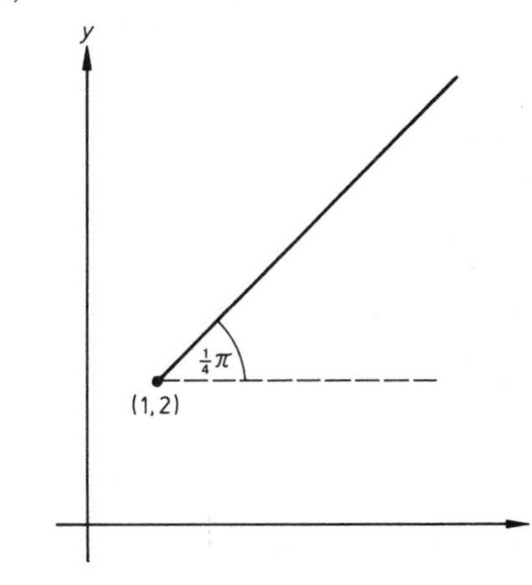

$z - 1 - 2i = z - (1 + 2i)$
Thus, if A is the point representing $1 + 2i$, $\arg(z - 1 - 2i)$ is the angle which $\overrightarrow{AP}$ makes with the real axis.
Hence the equation

$$\arg(z - 1 - 2i) = \tfrac{1}{4}\pi$$

represents the half-line with end-point $(1, 2)$, inclined at an angle $\frac{1}{4}\pi$ to the real axis.

(b)

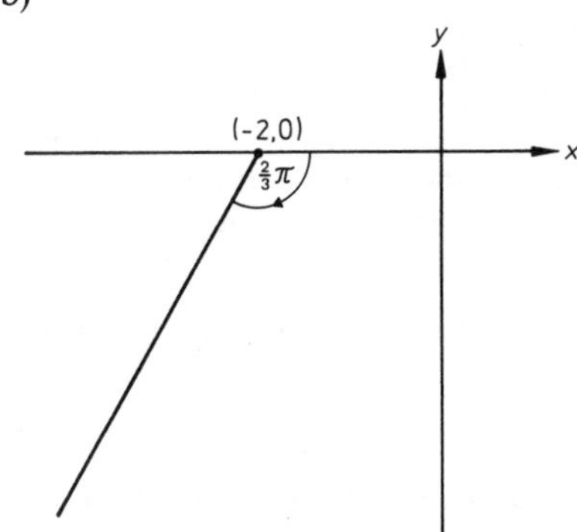

Using similar arguments the equation $\arg(z + 2) = -\frac{2}{3}\pi$ represents the half-line with end-point $(-2, 0)$, inclined to the real axis at an angle $\frac{2}{3}\pi$, measured clockwise.

Example 3 Sketch the locus of the point $P(x, y)$ representing the complex number $z = x + iy$, given that $|z - 1| = |z + i|$. Write down the Cartesian equation of the locus.

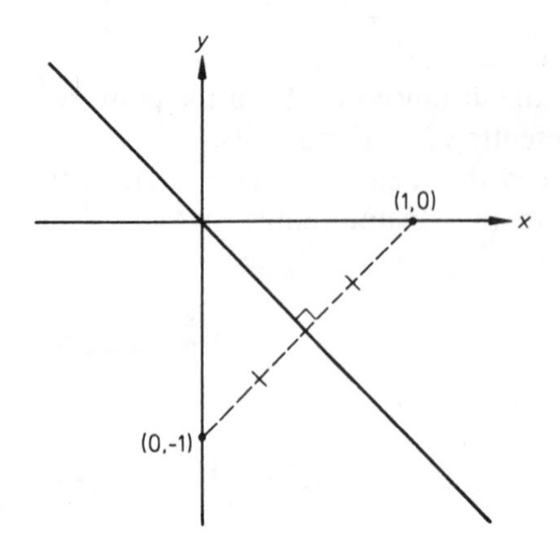

$|z - 1|$ and $|z + i|$ are the distances between the point P and the points representing 1 and $-i$ respectively. Hence the equation

$$|z - 1| = |z + i|$$

represents the locus of points equidistant from $(1, 0)$ and $(0, -1)$ i.e. the perpendicular bisector of the line joining the points $(1, 0)$ and $(0, -1)$. From the sketch it is clear that the Cartesian equation of the locus is $y = -x$.

Example 4 Find the Cartesian equation of the locus of the point $P(x, y)$ representing the complex number z, given that $2|z - 3i| = |z|$. Show that the locus is a circle, giving its centre and radius.

$$\begin{aligned} 2|z - 3i| = |z| \quad &\Rightarrow \quad 2|x + iy - 3i| = |x + iy| \\ &\Rightarrow \quad 2|x + i(y - 3)| = |x + iy| \\ &\Rightarrow \quad 4\{x^2 + (y - 3)^2\} = x^2 + y^2 \\ &\Rightarrow \quad 4x^2 + 4y^2 - 24y + 36 = x^2 + y^2 \\ &\Rightarrow \quad 3x^2 + 3y^2 - 24y + 36 = 0 \end{aligned}$$

Hence the Cartesian equation of the locus is

$$x^2 + y^2 - 8y + 12 = 0$$

Rearranging: $$x^2 + (y - 4)^2 = 4$$

This equation represents a circle with centre $(0, 4)$ and radius 2.

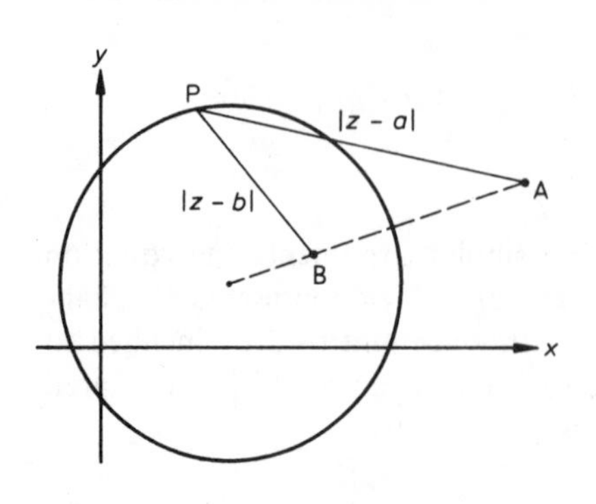

In general, if a point P moves so that the ratio of its distances from two fixed points A and B is constant, then the locus of P is a circle. This locus is referred to as Apollonius' circle and is represented in the Argand diagram by an equation of the form

$$|z - a| = k|z - b| \quad \text{or} \quad \left|\frac{z - a}{z - b}\right| = k,$$

$(k \neq 1)$.

As shown in Example 3, when $k = 1$, the equation represents the perpendicular bisector of the line joining A to B.

Equations involving $\arg\left(\dfrac{z-a}{z-b}\right)$ are more difficult to interpret.

If $\arg(z-a)=\alpha$, $\arg(z-b)=\beta$ and $\arg\left(\dfrac{z-a}{z-b}\right)=\gamma$,

then $$\gamma=\alpha-\beta(\pm 2\pi \text{ if necessary}).$$

Thus γ is the angle which the vector $\overrightarrow{AP}$ makes with the vector $\overrightarrow{BP}$. If the turn from $\overrightarrow{BP}$ to $\overrightarrow{AP}$ is anti-clockwise then γ is positive; if the turn is clockwise then γ is negative.

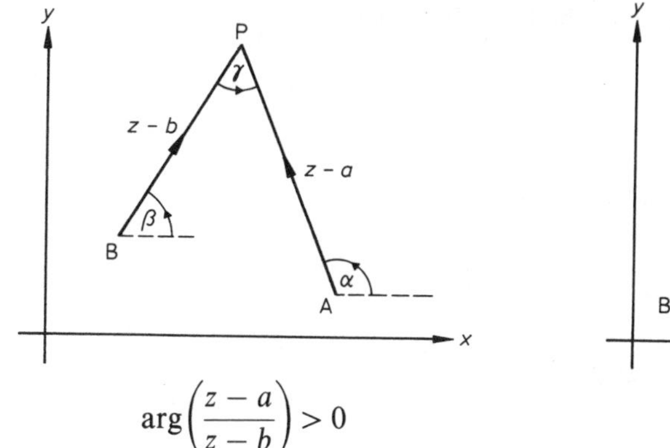

$\arg\left(\dfrac{z-a}{z-b}\right)>0$

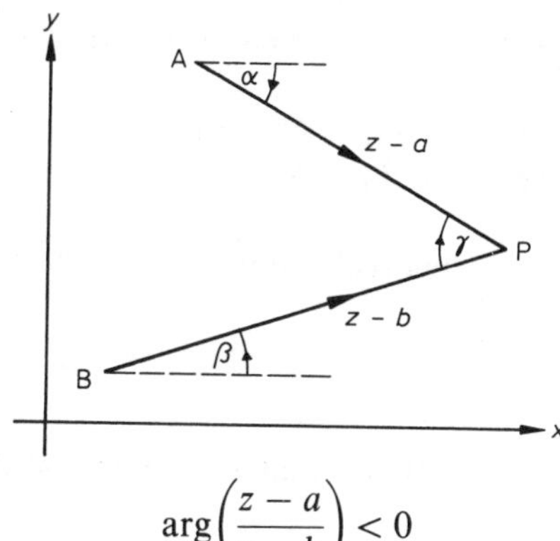

$\arg\left(\dfrac{z-a}{z-b}\right)<0$

It follows that the equation $\arg\left(\dfrac{z-a}{z-b}\right)=\lambda$, where λ is constant, represents a circular arc with end-points A and B.

For instance, if $\arg\left(\dfrac{z-3}{z-1}\right)=\frac{1}{4}\pi$, then the locus of P is a circular arc with end-points $A(3,0)$ and $B(1,0)$, such that $\angle APB=\frac{1}{4}\pi$.

Similarly, if $\arg\left(\dfrac{z+2}{z-i}\right)=\frac{1}{3}\pi$, then the locus of P is a circular arc with end-points $A(-2,0)$ and $B(0,1)$, such that $\angle APB=\frac{1}{3}\pi$. Since in both cases the given arguments are positive, the arcs must be drawn so that the turn from $\overrightarrow{BP}$ to $\overrightarrow{AP}$ is anti-clockwise.

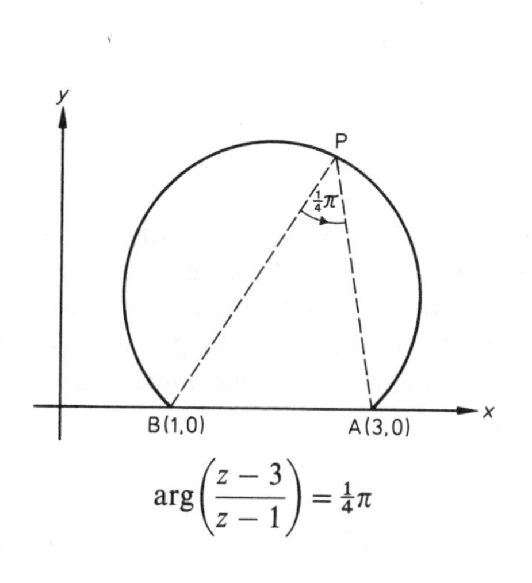

$\arg\left(\dfrac{z-3}{z-1}\right)=\frac{1}{4}\pi$

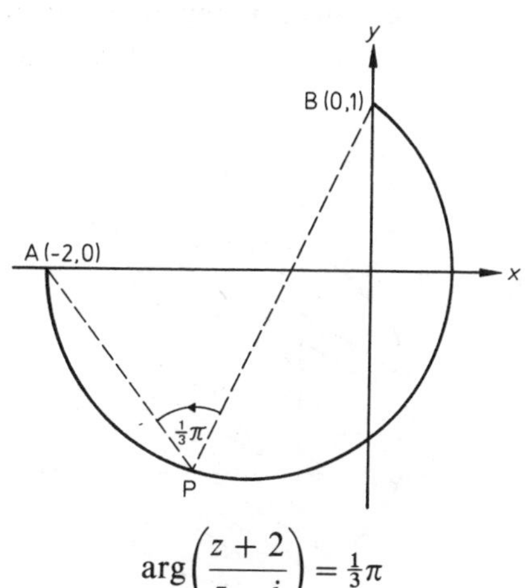

$\arg\left(\dfrac{z+2}{z-i}\right)=\frac{1}{3}\pi$

We use a worked example to illustrate the use of equations involving parameters to represent loci in the Argand diagram.

Example 5 If z is represented in an Argand diagram by the point P and $z - 6i = \lambda iz$, where λ is real, find the Cartesian equation of the locus of P as λ varies.

Let $z = x + iy$, then the equation of the locus becomes

$$x + iy - 6i = \lambda i(x + iy)$$

i.e. $x + i(y - 6) = -\lambda y + i\lambda x$

Equating real and imaginary parts,

$$x = -\lambda y, \qquad y - 6 = \lambda x$$

$\therefore$ provided that $x \neq 0$ and $y \neq 0$, $\dfrac{x}{-y} = \dfrac{y-6}{x} = \lambda$

i.e. $$x^2 = -y^2 + 6y$$

Hence the Cartesian equation of the locus of P is $x^2 + y^2 - 6y = 0$, which represents a circle with centre $(0, 3)$ and radius 3.

[Investigating the excluded values $x = 0, y = 0$, we find that there is no finite value of λ such that $z = 0$ satisfies the equation $z - 6i = \lambda iz$. This means that strictly speaking the origin $(0, 0)$ is not part of the locus.]

Inequalities involving the modulus or argument of a complex variable can be used to represent regions of an Argand diagram.

Example 6 Shade in separate Argand diagrams the regions in which
(a) $|z - 1 - i| < 3$, (b) $\frac{1}{3}\pi \leqslant \arg(z - 2) \leqslant \pi$.

(a)

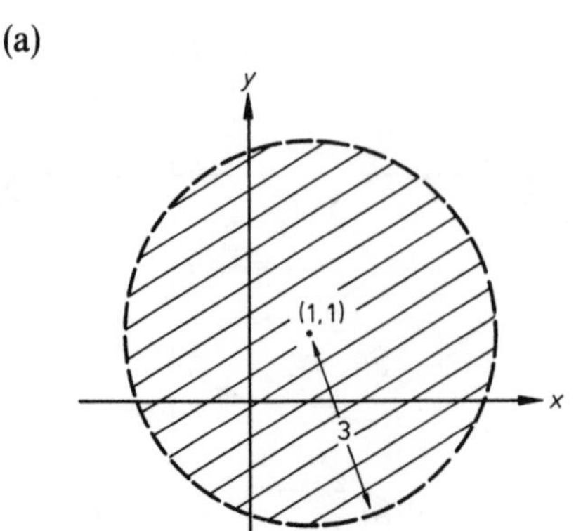

Since $z - 1 - i = z - (1 + i)$, the equation $|z - 1 - i| = 3$ represents a circle with centre $(1, 1)$ and radius 3. Hence the inequality $|z - 1 - i| < 3$ represents the interior of this circle, but not the circle itself.

(b)

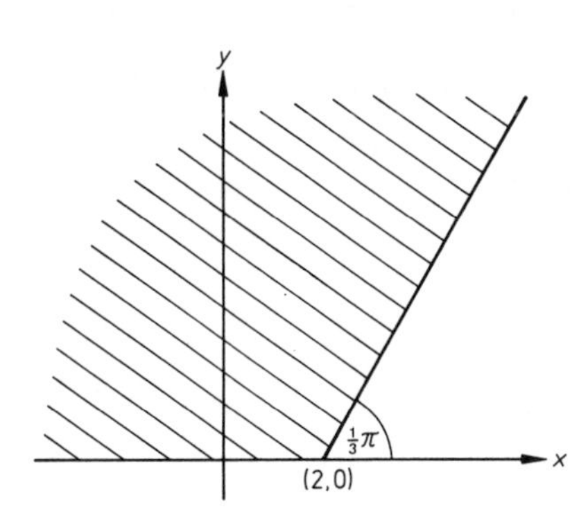

The equations $\arg(z-2) = \frac{1}{3}\pi$ and $\arg(z-2) = \pi$ represent half-lines with end-point $(2,0)$. Hence the inequality

$$\tfrac{1}{3}\pi \leqslant \arg(z-2) \leqslant \pi$$

represents these two lines and the region between them.

A geometrical approach is sometimes helpful when solving problems.

Example 7 Given that z is a complex number which varies such that $|z - i| = 1$, find the greatest and least values of $|z + 1|$.

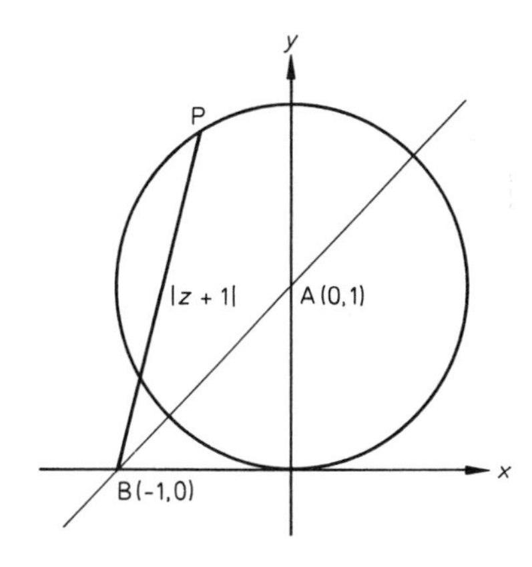

Let z be represented in an Argand diagram by the point P. Given that $|z - i| = 1$, the locus of P is a circle with centre $A(0,1)$ and radius 1. Since $|z + 1|$ is the distance of P from the point $B(-1,0)$, its greatest and least values occur when P lies on the straight line through A and B. As $AB = \sqrt{2}$ and the radius of the circle is 1, these greatest and least values must be $\sqrt{2} + 1$ and $\sqrt{2} - 1$.

Exercise 14.5

In questions 1 to 11 sketch in separate Argand diagrams the loci defined by the given equations.

1. (a) $|z| = 5$, (b) $|z - 2| = 3$.

2. (a) $|z - 3i| = 2$, (b) $|z + 4| = 4$.

3. (a) $|z - 1 + i| = 1$, (b) $|z - 2 - 3i| = 4$.

4. (a) $|z + 1 + 2i| = 3$, (b) $|z + 3 - 4i| = 5$.

5. (a) $\arg z = \frac{1}{4}\pi$, (b) $\arg(z - i) = \frac{1}{3}\pi$.

6. (a) $\arg(z + 1 - 3i) = -\frac{1}{6}\pi$, (b) $\arg(z - 3 + 2i) = \pi$.

7. (a) $\arg(z + 2 + i) = \frac{1}{2}\pi$, (b) $\arg(z - 1 - i) = -\frac{3}{4}\pi$.

8. (a) $|z + 1| = |z - 3|$, (b) $|z| = |z - 6i|$.

9. (a) $\left|\frac{z - i}{z - 1}\right| = 1$, (b) $\left|\frac{z - 2 - 3i}{z + 2 + i}\right| = 1$.

10. (a) $\arg\left(\frac{z - 1}{z + 1}\right) = \frac{1}{3}\pi$, (b) $\arg\left(\frac{z + 1}{z - 1}\right) = \frac{1}{3}\pi$.

11. (a) $\arg\left(\frac{z - 3}{z - 2i}\right) = \frac{1}{4}\pi$, (b) $\arg\left(\frac{z}{z - 4 + 2i}\right) = \frac{1}{2}\pi$.

In questions 12 to 24 find the Cartesian equation of the locus of the point P representing the complex number z. Sketch the locus of P in each case.

12. $2|z + 1| = |z - 2|$

13. $|z + 4i| = 3|z - 4|$

14. $\left|\frac{z}{z - 4}\right| = 5$

15. $\left|\frac{z + i}{z - 5 - 2i}\right| = 1$

16. $\left|\frac{z}{z + 6}\right| = 5$

17. $\left|\frac{z - 1}{z + 1 - i}\right| = \frac{2}{3}$

18. $z - 5 = \lambda i(z + 5)$, where λ is a real parameter.

19. $\frac{z + 2i}{z - 2} = \lambda i$, where λ is a real parameter.

20. $z = 3i + \lambda(2 + 5i)$, where λ is a real parameter.

21. $\operatorname{Im}(z^2) = 2$

22. $\operatorname{Re}(z^2) = 1$

23. $\operatorname{Re}\left(z - \frac{1}{z}\right) = 0$

24. $\operatorname{Im}\left(z + \frac{9}{z}\right) = 0$.

In questions 25 to 34 shade in separate Argand diagrams the regions represented by the given inequalities. Indicate in each case whether the boundaries are to be included in the region.

25. $|z| > 5$

26. $1 \leqslant |z| \leqslant 2$

27. $|z - i| \leqslant 3$

28. $|z - 4 + 3i| < 4$

29. $0 \leqslant \arg z \leqslant \frac{1}{3}\pi$

30. $\frac{1}{4}\pi < \arg z < \frac{3}{4}\pi$

31. $-\frac{1}{6}\pi < \arg(z - 1) < \frac{1}{6}\pi$

32. $-\frac{1}{2}\pi \leqslant \arg(z + i) \leqslant \frac{2}{3}\pi$

33. $|z| > |z + 2|$

34. $|z + i| \leqslant |z - 3i|$

35. Represent each of the following loci in an Argand diagram.
(a) $\arg(z - 1) = \arg(z + 1)$, (b) $\arg z = \arg(z - 1 - i)$,
(c) $\arg(z - 2) = \pi + \arg z$, (d) $\arg(z - 1) = \pi + \arg(z - i)$.

36. Find the least value of $|z + 4|$ for which
(a) $\text{Re}(z) = 5$, (b) $\text{Im}(z) = 3$, (c) $|z| = 1$, (d) $\arg z = \frac{1}{4}\pi$.

37. Given that the complex number z varies such that $|z - 7| = 3$, find the greatest and least values of $|z - i|$.

38. Given that the complex numbers w and z vary subject to the conditions $|w - 12| = 7$ and $|z - 5i| = 4$, find the greatest and least values of $|w - z|$.

39. In an Argand diagram, the point P represents the complex number z, where $z = x + iy$. Given that $z + 2 = \lambda i(z + 8)$, where λ is a real parameter, find the Cartesian equation of the locus of P as λ varies. If also $z = \mu(4 + 3i)$, where μ is real, prove that there is only one possible position for P. (JMB)

40. (i) Represent on the same Argand diagram the loci given by the equations $|z - 3| = 3$ and $|z| = |z - 2|$. Obtain the complex numbers corresponding to the points of intersection of these loci. (ii) Find a complex number z whose argument is $\pi/4$ and which satisfies the equation $|z + 2 + i| = |z - 4 + i|$. (L)

14.6 De Moivre's theorem for an integral index

In §14.4 it was shown that

$$(\cos\theta_1 + i\sin\theta_1)(\cos\theta_2 + i\sin\theta_2) = \cos(\theta_1 + \theta_2) + i\sin(\theta_1 + \theta_2).$$

Writing $\theta_1 = \theta_2 = \theta$, we find that

$$(\cos\theta + i\sin\theta)^2 = \cos 2\theta + i\sin 2\theta$$

$\therefore$ $$(\cos\theta + i\sin\theta)^3 = (\cos 2\theta + i\sin 2\theta)(\cos\theta + i\sin\theta)$$

Thus $$(\cos\theta + i\sin\theta)^3 = \cos 3\theta + i\sin 3\theta$$

These results suggest a general formula,

$$\boxed{(\cos\theta + i\sin\theta)^n = \cos n\theta + i\sin n\theta}$$

This statement, known as *De Moivre's*[†] *theorem*, may be proved for positive integral values of n by the method of induction.
Assuming that the theorem holds for $n = k$,

$$(\cos\theta + i\sin\theta)^k = \cos k\theta + i\sin k\theta$$

$$\therefore \quad (\cos\theta + i\sin\theta)^{k+1} = (\cos\theta + i\sin\theta)^k(\cos\theta + i\sin\theta)$$
$$= (\cos k\theta + i\sin k\theta)(\cos\theta + i\sin\theta)$$
$$= \cos(k\theta + \theta) + i\sin(k\theta + \theta)$$
$$= \cos(k+1)\theta + i\sin(k+1)\theta$$

$\therefore$ if the theorem holds for $n = k$, it also holds for $n = k + 1$.

When $n = 1$, $\quad (\cos\theta + i\sin\theta)^n = \cos\theta + i\sin\theta$
and $\quad \cos n\theta + i\sin n\theta = \cos\theta + i\sin\theta$

$\therefore$ the theorem holds for $n = 1$.
Hence, by induction, De Moivre's theorem holds when n is a positive integer.

Example 1 Find the value of $(\cos\frac{1}{4}\pi + i\sin\frac{1}{4}\pi)^{12}$.

By De Moivre's theorem,

$$(\cos\tfrac{1}{4}\pi + i\sin\tfrac{1}{4}\pi)^{12} = \cos(12 \times \tfrac{1}{4}\pi) + i\sin(12 \times \tfrac{1}{4}\pi)$$
$$= \cos 3\pi + i\sin 3\pi = -1.$$

Example 2 Express $(1 - i\sqrt{3})^4$ in the form $a + ib$.

$$|1 - i\sqrt{3}| = \sqrt{\{1^2 + (-\sqrt{3})^2\}} = \sqrt{\{1 + 3\}} = 2$$

If $\arg(1 - i\sqrt{3}) = \theta$, then $\tan\theta = \dfrac{-\sqrt{3}}{1} = -\sqrt{3}$

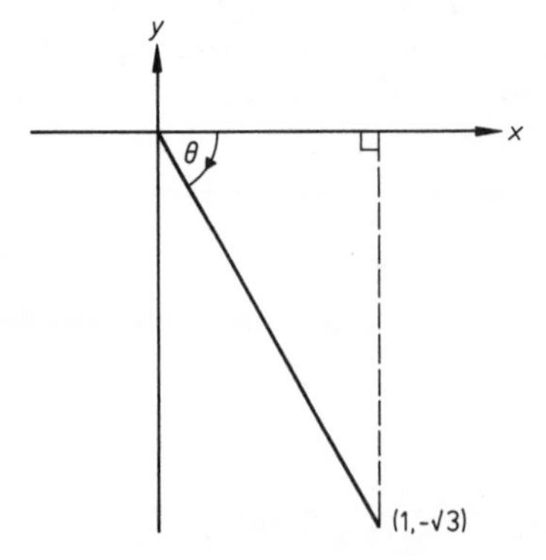

Since the point representing $1 - i\sqrt{3}$ lies in the fourth quadrant, $\arg(1 - i\sqrt{3}) = -\frac{1}{3}\pi$.

$$\therefore \quad 1 - i\sqrt{3} = 2\{\cos(-\tfrac{1}{3}\pi) + i\sin(-\tfrac{1}{3}\pi)\}$$

By De Moivre's theorem,

$$(1 - i\sqrt{3})^4 = 2^4\{\cos(-\tfrac{1}{3}\pi) + i\sin(-\tfrac{1}{3}\pi)\}^4$$
$$= 16\{\cos(-\tfrac{4}{3}\pi) + i\sin(-\tfrac{4}{3}\pi)\} = 16\left\{-\frac{1}{2} + \frac{\sqrt{3}}{2}i\right\}$$

[†] *De Moivre, Abraham* (1667–1754) French mathematician of Huguenot extraction who settled in England, and was a friend of Newton. He published his theorem in *Miscellanea Analytica* in 1730.

Hence $(1 - i\sqrt{3})^4 = -8 + 8\sqrt{3}i$.

To prove De Moivre's theorem when n is a negative integer, let $n = -m$ where m is a positive integer, then

$$(\cos\theta + i\sin\theta)^n = (\cos\theta + i\sin\theta)^{-m} = \frac{1}{(\cos\theta + i\sin\theta)^m}$$

Applying De Moivre's theorem for positive integral index,

$$\begin{aligned}(\cos\theta + i\sin\theta)^n &= \frac{1}{\cos m\theta + i\sin m\theta}\\ &= \cos m\theta - i\sin m\theta\\ &= \cos(-m\theta) + i\sin(-m\theta)\\ &= \cos n\theta + i\sin n\theta\end{aligned}$$

Hence De Moivre's theorem holds for all integral values of n.

Example 3 Evaluate $\dfrac{1}{(1 - i\sqrt{3})^3}$.

As shown in Example 2, $1 - i\sqrt{3} = 2\{\cos(-\tfrac{1}{3}\pi) + i\sin(-\tfrac{1}{3}\pi)\}$
By De Moivre's theorem,

$$\begin{aligned}(1 - i\sqrt{3})^{-3} &= 2^{-3}\{\cos(-\tfrac{1}{3}\pi) + i\sin(-\tfrac{1}{3}\pi)\}^{-3}\\ &= \tfrac{1}{8}\{\cos[(-3)(-\tfrac{1}{3}\pi)] + i\sin[(-3)(-\tfrac{1}{3}\pi)]\}\\ &= \tfrac{1}{8}\{\cos\pi + i\sin\pi\}\end{aligned}$$

Hence $\dfrac{1}{(1 - i\sqrt{3})^3} = -\tfrac{1}{8}$

De Moivre's theorem may be used to obtain certain types of trigonometric identity.

Example 4 Express $\cos 4\theta$ in terms of $\cos\theta$ and $\tan 4\theta$ in terms of $\tan\theta$.

By De Moivre's theorem, $\cos 4\theta + i\sin 4\theta = (\cos\theta + i\sin\theta)^4$

Using the binomial expansion,

$$\begin{aligned}\cos 4\theta + i\sin 4\theta &= \cos^4\theta + 4\cos^3\theta(i\sin\theta)\\ &\quad + 6\cos^2\theta(i\sin\theta)^2 + 4\cos\theta(i\sin\theta)^3 + (i\sin\theta)^4\\ &= (\cos^4\theta - 6\cos^2\theta\sin^2\theta + \sin^4\theta)\\ &\quad + i(4\cos^3\theta\sin\theta - 4\cos\theta\sin^3\theta)\end{aligned}$$

Equating real and imaginary parts,

$$\cos 4\theta = \cos^4\theta - 6\cos^2\theta\sin^2\theta + \sin^4\theta \tag{1}$$
$$\sin 4\theta = 4\cos^3\theta\sin\theta - 4\cos\theta\sin^3\theta \tag{2}$$

From (1):
$$\begin{aligned}\cos 4\theta &= \cos^4\theta - 6\cos^2\theta(1 - \cos^2\theta) + (1 - \cos^2\theta)^2\\ &= \cos^4\theta - 6\cos^2\theta + 6\cos^4\theta + 1 - 2\cos^2\theta + \cos^4\theta\end{aligned}$$

$$\therefore \quad \cos 4\theta = 8\cos^4\theta - 8\cos^2\theta + 1$$

Using (1) and (2): $\tan 4\theta = \dfrac{4\cos^3\theta\sin\theta - 4\cos\theta\sin^3\theta}{\cos^4\theta - 6\cos^2\theta\sin^2\theta + \sin^4\theta}$

Dividing numerator and denominator by $\cos^4\theta$,

$$\tan 4\theta = \frac{4\tan\theta - 4\tan^3\theta}{1 - 6\tan^2\theta + \tan^4\theta}$$

Expressions for powers of $\sin\theta$ and $\cos\theta$ in terms of sines and cosines of multiples of θ can be derived using the following results.

If $\quad z = \cos\theta + i\sin\theta,\quad$ then $\quad \dfrac{1}{z} = \cos\theta - i\sin\theta$

$$\therefore\quad z + \frac{1}{z} = 2\cos\theta \quad\text{and}\quad z - \frac{1}{z} = 2i\sin\theta$$

By De Moivre's theorem,

$$z^n = \cos n\theta + i\sin n\theta,\quad\text{so that}\quad \frac{1}{z^n} = \cos n\theta - i\sin n\theta$$

$$\boxed{z^n + \frac{1}{z^n} = 2\cos n\theta \quad\text{and}\quad z^n - \frac{1}{z^n} = 2i\sin n\theta}$$

Example 5 Express $\cos^4\theta$ in terms of cosines of multiples of θ.

$$\left(z + \frac{1}{z}\right)^4 = z^4 + 4z^3 . \frac{1}{z} + 6z^2 . \frac{1}{z^2} + 4z . \frac{1}{z^3} + \frac{1}{z^4}$$

$$= z^4 + 4z^2 + 6 + \frac{4}{z^2} + \frac{1}{z^4}$$

$$= \left(z^4 + \frac{1}{z^4}\right) + 4\left(z^2 + \frac{1}{z^2}\right) + 6$$

If $z = \cos\theta + i\sin\theta$, $z^n + \dfrac{1}{z^n} = 2\cos n\theta$

$\therefore\quad (2\cos\theta)^4 = 2\cos 4\theta + 4 \times 2\cos 2\theta + 6$

i.e. $\quad 16\cos^4\theta = 2\cos 4\theta + 8\cos 2\theta + 6$

Hence $\quad \cos^4\theta = \frac{1}{8}(\cos 4\theta + 4\cos 2\theta + 3)$

Exercise 14.6

1. Use De Moivre's theorem to simplify the following:

(a) $\left(\cos\dfrac{\pi}{5} + i\sin\dfrac{\pi}{5}\right)^{10}$ (b) $\left(\cos\dfrac{\pi}{8} + i\sin\dfrac{\pi}{8}\right)^{12}$

(c) $\left(\cos\frac{\pi}{9} + i\sin\frac{\pi}{9}\right)^{-3}$ (d) $\left\{\cos\left(-\frac{\pi}{6}\right) + i\sin\left(-\frac{\pi}{6}\right)\right\}^{-4}$

2. Express $\sqrt{3} + i$ in modulus-argument form. Hence find $(\sqrt{3} + i)^{10}$ and $1/(\sqrt{3} + i)^7$ in the form $a + ib$.

3. Express $-1 + i$ in modulus-argument form. Hence show that $(-1 + i)^{16}$ is real and that $1/(-1 + i)^6$ is purely imaginary, giving the value of each.

4. Simplify the following expressions:

(a) $\dfrac{\left(\cos\frac{2\pi}{7} - i\sin\frac{2\pi}{7}\right)^3}{\left(\cos\frac{2\pi}{7} + i\sin\frac{2\pi}{7}\right)^4}$ (b) $\dfrac{\left(\cos\frac{2\pi}{5} + i\sin\frac{2\pi}{5}\right)^8}{\left(\cos\frac{3\pi}{5} - i\sin\frac{3\pi}{5}\right)^3}$

5. Find expressions for $\cos 3\theta$ in terms of $\cos\theta$, $\sin 3\theta$ in terms of $\sin\theta$ and $\tan 3\theta$ in terms of $\tan\theta$.

6. Express $\sin 5\theta$ and $\cos 5\theta/\cos\theta$ in terms of $\sin\theta$.

7. Prove that $\tan 5\theta = \dfrac{5\tan\theta - 10\tan^3\theta + \tan^5\theta}{1 - 10\tan^2\theta + 5\tan^4\theta}$. By considering the equation $\tan 5\theta = 0$, show that $\tan^2(\pi/5) = 5 - 2\sqrt{5}$.

8. Find expressions for $\cos 6\theta$ and $\sin 6\theta/\sin\theta$ in terms of $\cos\theta$ and for $\tan 6\theta$ in terms of $\tan\theta$.

9. Express in terms of cosines of multiples of θ:
(a) $\cos^5\theta$ (b) $\cos^7\theta$ (c) $\sin^2\theta\cos^3\theta$

10. Express in terms of sines of multiples of θ:
(a) $\sin^3\theta$ (b) $\sin^7\theta$ (c) $\cos^4\theta\sin^3\theta$

11. Prove that $\cos^6\theta + \sin^6\theta = \frac{1}{8}(3\cos 4\theta + 5)$.

12. Evaluate (a) $\displaystyle\int_0^{\pi} \sin^4\theta\, d\theta$, (b) $\displaystyle\int_0^{\pi/2} \cos^4\theta\sin^2\theta\, d\theta$.

Exercise 14.7 (miscellaneous)

1. Let $z = x + iy$ be any non-zero complex number. Express $\dfrac{1}{z}$ in the form $u + iv$.

Given that $z + \dfrac{1}{z} = k$ with k *real*, prove that either $y = 0$ or $x^2 + y^2 = 1$. Show (i) that if $y = 0$ then $|k| \geqslant 2$, (ii) that if $x^2 + y^2 = 1$ then $|k| \leqslant 2$. (JMB)

2. In the quadratic equation $x^2 + (p + iq)x + 3i = 0$, p and q are real. Given that the sum of the squares of the roots is 8, find all possible pairs of values of p and q. (JMB)

3. Express the complex number $z_1 = \dfrac{11 + 2i}{3 - 4i}$ in the form $x + iy$ where x and y are real. Given that $z_2 = 2 - 5i$, find the distance between the points in the Argand diagram which represent z_1 and z_2. Determine the real numbers α, β such that $\alpha z_1 + \beta z_2 = -4 + i$. (JMB)

4. (i) Find two complex numbers z satisfying the equation $z^2 = -8 - 6i$.
   (ii) Solve the equation $z^2 - (3 - i)z + 4 = 0$ and represent the solutions on an Argand diagram by vectors $\vec{OA}$ and $\vec{OB}$, where O is the origin. Show that triangle OAB is right-angled.

5. If z and w are complex numbers, show that

$$|z - w|^2 + |z + w|^2 = 2\{|z|^2 + |w|^2\}.$$

Interpret your results geometrically. (AEB 1977)

6. A regular octagon is inscribed in the circle $|z| = 1$ in the complex plane and one of its vertices represents the number $\dfrac{1}{\sqrt{2}}(1 + i)$. Find the numbers represented by the other vertices. (AEB 1977)

7. (i) Two complex numbers z_1 and z_2 each have arguments between 0 and π. If $z_1 z_2 = i - \sqrt{3}$ and $z_1/z_2 = 2i$ find the values of z_1 and z_2 giving the modulus and argument of each.
   (ii) Obtain in the form $a + ib$ the solutions of the equation $z^2 - 2z + 5 = 0$, and represent the solutions on an Argand diagram by the points A and B.

   The equation $z^2 - 2pz + q = 0$ is such that p and q are real, and its solutions in the Argand diagram are represented by the points C and D. Find in the simplest form the algebraic relation satisfied by p and q in each of the following cases:
(a) $p^2 < q$, $p \neq 1$ and A, B, C, D are the vertices of a rectangle;
(b) $p^2 > q$ and $\angle CAD = \frac{1}{2}\pi$. (O&C)

8. (a) If $-\pi < \arg z_1 + \arg z_2 \leqslant \pi$ show that $\arg(z_1 z_2) = \arg z_1 + \arg z_2$. The complex numbers $a = 4\sqrt{3} + 2i$ and $b = \sqrt{3} + 7i$ are represented in the Argand diagram by points A and B respectively. O is the origin. Show that triangle OAB is equilateral and find the complex number c which the point C represents where $OABC$ is a rhombus. Calculate $|c|$ and $\arg c$. (You may leave answers in surd form.)
   (b) z is a complex number such that $z = \dfrac{p}{2 - i} + \dfrac{q}{1 + 3i}$ where p and q are real. If $\arg z = \pi/2$ and $|z| = 7$ find the values of p and q. (SU)

9. (i) Given that $z_1 = 3 + 4i$ and $z_2 = -1 + 2i$, represent z_1, z_2, $(z_1 + z_2)$ and $(z_2 - z_1)$ by vectors in the Argand diagram. Express $(z_1 + z_2)/(z_2 - z_1)$ in the form $a + ib$, where a and b are real. Find the magnitude of the angle between the vectors representing $(z_1 + z_2)$ and $(z_2 - z_1)$.
(ii) One root of the equation $z^3 - 6z^2 + 13z + k = 0$, where k is real, is $z = 2 + i$. Find the other roots and the value of k. (L)

10. (a) Show that $(1 + 3i)^3 = -(26 + 18i)$.
(b) Find the three roots z_1, z_2, z_3 of the equation $z^3 = -1$.
(c) Find, in the form $a + ib$, the three roots z'_1, z'_2, z'_3 of the equation $z^3 = 26 + 18i$.
(d) Indicate in the same Argand diagram the points represented by z_r and z'_r for $r = 1, 2, 3$, and prove that the roots of the equations may be paired so that $|z_1 - z'_1| = |z_2 - z'_2| = |z_3 - z'_3| = 3$. (O&C)

11. Write down, or obtain, the non-real cube roots of unity, ω_1 and ω_2, in the form $a + ib$, where a and b are real. A regular hexagon is drawn in an Argand diagram such that two adjacent vertices represent ω_1 and ω_2, respectively, and the centre of the circumscribing circle of the hexagon is the point $(1, 0)$. Determine, in the form $a + ib$, the complex numbers represented by the other four vertices of the hexagon and find the product of these four complex numbers. (JMB)

12. A complex number ω is such that $\omega^3 = 1$ and $\omega \neq 1$. Show that

(i) $\omega^2 + \omega + 1 = 0$, (ii) $(x + a + b)(x + \omega a + \omega^2 b)(x + \omega^2 a + \omega b)$

is real for real x, a and b, and simplify this product. Hence, or otherwise, find the three roots of the equation $x^3 - 6x + 6 = 0$, giving your answers in terms of ω, ω^2 and cube roots of integers. (JMB)

13. (i) Find, without the use of tables, the two square roots of $5 - 12i$ in the form $x + iy$, where x and y are real.
(ii) Represent on an Argand diagram the loci $|z - 2| = 2$ and $|z - 4| = 2$. Calculate the complex numbers corresponding to the points of intersection of these loci. (L)

14. (i) Given that $(1 + 5i)p - 2q = 3 + 7i$, find p and q when (a) p and q are real, (b) p and q are conjugate complex numbers.
(ii) Shade on the Argand diagram the region for which $3\pi/4 < \arg z < \pi$ and $0 < |z| < 1$. Choose a point in the region and label it A. If A represents the complex number z, label clearly the points B, C, D and E which represent $-z$, iz, $z + 1$ and z^2 respectively. (L)

15. (i) Show that $z = 1 + i$ is a root of the equation $z^4 + 3z^2 - 6z + 10 = 0$. Find the other roots of this equation.
(ii) Sketch the curve in the Argand diagram defined by $|z - 1| = 1$, $\operatorname{Im} z \geqslant 0$. Find the value of z at the point P in which this curve is cut by the line $|z - 1| = |z - 2|$. Find also the value of $\arg z$ and $\arg(z - 2)$ at P. (L)

16. (i) If $z = 1 + i\sqrt{3}$, find $|z|$ and $|z^5|$, and also the values of $\arg z$ and $\arg(z^5)$ lying between $-\pi$ and π. Show that $\text{Re}(z^5) = 16$ and find the value of $\text{Im}(z^5)$.
(ii) Draw the line $|z| = |z - 4|$ and the half-line $\arg(z - i) = \pi/4$ in the Argand diagram. Hence find the complex number that satisfies both equations. (L)

17. (i) Without using tables, simplify $\dfrac{\left(\cos\dfrac{\pi}{9} + i\sin\dfrac{\pi}{9}\right)^4}{\left(\cos\dfrac{\pi}{9} - i\sin\dfrac{\pi}{9}\right)^5}$.

(ii) Express $z_1 = \dfrac{7 + 4i}{3 - 2i}$ in the form $p + qi$, where p and q are real.
Sketch in an Argand diagram the locus of points representing complex numbers z such that $|z - z_1| = \sqrt{5}$. Find the greatest value of $|z|$ subject to this condition. (L)

18. (i) Given that $z = 1 - i$, find the values of $r(>0)$ and θ, $-\pi < \theta \leqslant \pi$, such that $z = r(\cos\theta + i\sin\theta)$. Hence, or otherwise, find $1/z$ and z^6, expressing your answers in the form $p + iq$, where $p, q \in \mathbb{R}$.
(ii) Sketch on an Argand diagram the set of points corresponding to the set A, where $A = \{z : z \in \mathbb{C},\ \arg(z - i) = \pi/4\}$. Show that the set of points corresponding to the set B, where $B = \{z : z \in \mathbb{C},\ |z + 7i| = 2|z - 1|\}$, forms a circle in the Argand diagram. If the centre of this circle represents the numbers z_1, show that $z_1 \in A$. (L)

19. Use De Moivre's theorem to show that

$$\cos 7\theta = 64\cos^7\theta - 112\cos^5\theta + 56\cos^3\theta - 7\cos\theta.$$

20. (i) If $(1 + 3i)z_1 = 5(1 + i)$, express z_1 and z_1^2 in the form $x + iy$, where x and y are real. Sketch in an Argand diagram the circle $|z - z_1| = |z_1|$ giving the coordinates of its centre.
(ii) If $z = \cos\theta + i\sin\theta$. show that

$$z - \frac{1}{z} = 2i\sin\theta. \qquad z^n - \frac{1}{z^n} = 2i\sin n\theta.$$

Hence, or otherwise, show that

$$16\sin^5\theta = \sin 5\theta - 5\sin 3\theta + 10\sin\theta.$$ (L)

21. (i) Given that $(1 + i)^n = x + iy$. where x and y are real and n is an integer, prove that $x^2 + y^2 = 2^n$.
(ii) Given that $\left|\dfrac{z - 1}{z + 1}\right| = 2$, find the cartesian equation of the locus of z and represent the locus by a sketch in the Argand diagram. Shade the region for

which the inequalities $\left|\dfrac{z-1}{z+1}\right| > 2$ and $0 < \arg z < 3\pi/4$ are both satisfied.

(L)

22. (i) Given that x and y are real, find the values of x and y which satisfy the equation $$\frac{2y+4i}{2x+y} - \frac{y}{x-i} = 0.$$
(ii) Given that $z = x + iy$, where x and y are real, (a) show that, when $\operatorname{Im}\left(\dfrac{z+i}{z+2}\right) = 0$, the point (x, y) lies on a straight line, (b) Show that, when $\operatorname{Re}\left(\dfrac{z+i}{z+2}\right) = 0$, the point (x, y) lies on a circle with centre $(-1, -\frac{1}{2})$ and radius $\frac{1}{2}\sqrt{5}$.

(L)

23. (i) Find $|z|$ and $\arg z$ for each of the complex numbers z given by (a) $12 - 5i$, (b) $(1 + 2i)/(2 - i)$, giving the argument in degrees (to the nearest degree) such that $-180° < \arg z \leqslant 180°$.
(ii) By expressing $\sqrt{3} - i$ in modulus-argument form, or otherwise, find the least positive integer n such that $(\sqrt{3} - i)^n$ is real and positive.
(iii) The point P in the Argand diagram lies outside or on the circle of radius 4 with centre at $(-1, -1)$. Write down in modulus form the condition satisfied by the complex number z represented by the point P. (L)

24. Sketch the circle C with Cartesian equation $x^2 + (y-1)^2 = 1$. The point P, representing the non-zero complex number z, lies on C. Express $|z|$ in terms of θ, the argument of z. Given that $z' = 1/z$, find the modulus and argument of z' in terms of θ. Show that, whatever the position of P on the circle C, the point P' representing z' lies on a certain line, the equation of which is to be determined.

(JMB)

25. (a) The sum of the infinite series $1 + z + z^2 + z^3 + \ldots$ for values of z such that $|z| < 1$ is $1/(1-z)$. By substituting $z = \frac{1}{2}(\cos\theta + i\sin\theta)$ in this result and using De Moivre's theorem, or otherwise, prove that
$$\frac{1}{2}\sin\theta + \frac{1}{2^2}\sin 2\theta + \ldots + \frac{1}{2^n}\sin n\theta + \ldots = \frac{2\sin\theta}{5 - 4\cos\theta}.$$
(b) Two variable complex numbers z and w are such that $|z - i| = |z - 1|$ and $|w - 4| = 2$. Show on the same Argand diagram the loci of the points representing z and w. Mark on the diagram the points A and B representing z and w respectively for which $|z - w|$ has its least value and find this value. (O&C)

26. The point representing the complex number z in an Argand diagram describes the circle $|z - 1| = 1$. Show that $z = 1 + \cos\theta + i\sin\theta$, where $-\pi < \theta \leqslant \pi$, and deduce that the point representing the complex number $1/z$ describes a straight line. Find the modulus and argument of z and hence, or otherwise, express

$(1 + \cos\theta + i\sin\theta)^n$ in the form $x + iy$, where n is a positive integer, and x and y are real. By writing $\cos\theta + i\sin\theta = \omega$, and using the binomial expansion of $(1 + \omega)^n$, prove that

$$1 + \binom{n}{1}\cos\theta + \binom{n}{2}\cos 2\theta + \ldots + \cos n\theta \equiv \left[2\cos\left(\frac{\theta}{2}\right)\right]^n \cos\left(\frac{n\theta}{2}\right). \qquad \text{(L)}$$

15 Probability distributions

15.1 Discrete probability distributions

The theory of probability is used to set up mathematical models of random experiments or trials. In its simplest form a *probability model* of an experiment consists of a description of the sample space (i.e. the set of possible outcomes) together with estimates of the probabilities associated with the outcomes. Let us consider, for instance, a trial in which a single unbiased die is thrown and the score noted. This can be described as a trial with a sample space containing 6 outcomes each with probability 1/6. We can use this model to make predictions about the behaviour of dice in experiments involving large numbers of such trials. The theory can be tested by comparing these predictions with the observed results of the experiments.

Since the score obtained in any particular throw of a single die cannot be predicted in advance, the trial is said to be *random* and the score obtained is called a *random variable*. In general, the values of a random variable are real numbers assigned to the elements of a sample space. This leads to the following formal definition.

> A random variable X is a function which assigns a real number x to each element s of a sample space S.

A random variable such as the height of students in a class, which may take any real value within a given interval, is said to be *continuous*. A random variable such as the score on a die, which takes only certain specified values, is described as *discrete*.

As we saw in Book 1, any subset A of a sample space S is called an event and the probability that event A occurs is denoted by $P(A)$. Thus the probability that the random variable X takes the value x may be written $P(X = x)$. We can now define a probability function on the set of values taken by X.

The probability function p of the discrete random variable X is defined by $p(x) = P(X = x)$, where $0 \leqslant p(x) \leqslant 1$ for all values of x.

The set of values $x_1, x_2, x_3, \ldots$ of a discrete random variable X, together with the corresponding values $p_1, p_2, p_3, \ldots$ of the probability function p, form a *discrete probability distribution.*

Since the sum of the probabilities of all possible outcomes of any experiment must be 1, all discrete probability distributions have the following property.

If a discrete random variable X takes the n values $x_1, x_2, \ldots x_n$ with probabilities $p_1, p_2, \ldots, p_n$ respectively, then $\sum_{r=1}^{n} p_r = 1$.

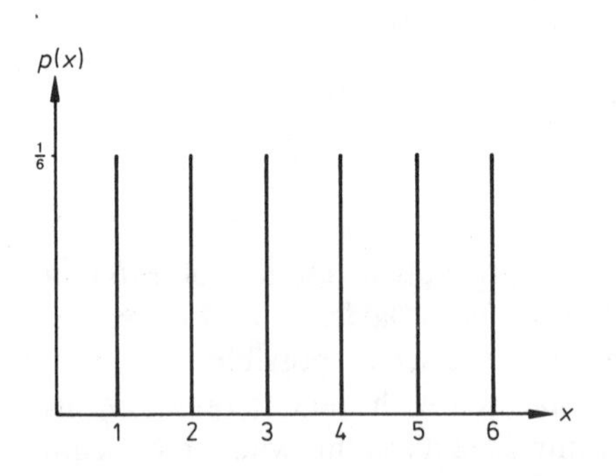

If X represents the score obtained when a single die is thrown, then $p(x) = 1/6$ for $x = 1, 2, 3, 4, 5$ and 6. A distribution such as this, in which all values of the random variable are equally likely, is said to be *rectangular* or *uniform.*

For a trial in which two dice are thrown and the total score noted the sample space S can be regarded as a set of 36 ordered pairs such as (1, 3), (4, 6) and (6, 4). The total score obtained in any throw is a discrete random variable defined on S and takes the values 2, 3, 4, ..., 12. The following diagrams show the sample space and corresponding probability distribution for this two dice trial.

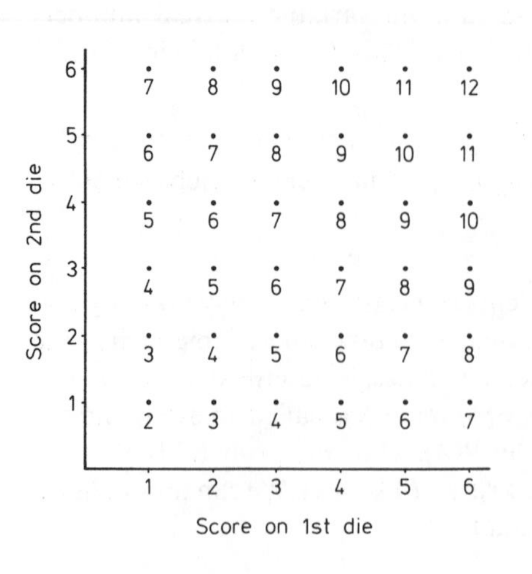

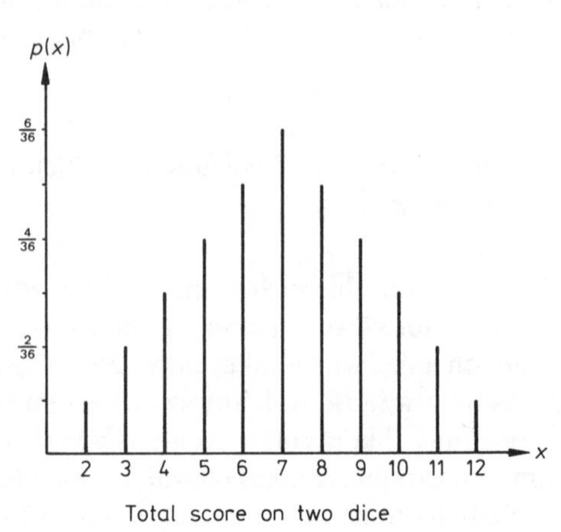

A further type of discrete probability distribution is obtained when a die is thrown repeatedly and the random variable X is the number of throws needed to obtain a six. We then find that

$$p(1) = \frac{1}{6}, p(2) = \frac{5}{6} \times \frac{1}{6}, p(3) = \left(\frac{5}{6}\right)^2 \times \frac{1}{6}, p(4) = \left(\frac{5}{6}\right)^3 \times \frac{1}{6}, \ldots$$

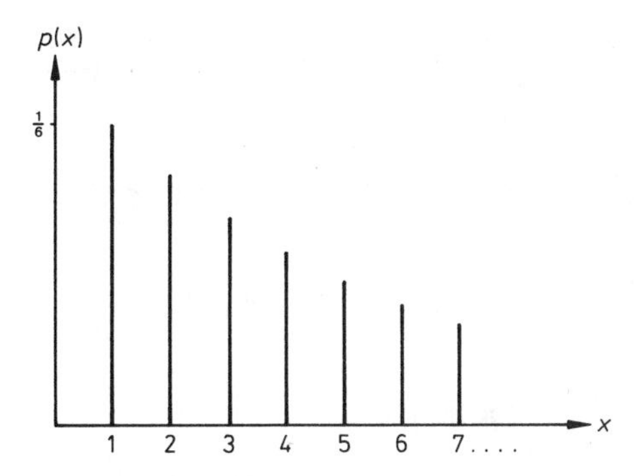

Clearly X may take any one of the infinite set of values 1, 2, 3, 4, ... Since the corresponding probabilities form a geometric progression, this distribution is described as *geometric.*

In order to discuss the properties of probability distributions in more detail we need to define quantities similar to the measures of location and spread used to describe frequency distributions.

For a frequency distribution in which a set of n observations $x_1, x_2, \ldots, x_n$ occur with frequencies $f_1, f_2, \ldots, f_n$,

$$\text{the mean } \bar{x} = \frac{1}{n}\sum f_r x_r, \text{ where } n = \sum f_r,$$

$$\text{the variance} = \frac{1}{n}\sum f_r(x_r - \bar{x})^2 = \frac{1}{n}\sum f_r x_r^2 - \bar{x}^2$$

If n is fairly large the relative frequency f_r/n is approximately equal to the probability $p(x_r)$. Thus we obtain suitable definitions of the mean and variance of a probability distribution by replacing f_r/n by $p(x_r)$ in the formulae used for frequency distributions.

The mean of a probability distribution is called the *expectation* or *expected value* of the random variable X and is denoted by $E(X)$ or μ. The *variance* is denoted by $\text{Var}(X)$ or σ^2. These quantities are defined as follows:

$$E(X) = \mu = \sum x_r p(x_r)$$
$$\text{Var}(X) = \sigma^2 = \sum (x_r - \mu)^2 p(x_r)$$

The expression for $\text{Var}(X)$ can be rearranged to give:

$$\text{Var}(X) = \sigma^2 = \sum x_r^2 p(x_r) - \mu^2$$

For the two dice trial discussed earlier:

$$E(X) = 2.\tfrac{1}{36} + 3.\tfrac{2}{36} + \ldots + 12.\tfrac{1}{36} = 7$$
$$\text{Var}(X) = 2^2.\tfrac{1}{36} + 3^2.\tfrac{2}{36} + \ldots + 12^2.\tfrac{1}{36} - 7^2 = \tfrac{35}{6} \approx 5{\cdot}83.$$

Example 1 A box contains 10 coloured discs of which 2 are red. A man pays 10 pence to play a game in which discs are drawn one at a time from the box, without replacement, until a red disc is drawn. He will receive 25 pence if the first disc drawn is red, 20 pence if the second disc is red, 5 pence if the third disc is red, but nothing otherwise. Find the man's expected profit or loss in any game.

If the random variable X is the amount in pence received in any game, then X takes the values 25, 20, 5 and 0.

$p(25) = \tfrac{2}{10} = \tfrac{1}{5}$, $p(20) = \tfrac{8}{10} \times \tfrac{2}{9} = \tfrac{8}{45}$ and $p(5) = \tfrac{8}{10} \times \tfrac{7}{9} \times \tfrac{2}{8} = \tfrac{7}{45}$.

$\therefore$ the expected value of X

$$= 25 \times p(25) + 20 \times p(20) + 5 \times p(5) + 0 \times p(0)$$
$$= 25 \times \tfrac{1}{5} + 20 \times \tfrac{8}{45} + 5 \times \tfrac{7}{45} + 0 = 9\tfrac{1}{3}$$

Since the man pays 10 pence per game, his expected loss in pence is $\tfrac{2}{3}$.

In the next example we make use of the following results obtained in §9.1.

$$\sum_{r=1}^{n} r = \tfrac{1}{2}n(n+1), \quad \sum_{r=1}^{n} r^2 = \tfrac{1}{6}n(n+1)(2n+1), \quad \sum_{r=1}^{n} r^3 = \tfrac{1}{4}n^2(n+1)^2.$$

Example 2 The probability distribution of a random variable X is given by

$$P(X = r) = k(n - r) \quad \text{for} \quad r = 1, 2, 3, \ldots, n.$$

Find k in terms of n. Find also the mean and variance of X.

Since X is a discrete random variable, $\sum_{1}^{n} p(r) = 1$.

$$\text{But} \quad \sum_{1}^{n} p(r) = \sum_{1}^{n} k(n-r) = k\sum_{1}^{n} n - k\sum_{1}^{n} r$$
$$= kn^2 - k.\tfrac{1}{2}n(n+1) = \tfrac{1}{2}kn(n-1)$$

Hence $k = 2/n(n-1)$.

$$E(X) = \sum_{1}^{n} rp(r) = \sum_{1}^{n} kr(n-r)$$
$$= kn\sum_{1}^{n} r - k\sum_{1}^{n} r^2$$
$$= kn.\frac{1}{2}n(n+1) - k.\frac{1}{6}n(n+1)(2n+1)$$
$$= \frac{2}{n(n-1)}.\frac{n(n+1)}{6}\{3n - (2n+1)\} = \frac{1}{3}(n+1)$$

$$\text{Var}(X) = \sum_1^n r^2 p(r) - \mu^2 = \sum_1^n kr^2(n-r) - \frac{1}{9}(n+1)^2$$

$$= kn\sum_1^n r^2 - k\sum_1^n r^3 - \frac{1}{9}(n+1)^2$$

$$= kn.\frac{1}{6}n(n+1)(2n+1) - k.\frac{1}{4}n^2(n+1)^2 - \frac{1}{9}(n+1)^2$$

$$= \frac{2}{n(n-1)}.\frac{n^2(n+1)}{12}\{2(2n+1) - 3(n+1)\} - \frac{1}{9}(n+1)^2$$

$$= \frac{1}{6}n(n+1) - \frac{1}{9}(n+1)^2$$

$$= \frac{1}{18}(n+1)\{3n - 2(n+1)\} = \frac{1}{18}(n+1)(n-2)$$

Thus the mean and variance of X are $\frac{1}{3}(n+1)$ and $\frac{1}{18}(n+1)(n-2)$ respectively.

Exercise 15.1

1. A student has a week-end job in a warehouse. The probability that he will be given 6 hours work is 1/6, the probability of 9 hours work is 1/3 and the probability of 12 hours work is 1/2. Find the expected number of hours the student will work.

2. A man pays x pence to play a game in which two unbiased dice are rolled. If the total score on the dice is an even number n, then the man receives n pence. Otherwise he receives nothing. Calculate the value of x which makes the game fair.

3. A bag contains one 50p coin, three 20p coins, seven 10p coins and several 5p coins. Given that when one coin is selected at random the expectation is 10p, find the number of 5p coins. Find also the expectation when two coins are selected at random without replacement.

4. Three discs are drawn at random, without replacement from a bag containing three black discs and five red discs. Find (a) the most probable number of black discs drawn, (b) the expected number of black discs drawn.

5. A bag X contains 3 red balls and 6 white balls and a second bag Y contains 5 red balls and 4 white balls. A ball is chosen at random from bag X and placed in bag Y. A ball is then chosen at random from bag Y and placed in bag X. If R is the number of red balls in bag X after these operations, find the probability distribution of R. Hence find the expected value of R.

6. A simple 'fruit machine' has three wheels on which there are pictures of fruit. When 10p is paid into the machine, the wheels spin and then come to rest independently, each showing one fruit. The probabilities that the three wheels show

a lemon are 0·4, 0·2 and 0·1 respectively. The machine pays out 10p for 1 lemon, 30p for 2 lemons and £1·60 for 3 lemons. Find the expected profit or loss per play.

7. In a certain gambling game a player nominates an integer x from 1 to 6 inclusive and he then throws three fair cubical dice. Calculate the probabilities that the number of x's thrown will be 0, 1, 2 and 3. The player pays 5 pence per play of the game and he receives 48 pence if the number of x's thrown is three, 15 pence if the number of x's thrown is two, 5 pence if only one x is thrown and nothing otherwise. Calculate the player's expected gain or loss per play of the game. (JMB)

8. In a certain snooker match the first man to win 3 frames wins the match. The winner of each frame receives £20 and the winner of the match receives an additional £40. Find the expected cost of the prizes in a match between players A and B if the probability that A wins a frame against B is (a) 0·5, (b) 0·2.

9. A player throws a die whose faces are numbered 1 to 6 inclusive. If the player obtains a six he throws the die a second time, and in this case his score is the sum of 6 and the second number; otherwise his score is the number obtained. The player has no more than two throws. Let X be the random variable denoting the player's score. Write down the probability distribution of X, and determine the mean of X. Show that the probability that the sum of two successive scores is 8 or more is 17/36. Determine the probability that the first of two successive scores is 7 or more, given that their sum is 8 or more. (C)

10. At a horse show there is a jumping competition in which each horse attempts to clear 4 successive fences, each fence being more difficult than the previous one. A particular horse has probability $(5 - i)/5$ of clearing the ith fence ($i = 1, 2, 3, 4$), independent of its performance over the other fences. (i) Find the probability that the horse clears all four fences. (ii) Find the probability that the horse does not clear any fence. (iii) Determine the expected value of the number of fences that the horse clears. (C)

11. A disc is drawn from a bag containing 10 discs numbered $0, 1, 2, \ldots 9$. The random variable X is defined as the square of the number drawn. Find $E(X)$ and $\text{Var}(X)$.

12. A box contains 6 cards all of different colours. A card is drawn at random from the box and its colour noted. If three cards are drawn, with replacement, find the mean and variance of the number of different colours obtained.

13. Let X denote the number of heads obtained when a fair coin is tossed three times. Find the mean and the variance of X. A pack of six cards consists of three cards labelled H and the other three labelled T. Three cards are drawn at random without replacement from the pack. Let Y denote the number of H's obtained. Calculate the probabilities that Y takes the values 0, 1, 2 and 3 respectively. Show that the mean of Y is equal to the mean of X and express the variance of Y as a percentage of the variance of X. (JMB)

14. In a certain field, each puffball which is growing in one year gives rise to a number, X, of new puffballs in the following year. None of the original puffballs is present in the following year. The probability distribution of the random variable X is as follows:

$$P(X = 0) = P(X = 2) = 0{\cdot}3, P(X = 1) = 0{\cdot}4.$$

Find the probability distribution of Y, the number of puffballs resulting from there being two puffballs in the previous year, and show that the variance of Y is 1·2. Hence, or otherwise, determine the probability distribution of the number, Z, of puffballs present in year 3, given that there was a single puffball present in year 1. Find also the mean and variance of Z. (C)

15. A cubical die is biased in such a way that the probability of scoring $n(n = 1$ to $6)$ is proportional to n. Determine the mean value and variance of the score obtained in a single throw. What would be the mean and variance if the score showing were doubled? (O&C)

16. A random variable R takes values $1, 2, \ldots, n$ with equal probabilities. Determine the expectation μ of R and show that the variance σ^2 is given by $12\sigma^2 = n^2 - 1$. Find $P(|R - \mu| > \sigma)$ in the case $n = 100$. (C)

17. The probability distribution of a random variable R is defined by $P(R = r) = kr$ for $r = 1, 2, \ldots, n$. Find the value of k. Hence find the mean and variance of the distribution.

18. Ten identically shaped discs are in a bag; two of them are black, the rest white. Discs are drawn at random from the bag in turn and NOT replaced. Let X be the number of discs drawn up to AND INCLUDING the first black one. List the values of X and the associated theoretical probabilities. Calculate the mean value of X and its standard deviation. What is the most likely value of X? If instead each disc is replaced before the next is drawn, construct a similar list of values and probabilities and point out the chief differences between the two lists. (SU)

19. (a) Expand $(1 - x)^{-2}$ in a series of ascending powers of x, giving the first three terms and the term in x^{r-1}.
(b) A boy is throwing stones at a tin can. The probability of hitting the can with any particular stone is p. Find the probability that the boy first hits the can with the rth stone. If the random variable R is the number of throws needed to hit the can, find the expected value of R in terms of p.

20. A faulty gas-lighter works successfully, on average, once in three attempts. When using it one goes on trying until it does work and the gas lights. What is the probability that the gas lights (i) first time, (ii) second time, (iii) at the fifth attempt? What is the probability that it will not have worked after five attempts? The sellers of the lighter promise that if one is returned, they will pay 25p for each time it fails to work before the gas is lit. What is one's expectation in taking up this offer? What is the probability of getting one's money back (or more) if the original price was £1? If

four people take up this offer, what is the probability that at least two get their money back (or more)? (AEB 1978)

21. An ordinary unbiased cubical die is to be thrown repeatedly. Show that the probability that a score of 6 will occur for the first time on the nth throw with the first $(n-1)$ throws resulting in one odd score and $(n-2)$ even scores (other than 6) is equal to $(n-1)(\frac{1}{4})(\frac{1}{3})^{n-1}$ for $n \geqslant 2$. If the die is to be thrown as many times as it is necessary for a score of 6 to occur, find, using the above result, or otherwise, the probability that only one of the throws will result in an odd score.

[Hint: For $|x| < 1$, $\sum_{r=0}^{\infty} (r+1)x^r = (1-x)^{-2}$] (JMB)

15.2 The binomial distribution

A *Bernouilli*[†] *trial* is one in which there are two possible outcomes, usually referred to as success and failure. If the probability of success is p and the probability of failure is q, then $p + q = 1$.

If a random variable X represents the number of successes in a sequence of n independent repetitions of a Bernouilli trial, then the resulting probability distribution is called a *binomial distribution.*

Consider, for example, a trial in which an unbiased die is rolled once and success is defined as obtaining a six. In one trial the probability of success is 1/6 and the probability of failure is 5/6. Suppose now that the random variable X is the number of sixes obtained in a sequence of 5 rolls of the die.

Clearly $$P(X = 0) = (\tfrac{5}{6})^5$$

The number of arrangements of a sequence of 4 failures and 1 success = 5.
The probability of each such arrangement $= (\frac{5}{6})^4(\frac{1}{6})$

$$\therefore \quad P(X = 1) = 5(\tfrac{5}{6})^4(\tfrac{1}{6})$$

The number of arrangements of a sequence of 3 failures and 2 successes $= \frac{5!}{3!2!} = 10$.
The probability of each such arrangement $= (\frac{5}{6})^3(\frac{1}{6})^2$

$$\therefore \quad P(X = 2) = 10(\tfrac{5}{6})^3(\tfrac{1}{6})^2$$

Using similar arguments it follows that

$$P(X = 3) = 10(\tfrac{5}{6})^2(\tfrac{1}{6})^3, P(X = 4) = 5(\tfrac{5}{6})(\tfrac{1}{6})^4, P(X = 5) = (\tfrac{1}{6})^5.$$

We see that the probabilities of 0, 1, 2, 3, 4, 5 sixes in a sequence of 5 rolls of a die are given by the terms in the binomial expansion of $(\frac{5}{6} + \frac{1}{6})^5$.

To obtain a more general result let us assume that the probability of success in a certain trial is p and that the probability of failure is q, where $q = 1 - p$.

[†]*Bernouilli, Jacques* (1654–1705) Swiss mathematician: one of the members of a large family who were all gifted mathematically. His *Ars Conjectandi* of 1713 contains a considerable quantity of work on probability.

The number of arrangements of a sequence of $(n - r)$ failures and r successes is $\frac{n!}{(n-r)!r!}$, written $\binom{n}{r}$ or ${}_nC_r$.
The probability of each such arrangement $= q^{n-r}p^r$.

Hence the probability of r successes in n independent trials is $\binom{n}{r}q^{n-r}p^r$.

Thus the probabilities that $r = 0, 1, 2, \ldots, n$ are the terms in the binomial expansion of $(q + p)^n$. This distribution is described as the binomial distribution with *parameters* n and p.

> A random variable X is said to have a binomial distribution $B(n, p)$, if
>
> $$P(X = r) = \binom{n}{r}q^{n-r}p^r \quad \text{for} \quad r = 0, 1, 2, \ldots, n,$$
>
> where $0 < p < 1$ and $q = 1 - p$.

This binomial distribution arises when X represents the number of successes in n *independent* trials, each trial having the *constant* probability p of success.

Example 1 A doctor estimates that his treatment of a particular illness is successful in $66\frac{2}{3}\%$ of cases. Find the probability that he will treat successfully exactly 4 out of the 6 patients currently in his care. Find also the probability of success in at least 2 out of the 6 cases.

If X is the number of successes, then X has a binomial distribution with parameters $n = 6$ and $p = \frac{2}{3}$.

$$P(X = 4) = \binom{6}{4}\left(\tfrac{1}{3}\right)^2\left(\tfrac{2}{3}\right)^4 = \frac{6!}{2!4!} \times \frac{2^4}{3^6} = \tfrac{80}{243} \approx 0{\cdot}329$$

$\therefore$ the probability of success in exactly 4 out of 6 cases is $0{\cdot}329$.

$P(X = 0) = \left(\tfrac{1}{3}\right)^6 = \tfrac{1}{729}$, $P(X = 1) = \binom{6}{1}\left(\tfrac{1}{3}\right)^5\left(\tfrac{2}{3}\right) = \tfrac{4}{243}$,

$\therefore$ the probability of at least 2 successes

$= 1 - \tfrac{1}{729} - \tfrac{4}{243} = \tfrac{716}{729} \approx 0{\cdot}982.$

If a random variable X has the binomial distribution $B(n, p)$, then

$$\sum_{r=0}^{n} P(X = r) = \sum_{r=0}^{n} \binom{n}{r}q^{n-r}p^r = (q + p)^n = 1.$$

Thus we have verified that for the distribution $B(n, p)$,

$$\sum_{r=0}^{n} p_r = 1, \quad \text{where} \quad p_r = P(X = r).$$

We now obtain the mean and variance of X.

$$E(X) = \sum_{r=0}^{n} r\binom{n}{r}q^{n-r}p^r$$

$$= \sum_{r=0}^{n} \frac{r(n!)}{(n-r)!r!}q^{n-r}p^r$$

$$= \sum_{r=1}^{n} \frac{n!}{(n-r)!(r-1)!}q^{n-r}p^r$$

$$= np\sum_{r=1}^{n} \frac{(n-1)!}{(n-r)!(r-1)!}q^{n-r}p^{r-1} = np(q+p)^{n-1} = np.$$

$$\text{Var}(X) = \sum_{r=0}^{n} r^2\binom{n}{r}q^{n-r}p^r - \{E(X)\}^2$$

Writing $r^2 = r(r-1) + r$, we obtain:

$$\text{Var}(X) = \sum_{r=0}^{n} \frac{r(r-1)n!}{(n-r)!r!}q^{n-r}p^r + \sum_{r=0}^{n} \frac{r(n!)}{(n-r)!r!}q^{n-r}p^r - n^2p^2$$

$$= \sum_{r=2}^{n} \frac{n!}{(n-r)!(r-2)!}q^{n-r}p^r + np - n^2p^2$$

$$= n(n-1)p^2\sum_{r=2}^{n} \frac{(n-2)!}{(n-r)!(r-2)!}q^{n-r}p^{r-2} + np - n^2p^2$$

$$= n(n-1)p^2(q+p)^{n-2} + np - n^2p^2$$
$$= n(n-1)p^2 + np - n^2p^2$$
$$= np - np^2 = np(1-p) = npq$$

$\therefore$ the mean and variance of the binomial distribution $B(n, p)$ are given by

$$\mu = np, \quad \sigma^2 = npq = np(1-p).$$

Example 2 The components produced by a particular machine are tested by taking samples containing 5 components and noting the number of rejects in each. The given table shows the results for 100 samples

Number of rejects (x)	0	1	2	3	4	5
Frequency (f)	10	27	31	20	9	3

(a) Calculate the mean and variance of the number of rejects.
(b) Estimate the probability p that a component selected at random is a reject.

(c) By assuming that the number of rejects in a sample of 5 components has a binomial distribution with parameters $n = 5$ and p, write down the theoretical mean and variance.

(d) Obtain the expected frequency distribution for 100 samples.

(a) $\sum f = 10 + 27 + 31 + 20 + 9 + 3 = 100$

$\sum fx = 0 \times 10 + 1 \times 27 + 2 \times 31 + 3 \times 20 + 4 \times 9 + 5 \times 3 = 200$

$\sum fx^2 = 0^2 \times 10 + 1^2 \times 27 + 2^2 \times 31 + 3^2 \times 20 + 4^2 \times 9 + 5^2 \times 3 = 550$

$$\therefore \quad \text{the mean } \bar{x} = \frac{\sum fx}{\sum f} = \frac{200}{100} = 2,$$

$$\text{the variance} = \frac{\sum fx^2}{\sum f} - \bar{x}^2 = \frac{550}{100} - 2^2 = 1{\cdot}5.$$

(b) Total number of rejects $= \sum fx = 200$
Total number of components $= 5 \times \sum f = 500$
$\therefore$ the probability p that any component is a reject $= \frac{200}{500} = \frac{2}{5}$.

(c) The mean and variance of the binomial distribution with parameters $n = 5$ and $p = 2/5$ are respectively:

$$\mu = np = 2, \qquad \sigma^2 = np(1 - p) = 1{\cdot}2.$$

(d) $$P(\text{no rejects}) = \left(\frac{3}{5}\right)^5 \approx 0{\cdot}08,$$

$$P(1 \text{ reject}) = \binom{5}{1}\left(\frac{3}{5}\right)^4\left(\frac{2}{5}\right) \approx 0{\cdot}26,$$

$$P(2 \text{ rejects}) = \binom{5}{2}\left(\frac{3}{5}\right)^3\left(\frac{2}{5}\right)^2 \approx 0{\cdot}35,$$

$$P(3 \text{ rejects}) = \binom{5}{3}\left(\frac{3}{5}\right)^2\left(\frac{2}{5}\right)^3 \approx 0{\cdot}23,$$

$$P(4 \text{ rejects}) = \binom{5}{4}\left(\frac{3}{5}\right)\left(\frac{2}{5}\right)^4 \approx 0{\cdot}08,$$

$$P(5 \text{ rejects}) = \left(\frac{2}{5}\right)^5 \approx 0{\cdot}01.$$

Multiplying these probabilities by 100 gives the expected frequencies in 100 samples (correct to the nearest integer).

Number of rejects	0	1	2	3	4	5
Expected frequency	8	26	35	23	8	1

[Note that because of rounding errors the sum of these expected frequencies is 101. This does not invalidate the results.]

We now consider an alternative way of finding the mean and variance of certain discrete probability distributions including the binomial distribution.

The *probability generating function* (p.g.f.) for the random variable X, which takes the values $0, 1, 2, \ldots, r, \ldots$ with probabilities $p_0, p_1, p_2, \ldots, p_r, \ldots$ is given by

$$G(t) = p_0 + p_1 t + p_2 t^2 + \ldots + p_r t^r + \ldots$$

Thus by using a "dummy" variable t, we are able to write down a single expression $G(t)$ representing a complete probability distribution. We note that since $\sum p_r = 1$, it follows that $G(1) = 1$.
For the binomial distribution with parameters n and p,

$$G(t) = q^n + \binom{n}{1} q^{n-1} pt + \binom{n}{2} q^{n-2} p^2 t^2 + \ldots + \binom{n}{r} q^{n-r} p^r t^r + \ldots$$

$\therefore$ the p.g.f. of the binomial distribution $B(n, p)$ is

$$G(t) = (q + pt)^n, \quad \text{where} \quad q = 1 - p.$$

If $G(t) = p_0 + p_1 t + p_2 t^2 + p_3 t^3 + \ldots + p_r t^r + \ldots$ then, differentiating with respect to t,

$$G'(t) = p_1 + 2p_2 t + 3p_3 t^2 + \ldots + rp_r t^{r-1} + \ldots$$
$$G''(t) = 2p_2 + 3.2p_3 t + \ldots + r(r-1)p_r t^{r-2} + \ldots$$
$$\therefore \quad G'(1) = p_1 + 2p_2 + 3p_3 + \ldots + rp_r + \ldots = \sum rp_r$$
$$G''(1) = 2p_2 + 3.2p_3 + \ldots + r(r-1)p_r + \ldots$$
$$= \sum r(r-1)p_r = \sum r^2 p_r - \sum rp_r$$

Hence $E(X) = \sum rp_r = G'(1)$
and $\text{Var}(X) = \sum r^2 p_r - [E(X)]^2 = G''(1) + G'(1) - [G'(1)]^2$

i.e. for a distribution with p.g.f. $G(t)$

$$\mu = G'(1), \quad \sigma^2 = G''(1) + G'(1) - [G'(1)]^2.$$

For the binomial distribution $B(n, p)$:

$$G(t) = (q + pt)^n$$
$$G'(t) = np(q + pt)^{n-1} \qquad \therefore \quad G'(1) = np$$
$$G''(t) = n(n-1)p^2(q + pt)^{n-2} \qquad \therefore \quad G''(1) = n(n-1)p^2$$

Hence $\mu = np$, $\sigma^2 = n(n-1)p^2 + np - n^2p^2 = np(1-p)$.
These results are the same as those obtained earlier by elementary methods.

Exercise 15.2

1. An unbiased die is thrown 9 times. Find the probability of (a) 2 sixes, (b) at least 2 sixes. Find also the mean and variance of the number of sixes thrown.

2. In a large batch of plastic mouldings 25% have faults. If a random sample of 10 mouldings is inspected, find the probability that the sample contains

(a) 3 faulty mouldings, (b) no more than 2 faulty mouldings. Find also the mean and variance of the number of faulty mouldings.

3. The probability that a certain darts player hits the "treble 20" with one dart is 0·3. Find the probability that the player scores at least one treble 20 with 3 darts. If the probability of scoring at least one treble 20 with n darts is greater than 0·9, find the least possible value of n.

4. One in eight of the torches assembled by a machine are imperfect. Eight torches are chosen at random. Find the probability that (a) one is imperfect, (b) at least 6 are imperfect. The probability that a box of n torches contains none that are imperfect is less than 0·1. Find the range of possible values of n.

5. Given that the binomial distribution $B(n, p)$ has mean 9·6 and standard deviation 2·4, find n and p. Hence obtain the probability generating function of the distribution.

6. Find the probability generating function of a binomial distribution whose mean is 1·8 and whose variance is 0·99. If the random variable X has this binomial distribution, (i) state the greatest possible value that X may take (with non-zero probability), (ii) write down a numerical expression for the probability that a randomly observed value of X will be equal to 3. (JMB)

7. A lake contains 1000 fish of species A, 2000 fish of species B and no other fish. Ten fish are taken from the lake at random. Explain why the probabilities of obtaining various numbers of fish of species A in the ten can be very well approximated by a binomial probability distribution, and hence estimate to three significant figures the probability of obtaining 4 fish of species A and six of species B. Estimate also to three significant figures the probability that in a catch of 5 fish altogether there are more fish of species A than there are of species B. (C)

8. A manufacturing company sends out invoices to its customers allowing 5% discount to those who settle their accounts within 15 days. In the past, 40% of the customers took advantage of the discount terms. On a particular day the company sent out 10 invoices. Examine critically the appropriateness of the binomial distribution model for this situation, discussing (a) the requirement of constant probability from trial to trial, (b) the independence of trials. Assuming the model to be appropriate, calculate the probability that, of the 10 invoices, less than 2 are settled within 15 days. For days when 10 invoices are sent out, calculate the mean and variance of the distribution of the number of invoices settled within 15 days. (L)

9. The numbers of males in a sample of 810 litters of 4 cats are recorded in the table below:

No. of males	0	1	2	3	4
No. of litters	156	326	242	74	12

Calculate the mean and variance of the number of males in this sample of litters. Obtain a corresponding table of the above form for a binomial distribution with $p = \frac{1}{3}$.

10. Six dice were thrown together 1000 times. At each throw, the number of ones obtained was noted and the frequency is shown in the table below:

No. of 1's in a single throw	0	1	2	3	4	5	6
No. of throws	370	380	150	90	9	1	0

Calculate the mean and variance of the number of ones obtained in the experiment. Produce a corresponding table of the expected frequencies, to the nearest integer in each case, based on the assumption that the six dice are unbiased and that the 1000 throws are independent. (L)

11. (a) Berg and Korner have a long rivalry in tennis: in the last 25 games, Berg has won 15 times. They start a new series of 8 games. Assuming that the binomial model may be applied, what is the probability that Berg will win at least 6 of these?
(b) Prove that the mean of the binomial distribution is np. A set of 100 pods, each containing 4 peas, was examined to see how many of the peas were good. The following were the results:

No. of good peas in pod	0	1	2	3	4
No. of pods	7	20	35	30	8

(i) What was the probability of getting a good pea? (ii) Calculate the theoretical frequencies of 0, 1, 2, 3, 4 good peas, using the associated theoretical binomial distribution. (SU)

12.

X	frequency
0	36
1	37
2	20
3	5
4	2

The number of wrong numbers (X) obtained in making a set of ten telephone calls is recorded in the given table which records the results obtained in making 100 sets of 10 telephone calls. Assuming that wrong numbers occur randomly, suggest a suitable statistical model to correspond to the making of 10 telephone calls and recording the number of wrong numbers obtained. Using this model estimate the percentage of sets of ten telephone calls in which 2 or more wrong numbers are obtained. What will this percentage be if the chance of getting a wrong number when making a call is doubled? (L)

13. (i) Find the probability of obtaining exactly two sixes in six throws of a single die [You may assume that $(5/6)^5 = 0{\cdot}402$.]
(ii) If two dice are to be thrown together and the sum of their scores is recorded, find the probability that (a) the sum in one throw will be eleven, (b) a score of

eleven will occur exactly twice in six throws [assume that $17^4/18^5 = 0{\cdot}044$], (c) in one throw the score will be less than eight, given that exactly one die shows a four. (L)

14. A crossword puzzle is published in *The Times* each day of the week, except Sunday. A man is able to complete, on average, 8 out of 10 of the crossword puzzles. (i) Find the expected value and the standard deviation of the number of completed crosswords in a given week. (ii) Show that the probability that he will complete at least 5 in a given week is 0·655 (to 3 significant figures). (iii) Given that he completes the puzzle on Monday, find, to three significant figures, the probability that he will complete at least 4 in the rest of the week. (iv) Find, to three significant figures, the probability that, in a period of four weeks, he completes 4 or less in only one of the four weeks. (C)

15. 4 girls and 3 boys plan to meet together on the following Saturday. The probability that each boy will be present is $\frac{2}{3}$ independently of the other boys. Find the probabilities that (a) 0, (b) 1, (c) 2, (d) 3 boys will be present. The probability that each girl will be present is $\frac{1}{2}$ independently of the other girls and of the boys. (e) Find the probability that the number of girls present will equal the number of boys. (f) Find the probability that both sexes will be present. (g) Afterwards it was reported that the gathering had included at least one boy and at least one girl. What is the probability that there were equal numbers of boys and girls in the light of this additional information? (O&C)

16. Each trial of a binomial experiment has a constant probability p of yielding a success. (i) Given that 7 successes occurred in 15 independent trials show that the conditional probability that the first and last trials were successes is 1/5. (ii) Find the value of p given that in 20 independent trials the probability of exactly 4 successes is twice that of exactly 6 successes. (JMB)

17. The random variable X has the binomial distribution defined by

$$P(X = r) = \binom{n}{r}(1 - p)^{n-r}p^r, \quad r = 0, 1, 2, \ldots, n.$$

Find an expression for $P(X = r)/P(X = r - 1)$. Hence deduce the most likely values of X when (a) $n = 20$, $p = 0{\cdot}4$, (b) $n = 14$, $p = 0{\cdot}2$.

18. Cards are drawn at random and with replacement from an ordinary pack of playing cards until three spades have been drawn. Find the probability that the number of draws required is (a) exactly six, (b) at least six. Find an expression for P_n, the probability that exactly n cards have to be drawn to get three spades ($n \geqslant 3$). Hence, or otherwise, find the most likely value (or values) of n. (L)

19. A random variable X has p.g.f. $G(t) = 1/(2 - t)^2$. Find the expressions for (a) $P(X = r)$, (b) $E(X)$, (c) $\mathrm{Var}(X)$.

20. A random variable X is defined as the number of throws of an unbiased die needed to obtain a six. Find the p.g.f. for X and hence determine $E(X)$ and $\mathrm{Var}(X)$.

15.3 Continuous probability distributions

If X is a continuous random variable which may take any real value within some given interval, $0 \leqslant x \leqslant 10$ say, then it is difficult to give a meaning to probabilities such as $P(X = 2)$, $P(X = 7{\cdot}32054)$, $P(X = 17/39)$. If $P(X = x)$ were non-zero for all values of x, the sum of all such probabilities would be infinite. This means that we must regard the value of $P(X = x)$ as zero. Thus, in order to set up a suitable probability model we clearly need to define a new type of function.

> The *probability density function* (p.d.f.) of the continuous random variable X is a function $f(x)$ such that
>
> $$P(a \leqslant X \leqslant b) = \int_a^b f(x)\,dx.$$

[Since $P(X = a) = P(X = b) = 0$, the values of $P(a < X < b)$, $P(a \leqslant X < b)$ and $P(a < X \leqslant b)$ are also given by this integral.]

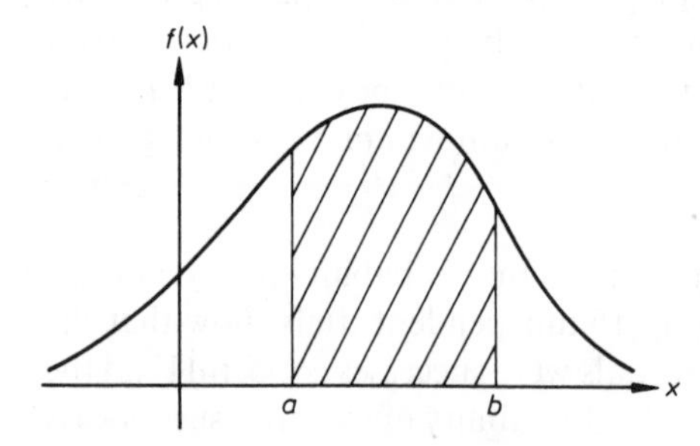

It follows that we can now think of probabilities as areas under the graph of $f(x)$. As probabilities are never negative, $f(x) \geqslant 0$ for all values of x. Since the sum of the probabilities of all possible outcomes of an experiment is 1, the total area under the graph must be 1.

i.e. $$\int_{-\infty}^{\infty} f(x)\,dx = 1.$$

For a discrete random variable X, $E(X) = \mu = \sum xp(x)$ and $\text{Var}(X) = \sigma^2 = \sum(x - \mu)^2 p(x) = \sum x^2 p(x) - \mu^2$. To obtain similar formulae for the mean and variance of a continuous random variable X, let us consider the probability that the value of X lies between x and $x + \delta x$. Provided that δx is small, this probability is given approximately by $f(x)\delta x$. Thus we let

$$E(X) = \lim_{\delta x \to 0} \sum x f(x)\,\delta x = \int_{-\infty}^{\infty} x f(x)\,dx.$$

Applying similar arguments to the variance, we have:

> For a continuous random variable X,
>
> $$E(X) = \mu = \int_{-\infty}^{\infty} xf(x)\,dx$$
>
> $$\text{Var}(X) = \sigma^2 = \int_{-\infty}^{\infty} (x - \mu)^2 f(x)\,dx = \int_{-\infty}^{\infty} x^2 f(x)\,dx - \mu^2.$$

One of the simplest types of continuous probability distribution is the *rectangular* or *uniform* distribution. A continuous random variable X is said to be *uniformly distributed* over the interval a to b if $f(x)$ has a constant non-zero value for $a \leqslant x \leqslant b$ and is zero otherwise.

Example 1 Find the mean and variance of the continuous random variable X which is uniformly distributed over the interval from 0 to 3.

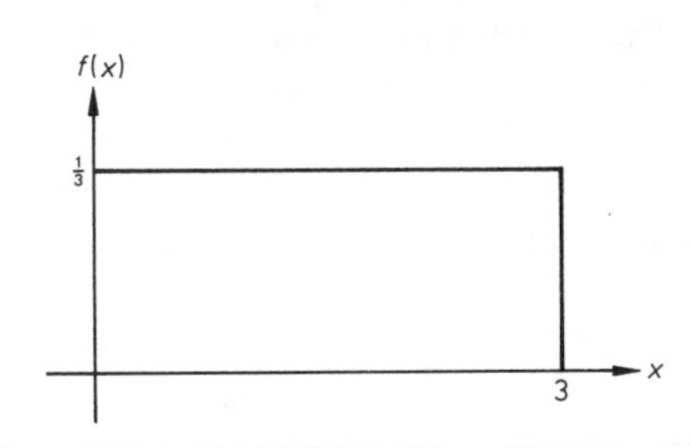

Since the total area under the graph of the probability density function $f(x)$ must be 1,

$$f(x) = \tfrac{1}{3} \quad \text{for} \quad 0 \leqslant x \leqslant 3$$
$$f(x) = 0 \quad \text{otherwise.}$$

$$\therefore \qquad E(X) = \int_0^3 x \times \frac{1}{3}\,dx = \left[\frac{1}{6}x^2\right]_0^3 = \frac{3}{2}$$

$$\text{Var}(X) = \int_0^3 x^2 \times \frac{1}{3}\,dx - \left(\frac{3}{2}\right)^2 = \left[\frac{1}{9}x^3\right]_0^3 - \frac{9}{4} = 3 - \frac{9}{4} = \frac{3}{4}$$

Hence the mean of the distribution is $1\frac{1}{2}$ and the variance is $\frac{3}{4}$.

A *mode* or *modal value* of a continuous random variable X with probability density function $f(x)$ is a value of x for which $f(x)$ takes a maximum value. Thus the modes of a distribution are the x-coordinates of the maximum points on the graph of $f(x)$.

The *median* of a continuous random variable X with probability density function $f(x)$ is the value m such that $\displaystyle\int_{-\infty}^{m} f(x)\,dx = 0{\cdot}5$.

Example 2 A continuous random variable X has probability density function $f(x) = kx(9 - x^2)$ for $0 \leqslant x \leqslant 3$ and $f(x) = 0$ otherwise. Find (a) the value of k, (b) the mean, (c) the mode, (d) the median.

$$\text{(a)} \quad \int_0^3 f(x)\,dx = k\int_0^3 (9x - x^3)\,dx = k\left[\frac{9x^2}{2} - \frac{x^4}{4}\right]_0^3 = \frac{81k}{4}$$

Since the total area under the graph of $f(x)$ must be 1, $k = 4/81$.

$$\text{(b)} \quad E(X) = \int_0^3 xf(x)\,dx = \frac{4}{81}\int_0^3 (9x^2 - x^4)\,dx$$

$$= \frac{4}{81}\left[3x^3 - \frac{x^5}{5}\right]_0^3 = 4\left(1 - \frac{3}{5}\right) = \frac{8}{5}$$

Hence the mean of the distribution is $1{\cdot}6$.

(c)

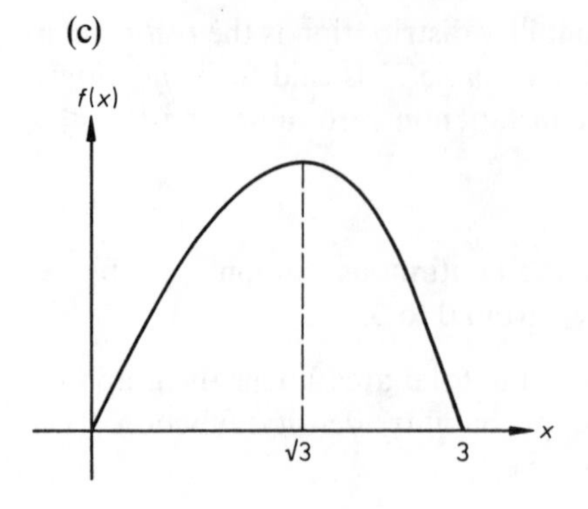

$f(x) = \frac{4}{81}(9x - x^3)$,

$f'(x) = \frac{4}{81}(9 - 3x^2) = \frac{4}{27}(3 - x^2)$

$\therefore$ in the interval $0 \leqslant x \leqslant 3$, $f'(x) = 0$ when $x = \sqrt{3}$.
The sketch shows that this gives a maximum value of $f(x)$.

Hence the mode of the distribution is 1·73 (to 3 s.f.)

(d) If m is the median, $\int_0^m f(x)\,dx = \frac{1}{2}$

$$\therefore \quad \frac{4}{81}\int_0^m (9x - x^3)\,dx = \frac{4}{81}\left[\frac{9x^2}{2} - \frac{x^4}{4}\right]_0^m = \frac{1}{81}(18m^2 - m^4) = \frac{1}{2}$$

$$36m^2 - 2m^4 = 81$$
$$2m^4 - 36m^2 + 81 = 0$$

$$\therefore \quad m^2 = \frac{36 \pm \sqrt{(36^2 - 8.81)}}{4} = \frac{36 \pm 18\sqrt{(4-2)}}{4} = 9 \pm \frac{9}{2}\sqrt{2}$$

Since $0 < m < 3$, we must have $m^2 = 9 - \frac{9}{2}\sqrt{2}$

$$\therefore \quad m = \sqrt{\left\{9 - \frac{9}{2}\sqrt{2}\right\}} \approx 1{\cdot}62$$

Hence the median of the distribution is 1·62 (to 3 s.f.).

Another function used when modelling continuous probability distributions is defined as follows:

> The *cumulative distribution function* (c.d.f.) of the continuous random variable X is the function $F(x)$ such that
>
> $$F(x) = P(X \leqslant x).$$

If X has probability density function $f(x)$, then

$$F(t) = P(X \leqslant t) = \int_{-\infty}^{t} f(x)\,dx.$$

It follows that: $\qquad F'(x) = f(x).$

Example 3 A random variable X has cumulative distribution function

$$F(x) = \begin{cases} 0 & \text{for } x \leqslant 0 \\ \frac{1}{6}x^2 & \text{for } 0 \leqslant x \leqslant 2 \\ -2 + 2x - \frac{1}{3}x^2 & \text{for } 2 \leqslant x \leqslant 3 \\ 1 & \text{for } x \geqslant 3. \end{cases}$$

Find (a) $P(X \leqslant 1{\cdot}5)$, (b) the median, (c) the probability density function $f(x)$.

(a) $P(X \leqslant 1{\cdot}5) = F(1{\cdot}5) = \frac{1}{6}(1{\cdot}5)^2 = 0.375.$

(b) The median is the value m such that $F(m) = \frac{1}{2}$.

Since $F(0) = 0$ and $F(2) = \frac{2}{3}$, m lies in the interval from 0 to 2 and thus $F(m) = \frac{1}{6}m^2$.

$\therefore \quad \frac{1}{6}m^2 = \frac{1}{2}$ which gives $m = \sqrt{3}$.

Hence the median of the distribution is 1·73 (to 3 s.f.).

(c) The probability density function $f(x) = F'(x)$.

$$\therefore \quad \text{for} \quad 0 \leqslant x \leqslant 2, \quad f(x) = \frac{d}{dx}(\tfrac{1}{6}x^2) = \tfrac{1}{3}x;$$

$$\text{for} \quad 2 \leqslant x \leqslant 3, \quad f(x) = \frac{d}{dx}(-2 + 2x - \tfrac{1}{3}x^2) = 2 - \tfrac{2}{3}x;$$

$$\text{otherwise,} \qquad f(x) = 0.$$

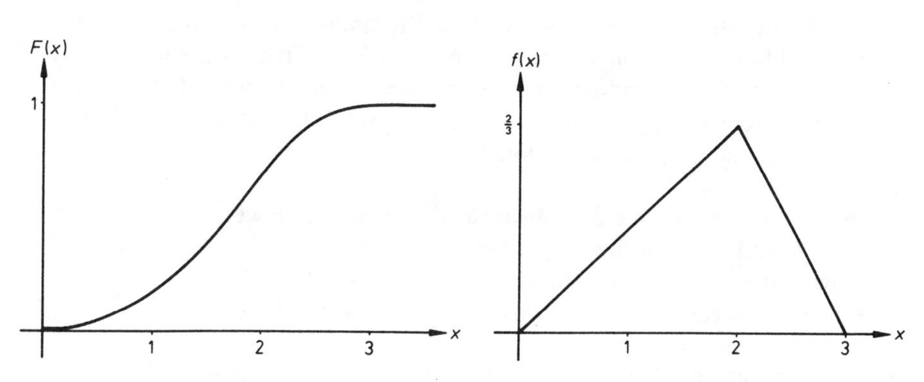

Example 4 A random variable X has probability density function $f(x) = \frac{1}{2}x$ for $0 \leqslant x \leqslant 2$ and $f(x) = 0$ otherwise. Find the cumulative distribution function $F(x)$ of the variable X. Hence find the cumulative distribution function $G(y)$ of the variable $Y = X^2$. Deduce the probability density function $g(y)$.

$$\text{For } 0 \leqslant t \leqslant 2, F(t) = \int_0^t \tfrac{1}{2}x\,dx = \left[\tfrac{1}{4}x^2\right]_0^t = \tfrac{1}{4}t^2.$$

$\therefore$ the cumulative distribution function of X is given by

$F(x) = 0\,(x \leqslant 0), \quad F(x) = \frac{1}{4}x^2\,(0 \leqslant x \leqslant 2), \quad F(x) = 1\,(x \geqslant 2).$

$G(y) = P(Y \leqslant y) = P(X^2 \leqslant y) = P(X \leqslant \sqrt{y}) = F(\sqrt{y})$

$\therefore$ the cumulative distribution function of Y is given by

$$G'(y) = 0\,(y \leqslant 0), \quad G(y) = \tfrac{1}{4}y\,(0 \leqslant y \leqslant 4), \quad G(y) = 1\,(y \geqslant 4).$$

Since $g(y) = G'(y)$, we find that $g(y) = \frac{1}{4}(0 \leqslant y \leqslant 4)$ and $g(y) = 0$ otherwise. [We see that Y is uniformly distributed over the interval from 0 to 4.]

Exercise 15.3

1. Find the mean and variance of the continuous random variable X which is uniformly distributed over the interval
(a) from 0 to 1, (b) from 1 to 5,
(c) from 0 to k, (d) from a to b, where $b > a$.

2. A continuous random variable X has probability density function

$$f(x) = k(3x^2 - x^3) \quad \text{for} \quad 0 \leqslant x \leqslant 3 \quad \text{and} \quad f(x) = 0 \text{ otherwise.}$$

Find the value of k. Find also the mean, mode and standard deviation of X.

3. (a) A continuous variable x is distributed at random between the values 2 and 3 and has a probability density function of $6/x^2$. Find the median value of x.
(b) A continuous random variable X takes values between 0 and 1, with a probability density function of $Ax(1 - x)^3$. Find the value of A, and the mean and standard deviation of X. (SU)

4. A continuous variable X is distributed at random between two values, $x = 0$ and $x = 2$, and has a probability density function of $ax^2 + bx$. The mean is 1·25. (i) Show that $b = \frac{3}{4}$, and find the value of a. (ii) Find the variance of X. (iii) Verify that the median value of X is approximately 1·3. (iv) Find the mode. (SU)

5. The random variable X has a probability density function given by

$$P(x) = \begin{cases} kx(1 - x^2) & (0 \leqslant x \leqslant 1), \\ 0 & \text{elsewhere,} \end{cases}$$

k being a constant. Find the value of k and find also the mean and variance of this distribution. Find the median of the distribution. (O&C)

6. The continuous random variable X has probability density function given by

$$f(x) = \begin{cases} k(1 + x^2) & \text{for} \quad -1 \leqslant x \leqslant 1, \\ 0 & \text{otherwise,} \end{cases}$$

where k is a constant. Find the value of k, and determine $E(X)$ and $\text{Var}(X)$. A is the event $X > \frac{1}{2}$; B is the event $X > \frac{3}{4}$. Find (i) $P(B)$, (ii) $P(B|A)$. (C)

7. Petrol is delivered to a garage every Monday morning. At this garage the weekly demand for petrol, in thousands of units, is a continuous random variable X distributed with a probability density function of the form

$$f(x) = \begin{cases} ax^2(b - x) & 0 \leqslant x \leqslant 1 \\ 0 & \text{otherwise.} \end{cases}$$

(i) Given that the mean weekly demand is 600 units, determine the values of a and b. (ii) If the storage tanks at this garage are filled to their total capacity of 900 units every Monday morning, what is the probability that in any given week the garage will be unable to meet the demand for petrol? (AEB 1978)

8. A continuous random variable X has the probability density function f defined by

$$\begin{aligned} f(x) &= \tfrac{1}{3}cx, & 0 &\leqslant x < 3, \\ f(x) &= c, & 3 &\leqslant x \leqslant 4, \\ f(x) &= 0, & &\text{otherwise,} \end{aligned}$$

where c is a positive constant. Find (i) the value of c, (ii) the mean of X, (iii) the value, a, for there to be a probability of 0·85 that a randomly observed value of X will exceed a. (JMB)

9. A continuous random variable X takes values between 0 and 1 only and its probability density function f is given by $f(x) = kx^2(1 - x)$. Prove that $k = 12$, and find $E(X)$. The cumulative distribution function F is such that $F(x) = P(X \leqslant x)$. Express $F(x)$ in terms of x, and sketch the graph of $y = F(x)$. Verify by calculation that the median of X lies between 0·60 and 0·62. (C)

10. The random variable X has the probability density function

$$f(x) = ae^{-ax} \, (0 < x < \infty).$$

Show that the cumulative distribution function of X is $F(x) = 1 - e^{-ax}$. The above distribution, with parameter $a = 0{\cdot}8$, is proposed as a model for the length of life, in years, of a species of bird. Find the expected frequencies, of a total of 50 birds, that would fall in the class intervals (years) 0 –, 1 –, 2 –, 3 –, 4 –, 5 and over. (O)

11. A random variable X takes values in the interval $0 < x < 3$ and has probability density function $f(x)$, where

$$f(x) = \begin{cases} ax & (0 < x \leqslant 1), \\ \tfrac{1}{2}a(3 - x) & (1 < x < 3). \end{cases}$$

(i) Find the value of a. (ii) Find the expected value and the standard deviation of X. (iii) Find the (cumulative) distribution function $F(x)$. (iv) Find $P(|X - 1| < \tfrac{1}{2})$. (C)

12. The continuous random variable X has probability density function $f(x)$ defined by

$$f(x) = \begin{cases} c/x^4 & (x < -1) \\ c(2 - x^2) & (-1 \leqslant x \leqslant 1) \\ c/x^4 & (x > 1). \end{cases}$$

(i) Show that $c = \frac{1}{4}$. (ii) Sketch the graph of $f(x)$. (iii) Determine the cumulative distribution function $F(x)$. (iv) Determine the expected value of X and the variance of X. (C)

13. The continuous random variable X has (cumulative) distribution function given by

$$F(x) = \begin{cases} (1 + x)/8 & (-1 \leqslant x \leqslant 0), \\ (1 + 3x)/8 & (0 \leqslant x \leqslant 2), \\ (5 + x)/8 & (2 \leqslant x \leqslant 3), \end{cases}$$

with $F(x) = 0$ for $x < -1$, and $F(x) = 1$ for $x > 3$.
(i) Sketch the graph of the probability density function $f(x)$.
(ii) Determine the expectation of X and the variance of X.
(iii) Determine $P(3 \leqslant 2X \leqslant 5)$. (C)

14. The length X of an offcut of wooden planking is a random variable which can take any value up to 0·5 m. It is known that the probability of the length being not more than x metres $(0 \leqslant x \leqslant 0{\cdot}5)$ is equal to kx. Determine (i) the value of k, (ii) the probability density function of X, (iii) the expected value of X, (iv) the standard deviation of X (correct to 3 s.f.). (C)

15. The (cumulative) distribution function for a continuous random variable X is given by

$$F(x) = \begin{cases} 0 & \text{for} \quad x < -2 \\ k(4x - \frac{1}{3}x^3 + \frac{16}{3}) & \text{for} \quad -2 \leqslant x \leqslant 2, \\ 1 & \text{for} \quad x > 2. \end{cases}$$

Find (i) the value of k, (ii) $P(-1 < X < 1)$, (iii) the probability density function for X, (iv) the mean and variance of X. (C)

16. A random variable X has probability density function $f(x) = \frac{2}{3}x$ for $1 \leqslant x \leqslant 2$ and $f(x) = 0$ otherwise. Find the cumulative distribution function $F(x)$ of the variable X. Hence find the cumulative distribution function $G(y)$ of the variable $Y = X^{1/2}$. Deduce the probability density function $g(y)$.

17. X is a continuous random variable which is uniformly distributed in the interval $-1 \leqslant x \leqslant 1$. A second random variable Y is given by $Y = X^2$. Show that, if $0 < y < 1$, then $P(Y \leqslant y) = \sqrt{y}$, and deduce the probability density function of Y. Two independent observations of Y are made, and Z denotes the larger of the two observations. By considering the probability that both values of Y are

less than or equal to z, where $0 \leqslant z \leqslant 1$, or otherwise, prove that z is uniformly distributed. Determine the probability density function of W, the smaller of the two values of Y. (C)

18. Write down the probability density function and the cumulative distribution function of a random variable which is uniformly distributed over the interval $[a, b]$.

A person drives to work, a distance of 12 miles. The time he takes, in minutes, to cover the journey is uniformly distributed over the interval $[20, 30]$. Let V denote the car's average speed in miles per hour on such a journey. Find an expression for $P(V \leqslant v)$, where v is an arbitrary value between 24 and 36. Hence, or otherwise, find (i) the probability density function of V, (ii) the probability that on such a journey the car's average speed will be in excess of 30 m.p.h., (iii) the median average speed for the journey. (W)

19. A continuous random variable X has probability density function defined by

$$\begin{aligned} f(x) &= 2x/a^2 \qquad 0 \leqslant x \leqslant a \\ f(x) &= 0, \qquad \text{otherwise.} \end{aligned}$$

(a) Find the variance of X. (b) Find the cumulative distribution function of X. (c) If $Y = 1 - (X/a)$, find (i) $P(Y > X)$, (ii) the probability density function of Y. (W)

20. A random variable X has probability density function $f(x) = |x|$ for $-1 \leqslant x \leqslant 1$ and $f(x) = 0$ otherwise. Find the probability density function $g(y)$ of the random variable $Y = X^2$.

15.4 The normal distribution

The *normal distribution* is the most important continuous probability distribution in theoretical statistics. It is found to provide a good probability model for the distributions of many variables such as the heights of people and other measurements of biological characteristics. One of the early uses of the normal distribution was as a model for the distribution of errors in experimental observations.

A continuous random variable X is said to be *normally distributed* with mean μ and variance σ^2 if it has probability density function.

$$f(x) = \frac{1}{\sigma\sqrt{(2\pi)}} e^{-(x-\mu)^2/2\sigma^2} \quad (\sigma > 0)$$

This distribution is denoted by $N(\mu, \sigma^2)$.

We cannot prove by elementary methods that this is a valid definition of a probability distribution. Clearly $f(x) \geqslant 0$ for all values of x, but for now we must assume that the total area under the graph of $f(x)$ is 1 and that the distribution has mean μ and variance σ^2.

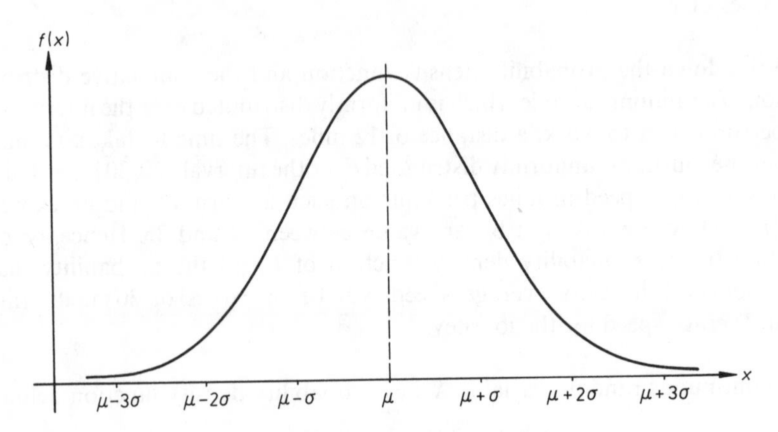

The graph of $f(x)$ is a bell-shaped curve called the *normal curve*. We see that
(1) the curve is symmetrical about the line $x = \mu$;
(2) the x-axis is an asymptote to the curve;
(3) when $x = \mu$, $f(x)$ takes its maximum value $1/\sigma\sqrt{(2\pi)}$ (i.e. $0{\cdot}4/\sigma$ approximately).
As described in the previous section, the probabilities associated with a continuous random variable X are represented by areas under the graph of the p.d.f. $f(x)$. Thus for the distribution $N(\mu, \sigma^2)$,

$$P(X \leqslant a) = \int_{-\infty}^{a} f(x)\,dx = \int_{-\infty}^{a} \frac{1}{\sigma\sqrt{(2\pi)}} e^{-(x-\mu)^2/2\sigma^2}\,dx.$$

This integral can be simplified by substituting

$$z = \frac{x-\mu}{\sigma} \quad \text{and} \quad dz = \frac{1}{\sigma}\,dx.$$

Writing $c = \dfrac{a-\mu}{\sigma}$, we have

$$P(X \leqslant a) = P(Z \leqslant c) = \int_{-\infty}^{c} \frac{1}{\sqrt{(2\pi)}} e^{-z^2/2}\,dz.$$

Hence the new random variable $Z = (X - \mu)/\sigma$ has probability density function $\phi(z) = \dfrac{1}{\sqrt{(2\pi)}} e^{-z^2/2}$. We deduce that Z is normally distributed with mean 0 and variance 1. We therefore call Z a *standardised* or *standard* normal variable. To summarise:

> If a normal variable X has the distribution $N(\mu, \sigma^2)$ then the random variable $Z = (X - \mu)/\sigma$ has the *standard normal distribution* $N(0, 1)$.

Using this change of variable any problem involving a normal distribution can be restated in terms of the standard distribution $N(0, 1)$.

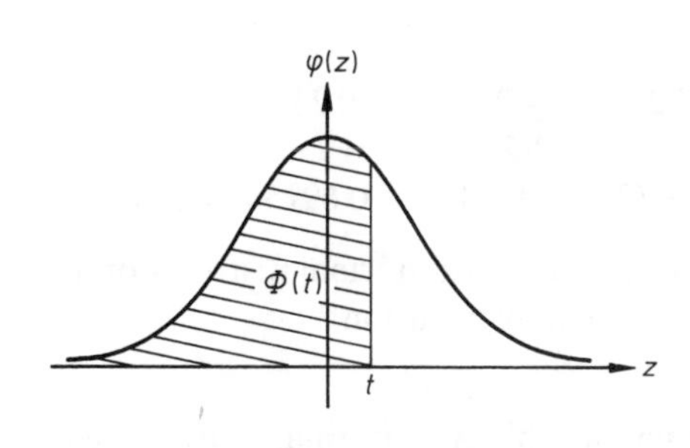

The (cumulative) *distribution function* for the standard normal variable Z is denoted by $\Phi(z)$,
where $\Phi(z) = P(Z \leqslant z)$,

$$\therefore \quad \Phi(t) = \int_{-\infty}^{t} \phi(z)\,dz,$$

where $\phi(z) = \dfrac{1}{\sqrt{(2\pi)}} e^{-z^2/2}$.

Since these integrals are difficult to evaluate, tables of values have been compiled. Table 1, which appears towards the end of this book, gives values of $\Phi(z)$ for $0 \leqslant z \leqslant 4$. For example, $P(Z \leqslant 2{\cdot}18) = \Phi(2{\cdot}18) = 0{\cdot}98537$. If intermediate values of $\Phi(z)$ are required linear interpolation can be used.

Example 1 Use Table 1 to estimate the value of $\Phi(0{\cdot}6354)$.

$\Phi(0{\cdot}64) - \Phi(0{\cdot}63) = 0{\cdot}7389 - 0{\cdot}7357 = 0{\cdot}0032$

$\therefore$ by linear interpolation,

$\Phi(0{\cdot}6354) = 0{\cdot}7357 + \frac{54}{100} \times 0{\cdot}0032 \approx 0{\cdot}7374$.

The following diagrams show how the symmetry of the normal distribution can be used to obtain $\Phi(z)$ for negative values of z.

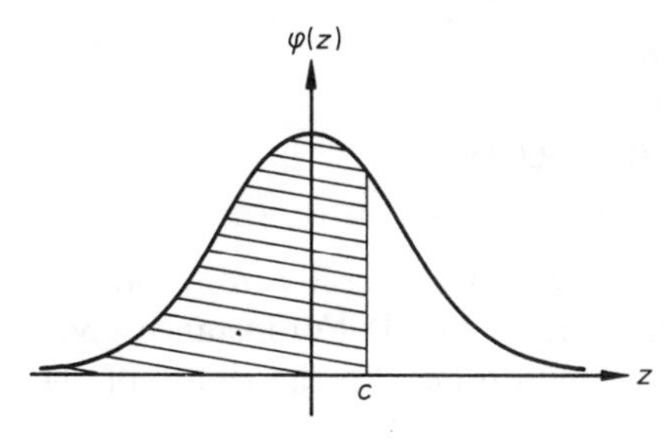

$P(Z \leqslant c) = \Phi(c)$

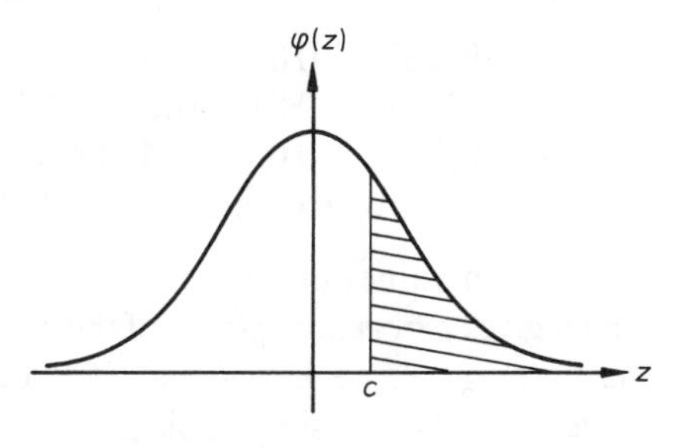

$P(Z \geqslant c) = 1 - \Phi(c)$

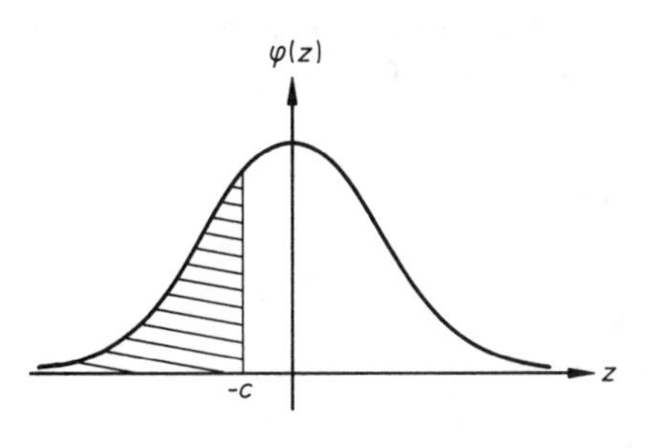

$P(Z \leqslant -c) = 1 - \Phi(c)$

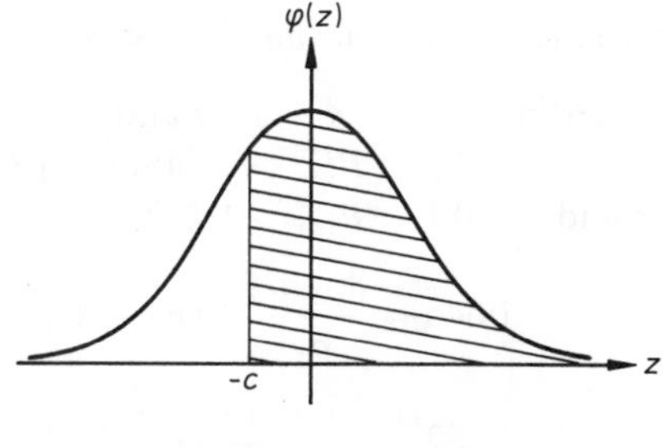

$P(Z \geqslant -c) = \Phi(c)$

We may apply these results as follows:

$$P(Z \leqslant -0.72) = 1 - \Phi(0{\cdot}72) = 1 - 0{\cdot}7642 = 0.2358$$
$$P(-1{\cdot}5 \leqslant Z \leqslant -0{\cdot}5) = P(Z \geqslant -1{\cdot}5) - P(Z \geqslant -0.5)$$
$$= \Phi(1{\cdot}5) - \Phi(0{\cdot}5) = 0{\cdot}9332 - 0{\cdot}6915 = 0{\cdot}2417$$

In the next example we use our knowledge of the standard normal distribution to find probabilities associated with a different normal distribution.

Example 2 Given that a continuous random variable X is normally distributed with mean 20 and variance 16, find $P(15 \leqslant X \leqslant 24)$.

X has mean $\mu = 20$ and standard deviation $\sigma = 4$,
$\therefore$ the corresponding standard normal variable is

$$Z = (X - \mu)/\sigma = (X - 20)/4$$

$$P(X \leqslant 24) = P\left(Z \leqslant \frac{24 - 20}{4}\right) = P(Z \leqslant 1) = \Phi(1) = 0{\cdot}8413$$

$$P(X \leqslant 15) = P\left(Z \leqslant \frac{15 - 20}{4}\right) = P(Z \leqslant -1{\cdot}25) = 1 - \Phi(1{\cdot}25) = 0{\cdot}1056$$

$\therefore$ $P(15 \leqslant X \leqslant 24) = 0{\cdot}8413 - 0{\cdot}1056 = 0{\cdot}7357.$

We now demonstrate a different way of using values of the normal distribution function.

Example 3 Given that $\Phi(t) = 0{\cdot}81$, use Table 1 to estimate the value of t.

$\Phi(0{\cdot}87) = 0{\cdot}8078$, $\Phi(0{\cdot}88) = 0{\cdot}8106$,
$\therefore$ if $\Phi(t) = 0{\cdot}81$, then $0{\cdot}87 < t < 0{\cdot}88$.
$\Phi(t) - \Phi(0{\cdot}87) = 0{\cdot}0022$, $\Phi(0{\cdot}88) - \Phi(0{\cdot}87) = 0{\cdot}0028$,
$\therefore$ by linear interpolation, $t = 0{\cdot}87 + \frac{22}{28} \times 0.01 \approx 0.878$.

For certain values of $\Phi(z)$, the values of z can be obtained directly from Table 2, which gives *percentage points* of the normal distribution $N(0, 1)$. For instance, given that $\Phi(t) = 0{\cdot}8$, we know that 20% of the distribution lies "to the right" of the value $z = t$. Thus, using Table 2, $t = 0{\cdot}842$.

Example 4 A random variable X is normally distributed with mean μ and variance σ^2. Given that $P(X < 8) = 0{\cdot}05$ and $P(X > 15) = 0{\cdot}1$, find μ and σ^2.

Letting $Z = (X - \mu)/\sigma$ and using tables,
$0{\cdot}05 = P(Z > 1{\cdot}645) = P(Z < -1{\cdot}645)$
and $0{\cdot}1 = P(Z > 1{\cdot}282)$.

$$\therefore \quad -1{\cdot}645 = \frac{8 - \mu}{\sigma} \quad \text{i.e.} \quad 1{\cdot}645\sigma = \mu - 8 \qquad (1)$$

$$1{\cdot}282 = \frac{15 - \mu}{\sigma} \quad \text{i.e.} \quad 1{\cdot}282\sigma = 15 - \mu \qquad (2)$$

Adding (1) and (2): $\quad 2{\cdot}927\sigma = 7$

$$\therefore \quad \sigma = \frac{7}{2{\cdot}927} \quad \text{and} \quad \sigma^2 \approx 5{\cdot}72$$

Substituting in (1): $\quad \mu = 8 + \dfrac{1{\cdot}645 \times 7}{2{\cdot}927} \approx 11{\cdot}9$

The normal distribution arises theoretically as a limit of the binomial distribution, as described in the following theorem:

> If R is a random variable with distribution $B(n, p)$, then the distribution of the variable $T = (R - \mu)/\sigma$, where $\mu = np$ and $\sigma = \sqrt{(npq)}$, tends to $N(0, 1)$ as n tends to infinity.

The proof of this statement is rather difficult, but we can demonstrate the limiting process involved by means of an example. The given set of histograms shows the distribution of the standardised binomial variable T as the value of n increases, but p remains fixed.

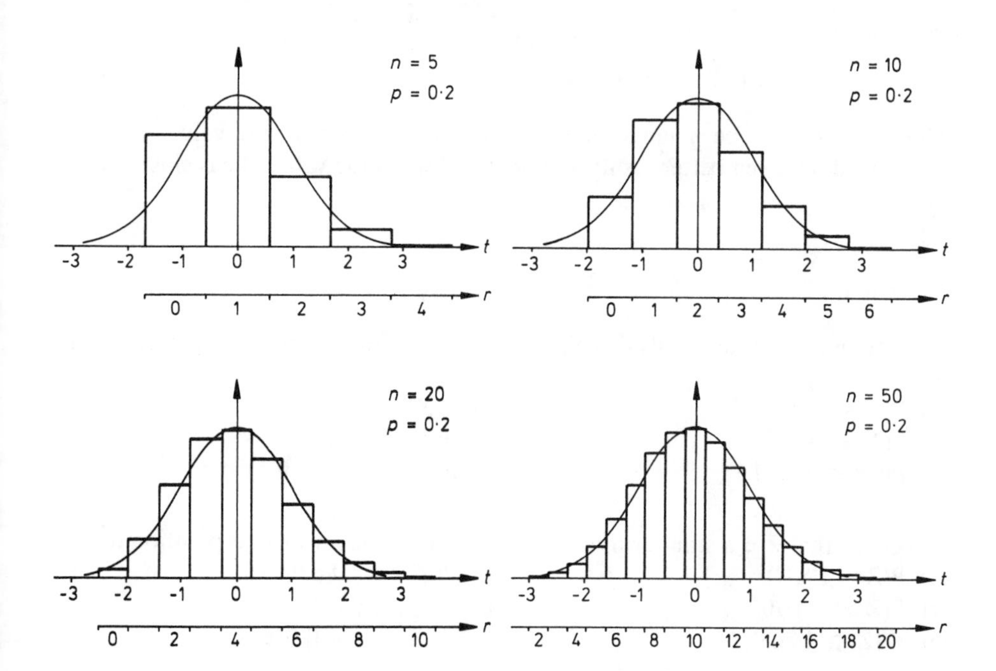

Transforming back to variables with mean $\mu = np$ and variance $\sigma^2 = npq$ we draw the following important conclusion.

> If R is a random variable with distribution $B(n, p)$, then for large n, the distribution of R can be approximated by the normal distribution $N(np, npq)$, where $q = 1 - p$.

This approximation will be close enough for most practical purposes if both np and nq are greater than 5. However, more accurate results will be obtained if both np and nq are greater than 20.

Before considering a numerical example we note that when using the distribution of a continuous variable X as an approximation to the distribution of a discrete variable R, it is necessary to make a *continuity correction*. For instance, if R takes only the integral values $0, 1, 2, \ldots$, then these are assumed to correspond to the intervals $X \leqslant 0{\cdot}5$, $0{\cdot}5 \leqslant X \leqslant 1{\cdot}5$, $1{\cdot}5 \leqslant X \leqslant 2{\cdot}5, \ldots$.

Let us now suppose that R has a binomial distribution with $n = 50$, $p = 0{\cdot}2$, then R will have mean $\mu = np = 10$ and variance $\sigma^2 = npq = 8$. Thus the corresponding normal variable X also has mean 10 and variance 8. Using $N(10, 8)$ as an approximation to $B(50, 0{\cdot}2)$:

$$\begin{aligned} P(12 \leqslant R \leqslant 15) &= P(11{\cdot}5 \leqslant X \leqslant 15{\cdot}5) \\ &= P\left(\frac{11{\cdot}5 - 10}{\sqrt{8}} \leqslant \frac{X - 10}{\sqrt{8}} \leqslant \frac{15{\cdot}5 - 10}{\sqrt{8}}\right) \\ &= P(0{\cdot}5303 \leqslant Z \leqslant 1{\cdot}9445) \\ &= \Phi(1{\cdot}9445) - \Phi(0{\cdot}5303) \\ &= 0{\cdot}9741 - 0{\cdot}7020 = 0{\cdot}2721 \\ \therefore \quad P(12 \leqslant R \leqslant 15) &\approx 0{\cdot}27. \end{aligned}$$

This turns out to be a fairly good approximation to the true value, which is 0·26 (to 2 d.p.). Even better results are obtained when the value of p is closer to 0·5.

Exercise 15.4

1. Given that X is normally distributed with mean 0 and variance 1, use Table 1 to find

(a) $P(X \leqslant 0{\cdot}3)$, (b) $P(X \geqslant 1{\cdot}2)$,
(c) $P(X > -2)$, (d) $P(X \leqslant -2{\cdot}23)$,
(e) $P(0{\cdot}8 \leqslant X \leqslant 1{\cdot}8)$, (f) $P(-1{\cdot}1 < X < 0{\cdot}5)$.

2. Given that Z is a standardised normal variable, use linear interpolation to find

(a) $P(Z \leqslant 1{\cdot}537)$, (b) $P(Z < -0.862)$,
(c) $P(Z > -1{\cdot}9035)$, (d) $P(Z \geqslant 2{\cdot}0067)$,
(e) $P(|Z| \geqslant 1{\cdot}263)$, (f) $P(|Z + 1| < 0{\cdot}616)$.

3. Given that X is normally distributed with mean 5 and standard deviation 1, find

(a) $P(X < 6{\cdot}7)$, (b) $P(X \leqslant 4{\cdot}2)$, (c) $P(3{\cdot}7 < X < 5{\cdot}7)$.

4. Given that X is normally distributed with mean 0·2 and standard deviation 0·01, find

(a) $P(X \geqslant 0{\cdot}205)$, (b) $P(X > 0{\cdot}198)$, (c) $P(0{\cdot}199 \leqslant X \leqslant 0{\cdot}201)$.

5. Given that X is normally distributed with mean 12 and variance 1·5, find
(a) $P(X \leqslant 14)$, (b) $P(X > 9{\cdot}6)$, (c) $P(9 \leqslant X \leqslant 10)$.

6. Use Table 1 to estimate the value of t given that
(a) $\Phi(t) = 0{\cdot}67$, (b) $\Phi(t) = 0{\cdot}72$, (c) $\Phi(t) = 0{\cdot}19$.

7. Use Table 2 to find the value of t such that
(a) $\Phi(t) = 0{\cdot}7$, (b) $\Phi(t) = 0{\cdot}99$, (c) $\Phi(t) = 0{\cdot}9995$.

8. The weights of a large number of cats are normally distributed with mean 1·55 kg and standard deviation 0·2 kg. Find the percentage of cats with weights (a) less than or equal to 1 kg, (b) between 1·2 and 1·3 kg, (c) between 1·5 and 1·75 kg, (d) greater than or equal to 2 kg. (L)

9. The weights of oranges are distributed normally with a mean of 96 g and standard deviation 25 g. For selling purposes, oranges are divided into three grades, I, II and III. Grade I oranges weigh more than 126 g, grade III oranges weigh less than 60 g. All remaining oranges are classed grade II. Calculate the percentage of oranges in each grade. Grade I, II and III oranges sell respectively for 12p, 8p and 4p each. Calculate the expected receipts from the sale of 80 000 oranges. (L)

10. Tests made on 2 types of electric light bulb show the following: Type A, lifetime distributed normally with an average life of 1150 hours and a standard deviation of 30 hours. Type B, long-life bulb, average lifetime of 1900 hours, with standard deviation of 50 hours. (i) What percentage of bulbs of type A could be expected to have a life of more than 1200 hours? (ii) What percentage of type B would you expect to last longer than 1800 hours? (iii) What lifetime limit would you estimate would contain the central 80% of the production of type A? (SU)

11. Batteries for a transistor radio have a mean life under normal usage of 160 hours, with a standard deviation of 30 hours. Assuming the battery life follows a normal distribution, (a) calculate the percentage of batteries which have a life between 150 hours and 180 hours; (b) calculate the range, symmetrical about the mean, within which 75% of the battery lives lie; (c) if a radio takes four of these batteries and requires all of them to be working, calculate the probability that the radio will run for at least 135 hours. (O&C)

12. Adult men in a certain country have heights which are normally distributed with a mean of 1·81 m and a standard deviation of 0.05 m.
(a) Estimate the number of men in a random sample of 100 men whose heights, when measured to the nearest 0·01 m, exceed 1·90 m.
(b) Find the probability that a man selected at random will have a height, when measured to the nearest 0·01 m, less than 1·72 m.
(c) Find the height which is exceeded by 90% of the men. (L)

13. Find the mean and variance of the normally distributed variable X in each of the following cases:
(a) $P(X > 3) = 0{\cdot}5$, $P(3 < X < 5) = 0{\cdot}1$.
(b) $P(X > 20) = P(X < 10) = 0{\cdot}05$.
(c) $P(X < 60) = 0{\cdot}01$, $P(X > 90) = 0{\cdot}3$.
(d) $P(X > 10) = 0{\cdot}1$, $P(9 < X < 10) = 0{\cdot}2$.

14. A machine is producing curtain rails whose lengths are normally distributed with mean 2·4 m. If 5% of the rails are longer than 2·42 m, find the standard deviation of the distribution. If the machine is adjusted so that 95% of the rails have lengths between 2·38 m and 2·42 m, find the new standard deviation of the distribution.

15. In an examination 30% of the candidates fail and 10% achieve distinction. Last year the pass mark (out of 200) was 84 and the minimum mark required for a distinction was 154. Assuming that the marks of the candidates were normally distributed, estimate the mean mark and the standard deviation. (O&C)

16. A random variable R has the binomial distribution $B(n, p)$. In each of the following cases find both the probability of r successes and the normal approximation to it.
(a) $n = 10, p = \frac{1}{2}, r = 6$,
(b) $n = 20, p = \frac{1}{2}, r = 12$,
(c) $n = 50, p = \frac{1}{2}, r = 30$,
(d) $n = 20, p = \frac{1}{4}, r = 4$,
(e) $n = 50, p = \frac{1}{4}, r = 10$,
(f) $n = 100, p = \frac{1}{4}, r = 20$.

17. Estimate the probability of obtaining more than 60 heads in 100 tosses of an unbiased coin.

18. In a large consignment of peaches 20% of the fruit are bruised. Estimate the probability of fewer than 10 bruised peaches in a box of 60.

19. Tutorial groups, each of 9 students, are picked at random from a large population of which two-thirds are men. (i) What is the probability that a group contains men only? (ii) Calculate the probability that the number of women in a group is either 1, 2, 3, 4 or 5. (iii) Compare this with the probability that a normal variable with the same mean and variance as the distribution appropriate to (ii) should lie between 0·5 and 5·5. (AEB 1979)

20. In a particular community the probability of a man having brown eyes is 2/5. If a random sample of 6 men is to be taken, find the probability that (a) exactly 3 men, (b) at least 3 men will have brown eyes. A second random sample of 100 men is to be taken. Use the normal distribution as an approximation to the binomial distribution, to estimate to 3 decimal places the probabilities that (a) at least half the men have brown eyes and (b) the number of men with brown eyes is between 30 and 45 inclusive. (L)

21. It is known that 30% of an apple crop has been attacked by insects. A random sample of 150 is selected from the crop. Assuming that the distribution of the

number of damaged apples in random samples of 150 may be approximated to by a normal distribution, estimate the probability that (a) more than half the sample is damaged, (b) less than 10% is damaged, (c) the number of damaged apples in the sample lies between 35 and 50, inclusive. (L)

22. In Urbania selection for the Royal Flying Corps (RFC) is by means of an aptitude test based on a week's intensive military training. It is known that the scores of potential recruits on this test follow a normal distribution with mean 45 and standard deviation 10. (i) What is the probability that a randomly chosen recruit will score between 40 and 60? (ii) What percentage of the recruits is expected to score more than 30? (iii) In a particular year 100 recruits take the test. Assuming that the pass mark is 50, calculate the probability that less than 35 recruits qualify for the RFC. (AEB 1978)

15.5 The Poisson distribution

The *Poisson*† *distribution* was originally derived as a limit of the binomial distribution. It is found to provide a useful model for distributions in which the variable is the number of occurrences of a particular event in a given interval of time or space. Some examples of variables which may have a Poisson distribution are: the number of customers entering a shop in an hour, the number of claims received by an insurance company in a day, the number of flaws in a given length of copper wire, the number of earthworms in a given area of land.

To demonstrate the way in which the Poisson distribution arises as a limit of the binomial distribution, we consider the number of telephone calls received in an hour by a small business. Assuming that the average number of calls received per hour is 4, we can construct a simple binomial model by dividing each hour into 12 five-minute "trials" and taking the probability of a call in any five-minute interval to be 1/3. However, since the firm may receive more than one call in a five-minute period, this model is not entirely satisfactory. It would be better to regard an hour as 60 one-minute trials, taking the probability of a call in any one-minute interval to be 1/15. Remembering the speed with which a modern switchboard can process calls, we should perhaps reduce the time interval to 10 seconds or even to 2 seconds. The given table shows that we are producing a sequence of binomial distributions with increasing numbers of trials but constant mean.

Interval	n	p	Mean	Variance
5 minutes	12	$\frac{1}{3}$	4	$4 \times \frac{2}{3} \approx 2{\cdot}67$
1 minute	60	$\frac{1}{15}$	4	$4 \times \frac{14}{15} \approx 3{\cdot}73$
10 seconds	360	$\frac{1}{90}$	4	$4 \times \frac{89}{90} \approx 3{\cdot}96$
2 seconds	1800	$\frac{1}{450}$	4	$4 \times \frac{449}{450} \approx 3{\cdot}99$

The limit of this sequence is the Poisson distribution with mean and variance 4.

†*Poisson, Siméon Denis* (1781–1840) French mathematician. His main contributions were to the theory of electricity and magnetism, but in 1837 he published *Recherches sur la probabilité des jugements*, dealing with probability distributions.

We obtain the general form of the Poisson distribution by considering the limit of the binomial distribution $B(n, p)$ as n tends to infinity in such a way that the mean $\mu = np$ remains constant.

$$P(X = r) = \binom{n}{r} p^r(1 - p)^{n-r}$$
$$= \frac{n(n-1)\dots(n-r+1)}{r!}\left(\frac{\mu}{n}\right)^r\left(1 - \frac{\mu}{n}\right)^{n-r}$$
$$= \frac{n}{n}\left(\frac{n-1}{n}\right)\dots\left(\frac{n-r+1}{n}\right)\frac{\mu^r}{r!}\left(1 - \frac{\mu}{n}\right)^n\left(1 - \frac{\mu}{n}\right)^{-r}$$
$$= \left(1 - \frac{1}{n}\right)\dots\left(1 - \frac{r-1}{n}\right)\frac{\mu^r}{r!}\left(1 - \frac{\mu}{n}\right)^n\bigg/\left(1 - \frac{\mu}{n}\right)^r$$

$$\therefore \qquad \lim_{n\to\infty} P(X = r) = \frac{\mu^r}{r!}\lim_{n\to\infty}\left(1 - \frac{\mu}{n}\right)^n$$

By the method of question 13, Exercise 9.4, it can be shown that

$$\lim_{n\to\infty}\left(1 - \frac{\mu}{n}\right)^n = e^{-\mu}$$

Thus, as $n \to \infty$, $P(X = r) \to \dfrac{\mu^r}{r!}e^{-\mu}$.

This result leads to the following definition:

> A discrete random variable X is said to have a Poisson distribution with parameter μ, where $\mu > 0$, if
>
> $$P(X = r) = \frac{\mu^r}{r!}e^{-\mu} \quad \text{for} \quad r = 0, 1, 2, \dots$$

Considering again the limit of the distribution $B(n, p)$ as $n \to \infty$, we see that if the mean $\mu = np$ remains constant, then $p \to 0$ and the variance $\sigma^2 = np(1 - p)$ tends to μ. This suggests that the Poisson distribution has mean μ and variance μ. These results can proved either directly or by using the appropriate probability generating function. For instance:

$$E(X) = \sum_{r=0}^{\infty} r\left(\frac{\mu^r}{r!}e^{-\mu}\right) = \mu e^{-\mu}\sum_{r=1}^{\infty}\frac{\mu^{r-1}}{(r-1)!}$$
$$= \mu e^{-\mu}\left(1 + \mu + \frac{\mu^2}{2!} + \frac{\mu^3}{3!} + \dots\right) = \mu e^{-\mu} \times e^{\mu} = \mu.$$

[For other proofs see Exercise 15.5 questions 12 and 13.]

It follows from this discussion that the Poisson distribution is a suitable model for the distribution of the number of events occurring in a specified interval of continuous time or space, provided that

(a) the events occur at random and independently, but may not occur simultaneously;
(b) the expected number of events in any interval is proportional to the size of the interval.

Example 1 A small business receives an average of 4 telephone calls per hour. Use the appropriate Poisson distribution to find the probability that (a) 7 calls are received in a two hour period, (b) up to 3 calls are received in an interval of 30 minutes.

(a) The expected number of calls in 2 hours is 8.
Thus if X is the number of calls received in 2 hours, X has a Poisson distribution with parameter 8

$$\therefore \quad P(X = 7) = \frac{8^7}{7!}e^{-8} \approx 0{\cdot}140.$$

Hence the probability that 7 calls are received in a two hour period is 0·140.
(b) The expected number of calls in 30 minutes is 2.
Thus if Y is the number of calls received in 30 minutes, Y has a Poisson distribution with parameter 2

$$\therefore \quad P(Y \leqslant 3) = P(Y = 0) + P(Y = 1) + P(Y = 2) + P(Y = 3)$$

$$= e^{-2} + 2e^{-2} + \frac{2^2}{2!}e^{-2} + \frac{2^3}{3!}e^{-2}$$

$$= (1 + 2 + 2 + \tfrac{4}{3})e^{-2} \approx 0{\cdot}857.$$

Hence the probability that up to 3 calls are received in 30 minutes is 0·857.

The following diagrams show the Poisson distributions used in Example 1.

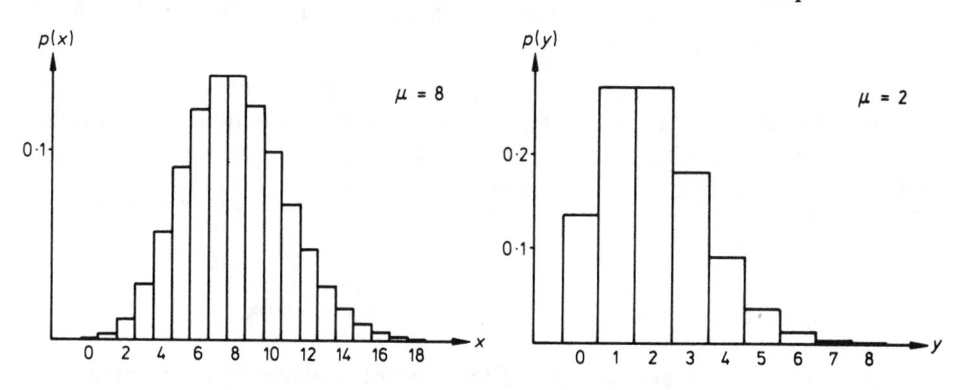

We have seen that the Poisson distribution with parameter $\mu = np$ is a limit of the binomial distribution $B(n, p)$. This means that under certain conditions a Poisson distribution can be used as an approximation to a binomial distribution with the same mean. In practice a Poisson approximation gives fairly accurate results when $n > 50$, $p < 0{\cdot}1$ and $np \leqslant 5$. It may also produce acceptable approximations to binomial distributions which do not satisfy these conditions. However, for a binomial distribution such that both np and $n(1-p)$ are greater than 5, a normal approximation is usually more appropriate.

It can be shown that as μ tends to infinity, the distribution of the standardised Poisson variable $(X-\mu)/\sqrt{\mu}$ approaches the standard normal distribution $N(0, 1)$. It follows that for large μ, the normal distribution $N(\mu, \mu)$ can be used as an approximation to the Poisson distribution with parameter μ. Fairly good approximations are obtained for $\mu > 5$, but for accurate results it is necessary to have $\mu > 20$.

Exercise 15.5

1. In a school laboratory an average of 8 test-tubes are broken in 20 working days. Find the probability that (a) 5 test-tubes will be broken in the next 10 days; (b) fewer than 2 test-tubes will be broken in the next 3 days.

2. The number of heavy vehicles driving past a certain historic building in an hour is assumed to have a Poisson distribution with mean 3. Find the probability that (a) in the next hour more than 3 heavy vehicles will drive by; (b) in the next two hours fewer than 5 heavy vehicles will drive by.

3. A textile manufacturer estimates that there is an average of 2 faults in every 100 metres of fabric he produces. Find the probability that there is at least 1 fault in a 10-metre length of fabric. Write down an expression for the probability that there is no fault in an n-metre length of fabric. What is the greatest integer n such that this probability is greater than 0·9?

4. The number of plants of a certain species found in one square metre of woodland is assumed to have a Poisson distribution with mean 0·003. Find the probability of observing at least 2 specimens in an area of $400\,\text{m}^2$. A botanist is more than 99% certain that he will find at least one specimen in an area of $n\ \text{m}^2$. Find the least possible value of the integer n.

In questions 5 to 8 draw up a table showing the values, correct to 3 d.p., of the probabilities $p(0), p(1), p(2), \ldots$ for (a) the binomial distribution $B(n, p)$; (b) the Poisson distribution with parameter $\mu = np$; (c) the normal approximation to (a); (d) the normal approximation to (b). Comment briefly on your results.

5. $n = 20, p = 0{\cdot}5$.

6. $n = 20, p = 0{\cdot}1$.

7. $n = 50, p = 0{\cdot}1$.

8. $n = 100, p = 0{\cdot}08$.

9. It is found that on average 0·5% of the watches produced by a certain factory are faulty. Estimate the probability that (a) in a batch of 200 watches there are no faulty watches; (b) in a batch of 500 watches more than 2 are faulty.

10. On average 1 out of 50 patients treated with a certain drug suffer side effects. Estimate the probability that (a) if 80 patients take the drug then exactly 2 will have side effects, (b) if 200 patients take the drug then at least 4 will have side effects.

11. In a certain experiment the probability of success in any trial is 0·03. Use appropriate approximations to find the probabilities of (a) at least 2 successes in 50 trials, (b) at least 4 successes in 100 trials, (c) at least 20 successes in 500 trials.

12. Use the identity $r^2 \equiv r(r-1) + r$ to show that

$$\sum_{r=0}^{\infty} \frac{r^2 x^r}{r!} = x^2 e^x + x e^x.$$

Hence show that the Poisson distribution with parameter μ has variance μ.

13. Show that the probability generating function for the Poisson distribution with parameter μ is given by $G(t) = e^{\mu(t-1)}$. Hence show that the distribution has mean μ and variance μ.

14. The number of road accidents at a certain traffic roundabout has been found to follow a Poisson distribution with a mean of 0·8 accidents per week. Calculate, correct to 2 significant figures, the probabilities that (i) there will be at least 2 accidents in a particular week, (ii) there will be exactly 3 accidents in a particular three-week period. After road works have been carried out, the traffic is observed for a period of 8 weeks and in that time only 1 accident occurs. Calculate the probability of at most 1 accident in 8 weeks if the mean is still 0·8 accidents per week. Comment briefly on this result. (C)

15. (a) I recorded the number of phone calls I received over a period of 150 days:

No. of calls	0	1	2	3	4
No. of days	51	54	36	6	3

(i) Find the average number of calls per day. (ii) Calculate the frequencies of the comparable Poisson distribution.

(b) A firm selling electrical components packs them in boxes of 60. On average 2% of the components are faulty. What is the chance of getting more than 2 defective components in a box? (Use the Poisson distribution.) (SU)

16. An aircraft has 116 seats. The airline has found, from long experience, that on average 2·5% of people with tickets for a particular flight do not arrive for that flight. If the airline sells 120 seats for a particular flight, determine, using a suitable approximation, the probability that more than 116 people arrive for that flight. Determine also the probability that there are empty seats on the flight. (C)

17. A number of different types of fungi are distributed at random in a field. Eighty per cent of these fungi are mushrooms, and the remainder are toadstools. Five per cent of the toadstools are poisonous. A man who cannot distinguish between mushrooms and toadstools, wanders across the field and picks a total of 100 fungi. Determine, correct to 2 significant figures, using approximations, the probability that the man has picked (i) at least 20 toadstools, (ii) exactly two poisonous toadstools. (C)

18. A new telephone directory is to be published, and before publication takes place the entries have to be checked for misprints and any necessary corrections made. Previous experience suggests that, on average, 0·1% of the entries will require correction, and that entries requiring correction will be randomly distributed. The directory will contain 240 000 entries altogether, printed on 800 pages with 300 entries per page. There are two methods of making corrections: for method A the costs are £C per page containing one entry requiring correction and £$3C$ per page containing two or more entries requiring correction; for method B the costs are £$2C$ per page containing one or more entries requiring correction. Calculate the expected costs of correction for each method. (C)

19. In a certain office the number of reams of typing paper used per month has a Poisson distribution with mean 25. Estimate the probability that more than 25 reams of paper will be used in the next month. Given that it is more than 95% certain that there is enough paper in the stationery cupboard to last a month, find the minimum number of reams of paper in the cupboard.

20. Show that the variance of a Poisson distribution is equal to the mean. The number of bacteria in 1 ml of inoculum has a Poisson distribution with mean 2·0. If at least 3 bacteria are needed for a dose to be infective, find the probability that a dose of 1 ml will cause infection. Find approximate limits, symmetrical about the mean, between which lies 95% of the distribution of the number of bacteria in 100 ml of inoculum. (O&C)

15.6 The algebra of expectation and variance

The expectation $E(X)$ and the variance $\text{Var}(X)$ of a random variable X were defined earlier in the chapter.

$$E(X) = \mu = \begin{cases} \sum x_r p(x_r), & \text{for discrete } X \\ \int_{-\infty}^{\infty} x f(x)\,dx, & \text{for continuous } X \end{cases}$$

$$\text{Var}(X) = \sigma^2 = \begin{cases} \sum (x_r - \mu)^2 p(x_r), & \text{for discrete } X \\ \int_{-\infty}^{\infty} (x - \mu)^2 f(x)\,dx, & \text{for continuous } X \end{cases}$$

If a new random variable $g(X)$ is defined in terms of X by a formula such as $g(X) = 2X + 3$ or $g(X) = X^2$, then the expectation of $g(X)$ is given by:

$$E[g(X)] = \begin{cases} \sum g(x_r) p(x_r), & \text{for discrete } X \\ \int_{-\infty}^{\infty} g(x) f(x)\,dx & \text{for continuous } X \end{cases}$$

If X is a discrete variable which takes the values $x_1, x_2, \ldots$ then we can show that this formula for $E[g(X)]$ is consistent with the definition of $E(X)$ by letting $Y = g(X)$.

(i) If Y takes distinct values $y_1 = g(x_1)$, $y_2 = g(x_2), \ldots,$ then

$$P(Y = y_r) = P(g(X) = g(x_r)) = P(X = x_r) = p(x_r),$$

$$\therefore \quad E[g(X)] = E(Y) = \sum_{\text{all } r} y_r P(Y = y_r) = \sum_{\text{all } r} g(x_r)p(x_r).$$

(ii) If the values $g(x_1), g(x_2), \ldots$ are not all distinct then we have relationships such as $y_k = g(x_i) = g(x_j) = \ldots$. In these cases

$$P(Y = y_k) = P(X = x_i) + P(X = x_j) + \ldots = p(x_i) + p(x_j) + \ldots$$

$$\therefore \quad y_k P(Y = y_k) = y_k p(x_i) + y_k p(x_j) + \ldots$$
$$= g(x_i)p(x_i) + g(x_j)p(x_j) + \ldots$$

$$\therefore \quad E[g(X)] = E(Y) = \sum_{\text{all } k} y_k P(Y = y_k) = \sum_{\text{all } r} g(x_r)p(x_r).$$

Thus the formula for $E[g(X)]$ is valid for discrete X. We will assume that it is also valid for continuous X.

Example 1 A discrete random variable X takes the values 0, 1, 2, 3 and $p(0) = 0{\cdot}1$, $p(1) = p(3) = 0{\cdot}2$, $p(2) = 0{\cdot}5$. Find $E(X^2 - X)$.

$$E(X^2 - X) = \sum (x^2 - x)p(x)$$
$$= (0^2 - 0)0{\cdot}1 + (1^2 - 1)0{\cdot}2 + (2^2 - 2)0{\cdot}5 + (3^2 - 3)0{\cdot}2$$
$$= 0 + 0 + 2 \times 0{\cdot}5 + 6 \times 0{\cdot}2 = 2{\cdot}2.$$

Example 2 Find $E(X^3)$ for the continuous random variable X which has probability density function $f(x) = 2(1 - x)$ for $0 \leqslant x \leqslant 1$ and $f(x) = 0$ otherwise.

$$E(X^3) = \int_0^1 x^3 f(x)\,dx = \int_0^1 x^3 . 2(1 - x)\,dx$$

$$= 2\int_0^1 (x^3 - x^4)\,dx = 2\left[\frac{x^4}{4} - \frac{x^5}{5}\right]_0^1 = 2\left(\frac{1}{4} - \frac{1}{5}\right) = 0{\cdot}1.$$

Two useful properties of expectation follow from the definition of $E[g(X)]$.

> If X is a random variable, discrete or continuous,
> (i) $E(aX + b) = aE(X) + b$, for constants a and b.
> (ii) $E[g(X) + h(X)] = E[g(X)] + E[h(X)]$.

We give here the proof of (i) for discrete X. The remaining proofs are left as an exercise.

$$E(aX + b) = \sum (ax_r + b)p(x_r)$$
$$= \sum \{ax_r p(x_r) + bp(x_r)\} = a\sum x_r p(x_r) + b\sum p(x_r)$$

But $\quad \sum x_r p(x_r) = E(X) \quad$ and $\quad \sum p(x_r) = 1$

Hence $\quad E(aX + b) = aE(X) + b.$

It is now possible to express the variance of a random variable in terms of expected values:

$$\text{Var}(X) = E[(X - \mu)^2], \quad \text{where} \quad \mu = E(X)$$

Thus, using results (i) and (ii):

$$\text{Var}(X) = E(X^2 - 2\mu X + \mu^2) = E(X^2) - 2\mu E(X) + \mu^2 = E(X^2) - \mu^2$$

i.e.

$$\boxed{\text{Var}(X) = E(X^2) - [E(X)]^2.}$$

Hence if a and b are constants,

$$\begin{aligned}\text{Var}(aX + b) &= E[(aX + b)^2] - [E(aX + b)]^2 \\ &= E(a^2X^2 + 2abX + b^2) - [aE(X) + b]^2 \\ &= a^2E(X^2) + 2abE(X) + b^2 - a^2[E(X)]^2 - 2abE(X) - b^2 \\ &= a^2E(X^2) - a^2[E(X)]^2 = a^2\{E(X^2) - [E(X)]^2\}\end{aligned}$$

i.e.

$$\boxed{\text{Var}(aX + b) = a^2 \text{Var}(X)}$$

To develop the theory of expectation further we must now consider the distribution of two discrete random variables X and Y with probability functions $p_X(x)$ and $p_Y(y)$ respectively. Let X take the values $x_1, x_2, \ldots, x_i, \ldots$ and let Y take the values $y_1, y_2, \ldots, y_j, \ldots$

The *joint probability distribution* of X and Y is the set of all probabilities $p(x_i, y_j)$, where $p(x_i, y_j) = P(X = x_i, Y = y_j)$.

For example if the possible values of X are 0, 1, 2 and the values of Y are 0, 1, 2, 3, then the joint distribution will be a set of 12 probabilities, which can be displayed in a two-way table as follows:

X \ Y	0	1	2	3	Total
0	$p(0,0)$	$p(0,1)$	$p(0,2)$	$p(0,3)$	$p_X(0)$
1	$p(1,0)$	$p(1,1)$	$p(1,2)$	$p(1,3)$	$p_X(1)$
2	$p(2,0)$	$p(2,1)$	$p(2,2)$	$p(2,3)$	$p_X(2)$
Total	$p_Y(0)$	$p_Y(1)$	$p_Y(2)$	$p_Y(3)$	1

As indicated in the table, the row totals give the values of $p_X(x)$. For instance

$$p(0,0) + p(0,1) + p(0,2) + p(0,3) = p_X(0).$$

Similarly the column totals are the values of $p_Y(y)$. Since these probabilities appear in the margins, they are called the *marginal distributions* of X and Y. Using this table we could obtain the probability distribution of a random variable such as $X + Y$. We see that $P(X + Y = 0) = p(0,0)$, $P(X + Y = 1) = p(1,0) + p(0,1)$, and so on. However, it is possible to obtain general formulae for the expected values of $X + Y$ and other functions of X and Y.

Taking into account all ordered pairs (x_i, y_j):

$$\begin{aligned}
E(X + Y) &= \sum_i \sum_j (x_i + y_j) p(x_i, y_j) \\
&= \sum_i \left[\sum_j x_i p(x_i, y_j) \right] + \sum_j \left[\sum_i y_j p(x_i, y_j) \right] \\
&= \sum_i \left[x_i \sum_j p(x_i, y_j) \right] + \sum_j \left[y_j \sum_i p(x_i, y_j) \right] \\
&= \sum_i x_i p_X(x_i) + \sum_i y_j p_Y(y_j) = E(X) + E(Y).
\end{aligned}$$

Similar arguments can be used to extend this result to any *linear combination* of random variables.

> If X and Y are random variables and a and b are constants, then
>
> $$E(aX + bY) = aE(X) + bE(Y).$$
>
> If $X_1, X_2, \ldots, X_n$ are random variables, then
>
> $$E(X_1 + X_2 + \ldots + X_n) = E(X_1) + E(X_2) + \ldots + E(X_n).$$

[These results are valid for both discrete and continuous variables, but in the continuous case the proofs require integration methods not covered in this course.]

If random variables X and Y are *independent*, then

$$P(X = x_i, Y = y_j) = P(X = x_i) \times P(Y = y_j)$$

i.e.

$$p(x_i, y_j) = p_X(x_i) p_Y(y_j)$$

$$\begin{aligned}
E(XY) &= \sum_i \sum_j x_i y_j p(x_i, y_j) \\
&= \sum_i \left[\sum_j x_i y_j p_X(x_i) p_Y(y_j) \right] \\
&= \sum_i \left[x_i p_X(x_i) \sum_j y_j p_Y(y_j) \right] \\
&= \left[\sum_i x_i p_X(x_i) \right] \left[\sum_j y_j p_Y(y_j) \right] = E(X)E(Y).
\end{aligned}$$

$\therefore$

> If X and Y are independent random variables, then
>
> $$E(XY) = E(X)E(Y).$$

We are now ready to consider the variance of a linear combination of random variables. In particular:

$$\begin{aligned}\text{Var}(X+Y) &= E[(X+Y)^2]-[E(X+Y)]^2\\ &= E(X^2+2XY+Y^2)-[E(X)+E(Y)]^2\\ &= E(X^2)+2E(XY)+E(Y^2)-[E(X)]^2-2E(X)E(Y)-[E(Y)]^2\\ &= \{E(X^2)-[E(X)]^2\}+2\{E(XY)-E(X)E(Y)\}\\ &\quad+\{E(Y^2)-[E(Y)]^2\}\end{aligned}$$

The quantity $E(XY)-E(X)E(Y)$ is called the *covariance* of X and Y, and is denoted by $\text{Cov}(X, Y)$. [We will see in Chapter 17 that it is a measure of the relationship between the distributions of X and Y.]
Thus, in general:

$$\text{Var}(X+Y) = \text{Var}(X) + 2\,\text{Cov}(X, Y) + \text{Var}(Y).$$

However, if the variables X and Y are independent,

$$\text{Cov}(X, Y) = E(XY) - E(X)E(Y) = 0$$

$$\therefore \quad \text{Var}(X+Y) = \text{Var}(X) + \text{Var}(Y).$$

As before, this basic result can be extended:

If X and Y are independent random variables and a and b are constants, then

$$\text{Var}(aX+bY) = a^2\,\text{Var}(X) + b^2\,\text{Var}(Y).$$

Two important special cases are given by:

$$\text{Var}(X+Y) = \text{Var}(X-Y) = \text{Var}(X) + \text{Var}(Y)$$

If $X_1, X_2, \ldots, X_n$ are independent random variables,

$$\text{Var}(X_1+X_2+\ldots+X_n) = \text{Var}(X_1) + \text{Var}(X_2) + \ldots + \text{Var}(X_n).$$

Example 3 In the ice-cream dessert served in a motorway cafeteria, the masses of the portions of ice-cream are distributed about a mean of 80 g with standard deviation 6 g. The fruit topping has mean mass 20 g and standard deviation 4 g. If the cost of 100 g of ice-cream is 5 p and the cost of 100 g of topping is 10 p, find the mean and standard deviation of the cost of a dessert.

Let the random variables X and Y represent the masses in grammes of portions of ice-cream and topping respectively. If Z is the total cost in pence of the dessert, then

$$Z = \tfrac{5}{100}X + \tfrac{10}{100}Y = \tfrac{1}{20}X + \tfrac{1}{10}Y$$

$$\therefore \quad E(Z) = \tfrac{1}{20}E(X) + \tfrac{1}{10}E(Y) = \tfrac{80}{20} + \tfrac{20}{10} = 6,$$

$$\text{Var}(Z) = (\tfrac{1}{20})^2 \text{Var}(X) + (\tfrac{1}{10})^2 \text{Var}(Y)$$
$$= \frac{6^2}{20^2} + \frac{4^2}{10^2} = \frac{9+16}{100} = \frac{1}{4}$$

Hence the cost of a dessert has mean 6 p and standard deviation $\frac{1}{2}$ p.

We now state without proof two important theorems concerning linear combinations of normal random variables.

> If X and Y are independent normal random variables with distributions $N(\mu_x, \sigma_x^2)$ and $N(\mu_y, \sigma_y^2)$ respectively, then $Z = aX + bY$, where a and b are constants, is also a normal random variable and has mean $a\mu_x + b\mu_y$, variance $a^2\sigma_x^2 + b^2\sigma_y^2$.

Sums and differences of independent normal variables are of particular interest:

$Z = X + Y$ has the distribution $N(\mu_x + \mu_y, \sigma_x^2 + \sigma_y^2)$,
$Z = X - Y$ has the distribution $N(\mu_x - \mu_y, \sigma_x^2 + \sigma_y^2)$.

> If $X_1, X_2, \ldots, X_n$ are independent observations of a random variable with distribution $N(\mu, \sigma^2)$, then $Z = X_1 + X_2 + \ldots + X_n$ is normally distributed with mean $n\mu$, variance $n\sigma^2$.

Example 4 Brand X beefburgers have masses normally distributed about a mean of 65 g with variance 15 g^2. The masses of Brand Y beefburgers are normally distributed with mean 100 g, variance 25 g^2. Find the probability that the mass of 3 Brand X beefburgers is greater than the mass of 2 Brand Y beefburgers.

[It is tempting to think that what we need to find here is $P(3X - 2Y > 0)$. However, in the expression $3X - 2Y$, $3X$ represents 3 times the mass of one beefburger, not the mass of three beefburgers.]

Let $D = (X_1 + X_2 + X_3) - (Y_1 + Y_2)$, where X_1, X_2, X_3 represent the masses (in grammes) of 3 Brand X beefburgers and Y_1, Y_2 represent the masses of 2 Brand Y beefburgers, then

$$E(D) = E(X_1) + E(X_2) + E(X_3) - E(Y_1) - E(Y_2)$$
$$= 3 \times 65 - 2 \times 100 = -5$$
$$\text{Var}(D) = \text{Var}(X_1) + \text{Var}(X_2) + \text{Var}(X_3) + \text{Var}(Y_1) + \text{Var}(Y_2)$$
$$= 3 \times 15 + 2 \times 25 = 95$$

Letting Z be the standardised normal variable $(D + 5)/\sqrt{95}$,

$$P(D > 0) = P(Z > 5/\sqrt{95})$$
$$= P(Z > 0{\cdot}5130)$$
$$= 1 - \Phi(0{\cdot}5130) = 1 - 0{\cdot}6961 = 0{\cdot}3039$$

$\therefore$ the required probability is approximately 0·304.

Two further results are sometimes useful.

> If independent random variables X and Y have distributions $B(m, p)$ and $B(n, p)$, then $X + Y$ has distribution $B(m + n, p)$.

> If independent random variables X and Y have Poisson distributions with means λ and μ, then $X + Y$ has a Poisson distribution with mean $\lambda + \mu$.

Methods of proof are indicated in Exercise 15.6 questions 23 and 25.

Exercise 15.6

1. A discrete random variable X takes the values 1, 2, 3, 4, 5 with probabilities $p(1) = p(2) = 0{\cdot}1, p(3) = 0{\cdot}15, p(4) = 0{\cdot}4, p(5) = 0{\cdot}25$. Find $E(X), E(X^2)$ and $E(1/X)$. Hence find $E(2X^2 - 1)$ and $E[(X + 1)^2/X]$.

2. A discrete random variable X takes the values $-1, 0, 1, 2$ with probabilities $p(-1) = 0{\cdot}2, p(0) = 0{\cdot}3, p(1) = 0{\cdot}4, p(2) = 0{\cdot}1$. Find $E(X), E(X^2)$ and $E(X^4)$. Hence find $\text{Var}(X)$ and $\text{Var}(X^2)$.

3. A continuous random variable X has probability density function

$$f(x) = \tfrac{1}{2}(3 - x) \text{ for } 1 \leqslant x \leqslant 3 \text{ and } f(x) = 0 \text{ otherwise.}$$

Find $E(X)$, $E(X^2)$ and $E(X^3)$. Hence find $E(3X^3 + 2X)$ and $\text{Var}(X)$.

4. A continuous random variable X has probability density function

$$f(x) = 6x(1 - x) \text{ for } 0 \leqslant x \leqslant 1 \text{ and } f(x) = 0 \text{ otherwise.}$$

Find $E(X)$, $E(X^2)$, and $E(\sqrt{X})$. Hence find $E[X(X - 1)]$ and $\text{Var}(\sqrt{X})$.

5. A random variable X takes the values $5, 8, 4, 7$ with equal probability and an independent variable Y takes the values 4, 7, 1 with equal probability. Find the mean and variance of the following:
(a) X (b) Y, (c) $4X + 3$, (d) $5 - 2Y$,
(e) $X + Y$, (f) $X - Y$, (g) $3X - Y$, (h) $2X + 3Y$.

6. Independent random variables X and Y are such that $E(X) = 10$, $\text{Var}(X) = 9$ and $E(Y) = 7$, $\text{Var}(Y) = 1$. Find (a) the mean and variance of $5X - 3Y + 2$, (b) the values of $E(X^2)$, $E(Y^2)$ and $E(X^2 + Y^2)$.

7. A factory produces nuts with masses distributed about a mean of 5 g with a standard deviation of 0·2 g. These are screwed on to bolts whose masses have mean

20 g and standard deviation 0·1 g. If each bolt is fitted with two nuts, find the mean and standard deviation of the resulting total mass.

8. In a can containing baked beans and 5 small sausages, the mass of each sausage is distributed about a mean of 20 g with standard deviation 2 g. The mean mass of the baked beans is 100 g with standard deviation 5 g. Find the mean mass of the total contents of a can and the standard deviation.

9.

X \ Y	0	1	2
1	0·06	0·12	0·02
2	0·15	0·3	0·05
3	0·09	0·18	0·03

The given table shows the joint probability distribution of two discrete random variables X and Y.

(a) Use the marginal distributions of X and Y to find $E(X)$, $\text{Var}(X)$ and $E(Y)$, $\text{Var}(Y)$.

(b) Are there any values of x and y for which

$$P(X = x, Y = y) \neq P(X = x) \times P(Y = y)?$$

Use your answer to determine whether or not X and Y are independent.

(c) Find the distribution of XY i.e. $P(XY = 0)$, $P(XY = 1)$, $P(XY = 2), \ldots$ Hence determine whether or not $E(XY) = E(X)E(Y)$.

(d) Find the distribution of $X + Y$. Hence determine whether or not

$$\text{Var}(X + Y) = \text{Var}(X) + \text{Var}(Y).$$

10. Repeat question 9 for the following joint distributions.

(i)

X \ Y	0	1	2	3
0	0·03	0·1	0.05	0·02
1	0·05	0	0·25	0·1
2	0·02	0·2	0·1	0·08

(ii)

X \ Y	1	2	3
0	0·05	0·2	0·05
1	0·2	0	0·2
2	0·05	0·2	0·05

11. A random variable X takes the values 0 and 2 each with probability 0·5. Find $E(X)$, $E(X^2)$, $E(X^3)$, $E(X^4)$. Use the formula $\text{Var}(Y) = E(Y^2) - [E(Y)]^2$ to verify that (a) if $Y = 3X - 5$, $\text{Var}(Y) = 9\,\text{Var}(X)$,

(b) if $Y = X^2 + X$, $\quad \text{Var}(Y) \neq \text{Var}(X^2) + \text{Var}(X)$.

12. The daily demand, x kilogrammes, for fresh fish at a certain fishmonger's shop is a continuous random variable with probability density function f, where f is given by $f(x) = (30 - x)/450, 0 \leqslant x \leqslant 30, f(x) = 0$, otherwise. The fishmonger receives a delivery of 20 kg of fresh fish each morning. He makes a profit of 40 pence on each kilogramme of fish sold on the day the fish is delivered; fish left unsold on the day it is

delivered is sold to a fertiliser manufacturer at a loss of 5 pence per kg. Show that on a day when x is less than 20, the fishmonger's profit is $(45x - 100)$ pence and that on a day when x is 20 or more, the fishmonger's profit is £8. Hence determine the fishmonger's expected daily profit from the sale of fresh fish. (JMB)

13 A sample of four mice is to be chosen at random from a litter of ten mice, of which six are male and four are female. Let X denote the number of male mice in the sample. Calculate the values of $P(X = r)$ for $r = 0, 1, 2, 3, 4$. Hence find the mean and the standard deviation of X. Deduce the mean and the standard deviation of the number of female mice in the sample. Independently of the above, a sample of four mice is to be chosen at random from another litter of ten mice, of which four are male and six are female. Calculate the mean and the standard deviation of the total number of male mice in the combined sample of eight mice. (W)

14. The joint probability distribution of two discrete random variables X and Y is displayed in the following table, the entries in the body of the table being $P(X = x, Y = y)$.

		x 0	1	2
y	0	0·1	0·1	0
	1	0·1	0·2	0·1
	2	0·1	0·2	0·1

(i) Find $E(X)$ and $E(Y)$. (ii) Determine the distribution of $Z = 2X - Y$ and verify that $E(Z) = 2E(X) - E(Y)$. (iii)Let (X_1, Y_1) and (X_2, Y_2) denote two independent observations of (X, Y). By using the distribution of Z, or otherwise, find the probability that

$$2(X_1 - X_2) = Y_1 - Y_2.$$

(W)

15. On a piece-work operation a company pays a bonus to any employee who processes in excess of 300 kg of raw material in a day. The daily amounts processed by two employees A and B are independent and normally distributed, the mean and standard deviation of the amounts processed by A being 291 kg and 9 kg, respectively, and of the amounts processed by B being 282 kg and 12 kg, respectively. Calculate, to two significant figures in each case, (i) the proportion of days on which both A and B will earn the bonus, (ii) the probability that the combined daily amount processed by A and B will exceed 600 kg, (iii) the proportion of days on which the amount processed by B will exceed that processed by A. (W)

16. The diameters of axles supplied by a factory have a mean value of 19·92 mm and a standard deviation of 0·05 mm. The inside diameters of bearings supplied by another factory have a mean of 20·04 mm and a standard deviation of 0·03 mm. What is the mean and standard deviation of the random variable defined to be the diameter of a bearing less the diameter of an axle? Assuming that both dimensions are normally distributed, what percentage of axles and bearings taken at random will not fit? (O&C)

17. The time of arrival of a bus at a bus stop varies in a normal distribution with a mean of 09.00 a.m. and a standard deviation of 2 minutes. Independently a second bus departs from this stop at a time which varies in a normal distribution with a mean of 09.01 a.m. (i.e. one minute past 09.00 a.m.) and a standard deviation of 1 minute. Find the probability that (a) the first bus arrives before the second bus leaves; (b) this happens on 5 given consecutive days. (O&C)

18. The weight of a large loaf of bread is a normal variable with mean 420 g and standard deviation 30 g. The weight of a small loaf of bread is a normal variable with mean 220 g and standard deviation 10 g. (i) Find the probability that 5 large loaves weigh more than 10 small loaves. (ii) Find the probability that the total weight of 5 large loaves and 10 small loaves lies between 4·25 kg and 4·4 kg. (C)

19. The mass of a cheese biscuit has a normal distribution with mean 6 g and standard deviation 0·2 g. Determine the probability that (i) a collection of twenty-five cheese biscuits has a mass of more than 149 g, (ii) a collection of thirty cheese biscuits has a mass of less than 180 g, (iii) twenty-five times the mass of a cheese biscuit is less than 149 g. The mass of a ginger biscuit has a normal distribution with mean 10 g and standard deviation 0·3 g. Determine the probability that a collection of seven cheese biscuits has a mass greater than a collection of four ginger biscuits. (C)

20. Four sprinters A, B, C and D each run 100 m. The times (in seconds) that they take may be regarded as independent observations from a normal distribution with mean 14 and standard deviation 0·2. The athlete E runs 400 m. The time (in seconds) that this athlete takes may be regarded as an observation from a normal distribution with mean 58 and standard deviation 1·0 and is independent of the times of the sprinters. Determine the probability that (i) the time taken by E is less than 3 seconds greater than the sum of the times taken by the four sprinters, (ii) the time taken by E is less than 4 times as great as the time taken by A. For the four sprinters A, B, C and D, find the probability that A is the fastest sprinter of the four. Find also the probability that C and D (in either order) are the two slowest sprinters. (C)

21. P is a fixed point on the circumference of a circle, centre O, of radius r and Q is a point on the circumference such that $\angle POQ = \theta$, where θ is a random variable with a rectangular distribution in $[0, 2\pi]$. Find the mean and median values for the length of the shorter arc PQ. The length of the chord PQ is X. Find $E(X)$ and show that $\text{Var}(X) = 2r^2(1 - 8/\pi^2)$. Find also $P(X > r\sqrt{3})$. (C)

22. X and Y are independent random variables with means and variances μ_x, σ_x^2 and μ_y, σ_y^2 respectively, and you may assume that for any such independent variables $E(XY) = E(X)E(Y)$. Find expressions for (i) the expectation of the area of a square of side X, (ii) the expectation and variance of the area of a rectangle of sides X and Y, (iii) the expectation of the volume of a cuboid with a square base of side X and with height Y. (C)

23. Two independent random variables X and Y have Poisson distributions with means λ and μ respectively. Obtain an expression for $P(X = r, Y = n - r)$. Hence express $P(X + Y = n)$ as the sum of a series. Deduce that $X + Y$ has a Poisson distribution with mean $\lambda + \mu$.

24. Telephone calls reach a secretary independently and at random, internal ones at a mean rate of 2 in any 5 minute period, and external ones at a mean rate of 1 in any 5 minute period. Calculate the probability that there will be more than 2 calls in any period of 2 minutes. (O&C)

25. (a) Verify that if a random variable X has probability generating function $G_X(t)$, then $G_X(t) = E(t^X)$.
(b) Two independent random variables X and Y have distributions $B(m, p)$ and $B(n, p)$ respectively. Write down expressions for $E(t^X)$ and $E(t^Y)$. Assuming that t^X and t^Y can be regarded as independent random variables, write down $E(t^X \times t^Y)$ and hence find the p.g.f. of the variable $Z = X + Y$. Deduce that Z has the distribution $B(m + n, p)$.

Exercise 15.7 (miscellaneous)

1. A random variable R takes the integer values $1, 2, \ldots, n$ each with probability $1/n$. Find the mean and variance of R. A pack of 15 cards bearing the numbers 1 to 15 is shuffled. Find the probability that the number on the top card is larger than that on the bottom card, giving reasons for your answer. If the sum of these two numbers is S, find (a) the probability that $S \leqslant 4$, (b) the expected value of S. (O&C)

2. The number X of radios sold per week by a certain shop is a binomially distributed random variable with $n = 10$ and $p = 0{\cdot}3$. The number Y of television sets sold per week by the same shop may be 0, 1 or 2 with $P(Y = 0) = 0{\cdot}6$, $P(Y = 1) = 0.3$, $P(Y = 2) = 0{\cdot}1$. The shopkeeper makes a profit of £2 on each radio sold and a profit of £20 on each television set sold. Find the shopkeeper's expected weekly profit from the sale of radios and television sets. (JMB)

3. (a) In a certain manufacturing process, it is known that approximately 10% of the items produced are defective. A quality control scheme is set up, by selecting 20 items out of a large batch, and rejecting the whole batch if 3 or more are defective. Find the probability that the batch is rejected.
(b) Two boys, John and David, play a game with a die. The die will be thrown 4 times. David will give John £x if there is an odd number of sixes, otherwise John will give David £1. If the game is to be a fair one to both John and David, find the value of x. (SU)

4. The probability of there being X unusable matches in a full box of Surelite matches is given by $P(X = 0) = 8k$, $P(X = 1) = 5k$, $P(X = 2) = P(X = 3) = k$,

$P(X \geqslant 4) = 0$. Determine the constant k and the expectation and variance of X. Two full boxes of Surelite matches are chosen at random and the total number Y of unusable matches is determined. Calculate $P(Y > 4)$, and state the values of the expectation and variance of Y. (C)

5. Fergus Lightfingers, an educated thief, has broken into a house and has come upon a large safe. Lying upon the table is a bunch of k similar keys, only one of which will open the safe. He considers two possible strategies: (i) randomly select keys, one at a time *without replacement*, until successful; (ii) randomly select keys, one at a time *with replacement*, until successful. Let N, a random variable, be the number of keys including the successful one tried by Fergus in order to open the safe. Derive the probability distribution of N for both strategies. Calculate the expected value of N in each case. Use your statistical judgment in order to select the strategy that Fergus should use. (You may assume that $\sum_{n=1}^{\infty} nx^{n-1} = \frac{1}{(1-x)^2}$ for $|x| < 1$.) (AEB 1978)

6. A batsman's performance is described by the following probability distribution. For each ball that he receives he has probability $\frac{1}{2}$ of scoring no runs, probability $1/5$ of scoring one run and otherwise he is equally likely to score two runs, four runs, or to be bowled out. In the last case he receives no further balls.
(i) Determine the probability that the batsman's total score is 0.
(ii) Given that the batsman is not bowled by a particular ball, calculate the expectation of his score from that ball.
(iii) Calculate the expectation of the number of balls that the batsman receives.
(iv) Deduce the expectation of the batsman's total score. (C)

7. In a series of N independent trials the probability of success in any trial is constant and equal to p. Show that the probability of r successes is given by

$$P(r) = \binom{N}{r} p^r (1-p)^{N-r}.$$

Find $P(3)$ if $N = 5$ and $p = 0{\cdot}2$. Write a flow diagram for a program to print values of $P(r)$ if $N = 25$ and $p = 0{\cdot}4$ for $r = 0(1)25$. (L)

8. In a certain factory 70% of the microprocessors produced are found to be imperfect and have to be discarded. Find the probability that, in a randomly chosen sample of 5 microprocessors produced, the number of perfect specimens will be (i) 0, (ii) 1, (iii) 2, (iv) 3, (v) 4, (vi) 5. Verify that the probability of more than 3 perfect specimens is less than 1/30. If a sample of 200 is taken from the same set of microprocessors, find the probability that there are at least 50 perfect specimens. (AEB 1980)

9. Derive the mean and variance of the binomial distribution. Mass production of miniature hearing aids is a particularly difficult process and so the quality of these products is monitored carefully. Samples of size six are selected regularly and tested for correct operation. The number of defectives in each sample is

recorded. During one particular week 140 samples are taken and the distribution of the number of defectives per sample is given in the following table.

Number of defectives per samples (x)	0	1	2	3	4	5	6
Number of samples with x defectives (f)	27	36	39	22	10	4	2

Find the frequencies of the number of defectives per sample given by a binomial distribution having the same mean and total as the observed distribution.

(AEB 1978)

10. Show that the probability generating function for the binomial distribution arising from n independent trials, in each of which the probability of a success is p, is $(1 - p + pt)^n$. Show that the probability generating function for the number of trials needed to obtain a success is $pt/(1 - t + pt)$, and find the mean and variance of this distribution.

When a gun first fires at a target the probability of a hit is 1/5. Firing continues until the first hit, after which the gun is resighted and the probability of a hit becomes 2/5. Find the probability generating function for the number of shots required to obtain 2 hits and hence find the probability that fewer than 5 shots will be needed.

(O&C)

11. A random variable X takes values in the intervals 0 to 1 and 2 to 3, and has probability density function given by

$$f(x) = \begin{cases} 1 - x & (0 \leqslant x \leqslant 1), \\ x - a & (2 \leqslant x \leqslant 3), \\ 0 & \text{elsewhere.} \end{cases}$$

Determine (i) the value of a, (ii) the expectation, μ, of X, (iii) the variance of X, (iv) the (cumulative) distribution function $F(x)$, (v) the probability that $|X - \mu|$ exceeds 1.

(C)

12. The random variable X is the distance, in metres, that an inexperienced tight-rope walker has moved along a given tight-rope before falling off. It is given that

$$P(X > x) = 1 - x^3/64, \quad 0 \leqslant x \leqslant 4.$$

(i) Show that $E(X) = 3$. (ii) Find the standard deviation, σ, of X. (iii) Show that $P(|X - 3| < \sigma) = \frac{69}{80}\sqrt{\frac{3}{5}}$.

(C)

13. Find the mean and the variance of the distribution whose probability density function f is given by

$$\begin{aligned} f(x) &= x, & 0 &\leqslant x < 1, \\ f(x) &= 2 - x, & 1 &\leqslant x \leqslant 2, \\ f(x) &= 0, & &\text{otherwise.} \end{aligned}$$

A rectangle is constructed with adjacent sides of lengths x cm and $(2 - x)$ cm where x is a random value from the above distribution. Find the probability that the area of the rectangle will exceed $0{\cdot}75\ \text{cm}^2$. Also find the expected value of the area of the rectangle. (JMB)

14. The operational lifetime in hundreds of hours of a battery-operated minicalculator may be regarded as a continuous random variable having probability density function

$$f(x) = cx(10 - x), \qquad 5 \leqslant x \leqslant 10,$$
$$f(x) = 0, \qquad \text{otherwise.}$$

(a) Find the value of c and of the expected operational lifetime of such a minicalculator.
(b) The purchase price of such a minicalculator is £20 and its running cost (for batteries) amounts to 20 pence per hundred hours operation. Thus, the overall average cost in pence per hundred hours operation of a minicalculator whose operational lifetime is X hundred hours is given by $Y = 20 + (2000/X)$. (i) Evaluate $E(Y)$, the expected overall average cost per hundred hours. (ii) Find the probability that the overall average cost per hundred hours will exceed £2.70. (W)

15. A random variable X has (cumulative) distribution function

$$F(x) = \begin{cases} 0 & (x \leqslant 0) \\ \frac{1}{4}x^2 & (0 \leqslant x \leqslant 1) \\ \alpha x + \beta & (1 \leqslant x \leqslant 2) \\ \frac{1}{4}(5 - x)(x - 1) & (2 \leqslant x \leqslant 3) \\ 1 & (3 \leqslant x) \end{cases}$$

Find (i) the values of the constants α and β, (ii) the probability that $1{\cdot}5 \leqslant X \leqslant 2{\cdot}5$, (iii) the probability density function $f(x)$. Sketch the graph of $f(x)$, and hence, or otherwise, deduce the mean of X. Determine also the variance of X. (C)

16. A failure time, X, has density function $\lambda e^{-\lambda x}$ for $x \geqslant 0$, $\lambda > 0$. Find (a) the distribution function; (b) the mean and variance; (c) the median; (d) the time below which it is 95% certain that there will be no failure. (O)

17. Henri de Lade regularly travels from his home in the suburbs to his office in Paris. He always tries to catch the same train, the 08.05 from his local station. He walks to the station from his home in such a way that his arrival times form a normal distribution with mean 08.00 hours and standard deviation 6 minutes. (i) Assuming that his train always leaves on time, what is the probability that on any given day Henri misses his train? (ii) If Henri visits his office in this way 5 days each week and if his arrival times at the station each day are independent, what is the probability that he misses his train once and only once in a given week? (AEB 1980)

18. The score, S, gained by an expert rifleman with a single shot, is a random variable with the following probability distribution:

$$P(S = 8) = 0{\cdot}01, \quad P(S = 9) = 0{\cdot}29, \quad P(S = 10) = 0{\cdot}70.$$

(i) Use a normal approximation to determine the probability that the rifleman obtains six or more scores of 8 in a series of 900 independent shots. (ii) Find the expectation and variance of S. (iii) Use a normal approximation to determine the probability that the rifleman scores less than 96 with ten independent shots. (C)

19. The prices of houses on an estate agent's books are approximately normally distributed with a mean price of £18 000. Given that 90% of the houses are priced at less than £30 000, find the standard deviation of the house prices and the percentage of the houses with a price of less than £10 000. Find the mean price and the standard deviation of the prices when the price of each house is increased by (a) £2000, (b) 10%. (L)

20. The weights of grade A oranges are normally distributed with mean 200 g and standard deviation 12 g. Determine, correct to 2 significant figures, the probability that (i) a grade A orange weighs more than 190 g but less than 210 g, (ii) a sample of 4 grade A oranges weighs more than 820 g. The weights of grade B oranges are normally distributed with mean 175 g and standard deviation 9 g. Determine, correct to 2 significant figures, the probability that (iii) a grade B orange weighs less than a grade A orange, (iv) a sample of 8 grade B oranges weighs more than a sample of 7 grade A oranges. (C)

21. Mass-produced right circular cylindrical pipes have internal diameters that are normally distributed with a mean of 10 cm and a standard deviation of 0·4 cm. (i) Find the probability that a randomly chosen pipe will have an internal diameter greater than 10·3 cm. (ii) Find the expected number of pipes in a random sample of 100 pipes that have internal diameters in the range from 9·7 cm to 10·3 cm. (iii) Find the expected value of the internal cross-sectional area of a randomly chosen pipe; give your answer correct to three significant figures. (iv) Mass-produced pistons have diameters that are normally distributed with a mean of 9·9 cm and a standard deviation of 0·3 cm. Find the probability that a randomly selected piston will have a diameter less than the internal diameter of a randomly selected pipe. (W)

22. Two gauges are used to test the thickness of manufactured metal sheets. Over a long period it is found that 1·25% of the sheets will pass through the smaller gauge of 1·4 mm while 95·4% will pass through the larger gauge of 1·6 mm. Assuming that the distribution of sheet thickness is normal, find its mean and standard deviation. At the next stage of manufacture 2 sheets are clamped together to make a thick plate. Find the mean and the standard deviation of the thickness of these plates. If a plate will pass through a gauge of 3·2 mm but not through one of 2·8 mm, find the probability that both the sheets of which it is composed will pass through the 1·6 mm gauge but not through the 1·4 mm gauge. Find also the size of a gauge which will allow 99·5% of these plates to pass through. (O&C)

23. Next May, an ornithologist intends to trap one male cuckoo and one female cuckoo. The mass M of the male cuckoo may be regarded as being a normal random variable with mean 116 g and standard deviation 16 g. The mass F of the female cuckoo may be regarded as being independent of M and as being a normal random

variable with mean 106 g and standard deviation 12 g. Determine (i) the probability that the mass of the two birds together will be more than 230 g. (ii) the probability that the mass of the male will be more than the mass of the female. By considering $X = 9M - 16F$, or otherwise, determine the probability that the mass of the female will be less than nine-sixteenths of that of the male. Suppose that one of the two trapped birds escapes. Assuming that the remaining bird will be equally likely to be the male or the female, determine the probability that its mass will be more than 118 g. (C)

24. (a) The number of accidents notified in a factory per day over a period of 200 days gave rise to the following table:

No. of accidents	0	1	2	3	4	5
No. of days	127	54	14	3	1	1

(i) Calculate the mean number of accidents per day. (ii) Assuming that this situation can be represented by a suitable Poisson distribution, calculate the corresponding frequencies.

(b) Of items produced by a machine, approximately 3% are defective, and these occur at random. What is the probability that, in a sample of 144 items, there will be at least two which are defective? (SU)

25. Define a Poisson distribution giving the general probability formula; state the relationship between its mean and variance. The distribution of the number of vehicles observed passing under a motorway bridge in 100 successive intervals of 12 seconds, at a time when the traffic flow is 720 vehicles per hour, may be considered to be Poissonian. During how many of the intervals would one expect (i) no cars to have passed? (ii) 2 cars to have passed, (iii) more than 2 cars to have passed? Find the smallest value of N such that the probability of at most N cars in an interval is greater than 0·999. (AEB 1978)

26. Each play of a one-armed bandit machine has probability 0·04 of yielding a win. It may be assumed that this probability applies independently for each play of the machine. (a) Find the probability that in ten plays of the machine there will be (i) at least two wins, (ii) exactly two wins. (b) For a sequence of n plays of the machine write down an expression for the probability that there will be no win. Hence find the smallest number of plays of the machine for there to be a probability of at least 0·95 of winning at least once. (c) Use an appropriate method to find an approximate value for the probability that there will be six or fewer wins in 80 plays of the machine. (W)

27. Gnat larvae are distributed at random in pond water so that the number of larvae contained in a random sample of 10 cm^3 of pond water may be regarded as a random variable having a Poisson distribution with mean 0·2. Ten independent random samples, each of 10 cm^3, of pond water, are taken by a zoologist. Determine (correct to three significant figures) (i) the probability that none of the samples contain larvae, (ii) the probability that one sample contains a single larva and the remainder contain no larvae, (iii) the probability that one sample contains two or

more larvae and the remainder contain no larvae, (iv) the expectation of the total number of larvae contained in the ten samples, (v) the expectation of the number of samples containing no larvae. (C)

28. A prospector for gold examines exactly 800 pans of material every month. The contents of each pan may be assumed to be independent of the others with the probability of a pan containing gold being 0·005 for each pan. The prospector has a good month when 4 or more of the pans that he examines contain gold. Show that, to three significant figures, the probability that a randomly chosen month is a good month is 0·567. Determine the probability that (i) a randomly chosen period of four months contains more than two good months, (ii) a randomly chosen period of 24 months contains more than 12 good months. (C)

29. The number of organic particles suspended in a volume $V\text{ cm}^3$ of a certain liquid follows a Poisson distribution with mean $0{\cdot}1V$. (a) Find the probabilities that a sample of 1 cm^3 of the liquid will contain (i) at least one organic particle, (ii) exactly one organic particle. (b) Use an appropriate approximate procedure to find the probability that a sample of 1000 cm^3 of the liquid will contain at least 90 organic particles. (c) The liquid is sold in vials, each vial containing 10 cm^3 of the liquid. The vials are dispatched for sale in boxes, each box containing 100 vials. Find the probability that a vial will contain at least one organic particle. Hence find the mean and the standard deviation of the number of vials per box of 100 vials that contain at least one organic particle. (W)

30. When a number is rounded off to its nearest integer value, the rounding off error X may be regarded as a random variable which is uniformly distributed over the interval $(-\frac{1}{2}, \frac{1}{2})$. Find the variance of X. If n numbers are rounded off to their nearest integer values and then summed, the total error in the sum is given by $Y = X_1 + X_2 + \ldots + X_n$, where $X_1, X_2, \ldots, X_n$ are the individual errors in the n numbers, which may be assumed to be independent and uniformly distributed over $(-\frac{1}{2}, \frac{1}{2})$. Write down the mean and the variance of Y in terms of n. Suppose n is large enough to justify approximating the distribution of Y by a normal distribution. Show that there is a probability in excess of 0·9 that Y will be numerically less than $\frac{1}{2}\sqrt{n}$. Find the largest integer value of n for which there is a probability of at least 0·6 that Y will be numerically less than 1·0. (W)

31. The joint probability function of two discrete random variables X and Y is given by $P(X = r,\ Y = s) = c|r + s|$ for $r = -1, 0, 1$ and $s = -2, -1, 0, 1, 2$,
$P(X = r,\ Y = s) = 0$, otherwise.
(a) Show that (i) $c = 0{\cdot}05$, (ii) $\text{Var}(Y) = 2 + \text{Var}(X)$.
(b) Evaluate $\text{Var}(Y - X)$. (W)

32. A continuous random variable X is distributed with probability density function $f(x) = 3x^{-4}$, $x \geqslant 1$, $f(x) = 0$, otherwise. Find (i) the cumulative distribution function of X, (ii) the mean and the variance of X. The random variable Y is defined by $Y = X^{-1}$. By first finding the cumulative distribution function of Y, or otherwise, determine the probability density function of Y. Verify that $E(X)E(Y)$

and $E(XY)$ are not equal and explain why they have different values in this particular example. (W)

33. A continuous random variable X has probability density function

$$f(x) = 4x(1 - x^2), \qquad 0 \leqslant x \leqslant 1,$$
$$f(x) = 0, \qquad \text{otherwise.}$$

(i) Verify that $f(x)$ satisfies the requirements to be a probability density function. (ii) Find the mean value of X. (iii) Find the cumulative distribution function of X. (iv) If X denotes the area of a square of side Y find the probability density function of Y and use it to verify that $E(Y^2) = E(X)$. (W)

34. The probability density function of a continuous random variable X is given by

$$f(x) = \tfrac{2}{3}x, \qquad 1 \leqslant x \leqslant 2,$$
$$f(x) = 0, \qquad \text{otherwise.}$$

(a) Find (i) $E(X^2)$, (ii) $E(X^4)$, (iii) the cumulative distribution function of X.
(b) Suppose X is the length of the side of a square. (i) Find the mean and the variance of the area of the square. (ii) Show that the area of the square has a uniform distribution over a certain interval which should be specified. (W)

16 Samples and significance

16.1 Sampling distributions

In this chapter we will be considering the interpretation of statistical data. This usually involves analysing the data, then setting up an appropriate probability model. The known properties of the theoretical distribution can then be used to draw conclusions or make predictions about the observed distribution. Later we will look at tests which help us to measure the significance of the available data. However, first we define some of the terms to be used.

A *population* is the set of all items under discussion in a particular survey or experiment, e.g. the ages of all the inhabitants of Bournemouth, the lengths of all components produced by a certain machine. A *sample* is a set of observations taken from a population, i.e. a subset of the population. For instance, a set of 10 scores obtained by tossing a single die is a sample of size 10 drawn from the theoretically infinite population of all possible tosses. A *random sample* is one in which each element of the population has an equal chance of being selected.

A *statistic* is a quantity calculated using the observations in a sample. The mean, the median and the variance of a sample are all examples of statistics. By considering all the samples of a given size which can be drawn from a particular population, it is possible to obtain the probability distribution of a statistic such as the sample mean. These theoretical distributions are called *sampling distributions.*

In order to investigate the properties of sampling distributions we will regard a *random sample of size n*, from a population with mean μ and variance σ^2, as a set of random variables $X_1, X_2, \ldots, X_n$ each with the same distribution as the parent population. We will assume that these variables are *independent* if there is little or no change in the population as elements of a sample are selected,

i.e. if (a) the population is infinite,
or (b) the population is finite, but very large,
or (c) the population is finite, but the sampling is *with replacement* (each item is replaced before the next is selected).

The variables $X_1, X_2, \ldots, X_n$ cannot be treated as independent if samples are taken *without replacement* from a relatively small population.

If $\bar{X}$ represents the sample mean $\frac{1}{n}(X_1 + X_2 + \ldots + X_n)$, then, using results obtained in §15.6:

$$E(\bar{X}) = E\left[\frac{1}{n}(X_1 + X_2 + \ldots + X_n)\right]$$
$$= \frac{1}{n}[E(X_1) + E(X_2) + \ldots + E(X_n)] = \frac{1}{n}(n\mu) = \mu.$$

Provided that the variables $X_1, X_2, \ldots, X_n$ are independent,

$$\text{Var}(\bar{X}) = \text{Var}\left[\frac{1}{n}(X_1 + X_2 + \ldots + X_n)\right]$$
$$= \frac{1}{n^2}[\text{Var}(X_1) + \text{Var}(X_2) + \ldots + \text{Var}(X_n)] = \frac{1}{n^2}(n\sigma^2) = \frac{\sigma^2}{n}$$

∴ If $\bar{X}$ is the mean of a random sample of size n from a population with mean μ, variance σ^2

$$E(\bar{X}) = \mu.$$

For a large population or sampling with replacement

$$\text{Var}(\bar{X}) = \sigma^2/n.$$

[It can be shown that for a random sample of size n drawn without replacement from a population of size ν, the appropriate formula is

$$\text{Var}(\bar{X}) = \left(\frac{\nu - n}{\nu - 1}\right)\frac{\sigma^2}{n}.$$

We see that as $\nu \to \infty$, $\text{Var}(\bar{X}) \to \sigma^2/n$.]

If the mean, $\bar{x}$, of a particular sample is regarded as an approximate value of the mean μ of the whole population, then the deviation $\bar{x} - \mu$ is the error in this approximation. For this reason the standard deviation of the sampling distribution of a statistic is often called the *standard error* of the statistic.

Thus, in general, the standard error of the mean of a random sample of size n is $\sigma/\sqrt{n}$.

Since linear combinations of normal random variables are also normally distributed we can extend these results as follows.

If $\bar{X}$ is the mean of a random sample of size n from a population which is normally distributed with mean μ, variance σ^2, then $\bar{X}$ is normally distributed with mean μ, variance σ^2/n.

It is more difficult to make general statements about the sampling distribution of the mean for populations which are not normally distributed. However, there is an important theorem which tells us a great deal about the behaviour of $\bar{X}$ when n is large.

The central limit theorem. Let X be a random variable with mean μ, variance σ^2. If $\bar{X}$ is the mean of a random sample of size n drawn from the distribution of X then the distribution of the standardised variable $(\bar{X} - \mu)/(\sigma/\sqrt{n})$ tends to the standard normal distribution $N(0, 1)$ as n tends to infinity.

In practice we use a less formal statement of the result.

If $\bar{X}$ is the mean of a random sample of size n from *any* population with mean μ, variance σ^2, then for large n, $\bar{X}$ has an approximately normal distribution with mean μ, variance σ^2/n.

If the population distribution is reasonably symmetrical, good approximations may be obtained for $n \geqslant 25$. However, when sampling from a rather skew (i.e. unsymmetrical) distribution, we may need $n > 100$ for good results.

Example 1 The metal discs produced by a certain machine have mean diameter 30·01 mm with standard deviation 0·08 mm. A sample of 100 discs is taken. Find the probability that the mean diameter of the discs in this sample is less than 30 mm.

Using the central limit theorem, the distribution of the means of samples of size 100 has an approximately normal distribution with mean 30·01 mm and standard deviation $0{\cdot}08/\sqrt{100}$ mm i.e. 0·008 mm.
Let $\bar{X}$ be the mean of a sample of size 100 and let Z be a standard normal variable, then

$$P(\bar{X} < 30) = P\left(Z < \frac{30 - 30{\cdot}01}{0{\cdot}008}\right)$$

$$= P(Z < -1{\cdot}25) = 1 - \Phi(1{\cdot}25) = 0{\cdot}1056$$

Hence the probability that the sample mean is less than 30 mm is 0·11 approximately.

We now consider briefly the sampling distribution of a *difference between two means.* Let us suppose that $\bar{X}$ and $\bar{Y}$ are the means of random samples of sizes n_x, n_y, taken from two populations with means μ_x, μ_y and variances σ_x^2, σ_y^2 respectively. The distributions of $\bar{X}$ and $\bar{Y}$ will then have means μ_x, μ_y and variances $\sigma_x^2/n_x, \sigma_y^2/n_y$.

$\therefore$ the sampling distribution of the difference $\bar{X} - \bar{Y}$ has mean $\mu_x - \mu_y$ and variance $\dfrac{\sigma_x^2}{n_x} + \dfrac{\sigma_y^2}{n_y}$.

If both parent populations are normally distributed then $\bar{X} - \bar{Y}$ will also have a normal distribution. If the populations are not normally distributed, then it follows from the central limit theorem that, for large values of n_x and n_y, $\bar{X} - \bar{Y}$ will have an approximately normal distribution.

One of the most useful statistics that can be obtained from a sample is a *proportion*. If r members of a sample of size n have a certain attribute, then the corresponding proportion p is given by r/n. For instance, if 4 students in a class of 25 are left-handed, the proportion of left-handed students is 0·16. If the proportion of the total population having the stated attribute is π, then the value π gives the probability that any one member of the population has the attribute. This means that if R is the number of "successes" in a random sample of size n, then R has the binomial distribution $B(n, \pi)$. Thus if P is the proportion of successes in the sample,

$$P(P = p) = P(R = r) = \binom{n}{r}\pi^r(1 - \pi)^{n-r}.$$

These probabilities form the *sampling distribution of a proportion P* and using our knowledge of the binomial distribution we have:

$$E(P) = E(R/n) = \frac{1}{n}E(R) = \frac{1}{n}(n\pi) = \pi$$

$$\text{Var}(P) = \text{Var}(R/n) = \frac{1}{n^2}\text{Var}(R) = \frac{1}{n^2}[n\pi(1 - \pi)] = \frac{\pi(1 - \pi)}{n}$$

For large n, R will have an approximately normal distribution,

∴ If a proportion π of the members of a population have a certain attribute, then the proportion P in a random sample of size n, where n is large, has an approximately normal distribution with mean π and variance $\pi(1 - \pi)/n$.

Elementary problems involving sample proportions can be solved using either the appropriate sampling distribution or the corresponding binomial distribution.

Example 2 It is estimated that in a certain large city the proportion of families that own a freezer is 0·35. Find the probability that in a random sample of n families over 40% own freezers, where (a) $n = 6$, (b) $n = 60$, (c) $n = 600$.

Let R be the number of families owning a freezer in a sample of size n, then R has the distribution $B(n, \pi)$, where $\pi = 0{\cdot}35$, and the proportion P of families with freezers is given by $P = R/n$.

(a) $P(P > 0{\cdot}4) = P(R > 2{\cdot}4)$

$$= 1 - \{P(R = 0) + P(R = 1) + P(R = 2)\}$$

$$= 1 - \left\{(0{\cdot}65)^6 + \binom{6}{1}(0{\cdot}65)^5(0{\cdot}35) + \binom{6}{2}(0{\cdot}65)^4(0{\cdot}35)^2\right\}$$

$$\approx 0{\cdot}353$$

Hence, when $n = 6$ the required probability is 0·353.

(b) Using a normal approximation to $B(60, 0{\cdot}35)$ and a suitable continuity correction,

$$P(P > 0{\cdot}4) = P(R > 24)$$

$$\approx P\left(Z > \frac{24{\cdot}5 - 21}{\sqrt{(60 \times 0{\cdot}35 \times 0{\cdot}65)}}\right)$$

$$= P(Z > 0{\cdot}9473) = 1 - \Phi(0{\cdot}9473) \approx 0{\cdot}172$$

Hence, when $n = 60$ the required probability is 0.172.

(c) When $n = 600$, the sampling distribution of the proportion P is approximately normal with mean $\mu_p = 0{\cdot}35$ and variance

$$\sigma_p^2 = (0{\cdot}35 \times 0{\cdot}65)/600.$$

$$\therefore \quad P(P > 0{\cdot}4) \approx P\left(Z > \frac{0{\cdot}4 - \mu_p}{\sigma_p}\right)$$

$$= P(Z > 2{\cdot}5678) = 1 - \Phi(2{\cdot}5678) \approx 0{\cdot}005$$

Hence, when $n = 600$ the required probability is $0{\cdot}005$.

[Note that when n has a very large value such as 600, applying a continuity correction would not produce a significant improvement in accuracy.]

Another important statistic is the *sample variance*, but the mathematics involved in studying its sampling distribution is rather difficult. All we can do for the moment is find the expected value of the sample variance.

Using the notation introduced earlier in the section, the variance V of a random sample of size n is given by

$$V = \frac{1}{n}\sum(X_r - \bar{X})^2 = \frac{1}{n}\sum X_r^2 - \bar{X}^2$$

$$\therefore \quad E(V) = E\left[\frac{1}{n}\sum X_r^2 - \bar{X}^2\right] = \frac{1}{n}\sum E(X_r^2) - E(\bar{X}^2)$$

But $\operatorname{Var}(X_r) = E(X_r^2) - [E(X_r)]^2$, so $E(X_r^2) = \sigma^2 + \mu^2$

and $\operatorname{Var}(\bar{X}) = E(\bar{X}^2) - [E(\bar{X})]^2$, so $E(\bar{X}^2) = \dfrac{\sigma^2}{n} + \mu^2$

$$\therefore \quad E(V) = \sigma^2 + \mu^2 - \left(\frac{\sigma^2}{n} + \mu^2\right) = \left(\frac{n-1}{n}\right)\sigma^2$$

i.e.

> If V is the variance of a random sample of size n from a population with mean μ, variance σ^2,
>
> $$E(V) = \left(\frac{n-1}{n}\right)\sigma^2.$$

[Note that this result is not valid for small populations sampled without replacement, since under those conditions, $\operatorname{Var}(\bar{X}) \neq \sigma^2/n$.]

Exercise 16.1

1. A random variable X takes the values $0, 1, 2, 3$ with probabilities $p(0) = 0{\cdot}3$, $p(1) = 0{\cdot}5$, $p(2) = p(3) = 0{\cdot}1$. Find the mean μ and the variance σ^2 of this distribution. Random samples of size 2 are drawn from the distribution. List all 10 possible samples and their probabilities. In each case find the sample mean $\bar{X}$ and the sample variance V. Hence obtain the sampling distributions of the mean and variance. Verify that $E(\bar{X}) = \mu$, $\text{Var}(\bar{X}) = \frac{1}{2}\sigma^2$ and $E(V) = \frac{1}{2}\sigma^2$.

2. Using the distribution given in question 1, obtain the sampling distributions of the mean $\bar{X}$ and variance V of random samples of size 3. Verify that in this case $E(\bar{X}) = \mu$, $\text{Var}(\bar{X}) = \frac{1}{3}\sigma^2$ and $E(V) = \frac{2}{3}\sigma^2$.

3.

Family	A	B	C	D	E
No. of children	2	0	0	6	2

The table shows the number of children in 5 families A, B, C, D, E. Find the mean μ and the variance σ^2 of the distribution. Random samples of size 2 are selected, *without replacement*, from the population of 5 families. List all 10 possible samples and find in each case the sample mean $\bar{X}$ and the sample variance V. Remembering that each sample occurs with probability $0{\cdot}1$, obtain the sampling distributions of $\bar{X}$ and V. Hence find $E(\bar{X})$, $\text{Var}(\bar{X})$ and $E(V)$. Write down the values of $E(\bar{X})$, $\text{Var}(\bar{X})$ and $E(V)$ for samples of size 2 selected *with replacement*.

4. Using the distribution given in question 3, obtain the sampling distributions of the mean $\bar{X}$ and variance V of random samples of size 3 selected without replacement. Hence find $E(\bar{X})$, $\text{Var}(\bar{X})$ and $E(V)$. Write down the values of $E(\bar{X})$, $\text{Var}(\bar{X})$ and $E(V)$ for samples of size 3 selected with replacement.

5. The masses of a large consignment of melons have an approximately normal distribution with mean $1{\cdot}5$ kg and standard deviation $0{\cdot}3$ kg. Find the probability that (a) the mass of one melon differs from $1{\cdot}5$ kg by more than $0{\cdot}5$ kg, (b) the mean mass of 25 melons differs from $1{\cdot}5$ kg by more than $0{\cdot}1$ kg.

6. A random variable X is normally distributed with mean 20 and variance 32. Find the probability that X lies between 15 and 25. If a random sample of size 50 is drawn from this distribution find the probability that the sample mean lies between 19 and 21.

7. The diameters x of 110 steel rods were measured in centimetres and the results were summarised as follows:

$$\sum x = 36{\cdot}5, \quad \sum x^2 = 12{\cdot}49.$$

Find the mean and standard deviation of these measurements. Assuming these measurements are a sample from a normal distribution with this mean and this variance, find the probability that the mean diameter of a sample of size 110 is greater than 0·345 cm (O&C)

8. A class of 18-year-old pupils contains 18 boys and 15 girls. The measured heights in metres of the boys are $x_i (i = 1, \ldots, 18)$, and those of the girls are $y_j (j = 1, \ldots, 15)$. It is found that $\sum x_i = 31{\cdot}50$, $\sum y_j = 24{\cdot}75$, $\sum x_i^2 = 55{\cdot}305$, $\sum y_j^2 = 41{\cdot}175$.
(a) Find the means and the sample variances of the heights of (i) the boys, (ii) the girls.
(b) Assuming that the heights of boys and of girls of age 18 are independent and both are normally distributed with the same means and variances as in these samples, find the difference in mean heights of boys and girls. Find also the standard deviation of this difference in a large population of 18-year-olds. (O&C)

9. Two random variables X and Y have independent normal distributions $N(3, 7)$ and $N(5, 11)$ respectively. A random sample of 12 observations is selected from each distribution. Find the sampling distribution of $\bar{X} - \bar{Y}$, the difference between the sample means. Hence find the probability that (a) $\bar{X} < \bar{Y}$, (b) $|\bar{X} - \bar{Y}| < 1$.

10. A random variable X has the distribution $N(5, 5)$ and $\bar{X}$ is the mean of a random sample of size n drawn from the distribution of X. Find the least value of n such that (a) $P(\bar{X} > 6) < 0{\cdot}01$, (b) $P(\bar{X} > 4{\cdot}9) > 0{\cdot}9$.

11. The random variables $X_1, X_2, \ldots, X_n$ are independent and each has a normal distribution with mean μ and variance 1. The random variable $\bar{X}$ is defined to be $(X_1 + X_2 + \ldots + X_n)/n$. Determine, in terms of n, the value v which is such that, when $\mu = 0$, the probability of $\bar{X}$ exceeding v is 0·05. For this value of v it is desired that the probability of $\bar{X}$ being less than v when $\mu = 0{\cdot}2$ should be at most 0·10. Calculate the smallest possible value of n which satisfies this requirement. (C)

12. In a presidential election the proportion of voters who voted for the socialist candidate was 0·48. Find the probability that in a random sample of n voters over 50% voted for the socialist, where (a) $n = 5$, (b) $n = 50$, (c) $n = 500$, (d) $n = 5000$.

13. It is known that in a large consignment of hyacinth bulbs 24% are expected to produce pink blooms. For each of the following samples find the probability that at least 20% of the bulbs will produce pink blooms. (a) 8 bulbs in a window box, (b) 80 bulbs in a rockery, (c) 800 bulbs in a municipal park.

16.2 Unbiased estimators

Having studied the properties of various sampling distributions we are now ready to consider the use of sample statistics to estimate population parameters, such as the mean μ, the variance σ^2 or a proportion π. When producing such an estimate we can calculate either (i) a single value or *point estimate* for the parameter, or (ii) two values which define a *confidence interval* within which the parameter is expected to lie. Thus there are two basic types of estimation, *point estimation* and *interval estimation*. In this section we concentrate on the former.

A function of the observations in a sample, which is used to estimate a population parameter θ is called an *estimator* of θ. The value taken by this estimator in a particular sample is an *estimate* of θ. For instance, the sample mean $\bar{X}$ is an estimator of the population mean μ and the mean $\bar{x}$ of a particular sample is an estimate of μ. To determine whether a statistic T is likely to be a "good" estimator of a parameter θ, we need to know something about its sampling distribution. Clearly the distribution of T should be centred on the true value of θ. An estimator with this property is said to be *unbiased*.

> A statistic T obtained from a random sample and used to estimate a population parameter θ is an *unbiased estimator* of θ if $E(T) = \theta$.

It is also desirable that the spread of values produced by an estimator should be as small as possible.

> If T_1 and T_2 are two unbiased estimators of a population parameter θ, both obtained from samples of size n, then if $\text{Var}(T_1) < \text{Var}(T_2)$, T_1 is said to be a *better* or *more efficient* estimator of θ than T_2.

It was shown in §16.1 that if $\bar{X}$ is the mean of a random sample of size n taken from a population with mean μ, then $E(\bar{X}) = \mu$.

$\therefore$ $\bar{X}$ is an unbiased estimator of μ.

Similarly, if a proportion P of the elements of a random sample of size n have a certain attribute and π is the proportion of the population having that attribute, then $E(P) = \pi$.

$\therefore$ P is an unbiased estimator of π.

However, if V is the variance of a random sample of size n from a population with variance σ^2, $E(V) \neq \sigma^2$. Thus V is a biased estimator of σ^2. Under most conditions,

$$E(V) = \left(\frac{n-1}{n}\right)\sigma^2$$

$$\therefore \quad E\left[\left(\frac{n}{n-1}\right)V\right] = \left(\frac{n}{n-1}\right)E(V) = \sigma^2$$

$\therefore$ $\left(\frac{n}{n-1}\right)V$ is an unbiased estimator of σ^2.

The standard notation for a typical value of this estimator is s^2.

Thus $s^2 = \left(\frac{n}{n-1}\right)\frac{1}{n}\sum(x_r - \bar{x})^2$ or $\left(\frac{n}{n-1}\right)\left(\frac{1}{n}\sum x_r^2 - \bar{x}^2\right)$

$\therefore$ an unbiased estimate of population variance from a random sample of size n with mean $\bar{x}$ is given by

$$s^2 = \frac{1}{n-1}\sum(x_r - \bar{x})^2 \quad \text{or} \quad \frac{1}{n-1}\left(\sum x_r^2 - n\bar{x}^2\right).$$

The corresponding formulae for data given in the form of a frequency distribution are

$$s^2 = \frac{1}{n-1}\sum f_r(x_r - \bar{x})^2 \quad \text{or} \quad \frac{1}{n-1}(\sum f_r x_r^2 - n\bar{x}^2).$$

Example 1 A random sample drawn from a large population contains 10 observations $x_1, x_2, \ldots, x_{10}$ such that $\sum x = 24{\cdot}5$ and $\sum x^2 = 61{\cdot}93$. Obtain unbiased estimates for the population mean and variance.

$$\bar{x} = \frac{1}{n}\sum x = \tfrac{1}{10} \times 24{\cdot}5 = 2{\cdot}45$$

$$s^2 = \frac{1}{n-1}\left[\sum x^2 - n\bar{x}^2\right] = \tfrac{1}{9}[61{\cdot}93 - 10(2{\cdot}45)^2] \approx 0{\cdot}212$$

Hence unbiased estimates of the mean and variance of the population are 2·45 and 0·212 respectively.

Example 2 A continuous random variable X has probability density function $f(x) = 2x/\theta^2$ for $0 \leqslant x \leqslant \theta$ and $f(x) = 0$ otherwise. Find the mean μ of the distribution of X. Given that $\bar{X}$ denotes the mean of a random sample of size n drawn from this distribution, show that $3\bar{X}/2$ is an unbiased estimator of θ.

$$\mu = E(X) = \int_0^\theta x f(x)\,dx = \int_0^\theta \frac{2x^2}{\theta^2}\,dx = \left[\frac{2x^3}{3\theta^2}\right]_0^\theta = \tfrac{2}{3}\theta$$

$$\therefore \quad E\left(\frac{3\bar{X}}{2}\right) = \tfrac{3}{2}E(\bar{X}) = \tfrac{3}{2}\mu = \tfrac{3}{2}(\tfrac{2}{3}\theta) = \theta$$

Hence $3\bar{X}/2$ is an unbiased estimator of θ.

Finally we show that data from two samples can be *pooled* to produce unbiased *two-sample estimates* of population parameters. We will suppose that two random samples of sizes m and n are drawn from the same population. These samples can be combined to form a single sample of size $m + n$. If the means of the original samples are $\bar{X}$ and $\bar{Y}$ respectively, then the mean $\bar{Z}$ of the combined sample is given by

$$\bar{Z} = \frac{m\bar{X} + n\bar{Y}}{m + n}$$

If the mean of the parent population is μ, then

$$E(\bar{Z}) = \frac{mE(\bar{X}) + nE(\bar{Y})}{m+n} = \frac{m\mu + n\mu}{m+n} = \mu$$

$\therefore$ If $\bar{X}$ and $\bar{Y}$ are the means of random samples of sizes m and n respectively, drawn from a population with mean μ, then an unbiased estimator of μ is given by:

$$\frac{m\bar{X} + n\bar{Y}}{m+n}$$

The next result is obtained using similar arguments:

Random samples of sizes m and n are drawn from the same population and proportions P_x and P_y of these samples have a certain attribute. If π is the proportion of the population having that attribute, then an unbiased estimator of π is given by:

$$\frac{mP_x + nP_y}{m+n}$$

If V_x and V_y are the variances of the samples under consideration and σ^2 is the population variance, then

$$\begin{aligned} E(mV_x + nV_y) &= mE(V_x) + nE(V_y) \\ &= (m-1)\sigma^2 + (n-1)\sigma^2 = (m+n-2)\sigma^2 \end{aligned}$$

$\therefore$ If V_x and V_y are the variances of random samples of sizes m and n respectively, drawn from a population with variance σ^2, then an unbiased estimator of σ^2 is given by:

$$\frac{mV_x + nV_y}{m+n-2}$$

In terms of the unbiased variance estimates, s_x^2 and s_y^2, obtained from random samples of sizes m and n, the two-sample unbiased estimate of σ^2 is:

$$\frac{(m-1)s_x^2 + (n-1)s_y^2}{m+n-2}$$

[The results obtained in §§16.1 and 16.2 are summarised in the Formulae for Reference section towards the end of the book.]

Exercise 16.2

1. The following random sample of size 12 was drawn from a large population: 2·2, 3·1, 4·7, 6·3, 5·0, 4·4, 1·9, 6·8, 3·2, 5·7, 2·5, 3·4. Obtain unbiased estimates of

(a) the population mean, (b) the population variance, (c) the proportion of the population having values between 2 and 6.

2. The following random sample of size 8 was drawn from a large population: 42·0, 35·7, 52·3, 35·9, 45·2, 54·1, 48·6, 31·8. Obtain unbiased estimates of (a) the population mean, (b) the population variance, (c) the proportion of the population having values less than 40.

3. A random sample drawn from a large population contains 20 observations $x_1, x_2, \ldots, x_{20}$ such that $\sum x = 48{\cdot}2$ and $\sum x^2 = 160{\cdot}66$. Obtain unbiased estimates for the population mean and variance. Find also the expected mean and variance of samples of size 100.

4. A random sample drawn from a large population contains 25 observations $x_1, x_2, \ldots, x_{25}$ such that $\sum x = 1492$ and $\sum x^2 = 95\,030$. Obtain unbiased estimates for the population mean and variance. Find also the expected mean and variance of samples of size 10.

5. The given table shows the distribution of days lost through illness in a three-month period for a random sample of 60 workers in a large factory.

No. of days lost	0	1	2	3	4	5	6
No. of workers	8	12	19	14	4	2	1

Find unbiased estimates for the mean and variance of the numbers of days lost per worker for the whole work force. Estimate also the proportion of workers losing more than 2 days work.

6. A continuous random variable X is uniformly distributed over the interval from 0 to θ. Find the mean μ and the variance σ^2 of the distribution of X. If a random sample of size 9 drawn from this distribution has mean $\bar{X}$ and variance V, show that (a) $2\bar{X}$ is an unbiased estimator of θ, (b) $13{\cdot}5V$ is an unbiased estimator of θ^2. Given that one random sample of size 9 contains the observations 5·7, 0·2, 1·7, 3·8, 2·5, 4·2, 5·1, 1·4, 3·3, calculate unbiased estimates of θ and θ^2.

7. A continuous random variable X has probability density function $f(x) = 3x^2/\theta^3$ for $0 \leqslant x \leqslant \theta$ and $f(x) = 0$ otherwise. Find the mean μ and the variance σ^2 of the distribution of X. If a random sample of size 6 drawn from this distribution has mean $\bar{X}$ and variance V, show that (a) $4\bar{X}/3$ is an unbiased estimator of θ, (b) $32V$ is an unbiased estimator of θ^2. Given that one random sample of size 6 contains the observations 3·7, 5·9, 7·1, 8·1, 9·0, 9·4, calculate unbiased estimates of θ and θ^2.

8. Two random sample of size 8 drawn from the same large population contain the following observations:

Sample I:	13,	37,	66,	72,	27,	33,	47,	61.
Sample II:	29,	46,	40,	68,	19,	60,	30,	44.

(a) Find the sample means and variances. Hence obtain unbiased two-sample estimates of the population mean and variance.

(b) Treating the observations in the two samples as a single sample of size 16, find further unbiased estimates of the population mean and variance.

9. Two random samples drawn from the same large population are of sizes 12 and 6.

Sample I:	3, 8, 10, 12, 6, 5, 9, 13, 11, 15, 6, 13.
Sample II:	14, 4, 11, 6, 9, 10.

(a) Find the sample means and variances. Hence obtain unbiased two-sample estimates of the population mean and variance.
(b) Treating the observations in the two samples as a single sample of size 18, find further unbiased estimates of the population mean and variance.

10. A discrete random variable X can take the values 0, 1 and 2 only, with respective probabilities $\frac{1}{2}\theta$, $1 - \theta$, and $\frac{1}{2}\theta$, where θ is an unknown number between 0 and 1. Let X_1 and X_2 denote two randomly observed values of X. List the possible values of $\{X_1, X_2\}$ that may arise and calculate the probability of each possibility; verify that your probabilities sum to unity. By calculating the value of $(X_1 - X_2)^2$ for each possible $\{X_1, X_2\}$ determine the sampling distribution of $(X_1 - X_2)^2$. Hence show that $Y = \frac{1}{2}(X_1 - X_2)^2$ is an unbiased estimator of θ and express its sampling variance in terms of θ. Since θ is the probability that X will not take the value 1, another possible estimator of θ is the proportion of sample values not equal to 1; for a sample of two observations this estimator is given by $Z = \frac{1}{2}N$, where N is the number of observations (0, 1 or 2) not equal to 1. State, giving your reasons, which of Y and Z you would prefer as the estimator of θ. (W)

11. Each trial of a random experiment has probability p, $(0 < p < 1)$, of yielding a success. In n_1 independent trials of the experiment the number of successes obtained was r_1. Write down an unbiased estimate, $\hat{p}_1$, of p and find its standard error in terms of n_1 and p. In further n_2 independent trials of the same experiment the number of successes obtained was r_2. Let $\hat{p}_2$ denote the unbiased estimate of p from these n_2 trials. Verify that $\frac{1}{2}(\hat{p}_1 + \hat{p}_2)$ is an unbiased estimate of p, and find its standard error in terms of n_1, n_2 and p. Determine the range of values of the ratio n_1/n_2 for which the estimate $\frac{1}{2}(\hat{p}_1 + \hat{p}_2)$ is to be preferred to each of $\hat{p}_1$ and $\hat{p}_2$. (W)

16.3 Confidence intervals

In the previous section we saw that a sample statistic, such as a sample mean, can be used to obtain estimates of a population parameter. However, since these point estimates may vary widely from sample to sample, it is important to know how close a particular estimate is likely to be to the true value of the parameter. This information is often given in the form of an interval estimate. Using the sampling distribution of a statistic we can determine a *confidence interval* for the corresponding population parameter. As the following definition shows, the probability that the chosen interval contains this parameter is expressed as a percentage.

If a and b are two quantities obtained using the observations in a random sample and θ is an unknown population parameter, such that

$$P(a \leqslant \theta \leqslant b) = 0{\cdot}95,$$

then the interval $[a, b]$ is said to be a *95% confidence interval* for θ. The numbers a and b are called *95% confidence limits* for θ.

Let us suppose that $\bar{X}$ is the mean of a random sample of size n from a population which is normally distributed with mean μ and variance σ^2. As shown in §16.1, the sampling distribution of $\bar{X}$ is $N(\mu, \sigma^2/n)$. Hence we can produce a variable Z with the standard normal distribution $N(0, 1)$, by letting $Z = \dfrac{\bar{X} - \mu}{\sigma/\sqrt{n}}$

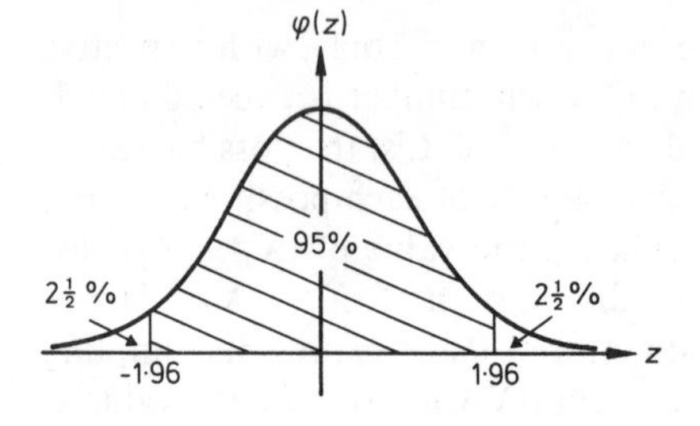

Using the 2·5% value given in Table 2 (percentage points of the normal distribution), we find that

$$P(-1{\cdot}96 \leqslant Z \leqslant 1{\cdot}96) = 0{\cdot}95$$

$$\therefore \quad P\left(-1{\cdot}96\frac{\sigma}{\sqrt{n}} \leqslant \bar{X} - \mu \leqslant 1{\cdot}96\frac{\sigma}{\sqrt{n}}\right) = 0{\cdot}95$$

$$\therefore \quad P\left(-1{\cdot}96\frac{\sigma}{\sqrt{n}} \leqslant \mu - \bar{X} \leqslant 1{\cdot}96\frac{\sigma}{\sqrt{n}}\right) = 0{\cdot}95$$

$$\therefore \quad P\left(\bar{X} - 1{\cdot}96\frac{\sigma}{\sqrt{n}} \leqslant \mu \leqslant \bar{X} + 1{\cdot}96\frac{\sigma}{\sqrt{n}}\right) = 0{\cdot}95$$

Thus the (symmetric) 95% confidence interval for μ is $\left[\bar{x} - 1{\cdot}96\dfrac{\sigma}{\sqrt{n}}, \bar{x} + 1{\cdot}96\dfrac{\sigma}{\sqrt{n}}\right]$, that is:

If $\bar{x}$ is the mean of a random sample of size n from a normal population with *unknown* mean μ and *known* variance σ^2, then the 95% confidence limits for μ are $\bar{x} \pm 1{\cdot}96\dfrac{\sigma}{\sqrt{n}}$.

Using the 0·5% value given in Table 2, the 99% confidence limits for μ are $\bar{x} \pm 2{\cdot}576\dfrac{\sigma}{\sqrt{n}}$.

It follows from the central limit theorem that these results can also be used to give approximate confidence limits for the mean of a population which is not normal, provided that n is fairly large.

Example 1 The diameters of a large consignment of ball-bearings are normally distributed with standard deviation 0·3 mm. The mean diameter of a random sample of size 16 is found to be 5·4 mm. Find (a) 95% confidence limits for the mean diameter of the whole consignment, (b) the smallest sample size needed to give a 99% confidence interval of width less than 0·2 mm.

(a) The 95% confidence limits for the mean diameter are (in mm):

$$\bar{x} \pm 1{\cdot}96\frac{\sigma}{\sqrt{n}} = 5{\cdot}4 \pm 1{\cdot}96 \times \frac{0{\cdot}3}{4} \approx 5{\cdot}4 \pm 0.15$$

Hence the required limits are 5·25 mm and 5·55 mm.

(b) The 99% confidence interval for the mean diameter has limits:

$$\bar{x} \pm 2{\cdot}576\frac{\sigma}{\sqrt{n}} \quad \text{(in mm)}$$

Hence this interval will have width less than 0·2 mm if

$$2{\cdot}576 \times \frac{0.3}{\sqrt{n}} < 0{\cdot}1$$

i.e. $$\frac{2.576 \times 0{\cdot}3}{0{\cdot}1} < \sqrt{n}$$

i.e. $$n > (2{\cdot}576 \times 3)^2 \approx 59{\cdot}7$$

Thus the smallest sample size needed is 60.

In practice it is often necessary to find confidence intervals for the mean of an approximately normal population whose variance is *unknown.* If in our earlier working we replace σ^2 by its estimated value s^2, then the standard normal variable $Z = (\bar{X} - \mu)/(\sigma/\sqrt{n})$ is replaced by a new variable which is not simply a linear function of $\bar{X}$, but a function of both $\bar{X}$ and the sample variance. In general, this new variable has what is called a *t-distribution,* which is not covered in this course. However, as n increases this t-distribution approaches the standard normal distribution. Thus, for large n we can again use percentage points of $N(0, 1)$ to obtain confidence intervals.

> If $\bar{x}$ is the mean of a random sample of size n from a normal population with mean μ and variance σ^2 *both unknown,* then for large n the approximate 95% confidence for μ are
>
> $$\bar{x} \pm 1{\cdot}96\frac{s}{\sqrt{n}},$$
>
> where s^2 is the unbiased estimate of σ^2.

[This result usually provides good approximations for $n \geqslant 30$. For $n \geqslant 200$ the sample variance $\frac{1}{n}\sum(x - \bar{x})^2$ can be used as an estimate of σ^2 with little loss of accuracy.]

More general results can be obtained by considering the $C\%$ confidence interval for a population parameter λ. Let us assume that a sample statistic L is an unbiased estimator of λ with standard error σ_l, and that L has a normal (or nearly normal) distribution. For a particular unbiased estimate l the corresponding $C\%$ confidence limits for λ are given by $l \pm z\sigma_l$, where z is the value of the standard normal variable Z such that $P(-z \leqslant Z \leqslant z) = C/100$. We have already seen that $C = 95$, $z = 1{\cdot}96$ and for $C = 99$, $z = 2{\cdot}576$.

For large samples drawn from a normal population it is usually possible to replace σ_l by an estimate s_l to give the approximate $C\%$ confidence limits $l \pm zs_l$.

Thus if p is the estimate of a proportion obtained using a large sample of size n, then approximate 95% confidence limits for the population proportion π are given by:

$$p \pm 1{\cdot}96\sqrt{\left\{\frac{p(1-p)}{n}\right\}}$$

If $\bar{x}_1, \bar{x}_2$ are the means of samples of sizes n_1, n_2 drawn from normal populations with means μ_1, μ_2 respectively, but the same variance σ^2, the 95% confidence limits for the difference between means, $\mu_1 - \mu_2$, are given by:

$$(\bar{x}_1 - \bar{x}_2) \pm 1{\cdot}96\sqrt{\left(\frac{\sigma^2}{n_1} + \frac{\sigma^2}{n_2}\right)} \quad \text{i.e.} \quad \bar{x}_1 - \bar{x}_2 \pm 1{\cdot}96\sigma\sqrt{\left(\frac{1}{n_1} + \frac{1}{n_2}\right)}$$

If s_1^2, s_2^2 are unbiased estimates of σ^2 obtained from large samples of sizes n_1, n_2, then approximate 95% confidence limits are given by:

$$\bar{x}_1 - \bar{x}_2 \pm 1{\cdot}96s\sqrt{\left(\frac{1}{n_1} + \frac{1}{n_2}\right)}, \quad \text{where} \quad s^2 = \frac{(n_1 - 1)s_1^2 + (n_2 - 1)s_2^2}{n_1 + n_2 - 2}.$$

[This two-sample estimate of variance was introduced in §16.2.]

Exercise 16.3

1. Over a long period it is found that the mass of soap powder in certain packets is normally distributed with standard deviation 15 g. Find 95% confidence limits for the mean of the distribution given that (a) the mean of a random sample of size 25 is 842 g, (b) the mean of a random sample of size 100 is 843 g.

2. A set of 50 observations $x_1, x_2, \ldots, x_{50}$ is a random sample from a normal distribution with mean μ. Given that $\sum x = 163$ and $\sum x^2 = 548$ find (a) approximate 95% confidence limits for μ, (b) approximate 99% confidence limits for μ.

3. The masses of a certain type of sausage are normally distributed with standard deviation 4 g. The masses of packs containing 8 sausages have a distribution with mean μ. Find 99% confidence limits for μ, given that (a) the mass of one pack selected at random is 468 g, (b) the total mass of a random sample of 10 packs is 4605 g.

4. In a certain survey 26 shoppers out of a random sample of 200 stated that they preferred brown bread to white. If π is the proportion of all shoppers having a

preference for brown bread find (a) approximate 95% confidence limits for π, (b) approximate 98% confidence limits for π.

5. The heights of young trees in a large plantation have an approximately normal distribution with mean μ and the proportion of trees damaged by deer is π. A random sample of size 300 is selected and the height, x m, of each tree recorded. It is found that $\sum x = 1185$ and $\sum x^2 = 5165$. The number of damaged trees in the sample is 28. Find (a) 95% confidence limits for μ, (b) 98% confidence limits for μ, (c) 96% confidence limits for π, (d) 99% confidence limits for π.

6. It is known that 78 of the 90 eleven year old children entering a certain school have already had chickenpox. Treating these children as a random sample of all eleven year old children in a certain city, obtain 99% confidence limits for the proportion who have had chickenpox. Find also the minimum sample size that would be needed to give a 95% confidence interval of width less than 0·1.

7. A random sample of size n is to be drawn from a normal population with unknown mean μ and known standard deviation 1·8. Find the smallest value of n which will give (a) a 90% confidence interval for μ for width less than 1, (b) a 95% confidence interval for μ of width less than 1.

8. Machines X and Y fill cartons with orange drink. The amounts of orange drink in 10 randomly selected cartons filled by X were measured in ml and recorded as follows: 223, 247, 229, 252, 243, 254, 238, 241, 232, 251. The amounts in 8 randomly selected cartons filled by Y were as follows: 226, 248, 236, 257, 242, 238, 253, 244. Given that both machines dispense quantities of orange drink normally distributed with a standard deviation of 10 ml, find (a) 90% confidence limits for the mean amount in a carton filled by X, (b) 95% confidence limits for the mean amount in a carton filled by Y, (c) 99% confidence limits for the difference, $\mu_Y - \mu_X$, between the mean amounts dispensed by the two machines.

9. A fruit farmer is comparing two methods of cultivation. Using method I, 60 fruit bushes yield crops with mean mass 0·81 kg and variance 0·042 kg^2. Using method II, 80 fruit bushes yield crops with mean mass 0·89 kg and variance 0·036 kg^2. Find 99% confidence limits for the mean mass of fruit per bush (a) for method I, (b) for method II. Treating both crops as random samples from normal populations with common variance σ^2, find (c) a two-sample unbiased estimate of σ^2, (d) 95% confidence limits for the difference, $\mu_2 - \mu_1$, between the mean crops for methods II and I respectively.

10. In a sample of 400 shops taken in 1972, it was discovered that 136 of them sold carpets at below the list prices which had been recommended by manufacturers (i) Estimate the percentage of all carpet selling shops selling below list price. (ii) Calculate the 95% confidence limits for this estimate, and explain briefly what these mean. (iii) What size sample would have to be taken in order to estimate the percentage to within $\pm 2\%$? (SU)

11. The table below gives the distribution of the age in years at marriage of 200 men.

Age (centre of interval)	17·5	22·5	27·5	32·5	37·5	42·5	47·5	52·5	57·5	62·5
Frequency	31	73	48	21	12	6	3	2	1	3

Calculate estimates of the mean and standard deviation of these ages. If these ages may be assumed to be a random sample from a large population, obtain 95% confidence limits for the population mean, and explain carefully what your limits signify. (C)

12. Two random samples, each consisting of 50 Bramley apples, are selected from a very large collection of these apples. It is known that the standard deviation of the mass of a Bramley apple is 15 g. Give a symmetrical 95% probability interval for X, where X is the mass of the first sample minus that of the second sample, stating your assumptions. (C)

16.4 Introduction to hypothesis testing

In this section we consider an elementary way of measuring the significance of statistical evidence. The principle of the method is that if we have only a small number of observations, then it is easier to reject a theory than to support one. For example, one black Manx cat is sufficient evidence to reject the theory that all cats have tails, but cannot be considered strong evidence that all cats are black. Thus, to test the significance of data, a theory is formulated and the evidence used to decide whether or not the theory can be rejected. This theory is called the *null hypothesis*, often denoted by H_0. We now look at the test procedure in more detail.

Illustration A A new drug is to be used to treat 10 patients suffering from a serious illness. It is known that with established forms of treatment the probability of recovery is 0·3.

Let the null hypothesis, H_0, be that using the new drug the probability of recovery is still 0·3. The research team hopes to reject this theory in favour of the *alternative hypothesis*, H_1, that using the new treatment the probability of a cure is greater than 0·3.

Assuming that H_0 is true, the random variable R, representing the number of patients who recover, will have the binomial distribution $B(10, 0{\cdot}3)$. The values of $p(r)$ for $r = 0, 1, \ldots, 10$ are given correct to 3 d.p. in the table below.

r	0	1	2	3	4	5	6	7	8	9	10
$p(r)$	0·028	0·121	0·233	0·267	0·200	0·103	0·037	0·009	0·001	0·000	0·000

The problem now is to determine which values of the *test statistic* R can be regarded as significant evidence that H_0 should be rejected. Even the recovery of all 10 patients, although extremely unlikely, could still be consistent with the null hypothesis. To decide where to draw the line between acceptance and rejection of H_0 we must consider the probabilities $P(R > r)$. In practice events with a probability of 0·05 or less are regarded as *unlikely*. Thus when testing a hypothesis such events are said to provide *significant* evidence against the null hypothesis H_0, or more formally, to provide sufficient evidence *at the 5% level* to reject H_0. The "jargon" of significance tests can be summarised as follows:

Probability	Event	Evidence	Significance
0·05 or less	unlikely	significant	5% level
0·01 or less	very unlikely	strong or highly significant	1% level
0·001 or less	extremely unlikely	very strong or very highly significant	0·1% level

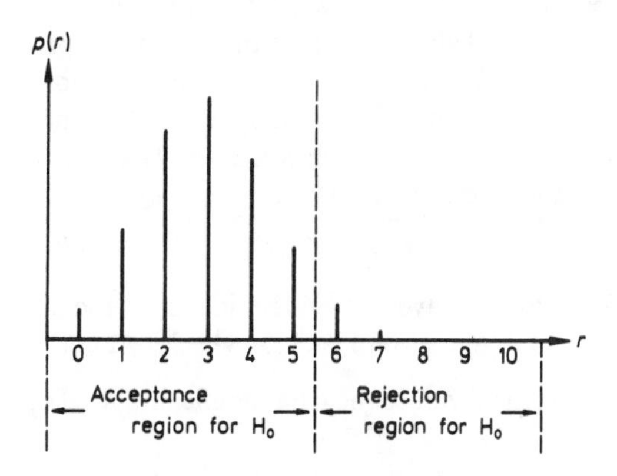

In our example the research team decide to look for evidence significant at the 5% level.

Since $P(R > 5) \approx 0{\cdot}047$
and $P(R > 4) \approx 0{\cdot}150$,

the appropriate *decision rule* is: "reject H_0 if $R > 5$".

Thus, given that H_0 is true, the probability of rejecting it in favour of H_1 is 0·047 or 4·7%. This quantity, the probability of wrongly rejecting the null hypothesis, is called the *significance level* of the test.

Hence we have constructed a test with a significance level of 4·7%, in which 6 or more recoveries will be regarded as evidence that the new treatment is more effective than the old.

Illustration B A production line is designed to produce bicycle wheels with mean diameter 42 cm. It is known that the diameters are normally distributed with standard deviation 1·5 cm. In order to test the hypothesis that the mean diameter is indeed 42 cm, a random sample of 25 wheels is selected. The sample mean is found to be 41·5 cm.

Using standard notation, we suggest a suitable test:

$$H_0\colon \mu = 42. \quad H_1\colon \mu \neq 42. \quad \text{Significance level: } 5\%.$$

Test statistic: $\bar{X}$, the mean of a sample of size 25.

It was shown in §16.1 that if a random sample of size n is drawn from the distribution $N(\mu, \sigma^2)$, then the sample mean has distribution $N(\mu, \sigma^2/n)$. Hence, on the assumption that H_0 is true, $\bar{X}$ is normally distributed with mean 42 and standard deviation 0·3.

Since we can reject H_0 if there is significant evidence that $\mu > 42$ or that $\mu < 42$, we require a test with a rejection region that covers *both* ends of the distribution of $\bar{X}$. This type of test is called a *two-tailed test*. (In our earlier example we used a *one-tailed test* which involved only *one* end of the distribution of the test statistic R.) For a test with a 5% significance level there is a 5% probability that the observed value of the test statistic will lie in the rejection region. In a two-tailed test the probability of rejection is divided equally between the two "tails".

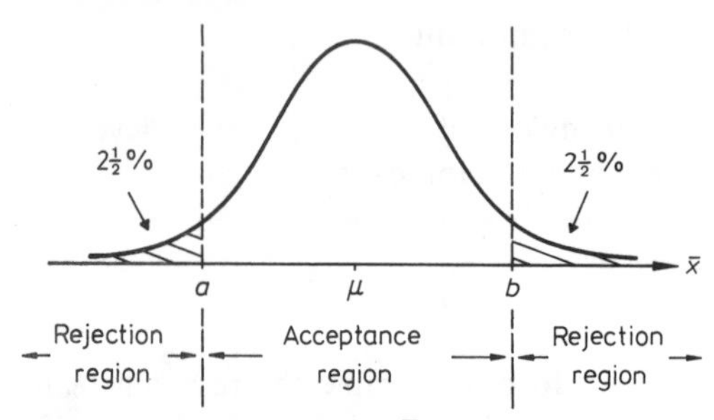

In our present example we now need to find critical values a, b, which lie at the boundaries of the rejection region for H_0.

Using the $2\frac{1}{2}\%$ value given in Table 2, we find that the appropriate critical values for the distribution $N(0, 1)$ would be $\pm 1{\cdot}96$. Hence for the distribution of $\bar{X}$ the critical values are $42 \pm 1{\cdot}96 \times 0{\cdot}3$ i.e. 41·412 and 42·588. It follows that for this test a suitable decision rule is:

$$\text{"reject } H_0 \text{ if } \bar{X} < 41{\cdot}4 \quad \text{or} \quad \text{if } \bar{X} > 42{\cdot}6\text{".}$$

Since the observed value of $\bar{X}$ is 41·5 we do not have sufficient evidence at the 5% level to reject the hypothesis that the mean diameter of a bicycle wheel is 42 cm.

[Note that because we have rounded the critical values the true significance level of this test is a little less than 5%.]

We now summarise the steps that must be taken when conducting a statistical test of this kind.

(1) Formulate a null hypothesis H_0.
(2) Formulate an alternative hypothesis H_1 and hence decide whether the test is to be one-tailed or two-tailed.
(3) Choose the significance level of the test.
(4) Choose a suitable test statistic and describe its sampling distribution on the assumption that H_0 is true.
(5) Formulate a decision rule either by defining a suitable rejection region or by obtaining the critical value(s) of the test statistic.
(6) Carry out the test.

Example 1 A die is rolled 150 times and 35 sixes are obtained. Test, at the the 1% level, the hypothesis that the die is unbiased.

Let p be the probability of obtaining a six in one roll of a die.
$H_0: p = 1/6$. $H_1: p \neq 1/6$. Two-tailed test.

Significance level: 1%.
Test statistic: R, the number of sixes obtained in 150 rolls of the die.
Assuming that H_0 is true, R has a binomial distribution with parameters $n = 150$, $p = 1/6$. Since n is large, this can be approximated by the normal distribution with mean $np = 25$ and variance $np(1 - p) = 125/6$.
In a two-tailed test with 1% significance level the critical values for the distribution $N(0, 1)$ are $\pm 2{\cdot}576$.
Hence the critical values of R are $25 \pm 2{\cdot}576 \times \sqrt{(125/6)}$ i.e. 13 and 37 (to the nearest integer).
Decision rule: reject H_0 if the number of sixes is either less than 13 or greater than 37.
Since 35 sixes were obtained, we do not have significant evidence (at the 1% level) that the die is biased.

Example 2 Two schools entered candidates for the same public examination. The marks obtained by 80 candidates in school A had mean 59·7 and standard deviation 12. The marks obtained by 60 candidates in school B had mean 56·5 and standard deviation 15. Regarding the sets of marks as large random samples from normal populations with the same variance σ^2, determine whether the performance of school A was significantly better than that of school B.

Let μ_1, μ_2 be the population means for marks obtained by students from schools A and B respectively.
H_0: $\mu_1 = \mu_2$. H_1: $\mu_1 > \mu_2$. One-tailed test.
Significance level: 5%.
Test statistic: $D = \bar{X}_1 - \bar{X}_2$, where $\bar{X}_1$ is the mean mark of 80 candidates from school A and $\bar{X}_2$ is the mean mark of 60 candidates from school B.
The sampling distributions of $\bar{X}_1$ and $\bar{X}_2$ are $N(\mu_1, \sigma^2/80)$ and $N(\mu_2, \sigma^2/60)$.
However, since the value of σ^2 is unknown we will have to use estimated variances based on the two-sample unbiased estimate $s^2 = \dfrac{80 \times 12^2 + 60 \times 15^2}{80 + 60 - 2}$.
Thus, on the assumption that H_0 is true, D is normally distributed with mean 0 and estimated variance $(s^2/80 + s^2/60)$
i.e. $7s^2/240$.

[Note that for large samples the statistic $D/\sqrt{(7s^2/240)}$ has an approximately normal distribution. This means that we can use values obtained from normal distribution tables to calculate the critical value of D.]

Using Table 2, the 5% value in the distribution $N(0, 1)$ is 1·645.
Hence the critical value of D is given approximately by

$$1{\cdot}645 \times \sqrt{\left(\frac{7s^2}{240}\right)} = 1{\cdot}645 \times \sqrt{\left(\frac{7(80 \times 12^2 + 60 \times 15^2)}{240 \times 138}\right)} \approx 3{\cdot}78.$$

Decision rule: reject H_0 if the difference $\bar{X}_1 - \bar{X}_2$ exceeds 3·78.
The observed value of $D = 59{\cdot}7 - 56{\cdot}5 = 3{\cdot}2$.
Thus we do not have sufficient evidence at the 5% level to suggest that the performance of school A was significantly better than that of school B.

Finally, we consider briefly the two types of error which may arise when testing a hypothesis.

A *Type I error* is the rejection of the null hypothesis when it is true. Since the probability of a Type I error is the probability of wrongly rejecting the null hypothesis, it must always be equal to the significance level of the test. For instance in Illustration A the probability of a Type I error was 0·047 or 4·7%.

A *Type II error* is the acceptance of a null hypothesis which is false. The probability of a Type II error is the probability of accepting the null hypothesis given that the alternative hypothesis is true. The value of this probability depends on the form of the alternative hypothesis and it is difficult to estimate for the tests discussed in this section.

Exercise 16.4

1. Anne thinks that the probability of growing a healthy plant from an avocado pear stone is $\frac{1}{3}$. Barbara has succeeded with 4 stones out of 5. Assuming that Anne is right, find the probability of succeeding with 4 stones or more. Hence decide whether Barbara's success provides sufficient evidence at the 5% level to suggest that the probability of growing a healthy plant is (a) not equal to $\frac{1}{3}$, (b) greater than $\frac{1}{3}$.

2. A driving school claims that 90% of its pupils pass the driving test at the first or second attempt. This hypothesis is to be tested against the alternative hypothesis that the success rate is less than 90%. Given that a random sample of 8 pupils will be taken, devise tests with significance levels of approximately (a) 5%, (b) 1%, (c) 0·1%. In each case state the decision rule and calculate the true significance level of the test correct to 2 d.p.

3. A cricketer estimates that the probability p that he will score more than 30 runs in an innings is 5/12. He intends to test this theory against the alternative hypothesis that $p \neq 5/12$ by noting the number of runs he scores in his next 7 innings. Obtain a suitable decision rule on the assumption that the significance level of the test is approximately 5%. Calculate the probability of a Type I error giving your answer correct to 2 s.f.

4. Just before an election the Conservative party claims that 60% of the voters support their candidate. In an opinion poll 281 people out of a random sample of 500 state that they will vote for the Conservative candidate. Decide whether there is sufficient evidence at the 5% level to suggest that less than 60% of the electorate will vote Conservative.

5. A random sample of size 80 is drawn from a normal distribution with unknown mean and variance 16. The sample mean is found to be 63·1. Does this result provide (a) significant evidence at the 1% level that the population mean is not equal to 62, (b) significant evidence at the 3% level that the population mean is less than 64.

6. The mean height of a large population of women is 1·65 m with a standard deviation of 0·07 m. Random samples of 100 are taken from this population.

Describe the distribution of the means of these samples, giving the mean and the standard deviation of this distribution. A particular group of 100 women has a mean height of 1·68 m. Should this be considered significantly different from the expected mean of a random sample? Give reasons for your answer. (AEB 1980)

7. A machine is designed to produce articles of mass 2·4 kg. A random sample of size 120 is found to have mean 2·52 kg and variance 0·8 kg^2. Treating this as a large sample from a normal population, decide whether the data suggests that the population mean is greater than 2·4 kg.

8. A keen gardener wishes to compare two types of seed potato. He finds that 50 plants of variety C (Champion Chipper) yield crops of potatoes with a mean mass of 1·59 kg and 60 plants of variety F (Fancy Fryer) have yields with a mean mass of 1·71 kg. Given that the population variances for varieties C and F are 0·060 kg^2 and 0·084 kg^2 respectively, test the hypothesis that there is no difference between their mean yields (a) at the 5% level, (b) at the 1% level.

9. In an experiment to investigate extra-sensory perception one card is drawn at random from an ordinary pack of 52 by the "sender"; he then concentrates on the suit of the card drawn, trying to communicate the name of the suit to a second person, the "receiver", who is in another room. The card is returned to the pack and the pack is shuffled after each trial. In an extended series of trials the receiver correctly identifies the suit of the card in 291 trials out of 1000. Explain carefully, with supporting calculations, whether you regard this result as contradicting the supposition that results are obtained entirely by guesswork. (C)

10. (a) After a survey a market research company asserted that 75% of T.V. viewers watched a certain programme. Another company interviewed 75 viewers and found that 51 watched the programme and 24 did not. Does this provide evidence at the 5% level of significance that the first company's figure of 75% was incorrect?

(b)

Tree	A	B
Sample size	60	80
Mean	11·4	10·7
S.D.	2·6	3·1

Samples of leaves were collected from two oak trees A and B. The number of galls was counted on each leaf and the mean and standard deviation of the number of galls per leaf was calculated with the given results. Assuming normal distributions, do the data provide evidence at the 5% significance level of different population means for the two trees? (SU)

11.

	number	$\sum x$	$\sum x^2$
males	120	198	327
females	160	248	385

From a large population of students, 120 males and 160 females are chosen at random.

Their heights x in metres are summarised in the given table. The males and females may be treated as random samples from two independent populations. (a) Find the sample means and variances. (b) Assuming that in both populations the heights are normally distributed with these means and variances, find the probability that a randomly-chosen female will be taller than a randomly-chosen male. (c) Assuming only that height is distributed normally in both populations, test the hypothesis that the mean height of the population of male students exceeds the mean height of the population of female students by less than 0·08 metres. (O&C)

12. A large number of tomato plants are grown under controlled conditions. Half of the plants, chosen at random, are treated with a new fertilizer, and the other half of the plants are treated with a standard fertilizer. Random samples of 100 plants are selected from each half, and records are kept of the total crop mass of each plant. For those treated with the new fertilizer, the crop masses (in suitable units) are summarised by the figures $\sum x = 1030{\cdot}0$, $\sum x^2 = 11\,045{\cdot}59$. Obtain an unbiased estimate of the population variance, and, treating the sample as a large sample from a normal distribution, obtain a symmetric 96% confidence interval for the mean crop mass.

The corresponding figures for those plants treated with the standard fertilizer are $\sum y = 990{\cdot}0$, $\sum y^2 = 10079{\cdot}19$. Treating the sample as a large sample from a normal distribution, and assuming that the population variances of both distributions are equal, obtain a two-sample pooled estimate of the common population variance. Assuming that it is impossible for the new fertilizer to be less efficacious than the old fertilizer and assuming that both distributions are normal, test whether the results provide significant evidence (at the 3% level) that the new fertilizer is associated with a greater mean crop mass, stating clearly your null and alternative hypothesis. (C)

13. A large group of sunflowers is growing in the shady side of a garden. A random sample of 36 of these sunflowers is measured. The sample mean height is found to be 2·86 m, and the sample standard deviation is found to be 0·60 m. Treating the sample as a large sample and assuming the heights to be normally distributed, give a symmetric 99% confidence interval for the mean height of the sunflowers in the shady side of the garden.

A second group of sunflowers is growing in the sunny side of the garden. A random sample of 26 of these sunflowers is measured. The sample mean height is found to be 3·29 m and the sample standard deviation is found to be 0·90 m. Treating the samples as large samples from normal distributions having the same variance but possibly different means, obtain a pooled estimate of the variance, and test whether the results provide significant evidence (at the 5% level) that the sunny-side sunflowers grow taller, on average, than the shady-side sunflowers. (C)

14. According to some recent accident statistics, a random sample of 800 car drivers injured in road accidents comprised 250 who were wearing seat belts and 550 who were not wearing seat belts. Determine a symmetric 99% confidence interval for the proportion of injured drivers wearing seat belts.

The injuries of the car drivers were described as either slight or serious. Of those wearing seat belts, 50 were seriously injured; of those not wearing seat belts, 150 were seriously injured. Determine an unbiased estimate of the overall proportion of serious injuries amongst the injured drivers. Test, at the 5% significance level, the hypothesis that the proportion of serious injuries is greater amongst those injured drivers not wearing seat belts than amongst those injured drivers wearing seat belts. (C)

15. Countries A and B contain large numbers of elm trees, many of which are affected by Dutch elm disease. It is found that in country A, out of a random sample of 100 elm trees, 67 are affected by the disease and that in country B, out of a random sample of 150 elm trees, 93 are affected by the disease. Tree experts have a theory that the proportions of affected trees in the two countries are the same, although there is a possibility that, since the disease affected the trees in country A before those in country B, the proportion of trees affected in country A may be greater than the proportion affected in country B. Using the sample results, test, at the 10% significance level, the theory of the experts. For the test that you perform, state clearly the hypotheses under comparison. Assuming that the proportions of affected trees in the two countries are the same, give an approximate symmetric 98% confidence interval for this common proportion. (C)

16. A person claims that he has an almost foolproof system for identifying the suit of a playing card placed face downwards. To test whether the person is merely guessing he is asked to identify the suits of 4 playing cards placed face downwards. It is agreed to accept his claim only if he correctly identifies at least 3 of the 4 cards. Viewing this as a hypothesis testing problem, write down an appropriate null hypothesis and an appropriate alternative hypothesis. Calculate the significance level of the test and the probability that the person will be regarded as merely guessing when in fact his system has a probability of 0·8 of correctly identifying the suit of a card. (JMB)

17. Two types of seed, X and Y, appear identical but have different probabilities of germination, these being 50% for type X seeds and 60% for type Y seeds. The germination, or otherwise, of each seed may be regarded as being independent of the germination, or otherwise, of all other seeds. A packet contains 100 type X seeds, which are sown. Show, using a normal approximation, that the smallest number n, which is such that $P(n$ or more germinate$) < 0{\cdot}05$, is 59. A second packet also contains 100 seeds, which are sown. It is not known whether these seeds (which are all of the same type) are type X or type Y. There are two hypotheses, H_0: the seeds are of type X, H_1: the seeds are of type Y. Construct a test based on n, which is such that $P(\text{Accept } H_1 | H_0 \text{ true}) < 0{\cdot}05$. State this test clearly and determine $P(\text{Accept } H_0 | H_1 \text{ true})$. (C)

Exercise 16.5 (miscellaneous)

1. What is meant by the 'sampling distribution of a statistic'? If random samples of size n are drawn from an unspecified distribution with mean μ and variance σ^2, what is the sampling distribution of the mean as n tends to infinity?

The distribution of the energy provided by certain fuel cells has mean 20 units and standard deviation 2 units. A power pack made up of 64 cells is assumed to have an output which is equal to the total energy supplied by all the cells in it. Find the probability that such a pack will provide 1300 units of energy. The manufacturers wish to quote an energy output for such packs that will be satisfied by 99% of the packs. Determine what energy level should be quoted.

2. (a) Explain briefly, in your own words, the idea of the Central Limit Theorem.
   Samples of 100 pebbles are taken from a large number of pebbles on a certain beach. The mean mass of the pebbles overall is known to be 283 g with a standard deviation of 23 g. Describe the distribution of the means of such samples, giving its mean and standard deviation. What effect, if any, is the type of distribution of the pebbles on the beach likely to have on the distribution of means?
   (b) How does the variance of the means of small samples of n elements compare with the variance of the population from which it is taken. A random sample of 10 lengths of straw was picked from a field. The lengths, in metres, were 0·8, 0·8, 0·8, 0·9, 1, 1, 1·1, 1·1, 1·2, 1·3. Give an estimate of the mean length of straw in the field and the standard error of this estimate. (AEB 1979)

3. In a certain examination with a very large entry, the percentage marks obtained by the male candidates were found to follow a normal distribution with a mean of 54 and a standard deviation of 16. Let $\bar{X}$ denote the mean of the percentage marks scored by a random sample of 4 male candidates. What is the sampling distribution of $\bar{X}$? Calculate the probability that $\bar{X}$ will exceed 70 and the value c such that there is a probability of 0·95 that $\bar{X}$ will be within c marks of the mean mark of 54. In the same examination the percentage marks obtained by the female candidates were found to follow a normal distribution with a mean of 59 and a standard deviation of 20. Let $\bar{Y}$ denote the mean of the percentage marks scored by a random sample of 5 female candidates. What is the sampling distribution of $\bar{Y} - \bar{X}$? Calculate the probability that the value of $\bar{Y}$ will be greater than the value of $\bar{X}$. (W)

4. A carton contains 10 boxes of matches. It was found that 3 of the boxes contained 46 matches, 6 contained 47 matches, and the remaining box contained 48 matches. Calculate the mean number of matches per box in the carton. Suppose a random sample of 3 boxes are chosen without replacement from the carton. Let m denote the mean number of matches per box in the sample. By considering the contents of all possible samples of 3 boxes and their corresponding probabilities, obtain the sampling probability distribution of m. Verify that m is an unbiased estimate of the mean number of matches per box in the carton and calculate its standard error. [The arithmetic can be reduced substantially by operating on a suitably chosen linear function of m and using properties of expectation and variance.] (W)

5. Show that, if $X_1, \ldots, X_n$ are n independent random variables with mean μ and variance σ^2, then $\sum X_i/n$ has mean μ and variance σ^2/n. The diameters of 25 steel rods are measured and found to have a mean of 0·980 cm and a standard deviation of 0·015 cm. Assuming this is a sample from a normal distribution with the same variance, find 99% confidence limits for the population mean. (O&C)

6.

Time	Frequency
$0 < t \leqslant 50$	31
$50 < t \leqslant 100$	24
$100 < t \leqslant 150$	20
$150 < t \leqslant 200$	13
$200 < t \leqslant 300$	11
$300 < t \leqslant 500$	1
	100

On 1 January 1979, 100 new "Eternity" light-bulbs were installed in a certain building, together with a device which records how long each light-bulb is in use. By 1 March all 100 bulbs had failed, and the recorded life-times, t, (in hours of use since 1 January) are summarised in the given table. Obtain estimates of the sample mean and the sample standard deviation for this set of data. Assuming that the bulbs constituted a random sample of "Eternity" light-bulbs, obtain a symmetric 99% confidence interval for the mean life-time of "Eternity" light-bulbs. (C)

7. If $\bar{X}$ is the mean of a random sample of size n_1 from a population distribution having mean μ_1 and variance $\sigma_1{}^2$, and n_1 is very large, what can you say about the sampling distribution of $\bar{X}$? If $\bar{Y}$ is the mean of a random sample of size n_2 from another population distribution having mean μ_2 and variance $\sigma_2{}^2$, and n_2 is also very large, what can you say about the sampling distribution of $\bar{Y} - \bar{X}$?

The heights, measured in metres, of 100 women drawn at random from a certain tribe had a sum of 160 and the sum of their squares was 265. Use this information to construct a 90% confidence interval for the mean height of women of this tribe. The heights, measured in metres, of 200 men drawn at random from the same tribe, had a sum of 360 and the sum of their squares was 712. Use the information contained in the two samples to construct a 95% confidence interval for the amount by which the average height of the men in the tribe exceeds that of the women. (W)

8. Let p denote the probability of obtaining a head when a certain coin is tossed. (i) If $p = 0{\cdot}4$, find the probability of obtaining at least 3 heads in 10 independent tosses of the coin. (ii) If $p = 0{\cdot}6$, find the probability of obtaining exactly 12 heads in 20 independent tosses of the coin. (iii) Write down an appropriate null hypothesis and an appropriate alternative hypothesis for testing whether the coin is unbiased. To carry out this test 20 independent tosses of the coin are made and the number of heads that occurs is observed. Given that 15 heads occurred, carry out the test, assuming a 5% significance level. Write down a statement of the conclusion you draw about the value of p for this coin. (JMB)

9. The masses of loaves from a certain bakery are normally distributed with mean 500 g and standard deviation 20 g. (a) Determine what percentage of the output would fall below 475 g and what percentage would be above 530 g. (b) The bakery produces 1000 loaves daily at a cost of 8 p per loaf and can sell all those above 475 g for 20 p each but is not allowed to sell the rest. Calculate the expected daily profit. (c) A sample of 25 loaves yielded a mean mass of 490 g. Does this provide evidence of of a reduced population mean? Use the 5% level of significance and state whether the test is one-tailed or two. (SU)

10. A man, Mr Knowall, claims to be able to recognize a well known brand, A, of whisky blindfolded. On 8 occasions he is given two glasses in random order, one containing brand A and the other containing brand X. On 7 occasions he names the brands correctly. Would you consider his claim justified? The same experiment is repeated the following day, and this time he names the brand correctly in 6 out of 8 trials. How does this affect your opinion of his claim? (AEB 1977)

11. Experimental data concerning a variable X, which measures the reliability of a certain electric component, is as follows: $\sum x = 1164{\cdot}2$, $\sum x^2 = 13911{\cdot}6$, $n = 100$. Calculate the sample mean and standard deviation from these figures. Explain whether, on the evidence of this sample, you would reject the hypothesis that the mean value of X is 12. Figures collected over a long period have established that the mean and standard deviation of X are 12 and 2 respectively. After a change in the manufacturing process it is expected that the mean will have been *increased*, but it may be assumed that the standard deviation remains equal to 2. A sample of n values of X is taken, with sample mean m; if m is greater than some critical value it will be accepted that the mean has in fact increased, but if m is less than this critical value the increase is not established. State carefully appropriate null and alternative hypotheses for this situation, and find, in terms of n, the critical value for a 1% significance level. (C)

12. The current medical treatment for a particular ailment has a cure rate of 80% (i.e. a patient with this ailment has a probability of 0·8 of being cured by the treatment). (i) Write down the mean and the variance of the number of patients that will be cured per 100 patients treated. (ii) Calculate the probability that, of 4 patients treated, at least 3 will be cured. (iii) A pharmaceutical company claims that it has developed a new treatment for the ailment which has a higher cure rate than the current treatment. To test this claim the new treatment is to be given to 20 patients. It is decided to accept the new treatment as being better than the current treatment only if at least 19 of the 20 patients are cured. Write down an appropriate null hypothesis and an appropriate alternative hypothesis for this test. Calculate the significance level of the test. Also calculate the probability that the new treatment will not be accepted as being better than the current treatment when the new treatment has a cure rate of 90%. (JMB)

13. The lengths (in millimetres) of nine screws selected at random from a large consignment are found to be 7·99, 8·01, 8·00, 8·02, 8·03, 7·99, 8·00, 8·01, 8·01. Calculate unbiased estimates of the population mean and variance. Assuming a normal distribution with variance 0·0001, test, at the 5% level, the hypothesis that the population mean is 8·00 against the alternative hypothesis that the population mean is not 8·00. From a second large consignment, sixteen screws are selected at random, and their mean length (in millimetres) is found to be 7·992. Assuming a normal distribution with variance 0·0001, test, at the 5% level, the hypothesis that this population has the same mean as the first population, against the alternative hypothesis that this population has a smaller mean than the first population. (C)

14. (a) A sample of 200 similar packets of breakfast cereal was examined, and the mass of the contents in each packet was noted. The following results were

obtained: sample mean = 341·2 g. sample standard deviation = 0·92 g. Calculate a 95% confidence interval for the mean mass of contents in all packets of this type, stating any assumptions that you make.

(b) Samples of packets of cereal filled by two different machines were examined, and the following results about the mass (x grams) of the contents were obtained:

	Sample size	$\bar{x}$	$\sum(x-\bar{x})^2$
Machine 1	20	340·9	16·82
Machine 2	10	341·4	8·63

Test whether the mean masses of contents in packets filled by the two machines are significantly different at the 5% level. Assume that the masses of contents in packets filled by each machine are normally distributed, and that the two normal distributions have equal variances. (C)

15. (a) In one county in England, a random sample of 225 12-year-old boys and 250 12-year-old girls was given an arithmetic test. The average mark for the boys was 57 with a standard deviation of 12, whilst the average for the girls was 60 with a standard deviation of 15. Assuming that the distributions are normal, does this provide evidence at the 2% level that 12-year-old girls are superior to 12-year-old boys at arithmetic?

(b) An IQ test which had been standardised giving a mean of 100 and a standard deviation of 12 was given to a random sample of 50 children in one area. The average mark obtained was 105. Does this provide evidence, at the 5% level, that children from this area are generally more intelligent than other children? (SU)

16. An organisation interviews a randomly chosen sample of 1000 adults from the population of the United Kingdom, and 517 of those interviewed claim to support the Conservative party. A second organisation independently interviews a random sample of 2000 adults, of whom 983 claim to support the Conservative party. (i) Verify that the results of the two organisations do not differ significantly at the 5% level. (ii) Obtain a symmetric 99% confidence interval (based on the combined results) for the proportion of the population who claim to support the Conservative party. (C)

17. Determine the probability that a six-digit number (leading zeros being relevant) chosen at random from the range 000000 to 999999 inclusive has no two successive digits equal to one another. An electronic device is supposed to generate six-digit numbers (leading zeros being relevant) in the range 000000 to 999999 inclusive, so that each of the one million possibilities is equally probable. However, it is believed that the machine may be biased so that it under-represents numbers having at least one pair of successive digits equal to one another. A random selection of twenty numbers generated by the device is examined and it is found that precisely r of these numbers contain at least one pair of successive digits equal to one another. (i) Given that $r = 6$, does this provide significant evidence, at the 10% level, that the device is biased in the manner described? (ii) How small would r have to be in order that

there was significant evidence, at the 1% level, that the device was biased in the manner described? (C)

18. A machine produces nails whose lengths are normally distributed with mean μ cm and standard deviation 0·1 cm. When the machine is working correctly $\mu = 3{\cdot}0$, but occasionally the machine goes wrong, in which case $\mu = 3{\cdot}05$, the standard deviation remaining 0·1 cm. In order to decide whether the machine is working correctly, the lengths of the nails in a sample of n nails are measured, and the sample mean $\bar{x}$ is found. If the value of $\bar{x}$ exceeds a predetermined value v then it is concluded that the machine has gone wrong, otherwise the machine is presumed to be working correctly. It is required that there should be a probability of no more than 5% of presuming that the machine has gone wrong when in fact it is working correctly, and that there should also be a probability of no more than 10% of presuming the machine to be working correctly when it has gone wrong. Determine appropriate values of n and v, if n is to be as small as possible. (C)

19. The mass, in grams, of marmalade in a randomly chosen jar from a large batch filled by a machine is normally distributed. The mean, μ, for a batch can be adjusted but the standard deviation is always 4. Each batch produced is tested by determining the mean mass, $\bar{x}$, of the marmalade in a random sample of n jars. Under the null hypothesis $\mu = 454$, a batch is rejected if $\bar{x}$ is such that there is significant evidence, at the 1% level, that the value of μ is less than 454. The sample size, n, is chosen so that, in the case $\mu = 452$, the probability of rejecting the batch is at least 0·95. Find the smallest value of n to achieve this, and the corresponding critical value of $\bar{x}$ for rejection. (C)

20. Two hypotheses concerning the probability density function of a random variable are

$$H_0\colon f(x) = \begin{cases} 1 & 1 < x < 2, \\ 0 & \text{otherwise;} \end{cases} \qquad H_1\colon f(x) = \begin{cases} 3x^2/7 & 1 < x < 2, \\ 0 & \text{otherwise.} \end{cases}$$

Give a sketch of the probability density function for each case.
The following test procedure is decided upon:

A single observation of X is made and if X is less than a particular value a, where $1 < a < 2$, then H_0 is accepted; otherwise H_1 is accepted.

Find a such that, when H_0 is true, the test procedure leads, with probability 0·1, to the acceptance of H_1. With this value of a, find the probability that, when H_1 is true, the test procedure leads to the acceptance of H_0. (C)

17 Correlation and regression

17.1 Bivariate distributions

In most of our earlier work on probability and statistics we have considered distributions involving values of one variable. However, it is often possible to measure two or more characteristics of the members of a sample. For instance, we may record both the height and the weight of each child admitted to a certain school. The pairs of observations obtained form a *bivariate distribution.* This chapter is devoted to the study of such distributions.

The main purpose of collecting bivariate data is to investigate any relationship between the variables. A possible first step in this process is to represent the data graphically. This is done by plotting each pair of observations as a single point on a graph called a *scatter diagram.*

Illustration A Ten students seeking careers advice were given two aptitude tests. Their test scores and the corresponding scatter diagram were as follows:

Student	Verbal test	Spatial test
1	69	52
2	80	71
3	57	55
4	52	48
5	85	89
6	66	64
7	49	37
8	76	60
9	63	56
10	73	78

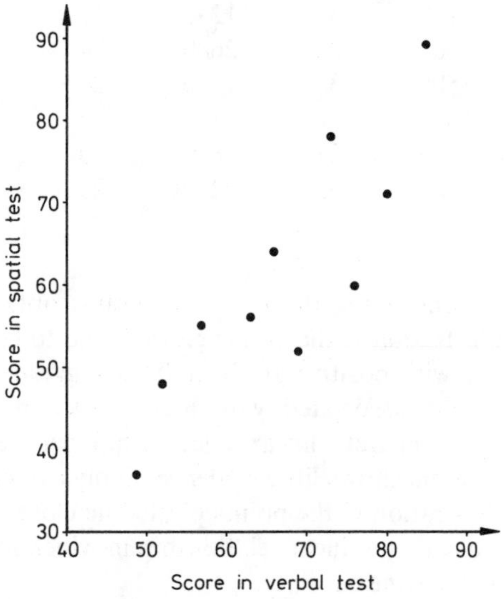

Illustration B The heights and batting averages of the members of a college cricket team were recorded. The resulting data and the scatter diagram are given below.

Player	Height in metres	Batting average
1	1·72	3
2	1·89	33
3	1·76	10
4	1·92	8
5	1·66	28
6	1·94	16
7	1·84	22
8	1·68	9
9	1·77	47
10	1·86	6
11	1·76	16

Illustration C Twenty types of car were tested by a motoring magazine. The engine capacity and fuel consumption of each car were recorded as follows:

Capacity in cc	Petrol in mpg	Capacity in cc	Petrol in mpg
1000	39	3440	17
1300	31	950	36
600	45	840	40
2000	27	1100	38
1170	36	1290	41
1600	32	2600	24
1500	33	1980	24
1580	30	1260	34
2800	21	1850	30
1600	35	2200	25

Considering the scatter diagrams obtained in these examples, we see that in Illustration A the points plotted tend to lie in a fairly narrow belt along a straight line with positive gradient. This suggests that high marks in the spatial test are usually associated with high marks in the verbal test and that there is an approximately linear relationship between the variables. In Illustration B the diagram shows little evidence of any association between the variables, whereas in Illustration C the points plotted lie close to a straight line with negative gradient, indicating a linear relationship in which low mileage per gallon is associated with high engine capacity.

Any relationship between the variables in a bivariate distribution is called a *correlation.* The *linear* correlation observed in Illustration A is described as a *positive* or *direct* correlation. Similarly the relationship in Illustration C is a *negative* or *inverse* correlation. In Illustration B the variables appear to be *uncorrelated.*

The following diagrams illustrate some further types of relationship which may rise.

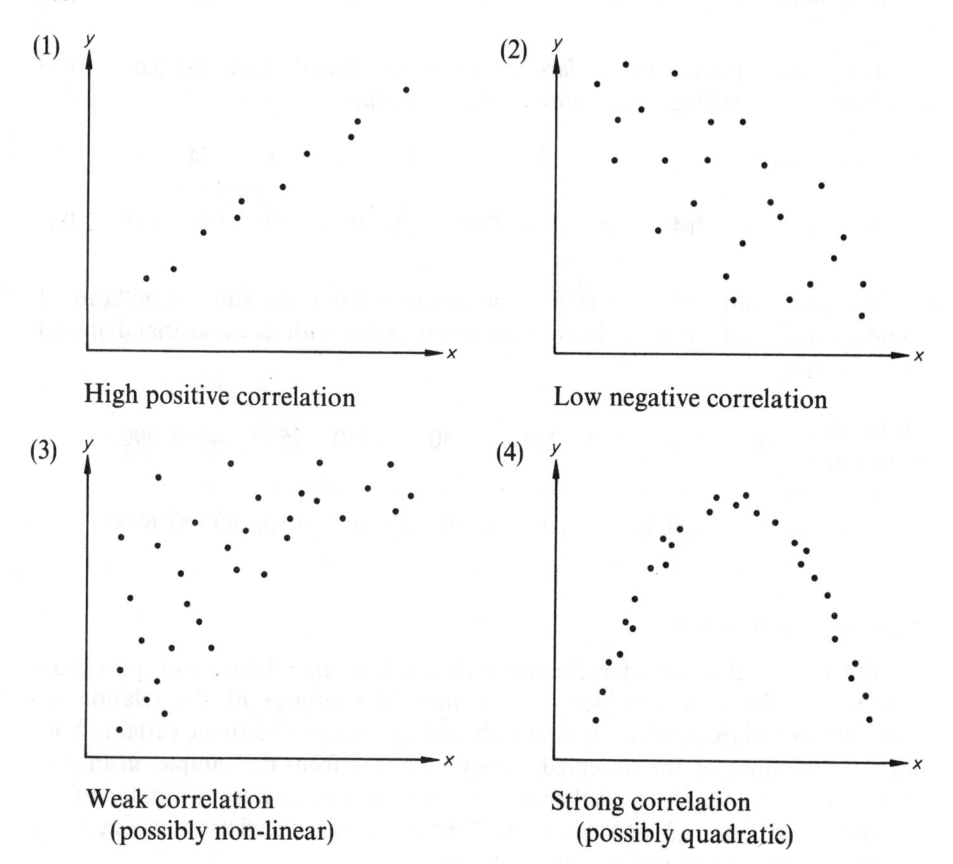

Exercise 17.1

In each of the following questions use the given data to plot a scatter diagram. Describe any apparent relationship between the variables.

1.

x	46	61	53	55	32	66	42	31	49	73
y	17	23	19	21	11	26	14	8	18	29

2.

x	14	11	5	12	9	15	8	15	10	6
y	11	9	16	11	12	8	14	4	10	12

3. The given table shows the oil production and oil consumption of various parts of the world in millions of metric tons per annum.

Oil production	68	538	277	23	299	690	390	0	67
Oil consumption	72	697	135	628	43	51	346	200	116

4. The given table shows the face values of the British coins minted in 1953 together with the selling prices quoted by a coin dealer.

Face value (in d.)	$\frac{1}{4}$	$\frac{1}{2}$	1	3	6	12	12	24	30	60
Selling price (in £)	0·40	1·00	1·80	0·80	0·50	0·60	0·90	1·30	1·60	2·00

5. The given table shows the average distance from the sun (in millions of kilometres) of each planet in the solar system together with its equatorial diameter (in kilometres).

Distance from sun	60	110	150	230	780	1430	2870	4500	5900
Diameter	4880	12 100	12 760	6790	142 800	120 000	51 800	49 500	6000

17.2 Linear correlation

In §17.1 we saw that a scatter diagram indicates the general nature of a bivariate distribution. We now consider some numerical measures of the relationship between two variables. When dealing with the distribution of a single variable X we use the deviations of the observed values $x_1, x_2, \ldots$ from the sample mean $\bar{x}$ to obtain useful measures such as the variance and standard deviation. Similarly for a bivariate distribution of variables X and Y the values of the deviations $x - \bar{x}$, $y - \bar{y}$ can be used to obtain measures of correlation.

Let us consider again the data given in Illustration A.

Verbal test score, x	69	80	57	52	85	66	49	76	63	73
Spatial test score, y	52	71	55	48	89	64	37	60	56	78

$$\bar{x} = \tfrac{1}{10}\sum x = 67, \quad \bar{y} = \tfrac{1}{10}\sum y = 61$$

We now calculate the values of $u = x - \bar{x}$, $v = y - \bar{y}$ and plot the corresponding scatter diagram.

u	2	13	-10	-15	18	-1	-18	9	-4	6
v	-9	10	-6	-13	28	3	-24	-1	-5	17

(A)

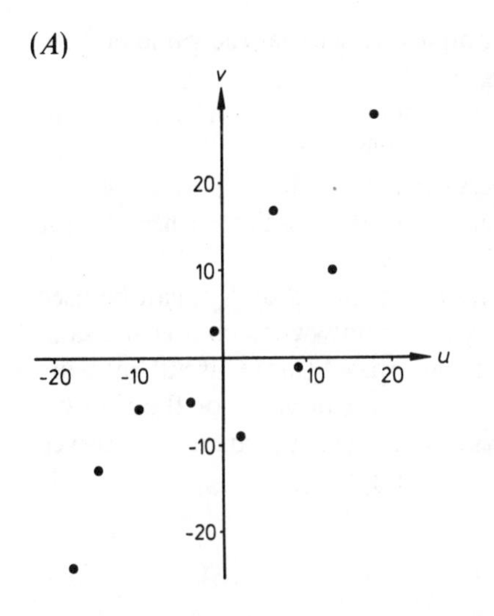

Because of the positive correlation between x and y, most of the points in this new scatter diagram lie in the 1st and 3rd quadrants. For these points the product uv takes positive values, whereas for the remaining points uv takes negative values. Thus it is not surprising to find that $\sum uv$ has the fairly large positive value 1413.

(B)

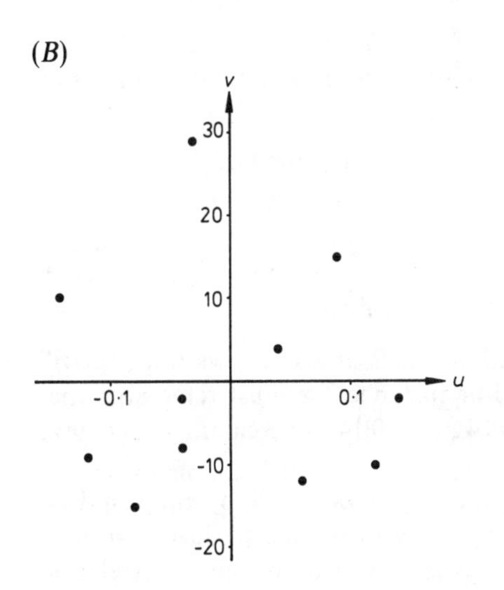

Performing similar calculations for the data in Illustration B we find that there is no definite pattern in the u, v scatter diagram and that the product uv takes a set of positive and negative values which tend to cancel each other out. In this case $\sum uv = -0{\cdot}28$.

(C)

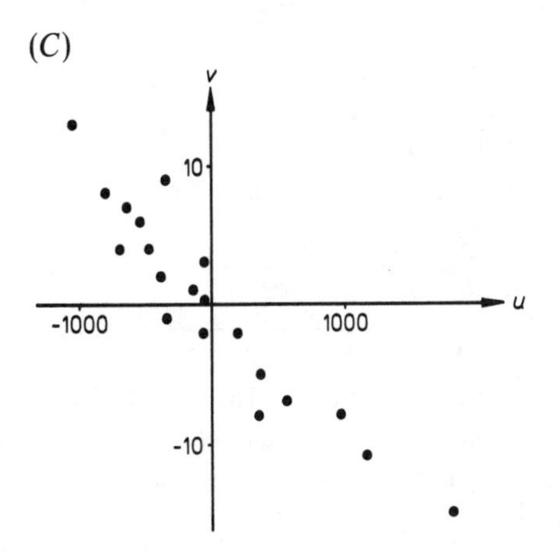

In illustration C there is a high negative correlation between the variables and nearly all the points in the u, v scatter diagram lie in the 2nd and 4th quadrants. It follows that most of the values of the product uv will be negative and thus we find that $\sum uv$ takes the negative value $-92\,304$.

Considering the results obtained in these examples we see that the value of $\sum uv$, i.e. $\sum (x - \bar{x})(y - \bar{y})$, is influenced by three main factors:

(i) the strength of the linear relationship or linear correlation between the variables,
(ii) the number of pairs of observations in the sample,
(iii) the spread or variation in the values of each variable (indicated here by the scales used on the u- and v-axes.)

We conclude that the sum $\sum (x - \bar{x})(y - \bar{y})$, often denoted by S_{xy}, can be used as a measure of the linear correlation between x and y only when comparing samples of equal size from similar bivariate distributions. A more useful measure is the *covariance* of x and y, which is defined as the mean value of the product $(x - \bar{x})(y - \bar{y})$. More formally, if $(x_1, y_1), (x_2, y_2), \ldots, (x_n, y_n)$ are the observed values of variables X and Y in a sample of size n, then:

$$\text{sample covariance} = \frac{1}{n} S_{xy} = \frac{1}{n} \sum (x - \bar{x})(y - \bar{y})$$

We now have a measure of correlation that can be used to compare samples of different sizes.

	S_{xy}	Covariance (to 3 s.f.)
Illustration A	1413	141
Illustration B	$-0{\cdot}28$	$-0{\cdot}0255$
Illustration C	$-92\,304$	-4620

However, further consideration of Illustrations A, B and C shows that we still need to allow for differences in "spread". For instance, we must remember that in B the values of $x - \bar{x}$ range from $-0{\cdot}14$ to $+0{\cdot}14$, whereas in C we have $-1033 \leqslant x - \bar{x} \leqslant +1807$.

We can solve this problem by dividing the values of $x - \bar{x}$ by the standard deviation of x and dividing the values of $y - \bar{y}$ by the standard deviation of y. The covariance of these standardised variables is a measure of linear correlation independent of both sample size and spread. This new measure, the *correlation coefficient* r, is defined as follows:

$$r = \frac{1}{n} \sum \left(\frac{x - \bar{x}}{\text{s.d. of } x} \right) \left(\frac{y - \bar{y}}{\text{s.d. of } y} \right)$$

i.e.

$$\text{correlation coefficient } r = \frac{\text{covariance of } x \text{ and } y}{\sqrt{(\text{variance of } x)}\sqrt{(\text{variance of } y)}}$$

(To avoid confusion with other measures of correlation r may also be called the *product moment correlation coefficient* or the *coefficient of linear correlation.*)

We obtain a simpler formula for r by extending the notation introduced to deal with covariance.

$$\text{Covariance} = \frac{1}{n}S_{xy} = \frac{1}{n}\sum(x-\bar{x})(y-\bar{y})$$

$$\text{Variance of } x = \frac{1}{n}S_{xx} = \frac{1}{n}\sum(x-\bar{x})^2$$

$$\text{Variance of } y = \frac{1}{n}S_{yy} = \frac{1}{n}\sum(y-\bar{y})^2$$

$$\therefore \qquad r = \frac{\frac{1}{n}S_{xy}}{\sqrt{\left(\frac{1}{n}S_{xx}\right)}\sqrt{\left(\frac{1}{n}S_{yy}\right)}} = \frac{S_{xy}}{\sqrt{(S_{xx}S_{yy})}}$$

We are already familiar with the results:

$$S_{xx} = \sum(x-\bar{x})^2 = \sum x^2 - n\bar{x}^2, \quad S_{yy} = \sum(y-\bar{y})^2 = \sum y^2 - n\bar{y}^2.$$

It can similarly be shown that:

$$S_{xy} = \sum(x-\bar{x})(y-\bar{y}) = \sum xy - n\bar{x}\bar{y}$$

$$\therefore \qquad r = \frac{S_{xy}}{\sqrt{(S_{xx}S_{yy})}} = \frac{\sum xy - n\bar{x}\bar{y}}{\sqrt{(\sum x^2 - n\bar{x}^2)}\sqrt{(\sum y^2 - n\bar{y}^2)}}$$

For Illustration A the calculation of r using this formula may be set out as follows:

x	y	xy	x^2	y^2
69	52	3588	4761	2704
80	71	5680	6400	5041
57	55	3135	3249	3025
52	48	2496	2704	2304
85	89	7565	7225	7921
66	64	4224	4356	4096
49	37	1813	2401	1369
76	60	4560	5776	3600
63	56	3528	3969	3136
73	78	5694	5329	6084
670	610	42 283	46 170	39 280

$$n = 10, \quad \bar{x} = \frac{1}{n}\sum x = \tfrac{670}{10} = 67, \quad \bar{y} = \frac{1}{n}\sum y = \tfrac{610}{10} = 61$$

$$\therefore \qquad S_{xy} = \sum xy - n\bar{x}\bar{y} = 42\,283 - 10 \times 67 \times 61 = 1413$$

$$S_{xx} = \sum x^2 - n\bar{x}^2 = 46\,170 - 10 \times 67^2 = 1280$$

$$S_{yy} = \sum y^2 - n\bar{y}^2 = 39\,280 - 10 \times 61^2 = 2070$$

Hence $r = \dfrac{S_{xy}}{\sqrt{(S_{xx}S_{yy})}} = \dfrac{1413}{\sqrt{(1280 \times 2070)}} \approx 0{\cdot}868.$

For Illustrations B and C the approximate values of r turn out to be $-0{\cdot}022$ and $-0{\cdot}939$ respectively. These results tend to confirm that the correlation coefficient is a satisfactory measure of the linear relationship between two variables.

It can be shown that r always lies between -1 and $+1$. (Methods of proof are suggested in Exercise 17.2, questions 12 and 13.)

$r = +1$ means perfect positive correlation.
$r = 0$ means that there is no linear correlation.
$r = -1$ means perfect negative correlation.

It is difficult to lay down more general rules for the interpretation of r as the significance of any particular value depends on the size and nature of the sample taken.

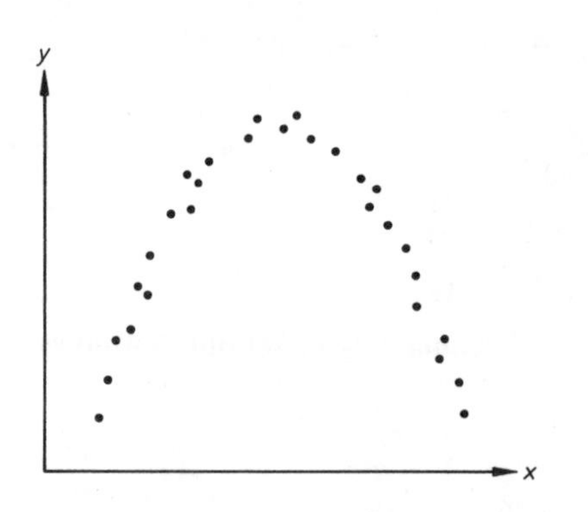

However, it must be emphasised that r is a measure of *linear* correlation. The given scatter diagram shows that it is possible to find a distribution in which there is a strong relationship between the variables, but a value of r close to zero.

We must also point out that a value of r close to -1 or to $+1$ does not necessarily imply a *causal* relationship between the variables. For instance, suppose that over a period of several years we had selected random samples of 10 year old children and recorded their consumption of frozen food and their levels of tooth decay. We would probably have found a negative correlation between these variables but it would clearly be wrong to assume that eating frozen food prevents tooth decay. Even if the variables measured had been use of "fluoride" toothpaste and levels of tooth decay, it would have been unwise to assume a connection without further evidence.

When working with *bivariate frequency distributions* a modified version of the formula for the correlation coefficient can be used. If a typical pair of observations (x, y) occurs with frequency f then:

$$S_{xx} = \sum fx^2 - n\bar{x}^2, \quad S_{yy} = \sum fy^2 - n\bar{y}^2 \quad \text{and}$$
$$S_{xy} = \sum fxy - n\bar{x}\bar{y}, \quad \text{where} \quad \sum f = n,$$

$$\therefore \quad \boxed{r = \frac{S_{xy}}{\sqrt{(S_{xx}S_{yy})}} = \frac{\sum fxy - n\bar{x}\bar{y}}{\sqrt{(\sum fx^2 - n\bar{x}^2)}\sqrt{(\sum fy^2 - n\bar{y}^2)}}}$$

Finally, we extend the work of Chapter 15 by considering briefly linear correlation between random variables. To set up a probability model of a bivariate experiment it is necessary to construct a joint probability distribution for two suitable random variables X and Y. As shown in §15.6:

$$\text{Var}(X) = E[\{X - E(X)\}^2] = E(X^2) - [E(X)]^2$$
$$\text{Var}(Y) = E[\{Y - E(Y)\}^2] = E(Y^2) - [E(Y)]^2$$

It therefore seems reasonable to define the covariance of X and Y as follows:

$$\text{Cov}(X, Y) = E[\{X - E(X)\}\{Y - E(Y)\}]$$

It can then be shown that:

$$\text{Cov}(X, Y) = E(XY) - E(X)E(Y)$$

We now define the correlation coefficient ρ of random variables X and Y by writing:

$$\boxed{\rho = \frac{\text{Cov}(X, Y)}{\sqrt{\{\text{Var}(X)\text{Var}(Y)\}}}}$$

It was further shown in §15.6 that if X and Y are independent random variables then $E(XY) = E(X)E(Y)$. Thus for independent variables $\rho = 0$. However, it is also possible for the value of ρ to be zero when X and Y are *not* independent.

Exercise 17.2

Questions 1 to 5. Find the correlation coefficients for the data given in Exercise 17.1 questions 1 to 5.

6. From 12 pairs of values of x and y the following were calculated:

$$\sum x = 112, \quad \sum y = 94, \quad \sum xy = 784, \quad \sum x^2 = 1116, \quad \sum y^2 = 1020.$$

Find (a) the covariance of x and y, (b) the correlation coefficient.

7.

		x 12	13	14	15	16
	4	0	0	0	1	1
	5	0	1	3	6	1
y	6	0	3	8	4	0
	7	0	5	2	0	0
	8	3	1	0	1	0

For the bivariate frequency distribution in the given table calculate (a) the mean and variance of each of x and y, (b) the covariance of x and y, (c) the correlation coefficient.

8. Claire Voyant claims that she possesses "extra-sensory perception". A small room is divided by a metal screen, with Claire seated on one side and an experimental assistant on the other. Each is given a set of cards numbered 1 to 5. At each trial of the experiment the assistant selects one of these cards and thinks of that number; Claire than records her "guess" by showing the appropriate card.

		Claire's "guess" 1	2	3	4	5
Assistant's choice	1	11	4	0	0	0
	2	1	16	8	1	0
	3	0	2	18	6	1
	4	0	0	2	13	5
	5	0	0	0	4	8

100 trials are completed and the results are given in the form of a bivariate table. Calculate the product-moment correlation coefficient for these data. (AEB 1978)

9. A set of pairs of values of variables x and y satisfy the linear relationship $y = mx + c$. Show that (a) if $m > 0$ then $r = +1$, (b) if $m < 0$ then $r = -1$.

10. A sample is drawn from a certain population and three characteristics with values represented by x, y and z are measured. The three coefficients of linear correlation between these variables are $r(x, y)$, $r(y, z)$ and $r(x, z)$. What values are possible for $r(x, z)$ given that (a) $r(x, y) = r(y, z) = +1$, (b) $r(x, y) = r(y, z) = -1$, (c) $r(x, y) = r(y, z) = 0$?

11. On a certain island there are large numbers of each of two clans, the Fatties and the Thinnies. Two random samples of 50 adult males are taken, one sample from each clan. Each of the clansmen is weighed and measured. For each clan, the value of the correlation coefficient between the heights and weights of the clansmen is found to be near $+1$. However, for the combined sample of 100 adult males, the value of the correlation coefficient is found to be near -1. Show, by a sketch of the scatter diagram for the combined sample, how this could arise. (C)

12. Using the definitions of u, v, S_{xx} and S_{yy} given in the text, let $U = u/\sqrt{S_{xx}}$ and $V = v/\sqrt{S_{yy}}$. Show that $\sum(U + V)^2 = 2n(1 + r)$ and hence prove that $r \geqslant -1$. By considering both $\sum(U + V)^2$ and $\sum(U - V)^2$, prove that $-1 \leqslant r \leqslant 1$.

13. Express $\sum(\lambda u + v)^2$ as a quadratic function of λ. Using the fact that this expression is never negative, prove that $(S_{xy})^2 \leqslant S_{xx}S_{yy}$. Deduce that $-1 \leqslant r \leqslant 1$.

Questions 14 to 16. Find the values of $\mathrm{Cov}(X, Y)$ and the correlation coefficient ρ for the joint probability distributions given in Exercise 15.6 questions 9, 10(i) and 10(ii).

17.3 Rank correlation

Let us suppose that two gardening experts, Mrs. Stone and Mr. Turvy are judging 8 entries in a prize cabbage competition at a village fair. They arrange the entries in order of merit and the positions or *ranks* allotted to the cabbages by each judge are given in the following table.

Cabbage	A	B	C	D	E	F	G	H
Mrs. Stone	3	7	4	5	8	1	2	6
Mr. Turvy	4	7	2	6	5	3	1	8

We can see that the rankings produced by the judges are broadly similar. However, to make a more precise statement about the level of agreement we need a numerical measure of *rank correlation.*

One such measure, *Spearman's*[†] *rank correlation coefficient* r_S, is simply the correlation coefficient introduced in the previous section calculated using the ranks assigned to sample members as data.

Mrs. Stone's rank, x	Mr. Turvy's rank, y	xy	x^2	y^2
3	4	12	9	16
7	7	49	49	49
4	2	8	16	4
5	6	30	25	36
8	5	40	64	25
1	3	3	1	9
2	1	2	4	1
6	8	48	36	64
36	36	192	204	204

Thus we have $n = 8$ and $\bar{x} = \bar{y} = 4{\cdot}5$

$$\therefore \qquad r_S = \frac{\sum xy - n\bar{x}\bar{y}}{\sqrt{(\sum x^2 - n\bar{x}^2)}\sqrt{(\sum y^2 - n\bar{y}^2)}} \approx 0{\cdot}714$$

Considering the values at the foot of the table we realise that for any ranking of 8 objects with *no tied ranks,* i.e. no equal placings, we will have $\sum x = 36, \sum x^2 = 204$. Similarly, for a ranking of n objects we can express $\sum x$ and $\sum x^2$ in terms of n and hence obtain a simpler formula for r_S.

Let us assume that ranks x and y take the integral values $1, 2, \ldots, n$ in some order. Using results obtained in §9.1 we have:

$$\sum x = \sum y = \tfrac{1}{2}n(n+1), \quad \sum x^2 = \sum y^2 = \tfrac{1}{6}n(n+1)(2n+1)$$

$$\therefore \qquad \bar{x} = \bar{y} = \tfrac{1}{2}(n+1)$$

If $d = x - y$, then $\sum d^2 = \sum (x-y)^2 = \sum x^2 + \sum y^2 - 2\sum xy$

$$\therefore \qquad \sum xy = \tfrac{1}{6}n(n+1)(2n+1) - \tfrac{1}{2}\sum d^2$$

[†] *Spearman, Charles Edward* (1863–1945) English psychologist. He was one of the first to apply correlation techniques to the analysis of data obtained from testing human abilities. He proposed his rank correlation method in 1906.

Hence
$$r_S = \frac{\frac{1}{6}n(n+1)(2n+1) - \frac{1}{2}\sum d^2 - \frac{1}{4}n(n+1)^2}{\frac{1}{6}n(n+1)(2n+1) - \frac{1}{4}n(n+1)^2}$$
$$= 1 - \frac{6\sum d^2}{n(n+1)[2(2n+1) - 3(n+1)]}$$
$$= 1 - \frac{6\sum d^2}{n(n+1)(n-1)}$$

For two rankings of n objects, with no tied ranks, Spearman's rank correlation coefficient is given by

$$r_S = 1 - \frac{6\sum d^2}{n(n^2-1)}$$

where d is the difference $x - y$ between ranks.

Applying this result to the cabbage rankings:

Cabbage	A	B	C	D	E	F	G	H
d	-1	0	2	-1	3	-2	1	-2

Thus $\sum d^2 = 1 + 4 + 1 + 9 + 4 + 1 + 4 = 24$

$$\therefore \qquad r_S = 1 - \frac{6 \times 24}{8(64-1)} = 1 - \tfrac{2}{7} \approx 0{\cdot}714$$

We now calculate r_S for Illustration A of §17.1, denoting the ranks of the x and y values by R_x and R_y respectively.

x	y	R_x	R_y	$d = R_x - R_y$	d^2
69	52	5	8	-3	9
80	71	2	3	-1	1
57	55	8	7	1	1
52	48	9	9	0	0
85	89	1	1	0	0
66	64	6	4	2	4
49	37	10	10	0	0
76	60	3	5	-2	4
63	56	7	6	1	1
73	78	4	2	2	4

$n = 10$ and $\sum d^2 = 24$,

$$\therefore \qquad r_S = 1 - \frac{6\sum d^2}{n(n^2-1)} = 1 - \frac{6 \times 24}{10 \times 99} \approx 0{\cdot}855$$

Strictly speaking, when there are *tied ranks*, the simplified formula for r_S is not valid and the original formula should be used. However, if the number of ties is small the "difference" formula can be used with little loss of accuracy provided that we adopt the usual ranking convention. For instance, if two objects are placed equal 3rd, both receive the rank 3·5 i.e. the mean of the ranks 3 and 4. We demonstrate the method using Illustration B.

x	y	R_x	R_y	$d = R_x - R_y$	d^2
1·72	3	9	11	−2	4
1·89	33	3	2	1	1
1·76	10	7·5	7	0·5	0·25
1·92	8	2	9	−7	49
1·66	28	11	3	8	64
1·94	16	1	5·5	−4·5	20·25
1·84	22	5	4	1	1
1·68	9	10	8	2	4
1·77	47	6	1	5	25
1·86	6	4	10	−6	36
1·76	16	7·5	5·5	2	4

$$n = 11 \text{ and } \sum d^2 = 208{\cdot}5,$$

$$\therefore \qquad r_S \approx 1 - \frac{6\sum d^2}{n(n^2-1)} = 1 - \frac{6 \times 208{\cdot}5}{11 \times 120} \approx 0{\cdot}052$$

[Note that the true correlation coefficient of R_x and R_y is 0·048 (to 3 d.p.).]

Kendall's[†] *rank correlation coefficient* r_K is a different measure of rank correlation. It is also designed to take values ranging from +1 for perfect agreement to −1 for total disagreement. To calculate r_K for two rankings of a sample, every pair of objects that can be selected from the sample is considered. For instance, in the case of the cabbage competition we must take into account the ranks given to AB, AC, AD, . . . , BC, BD, BE, . . . i.e. 28 pairs in all. Each pair is given a score of +1 if the judges have ranked the two cabbages in the same order, but −1 if the judges disagree. Kendall's coefficient is obtained by dividing the total score by the total number of pairs. Referring to the table of ranks we see that AB scores +1, AC scores −1 and so on. Clearly, determining all 28 scores one at a time would be rather tedious. A better method involves rearranging the table so that for the 1st judge the cabbages are in rank order.

Cabbage	Mrs. Stone	Mr. Turvy	Agreements	Disagreements
F	1	3	5	2
G	2	1	6	0
A	3	4	4	1
C	4	2	4	0
D	5	6	2	1
H	6	8	0	2
B	7	7	0	1
E	8	5		
			21	7

[†]*Kendall, Sir Maurice George* (1907–1983) English statistician. He proposed his rank correlation coefficient in 1938. *Rank Correlation Methods* was published in 1948.

The first entry in Mr. Turvy's column is 3. Below this 3 there are 5 ranks greater than 3, which correspond to 5 pairs FA, FD, FH, FB, FE, ranked in the same order by both judges. Similarly, below the 3 there are 2 ranks less than 3, which correspond to 2 disagreements. The second entry in Mr. Turvy's column is 1. Below the 1 there are 6 ranks greater than 1, but no ranks less than 1. Thus we have 6 further agreements, but no more disagreements. Working down the table in this way we eventually obtain the total numbers of agreements and disagreements.

Hence $$r_K = \frac{\text{no. of agreements} - \text{no. of disagreements}}{\text{total no. of pairs}} = \frac{21 - 7}{28} = 0{\cdot}5$$

For a set of n objects the total number of pairs is $\frac{1}{2}n(n-1)$. Thus the general formula for Kendall's rank correlation coefficient is

$$r_K = \frac{P - Q}{\frac{1}{2}n(n-1)}$$

where P is the number of agreements and Q is the number of disagreements.

This formula can also be used when there are tied ranks, by giving a pair tied in one or both rankings a score of zero. However, results obtained in this way must be treated with caution if the number of ties is large.

We can now calculate r_K for Illustration B using the ranks obtained earlier.

R_x	R_y	Agreements	Disagreements	Ties
1	5·5	5	4	1 (R_y col.)
2	9	2	7	0
3	2	7	1	0
4	10	1	6	0
5	4	4	2	0
6	1	5	0	0
7·5	7	2	1	1 (R_x col.)
7·5	5·5	2	1	0
9	11	0	2	0
10	8	0	1	0
11	3			
		28	25	

$$\therefore \quad r_K = \frac{P - Q}{\frac{1}{2}n(n-1)} = \frac{28 - 25}{\frac{1}{2} \times 11 \times 10} = \tfrac{3}{55} \approx 0{\cdot}055$$

Exercise 17.3

1. In a cookery competition three judges X, Y and Z place 6 sponge cakes in order of merit as follows.

	A	B	C	D	E	F
X	2	6	1	3	5	4
Y	1	5	4	2	3	6
Z	4	3	5	1	2	6

Calculate Spearman's rank correlation coefficient for the placings awarded by (a) judges X and Y, (b) judges Y and Z, (c) judges X and Z.

2. Calculate Kendall's rank correlation coefficient for the pairs of rankings given in question 1.

3. Three persons, P, Q and R, were asked to place in order of importance nine features of a house (A, B, C, ..., I). Calculate Spearman's rank correlation coefficients between the pairs of preferences as shown in the following table:

	A	B	C	D	E	F	G	H	I
P	1	2	4	8	9	7	6	3	5
Q	1	4	5	8	7	9	2	3	6
R	1	9	6	8	7	4	2	3	5

How far does this help to decide which pair from the three would be most likely to be able to compromise on a suitable house? (AEB 1978)

4. Calculate Kendall's rank correlation coefficient for the pairs of preferences given in question 3. Comment briefly on your results.

Questions 5 to 9. Find r_S and r_K for the data given in Exercise 17.1 questions 1 to 5.

10.

x	24	15	8	12	5	16	9
y	26	24	31	39	14	15	36

Use the given data to plot a scatter diagram. Find the product moment correlation coefficient r. Find also the rank correlation coefficients r_S and r_K. Comment briefly on your results.

11. Repeat question 10 for the following data.

x	38	7	14	21	19	4	11	24
y	4	27	11	6	7	43	13	5

12. Find r_S and r_K for the data given in Illustration C §17.1.

17.4 Regression lines

In §17.1 we found that the points in a scatter diagram representing a bivariate distribution sometimes tend to lie in a fairly narrow band along a straight line.

Such lines are called *regression*† *lines.* The equation of a regression line forms a linear approximation to any relationship which may exist between the variables.

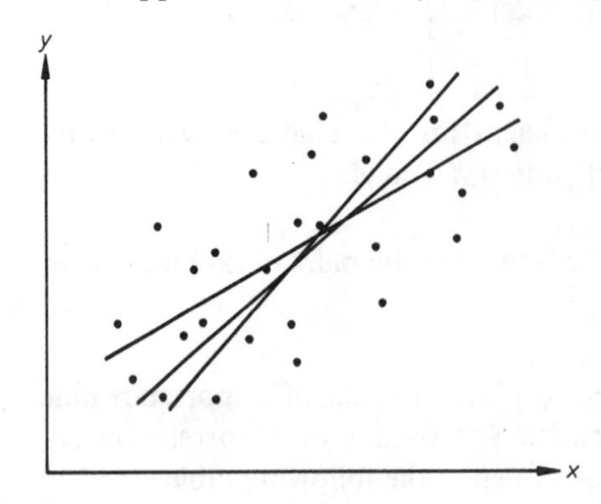

For the scatter diagram shown here it is difficult to decide which line best fits the data. All three lines illustrated might be suitable. We clearly need a numerical procedure for finding the equations of regression lines.

Let us consider a set of n pairs of observations, $(x_1, y_1), (x_2, y_2), \ldots, (x_n, y_n)$. We will assume that we need to find the regression line best suited to estimating unknown values of y for given values of x. This line is called the *regression line of y on x* and its equation will be denoted by $y = \alpha + \beta x$.

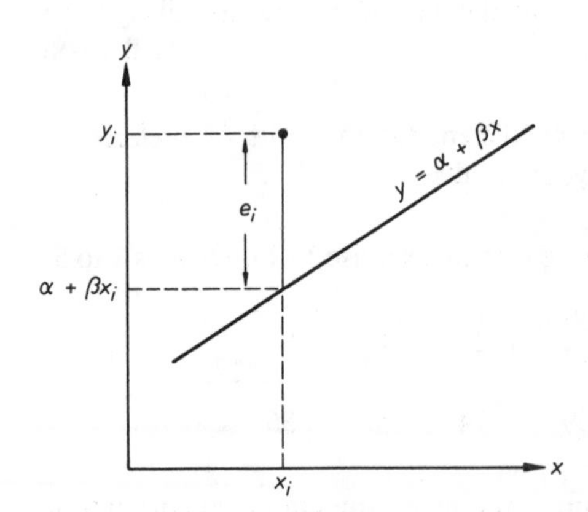

Using this line the estimated value of y corresponding to a typical observation x_i would be $\alpha + \beta x_i$. The deviation of the observed value y_i from this predicted value is called the *error of estimation* (or prediction) and is denoted by e_i. Thus we have:

$$y_i = \alpha + \beta x_i + e_i.$$

Our aim now is to obtain estimates for α and β such that the errors $e_1, e_2, \ldots, e_n$ (also called *residuals*) are as small as possible. The most satisfactory way of doing this is to choose the values which minimise the sum of the squares of the errors. This procedure is known as the *method of least squares.*

Let S be the sum of the squares of the (vertical) deviations of the points $(x_1, y_1), (x_2, y_2), \ldots, (x_n, y_n)$ from a variable line $y = a + bx$,
i.e. let $S = \sum\{y - (a + bx)\}^2$.

[Since S is a function of two variables, namely a and b, we cannot use elementary calculus to find its minimum value but must use an algebraic method. Readers with a knowledge of advanced calculus may prefer to use partial differentiation.]

† *Regression* Sir Francis Galton (1822–1911), English anthropologist, published *Natural Inheritance* in 1889. In this he called attention to the tendency of naturally inherited characteristics to revert to the mean. This tendency was later called regression.

We can simplify the algebra by writing:

$$y - (a + bx) = (y - \bar{y}) - b(x - \bar{x}) + (\bar{y} - a - b\bar{x}),$$

then letting $u = x - \bar{x}$, $v = y - \bar{y}$ and $c = \bar{y} - a - b\bar{x}$ to give:

$$\begin{aligned} S &= \sum (v - bu + c)^2 \\ &= \sum (v^2 + b^2u^2 + c^2 - 2buv - 2bcu + 2cv) \\ &= \sum v^2 + b^2 \sum u^2 + nc^2 - 2b\sum uv - 2bc\sum u + 2c\sum v. \end{aligned}$$

But $\sum u = \sum v = 0$ and in the notation of §17.2

$$\sum u^2 = S_{xx}, \quad \sum v^2 = S_{yy}, \quad \sum uv = S_{xy},$$

$$\therefore \qquad S = b^2 S_{xx} - 2bS_{xy} + S_{yy} + nc^2$$

"Completing the square" for the terms involving b:

$$S = S_{xx}\left(b - \frac{S_{xy}}{S_{xx}}\right)^2 + S_{yy} - \frac{(S_{xy})^2}{S_{xx}} + nc^2$$

Hence S takes its minimum value $S_{yy} - \dfrac{(S_{xy})^2}{S_{xx}}$ when $b - \dfrac{S_{xy}}{S_{xx}} = 0$ and $c = 0$

i.e. $\bar{y} - a - b\bar{x} = 0$.

$\therefore$ the *least squares* regression line of y on x has equation $y = a + bx$ where

$$b = \frac{S_{xy}}{S_{xx}} = \frac{\sum xy - n\bar{x}\bar{y}}{\sum x^2 - n\bar{x}^2} \quad \text{and} \quad a = \bar{y} - b\bar{x}.$$

We deduce that this regression line has gradient S_{xy}/S_{xx} and passes through the point $(\bar{x}, \bar{y})$. Its equation can also be expressed in the form $y - \bar{y} = b(x - \bar{x})$. The quantity b is referred to as the coefficient of regression of y on x or simply the *regression coefficient.*

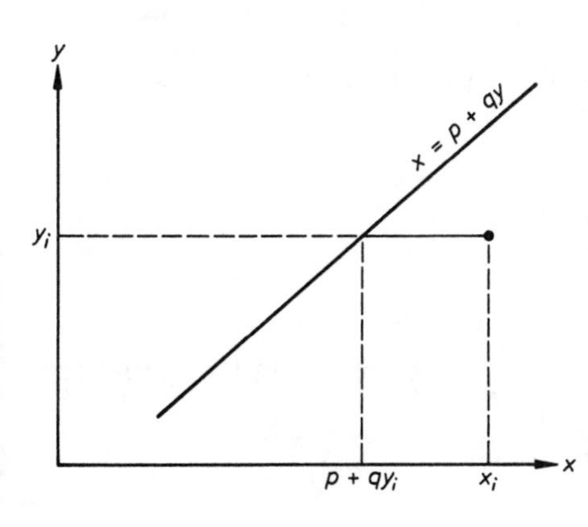

Provided that we are considering a bivariate experiment in which both x and y are allowed to vary freely, we can obtain a second least squares regression line called the *regression line of x on y.* This line is used to estimate values of x for given values of y. If we let its equation be $x = p + qy$, then the estimation errors take the form $x_i - (p + qy_i)$ and are represented graphically by horizontal deviations from the line. The values of p and q which minimise the sum of the squares of these errors are given by:

$$q = \frac{S_{xy}}{S_{yy}} \quad \text{and} \quad p = \bar{x} - q\bar{y}$$

Thus this regression line also passes through the point $(\bar{x}, \bar{y})$. In this case the gradient of the line is $1/q$ i.e. S_{yy}/S_{xy}.

Note that where less precision is required long calculations can be avoided by sketching regression lines on a scatter diagram "by eye". When doing this remember that

(i) *both* lines pass through the point $(\bar{x}, \bar{y})$,
(ii) the "y on x" line minimises *vertical* deviations,
(iii) the "x on y" line minimises *horizontal* deviations.

We now obtain the equations of both regression lines for the data given in Illustration A of §17.1. When in §17.2 we calculated the correlation coefficient r for this distribution we obtained the following results:

$$\bar{x} = 67, \quad \bar{y} = 61, \quad S_{xy} = 1413, \quad S_{xx} = 1280, \quad S_{yy} = 2070,$$

$$\therefore \qquad b = \tfrac{1413}{1280} \approx 1{\cdot}10, \quad a = 61 - (\tfrac{1413}{1280})67 \approx -12{\cdot}96.$$

Hence the regression line of y on x has equation

$$y = 1{\cdot}10x - 12{\cdot}96 \tag{1}$$

Similarly, $q = \frac{1413}{2070} \approx 0{\cdot}68$, $p = 67 - (\frac{1413}{2070})61 \approx 25{\cdot}36$.
Hence the regression line of x on y has equation

$$x = 0{\cdot}68y + 25{\cdot}36 \tag{2}$$

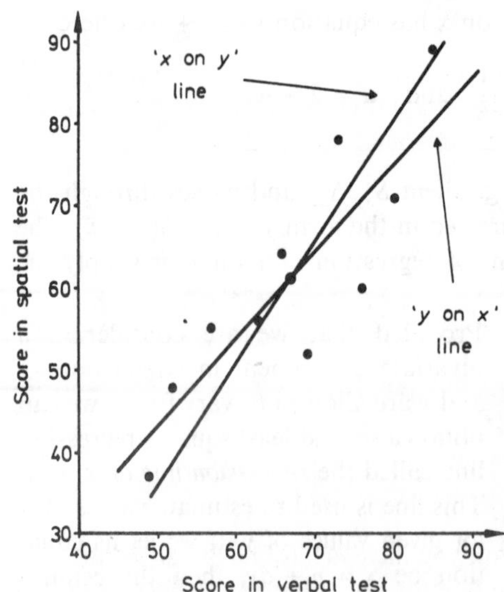

If a further student was given the verbal test and scored 61, we could predict his score in the spatial test using the "y on x" line.
From (1), if $x = 61$,

$$y = 1{\cdot}10 \times 61 - 12{\cdot}96 = 54{\cdot}14$$

Thus his predicted score in the spatial test is 54.

To estimate the verbal test score of a student who scored 74 in the spatial test we must use the "x on y" line.
From (2), if $y = 74$, $x = 0{\cdot}68 \times 74 + 25{\cdot}36 = 75{\cdot}68$.
Hence the predicted verbal test score is 76.

Finally we consider briefly the sampling distribution of the regression coefficient. For this purpose we will regard each value x_i as a given constant and the corresponding value y_i as an observed value of a random variable Y_i, such that $E(Y_i) = \alpha + \beta x_i$. It then follows that $\bar{x}$ is also a constant and that $E(\bar{Y}) = \alpha + \beta\bar{x}$. The random variable B representing the regression coefficient is now defined by writing:

$$B = \frac{\sum(x_i - \bar{x})(Y_i - \bar{Y})}{\sum(x_i - \bar{x})^2}$$

Using the rules of expectation algebra, (see §15.6),

$$\begin{aligned} E(B) &= \frac{\sum(x_i - \bar{x})E(Y_i - \bar{Y})}{\sum(x_i - \bar{x})^2} \\ &= \frac{\sum(x_i - \bar{x})\{(\alpha + \beta x_i) - (\alpha + \beta\bar{x})\}}{\sum(x_i - \bar{x})^2} \\ &= \frac{\sum\beta(x_i - \bar{x})^2}{\sum(x_i - \bar{x})^2} = \beta \end{aligned}$$

Thus the regression coefficient b is an unbiased estimate of the population parameter β.

If we also assume that each error $e_i = y_i - (\alpha + \beta x_i)$ is drawn at random from a population with variance σ^2, then $\text{Var}(Y_i) = \sigma^2$ for all values of i. It can then be shown (see Exercise 17.4 question 10) that

$$\text{Var}(B) = \sigma^2/S_{xx}, \quad \text{where} \quad S_{xx} = \sum(x_i - \bar{x})^2.$$

The quantity σ^2 is called the *residual variance* and when its value is known it is possible to obtain confidence limits for β by the methods of §16.3. We do not intend to discuss the estimation of σ^2, but give for reference the formula used.

$$\text{Estimated residual variance} = \frac{1}{n-2}\left\{S_{yy} - \frac{(S_{xy})^2}{S_{xx}}\right\} = \frac{S_{yy}(1 - r^2)}{n-2}.$$

Exercise 17.4

1. Calculate the equation of the regression line of y on x for the following distribution:

x	25	30	35	40	45	50
y	78	70	65	58	48	42

Is it possible to calculate from the equation you have just found (a) an estimate for the value of x when $y = 54$? (b) an estimate for the value of y when $x = 37$? In each case, if the answer is "Yes", calculate the estimate. If the answer is "No", say why not. (SU)

2. Fifteen boys took two exam papers in the same subject and the marks as percentages were as follows, where each boy's marks are in the same column.

Paper I (x)	65	73	42	52	84	60	70	79	60	83	57	77	54	66	89
Paper II (y)	78	88	60	73	92	77	84	89	70	99	73	88	70	85	89

Given that $\sum xy = 83\,734$, calculate the equation of the line of regression of y on x. Two boys were each absent for one paper. One scored 63 on Paper I, the other scored 81 on Paper II. In which case can you use your regression line to estimate the mark that the boy should be allocated for the paper he did not take, and what is that mark? (SU)

3. The 1962 and 1982 values, in pounds, of some "first editions" were as follows:

1962 value (x)	10	15	30	20	35
1982 value (y)	200	240	470	370	620

(i) Plot these values on a scatter diagram. (ii) Find the coordinates of a point through which the regression line of y on x must pass. (iii) Fit this regression line by eye. (iv) Find the equation of your line in the form $y = a + bx$, giving the constants a and b to two significant figures. Use the equation to estimate the 1982 value of a book valued at £25 in 1962.

Questions 4 to 8. Find the equations of the regression line of y on x and the regression line of x on y for the data given in Exercise 17.1 questions 1 to 5.

9. Using the usual notation, express in terms of S_{xx}, S_{yy} and r (a) the gradient of the regression line of y on x, (b) the gradient of the regression line of x on y. Use your results to prove that when $r = \pm 1$ the regression lines coincide. What form do the regression lines take when $r = 0$?

10. Using the definition of the random variable B given in the text, prove that

(a) $B = \dfrac{\sum (x_i - \bar{x}) Y_i}{S_{xx}}$, (b) $\text{Var}(B) = \dfrac{\text{residual variance}}{S_{xx}}$.

Exercise 17.5 (miscellaneous)

1. (a) Find the product moment correlation coefficient for the following distribution:

x	18	33	26	25	28	22	32	24
y	53	40	45	44	40	51	37	50

(b) Two judges in a baby competition rank the 8 babies as follow:

Baby	A	B	C	D	E	F	G	H
Judge 1	2	6	1	5	7	4	3	8
Judge 2	3	1	5	2	7	8	4	6

Calculate a rank correlation coefficient between the two judges. (SU)

2. (a) The marks of eight candidates in English and Mathematics are:

Candidate	1	2	3	4	5	6	7	8
English (x)	50	58	35	86	76	43	40	60
Mathematics (y)	65	72	54	82	32	74	40	53

Rank the results and hence find a rank correlation coefficient between the two sets of marks.

(b) Using the data in part (a), obtain the product-moment correlation coefficient. (SU)

3. (a) The following marks were awarded by two judges at a music competition:

	Child 1	Child 2	Child 3	Child 4	Child 5
Judge 1	10	5	8	7	9
Judge 2	9	6	10	5	8

Calculate a coefficient of rank correlation.

(b) Determine, by calculation, the equation of the regression line of x on y based on the following information about 8 children:

Child	1	2	3	4	5	6	7	8
Arithmetic mark (x)	45	33	27	23	18	14	8	0
English mark (y)	31	33	18	20	19	9	13	1

(SU)

4. 12 students were given a prognostic test at the beginning of a course and their scores X_i in the test were compared with their scores Y_i obtained in an examination at the end of the course ($i = 1, 2, \ldots, 12$). The results were as follows:

X_i	1	2	2	4	5	5	6	7	8	8	9	9
Y_i	3	4	5	5	4	8	6	6	6	7	8	10

Find the equation of the regression line of Y on X and determine the correlation coefficient between X and Y. (SU)

5. The heights h, in cm, and weights W, in kg, of 10 people are measured. It is found that $\sum h = 1710$, $\sum W = 760$, $\sum h^2 = 293\,162$, $\sum hW = 130\,628$ and $\sum W^2 = 59\,390$. Calculate the correlation coefficient between the values of h and W. What is the equation of the regression line of W on h? (O&C)

6. Explain briefly how the principle of least squares is used to find a regression line based on a sample of size n. Illustrate on a rough sketch the distances whose squares are minimised, taking care to distinguish the dependent and independent variates.

The table below shows the yields (y) obtained in an agricultural experiment in tonnes per hectare after using x tonnes of fertiliser per hectare.

Fertiliser (x)	0	0·2	0·4	0·6
Yield (y)	1·26	1·47	1·87	2·00

(a) Find the means of x and y. (b) Plot the measurements and their means on a graph of yield (y) against fertiliser (x). (c) Calculate the regression line of y on x in the form $y = a + bx$. (d) Draw the regression line you have calculated on the graph. (e) Estimate the total yield in tonnes from a 2·2 hectare field if 0·5 tonne of fertiliser is used altogether. (O&C)

7. Each of the variables x and y takes the values $1, 2, \ldots, n$ but not necessarily in the same order as each other. Prove that the covariance of x and y is

$$\text{cov}(x, y) = \frac{n^2 - 1}{12} - \frac{1}{2n}\sum_{i=1}^{n}(x_i - y_i)^2.$$

Hence show that Spearman's coefficient of rank correlation between x and y may be written as

$$1 - \frac{6}{n(n^2 - 1)}\sum_{i=1}^{n}(x_i - y_i)^2.$$

Seven army recruits (A, B, ..., G) were given two separate aptitude tests. Their orders of merit in each test were

Order of Merit	1st	2nd	3rd	4th	5th	6th	7th
1st test	G	F	A	D	B	C	E
2nd test	D	F	E	B	G	C	A

Find Spearman's coefficient of rank correlation between the two orders and comment briefly on the correlation obtained. (O&C)

8. The numbers of pupils, in a class of 20, gaining various scores in two tests are shown in the table below.

		Scores x in test A 1	2	3	4	Row totals
Score y in test B	1	1	—	—	—	1
	2	2	5	3	—	10
	3	—	2	4	2	8
	4	—	—	—	1	1
Column totals		3	7	7	3	20

(Thus, for example, 3 pupils gained a score of 3 in test A and a score of 2 in test B.) Find (a) the average score in each test, (b) the variances of x and y, (c) the covariance of x and y, (d) the equation of the regression line of y on x in the form $y = a + bx$. (e) If a new pupil scores 3·5 in test A, predict his score in test B. (O&C)

9. Explain the meaning of Spearman's rank correlation coefficient, giving particular attention to the cases when this coefficient takes the values (i) $+1$, (ii) -1.

The scores of ten batsmen in two successive innings were

Batsman	A	B	C	D	E	F	G	H	I	J
First innings	1	0	52	73	21	5	18	10	2	3
Second innings	36	0	5	21	46	32	11	2	6	1

Calculate Spearman's rank correlation coefficient. (AEB)

10. The body and heart masses of fourteen 10-month-old male mice are tabulated below:

Body mass (x) (grams)	27	30	37	38	32	36	32	32	38	42	36	44	33	38
Heart mass (y) (milligrams)	118	136	156	150	140	155	157	114	144	159	149	170	131	160

(i) Draw a scatter diagram of these data. (ii) Calculate the equation of the regression line of y on x and draw this line on the scatter diagram. (iii) Calculate the product-moment coefficient of correlation. (AEB 1977)

11. Explain clearly what is meant by the statistical term "correlation".

Vegboost Industries, a small chemical firm specializing in garden fertilizers, set up an experiment to study the relationship between a new fertilizer compound and the yield from tomato plants. Eight similar plants were selected and treated regularly throughout their life with x grams of fertilizer diluted in a standard volume of water. The yield y, in kilograms, of good tomatoes was measured for each plant. The following table summarises the results.

Plant	A	B	C	D	E	F	G	H
Amount of fertilizer x (g)	1·2	1·8	3·1	4·9	5·7	7·1	8·6	9·8
Yield y (kg)	4·5	5·9	7·0	7·8	7·2	6·8	4·5	2·7

(i) Calculate the product-moment correlation coefficient for these data. (ii) Calculate Spearman's rank correlation coefficient for these data. (iii) Is there any evidence of a relationship between these variables? Justify your answer. (No formal test is required.) (AEB 1978)

12. A scientist, working in an agricultural research station, believes there is a relationship between the hardness of the shells of eggs laid by chickens and the amount of a certain food supplement put into the diet of the chickens. He selects ten chickens of the same breed and collects the following data:

Chicken	A	B	C	D	E	F	G	H	I	J
Amount of food supplement x (g)	7·0	9·8	11·6	17·5	7·6	8·2	12·4	17·5	9·5	19·5
Hardness of shells y	1·2	2·1	3·4	6·1	1·3	1·7	3·4	6·2	2·1	7·1

(Hardness is measured on a 0 – 10 scale, 10 being the hardest. There are no units attached.)

(i) Calculate the equation of the regression line of y on x.
(ii) Calculate the product-moment correlation coefficient.
(iii) Do you believe that this linear model will continue to be appropriate no matter how large or small x becomes? Justify your reply. (AEB 1978)

13. The state of Tempora demands that every household in the country shall have a reliable clock; inspectors are being introduced throughout the country to

implement the policy. The Chief Inspector has the following data on the population size of towns, where Inspection Units have been set up, and the number of man-hours spent on inspection.

Population (thousands)	3	4	5	9	13	15	18	20	21	22
Man-hours (thousands)	8	11	13	18	24	26	31	32	34	33

(i) Calculate the regression line for predicting the number of man-hours from the population size (note that the mean value of each variate is a whole number). (ii) Predict the manpower required (in man-hours) for a new Inspection Unit to be installed in a town with a population of 17 000. (O)

14. In each of the following sections, which purport to be extracts from reports, the second sentence is an inference from the statement made in the first sentence. State whether the inferences are valid or invalid, and give the reasons for your decisions.

(a) The amount of fertilizer applied was varied from plot to plot and the correlation coefficient between the yield of corn from a plot and the amount of fertilizer applied was found to be 0·02. This shows that there was no relation between yield and amount of fertilizer applied

(b) Inspection of the police reports of car accidents in the town during 1975 revealed that, when the number of accidents involving drivers of a particular age was correlated with that age, the correlation coefficient was $-0{\cdot}72$. We conclude that older drivers are less likely to have an accident than younger drivers.

(c) The correlation coefficient between the sugar content s of the peas and the length of time t they have been in the greengrocer's shop was negative. It follows that the regression lines of s on t and of t on s have negative slopes.

(d) The correlation coefficient of percentage of children over sixteen at school and the Gross National Product, over the years for which we have records, is 0·91, which is a high correlation. It is obvious that if many more children can be persuaded to stay on at school after sixteen then the Gross National Product will increase substantially. (O)

15. A response variable y is known to be linearly related to the value of a variable x. To estimate the precise form of this relationship the response y is measured for each of six specified values of x, the results being as shown in the following table:

$x =$	10	20	30	40	50	60
$y =$	19	27	26	32	35	41

Whereas the specified values of x are known to be accurate, the observed values of y are subject to independent normally distributed errors with mean zero and standard deviation 1·5.

(i) Calculate the least squares estimate of the linear relationship connecting x and y.

(ii) Calculate 95 per cent confidence limits for the slope of the relationship. Hence, or otherwise, find 95 per cent confidence limits for the increase in the value of y when x is increased by 20 units. (W)

18 Matrices and determinants

18.1 Linear transformations in three dimensions

In this section we consider the effect of linear transformations on three-dimensional space. The effects of such transformations on two-dimensional space were examined in Chapter 19 of Book 1. We recall that a linear transformation is a mapping under which linear relationships between position vectors remain unchanged. Let us consider a linear transformation such that the images of the points $A(1, 0, 0)$, $B(0, 1, 0)$, $C(0, 0, 1)$ with position vectors **i**, **j**, **k** are the points A', B', C' with position vectors **a**, **b**, **c** respectively. Since the transformation preserves linear relationships, the image of a point $P(x, y, z)$ with position vector $x\mathbf{i} + y\mathbf{j} + z\mathbf{k}$ will be the point $P'(x', y', z')$ with position vector $x\mathbf{a} + y\mathbf{b} + z\mathbf{c}$. Thus, writing $\mathbf{a} = \begin{pmatrix} a_1 \\ a_2 \\ a_3 \end{pmatrix}$, $\mathbf{b} = \begin{pmatrix} b_1 \\ b_2 \\ b_3 \end{pmatrix}$, $\mathbf{c} = \begin{pmatrix} c_1 \\ c_2 \\ c_3 \end{pmatrix}$, we have:

$$\begin{pmatrix} x' \\ y' \\ z' \end{pmatrix} = x\begin{pmatrix} a_1 \\ a_2 \\ a_3 \end{pmatrix} + y\begin{pmatrix} b_1 \\ b_2 \\ b_3 \end{pmatrix} + z\begin{pmatrix} c_1 \\ c_2 \\ c_3 \end{pmatrix} = \begin{pmatrix} a_1x + b_1y + c_1z \\ a_2x + b_2y + c_2z \\ a_3x + b_3y + c_3z \end{pmatrix}$$

$$= \begin{pmatrix} a_1 & b_1 & c_1 \\ a_2 & b_2 & c_2 \\ a_3 & b_3 & c_3 \end{pmatrix}\begin{pmatrix} x \\ y \\ z \end{pmatrix}$$

$\therefore$ The matrix $\mathbf{M} = \begin{pmatrix} a_1 & b_1 & c_1 \\ a_2 & b_2 & c_2 \\ a_3 & b_3 & c_3 \end{pmatrix}$ represents the the linear transformation in which $\begin{pmatrix} 1 \\ 0 \\ 0 \end{pmatrix} \to \begin{pmatrix} a_1 \\ a_2 \\ a_3 \end{pmatrix}$, $\begin{pmatrix} 0 \\ 1 \\ 0 \end{pmatrix} \to \begin{pmatrix} b_1 \\ b_2 \\ b_3 \end{pmatrix}$, $\begin{pmatrix} 0 \\ 0 \\ 1 \end{pmatrix} \to \begin{pmatrix} c_1 \\ c_2 \\ c_3 \end{pmatrix}$.

To interpret this result geometrically we regard the origin O and the points A, B, C as vertices of a unit cube. In general, the transformation with matrix $\mathbf{M}$ maps this unit cube onto a parallelepiped with O, A', B', C' among its vertices.

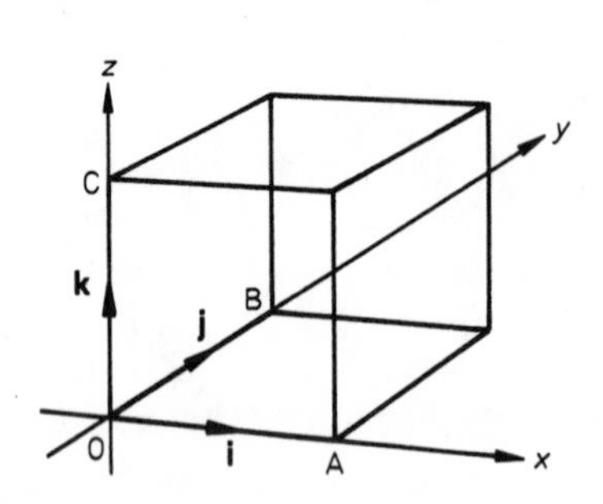

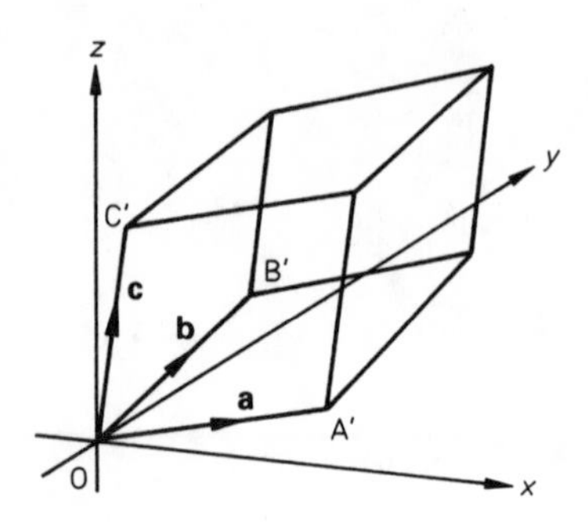

Example 1 Describe the transformations represented by the matrices

(a) $\begin{pmatrix} 1 & 0 & 0 \\ 0 & -1 & 0 \\ 0 & 0 & -1 \end{pmatrix}$ (b) $\begin{pmatrix} 0 & 0 & 1 \\ 1 & 0 & 0 \\ 0 & 1 & 0 \end{pmatrix}$.

(a)

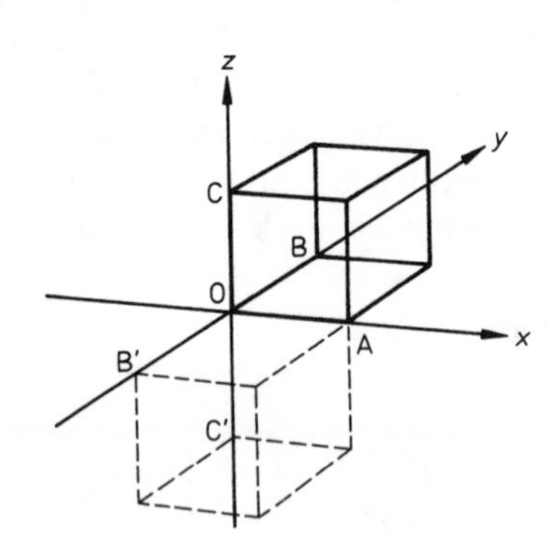

$$\begin{pmatrix} 1 \\ 0 \\ 0 \end{pmatrix} \to \begin{pmatrix} 1 \\ 0 \\ 0 \end{pmatrix}, \begin{pmatrix} 0 \\ 1 \\ 0 \end{pmatrix} \to \begin{pmatrix} 0 \\ -1 \\ 0 \end{pmatrix}, \begin{pmatrix} 0 \\ 0 \\ 1 \end{pmatrix} \to \begin{pmatrix} 0 \\ 0 \\ -1 \end{pmatrix}$$

$\therefore$ the images of the points $A(1,0,0)$, $B(0,1,0)$ and $C(0,0,1)$ are $A(1,0,0)$, $B'(0,-1,0)$ and $C'(0,0,-1)$ respectively.

The diagram shows that the matrix represents a rotation through 180° about the x-axis.

(b) $$\begin{pmatrix} 1 \\ 0 \\ 0 \end{pmatrix} \to \begin{pmatrix} 0 \\ 1 \\ 0 \end{pmatrix}, \begin{pmatrix} 0 \\ 1 \\ 0 \end{pmatrix} \to \begin{pmatrix} 0 \\ 0 \\ 1 \end{pmatrix}, \begin{pmatrix} 0 \\ 0 \\ 1 \end{pmatrix} \to \begin{pmatrix} 1 \\ 0 \\ 0 \end{pmatrix}$$

This means that the transformation maps A to B, B to C and C to A. Hence the matrix represents a rotation about an axis through the origin and the point $(1, 1, 1)$. Since three such rotations will return A, B and C to their original positions, the angle of rotation must be 120°.

We now consider some other basic transformations and their matrices.

(1) *Reflection in the y, z plane* (i.e. the plane $x = 0$).

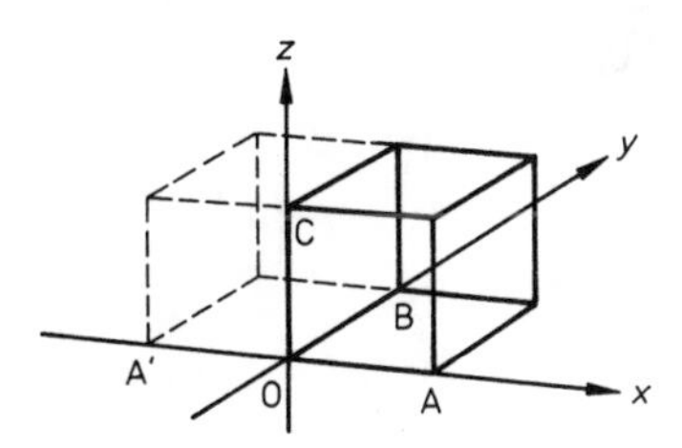

$$\begin{pmatrix}1\\0\\0\end{pmatrix} \to \begin{pmatrix}-1\\0\\0\end{pmatrix}, \quad \begin{pmatrix}0\\1\\0\end{pmatrix} \to \begin{pmatrix}0\\1\\0\end{pmatrix}, \quad \begin{pmatrix}0\\0\\1\end{pmatrix} \to \begin{pmatrix}0\\0\\1\end{pmatrix}$$

$\therefore$ the matrix of the transformation is $\begin{pmatrix}-1 & 0 & 0\\0 & 1 & 0\\0 & 0 & 1\end{pmatrix}$

(2) *Rotation* through 90° about the z-axis, so that the image of A is B.

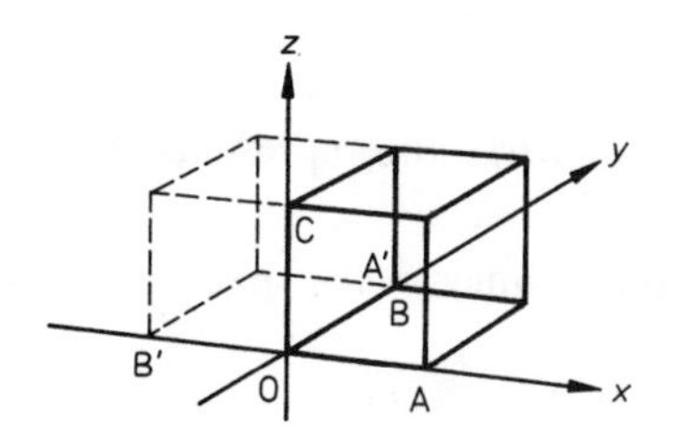

$$\begin{pmatrix}1\\0\\0\end{pmatrix} \to \begin{pmatrix}0\\1\\0\end{pmatrix}, \quad \begin{pmatrix}0\\1\\0\end{pmatrix} \to \begin{pmatrix}-1\\0\\0\end{pmatrix}, \quad \begin{pmatrix}0\\0\\1\end{pmatrix} \to \begin{pmatrix}0\\0\\1\end{pmatrix}$$

$\therefore$ the matrix of the transformation is $\begin{pmatrix}0 & -1 & 0\\1 & 0 & 0\\0 & 0 & 1\end{pmatrix}$

(3) *Enlargement* with scale factor k.
The matrix of the transformation is

$$\begin{pmatrix}k & 0 & 0\\0 & k & 0\\0 & 0 & k\end{pmatrix}$$

When $k = 1$ this matrix represents the identity transformation, which maps every point onto itself. If $k < 0$ the transformation reverses the direction of every position vector. This effect is sometimes called *central inversion.*

(4) *Orthogonal projection* on the plane $z = 0$.
Under this transformation the image of the point $P(x, y, z)$ is the point $P'(x, y, 0)$, which lies in the plane $z = 0$. This is called an orthogonal projection, because PP' is perpendicular (orthogonal) to the given plane.

$$\begin{pmatrix}1\\0\\0\end{pmatrix} \to \begin{pmatrix}1\\0\\0\end{pmatrix}, \quad \begin{pmatrix}0\\1\\0\end{pmatrix} \to \begin{pmatrix}0\\1\\0\end{pmatrix}, \quad \begin{pmatrix}0\\0\\1\end{pmatrix} \to \begin{pmatrix}0\\0\\0\end{pmatrix}$$

$\therefore$ the matrix of the transformation is $\begin{pmatrix}1 & 0 & 0\\0 & 1 & 0\\0 & 0 & 0\end{pmatrix}$

(5) *Shear* with plane $z = 0$ invariant, $(0, 0, 1) \to (1, 0, 1)$.

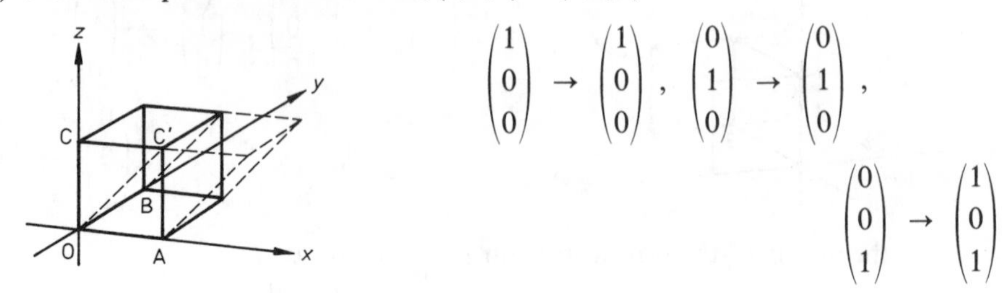

$$\begin{pmatrix}1\\0\\0\end{pmatrix} \to \begin{pmatrix}1\\0\\0\end{pmatrix}, \quad \begin{pmatrix}0\\1\\0\end{pmatrix} \to \begin{pmatrix}0\\1\\0\end{pmatrix}, \quad \begin{pmatrix}0\\0\\1\end{pmatrix} \to \begin{pmatrix}1\\0\\1\end{pmatrix}$$

$\therefore$ the matrix of the transformation is $\begin{pmatrix}1 & 0 & 1\\0 & 1 & 0\\0 & 0 & 1\end{pmatrix}$

The matrix $\mathbf{I} = \begin{pmatrix}1 & 0 & 0\\0 & 1 & 0\\0 & 0 & 1\end{pmatrix}$ is the 3×3 *identity matrix* and represents the identity transformation under which every point of three-dimensional space is mapped onto itself.

The *inverse* of a 3×3 matrix $\mathbf{M}$ is the matrix $\mathbf{M}^{-1}$ which has the property that

$$\mathbf{M}\mathbf{M}^{-1} = \mathbf{M}^{-1}\mathbf{M} = \mathbf{I}.$$

If it is impossible to find such a matrix, $\mathbf{M}$ is said to be *singular*. A matrix $\mathbf{M}$ which has an inverse is *non-singular*.

If the linear transformation under which the point $P(x, y, z)$ has image $P'(x', y', z')$ is represented by the matrix $\mathbf{M}$, then the inverse transformation, if it exists, maps P' to P and has matrix $\mathbf{M}^{-1}$. General methods for finding the inverses of 3×3 matrices will be discussed later in the chapter. However, in some cases the inverse of a matrix $\mathbf{M}$ can be found by considering the inverse of the transformation represented by $\mathbf{M}$.

We illustrate the method using the matrices

$$\mathbf{A} = \begin{pmatrix}-1 & 0 & 0\\0 & 1 & 0\\0 & 0 & 1\end{pmatrix}, \quad \mathbf{B} = \begin{pmatrix}0 & -1 & 0\\1 & 0 & 0\\0 & 0 & 1\end{pmatrix}, \quad \mathbf{C} = \begin{pmatrix}2 & 0 & 0\\0 & 2 & 0\\0 & 0 & 2\end{pmatrix}.$$

As shown in (1), $\mathbf{A}$ represents a reflection in the y, z plane. However, any reflection is its own inverse

$$\therefore \qquad \mathbf{A}^{-1} = \mathbf{A} = \begin{pmatrix}-1 & 0 & 0\\0 & 1 & 0\\0 & 0 & 1\end{pmatrix}.$$

As shown in (2), $\mathbf{B}$ represents a rotation through 90° about the z-axis. The inverse of this transformation is an equal rotation about the z-axis in the opposite sense. Hence $\mathbf{B}^{-1}$ represents the transformation under which the images of the points (1, 0, 0), (0, 1, 0), (0, 0, 1) are (0, −1, 0), (1, 0, 0), (0, 0, 1) respectively.

$$\therefore \qquad \mathbf{B}^{-1} = \begin{pmatrix} 0 & 1 & 0 \\ -1 & 0 & 0 \\ 0 & 0 & 1 \end{pmatrix}$$

Matrix **C** represents an enlargement, scale factor 2. Thus $\mathbf{C}^{-1}$ must represent an enlargement, scale factor $\frac{1}{2}$.

$$\text{i.e.} \qquad \mathbf{C}^{-1} = \begin{pmatrix} \frac{1}{2} & 0 & 0 \\ 0 & \frac{1}{2} & 0 \\ 0 & 0 & \frac{1}{2} \end{pmatrix}.$$

As in the case of transformations of the x, y plane, if two linear transformations with matrices **P** and **Q** are performed in succession using first **P** then **Q**, the matrix of the composite transformation is **QP**.
Considering the matrices **A** and **B** defined earlier we have:

$$\begin{pmatrix} 1 \\ 0 \\ 0 \end{pmatrix} \overset{\mathbf{B}}{\rightarrow} \begin{pmatrix} 0 \\ 1 \\ 0 \end{pmatrix} \overset{\mathbf{A}}{\rightarrow} \begin{pmatrix} 0 \\ 1 \\ 0 \end{pmatrix}, \quad \begin{pmatrix} 0 \\ 1 \\ 0 \end{pmatrix} \overset{\mathbf{B}}{\rightarrow} \begin{pmatrix} -1 \\ 0 \\ 0 \end{pmatrix} \overset{\mathbf{A}}{\rightarrow} \begin{pmatrix} 1 \\ 0 \\ 0 \end{pmatrix}, \quad \begin{pmatrix} 0 \\ 0 \\ 1 \end{pmatrix} \overset{\mathbf{B}}{\rightarrow} \begin{pmatrix} 0 \\ 0 \\ 1 \end{pmatrix} \overset{\mathbf{A}}{\rightarrow} \begin{pmatrix} 0 \\ 0 \\ 1 \end{pmatrix}$$

$$\text{and } \mathbf{AB} = \begin{pmatrix} -1 & 0 & 0 \\ 0 & 1 & 0 \\ 0 & 0 & 1 \end{pmatrix} \begin{pmatrix} 0 & -1 & 0 \\ 1 & 0 & 0 \\ 0 & 0 & 1 \end{pmatrix} = \begin{pmatrix} 0 & 1 & 0 \\ 1 & 0 & 0 \\ 0 & 0 & 1 \end{pmatrix}.$$

Thus the matrix **AB** has the same effect as **B** followed by **A** and represents a reflection in the plane $y = x$.

Certain linear transformations map every point in three-dimensional space onto a single straight line through the origin. There are other linear transformations, such as orthogonal projections, which map every point onto a single plane through the origin.

For instance under the transformation with matrix $\begin{pmatrix} 4 & -2 & 6 \\ -6 & 3 & -9 \\ 2 & -1 & 3 \end{pmatrix}$ the position vectors of the images of the points $A(1, 0, 0)$, $B(0, 1, 0)$, $C(0, 0, 1)$ are all multiples of the vector $\begin{pmatrix} 2 \\ -3 \\ 1 \end{pmatrix}$. Hence this transformation maps every point onto the straight line which passes through the origin and the point (2, −3, 1). The vector equation of this line is $\mathbf{r} = \lambda(2\mathbf{i} - 3\mathbf{j} + \mathbf{k})$.

Under the transformation with matrix $\begin{pmatrix} 2 & -1 & 3 \\ -1 & 3 & 1 \\ 1 & 2 & 4 \end{pmatrix}$ the position vectors of the images A', B', C' of the points A, B, C are

$$\mathbf{a} = \begin{pmatrix} 2 \\ -1 \\ 1 \end{pmatrix}, \quad \mathbf{b} = \begin{pmatrix} -1 \\ 3 \\ 2 \end{pmatrix}, \quad \mathbf{c} = \begin{pmatrix} 3 \\ 1 \\ 4 \end{pmatrix}$$

respectively. Since $\mathbf{c} = 2\mathbf{a} + \mathbf{b}$, C' must lie in the plane through the origin which contains A' and B'. Hence this transformation maps every point onto the plane through O and the points $A'(2, -1, 1)$ and $B'(-1, 3, 2)$. The vector equation of the plane is $\mathbf{r} = \lambda(2\mathbf{i} - \mathbf{j} + \mathbf{k}) + \mu(-\mathbf{i} + 3\mathbf{j} + 2\mathbf{k})$. The corresponding Cartesian equation is found to be $x + y - z = 0$.

Exercise 18.1

1. Find the images of the points $(1, 0, -2)$ and $(1, 1, 1)$ under the linear transformations with matrices

(a) $\begin{pmatrix} 1 & 0 & -1 \\ 0 & 2 & 0 \\ 0 & 1 & -1 \end{pmatrix}$ (b) $\begin{pmatrix} 0 & 1 & -1 \\ 1 & 0 & 1 \\ -1 & 1 & 0 \end{pmatrix}$ (c) $\begin{pmatrix} 1 & 3 & -2 \\ 0 & 0 & 5 \\ 0 & -4 & 0 \end{pmatrix}$

2. Write down the 3×3 matrices representing the following transformations:
(a) Reflection in the x, y plane,
(b) Enlargement, scale factor 5,
(c) Rotation through 180° about the y-axis,
(d) One way stretch parallel to the z-axis, $\times 2$,
(e) Rotations through 90° about the x-axis.

3. Describe geometrically the transformations represented by the following matrices:

(a) $\begin{pmatrix} 0 & 1 & 0 \\ -1 & 0 & 0 \\ 0 & 0 & 1 \end{pmatrix}$ (b) $\begin{pmatrix} -1 & 0 & 0 \\ 0 & -1 & 0 \\ 0 & 0 & -1 \end{pmatrix}$ (c) $\begin{pmatrix} 1 & 0 & 0 \\ 0 & 0 & 0 \\ 0 & 0 & 1 \end{pmatrix}$

(d) $\begin{pmatrix} 1 & 0 & 0 \\ 0 & -1 & 0 \\ 0 & 0 & 1 \end{pmatrix}$ (e) $\begin{pmatrix} 0 & 1 & 0 \\ 0 & 0 & 1 \\ 1 & 0 & 0 \end{pmatrix}$ (f) $\begin{pmatrix} 3 & 0 & 0 \\ 0 & 3 & 0 \\ 0 & 0 & 1 \end{pmatrix}$

4. Describe geometrically the transformations represented by

(a) $\begin{pmatrix} \cos\theta & -\sin\theta & 0 \\ \sin\theta & -\cos\theta & 0 \\ 0 & 0 & 1 \end{pmatrix}$ (b) $\begin{pmatrix} 1 & 0 & 0 \\ 0 & 0 & 1 \\ 0 & 1 & 0 \end{pmatrix}$ (c) $\begin{pmatrix} 1 & 0 & 0 \\ 0 & 1 & 0 \\ 1 & 0 & 1 \end{pmatrix}$

(d) $\begin{pmatrix} 0 & 0 & -1 \\ -1 & 0 & 0 \\ 0 & 1 & 0 \end{pmatrix}$ (e) $\begin{pmatrix} 0 & 0 & 1 \\ 0 & -1 & 0 \\ 1 & 0 & 0 \end{pmatrix}$ (f) $\begin{pmatrix} 1 & 1 & 0 \\ 1 & 1 & 0 \\ 0 & 0 & 1 \end{pmatrix}$

5. Write down the 3×3 matrices representing the following transformations:
(a) Reflection in the plane $x + y = 0$,
(b) Orthogonal projection on the plane $x = y$,

(c) Rotation through θ about the y-axis,
(d) Shear with x, y plane invariant, $(0,0,1) \to (1,1,1)$
(e) Rotation through 180° about the line $x + y = z = 0$ (i.e. the line through the origin and the point $(1,-1,0)$.)

6. If $\mathbf{M} = \begin{pmatrix} 0 & 0 & -1 \\ 0 & 1 & 0 \\ 1 & 0 & 0 \end{pmatrix}$, find $\mathbf{M}^2$ and $\mathbf{M}^4$. Describe geometrically the transformations represented by $\mathbf{M}$, $\mathbf{M}^2$ and $\mathbf{M}^4$.

7. If $\mathbf{P} = \begin{pmatrix} -1 & 0 & 0 \\ 0 & 1 & 0 \\ 0 & 0 & 1 \end{pmatrix}$, $\mathbf{Q} = \begin{pmatrix} 0 & 0 & 1 \\ 0 & 1 & 0 \\ -1 & 0 & 0 \end{pmatrix}$, find $\mathbf{QP}$ and $\mathbf{PQ}$.
Describe geometrically the transformations represented by $\mathbf{P}$, $\mathbf{Q}$, $\mathbf{QP}$ and $\mathbf{PQ}$. Verify that $\mathbf{QP}$ has the same effect as $\mathbf{P}$ followed by $\mathbf{Q}$ and that $\mathbf{PQ}$ is equivalent to $\mathbf{Q}$ then $\mathbf{P}$.

8. Find, where possible, the inverses of the matrices given in question 3. Verify for each matrix $\mathbf{A}$ which has an inverse $\mathbf{A}^{-1}$ that $\mathbf{AA}^{-1} = \mathbf{A}^{-1}\mathbf{A} = \mathbf{I}$, where $\mathbf{I}$ is the 3×3 identity matrix.

9. Find, where possible, the inverses of the matrices given in question 4.

10. Show that the linear transformation with matrix $\begin{pmatrix} 2 & -5 & 3 \\ -2 & 5 & -3 \\ 4 & -10 & 6 \end{pmatrix}$ maps every point of three-dimensional space onto a single line and write down a vector equation of this line. Describe the set of points mapped onto the origin under this transformation.

11. Show that the vectors $\begin{pmatrix} 1 \\ -3 \\ 4 \end{pmatrix}$, $\begin{pmatrix} 3 \\ 5 \\ -2 \end{pmatrix}$, $\begin{pmatrix} 14 \\ 0 \\ 14 \end{pmatrix}$ are coplanar. Deduce that the transformation with matrix $\begin{pmatrix} 1 & 3 & 14 \\ -3 & 5 & 0 \\ 4 & -2 & 14 \end{pmatrix}$ maps all the points of three-dimensional space onto a plane. Write down vector and Cartesian equations of this plane.

12. If $\mathbf{M} = \begin{pmatrix} 0 & 1 & -2 \\ 2 & 0 & -2 \\ 2 & -1 & 0 \end{pmatrix}$, find $\mathbf{M}^2$ and $\mathbf{M}^3$. Describe geometrically the transformations represented by $\mathbf{M}$, $\mathbf{M}^2$ and $\mathbf{M}^3$.

18.2 3 × 3 matrices and determinants

The basic properties of matrix addition and multiplication were established in Book 1. We now consider the application of this theory to 3 × 3 matrices.

Example 1 If $\mathbf{A} = \begin{pmatrix} 1 & 0 & -2 \\ 3 & 1 & -1 \\ 0 & 2 & 4 \end{pmatrix}$ and $\mathbf{B} = \begin{pmatrix} 2 & -1 & 3 \\ 0 & 0 & 1 \\ 4 & -2 & 1 \end{pmatrix}$,

find $3\mathbf{A} - 2\mathbf{B}$ and $\mathbf{A}^2\mathbf{B}$.

$$3\mathbf{A} - 2\mathbf{B} = 3\begin{pmatrix} 1 & 0 & -2 \\ 3 & 1 & -1 \\ 0 & 2 & 4 \end{pmatrix} - 2\begin{pmatrix} 2 & -1 & 3 \\ 0 & 0 & 1 \\ 4 & -2 & 1 \end{pmatrix} = \begin{pmatrix} -1 & 2 & -12 \\ 9 & 3 & -5 \\ -8 & 10 & 10 \end{pmatrix}$$

$$\mathbf{A}^2\mathbf{B} = \begin{pmatrix} 1 & 0 & -2 \\ 3 & 1 & -1 \\ 0 & 2 & 4 \end{pmatrix}\begin{pmatrix} 1 & 0 & -2 \\ 3 & 1 & -1 \\ 0 & 2 & 4 \end{pmatrix}\begin{pmatrix} 2 & -1 & 3 \\ 0 & 0 & 1 \\ 4 & -2 & 1 \end{pmatrix}$$

$$= \begin{pmatrix} 1 & -4 & -10 \\ 6 & -1 & 11 \\ 6 & 10 & 14 \end{pmatrix}\begin{pmatrix} 2 & -1 & 3 \\ 0 & 0 & 1 \\ 4 & -2 & 1 \end{pmatrix} = \begin{pmatrix} -38 & 19 & -11 \\ -32 & 16 & 6 \\ 68 & -34 & 42 \end{pmatrix}$$

Example 2 If $\mathbf{P} = \begin{pmatrix} 4 & 4 & -2 \\ -1 & 0 & 1 \\ 3 & 6 & -1 \end{pmatrix}$, show that $\mathbf{P}^2 = \mathbf{P} + k\mathbf{I}$ for some constant k, where $\mathbf{I}$ is the 3 × 3 unit matrix. Hence find the inverse matrix $\mathbf{P}^{-1}$.

$$\mathbf{P}^2 = \begin{pmatrix} 4 & 4 & -2 \\ -1 & 0 & 1 \\ 3 & 6 & -1 \end{pmatrix}\begin{pmatrix} 4 & 4 & -2 \\ -1 & 0 & 1 \\ 3 & 6 & -1 \end{pmatrix} = \begin{pmatrix} 6 & 4 & -2 \\ -1 & 2 & 1 \\ 3 & 6 & 1 \end{pmatrix}$$

$$= \begin{pmatrix} 4 & 4 & -2 \\ -1 & 0 & 1 \\ 3 & 6 & -1 \end{pmatrix} + \begin{pmatrix} 2 & 0 & 0 \\ 0 & 2 & 0 \\ 0 & 0 & 2 \end{pmatrix} = \mathbf{P} + 2\mathbf{I}$$

Rearranging: $\mathbf{P}^2 - \mathbf{P} = 2\mathbf{I}$

$\therefore$ $(\mathbf{P} - \mathbf{I})\mathbf{P} = 2\mathbf{I}$

$\therefore$ $\frac{1}{2}(\mathbf{P} - \mathbf{I}) \times \mathbf{P} = \mathbf{I}$

$$\text{Hence}\quad \mathbf{P}^{-1} = \tfrac{1}{2}(\mathbf{P} - \mathbf{I}) = \tfrac{1}{2}\left\{\begin{pmatrix} 4 & 4 & -2 \\ -1 & 0 & 1 \\ 3 & 6 & -1 \end{pmatrix} - \begin{pmatrix} 1 & 0 & 0 \\ 0 & 1 & 0 \\ 0 & 0 & 1 \end{pmatrix}\right\}$$

$$= \tfrac{1}{2}\begin{pmatrix} 3 & 4 & -2 \\ -1 & -1 & 1 \\ 3 & 6 & -2 \end{pmatrix} \text{ or } \begin{pmatrix} \frac{3}{2} & 2 & -1 \\ -\frac{1}{2} & -\frac{1}{2} & \frac{1}{2} \\ \frac{3}{2} & 3 & -1 \end{pmatrix}$$

Consider now two 3 × 3 matrices **A** and **B** with inverses $\mathbf{A}^{-1}$ and $\mathbf{B}^{-1}$ respectively.

$$(\mathbf{AB})(\mathbf{B}^{-1}\mathbf{A}^{-1}) = [\mathbf{A}(\mathbf{BB}^{-1})]\mathbf{A}^{-1} = (\mathbf{AI})\mathbf{A}^{-1} = \mathbf{AA}^{-1} = \mathbf{I}$$

i.e. $(\mathbf{AB})(\mathbf{B}^{-1}\mathbf{A}^{-1}) = \mathbf{I}$ and similarly $(\mathbf{B}^{-1}\mathbf{A}^{-1})(\mathbf{AB}) = \mathbf{I}$

Hence

$$\boxed{(\mathbf{AB})^{-1} = \mathbf{B}^{-1}\mathbf{A}^{-1}}$$

It can also be shown that if the transposes of **A** and **B** are $\mathbf{A}^T$ and $\mathbf{B}^T$ respectively, then

$$\boxed{(\mathbf{AB})^T = \mathbf{B}^T\mathbf{A}^T}$$

The *determinant* of a square matrix **A** is denoted by det **A**, $|\mathbf{A}|$ or simply Δ.

If
$$\mathbf{A} = \begin{pmatrix} a & b \\ c & d \end{pmatrix}, \quad \det \mathbf{A} = \begin{vmatrix} a & b \\ c & d \end{vmatrix} = ad - bc.$$

In Book 1 we saw that if **A** represents a linear transformation of the x, y plane, then under this transformation the image of a unit square is a parallelogram of area $|\det \mathbf{A}|$.

The determinant of a 3 × 3 matrix $\mathbf{A} = \begin{pmatrix} a_1 & b_1 & c_1 \\ a_2 & b_2 & c_2 \\ a_3 & b_3 & c_3 \end{pmatrix}$ is defined as

$$\boxed{\Delta = \begin{vmatrix} a_1 & b_1 & c_1 \\ a_2 & b_2 & c_2 \\ a_3 & b_3 & c_3 \end{vmatrix} = a_1\begin{vmatrix} b_2 & c_2 \\ b_3 & c_3 \end{vmatrix} - b_1\begin{vmatrix} a_2 & c_2 \\ a_3 & c_3 \end{vmatrix} + c_1\begin{vmatrix} a_2 & b_2 \\ a_3 & b_3 \end{vmatrix}}$$

It can be shown that under the linear transformation of three-dimensional space with matrix **A** the image of a unit cube is a parallelepiped of volume $|\det \mathbf{A}|$.

By considering the volume changes produced by the composite transformation with matrix **AB** we obtain the result:

$$\boxed{\det(\mathbf{AB}) = (\det \mathbf{A}).(\det \mathbf{B})}$$

Example 3 Evaluate $\begin{vmatrix} 1 & 2 & -1 \\ -1 & 3 & 0 \\ 2 & 1 & 2 \end{vmatrix}$

$$\begin{vmatrix} 1 & 2 & -1 \\ -1 & 3 & 0 \\ 2 & 1 & 2 \end{vmatrix} = 1\begin{vmatrix} 3 & 0 \\ 1 & 2 \end{vmatrix} - 2\begin{vmatrix} -1 & 0 \\ 2 & 2 \end{vmatrix} + (-1)\begin{vmatrix} -1 & 3 \\ 2 & 1 \end{vmatrix}$$

$$= 1 \times 6 - 2(-2) + (-1)(-7) = 17.$$

Let us consider now the effect produced when each element of a row or column of a determinant is multiplied by a constant.

$$\begin{vmatrix} ka & kb \\ c & d \end{vmatrix} = kad - kbc = k(ad - bc) = k\begin{vmatrix} a & b \\ c & d \end{vmatrix}$$

$$\begin{vmatrix} a_1 & kb_1 & c_1 \\ a_2 & kb_2 & c_2 \\ a_3 & kb_3 & c_3 \end{vmatrix} = a_1\begin{vmatrix} kb_2 & c_2 \\ kb_3 & c_3 \end{vmatrix} - kb_1\begin{vmatrix} a_2 & c_2 \\ a_3 & c_3 \end{vmatrix} + c_1\begin{vmatrix} a_2 & kb_2 \\ a_3 & kb_3 \end{vmatrix}$$

$$= ka_1\begin{vmatrix} b_2 & c_2 \\ b_3 & c_3 \end{vmatrix} - kb_1\begin{vmatrix} a_2 & c_2 \\ a_3 & c_3 \end{vmatrix} + kc_1\begin{vmatrix} a_2 & b_2 \\ a_3 & b_3 \end{vmatrix}$$

$$= k\begin{vmatrix} a_1 & b_1 & c_1 \\ a_2 & b_2 & c_2 \\ a_3 & b_3 & c_3 \end{vmatrix}$$

In general, if each element of a row or column of a determinant is multiplied by a constant k then the value of the determinant is multiplied by k.

Example 4 Evaluate $\begin{vmatrix} 12 & 6 & 10 \\ 2 & 3 & 15 \\ 6 & 0 & 5 \end{vmatrix}$

$$\begin{vmatrix} 12 & 6 & 10 \\ 2 & 3 & 15 \\ 6 & 0 & 5 \end{vmatrix} = 2\begin{vmatrix} 6 & 3 & 5 \\ 2 & 3 & 15 \\ 6 & 0 & 5 \end{vmatrix} = 2.2.3.5\begin{vmatrix} 3 & 1 & 1 \\ 1 & 1 & 3 \\ 3 & 0 & 1 \end{vmatrix}$$

$$= 60\left\{3\begin{vmatrix} 1 & 3 \\ 0 & 1 \end{vmatrix} - 1\begin{vmatrix} 1 & 3 \\ 3 & 1 \end{vmatrix} + 1\begin{vmatrix} 1 & 1 \\ 3 & 0 \end{vmatrix}\right\}$$

$$= 60(3 + 8 - 3) = 480.$$

In the expression $\Delta = \begin{vmatrix} a_1 & b_1 & c_1 \\ a_2 & b_2 & c_2 \\ a_3 & b_3 & c_3 \end{vmatrix} = a_1\begin{vmatrix} b_2 & c_2 \\ b_3 & c_3 \end{vmatrix} - b_1\begin{vmatrix} a_2 & c_2 \\ a_3 & c_3 \end{vmatrix} + c_1\begin{vmatrix} a_2 & b_2 \\ a_3 & b_3 \end{vmatrix}$ the determinants $\begin{vmatrix} b_2 & c_2 \\ b_3 & c_3 \end{vmatrix}$, $\begin{vmatrix} a_2 & c_2 \\ a_3 & c_3 \end{vmatrix}$ and $\begin{vmatrix} a_2 & b_2 \\ a_3 & b_3 \end{vmatrix}$ are called the *minors* of the elements a_1, b_1 and c_1 respectively.

$$\begin{vmatrix} a_1 & b_1 & c_1 \\ a_2 & b_2 & c_2 \\ a_3 & b_3 & c_3 \end{vmatrix}$$

In general the minor of an element of a determinant is obtained by deleting the row and column in which the element lies. For example the minor of c_2 is obtained by crossing out the second row and third column to give $\begin{vmatrix} a_1 & b_1 \\ a_3 & b_3 \end{vmatrix}$. If the

$$\begin{vmatrix} + & - & + \\ - & + & - \\ + & - & + \end{vmatrix}$$

minors are allotted signs alternately + and −, as shown in the diagram, the resulting quantities are the *cofactors* of the corresponding elements of the determinant. Thus the cofactor of c_2 is $-\begin{vmatrix} a_1 & b_1 \\ a_3 & b_3 \end{vmatrix}$

Denoting the cofactors of $a_1, b_1, \ldots, b_3, c_3$ by $A_1, B_1, \ldots, B_3, C_3$ respectively, the definition of Δ becomes

$$\Delta = \begin{vmatrix} a_1 & b_1 & c_1 \\ a_2 & b_2 & c_2 \\ a_3 & b_3 & c_3 \end{vmatrix} = a_1 A_1 + b_1 B_1 + c_1 C_1.$$

But $\Delta = a_1(b_2c_3 - c_2b_3) - b_1(a_2c_3 - c_2a_3) + c_1(a_2b_3 - b_2a_3)$

$$= a_1b_2c_3 - a_1b_3c_2 - a_2b_1c_3 + a_3b_1c_2 + a_2b_3c_1 - a_3b_2c_1$$

$$= -a_2b_1c_3 + a_3b_1c_2 + a_1b_2c_3 - a_3b_2c_1 - a_1b_3c_2 + a_2b_3c_1$$

$$= -b_1\begin{vmatrix} a_2 & c_2 \\ a_3 & c_3 \end{vmatrix} + b_2\begin{vmatrix} a_1 & c_1 \\ a_3 & c_3 \end{vmatrix} - b_3\begin{vmatrix} a_1 & c_1 \\ a_2 & c_2 \end{vmatrix}$$

$$= b_1B_1 + b_2B_2 + b_3B_3$$

Using this and other rearrangements of the original expression for Δ, it is found that Δ can be evaluated using the elements of any row or column together with the corresponding cofactors:

$$\begin{aligned} \Delta &= a_1A_1 + b_1B_1 + c_1C_1, & \Delta &= a_1A_1 + a_2A_2 + a_3A_3, \\ \Delta &= a_2A_2 + b_2B_2 + c_2C_2, & \Delta &= b_1B_1 + b_2B_2 + b_3B_3, \\ \Delta &= a_3A_3 + b_3B_3 + c_3C_3, & \Delta &= c_1C_1 + c_2C_2 + c_3C_3. \end{aligned}$$

These alternative expressions for Δ are particularly useful when one or more elements of the determinant are zero.

Example 5 Evaluate (a) $\begin{vmatrix} 2 & 4 & 3 \\ 0 & 5 & 4 \\ 0 & 2 & 3 \end{vmatrix}$ (b) $\begin{vmatrix} 2 & 4 & 5 \\ 0 & 1 & 0 \\ 3 & 9 & 7 \end{vmatrix}$

(a) Expanding down the first column

$$\begin{vmatrix} 2 & 4 & 3 \\ 0 & 5 & 4 \\ 0 & 2 & 3 \end{vmatrix} = 2\begin{vmatrix} 5 & 4 \\ 2 & 3 \end{vmatrix} - 0 + 0 = 2(15 - 8) = 14$$

(b) Expanding along the second row

$$\begin{vmatrix} 2 & 4 & 5 \\ 0 & 1 & 0 \\ 3 & 9 & 7 \end{vmatrix} = -0 + 1\begin{vmatrix} 2 & 5 \\ 3 & 7 \end{vmatrix} - 0 = 14 - 15 = -1$$

If two rows or two columns of a determinant are equal then the cofactors of the remaining row or column must all be zero. For instance, using the second column,

$$\begin{vmatrix} p & x & p \\ q & y & q \\ r & z & r \end{vmatrix} = -x\begin{vmatrix} q & q \\ r & r \end{vmatrix} + y\begin{vmatrix} p & p \\ r & r \end{vmatrix} - z\begin{vmatrix} p & p \\ q & q \end{vmatrix} = 0$$

Thus if two rows or columns of a determinant are equal then the value of the determinant is zero.

Example 6 Evaluate $\begin{vmatrix} -1 & 3 & 4 \\ 2 & -6 & -8 \\ 3 & 4 & 1 \end{vmatrix}$

$$\begin{vmatrix} -1 & 3 & 4 \\ 2 & -6 & -8 \\ 3 & 4 & 1 \end{vmatrix} = -2\begin{vmatrix} -1 & 3 & 4 \\ -1 & 3 & 4 \\ 3 & 4 & 1 \end{vmatrix} = 0$$

Exercise 18.2

1. If $\mathbf{P} = \begin{pmatrix} 2 & 1 & -4 \\ 0 & -3 & 1 \\ 5 & 0 & -1 \end{pmatrix}$ and $\mathbf{Q} = \begin{pmatrix} -1 & 0 & -1 \\ 2 & 0 & 3 \\ 0 & 1 & -2 \end{pmatrix}$, find (a) $\mathbf{P} - \mathbf{Q}$, (b) $2\mathbf{P} + 3\mathbf{Q}$, (c) $\mathbf{PQ}$, (d) $\mathbf{QP}$, (e) $\mathbf{Q}^2$.

2. If $\mathbf{A} = \begin{pmatrix} 1 & 3 & -9 \\ 5 & -1 & 3 \\ 1 & 3 & 7 \end{pmatrix}$ and $\mathbf{B} = \begin{pmatrix} 1 & 3 & 0 \\ 2 & -1 & 3 \\ -1 & 0 & 1 \end{pmatrix}$, find $\mathbf{AB}$ and hence write down $\mathbf{A}^{-1}$ and $\mathbf{B}^{-1}$.

3. If $\mathbf{M} = \begin{pmatrix} 0 & -1 & 0 \\ 0 & 0 & -1 \\ 1 & 0 & 0 \end{pmatrix}$, find $\mathbf{M}^2$ and $\mathbf{M}^3$. Hence find $\mathbf{M}^{-1}$.

4. If $\mathbf{A} = \begin{pmatrix} 0 & 2 & -1 \\ -1 & -3 & 1 \\ 1 & 2 & -2 \end{pmatrix}$, calculate $\mathbf{A}^2 + 4\mathbf{A}$ and hence find $\mathbf{A}^{-1}$.

5. If $\mathbf{A} = \begin{pmatrix} 1 & -2 & 0 \\ -3 & 0 & 2 \\ 2 & 1 & 3 \end{pmatrix}$ and $\mathbf{B} = \begin{pmatrix} 3 & -1 & 2 \\ 0 & 4 & -1 \\ -2 & 1 & 0 \end{pmatrix}$, write down $\mathbf{A}^T$ and $\mathbf{B}^T$. Verify that $(\mathbf{AB})^T = \mathbf{B}^T\mathbf{A}^T$ and that $(\mathbf{BA})^T = \mathbf{A}^T\mathbf{B}^T$.

6. If $\mathbf{P} = \begin{pmatrix} 1 & 2 & -2 \\ 2{\cdot}8 & -1 & 0{\cdot}4 \\ 0{\cdot}4 & 2 & 2{\cdot}2 \end{pmatrix}$, find $\mathbf{PP}^T$ and hence write down $\mathbf{P}^{-1}$ in terms of $\mathbf{P}^T$.

7. Evaluate the following determinants:

(a) $\begin{vmatrix} 2 & -1 & 1 \\ 0 & 3 & -2 \\ 1 & -1 & 0 \end{vmatrix}$ (b) $\begin{vmatrix} 0 & 2 & -1 \\ 1 & 0 & -3 \\ -2 & 1 & 0 \end{vmatrix}$ (c) $\begin{vmatrix} 1 & 2 & 0 \\ 4 & 0 & 1 \\ 3 & 1 & 2 \end{vmatrix}$

(d) $\begin{vmatrix} 4 & 10 & 7 \\ 7 & 9 & 8 \\ 5 & 9 & 7 \end{vmatrix}$ (e) $\begin{vmatrix} 8 & -2 & 7 \\ -2 & 2 & 7 \\ 5 & -3 & 6 \end{vmatrix}$ (f) $\begin{vmatrix} 6 & -8 & 11 \\ 2 & 1 & -2 \\ 3 & -6 & 7 \end{vmatrix}$

8. Evaluate the following determinants:

(a) $\begin{vmatrix} 2 & 3 & -1 \\ 5 & -4 & 3 \\ 2 & 3 & -1 \end{vmatrix}$ (b) $\begin{vmatrix} 12 & -8 & 10 \\ -9 & 0 & 2 \\ 15 & 1 & 0 \end{vmatrix}$ (c) $\begin{vmatrix} 2 & 1 & -4 \\ -1 & 0 & 2 \\ 1 & 3 & -2 \end{vmatrix}$

(d) $\begin{vmatrix} 6 & -2 & 0 \\ -3 & 2 & -7 \\ 30 & 5 & 35 \end{vmatrix}$ (e) $\begin{vmatrix} -21 & 0 & 14 \\ 24 & 40 & 0 \\ 0 & -15 & 18 \end{vmatrix}$ (f) $\begin{vmatrix} 9 & -3 & 6 \\ 15 & 7 & 20 \\ -6 & 2 & -4 \end{vmatrix}$

9. Evaluate the following determinants using the elements a, b, c and their corresponding cofactors.

(a) $\begin{vmatrix} a & 2 & 1 \\ b & 4 & 3 \\ c & 3 & 2 \end{vmatrix}$ (b) $\begin{vmatrix} -1 & 2 & -3 \\ a & b & c \\ 2 & -5 & 1 \end{vmatrix}$ (c) $\begin{vmatrix} 1 & 6 & a \\ 2 & 7 & b \\ 1 & 4 & c \end{vmatrix}$

10. Evaluate the following determinants

(a) $\begin{vmatrix} 0 & 4 & 1 \\ -1 & 2 & -3 \\ 0 & 3 & 2 \end{vmatrix}$ (b) $\begin{vmatrix} 2 & -4 & 6 \\ 5 & -3 & -7 \\ 0 & 0 & 2 \end{vmatrix}$ (c) $\begin{vmatrix} 4 & 0 & -1 \\ -5 & 0 & 2 \\ 1 & 3 & -5 \end{vmatrix}$

11. $\mathbf{A} = \begin{pmatrix} 2 & -1 & 0 \\ 0 & 3 & -2 \\ 6 & -3 & 1 \end{pmatrix}$, $\mathbf{B} = \begin{pmatrix} 0 & 4 & -2 \\ 0 & -1 & 1 \\ 3 & 2 & -1 \end{pmatrix}$, $\mathbf{C} = \begin{pmatrix} -1 & 0 & 3 \\ 0 & 2 & 0 \\ 2 & 4 & -5 \end{pmatrix}$

Find $\det \mathbf{A}$, $\det \mathbf{B}$ and $\det \mathbf{C}$. Verify that $\det(\mathbf{AB}) = (\det \mathbf{A})(\det \mathbf{B}) = \det(\mathbf{BA})$ and that $\det(\mathbf{ABC}) = (\det \mathbf{A})(\det \mathbf{B})(\det \mathbf{C})$.

18.3 Inverse matrices: adjoint method

We consider again the determinant $\Delta = \begin{vmatrix} a_1 & b_1 & c_1 \\ a_2 & b_2 & c_2 \\ a_3 & b_3 & c_3 \end{vmatrix}$. As we saw in the previous section, when the elements in any row or column are combined with the corresponding cofactors then the value of Δ is obtained e.g. $a_3A_3 + b_3B_3 + c_3C_3 = \Delta$, $c_1C_1 + c_2C_2 + c_3C_3 = \Delta$. However, if the elements of any row (or column) are combined with *alien cofactors*, i.e. with the cofactors of a different row (or column), the result is zero. For instance,

$$a_2A_1 + b_2B_1 + c_2C_1 = \begin{vmatrix} a_2 & b_2 & c_2 \\ a_2 & b_2 & c_2 \\ a_3 & b_3 & c_3 \end{vmatrix}$$

but since this determinant has two equal rows,

$$a_2A_1 + b_2B_1 + c_2C_1 = 0.$$

Using these results we find that

$$\begin{pmatrix} a_1 & b_1 & c_1 \\ a_2 & b_2 & c_2 \\ a_3 & b_3 & c_3 \end{pmatrix} \begin{pmatrix} A_1 & A_2 & A_3 \\ B_1 & B_2 & B_3 \\ C_1 & C_2 & C_3 \end{pmatrix} = \begin{pmatrix} A_1 & A_2 & A_3 \\ B_1 & B_2 & B_3 \\ C_1 & C_2 & C_3 \end{pmatrix} \begin{pmatrix} a_1 & b_1 & c_1 \\ a_2 & b_2 & c_2 \\ a_3 & b_3 & c_3 \end{pmatrix} = \begin{pmatrix} \Delta & 0 & 0 \\ 0 & \Delta & 0 \\ 0 & 0 & \Delta \end{pmatrix}$$

If $\mathbf{A} = \begin{pmatrix} a_1 & b_1 & c_1 \\ a_2 & b_2 & c_2 \\ a_3 & b_3 & c_3 \end{pmatrix}$ then the matrix $\begin{pmatrix} A_1 & A_2 & A_3 \\ B_1 & B_2 & B_3 \\ C_1 & C_2 & C_3 \end{pmatrix}$, obtained by replacing each element of **A** by its cofactor and then *transposing*, is called the *adjoint* (or adjugate) of **A**, written adj **A**. Thus we have shown that

$$\boxed{\mathbf{A}(\text{adj}\,\mathbf{A}) = (\text{adj}\,\mathbf{A})\mathbf{A} = \Delta\mathbf{I}, \quad \text{where} \quad \Delta = \det\mathbf{A}}$$

Example 1 If $\mathbf{A} = \begin{pmatrix} 1 & 2 & -1 \\ 2 & 1 & 1 \\ -2 & 0 & 2 \end{pmatrix}$, find adj **A** and verify that

$$\mathbf{A}(\text{adj}\,\mathbf{A}) = (\text{adj}\,\mathbf{A})\mathbf{A} = (\det\mathbf{A})\mathbf{I}.$$

Using the minors with their appropriate signs, the cofactors of the elements of **A** are found to be:

1st row: $+\begin{vmatrix} 1 & 1 \\ 0 & 2 \end{vmatrix} = 2, \quad -\begin{vmatrix} 2 & 1 \\ -1 & 2 \end{vmatrix} = -5, \quad +\begin{vmatrix} 2 & 1 \\ -1 & 0 \end{vmatrix} = 1,$

2nd row: $-\begin{vmatrix} 2 & -1 \\ 0 & 2 \end{vmatrix} = -4, \quad +\begin{vmatrix} 1 & -1 \\ -1 & 2 \end{vmatrix} = 1, \quad -\begin{vmatrix} 1 & 2 \\ -1 & 0 \end{vmatrix} = -2,$

3rd row: $+\begin{vmatrix} 2 & -1 \\ 1 & 1 \end{vmatrix} = 3, \quad -\begin{vmatrix} 1 & -1 \\ 2 & 1 \end{vmatrix} = -3, \quad +\begin{vmatrix} 1 & 2 \\ 2 & 1 \end{vmatrix} = -3$

$$\text{Hence adj }\mathbf{A} = \begin{pmatrix} 2 & -4 & 3 \\ -5 & 1 & -3 \\ 1 & -2 & -3 \end{pmatrix}$$

$$\mathbf{A}(\text{adj }\mathbf{A}) = \begin{pmatrix} 1 & 2 & -1 \\ 2 & 1 & 1 \\ -1 & 0 & 2 \end{pmatrix}\begin{pmatrix} 2 & -4 & 3 \\ -5 & 1 & -3 \\ 1 & -2 & -3 \end{pmatrix} = \begin{pmatrix} -9 & 0 & 0 \\ 0 & -9 & 0 \\ 0 & 0 & -9 \end{pmatrix} = -9\mathbf{I}$$

$$(\text{adj }\mathbf{A})\mathbf{A} = \begin{pmatrix} 2 & -4 & 3 \\ -5 & 1 & -3 \\ 1 & -2 & -3 \end{pmatrix}\begin{pmatrix} 1 & 2 & -1 \\ 2 & 1 & 1 \\ -1 & 0 & 2 \end{pmatrix} = \begin{pmatrix} -9 & 0 & 0 \\ 0 & -9 & 0 \\ 0 & 0 & -9 \end{pmatrix} = -9\mathbf{I}$$

Hence $\det \mathbf{A} = -9$ and $\mathbf{A}(\text{adj }\mathbf{A}) = (\text{adj }\mathbf{A})\mathbf{A} = (\det \mathbf{A})\mathbf{I}$.

Suppose now that $\mathbf{A}$ is a matrix with zero determinant, then for any matrix $\mathbf{B}$,

$$\det(\mathbf{AB}) = (\det \mathbf{A})(\det \mathbf{B}) = 0.$$

This means that it is impossible to find a matrix $\mathbf{B}$ such that $\mathbf{AB} = \mathbf{I}$.

$\therefore$ if $\det \mathbf{A} = 0$, then $\mathbf{A}$ is a singular matrix with no inverse and $\mathbf{A}(\text{adj }\mathbf{A}) = (\text{adj }\mathbf{A})\mathbf{A} = 0$.

However, if $\mathbf{A}$ is a non-singular matrix, then

$$\mathbf{A}(\text{adj }\mathbf{A}) = (\text{adj }\mathbf{A})\mathbf{A} = \Delta\mathbf{I}, \quad \text{where} \quad \Delta \neq 0$$

$$\therefore \qquad \mathbf{A}\left(\frac{1}{\Delta}\text{adj }\mathbf{A}\right) = \left(\frac{1}{\Delta}\text{adj }\mathbf{A}\right)\mathbf{A} = \mathbf{I}$$

Hence, provided that $\Delta \neq 0$, the inverse of $\mathbf{A}$ is the matrix $\frac{1}{\Delta}\text{adj }\mathbf{A}$.

i.e. if $\det \mathbf{A} \neq 0$, then $\mathbf{A}^{-1} = \dfrac{\text{adj }\mathbf{A}}{\det \mathbf{A}}$.

Example 2 Find the inverse of $\mathbf{A} = \begin{pmatrix} -1 & 3 & 2 \\ 2 & -6 & 0 \\ 1 & -2 & -4 \end{pmatrix}$

$$\text{adj }\mathbf{A} = \begin{pmatrix} +\ 24 & -(-8) & +\ 2 \\ -(-8) & +\ 2 & -(-1) \\ +\ 12 & -(-4) & +\ 0 \end{pmatrix}^T = \begin{pmatrix} 24 & 8 & 12 \\ 8 & 2 & 4 \\ 2 & 1 & 0 \end{pmatrix}$$

$$\det \mathbf{A} = (-1)24 + 3 \times 8 + 2 \times 2 = 4$$

$$\text{Hence } \mathbf{A}^{-1} = \frac{\text{adj }\mathbf{A}}{\det \mathbf{A}} = \frac{1}{4}\begin{pmatrix} 24 & 8 & 12 \\ 8 & 2 & 4 \\ 2 & 1 & 0 \end{pmatrix} = \begin{pmatrix} 6 & 2 & 3 \\ 2 & \frac{1}{2} & 1 \\ \frac{1}{2} & \frac{1}{4} & 0 \end{pmatrix}.$$

Exercise 18.3

In questions 1 to 12 find the adjoint and, where possible, the inverse of the given matrix.

1. $\begin{pmatrix} 2 & -1 & -2 \\ 0 & 2 & 3 \\ -3 & 0 & 1 \end{pmatrix}$ 2. $\begin{pmatrix} 0 & -1 & 2 \\ 3 & 1 & 0 \\ -4 & 1 & -5 \end{pmatrix}$ 3. $\begin{pmatrix} 3 & 0 & -2 \\ 1 & -2 & 4 \\ 5 & -4 & 6 \end{pmatrix}$

4. $\begin{pmatrix} 1 & 2 & 1 \\ 3 & 0 & 4 \\ 2 & 1 & 1 \end{pmatrix}$ 5. $\begin{pmatrix} -2 & -3 & 1 \\ 4 & 1 & -2 \\ 1 & -3 & -2 \end{pmatrix}$ 6. $\begin{pmatrix} 2 & -1 & 3 \\ 0 & -3 & 2 \\ -1 & 0 & -2 \end{pmatrix}$

7. $\begin{pmatrix} -6 & 9 & 3 \\ 10 & -15 & -5 \\ -8 & 12 & 4 \end{pmatrix}$ 8. $\begin{pmatrix} 2 & 1 & 2 \\ 3 & 1 & 2 \\ 1 & 6 & 3 \end{pmatrix}$ 9. $\begin{pmatrix} 2 & -1 & 0 \\ -1 & 3 & 1 \\ 2 & 2 & -3 \end{pmatrix}$

10. $\begin{pmatrix} 1 & -2 & 3 \\ 1 & 0 & 1 \\ 1 & 2 & 3 \end{pmatrix}$ 11. $\begin{pmatrix} 5 & -2 & 3 \\ 1 & 1 & -5 \\ -10 & 4 & -6 \end{pmatrix}$ 12. $\begin{pmatrix} 1 & -6 & 2 \\ -2 & 5 & 0 \\ 4 & -1 & -5 \end{pmatrix}$

13. If $\mathbf{A} = \begin{pmatrix} 0 & -1 & 1 \\ -1 & 0 & 1 \\ 0 & 0 & 1 \end{pmatrix}$ and $\mathbf{B} = \begin{pmatrix} 1 & 0 & 3 \\ 7 & 5 & 0 \\ 3 & 2 & 1 \end{pmatrix}$, find $\mathbf{A}^{-1}$ and $\mathbf{B}^{-1}$.
Verify that (a) $(\mathbf{A}^T)^{-1} = (\mathbf{A}^{-1})^T$ and $(\mathbf{B}^T)^{-1} = (\mathbf{B}^{-1})^T$, (b) $\det(\mathbf{AB}) = (\det \mathbf{A})(\det \mathbf{B})$, (c) $(\mathbf{AB})^{-1} = \mathbf{B}^{-1}\mathbf{A}^{-1}$ and $(\mathbf{BA})^{-1} = \mathbf{A}^{-1}\mathbf{B}^{-1}$.

[For further practice in finding inverses see Exercise 18.5.]

18.4 Sets of linear equations

In this section we consider sets of simultaneous linear equations of the form

$$\begin{aligned} a_1x + b_1y + c_1z &= d_1 \\ a_2x + b_2y + c_2z &= d_2 \\ a_3x + b_3y + c_3z &= d_3 \end{aligned}$$

Writing $\mathbf{A} = \begin{pmatrix} a_1 & b_1 & c_1 \\ a_2 & b_2 & c_2 \\ a_3 & b_3 & c_3 \end{pmatrix}$, $\mathbf{r} = \begin{pmatrix} x \\ y \\ z \end{pmatrix}$ and $\mathbf{d} = \begin{pmatrix} d_1 \\ d_2 \\ d_3 \end{pmatrix}$, we obtain a single matrix equation:

$$\mathbf{Ar} = \mathbf{d}$$

If $\det \mathbf{A} \neq 0$, the matrix $\mathbf{A}$ has an inverse $\mathbf{A}^{-1}$,

$$\therefore \quad \begin{aligned} \mathbf{A}^{-1}(\mathbf{A}\mathbf{r}) &= \mathbf{A}^{-1}\mathbf{d} \\ (\mathbf{A}^{-1}\mathbf{A})\mathbf{r} &= \mathbf{A}^{-1}\mathbf{d} \\ \mathbf{I}\mathbf{r} &= \mathbf{A}^{-1}\mathbf{d} \end{aligned}$$

Thus

$$\mathbf{r} = \mathbf{A}^{-1}\mathbf{d}$$

This result suggests one way of solving simultaneous linear equations.

Example 1 Solve the equations

$$\begin{aligned} x + 3y - 2z &= 1 \\ 2x \qquad + z &= 5 \\ - y + 2z &= 4 \end{aligned}$$

In matrix form these equations become:

$$\mathbf{A}\begin{pmatrix} x \\ y \\ z \end{pmatrix} = \begin{pmatrix} 1 \\ 5 \\ 4 \end{pmatrix}, \quad \text{where} \quad \mathbf{A} = \begin{pmatrix} 1 & 3 & -2 \\ 2 & 0 & 1 \\ 0 & -1 & 2 \end{pmatrix}$$

$$\operatorname{adj}\mathbf{A} = \begin{pmatrix} +1 & -4 & +(-2) \\ -4 & +2 & -(-1) \\ +3 & -5 & +(-6) \end{pmatrix}^T = \begin{pmatrix} 1 & -4 & 3 \\ -4 & 2 & -5 \\ -2 & 1 & -6 \end{pmatrix}$$

$$\det \mathbf{A} = 1 \times 1 + 3(-4) + (-2)(-2) = -7$$

$$\therefore \quad \mathbf{A}^{-1} = \frac{\operatorname{adj}\mathbf{A}}{\det \mathbf{A}} = \frac{1}{7}\begin{pmatrix} -1 & 4 & -3 \\ 4 & -2 & 5 \\ 2 & -1 & 6 \end{pmatrix}$$

$$\therefore \quad \begin{pmatrix} x \\ y \\ z \end{pmatrix} = \mathbf{A}^{-1}\begin{pmatrix} 1 \\ 5 \\ 4 \end{pmatrix} = \frac{1}{7}\begin{pmatrix} -1 & 4 & -3 \\ 4 & -2 & 5 \\ 2 & -1 & 6 \end{pmatrix}\begin{pmatrix} 1 \\ 5 \\ 4 \end{pmatrix} = \frac{1}{7}\begin{pmatrix} 7 \\ 14 \\ 21 \end{pmatrix} = \begin{pmatrix} 1 \\ 2 \\ 3 \end{pmatrix}$$

Hence the solution of the given set of equations is $x = 1$, $y = 2$, $z = 3$.

Since the method of Example 1 involves a great deal of numerical calculation it is not regarded as an efficient way of solving simultaneous equations. Elementary methods involving elimination of variables are often better.

$$\begin{aligned} x + 3y - 2z &= 1 && (1) \\ 2x \qquad + z &= 5 && (2) \\ - y + 2z &= 4 && (3) \\ (1) + 3 \times (3): \quad x \qquad + 4z &= 13 && (4) \\ 2 \times (4) - (2): \qquad 7z &= 21 \\ \therefore \qquad z &= 3 \end{aligned}$$

Substituting $z = 3$ in (2) and (3) we find that, as before, $x = 1$, $y = 2$, $z = 3$. This approach can be developed into a systematic elimination process which can be applied to any set of linear equations in three or more variables.

Suppose that

$$2x + y - 3z = 3 \quad (1)$$
$$x + 3y - 5z = 1 \quad (2)$$
$$6x - 2y + z = 9 \quad (3)$$

Using (1) to eliminate x from (2) and (3):

(1): $\quad 2x + y - 3z = 3 \quad (4)$

$2 \times (2) - (1)$: $\quad 5y - 7z = -1 \quad (5)$

$(3) - 3 \times (1)$: $\quad -5y + 10z = 0 \quad (6)$

Using (5) to eliminate y from (6) we produce the following set of equations:

(4): $\quad 2x + y - 3z = 3$

(5): $\quad 5y - 7z = -1$

(5) + (6): $\quad 3z = -1$

This new set of three equations is said to be in *echelon form*. In this context echelon means stepwise arrangement or ladder. To solve the equations we now work up the "ladder" obtaining $z = -\frac{1}{3}$, then $y = -\frac{2}{3}$ and $x = \frac{4}{3}$.

In reducing these equations to echelon form we have changed the matrix of coefficients from

$$\begin{pmatrix} 2 & 1 & -3 \\ 1 & 3 & -5 \\ 6 & -2 & 1 \end{pmatrix} \text{ to } \begin{pmatrix} 2 & 1 & -3 \\ 0 & 5 & -7 \\ 0 & 0 & 3 \end{pmatrix}$$

The new matrix has non-zero elements on or above the *leading diagonal* and zeros below it. Any matrix which either has zeros below the leading diagonal or has zeros above the leading diagonal is called a *triangular matrix*. In general, a set of linear equations is said to be in echelon form if the corresponding matrix is triangular.

Any systematic reduction method can be set out in table form. Each stage in the process is described by expressing the new rows r'_1, r'_2, r'_3 in terms of the previous rows r_1, r_2, r_3.

	Coefficients of x	y	z	Constant term
	2	1	−3	3
	1	3	−5	1
	6	−2	1	9
	2	1	−3	3
$r'_2 = 2r_2 - r_1$	0	5	−7	−1
$r'_3 = r_3 - 3r_1$	0	−5	10	0
	2	1	−3	3
	0	5	−7	−1
$r'_3 = r_2 + r_3$	0	0	3	−1

$\therefore$ as before, $z = -\frac{1}{3}$, $y = -\frac{2}{3}$, $x = \frac{4}{3}$.

We now consider a geometrical interpretation of the equations

$$\begin{aligned} a_1x + b_1y + c_1z &= d_1 \\ a_2x + b_2y + c_2z &= d_2 \\ a_3x + b_3y + c_3z &= d_3 \end{aligned}$$

These can be regarded as the equations of three planes in three-dimensional space.

If $\det \mathbf{A} \neq 0$, the equations have a unique solution and thus the planes must intersect in a single point. As we have seen, the coordinates of this point can be found either by using the inverse matrix $\mathbf{A}^{-1}$ or by reducing the equations to echelon form.

If $\det \mathbf{A} = 0$, the nature of the solution set of the equations depends on the way in which the three planes intersect.

Case (i): The planes have no common point. This situation arises when at least two of the planes are parallel (but not coincident), or when one plane is parallel to the line of intersection of the other two. The equations are then said to be inconsistent and have no solution.

Case (ii): The planes have a common line. This means that the equations will have an infinite solution set corresponding to the set of points on the line of intersection of the planes.

Case (iii): The planes are coincident. Again the equations have an infinite solution set, but this time the solutions are the coordinates of the points in the plane represented by all three equations.

Example 2 Discuss the solutions of the given equations when (a) $k = 5$, (b) $k = 6$.

$$\begin{aligned} 2x - 3y + z &= 1 \quad &(1) \\ x + 2y - z &= 3 \quad &(2) \\ 2x + 4y - 2z &= k \quad &(3) \end{aligned}$$

(a) When $k = 5$, equations (2) and (3) represent parallel planes. Hence the equations are inconsistent i.e. have no solution.

(b) When $k = 6$, equations (2) and (3) represent the same plane. Hence any solution of equations (1) and (2) also satisfies (3). Substituting $x = \lambda$ in (1) and (2) we have:

$$\begin{aligned} 2\lambda - 3y + z &= 1 \quad &(4) \\ \lambda + 2y - z &= 3 \quad &(5) \end{aligned}$$

(4) + (5): $\quad 3\lambda - y = 4 \quad \therefore \quad y = 3\lambda - 4$

From (5): $\quad \lambda + 6\lambda - 8 - z = 3 \quad \therefore \quad z = 7\lambda - 11$

Hence the solutions are $x = \lambda$, $y = 3\lambda - 4$, $z = 7\lambda - 11$, where λ may take any real value.

Example 3 Discuss the solutions of the given equations when (a) $k = 0$, (b) $k = 3$.

$$\begin{aligned} 2x - y + z &= 1 \quad &(1) \\ 2x - 3y - z &= 3 \quad &(2) \\ 4x - 3y + z &= k \quad &(3) \end{aligned}$$

	x	y	z	constant
	2	-1	1	1
	2	-3	-1	3
	4	-3	1	k
	2	-1	1	1
$r'_2 = r_1 - r_2$	0	2	2	-2
$r'_3 = 2r_1 - r_3$	0	1	1	$2-k$
	2	-1	1	1
$r'_2 = \frac{1}{2}r_2$	0	1	1	-1
$r'_3 = r_3 - \frac{1}{2}r_2$	0	0	0	$3-k$

(a) If $k = 0$, there is a contradiction in the new third row. Hence the equations are inconsistent.
(b) If $k = 3$, we have a line of zeros in the new third row together with the equations

$$\begin{aligned} 2x - y + z &= 1 \\ y + z &= -1 \end{aligned}$$

Hence in parametric form the solutions are $x = \lambda$, $y = \lambda - 1$, $z = -\lambda$.
This set of solutions represents the line of intersection of the planes with equations (1), (2) and (3).

Example 4 Discuss the solutions of the equations

$$\begin{aligned} 4x - 2y + 6z &= 2 \\ -2x + y - 3z &= -1 \\ 6x - 3y + 9z &= 3 \end{aligned}$$

Since all three equations represent the same plane their solution set represents the points in this plane. The solutions may be expressed in parametric form by writing:

$$x = \lambda, \quad y = \mu, \quad z = \tfrac{1}{3}(1 - 2\lambda + \mu).$$

Exercise 18.4

1. Find the inverse of the matrix $\begin{pmatrix} 1 & -1 & -2 \\ 0 & 1 & 1 \\ 2 & 0 & -1 \end{pmatrix}$ and hence solve the equations:

(a) $\begin{aligned} x - y - 2z &= 1 \\ y + z &= 1 \\ 2x - z &= 1 \end{aligned}$ (b) $\begin{aligned} p - q - 2r &= 2 \\ q + r &= -1 \\ 2p - r &= 3 \end{aligned}$ (c) $\begin{aligned} a - b - 2c &= 0 \\ b + c &= 0 \\ 2a - c &= 0 \end{aligned}$

2. Find the inverse of the matrix $\begin{pmatrix} 2 & -1 & 1 \\ 1 & 2 & -1 \\ 3 & 1 & 2 \end{pmatrix}$ and hence solve the equations:

(a) $\begin{aligned} 2x - y + z &= 1 \\ x + 2y - z &= 1 \\ 3x + y + 2z &= -2 \end{aligned}$

(b) $\begin{aligned} 2p - q + r &= 0 \\ p + 2q - r &= 3 \\ 3p + q + 2r &= -1 \end{aligned}$

(c) $\begin{aligned} 2a - b + c &= -1 \\ a + 2b - c &= 3 \\ 3a + b - 2c &= 2 \end{aligned}$

3. Use any method to solve for x, y and z the equations

(a) $\begin{aligned} 2x + 6y + z &= 0 \\ -x + 2y - z &= 10 \\ 4x + 3y + z &= 1 \end{aligned}$

(b) $\begin{aligned} 3x - 2y + 4z &= -7 \\ x + y - 6z &= -10 \\ 2x + 3y + 2z &= 3 \end{aligned}$

(c) $\begin{aligned} 4x + y + z &= 3 \\ x + 4y + z &= -1 \\ x + y + 4z &= -2 \end{aligned}$

(d) $\begin{aligned} x + 2y + z &= 0 \\ 3x - 2y - 5z &= 8 \\ 2x + 5y - z &= 6 \end{aligned}$

(e) $\begin{aligned} 3x - 2y + z &= -1 \\ 4x + y + 3z &= 2 \\ -x + 5y + 2z &= 5 \end{aligned}$

(f) $\begin{aligned} 9x - 10y + 2z &= -6 \\ x + 3y + 6z &= 13 \\ 4x - y + 8z &= 13 \end{aligned}$

4. Discuss the solutions of the given equations when (a) $k = 4$, (b) $k = 8$.

$$\begin{aligned} 3x + 3y + z &= 10 \\ x - y - z &= -4 \\ -2x + 2y + 2z &= k \end{aligned}$$

5. Discuss the solutions of the given equations when (a) $k = 9$, (b) $k = -9$.

$$\begin{aligned} x + y - z &= k \\ 3x - 2y + z &= 16 \\ 9x - y - z &= 5 \end{aligned}$$

6. Discuss the solutions of the given equations when (a) $k = -15$, (b) $k = -10$

$$\begin{aligned} 12x - 9y + 3z &= k \\ -4x + 3y - z &= 5 \\ 8x - 6y + 2z &= -10 \end{aligned}$$

7. Discuss the solutions of the given equations when (a) $k = -1$, (b) $k = +1$.

$$\begin{aligned} -4x + 12y + 8z &= -16 \\ kx - 3y - (k+1)z &= 4 \\ 2x - 6y - 4z &= 8 \end{aligned}$$

8. Show that the equations

$$\begin{aligned} 2x - y + z &= 3 \\ x + 2y - 3z &= 1 \\ 3x + y - 2z &= k \end{aligned}$$

are inconsistent when $k = 2$, but have an infinite set of solutions when $k = 4$. Obtain these solutions in parametric form.

9. Find the values of k for which the following sets of equations have solutions and find these solutions

(a) $$\begin{aligned} 2x + y - z &= 5 \\ x - y + z &= 1 \\ x - 2y + 2z &= k \end{aligned}$$

(b) $$\begin{aligned} 3x + y - z &= 1 \\ x + y + 3z &= 4 \\ x \qquad - 2z &= k \end{aligned}$$

10. Evaluate, in terms of k, the determinant $\begin{vmatrix} 1 & 1 & -3 \\ 3 & -1 & -1 \\ 5 & -3 & k \end{vmatrix}$.

Hence find the value of k such that the equations

$$\begin{aligned} x + y - 3z &= 1 \\ 3x - y - z &= 7 \\ 5x - 3y + kz &= 13 \end{aligned}$$

do not have a unique solution and obtain the general solution in this case.

11. Show that the equations

$$\begin{aligned} 2x + y + 2z &= 1 \\ 3x \qquad - z &= 3 \\ 4x - y + az &= b \end{aligned}$$

(i) have a unique solution unless $a = -4$; (ii) are inconsistent when $a = -4$ and $b = 2$; (iii) have an infinite number of solutions when $a = -4$ and $b = 5$. Interpret each case geometrically.

12. Solve the simultaneous equations

$$\begin{aligned} x + 2y - z &= 6, \\ 2x - y - z &= 1, \\ y + z &= 1. \end{aligned}$$

Find a if the set of equations

$$\begin{aligned} x + 2y - z &= 6, \\ 2x - y - z &= 1, \\ ay + z &= 1, \end{aligned}$$

has no solutions. Find a and b if the set of equations

$$\begin{aligned} x + 2y - z &= 6, \\ 2x - \;\; y - z &= 1, \\ ay + z &= b, \end{aligned}$$

has more than one solution. Find the solution for which $y = \lambda$. (AEB 1980)

18.5 Inverse matrices: row reduction method

The systematic reduction of a set of linear equations by means of row operations on the matrix of coefficients suggests another way of obtaining the inverse of a matrix. We illustrate the method by finding the inverse of the matrix

$$\mathbf{A} = \begin{pmatrix} 1 & -1 & 0 \\ -1 & 2 & 1 \\ 2 & -1 & 2 \end{pmatrix}$$

Let

$$\begin{pmatrix} 1 & -1 & 0 \\ -1 & 2 & 1 \\ 2 & -1 & 2 \end{pmatrix}\begin{pmatrix} x \\ y \\ z \end{pmatrix} = \begin{pmatrix} p \\ q \\ r \end{pmatrix}$$

then

$$x - y = p \qquad (1)$$
$$-x + 2y + z = q \qquad (2)$$
$$2x - y + 2z = r \qquad (3)$$

Using (1) to eliminate x from (2) and (3):

(1): $x - y = p$ (4)
(2) + (1): $y + z = p + q$ (5)
(3) − (2) × (1): $y + 2z = -2p + r$ (6)

Using (5) to eliminate y from (4) and (6):

(4) + (5): $x + z = 2p + q$ (7)
(5): $y + z = p + q$ (8)
(6) − (5): $z = -3p - q + r$ (9)

Using (9) to eliminate z from (7) and (8):

(7) − (9): $x = 5p + 2q - r$
(8) − (9): $y = 4p + 2q - r$
(9): $z = -3p - q + r$

Thus

$$\begin{pmatrix} x \\ y \\ z \end{pmatrix} = \begin{pmatrix} 5 & 2 & -1 \\ 4 & 2 & -1 \\ -3 & -1 & 1 \end{pmatrix}\begin{pmatrix} p \\ q \\ r \end{pmatrix}$$

Hence
$$\mathbf{A}^{-1} = \begin{pmatrix} 5 & 2 & -1 \\ 4 & 2 & -1 \\ -3 & -1 & 1 \end{pmatrix}$$

Using the notation of the previous section, it is quicker to set out the working as follows:

$$\mathbf{A} = \begin{pmatrix} 1 & -1 & 0 \\ -1 & 2 & 1 \\ 2 & -1 & 2 \end{pmatrix} \quad \Bigg| \quad \mathbf{I} = \begin{pmatrix} 1 & 0 & 0 \\ 0 & 1 & 0 \\ 0 & 0 & 1 \end{pmatrix}$$

$$\begin{array}{l} r_2' = r_2 + r_1 \\ r_3' = r_3 - 2r_1 \end{array} \quad \begin{pmatrix} 1 & -1 & 0 \\ 0 & 1 & 1 \\ 0 & 1 & 2 \end{pmatrix} \quad \Bigg| \quad \begin{pmatrix} 1 & 0 & 0 \\ 1 & 1 & 0 \\ -2 & 0 & 1 \end{pmatrix}$$

$$\begin{array}{l} r_2' = r_1 + r_2 \\ \\ r_3' = r_3 - r_2 \end{array} \quad \begin{pmatrix} 1 & 0 & 1 \\ 0 & 1 & 1 \\ 0 & 0 & 1 \end{pmatrix} \quad \Bigg| \quad \begin{pmatrix} 2 & 1 & 0 \\ 1 & 1 & 0 \\ -3 & -1 & 1 \end{pmatrix}$$

$$\begin{array}{l} r_1' = r_1 - r_3 \\ r_2' = r_2 - r_3 \end{array} \quad \begin{pmatrix} 1 & 0 & 0 \\ 0 & 1 & 0 \\ 0 & 0 & 1 \end{pmatrix} = \mathbf{I} \quad \Bigg| \quad \begin{pmatrix} 5 & 2 & -1 \\ 4 & 2 & -1 \\ -3 & -1 & 1 \end{pmatrix} = \mathbf{A}^{-1}$$

In the following more typical example a further step in the process is needed.

Example 1 Find the inverse of the matrix $\mathbf{A} = \begin{pmatrix} 1 & 2 & -1 \\ 2 & 1 & 1 \\ -1 & 0 & 2 \end{pmatrix}$

$$\begin{pmatrix} 1 & 2 & -1 \\ 2 & 1 & 1 \\ -1 & 0 & 2 \end{pmatrix} \quad \Bigg| \quad \begin{pmatrix} 1 & 0 & 0 \\ 0 & 1 & 0 \\ 0 & 0 & 1 \end{pmatrix}$$

$$\begin{array}{l} r_2' = r_2 - 2r_1 \\ r_3' = r_3 + r_1 \end{array} \quad \begin{pmatrix} 1 & 2 & -1 \\ 0 & -3 & 3 \\ 0 & 2 & 1 \end{pmatrix} \quad \Bigg| \quad \begin{pmatrix} 1 & 0 & 0 \\ -2 & 1 & 0 \\ 1 & 0 & 1 \end{pmatrix}$$

$$\begin{array}{l} r_1' = 3r_1 + 2r_2 \\ \\ r_3' = 3r_3 + 2r_2 \end{array} \quad \begin{pmatrix} 3 & 0 & 3 \\ 0 & -3 & 3 \\ 0 & 0 & 9 \end{pmatrix} \quad \Bigg| \quad \begin{pmatrix} -1 & 2 & 0 \\ -2 & 1 & 0 \\ -1 & 2 & 3 \end{pmatrix}$$

$$\begin{array}{l} r_1' = 3r_1 - r_3 \\ r_2' = 3r_2 - r_3 \end{array} \quad \begin{pmatrix} 9 & 0 & 0 \\ 0 & -9 & 0 \\ 0 & 0 & 9 \end{pmatrix} \quad \Bigg| \quad \begin{pmatrix} -2 & 4 & -3 \\ -5 & 1 & -3 \\ -1 & 2 & 3 \end{pmatrix}$$

$$\begin{array}{l} r_1' = r_1/9 \\ r_2' = -r_2/9 \\ r_3' = r_3/9 \end{array} \quad \begin{pmatrix} 1 & 0 & 0 \\ 0 & 1 & 0 \\ 0 & 0 & 1 \end{pmatrix} \quad \Bigg| \quad \begin{pmatrix} -\frac{2}{9} & \frac{4}{9} & -\frac{3}{9} \\ \frac{5}{9} & -\frac{1}{9} & \frac{3}{9} \\ -\frac{1}{9} & \frac{2}{9} & \frac{3}{9} \end{pmatrix}$$

$$\text{Hence } \mathbf{A}^{-1} = \begin{pmatrix} -\frac{2}{9} & \frac{4}{9} & -\frac{1}{3} \\ \frac{5}{9} & -\frac{1}{9} & \frac{1}{3} \\ -\frac{1}{9} & \frac{2}{9} & \frac{1}{3} \end{pmatrix} \text{ or } \frac{1}{9}\begin{pmatrix} -2 & 4 & -3 \\ 5 & -1 & 3 \\ -1 & 2 & 3 \end{pmatrix}$$

In some examples it is necessary to interchange rows to avoid zeros on the main diagonal.

Example 2 Find the inverse of the matrix $\mathbf{A} = \begin{pmatrix} -1 & 3 & 2 \\ 2 & -6 & 0 \\ 1 & -2 & -4 \end{pmatrix}$

$$\begin{pmatrix} -1 & 3 & 2 \\ 2 & -6 & 0 \\ 1 & -2 & -4 \end{pmatrix} \;\Bigg|\; \begin{pmatrix} 1 & 0 & 0 \\ 0 & 1 & 0 \\ 0 & 0 & 1 \end{pmatrix}$$

$$\begin{matrix} r_2' = r_2 + 2r_1 \\ r_3' = r_3 + r_1 \end{matrix} \quad \begin{pmatrix} -1 & 3 & 2 \\ 0 & 0 & 4 \\ 0 & 1 & -2 \end{pmatrix} \;\Bigg|\; \begin{pmatrix} 1 & 0 & 0 \\ 2 & 1 & 0 \\ 1 & 0 & 1 \end{pmatrix}$$

$$\begin{matrix} r_2' = r_3 \\ r_3' = r_2 \end{matrix} \quad \begin{pmatrix} -1 & 3 & 2 \\ 0 & 1 & -2 \\ 0 & 0 & 4 \end{pmatrix} \;\Bigg|\; \begin{pmatrix} 1 & 0 & 0 \\ 1 & 0 & 1 \\ 2 & 1 & 0 \end{pmatrix}$$

$$r_1' = r_1 - 3r_2 \quad \begin{pmatrix} -1 & 0 & 8 \\ 0 & 1 & -2 \\ 0 & 0 & 4 \end{pmatrix} \;\Bigg|\; \begin{pmatrix} -2 & 0 & -3 \\ 1 & 0 & 1 \\ 2 & 1 & 0 \end{pmatrix}$$

$$\begin{matrix} r_1' = r_1 - 2r_3 \\ r_2' = 2r_2 + r_3 \end{matrix} \quad \begin{pmatrix} -1 & 0 & 0 \\ 0 & 2 & 0 \\ 0 & 0 & 4 \end{pmatrix} \;\Bigg|\; \begin{pmatrix} -6 & -2 & -3 \\ 4 & 1 & 2 \\ 2 & 1 & 0 \end{pmatrix}$$

$$\begin{matrix} r_1' = -r_1 \\ r_2' = \frac{1}{2}r_2 \\ r_3' = \frac{1}{4}r_3 \end{matrix} \quad \begin{pmatrix} 1 & 0 & 0 \\ 0 & 1 & 0 \\ 0 & 0 & 1 \end{pmatrix} \;\Bigg|\; \begin{pmatrix} 6 & 2 & 3 \\ 2 & \frac{1}{2} & 1 \\ \frac{1}{2} & \frac{1}{4} & 0 \end{pmatrix}$$

$$\mathbf{A} = \begin{pmatrix} 6 & 2 & 3 \\ 2 & \frac{1}{2} & 1 \\ \frac{1}{2} & \frac{1}{4} & 0 \end{pmatrix} \quad \text{or} \quad \frac{1}{4}\begin{pmatrix} 24 & 8 & 12 \\ 8 & 2 & 4 \\ 2 & 1 & 0 \end{pmatrix}$$

The next example shows how the method breaks down for a matrix which has no inverse.

Example 3 Find, if possible, the inverse of the matrix $\mathbf{A} = \begin{pmatrix} 2 & -1 & 3 \\ -1 & 3 & 1 \\ 1 & 2 & 4 \end{pmatrix}$.

$$\left(\begin{array}{rrr|rrr} 2 & -1 & 3 & 1 & 0 & 0 \\ -1 & 3 & 1 & 0 & 1 & 0 \\ 1 & 2 & 4 & 0 & 0 & 1 \end{array}\right)$$

$$\begin{array}{l} \\ r_2' = 2r_2 + r_1 \\ r_3' = 2r_3 - r_1 \end{array} \quad \left(\begin{array}{rrr|rrr} 2 & -1 & 3 & 1 & 0 & 0 \\ 0 & 5 & 5 & 1 & 2 & 0 \\ 0 & 5 & 5 & -1 & 0 & 2 \end{array}\right)$$

$$\begin{array}{l} r_1' = 5r_1 + r_2 \\ \\ r_3' = r_3 - r_2 \end{array} \quad \left(\begin{array}{rrr|rrr} 10 & 0 & 20 & 6 & 2 & 0 \\ 0 & 5 & 5 & 1 & 2 & 0 \\ 0 & 0 & 0 & -2 & -2 & 2 \end{array}\right)$$

Since a complete row of zeros has been obtained, no inverse of the matrix **A** can be found.

[Note that the method of finding inverses using elementary row operations can be extended to square matrices of any order. For large matrices this process is usually more efficient than the adjoint method introduced in §18.3.]

Exercise 18.5

In questions 1 to 6 find, if possible, the inverses of the given matrices.

1. (a) $\begin{pmatrix} 1 & 0 & 0 \\ 0 & 1 & 1 \\ 1 & 1 & 2 \end{pmatrix}$ (b) $\begin{pmatrix} 1 & -1 & 1 \\ -1 & 2 & -2 \\ 0 & 0 & 1 \end{pmatrix}$

   (c) $\begin{pmatrix} 1 & -2 & 1 \\ 0 & 1 & -2 \\ 0 & 0 & 1 \end{pmatrix}$ (d) $\begin{pmatrix} 1 & 0 & 0 \\ -3 & 1 & 0 \\ 5 & -2 & 1 \end{pmatrix}$

2. (a) $\begin{pmatrix} 1 & -1 & 1 \\ -3 & 4 & -2 \\ 2 & 0 & 5 \end{pmatrix}$ (b) $\begin{pmatrix} 1 & -3 & 1 \\ -1 & 4 & -2 \\ 2 & -1 & -2 \end{pmatrix}$

   (c) $\begin{pmatrix} 1 & 1 & -1 \\ 1 & 0 & 0 \\ -1 & 0 & 1 \end{pmatrix}$ (d) $\begin{pmatrix} 1 & -2 & 0 \\ -2 & 3 & -1 \\ 1 & 1 & 2 \end{pmatrix}$

3. (a) $\begin{pmatrix} 1 & 1 & 1 \\ 1 & 2 & 3 \\ 2 & -1 & -3 \end{pmatrix}$ (b) $\begin{pmatrix} 1 & -2 & 0 \\ 3 & 1 & 1 \\ 0 & 1 & 0 \end{pmatrix}$

   (c) $\begin{pmatrix} 1 & -5 & -3 \\ -3 & 2 & -4 \\ 4 & 1 & 9 \end{pmatrix}$ (d) $\begin{pmatrix} -3 & 0 & 1 \\ 2 & 1 & 3 \\ 1 & 0 & 0 \end{pmatrix}$

4. (a) $\begin{pmatrix} -1 & 0 & 1 \\ -2 & 1 & 3 \\ 0 & 0 & 2 \end{pmatrix}$ (b) $\begin{pmatrix} 3 & -2 & 1 \\ -1 & 0 & -3 \\ 4 & -3 & 0 \end{pmatrix}$

(c) $\begin{pmatrix} 3 & 2 & 1 \\ 0 & 1 & 4 \\ 0 & 0 & 2 \end{pmatrix}$ (d) $\begin{pmatrix} 4 & 2 & 3 \\ 0 & -1 & -2 \\ 2 & 4 & 7 \end{pmatrix}$

5. (a) $\begin{pmatrix} 1 & 2 & 3 \\ 1 & 2 & 2 \\ 1 & 1 & 1 \end{pmatrix}$ (b) $\begin{pmatrix} 2 & 0 & -3 \\ 0 & 0 & -1 \\ 2 & 2 & 0 \end{pmatrix}$

(c) $\begin{pmatrix} 0 & 1 & 0 \\ 5 & 2 & 1 \\ 3 & 4 & 0 \end{pmatrix}$ (d) $\begin{pmatrix} 0 & 2 & -10 \\ 4 & 2 & 2 \\ 7 & 3 & 6 \end{pmatrix}$

6. (a) $\begin{pmatrix} 0 & 0 & 1 \\ 1 & 0 & 0 \\ 0 & 1 & 0 \end{pmatrix}$ (b) $\begin{pmatrix} 1 & 0 & 0 \\ 0 & 2 & 0 \\ 0 & 0 & 3 \end{pmatrix}$

(c) $\begin{pmatrix} 0 & 0 & 1 \\ 0 & 2 & 0 \\ 3 & 0 & 0 \end{pmatrix}$ (d) $\begin{pmatrix} 1 & 0 & 0 \\ 0 & 0 & -1 \\ 0 & -1 & 0 \end{pmatrix}$

7. Given that $\mathbf{A} = \begin{pmatrix} 1 & 0 & 0 \\ 4 & 1 & 0 \\ -3 & -2 & 1 \end{pmatrix}$ and $\mathbf{B} = \begin{pmatrix} 1 & 2 & 0 \\ 1 & 3 & 5 \\ 0 & -1 & -1 \end{pmatrix}$, find $\mathbf{A}^{-1}$ and $\mathbf{B}^{-1}$.

Use your answers to verify that
(a) $(\mathbf{AB})^{-1} = \mathbf{B}^{-1}\mathbf{A}^{-1}$, (b) $(\mathbf{A}^T)^{-1} = (\mathbf{A}^{-1})^T$, (c) $(\mathbf{B}^2)^{-1} = (\mathbf{B}^{-1})^2$.

8. Find the inverses of the matrices $\mathbf{A} = \begin{pmatrix} 1 & -1 & 3 \\ 1 & 0 & 4 \\ 2 & -4 & 3 \end{pmatrix}$, $\mathbf{B} = \begin{pmatrix} 1 & 1 & 1 \\ 0 & 1 & 1 \\ 0 & 0 & 1 \end{pmatrix}$.

Hence find $\mathbf{P}^{-1}$ and $\mathbf{Q}^{-1}$, where $\mathbf{P} = \mathbf{AB}$ and $\mathbf{Q} = \mathbf{B}^2\mathbf{A}$.

[For further practice in finding inverses see Exercise 18.3.]

Exercise 18.6 (miscellaneous)

1. Describe geometrically the transformations represented by the following matrices.

(a) $\begin{pmatrix} 1 & 0 & 0 \\ 0 & 0 & -1 \\ 0 & -1 & 0 \end{pmatrix}$ (b) $\begin{pmatrix} -2 & 0 & 0 \\ 0 & -2 & 0 \\ 0 & 0 & 2 \end{pmatrix}$ (c) $\begin{pmatrix} \frac{1}{2}\sqrt{3} & 0 & -\frac{1}{2} \\ 0 & 1 & 0 \\ \frac{1}{2} & 0 & \frac{1}{2}\sqrt{3} \end{pmatrix}$

2. Write down the 3×3 matrices representing the following transformations of three-dimensional space. (a) Reflection in the plane $y = x\tan\theta$, (b) $180°$ rotation about the line $y = x\tan\theta$ in the x, y plane.

3. T_1, T_2, T_3 are linear transformations of three-dimensional space. T_1 is reflection in the xy-plane; T_2 is rotation about the z-axis through $90°$ so that the positive x-axis is turned into the positive y-axis; T_3 is rotation through $180°$ about the line which lies in the xy-plane and has equation $y = x$. Give the coordinates of the images of the point (x, y, z) under each of these transformations, and hence find the matrix of each transformation. Find the matrices of the combined transformations T_3T_2 and T_2T_3 and describe each of these transformations geometrically. Describe also the transformations $T_3T_2T_1$ and $T_1T_2T_3$. (L)

4. Show that the linear transformation with matrix $\begin{pmatrix} 1 & -3 & 0 \\ 2 & -4 & 3 \\ 1 & 1 & 6 \end{pmatrix}$ maps every point of three-dimensional space onto a single plane and give the Cartesian equation of this plane. Describe the set of points mapped onto the origin under this transformation.

5. The transformations $\begin{pmatrix} x' \\ y' \\ 1 \end{pmatrix} = \begin{pmatrix} -1 & 0 & 2 \\ 0 & 1 & 0 \\ 0 & 0 & 1 \end{pmatrix}\begin{pmatrix} x \\ y \\ 1 \end{pmatrix}$ and

$\begin{pmatrix} x' \\ y' \\ 1 \end{pmatrix} = \begin{pmatrix} 0 & -1 & 0 \\ -1 & 0 & 0 \\ 0 & 0 & 1 \end{pmatrix}\begin{pmatrix} x \\ y \\ 1 \end{pmatrix}$ represent reflections of the plane; find the axes of these two reflections. Denote the matrices of reflections by R and S respectively; what is the transformation of the plane determined by the matrix SR? (W)

6. If $\mathbf{A} = \begin{pmatrix} 2 & 1 & 1 \\ 1 & 0 & 1 \\ 0 & -1 & 0 \end{pmatrix}$ and $\mathbf{B} = \begin{pmatrix} 1 & -1 & 1 \\ 0 & 0 & -1 \\ -1 & 2 & -1 \end{pmatrix}$, find (a) $\mathbf{AB}$, (b) a matrix $\mathbf{X}$ such that $\mathbf{AX} + \mathbf{B} = \mathbf{A}$. (O&C)

7. $\mathbf{P}$ is a 3×3 matrix such that $\mathbf{P}^2 = \mathbf{P} - \mathbf{I}$. (a) Find $\mathbf{P}^{-1}$ in terms of $\mathbf{P}$ and $\mathbf{I}$. (b) Show that $\mathbf{P}^3 + \mathbf{I} = \mathbf{0}$. (c) If $\mathbf{PQ} = 2\mathbf{I} - \mathbf{P}$, find $\mathbf{Q}$ in the form $\lambda\mathbf{I} + \mu\mathbf{P}$, where λ and μ are constants to be determined.

8. If $\mathbf{A}$ is a 3×3 matrix with transpose $\mathbf{A}^T$, then $\mathbf{A}$ is said to be *symmetric* if $\mathbf{A}^T = \mathbf{A}$ and *skew-symmetric* if $\mathbf{A}^T = -\mathbf{A}$. Prove that
(a) if $\mathbf{A}$ is both symmetric and skew-symmetric, then $\mathbf{A}$ is the zero matrix;
(b) if $\mathbf{A}$ is skew-symmetric then $\mathbf{A}^2$ is symmetric;
(c) if $\mathbf{A}$, $\mathbf{B}$ and $\mathbf{AB}$ are symmetric, then $\mathbf{AB} = \mathbf{BA}$. [You may assume that $(\mathbf{AB})^T = \mathbf{B}^T\mathbf{A}^T$.]

9. (a) Find the inverse of the matrix $A = \begin{pmatrix} 2 & 3 & -5 \\ 1 & 2 & -1 \\ -1 & -1 & 5 \end{pmatrix}$.

(b) Calculate the matrix product BC, where

$$B = \begin{pmatrix} 1 & 1 & -2 \\ 2 & -1 & 1 \\ 2 & -2 & 3 \end{pmatrix}, \quad C = \begin{pmatrix} 1 & -1 & 1 \\ 4 & -7 & 5 \\ 2 & -4 & 3 \end{pmatrix},$$

and hence or otherwise find the matrix X such that $CX = \begin{pmatrix} 0 & 1 & 0 \\ 0 & 0 & -1 \\ 1 & 0 & 0 \end{pmatrix}$. (W)

10. Obtain the multiplicative inverse, where it exists, of each of the following matrices. If there is no inverse, explain briefly why this is so.

(i) $\begin{pmatrix} 1 & 2 \\ -3 & -4 \end{pmatrix}$ (ii) $\begin{pmatrix} 1 & -2 \\ -3 & 6 \end{pmatrix}$ (iii) $\begin{pmatrix} 2 & 1 & 3 \\ 1 & 0 & 2 \\ 1 & 2 & 0 \end{pmatrix}$

(iv) $\begin{pmatrix} 1 & 2 \\ 2 & 1 \\ 3 & 4 \end{pmatrix}$ (v) $\begin{pmatrix} 1 & 1 & 0 \\ 0 & 1 & 1 \\ 0 & 0 & 1 \end{pmatrix}$ (AEB 1979)

11. Find the inverse of the matrix $\begin{pmatrix} 1 & -3 & 0 \\ 2 & 0 & 1 \\ 4 & 1 & 3 \end{pmatrix}$. Hence solve the equations $x - 3y - a = 0, 2x + z - b = 0$ and $4x + y + 3z - c = 0$ for x, y, z in terms of a, b, c. (O&C)

12. Find the inverse of the matrix $\mathbf{A} = \begin{pmatrix} 1 & 1 & -1 \\ 3 & 4 & -2 \\ -1 & 1 & 4 \end{pmatrix}$ by row reduction, or otherwise. Hence solve the simultaneous equations

$$\begin{aligned} x + y - z &= 1 \\ 3x + 4y - 2z &= 3 \\ -x + y + 4z &= 2, \end{aligned}$$

and find the matrix $\mathbf{X}$ such that $\mathbf{XA} = \begin{pmatrix} 1 & 2 & 1 \\ 1 & 0 & -2 \\ 0 & 1 & 1 \end{pmatrix}$. (W)

13. Find the inverse of the matrix $A = \begin{pmatrix} 2 & -1 & 1 \\ 2 & -2 & 3 \\ 1 & 1 & -2 \end{pmatrix}$ by row reduction, or

otherwise. Hence solve the simultaneous equations

$$\begin{aligned} 2x - y + z &= 1 \\ 2x - 2y + 3z &= -1 \\ x + y - 2z &= 3. \end{aligned}$$

Calculate the matrix $B = A^{-1}\begin{pmatrix} 0 & 0 & 1 \\ 0 & 1 & 0 \\ 1 & 0 & 0 \end{pmatrix} A$. *Without calculating B^2 directly,* show that $B^2 = I_3$ (the 3×3 identity matrix). (W)

14. Find the values of α for which the equations

$$\begin{aligned} x + 2y + 3z &= 1 \\ 5x + y + 3z &= \alpha \\ 3x + 9y + 13z &= \alpha^2 \end{aligned}$$

have solutions. Find all the solutions for each value of α. (AEB 1977)

15. Matrices **A**, **B** and vectors **X**, **K** are given by

$$\mathbf{A} = \begin{pmatrix} 1 & -3 & 1 \\ 2 & -4 & 3 \\ 3 & -1 & 4 \end{pmatrix}, \quad \mathbf{X} = \begin{pmatrix} x \\ y \\ z \end{pmatrix}, \quad \mathbf{B} = \begin{pmatrix} 1 & -3 & -4 \\ -1 & 3 & 5 \\ 2 & -3 & -5 \end{pmatrix}, \quad \mathbf{K} = \begin{pmatrix} 2 \\ 4 \\ 4 \end{pmatrix}.$$

(i) Use a *matrix* method to solve the system of equations $\mathbf{AX} = \mathbf{K}$.
(ii) Find values of the constants λ, μ such that the rows of matrix **B** are connected by the relation $R_3 = \lambda R_1 + \mu R_2$. Hence explain why the system of equations $\mathbf{BX} = \mathbf{K}$ would be inconsistent (i.e. they do not possess a solution set.) (W)

16. Find the matrix of the linear transformation which maps $U(2, 1, 1)$ to $U'(4, 2, 2)$, $V(3, 2, 1)$ to $V'(-3, -2, 1)$ and $W(0, 0, 1)$ to $W'(0, 0, -1)$.

17. For the matrix equation $\mathbf{M}\begin{pmatrix} x \\ y \\ z \end{pmatrix} = \begin{pmatrix} 3 \\ 6 \\ 4+k \end{pmatrix}$, where $\mathbf{M} = \begin{pmatrix} 1 & 1 & 1 \\ 1 & 2 & 3 \\ 1 & 3 & k \end{pmatrix}$, find the value of k for which the equation does not have a unique solution. For this value of k, solve the equation and interpret the solution geometrically in relation to the linear transformation with matrix **M**.

18. The linear transformation $\begin{pmatrix} x' \\ y' \\ z' \end{pmatrix} = \mathbf{M}\begin{pmatrix} x \\ y \\ z \end{pmatrix}$, where **M** is a 3×3 matrix, maps the points with position vectors $\begin{pmatrix} 1 \\ 0 \\ 0 \end{pmatrix}, \begin{pmatrix} 0 \\ 1 \\ 0 \end{pmatrix}, \begin{pmatrix} 0 \\ 0 \\ 1 \end{pmatrix}$ to the points with position

vectors $\begin{pmatrix}3\\2\\1\end{pmatrix}$, $\begin{pmatrix}2\\1\\0\end{pmatrix}$, $\begin{pmatrix}1\\0\\0\end{pmatrix}$ respectively. Write down the matrix $\mathbf{M}$ and find the inverse matrix $\mathbf{M}^{-1}$. Show that the transformation with matrix $\mathbf{M}$ maps points of the plane $x + y + z = 0$ to points of the plane $x = y$, and verify that the inverse transformation with matrix $\mathbf{M}^{-1}$ maps points of the plane $x = y$ to points of the plane $x + y + z = 0$. (L)

19. $\mathbf{M}$ is the matrix $\begin{pmatrix}3 & 1 & -3\\1 & 2a & 1\\0 & 2 & a\end{pmatrix}$.

(i) Find the two values of a for which $\mathbf{M}$ is singular.

(ii) Solve the equation $\mathbf{M}\begin{pmatrix}x\\y\\z\end{pmatrix} = \begin{pmatrix}-3\frac{1}{2}\\5\frac{1}{2}\\5\end{pmatrix}$ in the case $a = 2$, and also determine whether or not solutions exist for each of the two values of a found in (i).

(C)

19 Binary operations and groups

19.1 Definition of a group

A *binary operation* $*$ is a rule for combining two objects, a and b, to give a third object $c = a * b$.

We have already met many examples of binary operations e.g. addition of real numbers, intersection of sets, multiplication of matrices, composition of functions.

A set S is said to be *closed* under a binary operation $*$, if $a * b \in S$ for all elements $a, b \in S$.

For instance, if a and b are integers, then $a - b$ is also an integer
$\therefore$ the set of integers $\mathbb{Z}$ is closed under subtraction.
However, although 2 and 7 are positive integers, $2 - 7$ is not a positive integer,
$\therefore$ the set of positive integers $\mathbb{Z}^+$ is not closed under subtraction.

An element e of a set S is said to be an *identity element* for the binary operation $*$ if $e * a = a * e = a$ for all $a \in S$.

For example, since $\varnothing \cup A = A \cup \varnothing = A$ for any set A, the empty set $\varnothing$ is an identity element for the set union operation.
However, we can find no real number e such that $e \div a = a \div e = a$ for all $a \in \mathbb{R}$. Thus there is no identity element for division of real numbers.

[It will be shown in §19.4 that a set contains at most one identity element for any given binary operation.]

An *inverse* of an element a with respect to a binary operation $*$ is an element a^{-1} such that $a * a^{-1} = a^{-1} * a = e$.

We already know that the inverse of a non-zero real number a with respect to multiplication is $1/a$ and that the inverse of a non-singular matrix $\begin{pmatrix} a & b \\ c & d \end{pmatrix}$ with respect to matrix multiplication is $\frac{1}{\Delta}\begin{pmatrix} d & -b \\ -c & a \end{pmatrix}$, where $\Delta = ad - bc$.
However, for a non-empty set A there is no set B such that $A \cup B = B \cup A = \varnothing$. Therefore the set A has no inverse with respect to the operation $\cup$.

> A binary operation $*$ on a set S is *associative* if $a*(b*c) = (a*b)*c$ for all $a, b, c \in S$. The operation is *commutative* if $a*b = b*a$ for all $a, b \in S$.

For instance, union of sets is both associative and commutative; multiplication of matrices is associative but not commutative; subtraction of integers is neither associative nor commutative.

When investigating the properties of binary operation, it is often helpful to draw up an *operation table*. The following tables show two operations on the set $S = \{0, 1, 2, 3\}$, subtraction $-$ and the operation $*$, where $a*b = |a-b|$.

(1) Second element

	$-$	0	1	2	3
	0	0	-1	-2	-3
First	1	1	0	-1	-2
element	2	2	1	0	-1
	3	3	2	1	0

(2) Second element

	$*$	0	1	2	3
	0	0	1	2	3
First	1	1	0	1	2
element	2	2	1	0	1
	3	3	2	1	0

Since table (1) contains elements such as -1 which do not belong to S, the set S is not closed under $-$. All the entries in table (2) are elements of S, so S is closed under $*$.

There is no identity element for $-$ and hence the elements of S do not have inverses with respect to $-$. However, 0 is an identity element for $*$ and table (2) shows that every element of S is self-inverse.

For a commutative operation the operation table is symmetrical about the main diagonal. Thus we see that table (2) has this symmetry, but table (1) does not.

We cannot use the tables directly to decide whether or not the operations $-$ and $*$ are associative. In each case we must either prove the associative law or produce a counter-example to show that it does not hold.

Since $1-(2-3) = 2$, whereas $(1-2)-3 = -4$
and $1*(2*3) = 0$, whereas $(1*2)*3 = 2$,

we must conclude that neither operation is associative.

When studying the properties of a set and its associated binary operations it is useful to be able to identify patterns which can also be seen in other algebraic systems. One structure which occurs in many branches of mathematics is called a group.

A *group* $(G, *)$ is a set G with binary operation $*$ such that,
(1) G is closed under $*$ i.e. $a * b \in G$ for all $a, b \in G$.
(2) $*$ is associative i.e. $a * (b * c) = (a * b) * c$ for all $a, b, c \in G$.
(3) There is an identity element e in G such that $e * a = a * e = a$ for all $a \in G$.
(4) Every element a of G has an inverse a^{-1} in G such that $a * a^{-1} = a^{-1} * a = e$.

One familiar example of a group is $(\mathbb{Z}, +)$, the set of integers under addition.
$\mathbb{Z}$ is closed under addition.
Addition of integers is associative.
0, the identity element for addition, belongs to $\mathbb{Z}$.
Every integer a has an additive inverse $(-a)$, which is also an integer.
Since $\mathbb{Z}$ is an infinite set, $(\mathbb{Z}, +)$ is called an *infinite group.*

In the worked examples which follow we consider some finite groups. The *order* of a *finite group* is the number of elements it contains.

Example 1 Show that the set $A = \{1, i, -1, -i\}$, where $i^2 = -1$, forms a group under multiplication.

$\times$	1	i	-1	$-i$
1	1	i	-1	$-i$
i	i	-1	$-i$	1
-1	-1	$-i$	1	i
$-i$	$-i$	1	i	-1

(1) The operation table shows that A is closed under multiplication.
(2) Multiplication of complex numbers is associative.
(3) A contains 1, the identity element for multiplication.
(4) Since 1 appears in every row and column of the table, every element of A has an inverse in A. [1 and -1 are self-inverse, the inverse of i is $-i$ and the inverse of $-i$ is i.]

Hence A forms a group under multiplication.

Example 2 Given that $f(x) = x$, $g(x) = 1 - \dfrac{1}{x}$, $h(x) = \dfrac{1}{1-x}$, for all real x except $x = 0$, $x = 1$, show that the set $S = \{f, g, h\}$ forms a group under composition of functions.

$$f^2(x) = f(x), \quad fg(x) = gf(x) = g(x), \quad fh(x) = hf(x) = h(x),$$

$$g^2(x) = g\left(1 - \frac{1}{x}\right) = 1 - 1\Big/\left(1 - \frac{1}{x}\right) = 1 - \frac{x}{x-1} = \frac{1}{1-x} = h(x)$$

$$gh(x) = g\left(\frac{1}{1-x}\right) = 1 - 1\Big/\left(\frac{1}{1-x}\right) = 1 - (1-x) = x = f(x)$$

$$hg(x) = h\left(1 - \frac{1}{x}\right) = 1\Big/\left\{1 - \left(1 - \frac{1}{x}\right)\right\} = 1\Big/\left\{\frac{1}{x}\right\} = x = f(x)$$

$$h^2(x) = h\left(\frac{1}{1-x}\right) = 1\Big/\left\{1 - \left(\frac{1}{1-x}\right)\right\} = \frac{1-x}{-x} = 1 - \frac{1}{x} = g(x)$$

	f	g	h
f	f	g	h
g	g	h	f
h	h	f	g

(1) The table shows that the set S is closed under function composition.
(2) For any functions F, G, H,

$$F[GH](x) = F(GH(x)) = F(G(H(x))) = FG(H(x)) = [FG]H(x)$$
$$\therefore \quad F[GH] = [FG]H$$

Hence composition of functions is associative.
(3) The identity function f belongs to S.
(4) The table shows that $f^{-1} = f$, $g^{-1} = h$ and $h^{-1} = g$
$\therefore$ every element of S has an inverse in S.

> A group $(G, *)$ is said to be *Abelian*† if the operation $*$ is commutative in G i.e. if $a * b = b * a$ for all $a, b \in G$.

Since addition of integers is commutative, the group $(\mathbb{Z}, +)$ is Abelian. Similarly, the group A in Example 1 has a commutative binary operation and is therefore Abelian. In Example 2, although composition of functions is generally non-commutative, the symmetry of the operation table shows that within the set S the operation is commutative. Thus, this group is also Abelian.

For an example of a non-Abelian group, we consider a set of 2×2 matrices under multiplication.

Example 3 Show that the matrices $\mathbf{I} = \begin{pmatrix} 1 & 0 \\ 0 & 1 \end{pmatrix}$, $\mathbf{A} = \begin{pmatrix} 0 & -1 \\ 1 & 0 \end{pmatrix}$,

$\mathbf{B} = \begin{pmatrix} -1 & 0 \\ 0 & -1 \end{pmatrix}$, $\mathbf{C} = \begin{pmatrix} 0 & 1 \\ -1 & 0 \end{pmatrix}$, $\mathbf{D} = \begin{pmatrix} 1 & 0 \\ 0 & -1 \end{pmatrix}$, $\mathbf{E} = \begin{pmatrix} 0 & 1 \\ 1 & 0 \end{pmatrix}$, $\mathbf{F} = \begin{pmatrix} -1 & 0 \\ 0 & 1 \end{pmatrix}$,

$\mathbf{G} = \begin{pmatrix} 0 & -1 \\ -1 & 0 \end{pmatrix}$ form a group S under matrix multiplication.

First element \ Second element	**I**	**A**	**B**	**C**	**D**	**E**	**F**	**G**
I	**I**	**A**	**B**	**C**	**D**	**E**	**F**	**G**
A	**A**	**B**	**C**	**I**	**E**	**F**	**G**	**D**
B	**B**	**C**	**I**	**A**	**F**	**G**	**D**	**E**
C	**C**	**I**	**A**	**B**	**G**	**D**	**E**	**F**
D	**D**	**G**	**F**	**E**	**I**	**C**	**B**	**A**
E	**E**	**D**	**G**	**F**	**A**	**I**	**C**	**B**
F	**F**	**E**	**D**	**G**	**B**	**A**	**I**	**C**
G	**G**	**F**	**E**	**D**	**C**	**B**	**A**	**I**

(1) The operation table shows that S is closed under matrix multiplication.
(2) Matrix multiplication is associative.
(3) S contains the identity matrix **I**.
(4) Since **I** appears in every row and column of the table, every element of S has an inverse in S. [**I**, **B**, **D**, **E**, **F** and **G** are self-inverse, $\mathbf{A}^{-1} = \mathbf{C}$ and $\mathbf{C}^{-1} = \mathbf{A}$.]

Hence S forms a group under matrix multiplication.

[Since the operation table is not symmetrical about the main diagonal, the group is non-Abelian.]

† *Abel, Niels Henrik* (1802–1829) Norwegian mathematician. His work on the theory of equations (he proved that an algebraic solution of a quintic was impossible) helped to lay the foundations of modern group theory.

To prove that a set does *not* form a group under a given binary operation we must find a counter-example which demonstrates that one of the group axioms does not hold. It may not be necessary to consider all four group axioms.

Example 4 Show that the set $A = \{p, q, r, s\}$ does not form a group under either of the operations defined as follows:

(a)

$*$	p	q	r	s
p	q	p	s	r
q	p	q	r	s
r	s	r	s	p
s	r	s	p	q

(b)

$\circ$	p	q	r	s
p	p	q	r	s
q	q	p	q	r
r	r	q	p	q
s	s	r	q	p

(a) A is closed under $*$.
The identity element for $*$ is q.
However, since q does not appear in the third column of the table, r has no inverse.
$\therefore$ A does not form a group under $*$.

(b) A is closed under $\circ$.
The identity element for $\circ$ is p.
Each element is its own inverse under $\circ$.
Considering the associative law:

$$(q \circ r) \circ s = q \circ s = r, \quad q \circ (r \circ s) = q \circ q = p$$

$\therefore$ the associative law does not hold for $\circ$.
Hence A does not form a group under $\circ$.

Exercise 19.1

In questions 1 to 16 decide whether the given set forms a group under the stated binary operation, giving reasons for your answer. If the set does not form a group state which of the group axioms do not hold.

1. The set of even integers under addition.

2. The set of odd integers under multiplication.

3. $\mathbb{Q}$, the set of rational numbers, under multiplication.

4. The set $A = \{\frac{1}{4}, \frac{1}{2}, 1, 2, 4\}$ under multiplication.

5. The set of 2×3 matrices with real elements under matrix addition.

6. The set of numbers of the form p/q, where p and q are odd integers, under multiplication.

7. The set of all subsets of the set $S = \{a, b, c, d\}$ under the operation intersection $\cap$.

8. The set containing $\begin{pmatrix} 1 & 0 \\ 0 & 0 \end{pmatrix}, \begin{pmatrix} -1 & 0 \\ 0 & 0 \end{pmatrix}, \begin{pmatrix} i & 0 \\ 0 & 0 \end{pmatrix}, \begin{pmatrix} -i & 0 \\ 0 & 0 \end{pmatrix}$ under matrix multiplication.

9. The set of non-negative integers under $*$, where $a * b = |a - b|$.

10. The set $\mathbb{R}^+$, the positive real numbers, under $\circ$, where $a \circ b = ab/(a + b)$.

11. The set of integers $\mathbb{Z}$ under $*$, where $a * b = a$.

12. The set of real numbers $\mathbb{R}$ under $\oplus$, where $a \oplus b = a + b - 1$.

13. The set of non-negative rationals under $*$, where $a * b = \sqrt{(a^2 + b^2)}$.

14. The set $\mathbb{R}$ under $\circ$, where $a \circ b = 2(a + b)$.

15. The set of functions $e: x \to x$, $f: x \to -x$, $g: x \to \frac{1}{x}$, $h: x \to -\frac{1}{x}$ under composition of functions.

16. The set of matrices of the form $\begin{pmatrix} 1 & p \\ 0 & 1 \end{pmatrix}$, where $p \in \mathbb{Z}$, under matrix multiplication.

17. S is the set of all real numbers except 1 and the binary operation $*$ on S is defined by $a * b = ab - a - b + 2$. Show that $(S, *)$ is an Abelian group and explain why 1 must be excluded from the set S.

18. S is the set of ordered pairs (a, b), where $a, b \in \mathbb{R}$ and $a \neq 0$. The operation $*$ on S is defined by $(a, b) * (c, d) = (ac, bc + d)$. Show that $(S, *)$ is a non-Abelian group.

19. Given that, for all real x except $x = 0$ and $x = 1$, $e(x) = x$, $p(x) = \frac{x - 1}{x}$, $q(x) = \frac{1}{1 - x}$, $f(x) = 1 - x$, $g(x) = \frac{1}{x}$ and $h(x) = \frac{x}{x - 1}$, show that the set $A = \{e, p, q, f, g, h\}$ forms a group under composition of functions.

20. A set of six complex numbers forms a group under multiplication. Given that one element of the set is $\frac{1}{2}(1 + i\sqrt{3})$, find the other elements.

21. The following tables show a set $A = \{p, q, r, s\}$ under a binary operation $*$. In each case decide whether $(A, *)$ is a group. If $(A, *)$ does not form a group, state which of the group axioms do not hold.

(a)

$*$	p	q	r	s
p	p	q	r	s
q	q	t	p	r
r	r	p	s	q
s	s	r	q	t

(b)

$*$	p	q	r	s
p	q	r	s	p
q	p	q	r	s
r	s	p	q	r
s	r	s	p	q

(c)

$*$	p	q	r	s
p	r	s	p	q
q	s	s	q	p
r	p	q	r	s
s	q	p	s	r

(d)

$*$	p	q	r	s
p	r	s	q	p
q	s	r	p	q
r	q	p	s	r
s	p	q	r	s

22. For each of the systems $(A, *)$ defined in question 21, find the solution sets of the equations $p * x = q$ and $(x * r) * q = s$.

23. Prove that for any elements a, b, c of a group $(G, *)$,
(i) $(a * c = b * c) \Rightarrow (a = b)$,
(ii) $(a * c = c * b) \Rightarrow (a = c * b * c^{-1})$.

24.

$*$	I	A	B	C
I	I	A	B	C
A	A			I
B	B		I	
C	C	I		

Copy and complete the given table so that the set $\{I, A, B, C\}$ forms a group under $*$. Use the table to simplify
(a) $B * (C * A)$, (b) $(A * B) * (B * C)$.

25. (i) Prove that the set of non-zero real numbers forms a group under the operation $\circ$ defined by $x \circ y = 2xy$. State the identity element for this system, and give the inverse of x.
(ii) A set of four 2×2 matrices forms a group under matrix multiplication. Two members of the set are $\begin{pmatrix} -1 & 0 \\ 0 & 1 \end{pmatrix}$ and $\begin{pmatrix} 1 & 0 \\ 0 & -1 \end{pmatrix}$. Find the other members and write out the group table. (L)

26. Given that S is the set of all matrices of the form $\begin{pmatrix} a & b \\ 2b & a \end{pmatrix}$, where a and b are real, prove that S is closed under addition and multiplication. If a and b can take all real values, show that (i) S forms a group under addition, (ii) S does not form a group under multiplication. (The associative properties for addition and for multiplication may be assumed.) If a and b are rational and a is not zero, show that (iii) S does not form a group under addition, (iv) no member of S has zero determinant, (v) S forms a group under multiplication. (JMB)

19.2 The integers modulo *n*

The set of *integers modulo n*, often denoted by $\mathbb{Z}_n$, is the set of remainders or *residues* obtained when members of the set of integers are divided by n, where n is a positive integer. For instance, the set of integers modulo 5, $\mathbb{Z}_5 = \{0, 1, 2, 3, 4\}$. Arithmetic is performed in $\mathbb{Z}_n$ by replacing any sum or product by the appropriate remainder, e.g. $3 + 4 = 2 \pmod 5$ and $2 \times 4 = 3 \pmod 5$. The system consisting of the sets $\mathbb{Z}_n$ with their associated operations of addition and multiplication is given a variety of names, such as modulo arithmetic, modular arithmetic, finite arithmetic, and, in more abstract treatments, the algebra of residues.

[The elements of $\mathbb{Z}_n = \{0, 1, 2, \ldots, (n-1)\}$ are sometimes considered to represent infinite sets of integers. For instance, in $\mathbb{Z}_5$ the element 2 represents the set $\{\ldots, -8, -3, 2, 7, 12, 17, \ldots\}$ whose elements all leave remainder 2 when divided by 5.]

In this section we will be mainly concerned with group structures which arise in modulo arithmetic. Consider the addition and multiplication tables for the integers modulo 5.

$+_5$	0	1	2	3	4
0	0	1	2	3	4
1	1	2	3	4	0
2	2	3	4	0	1
3	3	4	0	1	2
4	4	0	1	2	3

$\times_5$	0	1	2	3	4
0	0	0	0	0	0
1	0	1	2	3	4
2	0	2	4	1	3
3	0	3	1	4	2
4	0	4	3	2	1

$\times_5$	1	2	3	4
1	1	2	3	4
2	2	4	1	3
3	3	1	4	2
4	4	3	2	1

$\mathbb{Z}_5$ is closed under both $+$ and $\times$.
Since addition and multiplication of integers is associative, these operations must also be associative in $\mathbb{Z}_5$.
$\mathbb{Z}_5$ contains the additive identity element 0 and the multiplicative identity element 1.
For addition, the inverses of 1, 2, 3 and 4 are 4, 3, 2 and 1 respectively.
For multiplication, the inverse of 2 is 3 and vice versa, 4 is self-inverse, but 0 has no inverse.
This means that $\mathbb{Z}_5$ forms a group under addition modulo 5 but not under multiplication. However, excluding 0, the set $\{1, 2, 3, 4\}$ does form a group under multiplication modulo 5.

Equations in $\mathbb{Z}_5$ can be solved using the additive and multiplicative inverses already obtained.

Example 1 Solve in $\mathbb{Z}_5$ the equation $2x + 3 = 1 \pmod 5$.

Performing all operations modulo 5:

$$
\begin{aligned}
2x + 3 = 1 \quad &\Rightarrow \quad (2x+3) + 2 = 1 + 2 && \text{Adding inverse of 3 for } + \\
&\Rightarrow \quad 2x + (3+2) = 1 + 2 && \text{Associativity of } + \\
&\Rightarrow \quad 2x + 0 = 3 && \\
&\Rightarrow \quad 2x = 3 && \\
&\Rightarrow \quad 3 \times 2x = 3 \times 3 && \text{Using inverse of 2 for } \times \\
&\Rightarrow \quad (3 \times 2)x = 3 \times 3 && \text{Associativity of } \times \\
&\Rightarrow \quad 1 \times x = 4 && \\
&\Rightarrow \quad x = 4 &&
\end{aligned}
$$

In general, the addition table for any set $\mathbb{Z}_n$ takes a similar form to the table for $\mathbb{Z}_5$. However, multiplication tables vary greatly in structure.

(1)

$\times_8$	1	2	3	4	5	6	7
1	1	2	3	4	5	6	7
2	2	4	6	0	2	4	6
3	3	6	1	4	7	2	5
4	4	0	4	0	4	0	4
5	5	2	7	4	1	6	3
6	6	4	2	0	6	4	2
7	7	6	5	4	3	2	1

(2)

$\times_8$	1	3	5	7
1	1	3	5	7
3	3	1	7	5
5	5	7	1	3
7	7	5	3	1

Table (1) shows that the set $\{1, 2, 3, 4, 5, 6, 7\}$ is not closed under multiplication modulo 8. However, from table (2) we see that the set $\{1, 3, 5, 7\}$ forms a group under this operation.

$\times_{10}$	2	4	6	8
2	4	8	2	6
4	8	6	4	2
6	2	4	6	8
8	6	2	8	4

It is more surprising to find that the set $\{2, 4, 6, 8\}$ forms a group under multiplication modulo 10. The identity element for the group is 6.

Exercise 19.2

1. Show that the set of integers $\{0, 1, 2, 3, 4, 5\}$ forms a group under addition modulo 6 but not under multiplication modulo 6. Decide whether the set $\{1, 2, 3, 4, 5\}$ forms a group under multiplication modulo 6.

2. Show that the set $\{0, 1, 2, 3, 4, 5, 6\}$ forms a group under addition modulo 7 and that the set $\{1, 2, 3, 4, 5, 6\}$ forms a group under multiplication modulo 7.

3. Solve the following equations in $\mathbb{Z}_5$ using the inverses of the elements with respect to addition and multiplication modulo 5.
(a) $x + 2 = 1$ (b) $4x = 3$ (c) $3x + 4 = 2$
(d) $4x + 3 = 0$ (e) $3(x + 1) = 0$ (f) $4x = 2x + 1$

4. Solve the following equations in $\mathbb{Z}_7$ by performing additions and multiplications modulo 7.
(a) $x + 4 = 3$ (b) $2x = 5$ (c) $3x + 2 = 3$
(d) $5x + 6 = 0$ (e) $6x + 3 = 5$ (f) $4x + 1 = 2x$

5. By investigating the subsets of the set $\{1, 2, 3, 4, 5, 6, 7, 8, 9\}$ which are closed under multiplication modulo 10, determine which of them form groups under this operation.

6. Find all possible groups of order 2 in $\mathbb{Z}_{12}$ under multiplication modulo 12.

7. Find all possible groups of order 4 in $\mathbb{Z}_{15}$ under multiplication modulo 15.

8. Show that the set $\mathbb{Z}_n$, where n is a positive integer, forms a group under addition modulo n.

9. Show that if $n = pq$, where both p and q are integers greater than 1, then the set $\mathbb{Z}_n$ does not form a group under multiplication modulo n.

19.3 Symmetries and permutations

The *symmetries* of a geometrical figure in two or three dimensional space are transformations of the figure in which distances between points and the region of space occupied by the figure remain unchanged. The two main types of symmetry are rotation about an axis of symmetry and reflection in a line or plane of symmetry.

The product AB of two symmetries A and B is defined as B followed by A. As composition of mappings is an associative operation, $A(BC) = (AB)C$ for all symmetries A, B, C. However, in general this operation is not commutative, $AB \neq BA$.

We demonstrate that the set of all symmetries of a given figure forms a group under composition by considering various examples.

(1) *The symmetries of a rectangle*

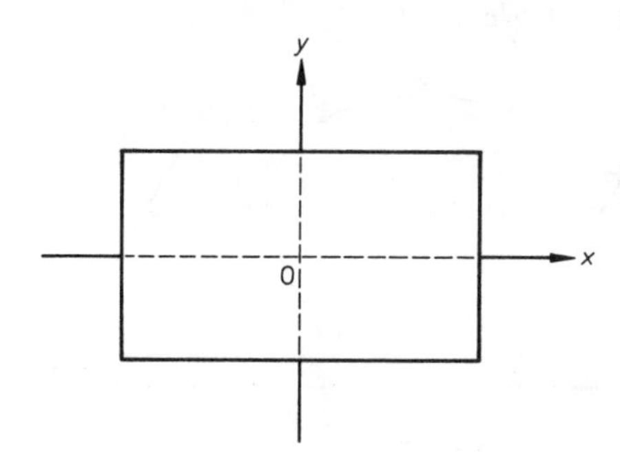

I: identity transformation
X: reflection in x-axis
Y: reflection in y-axis
H: half turn about O (i.e. about an axis through O perpendicular to the plane of the rectangle.)

Products of these symmetries can be obtained by labelling the corners of the rectangle then performing one transformation followed by another.

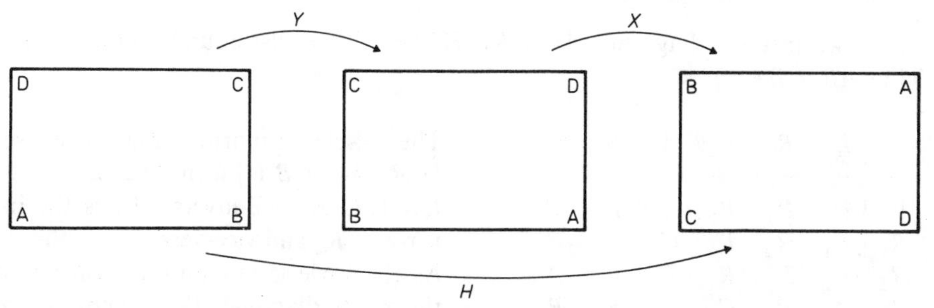

These diagrams show that Y followed by X has the same effect as H i.e. that $XY = H$.

	I	X	Y	H
I	I	X	Y	H
X	X	I	H	Y
Y	Y	H	I	X
H	H	Y	X	I

The operation table shows that the set $\{I, X, Y, H\}$ is closed under composition of symmetries.
The operation is associative.
The set contains the identity transformation. Each element is self-inverse.

Thus the symmetries of a rectangle form a group of order 4.

(2) *The symmetries of an equilateral triangle*

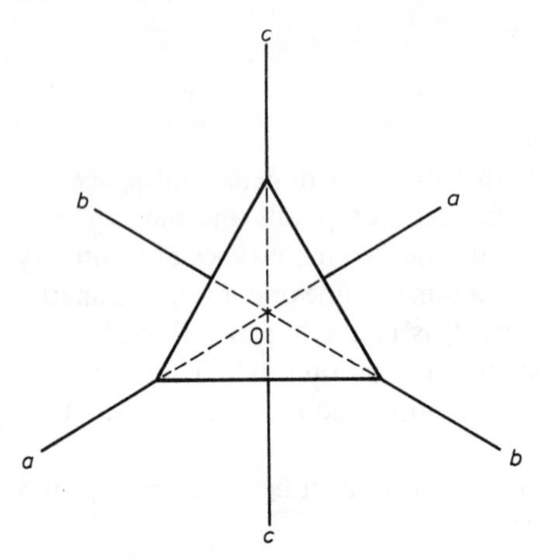

I: identity transformation
R_1: rotation through 120° anti-clockwise about O
R_2: rotation through 120° clockwise about O
A: reflection in a
B: reflection in b
C: reflection in c
[Note that the lines a, b, c are fixed in space and do not move with the triangle.]

In this more difficult example it may be helpful to make a cardboard copy of the triangle and use this to complete the operation table.

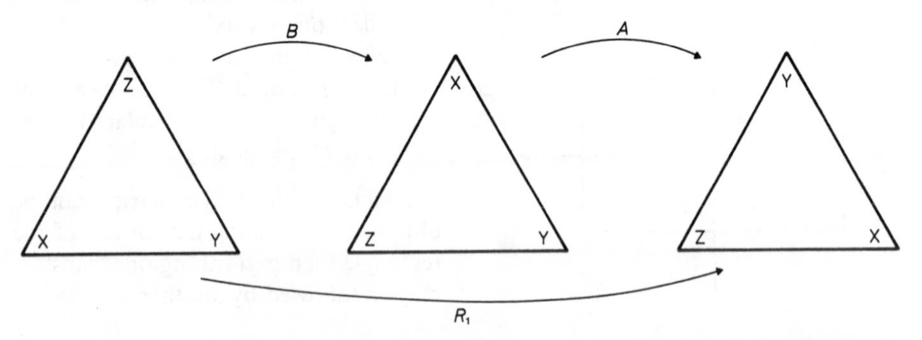

As shown in these diagrams, $AB = R_1$. However, it is also found that $BA = R_2$. Thus $AB \neq BA$.

	I	R_1	R_2	A	B	C
I	I	R_1	R_2	A	B	C
R_1	R_1	R_2	I	C	A	B
R_2	R_2	I	R_1	B	C	A
A	A	B	C	I	R_1	R_2
B	B	C	A	R_2	I	R_1
C	C	A	B	R_1	R_2	I

The table confirms that the set $\{I, R_1, R_2, A, B, C\}$ forms a group.
I, A, B, C are self-inverse. R_1 is the inverse of R_2 and vice-versa.
As the table is not symmetrical about the main diagonal, this group is not commutative.

(3) *The symmetries of the cuboid*

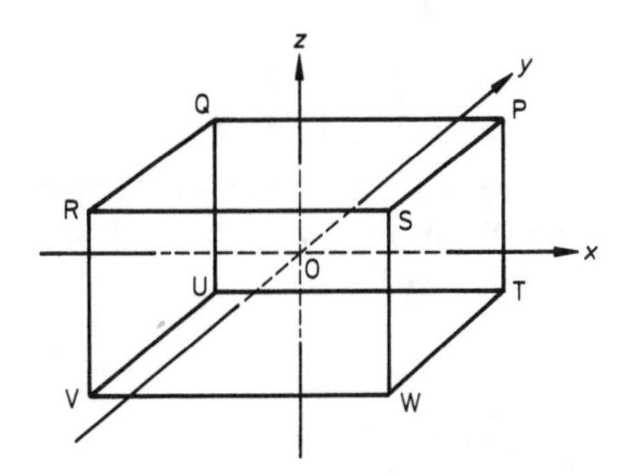

Initial position

I: identity transformation
X: reflection in y, z plane
Y: reflection in x, z plane
Z: reflection in x, y plane
H_1: half turn about x-axis
H_2: half turn about y-axis
H_3: half turn about z-axis
C: central inversion, (as shown in the second diagram).

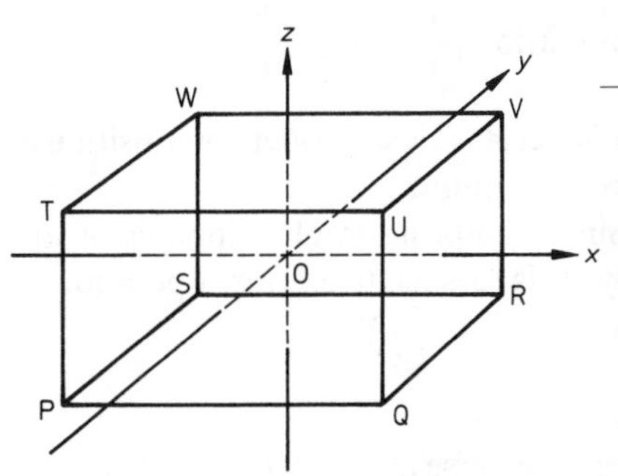

After central inversion

	I	H_1	H_2	H_3	X	Y	Z	C
I	I	H_1	H_2	H_3	X	Y	Z	C
H_1	H_1	I	H_3	H_2	C	Z	Y	X
H_2	H_2	H_3	I	H_1	Z	C	X	Y
H_3	H_3	H_2	H_1	I	Y	X	C	Z
X	X	C	Z	Y	I	H_3	H_2	H_1
Y	Y	Z	C	X	H_3	I	H_1	H_2
Z	Z	Y	X	C	H_2	H_1	I	H_3
C	C	X	Y	Z	H_1	H_2	H_3	I

The operation table shows that the symmetries of the cuboid form an Abelian group in which every element is self-inverse.

The symmetries of geometrical figures can be regarded as elements of a larger set of transformations called *isometries*. An isometry is a transformation of the points of two or three dimensional space in which distances between points remain unchanged. Thus under an isometry a geometrical figure retains its shape and size but may change its position in space. There are four types of *plane isometry*: rotation, reflection, translation and glide-reflection (i.e. reflection together with a translation in the direction of the mirror line). It can be shown that the set of all plane isometries form a group.

Any symmetry transformation of a geometrical figure results in a rearrangement of the vertices of the figure. These rearrangements form a subset of the set of all rearrangements or permutations of the vertices. We next consider a more abstract approach to transformations which rearrange the objects in a set.

A *permutation* is defined as a mapping of a finite set onto itself. If X is the permutation of the set $\{1, 2, 3, 4\}$ which maps $1 \to 3, 2 \to 4, 3 \to 2, 4 \to 1$, then X may be represented by the array $\begin{pmatrix} 1 & 2 & 3 & 4 \\ 3 & 4 & 2 & 1 \end{pmatrix}$.

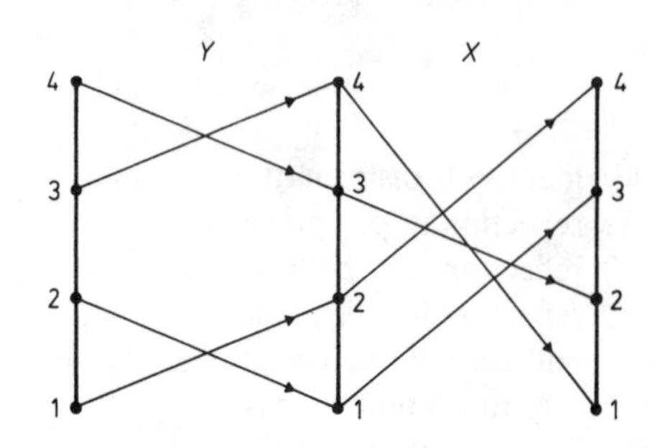

If $X = \begin{pmatrix} 1 & 2 & 3 & 4 \\ 3 & 4 & 2 & 1 \end{pmatrix}$ and

$$Y = \begin{pmatrix} 1 & 2 & 3 & 4 \\ 2 & 1 & 4 & 3 \end{pmatrix},$$

then the product XY, which means "first Y then X", maps $1 \to 2 \to 4$, $2 \to 1 \to 3$, $3 \to 4 \to 1$, $4 \to 3 \to 2$.

Hence $XY = \begin{pmatrix} 1 & 2 & 3 & 4 \\ 3 & 4 & 2 & 1 \end{pmatrix}\begin{pmatrix} 1 & 2 & 3 & 4 \\ 2 & 1 & 4 & 3 \end{pmatrix} = \begin{pmatrix} 1 & 2 & 3 & 4 \\ 4 & 3 & 1 & 2 \end{pmatrix}$.

Writing $Z = XY$, we can produce an operation table for the set of permutations $A = \{I, X, Y, Z\}$, where I is the identity permutation $\begin{pmatrix} 1 & 2 & 3 & 4 \\ 1 & 2 & 3 & 4 \end{pmatrix}$.

	I	X	Y	Z
I	I	X	Y	Z
X	X	Y	Z	I
Y	Y	Z	I	X
Z	Z	I	X	Y

The set A is closed under composition of permutations.
Since composition of mappings of any type is associative, this operation is associative.

A contains the identity I and each element has an inverse in A. Hence the set A forms a group.

It can also be shown that the set S_4 of *all* permutations of the set $\{1, 2, 3, 4\}$ forms a group. As before, we can assume the associative law for composition of mappings in S_4. Since S_4 contains all possible permutations, it contains the identity permutation and must also contain the product of any given pair of permutations. To demonstrate the existence of inverses we consider a typical element of S_4. If the permutation T maps $1 \to 2$, $2 \to 4$, $3 \to 3$, $4 \to 1$, then the inverse permutation T^{-1} maps $2 \to 1$, $4 \to 2$, $3 \to 3$, $1 \to 4$,

i.e. if $T = \begin{pmatrix} 1 & 2 & 3 & 4 \\ 2 & 4 & 3 & 1 \end{pmatrix}$, then $T^{-1} = \begin{pmatrix} 1 & 2 & 3 & 4 \\ 4 & 1 & 3 & 2 \end{pmatrix}$.

Clearly the inverse of any element can be found in this way. Since $XT = \begin{pmatrix} 1 & 2 & 3 & 4 \\ 4 & 1 & 2 & 3 \end{pmatrix}$ and $TX = \begin{pmatrix} 1 & 2 & 3 & 4 \\ 3 & 1 & 4 & 2 \end{pmatrix}$, composition of permutations is not usually commutative.

Thus the set S_4 forms a non-Abelian group. Since the number of different arrangements of 4 objects is 4!, the order of the permutation group S_4 is 24.

> In general, the set of all permutations of a set with n elements forms a permutation group of order $n!$ called the *symmetric group* S_n.

Exercise 19.3

1. Use the group table for the symmetries of an equilateral triangle to simplify

(a) ABC (b) AR_1BR_2

(c) $(R_1)^2R_2$ (d) $R_1^{-1}(R_2^{-1})^2$

2. Use the group table for the symmetries of a cuboid to verify that

(a) $X(YZ) = (XY)Z$ (b) $H_1(CX) = (H_1C)X$

3. Make a group table for the group of symmetries of a square using the following notation:

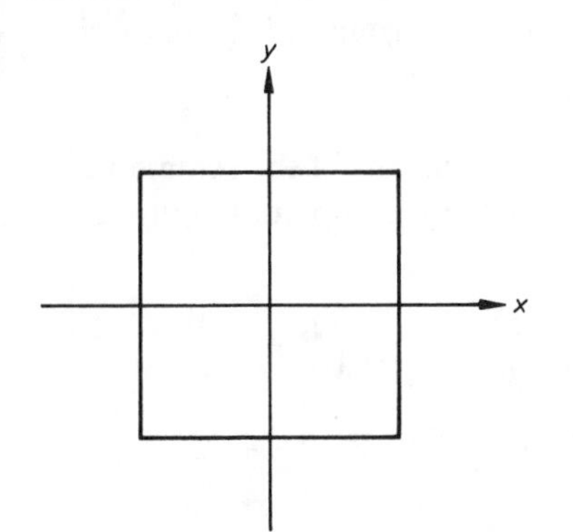

I: identity
R_1: rotation through 90° anti-clockwise
R_2: rotation through 180° anti-clockwise
R_3: rotation through 270° anti-clockwise
X: reflection in the x-axis
Y: reflection in the y-axis
A: reflection in the line $y = x$
B: reflection in the line $y = -x$

4. Discuss the symmetry groups of the following plane figures, stating the order of the group in each case.

(a) a parallelogram, (b) a rhombus,

(c) a regular pentagon, (d) a circle.

5. Discuss the symmetry groups of the following solids, stating the order of the group in each case.

(a) a pyramid with a square base,
(b) a prism whose cross-section is an isosceles triangle,
(c) a cylinder with an elliptical cross-section,
(d) a prism whose cross-section is an equilateral triangle.

6. Are the following statements about plane isometries true in all cases? If false, justify your answer.

(a) A reflection in one line followed by a reflection in another line results in a rotation.
(b) A rotation followed by a translation is equivalent to a rotation about a different centre.
(c) A reflection followed by a translation is equivalent to a glide-reflection.
(d) A rotation about A followed by a rotation about B results in a rotation about a third point C.
(e) A rotation followed by a reflection results in a reflection in a different mirror line.

7. Find the following products of permutations:

(a) $\begin{pmatrix} 1 & 2 & 3 & 4 \\ 1 & 3 & 2 & 4 \end{pmatrix}\begin{pmatrix} 1 & 2 & 3 & 4 \\ 3 & 2 & 4 & 1 \end{pmatrix}$ (b) $\begin{pmatrix} 1 & 2 & 3 & 4 \\ 2 & 3 & 1 & 4 \end{pmatrix}\begin{pmatrix} 1 & 2 & 3 & 4 \\ 4 & 2 & 1 & 3 \end{pmatrix}$

(c) $\begin{pmatrix} 1 & 2 & 3 & 4 \\ 3 & 4 & 2 & 1 \end{pmatrix}\begin{pmatrix} 1 & 2 & 3 & 4 \\ 3 & 2 & 4 & 1 \end{pmatrix}\begin{pmatrix} 1 & 2 & 3 & 4 \\ 3 & 1 & 4 & 2 \end{pmatrix}$

(d) $\begin{pmatrix} 1 & 2 & 3 & 4 \\ 1 & 4 & 2 & 3 \end{pmatrix}\begin{pmatrix} 1 & 2 & 3 & 4 \\ 2 & 3 & 4 & 1 \end{pmatrix}\begin{pmatrix} 1 & 2 & 3 & 4 \\ 4 & 3 & 2 & 1 \end{pmatrix}$

8. Write down the six arrays representing the elements of the set of permutations of the integers $\{1, 2, 3\}$. Prove that this set forms a group under composition of mappings.

9. Given below are arrays representing certain elements of the symmetric group S_5. In each case determine the order of the smallest group containing the given element(s).

(a) $\begin{pmatrix} 1 & 2 & 3 & 4 & 5 \\ 1 & 4 & 3 & 5 & 2 \end{pmatrix}$ (b) $\begin{pmatrix} 1 & 2 & 3 & 4 & 5 \\ 2 & 1 & 5 & 4 & 3 \end{pmatrix}$ (c) $\begin{pmatrix} 1 & 2 & 3 & 4 & 5 \\ 4 & 5 & 1 & 2 & 3 \end{pmatrix}$

(d) $\begin{pmatrix} 1 & 2 & 3 & 4 & 5 \\ 4 & 5 & 1 & 3 & 2 \end{pmatrix}$ (e) $\begin{pmatrix} 1 & 2 & 3 & 4 & 5 \\ 3 & 1 & 5 & 4 & 2 \end{pmatrix}, \begin{pmatrix} 1 & 2 & 3 & 4 & 5 \\ 3 & 5 & 1 & 4 & 2 \end{pmatrix}$

10. The symmetries of the square $ABCD$ are related to the permutations of its vertices. For instance, the permutation $\begin{pmatrix} A & B & C & D \\ B & C & D & A \end{pmatrix}$ is equivalent to a rotation of the square through 90°. Describe the transformations of the square $ABCD$ equivalent to the following:

(a) $\begin{pmatrix} A & B & C & D \\ C & D & A & B \end{pmatrix}$ (b) $\begin{pmatrix} A & B & C & D \\ C & B & A & D \end{pmatrix}$ (c) $\begin{pmatrix} A & B & C & D \\ D & C & B & A \end{pmatrix}$

Do all permutations of the vertices A, B, C, D correspond to symmetries of the square $ABCD$?

19.4 Properties of groups

First we use the group axioms to obtain two basic properties of groups. To simplify the working we will now write the product of elements a and b as ab instead of $a * b$.

Example 1 Prove that in any group G, (a) the identity element is unique, (b) the inverse of any element is unique.

(a) Let e_1 and e_2 be two identity elements in G,
then since e_1 is an identity element, $e_1 e_2 = e_2$
and since e_2 is an identity element, $e_1 e_2 = e_1$
Hence $e_1 = e_2$.

(b) Let an element a of G have two inverses b and c
then $ab = ba = e$ and $ac = ca = e$,
where e is the identity element in G.

$$\begin{aligned} ab = e \quad &\Rightarrow \quad c(ab) = ce \\ &\Rightarrow \quad (ca)b = c \\ &\Rightarrow \quad eb = c \\ &\Rightarrow \quad b = c \end{aligned}$$

Thus the element a has a unique inverse.

If a subset H of a group $(G, *)$ also forms a group under $*$, then H is said to be a *subgroup* of G.

Since the associative law for $*$ holds in G, it must hold in any subset of G. Thus to prove that a subset H is a subgroup, it is necessary to show that
(i) H is closed under $*$,
(ii) H contains the identity element,
(iii) if $x \in H$, then $x^{-1} \in H$.

The given operation table defines a group $(G, *)$. Suppose that we wish to list all the subgroups of G. To satisfy condition (ii) we need consider only subsets of G which contain the identity element e, namely: $\{e\}$, $\{e,a\}$, $\{e,b\}$, $\{e,c\}$, $\{e,a,b\}$, $\{e,a,c\}$, $\{e,b,c\}$ and G itself. From the table we see that the sets $\{e\}$, $\{e,a\}$, $\{e,b\}$, $\{e,c\}$ and G are closed under $*$, but that the remaining sets are not. Since every element of G is self-inverse, property (iii) is also satisfied in these five sets. Hence G has five subgroups, $\{e\}$, $\{e,a\}$, $\{e,b\}$, $\{e,c\}$ and G itself.

$*$	e	a	b	c
e	e	a	b	c
a	a	e	c	b
b	b	c	e	a
c	c	b	a	e

Every group G has two subgroups, $\{e\}$ and G. Any other subgroup is described as a *proper subgroup*. In this example, G has three proper subgroups: $\{e,a\}$, $\{e,b\}$, $\{e,c\}$.

Let us consider now the group formed by the set $\{1,2,3,4,5,6\}$ under multiplication modulo 7. Since 6 is self-inverse the set $\{1,6\}$ is a subgroup of order 2. The table also shows that the set $\{1,2,4\}$ forms a subgroup of order 3. Thus this group has two proper subgroups. This example illustrates a general result which is often useful when studying the subgroups of a finite group.

$\times_7$	1	2	3	4	5	6
1	1	2	3	4	5	6
2	2	4	6	1	3	5
3	3	6	2	5	1	4
4	4	1	5	2	6	3
5	5	3	1	6	4	2
6	6	5	4	3	2	1

Lagrange's theorem: The order of any subgroup of a finite group is a factor of the order of the group.

[A method of proof is indicated in Exercise 19.4, question 27.]

† *Lagrange, Joseph Louis* (1736–1813) French mathematician. His main work was *Mécanique Analytique* (1788). He also made advances in the theory of equations which helped in the formulation of group theory.

Example 2 A set S is defined as the set of all elements of a group G which commute with every element of G i.e. $a \in S$ if and only if $ax = xa$ for every element $x \in G$. Prove that S is a subgroup of G.

If $a, b \in S$, then $ax = xa$, $bx = xb$ for all $x \in G$
$\therefore \quad (ab)x = a(bx) = a(xb) = (ax)b = (xa)b = x(ab)$
i.e. $(ab)x = x(ab)$ for all $x \in G$
$\therefore \quad$ if a and b belong to S, ab also belongs to S.
Hence S is closed under the group operation.
If e is the identity element in G,

$$ex = xe = x \quad \text{for all} \quad x \in G$$

Hence the identity element e belongs to S.

$$\begin{aligned} ax = xa \quad &\Rightarrow \quad a^{-1}(ax)a^{-1} = a^{-1}(xa)a^{-1} \\ &\Rightarrow \quad (a^{-1}a)(xa^{-1}) = (a^{-1}x)(aa^{-1}) \\ &\Rightarrow \quad e(xa^{-1}) = (a^{-1}x)e \\ &\Rightarrow \quad xa^{-1} = a^{-1}x \end{aligned}$$

Hence if a belongs to S, a^{-1} also belongs to S.
Therefore the set S is a subgroup of the group G.

In a group G with identity element e, the *order* or *period* of any element x is the least positive integer n such that $x^n = e$.

The only element of order 1 in any group is the identity itself. If an element x has order 2, then $x^2 = e$. It follows that any element of order 2 is self-inverse.

A group whose elements can be expressed in the form $e, x, x^2, x^3, \ldots, x^{n-1}$ is called a *cyclic group* of order n and denoted by C_n. The element x is said to *generate* the group and is described as a *generator* of the group.

We now give two examples of cyclic groups.

(1) The rotational symmetries of an equilateral triangle.

	I	R_1	R_2
I	I	R_1	R_2
R_1	R_1	R_2	I
R_2	R_2	I	R_1

Using the notation of §19.3, the elements of the group can be expressed either as $\{I, R_1, R_1^2\}$ or as $\{I, R_2, R_2^2\}$.
Hence this group is a cyclic group of order 3, which may be generated by R_1 or R_2.

(2) The set $\{1, 2, 3, 4\}$ under multiplication modulo 5.

$\times_5$	1	2	3	4
1	1	2	3	4
2	2	4	1	3
3	3	1	4	2
4	4	3	2	1

The elements of the group can be expressed either as $\{1, 2, 2^2, 2^3\}$ or a $\{1, 3, 3^2, 3^3\}$.
Hence this group is a cyclic group of order 4, which may be generated by 2 or 3. Both 2 and 3 are elements of order 4, whereas the element 4 is of order 2 and generates the subgroup $\{1, 4\}$.

Example 3 Prove that every cyclic group is Abelian.

If a and b are elements of a cyclic group with generator x, then there are positive integers p and q such that $a = x^p$ and $b = x^q$.
Thus $ab = (x^p)(x^q) = x^{p+q} = x^{q+p} = (x^q)(x^p) = ba$.
Hence every cyclic group is Abelian.

Example 4 Prove that the cyclic group C generated by an element x of order n has n distinct elements.

The elements of C are all of the form x^p, where p is a positive integer. Since x is of order n, $x^n = e$, where e is the identity.
Considering the division of p by n, we see that there are integers q and r such that

$$p = nq + r, \quad \text{where} \quad q \geqslant 0, \quad 0 \leqslant r \leqslant n-1$$

If $r = 0$, $x^p = x^{nq} = (x^n)^q = e^q = e$.
If $r \neq 0$, $x^p = x^{nq+r} = (x^n)^q(x^r) = e(x^r) = x^r$.
Thus, for all values of p, x^p is an element of the set $\{e, x, x^2, x^3, \ldots, x^{n-1}\}$.
If $\quad x^s = x^t, \quad \text{where} \quad 0 < t < s \leqslant n$
then $\quad x^{s-t} = e$
$\therefore \quad s - t$ is a multiple of n.
Since this is impossible if s and t are both less than or equal to n, the n elements $e, x, x^2, x^3, \ldots, x^{n-1}$ must be distinct. Hence the group C generated by x is of order n.

It follows from Example 4 that:

> A group of order n is cyclic if and only if it contains an element of order n.

	I	X	Y	H
I	I	X	Y	H
X	X	I	H	Y
Y	Y	H	I	X
H	H	Y	X	I

Considering again the symmetries of a rectangle as defined in §19.3, we find that each of the elements X, Y, H is of order 2 and generates a cyclic subgroup of order 2. There is no element of order 4 which generates the whole group. Hence this group is not cyclic.

The elements of a group which is not cyclic can be expressed in terms of two or more generators. Although the group $\{I, X, Y, H\}$ has no single generator, the whole group is generated by the two elements X and Y. Referring back to Example 3, §19.1 we find that the elements of the group $\{\mathbf{I}, \mathbf{A}, \mathbf{B}, \mathbf{C}, \mathbf{D}, \mathbf{E}, \mathbf{F}, \mathbf{G}\}$ can be expressed as $\mathbf{I}, \mathbf{A}, \mathbf{A}^2, \mathbf{A}^3, \mathbf{D}, \mathbf{AD}, \mathbf{A}^2\mathbf{D}, \mathbf{A}^3\mathbf{D}$ respectively. In this case two elements $\mathbf{A}$ and $\mathbf{D}$ generate a group of order 8.

Exercise 19.4

1. Given that a and b are elements of a group G, show that
(i) $(ab)^{-1} = b^{-1}a^{-1}$, (ii) $(a^{-1})^{-1} = a$, (iii) $(a^2)^{-1} = (a^{-1})^2$.

2. Given that a, b, x and y belong to a group G, show that
(i) $ax = ay \;\Rightarrow\; x = y$, (ii) $ax = b \;\Rightarrow\; x = a^{-1}b$.

In questions 3 to 10 list the proper subgroups of the given groups.

3. The set containing the sets $\varnothing, X = \{1\}, Y = \{2\}, Z = \{1, 2\}$ under the operation symmetric difference Δ, where $A \,\Delta\, B = (A \cap B') \cup (A' \cap B)$.

4. The set of functions $\{e, f, g, h\}$ under function composition, where $e(x) = x$, $f(x) = 1 - \dfrac{1}{2x}$, $g(x) = \dfrac{1-x}{1-2x}$, $h(x) = \dfrac{1}{2-2x}$.

5. The set containing the matrices, $\mathbf{I} = \begin{pmatrix} 1 & 0 & 0 \\ 0 & 1 & 0 \\ 0 & 0 & 1 \end{pmatrix}$, $\mathbf{A} = \begin{pmatrix} 0 & 0 & 1 \\ 1 & 0 & 0 \\ 0 & 1 & 0 \end{pmatrix}$,

$\mathbf{B} = \begin{pmatrix} 0 & 1 & 0 \\ 0 & 0 & 1 \\ 1 & 0 & 0 \end{pmatrix}$, $\mathbf{C} = \begin{pmatrix} 0 & 1 & 0 \\ 1 & 0 & 0 \\ 0 & 0 & 1 \end{pmatrix}$, $\mathbf{D} = \begin{pmatrix} 1 & 0 & 0 \\ 0 & 0 & 1 \\ 0 & 1 & 0 \end{pmatrix}$, $\mathbf{E} = \begin{pmatrix} 0 & 0 & 1 \\ 0 & 1 & 0 \\ 1 & 0 & 0 \end{pmatrix}$, under matrix multiplication.

6. The symmetry group of the equilateral triangle as defined in §19.3.

7. The set $\{2, 4, 8, 10, 14, 16\}$ under multiplication modulo 18.

8. The set $\left\{\pm 1, \pm i, \pm\dfrac{1}{\sqrt{2}}(1+i), \pm\dfrac{1}{\sqrt{2}}(1-i)\right\}$ under multiplication.

9. The group $S = \{\mathbf{I}, \mathbf{A}, \mathbf{B}, \mathbf{C}, \mathbf{D}, \mathbf{E}, \mathbf{F}, \mathbf{G}\}$ defined in Example 3 §19.1.

10. The symmetry group of the cuboid as defined in §19.3.

11. Decide whether the following sets are subgroups of the group $(\mathbb{Z}, +)$.
(a) the set of even integers,
(b) the set of odd integers,
(c) the set $\{0, 10, 20, 30, 40, \ldots\}$
(d) the set $\{\ldots, -6, -3, 0, 3, 6, 9, \ldots\}$.

12. Decide whether the following sets are subgroups of the group formed by the set of non-zero real numbers under multiplication.
(a) the set of non-zero integers,
(b) the set of positive rational numbers,
(c) the set $\{x \in \mathbb{R}: 0 < x \leqslant 1\}$,
(d) the set of real numbers of the form $a\sqrt{3} + b$, where a and b are non-zero rational numbers.

In questions 13 to 20 determine whether the given group is cyclic and if so list the elements which are generators.

13. The set $\{1, 3, 5, 7\}$ under multiplication modulo 8.

14. The symmetries of a square.

15. The set $\{0, 1, 2, 3, 4\}$ under addition modulo 5.

16. The set $\{1, i, -1, -i\}$ under multiplication.

17. The functions which map x onto x, $1/x$, $1-x$, $1-1/x$, $1/(1-x)$, $x/(x-1)$ under composition of functions.

18. The rotations of a regular hexagon.

19. The set $\{1, 2, 3, 4, 5, 6\}$ under multiplication modulo 7.

20. The permutations of the set $\{1, 2, 3, 4\}$ in which 2 is mapped onto itself.

21. A cyclic group has elements $e, x, x^2, x^3, x^4, x^5, x^6, x^7$. Which of these elements generate the whole group? Find the orders of the remaining elements.

22. Repeat question 21 for the cyclic group with elements $e, a, a^2, a^3, a^4, a^5, a^6$.

23.

	I	A	B	C	D	E	F	G
I	I	A	B	C	D	E	F	G
A	A	B	C	I	E	F	G	D
B	B	C	I	A	F	G	D	E
C	C	I	A	B	G	D	E	F
D	D	G	F	E	B	A	I	C
E	E	D	G	F	C	B	A	I
F	F	E	D	G	I	C	B	A
G	G	F	E	D	A	I	C	B

Find the orders of the elements of the group defined in the given table. Deduce that the group is not cyclic. Show that the group has four proper subgroups, which are all cyclic.

24. The matrices $I = \begin{pmatrix} 1 & 0 \\ 0 & 1 \end{pmatrix}$, $A = \begin{pmatrix} i & 0 \\ 0 & -i \end{pmatrix}$, $B = \begin{pmatrix} -1 & 0 \\ 0 & -1 \end{pmatrix}$, $C = \begin{pmatrix} -i & 0 \\ 0 & i \end{pmatrix}$, $D = \begin{pmatrix} 0 & 1 \\ 1 & 0 \end{pmatrix}$, $E = \begin{pmatrix} 0 & i \\ -i & 0 \end{pmatrix}$, $F = \begin{pmatrix} 0 & -1 \\ -1 & 0 \end{pmatrix}$, $G = \begin{pmatrix} 0 & -i \\ i & 0 \end{pmatrix}$, where $i^2 = -1$, form a group of order 8 under matrix multiplication. (Do not prove this result.)

(i) Determine the order of each element.
(ii) Write down a subgroup of order 2.
(iii) Write down a cyclic subgroup of order 4 and explain why it is cyclic.
(iv) Write down a non-cyclic subgroup of order 4 and explain why it is non-cyclic.
(v) Prove that the given group is non-commutative.
(vi) Write down two elements which do commute with every element of the group.

(JMB)

25. Two elements a and b of a group G are such that $a^2 = (ab)^2 = b^4$. Deduce that $aba = b^3$ and $bab = a$. Given that a is of order 4, find the orders of the elements a^3b and b^3a.

26. Prove that if S and T are distinct subgroups of a group G then $S \cap T$ is also a subgroup of G. Illustrate this result by taking G to be the set $\{1, 2, 3, \ldots, 12\}$ under multiplication modulo 13 and letting S and T be the subgroups of order 4 and 6 respectively. Demonstrate that in this case $S \cup T$ is not a subgroup of G.

27. A finite group G of order n has a proper subgroup H of order p. The set xH, where $x \in G$, is defined as the set containing all products of the form xh, where $h \in H$. Prove that (a) xH has p distinct elements, (b) if xH and yH have a common element z, then $xH = yH$, (c) G can be expressed as the union of disjoint sets of the form xH, (d) p is a factor of n.

19.5 Isomorphisms

Two groups G and H which have precisely the same structure are said to be *isomorphic* and we write $G \cong H$.

One elementary way of showing that two groups are isomorphic is to demonstrate that their operation tables are equivalent. For instance, if $G = \{1, 2, 3, 4\}$ under multiplication modulo 5 and $H = \{2, 4, 6, 8\}$ under multiplication modulo 10, we obtain the tables:

$\times_5$	1	2	3	4
1	1	2	3	4
2	2	4	1	3
3	3	1	4	2
4	4	3	2	1

$\times_{10}$	2	4	6	8
2	4	8	2	6
4	8	6	4	2
6	2	4	6	8
8	6	2	8	4

We now rearrange the tables:

$\times_5$	1	2	4	3
1	1	2	4	3
2	2	4	3	1
4	4	3	1	2
3	3	1	2	4

$\times_{10}$	6	2	4	8
6	6	2	4	8
2	2	4	8	6
4	4	8	6	2
8	8	6	2	4

Since the tables now have exactly the same structure the groups G and H must be isomorphic.

More formally, two groups $(G, *)$ and $(G', \circ)$ are said to be isomorphic if there is a one-one correspondence between the elements $a, b, \ldots \in G$ and $a', b', \ldots \in G'$, such that if $a \leftrightarrow a'$ and $b \leftrightarrow b'$, then $a * b \leftrightarrow a' \circ b'$. This one-one correspondence between G and G' is called an *isomorphism.*

In the above example, the rearrangement of the tables was equivalent to setting up the isomorphism $1 \leftrightarrow 6, 2 \leftrightarrow 2, 4 \leftrightarrow 4, 3 \leftrightarrow 8$.

$\times_5$	1	3	4	2
1	1	3	4	2
3	3	4	2	1
4	4	2	1	3
2	2	1	3	4

A further rearrangement of the first table shows that there is a different isomorphism,

$$1 \leftrightarrow 6, 3 \leftrightarrow 2, 4 \leftrightarrow 4, 2 \leftrightarrow 8.$$

Thus it may be possible to set up an isomorphism between two groups in more than one way.

Consider now two further groups of order 4: the set of symmetries of a rectangle, $S = \{I, X, Y, H\}$, and the set of integers $\{1, 3, 5, 7\}$ under multiplication modulo 8.

	I	X	Y	H
I	I	X	Y	H
X	X	I	H	Y
Y	Y	H	I	X
H	H	Y	X	I

$\times_8$	1	3	5	7
1	1	3	5	7
3	3	1	7	5
5	5	7	1	3
7	7	5	3	1

The tables show that the mapping $I \leftrightarrow 1, X \leftrightarrow 3, Y \leftrightarrow 5, H \leftrightarrow 7$ is an isomorphism, $\therefore$ the groups S and T are isomorphic, $S \cong T$.

In any rearrangement of these group tables, the identity elements always appear along the main diagonal. It is therefore impossible to set up an isomorphism between S and T and the earlier groups of order 4, G and H. These pairs of isomorphic groups can be regarded as examples of two distinct abstract groups.

	e	a	b	c
e	e	a	b	c
a	a	b	c	e
b	b	c	e	a
c	c	e	a	b

	e	a	b	c
e	e	a	b	c
a	a	e	c	b
b	b	c	e	a
c	c	b	a	e

The first is the cyclic group of order 4, C_4, and consists of an identity element, one element of order 2 and two elements of order 4. The second is called the Klein[†] four-group and has an identity and three elements of order 2. These are the only possible groups of order 4.

This discussion of groups of order 4 suggests that we can test a pair of groups for isomorphism by comparing the orders of the elements in each group. It follows from the definition of an isomorphism that if $a \leftrightarrow a'$ then $a^n \leftrightarrow (a')^n$. Hence in any isomorphism between groups G and G', any element of G must correspond to an element of the same order in G'.

[†] *Klein, Felix* (1849–1925) German mathematician. In his *Erlangen Programm* he attempted a unification of mathematics in terms of group theory.

Example 1 Decide whether any of the groups, G_1, G_2, G_3 defined below are isomorphic.

G_1 is the set $\{0, 1, 2, 3, 4, 5\}$ under addition modulo 6.
G_2 is the set $\{1, 2, 3, 4, 5, 6\}$ under multiplication modulo 7.
G_3 is the set $\{I, R_1, R_2, A, B, C\}$ of symmetries of an equilateral triangle, as defined in §19.3.

$+_6$	0	1	2	3	4	5
0	0	1	2	3	4	5
1	1	2	3	4	5	0
2	2	3	4	5	0	1
3	3	4	5	0	1	2
4	4	5	0	1	2	3
5	5	0	1	2	3	4

$\times_7$	1	2	3	4	5	6
1	1	2	3	4	5	6
2	2	4	6	1	3	5
3	3	6	2	5	1	4
4	4	1	5	2	6	3
5	5	3	1	6	4	2
6	6	5	4	3	2	1

	I	R_1	R_2	A	B	C
I	I	R_1	R_2	A	B	C
R_1	R_1	R_2	I	C	A	B
R_2	R_2	I	R_1	B	C	A
A	A	B	C	I	R_1	R_2
B	B	C	A	R_2	I	R_1
C	C	A	B	R_1	R_2	I

From the group tables we see that G_1 and G_2 are both Abelian with one element of order 2, whereas G_3 is non-Abelian with three elements of order 2. Hence G_3 is isomorphic to neither G_1 nor G_2, but G_1 may be isomorphic to G_2.

In G_1 the element 1 is of order 6. To set up an isomorphism between G_1 and G_2 we must find an element of order 6 in G_2.

In G_2 1 is the identity and, since $2^3 = 1 \pmod 7$, the element 2 is of order 3. However, working modulo 7:

$$3^1 = 3, \qquad 3^2 = 2, \qquad 3^3 = 2 \times 3 = 6,$$
$$3^4 = 6 \times 3 = 4, \quad 3^5 = 4 \times 3 = 5, \quad 3^6 = 5 \times 3 = 1.$$

Thus in G_2 the element 3 is of order 6. Hence the required isomorphism between G_1 and G_2 is given by:

$$\begin{array}{lll}
1 & \leftrightarrow 3^1 \quad \text{i.e.} & 1 \leftrightarrow 3 \\
1+1 & \leftrightarrow 3^2 & 2 \leftrightarrow 2 \\
1+1+1 & \leftrightarrow 3^3 & 3 \leftrightarrow 6 \\
1+1+1+1 & \leftrightarrow 3^4 & 4 \leftrightarrow 4 \\
1+1+1+1+1 & \leftrightarrow 3^5 & 5 \leftrightarrow 5 \\
1+1+1+1+1+1 & \leftrightarrow 3^6 & 0 \leftrightarrow 1
\end{array}$$

$\times_7$	1	3	2	6	4	5
1	1	3	2	6	4	5
3	3	2	6	4	5	1
2	2	6	4	5	1	3
6	6	4	5	1	3	2
4	4	5	1	3	2	6
5	5	1	3	2	6	4

We can now rearrange the operation table for G_2 so that it has the same structure as the table for G_1.

Example 2 Show that the mapping $f: x \to 2^x$ from the set of integers $\mathbb{Z}$ to the set $S = \{\ldots, \frac{1}{4}, \frac{1}{2}, 1, 2, 4, 8, \ldots\}$ sets up an isomorphism between the groups $(\mathbb{Z}, +)$ and $(S, \times)$.

$f(x) = f(y) \quad \Rightarrow \quad 2^x = 2^y \quad \Rightarrow \quad x = y$

Hence f is a one-one correspondence between $\mathbb{Z}$ and S.

$f(x + y) = 2^{x+y} = (2^x) \times (2^y) = f(x) \times f(y)$

Hence f maps the sum of two elements in $(\mathbb{Z}, +)$ to the product of the corresponding elements in $(S, \times)$.
Thus f sets up an isomorphism between $(\mathbb{Z}, +)$ and $(S, \times)$.

Exercise 19.5

1. Determine which of the following groups of order 4 are isomorphic.
(a) The set of integers $\{0, 1, 2, 3\}$ under addition modulo 4.
(b) The set of matrices $\mathbf{P} = \begin{pmatrix} 1 & 1 \\ 0 & 0 \end{pmatrix}$, $\mathbf{Q} = \begin{pmatrix} -1 & -1 \\ 0 & 0 \end{pmatrix}$, $\mathbf{R} = \begin{pmatrix} i & i \\ 0 & 0 \end{pmatrix}$, $\mathbf{S} = \begin{pmatrix} -i & -i \\ 0 & 0 \end{pmatrix}$ under matrix multiplication.
(c) The functions $e(x) = x, f(x) = -\dfrac{1}{x}, g(x) = \dfrac{1 + x}{1 - x}, h(x) = \dfrac{x - 1}{x + 1}$ under function composition.
(d) The set of symmetries of a cube containing the identity transformation and the 180° rotations about the three axes of symmetry which pass through the mid-points of opposite faces.
(e) The set of integers $\{1, 3, 7, 9\}$ under multiplication modulo 10.
(f) The functions $a: x \to 2 - x$, $b: x \to x$, $c: x \to \dfrac{x - 2}{x - 1}$, $d: x \to \dfrac{x}{x - 1}$ under function composition.

2. Determine which of the following groups of order 6 are isomorphic.
(a) The symmetry group of the equilateral triangle.
(b) The set $\{1, 2, 4, 5, 7, 8\}$ under multiplication modulo 9.
(c) The matrices $\begin{pmatrix} -1 & 0 \\ 0 & -1 \end{pmatrix}$, $\begin{pmatrix} \frac{1}{2} & -\frac{1}{2}\sqrt{3} \\ \frac{1}{2}\sqrt{3} & \frac{1}{2} \end{pmatrix}$ and their products under matrix multiplication.

(d) The permutation group with elements $I = \begin{pmatrix} 1 & 2 & 3 & 4 & 5 \\ 1 & 2 & 3 & 4 & 5 \end{pmatrix}$,

$$A = \begin{pmatrix} 1 & 2 & 3 & 4 & 5 \\ 4 & 5 & 3 & 1 & 2 \end{pmatrix}, B = \begin{pmatrix} 1 & 2 & 3 & 4 & 5 \\ 3 & 2 & 4 & 1 & 5 \end{pmatrix}, C = \begin{pmatrix} 1 & 2 & 3 & 4 & 5 \\ 4 & 2 & 1 & 3 & 5 \end{pmatrix}$$

$$D = \begin{pmatrix} 1 & 2 & 3 & 4 & 5 \\ 3 & 5 & 1 & 4 & 2 \end{pmatrix}, E = \begin{pmatrix} 1 & 2 & 3 & 4 & 5 \\ 1 & 5 & 4 & 3 & 2 \end{pmatrix}.$$

(e) The set $\{2, 4, 6, 8, 10, 12\}$ under multiplication modulo 14.

(f) The matrices $\begin{pmatrix} -1 & 0 \\ 0 & 1 \end{pmatrix}$, $\begin{pmatrix} \frac{1}{2} & \frac{1}{2}\sqrt{3} \\ \frac{1}{2}\sqrt{3} & -\frac{1}{2} \end{pmatrix}$ and their products under matrix multiplication.

3. Show that the set $\{3, 6, 9, 12\}$ forms a group under multiplication modulo 15, which may be assumed to be associative. Show that this group is isomorphic to that formed by the set $\{1, -1, i, -i\}$ under multiplication of complex numbers. Name a geometrical figure which has a symmetry group of order four which is not isomorphic to these groups. By considering the orders of the elements, give a reason for this non-isomorphism. (JMB)

4. Given that $S = \{1, 2, 3, 4\}$, show that S forms a group under multiplication modulo 5 and that three elements of S form a group under multiplication modulo 7. Deduce that there are elements $x, y, z, \in S$ such that the set $T = \{(x, x), (x, y), (x, z), (y, x), (y, y), (y, z)\}$ forms a group under the operation $*$, where $(a, b) * (c, d) = (ac\ [\text{mod } 5], bd\ [\text{mod } 7])$.

Show that the set of permutations on S which map 1 onto itself forms a group G. Explain why the set of permutations on S which map 2 onto 3 does not form a group.

Show that the group T is not isomorphic to the group G. Show further that the symmetry group of the regular hexagon has both a subgroup isomorphic to T and a subgroup isomorphic to G.

5. Given that f is an isomorphism from a group $(G, *)$ to a group $(H, \circ)$, prove that (a) if e is the identity in G, then $f(e)$ is the identity in H, (b) if x^{-1} is the inverse of x in G, then $f(x^{-1})$ is the inverse of $f(x)$ in H.

6. Prove that the mapping $f: x \to \ln x$ from the set of positive real numbers $\mathbb{R}^+$ to the set of real numbers $\mathbb{R}$ sets up an isomorphism between the groups $(\mathbb{R}^+, \times)$ and $(\mathbb{R}, +)$.

7. Prove that the mapping $f: x \to 2x$ from the set of integers $\mathbb{Z}$ to the set of even integers E sets up an isomorphism between the groups $(\mathbb{Z}, +)$ and $(E, +)$.

8. Decide which of the following mappings sets up an isomorphism of the group of non-zero real numbers under multiplication with itself.

$$a: x \to \frac{1}{x}, \quad b: x \to -x, \quad c: x \to 3x, \quad d: x \to x^3.$$

9. Prove that the mapping $f: a \to a^{-1}$ is an isomorphism of a group G onto itself if and only if the group is Abelian.

Exercise 19.6 (miscellaneous)

1. Prove that the set of the form $\dfrac{1+2m}{1+2n}$, where m and n are integers $(0, \pm 1, \pm 2, \ldots)$, forms a group under the operation of multiplication.

(AEB 1976)

2. S is the subset of the real numbers given by $S = \{x : 0 \leqslant x < 1\}$. A binary operation $*$ is defined on S as follows: $x * y = \dfrac{x+y}{1+xy}$. Prove that $*$ is closed and associative. [Hint for closure: simplify $1 - (x*y)$.]

Find the identity element for $*$, and show that no element of S (other than the identity element) has an inverse in S. Determine the truth or falsity of the following proposition: given $x, y \in S$ then $((x * x) = (y * y)) \Rightarrow (x = y)$. (C)

3. The multiplication table for the set $\{e, a, b, c, d\}$ is given below.

	e	a	b	c	d
e	e	a	b	c	d
a	a	e	c	d	b
b	b	d	e	a	c
c	c	b	d	e	a
d	d	c	a	b	e

Using this table, determine $(ab)c$, $a(bc)$, $(bc)d$, $b(cd)$. Ascertain which group axioms are satisfied by the given set under the given multiplication. Find two subsets from the above set which form a group under the given multiplication.

(L)

4. S is the set of functions having the form $f: x \to \dfrac{x}{nx+1}$ where n is an integer, and the domain of f is all real x except numbers of the form $x = 1/m$ where m is an integer. Show that S forms a group under the operation of composition of functions. (The Associative Law for composition of functions should not be assumed.)

(JMB)

5. Show that H, the set of all non-singular 2×2 real matrices, forms a group under matrix multiplication. (It may be assumed that matrix multiplication is associative and that $\det \mathbf{A} \times \det \mathbf{B} = \det \mathbf{AB}$.) Given that S is the set of all non-singular 2×2 matrices which commute with the matrix $\begin{pmatrix} 1 & 0 \\ 1 & 1 \end{pmatrix}$, show that each member of S must be of the form $\begin{pmatrix} a & 0 \\ b & a \end{pmatrix}$, where $a \neq 0$. Show that under the operation of matrix multiplication, S is a commutative group. (L)

6. Prove that the set S of matrices of the form $\begin{pmatrix} 1 & p & q \\ 0 & 1 & r \\ 0 & 0 & 1 \end{pmatrix}$, where p, q, and r are integers, form a non-Abelian group under matrix multiplication. Show that the subset of S containing matrices in which $p = 0$ is an Abelian subgroup of S, but that the subset in which $q = 0$ is not a subgroup.

7. Prove that the set of integers $\{1, 3, 4, 5, 9\}$ forms a cyclic group under multiplication modulo 11. List the elements which generate the whole group.

8.

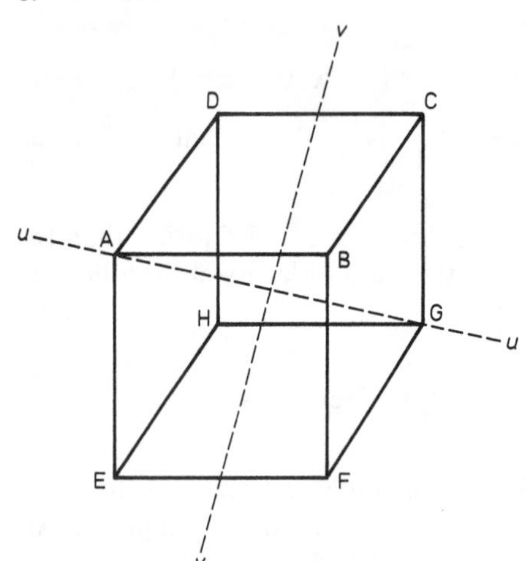

U and V are rotational symmetries of a cube. U is a rotation through 120° about the line u. V is a rotation through 180° about the line v, which passes through the mid-points of two opposite edges. Show that U and V generate a group of order 6. Find the order of the group containing all rotations of the cube.

9. Let S be the set of permutations of the numbers 1, 2 and 3. The elements of S may be listed as follows:

$$p_1 = \begin{pmatrix} 1 & 2 & 3 \\ 1 & 2 & 3 \end{pmatrix} \quad p_2 = \begin{pmatrix} 1 & 2 & 3 \\ 3 & 1 & 2 \end{pmatrix} \quad p_3 = \begin{pmatrix} 1 & 2 & 3 \\ 2 & 3 & 1 \end{pmatrix}$$

$$p_4 = \begin{pmatrix} 1 & 2 & 3 \\ 2 & 1 & 3 \end{pmatrix} \quad p_5 = \begin{pmatrix} 1 & 2 & 3 \\ 3 & 2 & 1 \end{pmatrix} \quad p_6 = \begin{pmatrix} 1 & 2 & 3 \\ 1 & 3 & 2 \end{pmatrix},$$

and $p_1 * p_2$ is defined to mean "p_2 followed by p_1". *It is given that* the elements of S under $*$ form a group. State the order of each element of S. Construct the group table for $(S, *)$. Show that the group $(S, *)$ is not commutative. Write down a cyclic sub-group of $(S, *)$ of order three. Name a generator of this sub-group. List all other non-trivial sub-groups of $(S, *)$. (JMB)

10. (i) Show that the set $\{2, 4, 6, 8, 10, 12\}$ forms a group under the operation of multiplication modulo 14. Name the identity element and the inverse of each of the other elements. Show that the set $\{1, 3, 5, 7, 9, 11, 13\}$ does NOT form a group under multiplication modulo 14. (Only one reason need be given, but the fact that one of the group postulates is not satisfied must be clearly demonstrated.)

(ii) The tables of two groups G_1, G_2 of order 4 are given.

G_1	a	b	c	d
a	a	b	c	d
b	b	a	d	c
c	c	d	a	b
d	d	c	b	a

G_2	p	q	r	s
p	p	q	r	s
q	q	r	s	p
r	r	s	p	q
s	s	p	q	r

Name a plane figure whose group of symmetries is isomorphic to G_1 and another plane figure whose group of symmetries contains a subgroup isomorphic to G_2. (L)

11. (a) (i) Show that the set of all matrices of the form

$$\begin{pmatrix} \exp(a) & b \\ 0 & \exp(-a) \end{pmatrix}, \quad a, b \in \mathbb{R}$$

forms a group under matrix multiplication. State the values of a and b which give the identity element and the inverse of a general element. (ii) Find a subgroup of the group in (i).

(b) Prove that the symmetry group of the equilateral triangle is not isomorphic to the cyclic group of order 6. (AEB 1975)

12. Construct operation tables for two *distinct* groups of order 4, explaining why they are not isomorphic.

Write down a table for the symmetry group of the square and show that this group contains subgroups of order 4 of each type. Show further that two such subgroups of different types contain two elements in common and that both of these elements commute with all other elements of the whole group. (AEB 1978)

13. **A** and **B** are square matrices of the same order, such that $\mathbf{A}^4 = \mathbf{I}$, $\mathbf{A}^2 = \mathbf{B}^2$, $\mathbf{AB} = \mathbf{BA}^3$, where **I** is the identity matrix. Prove that $\mathbf{A}^3\mathbf{B} = \mathbf{BA}$ and $\mathbf{A}^2\mathbf{B} = \mathbf{BA}^2$. Complete the combination table (where $\otimes$ denotes matrix multiplication, which may be assumed associative).

$\otimes$	$\mathbf{I}$	$\mathbf{A}$	$\mathbf{A}^2$	$\mathbf{A}^3$	$\mathbf{B}$	$\mathbf{BA}$	$\mathbf{BA}^3$	$\mathbf{BA}^3$
$\mathbf{I}$	$\mathbf{I}$	$\mathbf{A}$	$\mathbf{A}^2$	$\mathbf{A}^3$	$\mathbf{B}$	$\mathbf{BA}$	$\mathbf{BA}^2$	$\mathbf{BA}^3$
$\mathbf{A}$	$\mathbf{A}$	$\mathbf{A}^2$	$\mathbf{A}^3$	$\mathbf{I}$	$\mathbf{BA}^3$			
$\mathbf{A}^2$	$\mathbf{A}^2$	$\mathbf{A}^3$	$\mathbf{I}$	$\mathbf{A}$	$\mathbf{BA}^2$			
$\mathbf{A}^3$	$\mathbf{A}^3$	$\mathbf{I}$	$\mathbf{A}$	$\mathbf{A}^2$	$\mathbf{BA}$			
$\mathbf{B}$	$\mathbf{B}$	$\mathbf{BA}$	$\mathbf{BA}^2$	$\mathbf{BA}^3$	$\mathbf{A}^2$	$\mathbf{A}^3$	$\mathbf{I}$	$\mathbf{A}$
$\mathbf{BA}$	$\mathbf{BA}$				$\mathbf{A}$	$\mathbf{A}^2$	$\mathbf{A}^3$	$\mathbf{I}$
$\mathbf{BA}^2$	$\mathbf{BA}^2$				$\mathbf{I}$	$\mathbf{A}$	$\mathbf{A}^2$	$\mathbf{A}^3$
$\mathbf{BA}^3$	$\mathbf{BA}^3$				$\mathbf{A}^3$	$\mathbf{I}$	$\mathbf{A}$	$\mathbf{A}^2$

Hence prove that the set $\{\mathbf{I}, \mathbf{A}, \mathbf{A}^2, \mathbf{A}^3, \mathbf{B}, \mathbf{BA}, \mathbf{BA}^2, \mathbf{BA}^3\}$ forms a group under matrix multiplication. (L)

14. If ω is one of the complex cube roots of 1, prove that the set of numbers of the form $a + b\omega$ (where a and b are real numbers, not both zero), with the operation of multiplication, forms a group G. [You need not verify the associative law, because this is valid for complex numbers in general.] State, with reasons, whether the subsets of G, defined as follows, are subgroups:
(i) a and b are rational numbers (ii) a and b are integers. (O)

15. Show that the set of permutations of three symbols (a, b, c) forms a group under the operation of successive performance of permutations. By writing the symbols as a column matrix, $\begin{pmatrix} a \\ b \\ c \end{pmatrix}$, the permutation $(a, b, c) \to (b, a, c)$, for example, can be represented by a matrix A, where $A\begin{pmatrix} a \\ b \\ c \end{pmatrix} = \begin{pmatrix} b \\ a \\ c \end{pmatrix}$. Write down this matrix A and find the set of matrices which, under matrix multiplication, forms a group isomorphic to the permutation group. Find two subgroups of the permutation group which each contain two elements, and one subgroup which contains three elements. (AEB 1977)

16. Given the matrices

$$\mathbf{M}_1 = \begin{pmatrix} 1 & 0 \\ 0 & 1 \end{pmatrix}, \quad \mathbf{M}_2 = \begin{pmatrix} \omega & 0 \\ 0 & \omega^2 \end{pmatrix}, \quad \mathbf{M}_3 = \begin{pmatrix} \omega^2 & 0 \\ 0 & \omega \end{pmatrix},$$

$$\mathbf{M}_4 = \begin{pmatrix} 0 & 1 \\ 1 & 0 \end{pmatrix}, \quad \mathbf{M}_5 = \begin{pmatrix} 0 & \omega^2 \\ \omega & 0 \end{pmatrix}, \quad \mathbf{M}_6 = \begin{pmatrix} 0 & \omega \\ \omega^2 & 0 \end{pmatrix},$$

where $\omega^3 = 1$, $\omega \neq 1$, complete the multiplication table, where $\otimes$ denotes matrix multiplication.

$\otimes$	$\mathbf{M}_1$	$\mathbf{M}_2$	$\mathbf{M}_3$	$\mathbf{M}_4$	$\mathbf{M}_5$	$\mathbf{M}_6$
$\mathbf{M}_1$	$\mathbf{M}_1$	$\mathbf{M}_2$	$\mathbf{M}_3$	$\mathbf{M}_4$	$\mathbf{M}_5$	$\mathbf{M}_6$
$\mathbf{M}_2$	$\mathbf{M}_2$				$\mathbf{M}_4$	$\mathbf{M}_5$
$\mathbf{M}_3$	$\mathbf{M}_3$				$\mathbf{M}_6$	$\mathbf{M}_4$
$\mathbf{M}_4$	$\mathbf{M}_4$				$\mathbf{M}_2$	$\mathbf{M}_3$
$\mathbf{M}_5$	$\mathbf{M}_5$	$\mathbf{M}_6$	$\mathbf{M}_4$	$\mathbf{M}_3$	$\mathbf{M}_1$	$\mathbf{M}_2$
$\mathbf{M}_6$	$\mathbf{M}_6$	$\mathbf{M}_4$	$\mathbf{M}_5$	$\mathbf{M}_2$	$\mathbf{M}_3$	$\mathbf{M}_1$

Deduce that the above set of matrices forms a group under matrix multiplication, which may be assumed to be associative. Find a subset of three matrices which forms a group and is isomorphic to the group of mappings F, G, H of $S \to S$, where $S = \{x : x \in \mathbb{R}, x \neq 0, x \neq 1\}$, under composition of mappings where $F: x \to x$, $G: x \to 1 - 1/x$, $H: x \to 1/(1 - x)$. Indicate the corresponding elements of the isomorphism. (L)

17. Show that the set of matrices of the form $\begin{pmatrix} 1 & \ln a \\ 0 & 1 \end{pmatrix}$, where a is real and positive, forms an Abelian group under matrix multiplication. Find two groups, whose elements are real numbers, which are isomorphic to the given group.

18. Given the set S of six distinct elements, where $S = \{x_0, x_1, \ldots, x_5\}$, the law of composition is defined $\circ$ is defined by $x_i \circ x_j = x_k$, where $i + j \equiv k \pmod 6$. Prove that $(S, \circ)$ forms a group. Prove that the group has an element of order 6.

The group $(T, *)$ is given by the set T, where $T = \{0, 2, 3, 4, 5, 6\}$, and the law of composition $*$, where $a * b \equiv a + b - ab \pmod 7$. Determine whether or not $(S, \circ)$ and $(T, *)$ are isomorphic. (JMB)

19. $(G, .)$ is a group. The element $x . y$, where $x \in G$, $y \in G$, is denoted by xy and the inverse of x is denoted by x^{-1}.

(a) S is a non-empty subset of G such that

$$(a \in S \quad \text{and} \quad b \in S) \Rightarrow (ab^{-1} \in S).$$

(This includes the case where a and b are the same element.) Prove that S is a subgroup of G.

(b) If H is a subgroup of G and p is a *given* element of G, prove that the set $\{p^{-1}hp : h \in H\}$ is a subgroup of G isomorphic with H. (L)

20. G is a multiplicative group with identity element e. Every element $g \in G$ has the property that $g^2 = e$. Use this property applied to the element ab to prove that $ab = ba$ for any elements $a, b \in G$. H is the subset of G containing the three elements e, a and b. Prove that H cannot be a subgroup of G. Determine whether it is possible for a subset of G containing four distinct elements to be a subgroup of G. (C)

21. Let G be a group generated by elements a and b satisfying the relations $a^2 = b^3 = e$ and $ab = ba$, where e is the unit element of G. Show that G is isomorphic to the cyclic group of order 6. (C)

Formulae for reference

Logarithms

$$x = a^p \quad \Leftrightarrow \quad \log_a x = p; \quad \log_a x = \log_b x / \log_b a.$$

Series

Sum of A.P. $= \frac{1}{2}n\{2a + (n - 1)d\}$, sum of G.P. $= \dfrac{a(1 - r^n)}{1 - r}$

For $|r| < 1$, sum to infinity of G.P. $= a/(1 - r)$

$$(a + b)^n = a^n + na^{n-1}b + \frac{n(n-1)}{2!}a^{n-2}b^2 + \cdots + \binom{n}{r}a^{n-r}b^r + \cdots + b^n,$$

where n is a positive integer.

$$\sum_1^n r = \tfrac{1}{2}n(n + 1); \quad \sum_1^n r^2 = \tfrac{1}{6}n(n + 1)(2n + 1); \quad \sum_1^n r^3 = \tfrac{1}{4}n^2(n + 1)^2$$

Trigonometry

$$\cos^2\theta + \sin^2\theta = 1; \quad 1 + \tan^2\theta = \sec^2\theta; \quad \cot^2\theta + 1 = \operatorname{cosec}^2\theta.$$

$$\sin(A + B) = \sin A\cos B + \cos A\sin B$$
$$\sin(A - B) = \sin A\cos B - \cos A\sin B$$
$$\cos(A + B) = \cos A\cos B - \sin A\sin B$$
$$\cos(A - B) = \cos A\cos B + \sin A\sin B$$

$$\tan(A + B) = \frac{\tan A + \tan B}{1 - \tan A\tan B}$$

$$\tan(A - B) = \frac{\tan A - \tan B}{1 + \tan A\tan B}$$

$\sin 2A = 2\sin A\cos A$; $\tan 2A = 2\tan A/(1-\tan^2 A)$
$\cos 2A = \cos^2 A - \sin^2 A = 2\cos^2 A - 1 = 1 - 2\sin^2 A$
$\cos^2 A = \frac{1}{2}(1+\cos 2A)$, $\sin^2 A = \frac{1}{2}(1-\cos 2A)$

$$\sin\theta = \frac{2t}{1+t^2},\quad \cos\theta = \frac{1-t^2}{1+t^2},\quad \tan\theta = \frac{2t}{1-t^2},\quad \text{where}\quad t = \tan\tfrac{1}{2}\theta.$$

$$\begin{aligned}2\sin A\cos B &= \sin(A+B)+\sin(A-B)\\ 2\cos A\cos B &= \cos(A+B)+\cos(A-B)\\ 2\sin A\sin B &= -\{\cos(A+B)-\cos(A-B)\}\\ \sin X+\sin Y &= 2\sin\tfrac{1}{2}(X+Y)\cos\tfrac{1}{2}(X-Y)\\ \sin X-\sin Y &= 2\cos\tfrac{1}{2}(X+Y)\sin\tfrac{1}{2}(X-Y)\\ \cos X+\cos Y &= 2\cos\tfrac{1}{2}(X+Y)\cos\tfrac{1}{2}(X-Y)\\ \cos X-\cos Y &= -2\sin\tfrac{1}{2}(X+Y)\sin\tfrac{1}{2}(X-Y)\end{aligned}$$

$$\frac{a}{\sin A} = \frac{b}{\sin B} = \frac{c}{\sin C} = 2R,\quad a^2 = b^2 + c^2 - 2bc\cos A$$

Area of $\triangle ABC = \frac{1}{2}bc\sin A = \sqrt{\{s(s-a)(s-b)(s-c)\}}$,
where $s = \frac{1}{2}(a+b+c)$.
Length of circular arc $= r\theta$, area of sector $= \frac{1}{2}r^2\theta$ (θ in radians)

Calculus

$$\frac{d}{dx}(uv) = u\frac{dv}{dx} + v\frac{du}{dx},\qquad \frac{d}{dx}\left(\frac{u}{v}\right) = \frac{v\dfrac{du}{dx} - u\dfrac{dv}{dx}}{v^2}$$

y	$\sin x$	$\cos x$	$\tan x$	$\cot x$	$\sec x$	$\operatorname{cosec} x$
dy/dx	$\cos x$	$-\sin x$	$\sec^2 x$	$-\operatorname{cosec}^2 x$	$\sec x\tan x$	$-\operatorname{cosec} x\cot x$

$$\int u\frac{dv}{dx}dx = uv - \int v\frac{du}{dx}dx$$

$$\int e^{kx}dx = \frac{1}{k}e^{kx} + c,\qquad \int\frac{dx}{ax+b} = \frac{1}{a}\ln|ax+b| + c,$$

$$\int\frac{dx}{\sqrt{(a^2-x^2)}} = \sin^{-1}\frac{x}{a} + c,\quad \int\frac{dx}{a^2+x^2} = \frac{1}{a}\tan^{-1}\frac{x}{a} + c.$$

If $\dfrac{d^2y}{dx^2} + n^2y = k$, then $y = A\cos nx + B\sin nx + \dfrac{k}{n^2}$.

If $\dfrac{d^2y}{dx^2} - n^2y = k$, then $y = Ae^{nx} + Be^{-nx} - \dfrac{k}{n^2}$.

Expansions

$$(1+x)^n = 1 + nx + \frac{1}{2!}n(n-1)x^2 + \ldots + \binom{n}{r}x^r + \ldots \quad (|x| < 1)$$

$$e^x = 1 + x + \frac{x^2}{2!} + \ldots + \frac{x^r}{r!} + \ldots \quad (\text{all } x)$$

$$\ln(1+x) = x - \frac{x^2}{2} + \frac{x^3}{3} - \ldots + \frac{(-1)^{r+1}x^r}{r} + \ldots \quad (-1 < x \leqslant 1)$$

$$\sin x = x - \frac{x^3}{3!} + \frac{x^5}{5!} - \ldots, \quad \cos x = 1 - \frac{x^2}{2!} + \frac{x^4}{4!} - \ldots$$

Maclaurin's series: $f(x) = f(0) + f'(0)x + \dfrac{f''(0)}{2!}x^2 + \dfrac{f'''(0)}{3!}x^3 + \ldots$

Hyperbolic functions

$$\cosh x = \tfrac{1}{2}(e^x + e^{-x}), \quad \sinh x = \tfrac{1}{2}(e^x - e^{-x})$$

$$\cosh^2 x - \sinh^2 x = 1.$$

Numerical methods

Newton-Raphson process: $x_{n+1} = x_n - \dfrac{f(x_n)}{f'(x_n)}$

Trapezium rule:

$$\int_a^b f(x)\,dx \approx \tfrac{1}{2}h\{(y_0 + y_n) + 2(y_1 + y_2 + \ldots + y_{n-1})\}$$

Simpson's rule:

$$\int_a^b f(x)\,dx \approx \tfrac{1}{3}h\{(y_0 + y_n) + 4(y_1 + y_3 + \ldots) + 2(y_2 + y_4 + \ldots)\}$$

Coordinate geometry

Point dividing line joining (x_1, y_2) to (x_2, y_2) in ratio $\mu : \lambda$ is

$$\left(\frac{\lambda x_1 + \mu x_2}{\lambda + \mu}, \frac{\lambda y_1 + \mu y_2}{\lambda + \mu}\right)$$

Distance of (x_1, y_1) from $ax + by + c = 0$ is $\dfrac{|ax_1 + by_1 + c|}{\sqrt{(a^2 + b^2)}}$

Area of triangle $= \frac{1}{2}|x_1(y_2 - y_3) + x_2(y_3 - y_1) + x_3(y_1 - y_2)|$

Circle: $(x - h)^2 + (y - k)^2 = a^2$, centre (h, k), radius a;
$x^2 + y^2 + 2gx + 2fy + c = 0$, centre $(-g, -f)$, radius $\sqrt{(f^2 + g^2 - c)}$.

Rectangular hyperbola: $xy = c^2$, $(ct, c/t)$

Semicubical parabola: $ay^2 = x^3$, (at^2, at^3)

Parabola: $y^2 = 4ax$, $(at^2, 2at)$ focus $(a, 0)$, directrix $x = -a$.

Ellipse: $\dfrac{x^2}{a^2} + \dfrac{y^2}{b^2} = 1$, $(a\cos\theta, b\sin\theta)$, where $b^2 = a^2(1 - e^2)$, $e < 1$, foci $(\pm ae, 0)$ or $(\pm\sqrt{\{a^2 - b^2\}}, 0)$, directrices $x = \pm a/e$ or $\pm a^2/\sqrt{(a^2 - b^2)}$

Hyperbola: $\dfrac{x^2}{a^2} - \dfrac{y^2}{b^2} = 1$, $(a\sec\theta, b\tan\theta)$, where $b^2 = a^2(e^2 - 1)$, $e > 1$, foci $(\pm ae, 0)$, directrices $x = \pm a/e$, asymptotes $y = \pm bx/a$

Mechanics

Straight line motion with constant acceleration

$$v = u + at; \quad s = ut + \tfrac{1}{2}at^2; \quad s = \tfrac{1}{2}(u + v)t; \quad v^2 = u^2 + 2as.$$

Hooke's law: $T = \dfrac{\lambda}{l}x$; energy stored in string $= \dfrac{\lambda x^2}{2l}$.

Uniform body	Length, area or volume	Centre of mass
Triangular lamina, height h	$\frac{1}{2}h \times$ base	$\frac{1}{3}h$ from base
Circular arc, radius r, angle at centre 2α	$2r\alpha$	$\dfrac{r\sin\alpha}{\alpha}$ from centre
Sector of circle, radius r, angle at centre 2α	$r^2\alpha$	$\dfrac{2r\sin\alpha}{3\alpha}$ from centre
Solid cone or pyramid, height h	$\frac{1}{3}h \times$ base area	$\frac{1}{4}h$ from base
Solid hemisphere, radius r	$\frac{2}{3}\pi r^3$	$\frac{3}{8}r$ from centre
Hemispherical shell, radius r	$2\pi r^2$	$\frac{1}{2}r$ from centre

Probability

$P(A \cup B) = P(A) + P(B) - P(A \cap B), \quad P(A \cap B) = P(A) \times P(B \mid A)$

$$E(X) = \mu = \begin{cases} \sum x_r p(x_r) & X \text{ discrete} \\ \int_{-\infty}^{\infty} xf(x)\,dx & X \text{ continuous} \end{cases}$$

$$\operatorname{Var}(X) = \sigma^2 = \begin{cases} \sum (x_r - \mu)^2 p(x_r) = \sum x_r^2 p(x_r) - \mu^2, & X \text{ discrete} \\ \int_{-\infty}^{\infty} (x-\mu)^2 f(x)\,dx = \int_{-\infty}^{\infty} x^2 f(x)\,dx - \mu^2, & X \text{ continuous} \end{cases}$$

Binomial distribution $\mathrm{B}(n, p)$:

$$P(X = r) = \binom{n}{r} p^r (1-p)^{n-r} \quad (r = 0, 1, 2, \ldots, n)$$

$$E(X) = np, \quad \operatorname{Var}(X) = np(1-p)$$

Poisson distribution, parameter μ:

$$P(X = r) = \frac{\mu^r}{r!} e^{-\mu} \quad (r = 0, 1, 2, \ldots)$$

$$E(X) = \mu, \quad \operatorname{Var}(X) = \mu$$

Properties of expectation and variance:

$$E[g(X)] = \begin{cases} \sum g(x_r) p(x_r) & X \text{ discrete} \\ \int_{-\infty}^{\infty} g(x) f(x)\,dx & X \text{ continuous} \end{cases}$$

$$E(aX + b) = aE(X) + b, \qquad E(aX + bY) = aE(X) + bE(Y)$$

$$\operatorname{Var}(X) = E(X^2) - [E(X)]^2, \quad \operatorname{Var}(aX + b) = a^2 \operatorname{Var}(X)$$

For independent random variables $E(XY) = E(X)E(Y)$ and $\operatorname{Var}(aX + bY) = a^2 \operatorname{Var}(X) + b^2 \operatorname{Var}(Y)$

$\operatorname{Cov}(X, Y) = E[\{X - E(X)\}\{Y - E(Y)\}] = E(XY) - E(X)E(Y)$

Correlation coefficient $\rho = \dfrac{\operatorname{Cov}(X, Y)}{\sqrt{\{\operatorname{Var}(X)\operatorname{Var}(Y)\}}}$

Sampling distributions (large populations):

$$E(\bar{X}) = \mu, \qquad \operatorname{Var}(\bar{X}) = \sigma^2/n$$

$$E(\bar{X} - \bar{Y}) = \mu_x - \mu_y, \quad \operatorname{Var}(\bar{X} - \bar{Y}) = \frac{\sigma_x^2}{n_x} + \frac{\sigma_y^2}{n_y}$$

For sample variance V, $E(V) = \left(\dfrac{n-1}{n}\right)\sigma^2$

Statistics

Sample mean, $\bar{x} = \dfrac{1}{n}\sum x$

Sample variance, $\text{var}(x) = \dfrac{1}{n}\sum (x - \bar{x})^2 = \dfrac{1}{n}\sum x^2 - \bar{x}^2$

Sample covariance, $\text{cov}(x, y) = \dfrac{1}{n}\sum (x - \bar{x})(y - \bar{y}) = \dfrac{1}{n}\sum xy - \bar{x}\bar{y}$

Unbiased estimate of $\mu = \bar{x}$

Unbiased estimate of $\sigma^2 = s^2 = \dfrac{1}{n-1}\sum (x - \bar{x})^2$

Two sample estimate of $\mu = \dfrac{m\bar{x} + n\bar{y}}{m + n}$

Two sample estimate of $\sigma^2 = \dfrac{(m-1)s_x^2 + (n-1)s_y^2}{m + n - 2}$

$$S_{xy} = \sum (x - \bar{x})(y - \bar{y}) = \sum xy - n\bar{x}\bar{y}$$
$$S_{xx} = \sum (x - \bar{x})^2 = \sum x^2 - n\bar{x}^2$$
$$S_{yy} = \sum (y - \bar{y})^2 = \sum y^2 - n\bar{y}^2$$

Correlation coefficient $r = \dfrac{\text{cov}(x, y)}{\sqrt{\{\text{var}(x)\text{var}(y)\}}} = \dfrac{S_{xy}}{\sqrt{(S_{xx}S_{yy})}}$

Spearman's rank correlation coefficient $r_S = 1 - \dfrac{6\sum d^2}{n(n^2 - 1)}$

Kendall's rank correlation coefficient $r_K = \dfrac{P - Q}{\frac{1}{2}n(n-1)}$

Least squares regression line of y on x:

$$y = a + bx, \quad \text{where} \quad b = \frac{S_{xy}}{S_{xx}}, \quad a = \bar{y} - b\bar{x}.$$

Table 1 The normal distribution function $\Phi(z)$

z	·00	·01	·02	·03	·04	·05	·06	·07	·08	·09
0·0	·5000	·5040	·5080	·5120	·5160	·5199	·5239	·5279	·5319	·5359
01	·5398	·5438	·5478	·5517	·5557	·5596	·5636	·5675	·5714	·5753
0·2	·5793	·5832	·5871	·5910	·5948	·5987	·6026	·6064	·6103	·6141
0·3	·6179	·6217	·6255	·6293	·6331	·6368	·6406	·6443	·6480	·6517
0·4	·6554	·6591	·6628	·6664	·6700	·6736	·6772	·6808	·6844	·6879
0·5	·6915	·6950	·6985	·7019	·7054	·7088	·7123	·7157	·7190	·7224
0·6	·7257	·7291	·7324	·7357	·7389	·7422	·7454	·7486	·7517	·7549
0·7	·7580	·7611	·7642	·7673	·7704	·7734	·7764	·7794	·7823	·7852
0·8	·7881	·7910	·7939	·7967	·7995	·8023	·8051	·8078	·8106	·8133
0·9	·8159	·8186	·8212	·8238	·8264	·8289	·8315	·8340	·8365	·8389
1·0	·8413	·8438	·8461	·8485	·8508	·8531	·8554	·8577	·8599	·8621
1·1	·8643	·8665	·8686	·8708	·8729	·8749	·8770	·8790	·8810	·8830
1·2	·8849	·8869	·8888	·8907	·8925	·8944	·8962	·8980	·8997	·9015
1·3	·9032	·9049	·9066	·9082	·9099	·9115	·9131	·9147	·9162	·9177
1·4	·9192	·9207	·9222	·9236	·9251	·9265	·9279	·9292	·9306	·9319
1·5	·9332	·9345	·9357	·9370	·9382	·9394	·9406	·9418	·9429	·9441
1·6	·9452	·9463	·9474	·9484	·9495	·9505	·9515	·9525	·9535	·9545
1·7	·9554	·9564	·9573	·9582	·9591	·9599	·9608	·9616	·9625	·9633
1·8	·9641	·9649	·9656	·9664	·9671	·9678	·9686	·9693	·9699	·9706
1·9	·9713	·9719	·9726	·9732	·9738	·9744	·9750	·9756	·9761	·9767
2·0	·97725	·97778	·97831	·97882	·97932	·97982	·98030	·98077	·98124	·98169
2·1	·98214	·98257	·98300	·98341	·98382	·98422	·98461	·98500	·98537	·98574
2·2	·98610	·98645	·98679	·98713	·98745	·98778	·98809	·98840	·98870	·98899
2·3	·98928	·98956	·98983	·99010	·99036	·99061	·99086	·99111	·99134	·99158
2·4	·99180	·99202	·99224	·99245	·99266	·99286	·99305	·99324	·99343	·99361
2·5	·99379	·99396	·99413	·99430	·99446	·99461	·99477	·99492	·99506	·99520
2·6	·99534	·99547	·99560	·99573	·99585	·99598	·99609	·99621	·99632	·99643
2·7	·99653	·99664	·99674	·99683	·99693	·99702	·99711	·99720	·99728	·99736
2·8	·99744	·99752	·99760	·99767	·99774	·99781	·99788	·99795	·99801	·99807
2·9	·99813	·99819	·99825	·99831	·99836	·99841	·99846	·99851	·99856	·99861
3·0	·99865	·99869	·99874	·99878	·99882	·99886	·99889	·99893	·99896	·99900

z	3·1	3·2	3·3	3·4	3·5	3·6	3·7	3·8	3·9	4·0
Φ	·99903	·99931	·99952	·99966	·99977	·99984	·99989	·99993	·99995	·99997

Table 2 Percentage points of the normal distribution

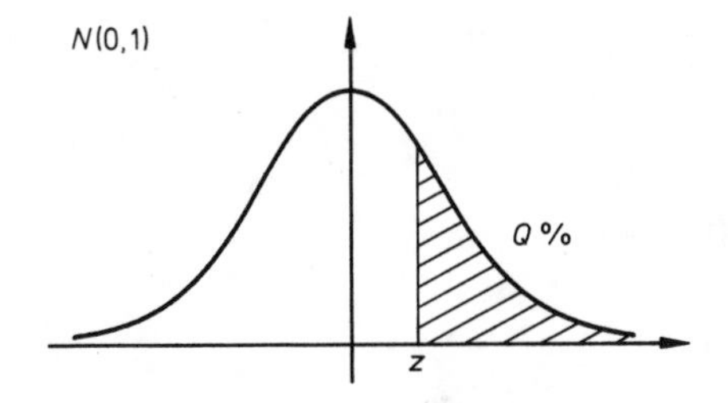

The shaded area represents $Q\%$ of the total distribution.

$$Q(z) = 100P(Z \geqslant z) = 100\{1 - \Phi(z)\}$$

Q	z	Q	z	Q	z
50	0·000	1·5	2·170	0·2	2·878
40	0·253	1·0	2·326	0·15	2·968
30	0·524	0·9	2·366	0·1	3·090
20	0·842	0·8	2·409	0·05	3·291
10	1·282	0·7	2·457	0·01	3·719
5	1·645	0·6	2·512	0·005	3·891
4	1·751	0·5	2·576	0·001	4·265
3	1·881	0·4	2·652	0·0005	4·417
2·5	1·960	0·3	2·748	0·0001	4·753
2	2·054	0·25	2·807	0·00005	4·892

Answers

Exercise 1.1

1. (a) Most people disapprove of the Knave of Hearts.
   (b) N is not prime.
   (c) Boycott took over as England cricket captain.
2. (a) Not valid. There could be swallow-tail butterflies in Norfolk, but there may not be.
   (b) Not valid. We are not told that absent-minded people do wear odd socks.
   (c) Not valid. $ABCD$ could be an isosceles trapezium.
3. Since $x - y = 0$, we cannot divide by $(x - y)$ to obtain $x + y = y$.
4. (1) Either, (2) Either, (3) Either, (4) True, (5) Either, (6) Either, (7) True, (8) Either, (9) Either, (10) Either, (11) Either, (12) True, (13) Either, (14) False, (15) False (16) True (17) False (18) Either.

Exercise 1.2

1. (a) $p \Rightarrow q, p \Rightarrow r, q \Rightarrow r$ (b) $p \Rightarrow r$ (c) $p \Rightarrow q, p \Rightarrow r$ (d) $p \Leftrightarrow q, p \Leftrightarrow r, q \Leftrightarrow r$
2. (a) $q \not\Rightarrow p, r \not\Rightarrow p, r \not\Rightarrow q$ (b) $p \not\Rightarrow q, q \not\Rightarrow p, q \not\Rightarrow r, r \not\Rightarrow p, r \not\Rightarrow q$ (c) $q \not\Rightarrow p, q \not\Rightarrow r, r \not\Rightarrow p, r \not\Rightarrow q$ (d) none
3. (a) $\sim q \Rightarrow \sim p, \sim r \Rightarrow \sim p, \sim r \Rightarrow \sim q$ (b) $\sim r \Rightarrow \sim p$ (c) $\sim q \Rightarrow \sim p, \sim r \Rightarrow \sim p$ (d) $\sim p \Leftrightarrow \sim q, \sim p \Leftrightarrow \sim r, \sim q \Leftrightarrow \sim r$
4. (a) $q \Rightarrow p, \sim p \Rightarrow \sim q$ (b) $p \Rightarrow \sim q, q \Rightarrow \sim p$ (c) $p \Leftrightarrow \sim q, \sim p \Leftrightarrow q$ (d) $p \Leftrightarrow \sim q, \sim p \Leftrightarrow q$
5. (a) $q \Rightarrow p$ (b) $\sim p \Leftrightarrow q$ (c) $\sim q \not\Rightarrow p$ (d) $\sim p \not\Leftrightarrow \sim q$ (e) $q \Rightarrow \sim p$ (f) $p \Leftrightarrow \sim q$
6. (a) $q \Rightarrow p$ (b) $q \Rightarrow p$ (c) $q \Rightarrow p$ (d) $q \Rightarrow p$ (e) $p \Rightarrow q$ (f) $p \Rightarrow q$
7. $AB = AC$ if $\triangle ABC$ is equilateral.
   $\triangle ABC$ is equilateral only if $AB = AC$.
   $\triangle ABC$ being equilateral is a sufficient condition for $AB = AC$.
   $AB = AC$ is a necessary condition for $\triangle ABC$ to be equilateral.

Exercise 1.3

1. If $\angle ABC = \angle CAT$, then AT must be a tangent to the circle.
3. (a) If a diameter of a circle is perpendicular to a chord then it bisects that chord; (i) True, (ii) True.
   (b) If $2x^2 - x - 3 = 0$, then $2x - 3 = 0$; (i) True, (ii) False.
   (c) If a pentagon is equilateral, it must be regular; (i) True, (ii) False.
   (d) If x^2 is greater than y^2, then x is greater than y; (i) False, (ii) False.
   (e) If the sum of the digits of a positive integer N is divisible by 9, then N is divisible by 9; (i) True, (ii) True.
   (f) If there are five pairs of socks in a drawer, then any set of six socks taken from the drawer will contain a pair; (i) False, (ii) True.
4. (a) True, (b) False. 5. (a) True, (b) False.
6. (a) False, (b) True.

Exercise 1.4

9. (a) 2870 (b) 19 270 (c) 11 480 (d) 10 660 10. $\frac{1}{4}$
11. (a) $(n, 2n - 1)$ (b) $(n + 1)y = (2n + 1)x$
12. $\dfrac{x}{2x+1}, \dfrac{x}{3x+1}, \dfrac{x}{nx+1}$

Exercise 1.5

1. If $(x - 3)^2 = (x + 1)^2$, then $x - 3 = \pm(x + 1)$.
2. Since $\lg(0{\cdot}2) < 0$, $2\lg(0{\cdot}2) < \lg(0{\cdot}2)$.
3. No value of x satisfies the initial statement.
4. $a + b > 0$ implies $a^2 + ab > 0$ only if $a > 0$.
5. The diagram is incorrect. P lies outside $\triangle ABC$.
6. There is no value of n for which a_n is divisible by 6.
7. It is necessary to prove that $r \Rightarrow q$ and that $s \Rightarrow q$.
8. It is necessary to prove that $P(1)$ and $P(2)$ are true.

14. $\begin{pmatrix}1 & 3\\ 0 & 4\end{pmatrix}, \begin{pmatrix}1 & 7\\ 0 & 8\end{pmatrix}, \begin{pmatrix}1 & 2^n - 1\\ 0 & 2^n\end{pmatrix}$ 15. 1.

18. (i) False, (ii) True, (iii) False. 19. One example is $f(x) \equiv x$.

20. One possibility is $\mathbf{P} = \begin{pmatrix}1 & 0\\ 0 & 2\end{pmatrix}, \mathbf{X} = \begin{pmatrix}1 & 1\\ 3 & 2\end{pmatrix}$; $\begin{pmatrix}-2 + 2^n.3 & -2 + 2^{n+1}\\ 3 - 2^n.3 & 3 - 2^{n+1}\end{pmatrix}$.

Exercise 2.1

1. (a) Even, (b) neither, (c) odd. 2. (a) Even, (b) even, (c) odd.
3. (a) Even, (b) odd, (c) neither. 4. (a) Neither, (b) neither, (c) even.

5. (a) Yes, $2\pi/3$; (b) yes, $\pi/2$; (c) yes, 8π.
6. (a) No; (b) yes, 2π; (c) yes, π.
7. (a) Yes, 2π; (b) no; (c) yes, 2π.
8. (a) Yes, π; (b) no; (c) yes, π.
9. f: even, $f(x) \geqslant -1$; g: neither, $\mathbb{R}$; fg: neither, $fg(x) \geqslant -1$; gf: even, $gf(x) \geqslant -1$.
10. f: neither, $\mathbb{R}$; g: odd, $-1 \leqslant g(x) \leqslant 1, 2\pi$; fg: neither, $\frac{1}{2}(\pi - 1) \leqslant fg(x) \leqslant \frac{1}{2}(\pi + 1), 2\pi$; gf: even, $-1 \leqslant gf(x) \leqslant 1, 4\pi$.
11. f: even, $-1 \leqslant f(x) \leqslant 1, 2\pi$; g: even, $g(x) \geqslant 0$; fg: even, $-1 \leqslant fg(x) \leqslant 1$; gf: even, $0 \leqslant gf(x) \leqslant 1, \pi$.
12. f: even, $f(x) \geqslant 0$; g: odd, $\mathbb{R}$; fg: even, $fg(x) \geqslant 0$; gf: even, $gf(x) \geqslant -2$.
17. (a) $-1, 1$, (b) $-1, 1$. 18. (a) none, (b) $-1, 1$.
19. (a) $x \in \mathbb{Z}$, (b) $x \in \mathbb{Z}$. 20. (a) 1, (b) $-1, 1$. 22. 3.

Exercise 2.2

7. $(-1, -1)$ min; $(-1, -1)$ max, $x = -2$, $x = 0$, $y = 0$.
8. $(-1, 0)$ min; $x = -1, y = 0$. 9. $(-1, 1)$ min; $(-1, 1)$ max, $y = 0$.
10. $(-3, 36)$ max, $(5/3, -400/27)$ min; $(-3, 1/36)$ min, $(5/3, -27/400)$ max, $x = -5, x = 0, x = 3, y = 0$.
11. $x = -2, y = 2$; $x = \frac{1}{2}, y = \frac{1}{2}$. 12. None, $(3/2, 0)$ min.
13. (0.4) max; $(-2, 0)$ min, $(0, 16)$ max, $(2, 0)$ min.
14. $(2n\pi + \frac{1}{2}\pi, 1)$ max, $(2n\pi - \frac{1}{2}\pi, -1)$ min; $(n\pi, 0)$ min, $(n\pi + \frac{1}{2}\pi, 1)$ max.
15. $x = 1, y = -2$; $(0, 0)$ min, $x = 1, y = 4$.
16. $(0, 0)$ min, $(2, 4)$ max; $(2, -2)$ min, $(2, 2)$ max.

Exercise 2.3

1. Yes; $\frac{1}{2}(x - 1)$; $\mathbb{R}$. 2. Yes; $5 - x$; $\mathbb{R}$. 3. No; e.g. $-2, 0$.
4. Yes; $\sqrt{(x + 2)}$; $\{x \in \mathbb{R} : x \geqslant -2\}$.
5. Yes; $1/(x - 1)$; $\mathbb{R} - \{1\}$.
6. Yes, $9 - 1/x$; $\mathbb{R} - \{0\}$. 7. Yes; $3 - x^2$; $\{x \in \mathbb{R} : x \geqslant 0\}$.
8. No; e.g. $-1, 1$. 9. Yes; $\log_3 x$; $\mathbb{R}^+$. 10. Yes, $10^x - 2$; $\mathbb{R}$.
11. No; e.g. $-1, 1$. 12. Yes; $x/(x - 2)$; $\mathbb{R} - \{2\}$.
13. No; e.g. $0, 2\pi$. 14. Yes; $\sin x$; $\{x \in \mathbb{R} : -\frac{1}{2}\pi \leqslant x \leqslant \frac{1}{2}\pi\}$
15. $1 + 4x^2$; $\frac{1}{2}\sqrt{(x - 1)}$. 16. $gf : x \to 3 - 9/x$.
17. (a) $f^{-1} : x \to 2 + \sqrt{x}, x \geqslant 0$, (b) $f^{-1} : x \to 1 + 10^{x-1}$,

$$\text{(c) } f^{-1} : x \to \frac{1 - x}{1 + x}, x \neq -1 \quad \text{(d) } f^{-1} : x \to \sqrt{\left(1 + \frac{1}{x^2}\right)}, x > 0.$$

18. $k = 2$, $\{x \in \mathbb{R} : x \geqslant 4\}$. 19. $k = 3$, $\{x \in \mathbb{R} : x \leqslant 9\}$.
20. $k = 4$, $\{x \in \mathbb{R} : x \geqslant 0\}$. 21. $k = 1$, $\{x \in \mathbb{R} : x \geqslant -1\}$.
22. $k = 1$, $\{x \in \mathbb{R} : x \geqslant 2\}$. 23. $k = \frac{1}{4}$, $\{x \in \mathbb{R} : x \geqslant -\frac{1}{4}\}$.
24. (a) No, (b) Yes, (c) Yes, (d) No, (e) No, (f) Yes.

Exercise 2.4

1. (a) 5, (b) 4, (c) -2, (d) $x-7$, (e) 0, (f) $-4x+5$.
2. $5, -12$. 3. $-3x+4$. 4. $2, -8$. 5. $4, -2, -5$.
6. $3(a+b)(b+c)(c+a)$. 7. $3(x-y)(y-z)(z-x)$.
8. $k=-7, (x+1)^2(2x-7)$; $k=20, (x-2)^2(2x+5)$.
9. $(x+2)^2(3x^2-4x+4)$. 10. $-2, 13; (x-1)^2(x-6)$.
11. $-3, -3, 12$. 12. $k=-5$: $-1/2, -1/2, 5/2$; $k=27$: $3/2, 3/2, -3/2$.
13. $-11; -3, -\frac{1}{2}, 2$. 14. $-14; 3, -2, 4/3$. 15. $1, 4, 2$.
16. $1, 3, 1, 6$. 17. $1, -3, 1, 0$. 19. $-3, \frac{1}{4}$. 20. $0, 2$.
21. $(x-a)^2g'(x)+2(x-a)g(x)+A$; $f'(a), f(a)-af'(a)$; $k=-1$: $-3, 0, 0$; $k=1$; $-1, 2, 2$.

Exercise 2.5

7. $y \leqslant 1/5$ or $y \geqslant 5$; $x=-4, x=1, y=0$; $(-1,5)$ min, $(11, 1/5)$ max.
8. $-\frac{1}{2} \leqslant y \leqslant 1$; $y=0$; $(-2, -\frac{1}{2})$ min, $(1,1)$ max.
9. All real values; $x=-3, x=0, y=0$; no turning points.
10. $y \leqslant 1/9$ or $y \geqslant 1$; $x=-3, x=0, y=0$; $(-1,1)$ min, $(3, 1/9)$ max.
11. $y \leqslant 1$ or $y \geqslant 9$; $x=1, x=2, y=0$; $(0,1)$ max, $(4/3, 9)$ min.
12. $-5/4 \leqslant y \leqslant 5$; $y=0$; $(4, -5/4)$ min, $(-1, 5)$ max.
13. $-\frac{1}{3} \leqslant y \leqslant 1$; $y=0$; $(0,1)$ max, $(2, -\frac{1}{3})$ min.
14. $y \leqslant -1$ or $y \geqslant -1/9$; $x=-5, x=-2, y=0$; $(-8, -1/9)$ min, $(-4, -1)$ max.
15. $y \leqslant 0$ or $y \geqslant 4/3$; $x=1, x=4, y=0$; $(5/2, 4/3)$ min.
16. All real values; $x=1, x=4, y=0$; no turning points.
20. $x=3, y=x$; $(1,-1)$ max, $(5,7)$ min.

Exercise 2.6

1. $\dfrac{1}{x+1}-\dfrac{1}{x+2}$. 2. $\dfrac{1}{x-1}-\dfrac{2}{x+2}$. 3. $\dfrac{2}{x-2}+\dfrac{1}{x-4}$.

4. $\dfrac{3}{x}+\dfrac{1}{x-3}$. 5. $\dfrac{3}{x-2}-\dfrac{2}{x+1}$. 6. $\dfrac{2}{x+1}+\dfrac{1}{2x+1}$.

7. $\dfrac{1}{2(x-1)}+\dfrac{1}{2(x+1)}$. 8. $\dfrac{1}{3+x}+\dfrac{1}{2-x}$. 9. $\dfrac{1}{4x-1}+\dfrac{2}{2x+1}$.

10. $\dfrac{1}{2(2x-1)}-\dfrac{3}{2(2x+5)}$. 11. $\dfrac{3}{x-2}-\dfrac{4}{x+3}+\dfrac{1}{x+1}$.

12. $\dfrac{6}{x-1}+\dfrac{2}{x+1}-\dfrac{3}{x-2}$. 13. $\dfrac{x+2}{x^2+1}-\dfrac{1}{x+1}$. 14. $\dfrac{2}{2x-1}-\dfrac{x}{x^2+1}$.

15. $\dfrac{1}{x} + \dfrac{3-2x}{2x^2+1}$. 16. $\dfrac{2x+1}{x^2+3} - \dfrac{3}{2x-1}$.

17. $\dfrac{2}{x-1} - \dfrac{4}{(x-1)^2} - \dfrac{1}{x+2}$. 18. $\dfrac{1}{x} + \dfrac{2}{x+1} + \dfrac{3}{(x+1)^2}$.

19. $\dfrac{5}{x} + \dfrac{2}{x^2} - \dfrac{3}{x-1}$. 20. $\dfrac{1}{x-2} - \dfrac{1}{x+1} - \dfrac{2}{(x+1)^2} - \dfrac{8}{(x+1)^3}$.

21. $\dfrac{x}{2x^2+5} - \dfrac{1}{4x-5}$. 22. $\dfrac{1}{x-2} - \dfrac{x}{x^2-x+2}$.

23. $1 + \dfrac{1}{x-1} - \dfrac{1}{x+1}$. 24. $1 + \dfrac{4}{3(x-2)} - \dfrac{1}{3(x+1)}$.

25. $\dfrac{1}{3} + \dfrac{5}{6(3x-1)} - \dfrac{1}{2(x-1)}$. 26. $x + \dfrac{2}{x-2} + \dfrac{2}{x+2}$.

27. $\dfrac{x+3}{x^2+3} - \dfrac{x+2}{x^2+2}$. 28. $\dfrac{2x-1}{x^2+2} + \dfrac{1}{x+1} - \dfrac{2}{(x+1)^2}$.

29. $\dfrac{x}{2} + 1 - \dfrac{1}{4(2x-1)} - \dfrac{1}{4(2x-1)^2} + \dfrac{2}{x-2}$.

30. $x^2 - x - 3 + \dfrac{3}{x^2+1} + \dfrac{3x}{x^2+2}$.

31. $\dfrac{2}{2x+3} + \dfrac{1}{x+1}; -\dfrac{4}{(2x+3)^2} - \dfrac{1}{(x+1)^2}; \dfrac{16}{(2x+3)^3} + \dfrac{2}{(x+1)^3}$.

32. (a) $(0, 0)$, (b) $(-7, -3/7)$.

Exercise 2.7

1. f: even, periodic, $\frac{1}{3}\pi$; g: neither, not periodic; h: even, not periodic; k: odd, periodic, 4π.
2. f: none; 2; g: x an even integer; x an even integer; h: $-1, 2$; $-1, 2$; k: none; 1.
3. $\pi, \pi; \pi; 0 \leqslant f(x).g(x) < 2$. 4. f: neither; g: even; h: even; 12π, 6π.
5. $4 \leqslant f(x) \leqslant 20$, f^{-1} exists; $-4 \leqslant g(x) \leqslant 0$, no inverse; $-5 \leqslant h(x) \leqslant 5$, no inverse; $k(x) \leqslant 1$, k^{-1} exists.
6. $a = 1$; $g^{-1}(x) = (2 - \sqrt{\{4 - x^2\}})/x$, $-2 \leqslant x \leqslant 2$, $-1 \leqslant g^{-1}(x) \leqslant 1$.
7. 1, 1/5.
8. $-5 < f(x) \leqslant 4$; $-2 < x \leqslant 1$: $f^{-1}(y) = \frac{1}{2}(y+1)$, $1 < x \leqslant 2$: $f^{-1}(y) = \sqrt{y}$, $2 \leqslant x < 3$: $f^{-1}(y) = \frac{1}{3}(10 - y)$.

9. $-\frac{1}{4}\pi \leqslant x \leqslant \frac{1}{4}\pi, f^{-1}(y) = \frac{1}{2}\sin^{-1} y; \frac{1}{4}\pi \leqslant x \leqslant \frac{3}{4}\pi, f^{-1}(y) = \frac{1}{4}\pi(2 - y)$.
10. (i) F: odd, periodic, 6π; G: even, periodic, 2π; f_3: odd; f_4: even, period $\frac{1}{3}\pi$, (ii) 0.
11. $\{y \in \mathbb{R} : y \geqslant 0, y \neq 2r + 1, r \in \mathbb{Z}\}$; not one-one; $g^{-1} : x \to x - \frac{1}{2}[x]$; $\{x \in \mathbb{R}^+ : 2r < x < 2r + 1, r \in \mathbb{Z}\}$.
12. $-1 < x < 2, 3 < x < 6$. 14. $x < -1$ or $x > 0$.
16. $-12\frac{1}{24}$; $-3/2, 4/3$; $-1{\cdot}56, -1{\cdot}44, 1{\cdot}27, 1{\cdot}39$.
17. $a = b = 1$; $x^3 - 2x^2 + 3$. 18. (a) $-6x + 2$, (b) $-2x$.
19. $2, \pm\sqrt{5}$.
20. (a) $(x + y + z)(x - y)(y - z)(z - x)$, (b) $3(a - b - c)(b - c - a)(c - a - b)$.
21. $a = -2, b = -1, c = 16$. 23. $-2, -2, 4 \pm 2\sqrt{2}$. 24. $0, -3, 125$.
26. $4, -5$.
27. (i) 2, (ii) HCF of $A(x)$ and $B(x)$ is $8x + 8$; $\alpha : g(x)$, $\beta : A(x)$ and $B(x)$ have no common factor.
28. $-\frac{1}{3} \leqslant y \leqslant 1, x = -1; -\frac{1}{3} \leqslant f(x) \leqslant 1; (-2, -\frac{1}{3}), (0, 1)$.
29. (a) $\mathbb{R} - \{\frac{1}{2}\}$; $x = -\frac{1}{2}$, $y = \frac{1}{2}$; none, (b) $\{y \in \mathbb{R} : y < 0$ or $y \geqslant 1\}$; $x = -3$, $x = 1$, $y = 0$; $(-1, 1)$, (c) $\{y \in \mathbb{R} : y \leqslant -3$ or $y \geqslant -\frac{1}{3}\}$; $x = 0$, $x = 3$, $y = 0$; $(-3, -\frac{1}{3}), (1, -3)$, (d) $\mathbb{R}$; $x = -2, x = 2, y = 0$; none.
30. (a) none, (b) $(0, 0)$ max, $(2, 4)$ min, (c) $(0, 0)$ min, (d) $(3, 1/8)$ max.
31. $\{x \in \mathbb{R} : x \neq -1, 3\}$; $\{y \in \mathbb{R} : y \leqslant -1$ or $y \geqslant \frac{1}{2}\}$; $x = -1, x = 3, y = 1$.
32. (a) max $(1, \frac{1}{2})$, min $(-1, -\frac{1}{2})$; inflexions $(-\sqrt{3}, -\frac{1}{4}\sqrt{3}), (0, 0), (\sqrt{3}, \frac{1}{4}\sqrt{3})$.
33. $\dfrac{3}{3x - 2} - \dfrac{1}{2x + 3}$. 34. $\dfrac{1}{x - 1} + \dfrac{1}{(x - 1)^2} - \dfrac{2}{2x - 1}$.
35. $\dfrac{1}{2(x - 1)} + \dfrac{1 - x}{2(x^2 + 1)}$. 36. $\dfrac{1}{3(x + 1)} - \dfrac{1}{x - 1} + \dfrac{2}{3(x - 2)}$.
37. $-\dfrac{1}{2} + \dfrac{3}{2(2x + 1)} + \dfrac{2}{3 - x}$. 38. $\dfrac{1}{1 - x} - \dfrac{1 - x}{1 + x + x^2}$.
39. $x - 2 - \dfrac{2}{3(x - 1)} + \dfrac{8}{3(x + 2)}$.
40. $\dfrac{1}{x - 1} + \dfrac{2}{(x - 1)^2} - \dfrac{1}{x + 1} + \dfrac{3}{(x + 1)^2}$.

Exercise 3.1

1. (a) $2e^{2x}$, (b) $-5e^{-x}$, (c) $3e^{3x+5}$.
2. (a) $6x^2e^{2x^3}$, (b) $\dfrac{1}{2\sqrt{x}}e^{\sqrt{x}}$, (c) $\dfrac{1}{x^2}e^{-1/x}$.
3. (a) $\cos xe^{\sin x}$, (b) $-2\sin 2xe^{\cos 2x}$, (c) $4\sec^2 xe^{4\tan x}$.
4. (a) $(2x^2 + 1)e^{x^2}$, (b) $(5x + 2)xe^{5x}$, (c) $(\cos x - x\sin x)e^{x\cos x}$.
5. (a) $e^{2x}(2\cos 3x - 3\sin 3x)$, (b) $e^{-x^2}(\cos x - 2x\sin x)$, (c) $\frac{1}{2}(x + 1)^2(x + 7)e^{x/2}$.
6. (a) $-2e^{4x}(1 - e^{4x})^{-\frac{1}{2}}$, (b) $e^{(x + e^x)}$, (c) $e^x\cos(e^x)$.
7. (a) $e^{2x} - e^{-2x}$, (b) $(2 + e^x)/(1 + e^{-x})^2$, (c) 0.

8. (a) $\frac{1}{3}e^{3x} + c$, (b) $-e^{-x} + c$, (c) $e^{x+4} + c$.
9. (a) $3(e^2 - 1)$, (b) $2(e - 1)$, (c) $e^6 - 1$.
10. (a) $\dfrac{1}{x}$, (b) $-\dfrac{1}{x}$, (c) $\dfrac{2}{x} - 1$. 12. $y = e(3x - 5)$. 13. $e - 1$.
15. $(1, e^{-1})$ max. 16. $(-2, -e^{-2})$ min.
17. $(-1, -e^{-\frac{1}{2}})$ min, $(1, e^{-\frac{1}{2}})$ max. 18. $(0, 1)$ max, $(1, 0)$ min.
19. $(1, e)$ min. 20. $(2k\pi, 1)$ min, $(\{2k + 1\}\pi, e^2)$ max, $k \in \mathbb{Z}$.
21. 4 m, -2 m s^{-2}. 22. $x = -1$, max; $x = 2$, min.
23. $x = \pi/4$, max; $x = 5\pi/4$, min.
24. $\frac{1}{2}(e^2 - 3 + 2e^{-1}) \approx 2{\cdot}56$; $\frac{1}{4}\pi(e^4 - 3 + 2e^{-2}) \approx 40{\cdot}7$.

Exercise 3.2

1. (a) $\dfrac{1}{x}$, (b) $\dfrac{1}{x + 3}$, (c) $\dfrac{2}{2x - 1}$.
2. (a) $\dfrac{3x^2}{x^3 + 4}$, (b) $2 \cot 2x$, (c) $\tan x$.
3. (a) $x(2 \ln x + 1)$, (b) $(1 - 2 \ln x)/x^3$, (c) $1/x \ln x$.
4. (a) $\dfrac{3}{x}$, (b) $\dfrac{2}{4x + 5}$, (c) $-1 - \ln x$.
5. (a) $\sec x$, (b) $\dfrac{4x}{x^2 + 4}$, (c) $(\log_{10} e)/x$.
6. (a) $\dfrac{3x - 4}{x(3x - 2)}$, (b) $\dfrac{1}{2x(x + 1)}$, (c) $\dfrac{x}{1 - x^2} - \tan x$.
7. (a) $2x$, (b) $\sec^2 x$, (c) $\dfrac{2}{x} - 1$. 8. $y = 3x - 6$. 9. $t/(1 + t)$.
11. (a) $2p + q + 1$, (b) p/q, (c) $1/(p + q)$, (d) e^{p+q}
12. (a) $t = 0$ or $\ln 2$, (b) $x = -\ln 2$.
13. (a) none, (b) none, (c) $(1, 0)$ min. 14. $(1, 1)$ min.
17. (a) $3^x \ln 3$, (b) $5^{2x}\, 2 \ln 5$, (c) $\ln 10 \cos x(10^{\sin x})$.
18. (a) $x^x(1 + \ln x)$, (b) $(2 \ln x)x^{(\ln x - 1)}$, (c) $(\ln x)^x\left\{\ln(\ln x) + \dfrac{1}{\ln x}\right\}$.
19. (a) $\dfrac{2(x - 2)}{(2x + 1)^3}$, (b) $\dfrac{(6 - 5x)x^2}{2(1 - x)^{3/2}}$, (c) $\dfrac{4x^2(x - 1)e^{4x}}{(x + 1)^3}$.
20. (a) $\frac{1}{2}(13 - 10x)\sqrt{\left\{\dfrac{x + 2}{(1 - 5x)^3}\right\}}$, (b) $\dfrac{4x}{(x - 2)^{2/3}(x + 4)^{4/3}}$.

(c) $\dfrac{4 \sin^2 x + 1}{\cos^3 x \sqrt{(\sin 2x)}}$.

Exercise 3.4

1. (a) $\frac{1}{2}\ln(2x+7)+c$, (b) $\frac{1}{5}\ln(5x-2)+c$.
2. (a) $-\ln(1-x)+c$, (b) $\frac{1}{2}\ln(1-2x)+c$.
3. (a) $\ln(2-3x)+c$, (b) $-2\ln(3-4x)+c$.
4. (a) $\ln|x+3|+c$, (b) $-3\ln|3-x|+c$, (c) $\frac{1}{2}\ln|4x+5|+c$.
5. (a) $x+\ln|x-1|+c$, (b) $x+2\ln|x-4|+c$, (c) $\frac{1}{2}x+\frac{1}{4}\ln|2x+1|+c$.
6. (a) $\frac{1}{2}x^2-\ln|x|+c$, (b) $x-\ln(1+x^2)+c$, (c) $-\frac{1}{2}x^2-\ln|1-x|+c$.
7. (a) $-\frac{1}{2}\ln|1-x^2|+c$, (b) $\frac{1}{3}\ln|x^3+8|+c$, (c) $\frac{1}{4}\ln|2x^2+4x-1|+c$.
8. (a) $\ln(1+\sin x)+c$, (b) $\ln|\sin x|+c$, (c) $x+\frac{1}{2}\ln|\sec 2x|+c$.
9. (a) $\ln(1+e^x)+c$, (b) $\ln|\ln x|+c$, (c) $2\ln(1+\sqrt{x})+c$.
10. (a) $5\ln 2$, (b) $\ln 4$, (c) $\ln 3$. 11. $\frac{1}{3}\ln\frac{5}{2}$, (b) $-\ln 4$, (c) $2+\ln\frac{2}{3}$.
12. (a) $\ln 2$, (b) $\ln 5$, (c) $\ln\frac{4}{3}$. 13. (a) 0·805, (b) 1·10.
14. $\dfrac{5(1-x)}{(2x-1)(3-x)}$; $-0{\cdot}167$. 15. $\frac{3}{4}-\ln 2$. 16. $\pi(8+\ln 3)$.

Exercise 3.5

1. (a) $2\cosh 2x$, (b) $3\sinh 3x$, (c) $2\sinh x\cosh x$, (d) $\dfrac{\sinh x}{2\sqrt{(\cosh x)}}$.
2. (a) $\cosh x+c$, (b) $\frac{1}{2}\sinh 2x+c$, (c) $\ln(\cosh x)+c$.
3. (a) $\operatorname{sech}^2 x$, (b) $-\operatorname{cosech}^2 x$, (c) $-\operatorname{sech} x\tanh x$, (d) $-\operatorname{cosech} x\coth x$.
5. (a) 0 or $\ln 5$, (b) $-\ln 2$.

Exercise 3.6

1. (a) $e^{3x}(3x+1)$, (b) $e^{2x}\left(\dfrac{1}{x}+2\ln 4x\right)$, (c) $a^x\ln a$.
2. (a) $\dfrac{1}{x}\cos(\ln x)$, (b) $\dfrac{1}{x}-x-2x\ln x$, (c) $\operatorname{cosec} x$.
3. (a) $\dfrac{8}{(x-3)(3x-1)}$, (b) $2\sec x$, (c) $\dfrac{3-4x^2}{1-x^2}$.
4. (a) $(\ln 2, 4)$ min, (b) $(0,0)$ inflexion, $(3, 27e^{-3})$ max.
5. $x=e$, max. 6. 24/7
9. $(-1, -2\sqrt{e})$ min, $(4, 18e^{-2})$ max.
10. (a) $\ln 3$, (b) $\frac{1}{2}\ln 6$, (c) $\frac{3}{2}\ln 3$.
11. (a) $\frac{1}{3}x-\frac{1}{9}\ln|3x+1|+c$, (b) $\ln|x|+\dfrac{2}{x^2}+c$,

    (c) $2x-\ln(x^2+x+1)+c$.
12. $\frac{1}{2}\ln(1+x)-\frac{1}{2}\ln(1-x)-x$; 0. 14. $2xe^{2x}$; 59·5.

16. $\frac{5}{6} + 2\ln 2$. 17. $2 - \dfrac{2}{e}$; $2\pi\left(1 - \dfrac{1}{e^2}\right)$.

18. $(2, 3), (3, 0)$; $y + 3x = \pm 6\sqrt{2}$.
19. $(-1, -\frac{1}{2})$ min, $(1, \frac{1}{2})$ max; $\frac{1}{2}\ln(1 + a^2)$.
20. Inflexion at $(-\{\ln b\}/c, a/2b)$.
21. Asymptote $x = 6$; $(4, 1)$ max, $(8, 9)$ min; $\frac{15}{2} - 8\ln 2$.
22. $\frac{1}{2}\pi(e^{2h} + 8e^h + 8h - 9)$; $50/\{\pi(e^{1\cdot 5} + 2)^2\} \approx 0{\cdot}379$.

Exercise 4.1

1. $\dfrac{3}{2x^2} - \dfrac{1}{x} + c$. 2. $\frac{1}{7}x^7 - \frac{1}{2}x^4 + x + c$.

3. $\frac{2}{5}x^{5/2} + \frac{2}{3}x^{3/2} + c$. 4. $\frac{1}{3}\sin 3x + c$. 5. $\frac{1}{4}e^{(4x-3)} + c$.

6. $\frac{1}{2}\cos(1 - 2x) + c$. 7. $-\frac{1}{5}(1 - x)^5 + c$. 8. $-\dfrac{1}{x + 4} + c$.

9. $-\dfrac{1}{4(2x - 5)^2} + c$. 10. $\frac{1}{2}\ln|2x - 5| + c$. 11. $\frac{1}{5}\tan 5x + c$.

12. $2\ln|\sin\frac{1}{2}x| + c$. 13. $\frac{1}{2}x - \frac{1}{4}\sin 2x + c$.
14. $\frac{1}{4}\cos 2x - \frac{1}{8}\cos 4x + c$. 15. $-x - \cot x + c$.
16. $-\frac{1}{3}\cos^3 x + c$. 17. $\sec x + c$. 18. $\frac{1}{3}\cos^3 x - \cos x + c$.
19. $x - \ln|x + 2| + c$. 20. $x + \frac{1}{2}x^2 + \ln|x - 1| + c$.
21. $\frac{1}{2}x^2 - 3\ln(x^2 + 5) + c$. 22. $\frac{2}{5}\sqrt{(5x + 3)} + c$.
23. $\frac{1}{4}\ln(1 + x^4) + c$. 24. $\frac{1}{16}(1 + x^4)^4 + c$. 25. $e - 1$.
26. $40\frac{1}{2}$. 27. $3/16$. 28. $-2/21$. 29. $(\pi + 2)/24$. 30. $2/15$.
31. $3/2$. 32. $1/4$. 33. $2 - \frac{1}{2}\pi$. 34. $\ln 2$. 35. 2.
36. $\frac{1}{2}\ln(4/3)$. 38. $\sec x$; $\frac{1}{2}\ln 3$. 39. $\tan\frac{1}{2}x + c$. 40. $3\pi/8$.

Exercise 4.2

1. $\ln\left|\dfrac{x - 1}{x + 2}\right| + c$. 2. $\frac{1}{2}\ln\left|\dfrac{1 + x}{1 - x}\right| + c$. 3. $\frac{1}{3}\ln\left|\dfrac{x - 3}{x}\right| + c$.

4. $\frac{1}{2}\ln|x^2 - 4| + c$. 5. $\ln|(x - 3)^3(x + 1)| + c$.

6. $\ln|(x - 2)(x - 3)| + c$. 7. $\ln|(x + 3)^2(x + 1)| + c$.

8. $\frac{1}{2}\ln\left|\dfrac{x^2}{2x - 1}\right| + c$. 9. $\ln\left|\dfrac{(x - 1)^3}{x - 4}\right| + c$. 10. $\frac{1}{2}\ln\left|\dfrac{1 + x}{1 - x}\right| - x + c$.

11. $\frac{1}{2}\ln|x^2 - 4x - 8| + c$. 12. $\frac{1}{2}x^2 + 2\ln|x^2 - 4| + c$.

13. $\ln|(x+1)(x-2)| + \dfrac{1}{x+1} + c.$ 14. $3\ln\left|\dfrac{x-1}{x}\right| + \dfrac{2}{x} + c.$

15. $\ln|x+3| + \dfrac{3}{x+3} + c.$ 16. $\ln\left|\dfrac{x^2+1}{x+1}\right| + c.$

17. $\frac{1}{2}\ln\dfrac{(x-1)^2}{x^2+1} + c.$ 18. $\frac{1}{4}\ln\dfrac{x^2+3}{x^2+5} + c.$

19. $\frac{1}{2}\ln(2x+1)^2(x^2+2) + c.$ 20. $x + \ln\dfrac{x^2+x+1}{(x-1)^2} + c.$

21. $\ln\frac{32}{27}.$ 22. $\ln\frac{12}{7}.$ 23. $\ln 6.$ 24. $\frac{3}{2}\ln\frac{9}{8}.$ 25. $\frac{1}{3}\ln\frac{27}{2}.$
26. $\ln\frac{4}{3}.$ 27. $\ln\frac{9}{10}.$ 28. $2 + \ln\frac{5}{9}.$ 29. $\frac{1}{2} + \ln\frac{3}{4}.$
30. $\frac{3}{2} + \ln\frac{8}{3}.$ 31. $\frac{13}{2} + \ln\frac{512}{3}.$ 32. $11 + 2\ln\frac{6}{5}.$
33. $\frac{1}{2}\ln\frac{40}{17}.$ 34. $1 + \ln\frac{5}{2}.$ 35. $\ln 2.$ 36. $\ln\frac{12}{7}.$

37. (a) $\ln\left(\dfrac{x-1}{x+3}\right) + c,$ (b) $\ln\left(\dfrac{1-x}{x+3}\right) + c,$ (c) $\ln\left(\dfrac{x-1}{x+3}\right) + c.$

38. (a) $\frac{1}{2}\ln\dfrac{(x+1)^2}{2x-3} + c,$ (b) $\frac{1}{2}\ln\dfrac{(x+1)^2}{3-2x} + c.$

Exercise 4.3

1. $\frac{1}{12}(3x-2)^4 + c.$ 2. $-\frac{1}{10}(1-2x)^5 + c.$
3. $-\frac{1}{42}(1-x)^6(1+6x) + c.$ 4. $\frac{1}{6}(x^3-1)^4 + c.$ 5. $\frac{1}{3}\tan^3 x + c.$
6. $-\frac{1}{6}\cos^6 x + c.$ 7. $\frac{1}{3}(x^2+4)^{3/2} + c.$
8. $-\frac{2}{15}(1-x)^{3/2}(2+3x) + c.$ 9. $\frac{1}{3}(x-1)\sqrt{(2x+1)} + c.$
10. $\ln(1+e^x) + c.$ 11. $\sin\theta - \frac{1}{3}\sin^3\theta + c.$
12. $\frac{1}{5}\cos^5\theta - \frac{1}{3}\cos^3\theta + c.$ 13. $\frac{1}{2}\sec\theta + c.$ 14. $2\tan(\sqrt{x}) + c.$

15. 12/5. 16. 16/105. 17. $\ln 2.$ 18. $\dfrac{e-1}{2(e+1)}.$ 19. $\pi/2.$

20. $\pi/18.$ 21. 4/5. 22. $\frac{1}{24}(2\pi + 3\sqrt{3}).$ 23. $\frac{1}{3}\ln 2.$
24. $\frac{1}{4}\ln 3.$ 25. $\frac{1}{120}(2x+3)^5(10x-3) + c.$
26. $\frac{1}{40}(x^2-1)^4(4x^2+1) + c.$ 27. $\frac{2}{9}(1+3x)^{3/2} + c.$

28. $\frac{2}{3}(x-6)\sqrt{(x+3)} + c.$ 29. $\dfrac{1}{3(1-x^3)} + c.$ 30. $-\dfrac{x}{(x-1)^2} + c.$

31. $\frac{1}{2}e^{x^2} + c.$ 32. $2\sqrt{(\ln x)} + c.$ 33. $\frac{1}{4}\ln\left|\dfrac{2+e^x}{2-e^x}\right| + c.$

34. 6. 35. 9. 36. $9\pi/4.$ 37. $\sqrt{3}.$ 38. $\frac{1}{6}(4\pi + 3\sqrt{3}).$
39. $\ln(4/3).$ 40. $23\frac{1}{5}.$ 41. $8; 5\pi.$ 42. $3\pi.$ 44. $\frac{1}{2}\pi.$

Exercise 4.4

1. $\sin x - x\cos x + c$. 2. $2x\sin\frac{1}{2}x + 4\cos\frac{1}{2}x + c$.
3. $-e^{-x}(x+1) + c$. 4. $\frac{1}{4}x^2(2\ln 2x - 1) + c$.
5. $\frac{1}{4}e^{2x}(2x-1) + c$. 6. $x\tan x + \ln(\cos x) + c$.
7. $\frac{1}{8}\sin 2x - \frac{1}{4}x\cos 2x + c$. 8. $x\tan x + \ln(\cos x) - \frac{1}{2}x^2 + c$.
9. $\frac{1}{4}x^2 + \frac{1}{4}x\sin 2x + \frac{1}{8}\cos 2x + c$. 10. $(x^2 - 2)\sin x + 2x\cos x + c$.
11. $e^x(x^3 - 3x^2 + 6x - 6) + c$. 12. $\frac{1}{16}x^4(4\ln x - 1) + c$.
13. $\frac{1}{2}x^2(\ln x)^2 - \frac{1}{2}x^2\ln x + \frac{1}{4}x^2 + c$. 14. $(x-1)\ln(x-1) - x + c$.
15. $-\frac{1}{2}e^{-x}(\sin x + \cos x) + c$. 16. -2. 17. $2/e$. 18. $\pi/9$.
19. 1. 20. $\frac{1}{4}\pi^2$. 21. $\frac{8}{3}\ln 2 - \frac{7}{9}$. 22. $10e^{-2} - 17e^{-3}$.
23. $\frac{1}{4} - \frac{3}{4}e^{-2}$. 24. $\pi - 2$. 25. $\frac{1}{2}(e^{\pi} + 1)$. 26. $\frac{5}{4}e^4 - \frac{1}{4}e^2$.
27. $\frac{1}{2}e^2$. 28. $\frac{1}{30}(x+1)^5(5x-1) + c$.

29. (a) $\frac{2}{15}(x-1)^{3/2}(3x+2) + c$, (b) $-\dfrac{(x+1)}{(x+2)^2} + c$.

Exercise 4.5

1. $\sin^{-1}\dfrac{x}{4} + c$. 2. $\frac{1}{5}\tan^{-1}\dfrac{x}{5} + c$. 3. $\frac{1}{2}\sin^{-1}2x + c$.

4. $\frac{1}{2}\ln(x^2+1) + \tan^{-1}x + c$. 5. $\frac{1}{5}\sin^{-1}\dfrac{5x}{2} + c$.

6. $\frac{1}{6}\tan^{-1}\dfrac{3x}{2} + c$. 7 $\ln|x+1| - \tan^{-1}x + c$.

8. $\ln\left(\dfrac{x^2+1}{(1-x)^2}\right) + 2\tan^{-1}x + c$. 9. $\ln\left(\dfrac{x^2+9}{x^2}\right) + 2\tan^{-1}\dfrac{x}{3} + c$.

10. $3\ln(x^2+4) - \ln|x+2| - 4\tan^{-1}\dfrac{x}{2} + c$. 11. $\frac{1}{2}\tan^{-1}\dfrac{(x+1)}{2} + c$.

12. $\sin^{-1}\dfrac{(x-2)}{3} + c$. 13. $\pi/18$. 14. $\pi/2$. 15. $\pi/8$.

16. $\pi/2$. 17. $\ln 2 + \pi/2$. 18. $\pi/4$. 19. $\pi/4$.
20. $\frac{1}{2}\ln(1 + e^{2x}) + \tan^{-1}(e^x) + c$. 21. $\tan^{-1}(e^x) + c$.
22. $\frac{1}{2}\sin^{-1}x + \frac{1}{2}x\sqrt{(1-x^2)} + c$. 23. $\sin^{-1}x + \sqrt{(1-x^2)} + c$.

24. $\ln(x^2 - 2x + 10) + \frac{2}{3}\tan^{-1}\dfrac{(x-1)}{3} + c$. 25. $\frac{1}{2}\tan^{-1}(x^2+1) + c$.

26. $-\sin^{-1}\left(\dfrac{1}{x}\right) + c$. 27. (a) $-0{\cdot}123$, (b) $0{\cdot}545$.

Exercise 4.6

1. (a) $2/5$, (b) 0, (c) π, (d) $\sqrt{3}$, (e) 0, (f) 0. 2. 0.
3. (a) -2, (b) 4, (c) 4. 4. (a) $-1\frac{1}{2}$, (b) $1\frac{1}{2}$, (c) $2\frac{1}{2}$.
5. (a) 0, (b) $8/3$. 7. $e-1$. 8. $\frac{2}{3}(2-\sqrt{2})$.
9. $I=\frac{1}{2}\pi$, $J=2(e-1)$. 10. (a) 2, (b) 1, (c) π.

Exercise 4.7

1. $2-\frac{1}{2}\ln 3$. 2. $\ln 2-1$. 3. $\frac{1}{4}\ln\frac{9}{5}$. 4. 2.

5. $1+\ln\frac{8}{3}$. 6. $\dfrac{64\sqrt{2}}{21}$. 7. $8/3$. 8. π. 9. $7/6$.

10. $\frac{1}{2}(1-\ln 2)$. 11. $1/24$. 12. $\ln 2$. 13. $\ln(4/3)$. 14. $\pi/16$.

15. $1/(\ln 2)$. 16. $\frac{1}{5}\ln\left(\dfrac{x-3}{x+2}\right)-\dfrac{1}{x-3}+c$. 18. $\pi/12$.

19. $(\pi-2)/4$. 20. $(\pi+2)/8$. 21. $\frac{1}{2}\ln 3$.
23. $(\frac{1}{4},\frac{8}{3})$ min; $\frac{3}{2}-\ln 4$. 24. $(-4,\frac{4}{3})$ max; $2\ln 3$.

25. $\dfrac{2}{x-1}-\dfrac{x}{x^2+1}$; $1{\cdot}04$, $-0{\cdot}23$.

27. (i) a^2+b^2, (ii) $2\log_{10}2-\log_{10}e$, (iii) $(2\pi+3\sqrt{3})/48$.
28. $y=x-a$, $x+y=5a$; $8a^2/3$.
29. $-3\sin t\cos t$; $y+3x\sin t_1\cos t_1=3\sin t_1-2\sin^3 t_1$.
31. $\frac{1}{2}x^2\sin 2x+\frac{1}{2}x\cos 2x-\frac{1}{4}\sin 2x+c$; π; $\pi^2(2\pi^2-3)/12$; $\pi^4/6$.

32. $\dfrac{1}{x+1}+\dfrac{x+2}{x^2+3}$; $\ln(\sqrt{2}+\sqrt{6})+\dfrac{\pi}{2\sqrt{3}}$; $\dfrac{1}{\sqrt{3}}\ln(\sqrt{2}+\sqrt{6})+\dfrac{\pi}{6}$.

33. $p=2$, $q=1$; $\dfrac{1}{4}\pi$; $\ln 2$; $\pi+\ln 2$.

34. (a) $\dfrac{1-t^2}{1+t^2}$ (b) $\dfrac{1}{1-t}+\dfrac{1}{1+t}+\dfrac{8}{1+9t^2}$.

35. (a) $x(\log_e x-1)+c$; $x\{(\log_e x)^2-2\log_e x+2\}+c$, (b) $\pi(e-2)$.

36. (i) $\frac{1}{2}\pi-1$, (ii) $\frac{1}{2}(1-e^{-2})$. 37. π.

38. $\dfrac{a^2}{12}(4\pi-3\sqrt{3})$; $\dfrac{5\pi a^3}{24}$.

39. (i) $\pi(2\ln 2-1)$, (ii) $\frac{1}{2}\pi-1$.

40. $\frac{22}{15}a^2$, $\frac{7}{12}\pi a^3$. 41. $\frac{16}{3}$, $\dfrac{64\pi}{15}$.

42. $x=\dfrac{\pi}{4},\dfrac{9\pi}{4}$: max, $x=\dfrac{5\pi}{4},\dfrac{13\pi}{4}$: min; $\frac{1}{2}(1+e^{-\pi})$; $\frac{1}{2}e^{-2\pi}(1+e^{-\pi})$.

Exercise 5.1

1. -42 N m, 14 N m, -21 N m, 0 N m.
2. 42 N m, -30 N m, -42 N m, 30 N m, 0 N m.
3. $(4+\sqrt{3})$ N m, $3\sqrt{3}$ N m, $(2+5\sqrt{3})$ N m.
4. (a) 6 N, 2 m, (b) 8 N, 3 m.
5. (a) 1 N, 10 m, (b) 1 N, 15 m.
6. 50 N m. 7. $F = G = 4$.
8. (a) $5\sqrt{2}$ N, 4 m, (b) 15 N, 3 m.
9. (a) $4\sqrt{2}$ N, 5 m, (b) 6·5 N, 12 m. 10. $6\sqrt{2}$ N along BD.
11. $20\sqrt{3}$ N m, $10\sqrt{3}$ N m, 0 N m, 0 N m, $10\sqrt{3}$ N m, $20\sqrt{3}$ N m; 10 N along CD.

Exercise 5.2

1. 4 N at 90° to AB, 4 m. 2. 2 N at 90° to AB, 5 m.
3. $6\sqrt{3}$ N at 60° to AB, 1 m. 4. 6 N at 60° to AB, 2 m.
5. (a) $5P$ at 53·1° to AB, (b) $6a$. 6. $3x - 7y + 7 = 0$.
7. $2x - 5y + 10 = 0$. 8. a, b. 9. c. 10. a, c. 11. b, c.
12. c. 13. c. 14. a, c. 15. b, c. 16. c. 17. a.
18. b. 19. a, c. 20. 6 N m clockwise; $y = x + 1$.
21. (a) 5, 13, (b) 2·5, 19·5. 22. $7\sqrt{5}$ N, 1 m; 28 N m in sense CBA.
23. (a) 10; 28 N along AD, (b) 18 N, $14\sqrt{3}$ N m in sense ABC.
24. (a) $25P$, $24y = 7x$, (b) (i) $S = -18P$, $T = 15P$, (ii) $72Pa$, (iii) clockwise.

Exercise 5.3

1. 18 cm from A. 2. 1·05 m from P. 3. 5 cm. 4. 3 cm, 4 cm.
5. (2, 1). 6. $(3, -\frac{1}{2})$. 7. $3\mathbf{i} + \mathbf{j}$. 8. $2\mathbf{i} + 2{\cdot}5\mathbf{j}$. 9. (3, 0).
10. (0·4, 1·6). 11. 10 cm, 3 cm. 12. 10 cm. 13. $1\frac{2}{3}$ cm from AB.
14. $6\frac{1}{3}$ cm from AD, $4\frac{4}{5}$ cm from AB. 15. 11 cm from both AB and AF.
16. $3\frac{1}{2}$ cm from AE, 4 cm from AB. 17. 4 cm from AD, 3·5 cm from AB.
18. $7a/5$ above base, $16a/11$ above base.

Exercise 5.4

3. $\dfrac{2a(14+3\pi)}{3(8+\pi)}$. 4. $\dfrac{2a(26-3\pi)}{3(16-\pi)}$. 5. $\dfrac{28r}{9\pi}$.

6. $\dfrac{27\sqrt{3}a}{3\sqrt{3}+4\pi}$. 7. $\dfrac{11h}{28}$.

8. (a) $\dfrac{7\sqrt{3}r}{8\pi+3\sqrt{3}}$, (b) $\dfrac{7\sqrt{3}r}{4\pi-3\sqrt{3}}$. 9. $28\pi a^3/3$, $12a/7$. 11. $3r$.

Exercise 5.5

1. 6, 103/15; (9/7, 103/70). 2. 153/4, 414/5; (17/8, 23/5).
3. ln 2, 7/48; (2 ln 2, 7/24). 4. 1/4, 2/5; (3/8, 3/5). 5. (20/7, 0).
6. (3/5, 0). 7. $(0, \pi/8)$. 8. (0, 7/3). 9. $\left(\frac{1}{e-1}, \frac{e+1}{4}\right)$.
10. $(1, \pi/8)$. 11. (2, 0). 12. (2/3, 0). 13. (2 ln 2, 0).
14. $\left(\frac{e^2-3}{2(e^2-1)}, 0\right)$. 15. (9/7, 0). 16. (0, 14/5).
17. $2\ln 2 - 1$, $2\ln 2 - \frac{3}{4}$, $2(\ln 2)^2 - 4\ln 2 + 2$; (1·65, 0·24).
18. 1/6; (1/2, 2/5).

Exercise 5.6

1. (a) $13P$, 12/5, (b) $3a$; $12Pa$, sense ABC. 2. $\lambda = 5/2$; $5a/2$; $5a/7$.
3. (c) (i) $3\sqrt{2}$ along AC, $\sqrt{2}$ along BD; $7a$;
(ii) 7 along AD, $4\sqrt{2}$ along CA, $6\sqrt{2}$ along BD.
4. (a) $4\sqrt{3}P, \frac{\sqrt{3}}{3}P; \frac{7\sqrt{3}}{3}P$; (b) $\frac{5\sqrt{3}}{3}P, \frac{\sqrt{3}}{3}P; \frac{7\sqrt{3}}{3}Pa$ in sense ABC.
5. $p = 11, q = 14$; (a) $2\sqrt{19}$ N, (b) $12\sqrt{3}$ N m.
6. (i) $m = 2, n = 6$, (iii) $x = 3a\sqrt{3}$; $\sqrt{31}$, $5Pa\sqrt{3}$.
7. $(7a/18, 4a/9)$.
8. 4·8 cm from OA, 8·8 cm from OC.
9. $\frac{11}{24}\pi a^3, \frac{5}{24}\pi a^3; \frac{21}{88}a, \frac{27}{40}a$ from centre of hemisphere.
11. $\frac{3a}{8}$. 12. $\frac{32}{15}\sqrt{2}; \frac{8\pi}{5}$. 13. 16/3; 8π; (0, 20/3).
14. (0·864, 0·333). 15. 4/3, $104\pi/15$; (15/13, 0). 16. $3\pi a^2$, $(\pi a, 5a/6)$.

Exercise 6.1

1. 360 N, 2·5 m. 2. 41 N at A, 33 N at B. 3. 1·6 m.
4. 0·6 m, 20 N. 5. (a) 1200 N (b) 3 m. 6. 0·4 m.
7. $5W/8$ at A, $3W/8$ at B. 8. 5. 9. $5a/3$, $5a/2$; $\tan^{-1} 10/21 \approx 25·5°$.
10. (a, a); 1/2. 11. $2/(\pi + 2)$. 12. $1/\pi$.
13. $\tan^{-1} 3/4 \approx 36·9°$; $3aW$. 14. $\tan^{-1} 11/13 \approx 40·2°$; $26aW/9$.
15. $\sqrt{2}$. 16. $2W/3$.
17. $h + \left(\frac{3a^2 - h^2}{8a + 4h}\right)$ from vertex of cone; (a) 17°, (b) $h = a\sqrt{3}$.
18. 30°; $1/\sqrt{3}$. 19. 36·9°. 20. (a) 53·1°, (b) 26·6°.

Exercise 6.2

1. (a) 10, (b) $10\sqrt{5}$ N at 26·6° to vertical.
2. $\frac{\sqrt{3}}{2}mg$ in AC, $\frac{1}{2}mg$ in BC; $\frac{1}{2}a$. 3. 2 N at A, $2\sqrt{17}$ N at C.
4. 30 N at 60° to vertical. 5. 45 cm, 16 N.
6. (a) $20\sqrt{3}$ N, 90°, (b) $15\sqrt{5}$ N, 63·4°. 7. 5/6; 5W/12, 13W/12.
8. (a) $W\sqrt{3}$ at both A and C, (b) $2a$.
9. (a) 15 N, (b) 9 N at 53·1° to vertical. 10. $5W$, 75 cm. 11. $\frac{1}{2}$; 30 cm.
12. 5/24. 13. (a) $\frac{1}{2}mg\cot\alpha$, (b) $\frac{1}{2}\cot\alpha$.
14. 45°, $\frac{1}{2}W$; half way up, all the way. 15. (a) 30°, (b) $1/\sqrt{3}$.
17. 1/3.

Exercise 6.3

1. (a) 24 N, (b) $12\sqrt{3}$ N, 13 N.
2. (a) 60 N, (b) 48·2 N at 41·6° to horizontal.
3. (a) 15, (b) 47·4 N at 71·6° to horizontal.
4. $4\sqrt{3}$ N perpendicular to rod; $2\sqrt{3}$ N, 10 N.
5. (a) 14 N, 30 N (b) 1·92 m. 6. (a) 4/7, (b) $\tan^{-1}\frac{2}{3} \approx 33{\cdot}7°$.
7. 5 N, 1/2. 8. 121 N, 17·5 N; 7/48. 9. $\frac{1}{2}\tan\alpha$, $2W\tan\alpha$.
10. $3W/10$, $W/10$; 0·8. 11. 1/5. 12. $8W/15$; $24W/75$, $43W/75$.
13. $(5\sqrt{3})/18$; $(W\sqrt{3})/2$. 14. $5a/3$. 15. $\tan^{-1}(4/3)$; 40, 900 N.
16. 10, 6/17. 17. $W/4$, 1/3. 18. $11W/4$. 19. 5 m.

Exercise 6.4

1. (a) $5\sqrt{10}$ N, $5\sqrt{5}$ N (b) 21·8°, 45°. 2. $\frac{1}{2}W\sqrt{5}$, 63·4°.
3. $\frac{1}{3}W\sqrt{3}$, $\frac{1}{2}W$.
4. (ii) $W\sqrt{(1+\mu^2)}$ at $\tan^{-1}\mu$ to vertical, (iii) $\mu W = \frac{1}{2}W\cot\theta$.
5. (a) 7/10, (b) $5W/2$, $3W/4$. 6. C; 67·4°. 7. C will slip.
8. (a) $9W/5$, $12W/5$, (b) 3·2°. 9. $\sqrt{3}/2$. 10. $2-\sqrt{3}$.
11. $\frac{3}{4}W$, $\frac{1}{4}W\tan\alpha$; $1/\sqrt{3}$.
13. (i) $\frac{9w}{25}$, (ii) $W+\frac{27w}{125}$, (iii) $\frac{36w}{125}$.

Exercise 6.5

1. AB: $\sqrt{3}$ kN tension; BC: $4\sqrt{3}$ kN tension; CD: $2\sqrt{3}$ kN thrust; AD: $2\sqrt{3}$ kN thrust; BD: $2\sqrt{3}$ kN tension.
2. AB and BC: 5 kN thrust; BD: no stress; CD: 4 kN tension; AD: 3 kN tension.
3. AB: 2 kN thrust; BC: $2\sqrt{2}$ kN tension; CD: 4 kN tension; AC: $2\sqrt{2}$ kN thrust.

4. AB: 8 kN thrust; BC: $4\sqrt{3}$ kN thrust; CD: $2\sqrt{3}$ kN tension; BD: 4 kN tension.
5. AB: $10\sqrt{2}$ kN thrust; BC: $5\sqrt{2}$ kN thrust; CD: 5 kN tension; DE: $5\sqrt{2}$ kN tension; AE: 10 kN tension; BD: 5 kN thrust; BE: 5 kN tension.
6. AB: 4 kN tension; BC: 3 kN thrust; CD: 3 kN thrust; AD: 4 kN tension; AC: 15 kN tension.
7. AB: $4\sqrt{3}$ kN tension; BC: $2\sqrt{3}$ kN tension; CD: $4\sqrt{3}$ kN thrust; DE: $3\sqrt{3}$ kN thrust; AE: $8\sqrt{3}$ kN thrust; BD: $2\sqrt{3}$ kN tension; BE: $2\sqrt{3}$ kN thrust.
8. AB and BC: 6 kN tension; CD and AE: $6\sqrt{3}$ kN thrust; BD and BE: $2\sqrt{3}$ kN tension; DE: $4\sqrt{3}$ kN thrust.
9. $P = \frac{1}{3}W\sqrt{2}$; AB: $\frac{2}{3}W$ thrust; BC: $\frac{1}{3}W\sqrt{2}$ tension; CD: $\frac{1}{3}W$ thrust; AC: $\frac{2}{3}W\sqrt{2}$ tension; AD: $\frac{2}{3}W$ thrust.
10. (ii) AB: 390 N; BC: 650 N; CD: 150 N; AD: 250 N; AC: $120\sqrt{5}$ N (iii) AC and CD.

Exercise 6.6

2. 4/11; 7:2. 4. $5a/2$, $5a/2$; $\tan^{-1}(1/6) \approx 9{\cdot}5°$; $W/6$.
7. (a) $12\frac{1}{2}$ N; $7\frac{1}{2}$ N in the direction CA, (b) $6\frac{1}{4}$ N; $6\frac{1}{4}$ N at 53·1° to CA produced; 8 N acting 0·625 m from A.
8. $\frac{1}{2}W\sqrt{5}$; $7l/12$.
9. (a) 15 N, (b) 8/3 m, (c) 42·7 N at 73·7° to the horizontal.
10. (ii) $\frac{1}{4}\sqrt{3}$, (iii) $24c/7$. 11. 3/46. 12. 3/8, $5l/3$.
14. $\frac{3}{2}mg\cot\alpha$; $mg\cot\alpha$. 15. $\cos^{-1}(2m/M)$.
16. $W(1 - \sin 2\theta)$; 18·4°.
17. $AX = \frac{2}{3}a$, $13W$ at 67·4° to XA; $CY = 11a/12$.
18. (i) $25W/37$, $20W/37$, (iii) 13/40.
20. $(8a/15, 3a/16)$; $128/45 \approx 2{\cdot}84$.
21. $h + \left(\dfrac{6l^2 - h^2}{4h + 12l}\right)$ from vertex of cone. 22. $\sin^{-1}\left(\dfrac{k+3}{3k-24}\right)$.
23. (i) $35\sqrt{2}$ N, (ii) $5\sqrt{2}$ N.
24. Force at the ground. 25. $\frac{1}{4}Wl\sqrt{3}$.
26. $\frac{1}{2}W/5$; $\frac{1}{2}W$, 0. 27. Slipping at A.
28. (i) $\frac{1}{2}W$, (ii) $\frac{1}{2}W$, W, (iv) $\frac{1}{2}$.
29. $2W\sqrt{3}$ along BA; $\frac{1}{3}W\sqrt{3}$, $3W/2$ at A. $\frac{1}{3}W\sqrt{3}$, $\frac{1}{2}W$ at C.
30. $\frac{2}{3}W\sqrt{3}$ between cylinders, $\frac{1}{3}W\sqrt{3}$ at both walls. 31. $1/\sqrt{3}$.
32. (i) $35W/4$, (ii) 0, $5W$ at A; $7W$, $5W/4$ at B, (iii) $21aW/2$.
33. (i) 2400 N, (ii) 2600 N at 67·4° to vertical, (iii) AB: 2400 N; BC: 700 N; CD: $700\sqrt{2}$ N; DE: 1400 N; BD: $700\sqrt{2}$ N; BE: $1000\sqrt{2}$ N; rods AB, BC, BD in compression.
34. (i) 14, (ii) AB: $4/\sqrt{3}$ kN tension; BC: $8/\sqrt{3}$ kN thrust; CD: $8/\sqrt{3}$ kN thrust; DE: $16/\sqrt{3}$ kN thrust; AE: $2/\sqrt{3}$ kN thrust; BD: $4/\sqrt{3}$ kN tension; BE: $4\sqrt{3}$ kN thrust, (iii) 2 kN vertically downwards.

Exercise 7.1

1. $\mathbf{a} - \mathbf{b} + \mathbf{c}$. 2. $\mathbf{a} - 2\mathbf{b} + 2\mathbf{c}$. 3. $60°$. 4. $7, \sqrt{19}$.
5. (a) $s = -2, t = 6$, (b) $s = 6/5, t = 3/5; s = -2, t = -1$.
6. (a) 9, (b) -1. 7. (a) $\frac{3}{8}\mathbf{a} + \frac{5}{8}\mathbf{b}$, (b) $-\frac{1}{2}\mathbf{a} + \frac{3}{2}\mathbf{b}$, (c) $2\mathbf{a} - \mathbf{b}$.
8. (a) $\mathbf{i} + 2\mathbf{j} - \mathbf{k}$, (b) $-7\mathbf{i} - 2\mathbf{j} - 5\mathbf{k}$.
9. $2\mathbf{i} - 5\mathbf{j} + 3\mathbf{k}, 7\mathbf{i} - 2\mathbf{k}; (4, -3, 1), (-1, -8, 6)$.
10. P is a vertex of parallelogram $ABCP$; Q divides AC internally in the ratio 4:3; R divides BC externally in the ratio 2:3.
11. When produced BA and CD meet a point Q such that $BQ:AQ = 7:4$ and $CQ:DQ = 5:2$.
12. $\mathbf{r} = 3\mathbf{q} - 2\mathbf{p}; 3:2$.

Exercise 7.2

1. (a) $\mathbf{r} = (1 + 3\lambda)\mathbf{i} + 2\mathbf{j} - (1 + \lambda)\mathbf{k}$, (b) $\mathbf{r} = (\lambda + 4)\mathbf{i} - 3\lambda\mathbf{j} + 3(\lambda - 1)\mathbf{k}$.
2. (a) $\mathbf{r} = 2(1 - 2\lambda)\mathbf{i} + (4\lambda - 3)\mathbf{j} + \mathbf{k}$, (b) $\mathbf{r} = (1 + 2\lambda)\mathbf{i} + (5 - 6\lambda)\mathbf{j} + 2\lambda\mathbf{k}$, (c) $\mathbf{r} = 5\lambda\mathbf{i} + (6 - 13\lambda)\mathbf{j} + 2(5\lambda - 4)\mathbf{k}$, (d) $\mathbf{r} = 5\lambda\mathbf{i} - 2\lambda\mathbf{j} + 3\lambda\mathbf{k}$.
3. $\mathbf{r} = (1 - \lambda)\mathbf{a} + \frac{1}{3}\lambda\mathbf{b} + \frac{2}{3}\lambda\mathbf{c}, \mathbf{r} = -\mu\mathbf{a} + (1 - \mu)\mathbf{b} + 2\mu\mathbf{c}; -\frac{1}{2}\mathbf{a} + \frac{1}{2}\mathbf{b} + \mathbf{c}$.
4. $\frac{1}{2}\mathbf{a}, \frac{1}{2}\mathbf{a} + \mathbf{b}, \frac{1}{3}\mathbf{a} + \frac{2}{3}\mathbf{b}$. 5. $\frac{3}{2}\mathbf{a}$. 6. 1:3.
7. (a) Intersect at $2\mathbf{i} - 3\mathbf{j} + 3\mathbf{k}$, (b) skew, (c) parallel, (d) intersect at $6\mathbf{i} + 13\mathbf{j} - 4\mathbf{k}$.
8. (a) A, B, (b) B, (c) C, (d) A, B, C.
9. (a) $\mathbf{r} = (\lambda + 1)\mathbf{i} + (3\lambda - \mu - 2)\mathbf{j} + 2\mu\mathbf{k}$,
   (b) $\mathbf{r} = (2\lambda - \mu)\mathbf{i} + (2\mu - \lambda)\mathbf{j} - 7\mu\mathbf{k}$,
   (c) $\mathbf{r} = (\mu + 3)\mathbf{i} + (\lambda + \mu + 1)\mathbf{j} + (\mu - 1)\mathbf{k}$.
10. $\mathbf{r} = (\lambda + 5\mu)\mathbf{i} + (2\mu - 3\lambda)\mathbf{j} + (2 - \lambda - 2\mu)\mathbf{k}$.
11. $-\mathbf{i} + 3\mathbf{k}, \mathbf{j}; \mathbf{r} = (2 - \lambda)\mathbf{i} + (5 + \mu)\mathbf{j} + 3\lambda\mathbf{k}$.
12. $\mathbf{r} = (2t + 1)\mathbf{i} + (2s - t)\mathbf{j} - (2 + s - 3t)\mathbf{k}; (5, 0, 3)$.

Exercise 7.3

1. $6\mathbf{i} - 8\mathbf{j}, 4\mathbf{i} + 3\mathbf{j}; 10\mathbf{i} - 5\mathbf{j}$.
2. $3\sqrt{2}$ N, $5\sqrt{2}$ N. 3. 27 N.
4. (a) $\mathbf{r} = (1 + \lambda)\mathbf{i} + (1 + 2\lambda)\mathbf{j}$, (b) $\mathbf{r} = -2\lambda\mathbf{i} + (5\lambda + 2)\mathbf{j} + (3\lambda - 1)\mathbf{k}$, (c) $\mathbf{r} = 3\mathbf{i} - 2\lambda\mathbf{j} + 3\lambda\mathbf{k}$.
5. $\mathbf{r} = (2 + 3\lambda)\mathbf{i} + (\lambda - 1)\mathbf{j} - 2\lambda\mathbf{k}, \mathbf{r} = (9 + \mu)\mathbf{i} - \mu\mathbf{j} + (3\mu - 1)\mathbf{k}; 8\mathbf{i} + \mathbf{j} - 4\mathbf{k}$; $\mathbf{r} = (8 + 4t)\mathbf{i} + \mathbf{j} + (t - 4)\mathbf{k}$.
6. 11 N, $\mathbf{r} = (3 + 7\lambda)\mathbf{i} + (5 + 6\lambda)\mathbf{j} + 6\lambda\mathbf{k}$.
7. $3x + y = 26$.
8. $3\sqrt{13}$ N, $\frac{2}{3}$; $2x - 3y + 2 = 0$.
9. (a) 15**YX**, where $BX : XC = 3:2, AY:YB = 2:1$,
   (b) 24**XY**, where $AX:XC = 1:3, AY:YB = 1:5$,
   (c) 6**YX**, where $AX:XC = 2:1$ and $AY = YB$,
   (d) 21**XY**, where $AX:XC = 1:2, AY:YB = 3:4$.

Exercise 7.4

1. $\mathbf{r}_A = (3 + 5t)\mathbf{i} + (4 + 2t)\mathbf{j}$, $\mathbf{r}_B = (5 + 2t)\mathbf{i} + (8 + t)\mathbf{j}$;
   $\mathbf{AB} = (2 - 3t)\mathbf{i} + (4 - t)\mathbf{j}$; $\sqrt{10}$ m.
2. $4\mathbf{i} - 6\mathbf{j}$; $(4t - 2)\mathbf{i} + (3 - 6t)\mathbf{j}$; $t = \frac{1}{2}$.
3. $(2 - 5t)\mathbf{i} + (4t - 18)\mathbf{j}$; $2\sqrt{41}$ km; $16\mathbf{i} + 19\mathbf{j}$, $8\mathbf{i} + 9\mathbf{j}$.
4. $8\cos\omega t\mathbf{i} + 8\sin\omega t\mathbf{j}$. 5. $\mathbf{i} - 2\mathbf{j}$, 3 m; 6 m s^{-1}.
6. (a) $\pi/4$, (b) $5\pi/4$. 7. (a) 1/3, (b) 3.
8. (a) $8t\mathbf{i} + (9t^2 - 2)\mathbf{j}$, $m(8\mathbf{i} + 18t\mathbf{j})$, (b) $-6t^2\mathbf{j} + 2t\mathbf{k}$, $m(-12t\mathbf{j} + 2\mathbf{k})$,
   (c) $(2 - 2t)\mathbf{i} + (6\cos 2t)\mathbf{j}$, $m\{-2\mathbf{i} - (12\sin 2t)\mathbf{j}\}$,
   (d) $e^{-t}(1 - t)\mathbf{i} - 2\mathbf{j}$, $me^{-t}(t - 2)\mathbf{i}$.
9. (a) $\sqrt{19}$ m s^{-1}, (b) $20\mathbf{i}$. 10. $12\mathbf{i} + 30\mathbf{j}$, $18\mathbf{i} + 45\mathbf{j}$.
11. 40 m s^{-1}, $18\mathbf{i} - 35\mathbf{j}$. 12. $5\mathbf{i} - 3\mathbf{j} + 2\mathbf{k}$.
13. $2\mathbf{i} + 6\mathbf{j}$; $9\sqrt{10}$ m.
14. $(16t - 5t^2)\mathbf{j}$, $12t\mathbf{i} + (25 - 5t^2)\mathbf{j}$; 15 m.

Exercise 7.5

1. (a) $p^2 + 2\mathbf{p}.\mathbf{q} + q^2$, (b) $p^2 - q^2$, (c) $2\mathbf{p}.\mathbf{r}$, (d) $p^2 - 2\mathbf{p}.\mathbf{q} + q^2$.
2. (a) x^2, (b) $2x^2 - 3y^2$. 3. 81, 49; 9, 7. 4. 13, 9. 8. $-3/4$.
9. $3\mathbf{a} - 2\mathbf{b}$. 14. (a) 68·3°, (b) 60°, (c) 90°, (d) 35·3°.
15. (a) 1/3, (b) 5/11. 16. (a) 3, (b) 2, (c) 1, (d) -1. 17. 65/7, 7/9.
18. (a) $4\mathbf{i} - 3\mathbf{j}$, (b) $2\mathbf{i} + 5\mathbf{j}$, (c) $3\mathbf{i} - 2\mathbf{j}$.
19. $\angle A = 90°$, $\angle B = 36·9°$, $\angle O = 53·1°$; 30 sq. units.
20. 11/21, $8\sqrt{5}$ sq. units. 21. $5\mathbf{i} + 2\mathbf{k}$.
22. (a) $(2, -1, 1)$, (b) 70·5°.
23. $\mathbf{i} - \mathbf{j}$, $\mathbf{j} + \mathbf{k}$; 60°. 24. -11, -7; 7; 7/57.

Exercise 7.6

1. $-2\mathbf{i} - 2\mathbf{j} + \mathbf{k}$; 3 units. 2. $\mathbf{i} - \mathbf{k}$; 9 sq. units.
3. $3\mathbf{i} - \mathbf{j} + 4\mathbf{k}$; $6\mathbf{i} - 2\mathbf{j} + 8\mathbf{k}$. 4. $5\mathbf{i} - \mathbf{j} + 6\mathbf{k}$; 13 units.
5. (a) $\mathbf{r}.(4\mathbf{i} - \mathbf{k}) = 3$, (b) $\mathbf{r}.(\mathbf{i} - \mathbf{j} - 2\mathbf{k}) = 1$.
6. (a) $\mathbf{r}.\mathbf{k} = 1$, (b) $\mathbf{r}.(3\mathbf{i} + 4\mathbf{j} - 2\mathbf{k}) = 0$.
7. (a) 5, (b) 3, (c) 4, (d) 0.
8. $2\mathbf{i} - 5\mathbf{j} - \mathbf{k}$, $\mathbf{r}.(2\mathbf{i} - 5\mathbf{j} - \mathbf{k}) = 7$. 9. $\mathbf{r}.(\mathbf{i} + 3\mathbf{j} - 2\mathbf{k}) = 1$.
10. $\mathbf{r}.(\mathbf{j} - 2\mathbf{k}) = -8$. 11. (a) 60°, (b) 48·2°, (c) 70·5°.
12. (a) 11/18, (b) 5/6, (c) 6/7. 13. $-\mathbf{i} - 2\mathbf{j} - 3\mathbf{k}$, $2\mathbf{i} - 2\mathbf{j} + 3\mathbf{k}$.
14. $\mathbf{r} = (3\lambda + 5)\mathbf{i} - (4\lambda + 3)\mathbf{j} + (\lambda + 4)\mathbf{k}$; $2\mathbf{i} + \mathbf{j} + 3\mathbf{k}$.
15. (a) $\dfrac{x-1}{2} = \dfrac{y+2}{3} = \dfrac{z-1}{4}$, (b) $\dfrac{x}{1} = \dfrac{y-2}{-1} = \dfrac{z-3}{2}$,

    (c) $\dfrac{x-2}{5} = \dfrac{y+3}{-6} = \dfrac{z-1}{3}$, (d) $\dfrac{x-5}{1} = \dfrac{y}{3} = \dfrac{z+4}{0}$
16. (a) $2x - 3y + z - 5 = 0$, (b) $3x + y - 5z - 9 = 0$.
17. $5x + y + z - 6 = 0$. 18. $2x - y + 5z - 3 = 0$.

19. $\mathbf{i} - 3\mathbf{j} + 2\mathbf{k}$; (a) $\mathbf{r}.(\mathbf{i} - 3\mathbf{j} + 2\mathbf{k}) = -5$, (b) $x - 3y + 2z + 5 = 0$.
20. $3x + y - 4z - 1 = 0$. 21. (a) 70·5°, (b) 65·9°, (c) 28·5°.
22. (a) Plane through O perpendicular to OA,
    (b) Plane through A perpendicular to OA,
    (c) Sphere with centre A, radius 1 unit,
    (d) Plane through O perpendicular to AB,
    (e) Sphere with diameter OA,
    (f) Sphere with diameter AB.

Exercise 7.7

1. 31·6°, $5\,\mathrm{m\,s^{-1}}$. 2. $\sqrt{41}\,\mathrm{N} \approx 6{\cdot}40\,\mathrm{N}$, 50·6°.
3. (a) 2 N, (b) −3 N, (c) $\sqrt{2}$ N, (d) −5/3 N.
4. (a) $8\mathbf{i} - 6\mathbf{k}$, $3\mathbf{i} + 4\mathbf{j} + 4\mathbf{k}$, (b) $-10\mathbf{j} + 24\mathbf{k}$, $5\mathbf{i} - 12\mathbf{j} - 5\mathbf{k}$,
   (c) $-6\mathbf{i} + 18\mathbf{j} - 27\mathbf{k}$, $24\mathbf{i} + 17\mathbf{j} + 6\mathbf{k}$, (d) $\frac{2}{3}(4\mathbf{i} - 8\mathbf{j} + \mathbf{k})$, $\frac{1}{3}(\mathbf{i} + 4\mathbf{j} + 28\mathbf{k})$.
5. (a) 2 J, (b) 0 J, (c) 21 J, (d) −11 J.
6. (a) $8m$ J, (b) $24m$ J, (c) $360m$ J, (d) 0 J.
7. (a) $4m\mathbf{i}$, $4t\mathbf{i} - 5\mathbf{k}$; $16mt$, (b) $6m(t\mathbf{j} - \mathbf{k})$, $3t^2\mathbf{j} - 6t\mathbf{k}$; $18mt(t^2 + 2)$,
   (c) $-me^{-t}\mathbf{j}$, $\mathbf{i} + e^{-t}\mathbf{j}$; $-me^{-2t}$.
   (d) $-m(\sin t\mathbf{j} + 3\cos t\mathbf{k})$, $\cos t\mathbf{j} - 3\sin t\mathbf{k}$; $4m\sin 2t$.
8. 14 J; $8\,\mathrm{m\,s^{-1}}$. 9. −114 J; 6; −96 W, −18 W.
10. $t^2\mathbf{i} + (1 - 3t)\mathbf{j} + \mathbf{k}$; 20 J; 31 W.
11. $(4t + 6)\mathbf{i} - 2\sin t\mathbf{j} + \frac{1}{2}t^2\mathbf{k}$, $t^3 + 32t + 48 + 4\sin 2t$. 12. 24 J.
13. $10t^4 + 6t^2 + 2t$; 544 J. 14. $\int_1^2 \mathbf{F}.(\mathbf{i} - 2\lambda\mathbf{j})\,d\lambda$; 12.

Exercise 7.8

1. $\mathbf{b} - \mathbf{a}$, $\frac{1}{2}(\mathbf{a} + \mathbf{c})$, $\mathbf{a} - \mathbf{b} + \mathbf{c}$; $2\mathbf{k}$, $2\mathbf{j}$, $2\mathbf{i} + 2\mathbf{j} + 2\mathbf{k}$, $\mathbf{i} + \mathbf{j} + 2\mathbf{k}$, $2\mathbf{j} + 2\mathbf{k}$; $(2\sqrt{2})/3$.
2. $\frac{1}{2}(\mathbf{a} + \mathbf{b})$, $\frac{1}{2}(\mathbf{b} + \mathbf{c})$, $\frac{1}{2}(\mathbf{c} + \mathbf{d})$, $\frac{1}{2}(\mathbf{d} + \mathbf{a})$, $\frac{1}{2}(\mathbf{a} + \mathbf{c})$, $\frac{1}{2}(\mathbf{b} + \mathbf{d})$; $(\mathbf{i} + \mathbf{j})/\sqrt{2}$.
3. Mid-point of XC, where X divides AB in ratio 1:2; $\frac{1}{3}\mathbf{a} + \frac{2}{3}(\frac{1}{4}\mathbf{b} + \frac{3}{4}\mathbf{c})$;
   R divides AY in ratio 2:1, where Y divides BC in ratio 3:1.
4. $\{(1 + \lambda)\mathbf{a} + \mathbf{b} + \lambda\mathbf{c}\}/2(1 + \lambda)$, $\{\lambda\mathbf{a} + (1 + \lambda)\mathbf{b} + \mathbf{c}\}/2(1 + \lambda)$,
   $\{\mathbf{a} + \lambda\mathbf{b} + (1 + \lambda)\mathbf{c}\}/2(1 + \lambda)$; $\frac{1}{2}(1 + \sqrt{5})$.
6. $[r(1 - p)\mathbf{c} + p(r - 1)\mathbf{a}]/(r - p)$, $[p(1 - q)\mathbf{a} + q(p - 1)\mathbf{b}]/(p - q)$.
7. $3\mathbf{i} + 2\mathbf{j} + 4\mathbf{k}$; $\mathbf{r} = 3\mathbf{i} + s(2\mathbf{i} + 5\mathbf{j} + 4\mathbf{k}) + t(\mathbf{j} + 2\mathbf{k})$.
8. $\mathbf{r} = (1 - \lambda)\mathbf{i} + (6\lambda + 3)\mathbf{j} + (4\lambda + 2)\mathbf{k}$. 10. 70·5°.
12. $\frac{1}{3}(\mathbf{a} + \mathbf{b} + \mathbf{c})$. 13. $\lambda:\mu = 1:6$; $\mathbf{r} = (1 + t)\mathbf{a} + 6t\mathbf{b}$; $-12\mathbf{i} - 6\mathbf{j} + 6\mathbf{k}$.
14. (i) $\mathbf{r} = (1 + \lambda + \mu)\mathbf{i} - 2\lambda\mathbf{j} - 3\mu\mathbf{k}$ or $\mathbf{r}.(6\mathbf{i} + 3\mathbf{j} + 2\mathbf{k}) = 6$,
    (ii) $\mathbf{r} = \lambda(6\mathbf{i} + 3\mathbf{j} + 2\mathbf{k})$, (iii) $6(6\mathbf{i} + 3\mathbf{j} + 2\mathbf{k})/49$.
15. (i) $(1, -2, 0)$, $(4, 4, -3)$, (ii) 90°, 90°,
    (iii) $x - y - z - 3 = 0$, $x + z = 1$, (iv) 90°.
16. $(p - \mathbf{a}.\mathbf{n})/(\mathbf{b}.\mathbf{n})$; $(2\frac{1}{3}, 2\frac{1}{3}, 2\frac{2}{3})$.
17. −1, 1; $\mathbf{i} + 5\mathbf{j} - \mathbf{k}$, $-3\mathbf{i} + 3\mathbf{j} + \mathbf{k}$; $2\sqrt{6}$.
18. (i) $-\mathbf{i} + \mathbf{j} + 3\mathbf{k}$, (ii) $3\mathbf{j} - \mathbf{k}$, (iii) 55°, (iv) 35°.

19. $61{\cdot}9°$, $5\sqrt{3}$ N; $\mathbf{r} = \lambda\mathbf{i} + (7\lambda + 1)\mathbf{j} + (5\lambda - 1)\mathbf{k}$, $\dfrac{x}{1} = \dfrac{y-1}{7} = \dfrac{z+1}{5}$.

20. (a) AD, (b) the straight line perpendicular to AB which divides AB in the ratio 1:5.
21. (i) $26{\cdot}7$ N in the direction of the vector $3\mathbf{i} + 13\mathbf{j}$; $-36/65$, (ii) $6\mathbf{i} + 26\mathbf{j}$.
22. $(2T^4 - 2)\mathbf{i} + (3T^2 + 3)\mathbf{j}$; $94\mathbf{i} + 30\mathbf{j}$.
23. $6\mathbf{i} + 8\mathbf{j}$, 6; $6\mathbf{i} + 8\mathbf{j} + 6(\mathbf{i}\cos(\pi t/12) + \mathbf{j}\sin(\pi t/12))$; $-\pi^2/24$, $-\pi^2\sqrt{3}/24$; $(9 - \sqrt{3}\pi)\mathbf{i} + (8 + 3\sqrt{3} + \pi)\mathbf{j}$.
24. $1/(1 + \cos\theta)$; towards O.
25. $2/\sqrt{13}$.
26. (i) $(10 - 2t)\mathbf{i} + (5 + 4t)\mathbf{j}$, (ii) 14.00 hrs, (iii) $4\sqrt{5}$ km.
27. $(260 - 35t)\mathbf{i} - 5\sqrt{3}t\mathbf{j}$; 62 km, 19.00 hrs.

28. (a) $\mathbf{v} - \mathbf{u}$, (b) $\mathbf{p} + t(\mathbf{v} - \mathbf{u})$; $\left\{p^2 - \dfrac{[\mathbf{p}.(\mathbf{u} - \mathbf{v})]^2}{|\mathbf{u} - \mathbf{v}|^2}\right\}^{1/2}$

29. (i) $\mathbf{F}.\mathbf{v}$, (ii) $\displaystyle\int_0^T \mathbf{F}.\mathbf{v}\,dt$; $-4a\sin 2t\mathbf{i} + 2a\cos 2t\mathbf{j}$,

$-8ma\cos 2t\mathbf{i} - 4ma\sin 2t\mathbf{j}$; $12ma^2\sin 4t$; $6ma^2$, $\frac{1}{4}\pi$.
30. $\frac{1}{2}ma^2(2 - 2\cos\theta)\dot{\theta}^2 + mga(1 + \cos\theta) = 2mga$.

Exercise 8.1

1. (a) $8\mathbf{i} - 28\mathbf{j}$, (b) $2\mathbf{i} - \frac{2}{3}\mathbf{j} + \mathbf{k}$, (c) 65, (d) 4.
2. (a) $10\mathbf{i} + 6\mathbf{j}$, (b) $2\mathbf{i} - 52\mathbf{j} + 8\mathbf{k}$, (c) $3\mathbf{j}$.
3. (a) 480 N s, (b) 30 000 N s, (c) 12 N s.
4. (a) $12\mathbf{j}$, (b) $-2\mathbf{i} + 6\mathbf{k}$, (c) $-60\mathbf{k}$.
5. 54 N s.
6. 4 N s.
7. 600 N.
8. (a) $16{\cdot}8\,\mathrm{m\,s^{-1}}$, (b) $7{\cdot}2\,\mathrm{m\,s^{-1}}$.
9. $7{\cdot}5$ s.
10. $12\mathbf{i} + 15\mathbf{j}$; $2\mathbf{i} + 2{\cdot}5\mathbf{j}$.
11. $8\mathbf{i} + 4\mathbf{j}$; 10 J
12. (a) $14\,\mathrm{m\,s^{-1}}$, 240 J gain, (b) $2\,\mathrm{m\,s^{-1}}$, 48 J loss, (c) $10\,\mathrm{m\,s^{-1}}$, 96 J gain.
13. $2{\cdot}5$ N s.
14. $20\sqrt{3}$ N s.
15. 600 N.
16. $12\,\mathrm{m\,s^{-1}}$.
17. 2 N at $53{\cdot}1°$ to horizontal
18. (a) $2\,\mathrm{m\,s^{-1}}$, $6\sqrt{3}$ N s, (b) $(4\sqrt{3})/3\,\mathrm{m\,s^{-1}}$, $4\sqrt{6}$ N s.
19. $1{\cdot}5\,\mathrm{m\,s^{-1}}$ along AB, $5{\cdot}22\,\mathrm{m\,s^{-1}}$ at $73{\cdot}3°$ to AB; $4{\cdot}5$ N s.
20. (a) $2\sqrt{2}\,\mathrm{m\,s^{-1}}$, $2\sqrt{2}\,\mathrm{m\,s^{-1}}$, $4\,\mathrm{m\,s^{-1}}$, (b) $2\,\mathrm{m\,s^{-1}}$, $2\sqrt{3}\,\mathrm{m\,s^{-1}}$, $4\,\mathrm{m\,s^{-1}}$.

Exercise 8.2

1. $5\,\mathrm{m\,s^{-1}}$.
2. $720\,\mathrm{m\,s^{-1}}$.
3. $7{\cdot}5\,\mathrm{m\,s^{-1}}$ towards A; $78{\cdot}75$ J.
4. $2{\cdot}5\,\mathrm{m\,s^{-1}}$, $2512{\cdot}5$ J.
5. $5\,\mathrm{m\,s^{-1}}$.
6. $7\,\mathrm{m\,s^{-1}}$.
7. $2\mathbf{i} - 5\mathbf{j}$.
8. $5\mathbf{i} + \frac{5}{2}\mathbf{j}$.
9. 28 J.
10. $3u$, $2u$.
11. $6\,\mathrm{m\,s^{-1}}$, 26 kJ.
12. $3{\cdot}5\,\mathrm{m\,s^{-1}}$, 2000 N.
13. $0{\cdot}6\,\mathrm{m\,s^{-1}}$, 720 N s.
14. 30.
15. (a) $14\,\mathrm{m\,s^{-1}}$, (b) $23{\cdot}8\,\mathrm{m\,s^{-1}}$.
16. $u/2g\sin\alpha$.

Exercise 8.3

1. 4800 N, 5 s.
2. 9000 N, 1/600 s.
3. $2\,\mathrm{m\,s^{-1}}$, 25 m.
4. 1.
5. (a) 70 N s, (b) 77 N s.
6. 10/7 s, $14\sqrt{2}\,\mathrm{m\,s^{-1}}$.

7. $4{\cdot}2\,\text{m s}^{-1}$, 78 400 N. 8. 90 160 N, 1/7 s. 9. 1·25 m.
10. $469\,\text{m s}^{-1}$. 11. $h, \frac{1}{4}h$. 12. 0·1 m. 13. $2{\cdot}1\,\text{m s}^{-1}$, 6·3 N s.
14. $7{\cdot}35\,\text{m s}^{-1}$, $2\frac{1}{4}$ s. 15. 3 s, $4{\cdot}2\,\text{m s}^{-1}$. 16. (a) $\frac{7}{3}h$, (b) $\frac{2}{3}m\sqrt{(\frac{2}{3}gh)}$.
17. $2{\cdot}8\,\text{m s}^{-1}$, 20·6 J. 18. (i) $\frac{3}{4}mgh$, (ii) $9m^2gh/4M$.
19. (ii) $\frac{12}{5}m\sqrt{(gc)}$, (iii) $\frac{33}{25}c$, (iv) $\frac{12}{5}mgc$.
20. $40\,\text{m s}^{-1}$, 80 J; 1200 J; $(d + 14)$ cm; 1, 8000 N.

Exercise 8.4

1. (a) 2·5 m, (b) 14·7 J, (c) 4·2 N s. 2. $\frac{3}{4}$, 4·05 m. 3. $\frac{2}{3}$.
4. $8\sqrt{5}\,\text{m s}^{-1}$, 60 N s. 5. $1/\sqrt{3}$; 45°. 6. 60°.
7. $4\frac{1}{6}\,\text{m s}^{-1}$ at 53·1° to horizontal. 8. $2u, 4u; 18mu^2$. 9. $\frac{1}{2}$.
10. (a) $1\,\text{m s}^{-1}$, $4\,\text{m s}^{-1}$, same direction.
(b) $\frac{1}{2}\,\text{m s}^{-1}$, $5\frac{1}{2}\,\text{m s}^{-1}$, opposite directions.
11. $\frac{1}{4}$, 2:3. 12. $5\,\text{m s}^{-1}$, $-4\,\text{m s}^{-1}$, $16\,\text{m s}^{-1}$; yes.
13. $u/4, 3u/16, 9u/16$; yes. 14. 2/9. 16. $u(1 - e), u$.
17. $\frac{1}{2}\sqrt{(2ag)}, \frac{1}{4}\sqrt{(2ag)}$ in same direction.
18. (a) $\frac{1}{3}V(1 - 2e), \frac{1}{3}V(1 + e), 0$ (b) $\frac{1}{3}V(1 - 2e), \frac{1}{9}V(1 + e)(2 - e), \frac{2}{9}V(1 + e)^2$.

Exercise 8.5

1. (i) $9mu$, (ii) $81mu^2/4$. 2. (a) $12\mathbf{i} + 4\mathbf{j}$, (b) $5\mathbf{i} - 2\mathbf{j}$, (c) 32 J.
3. $-\dfrac{mv}{2}\mathbf{i} + \dfrac{mv}{2\sqrt{3}}\mathbf{j}$; $\dfrac{v}{3\sqrt{3}}$ at 30° to the x-axis in a clockwise sense.
4. (i) 283 kg, (ii) 11 500 J, (iii) 39 kW, (iv) 56 kW; 2500 N.
5. (ii) $u/6, (u\sqrt{3})/4; mu/6$. 6. $|m_1u_1 - m_2u_2|/(m_1 + m_2)$.
8. $\frac{1}{2}mgh, m(1 + \sqrt{2})\sqrt{(gh)}$. 9. $1{\cdot}6\,\text{m s}^{-1}$, $0{\cdot}4\,\text{m s}^{-1}$; 131 mm, 8 mm.
10. (i) $5{\cdot}6\,\text{m s}^{-1}$, (ii) $3{\cdot}2\,\text{m s}^{-1}$, (iii) 13 440 J, (iv) 123·9 kN.
11. (i) $\frac{6}{5}m\sqrt{(2ga)}$, (ii) $\frac{6}{5}mga$, (iii) $\frac{2}{5}\sqrt{(17ga)}$.
12. $1{\cdot}96\,\text{m s}^{-2}$; $1{\cdot}12\,\text{m s}^{-1}$, 3·36 N s.
14. (i) $g/5, 12mg/5$, (iii) $9mv/8$, (iv) $15mv^2/16$, (v) $1/\sqrt{3}$.
15. $mMu^2/2a(m + M)$. 16. (b) 16°, (c) 5:6.
17. $\dfrac{u\sin\alpha}{\sin\beta}$; $\frac{1}{2}u\left|\dfrac{\sin(\alpha - \beta)}{\sin\beta}\right|$; $\dfrac{u^2}{g\sqrt{3}}$.
19. (a) $\cot\alpha\tan\beta$, (b) $u\cos\alpha\sec\beta$, (c) $mu\sin(\alpha + \beta)\sec\beta$.
20. (a) $\frac{1}{2}mu^2(1 - e^2)\cos^2\alpha$, (b) $u^2e\sin 2\alpha/g(1 + e)$.
21. $\frac{1}{2}$; 2:7; $2\,\text{m s}^{-1}$, $7\,\text{m s}^{-1}$ in opposite directions.
22. (i) u along $\overrightarrow{BA}$, (ii) 4/5, (iii) $3mu^2$,
(iv) u, $4u/3$ along $\overrightarrow{BC}$, (v) 1/9.
23. (i) $\frac{1}{2}$, (ii) $\frac{1}{4}mu^2$, (iii) $\frac{2}{3}$, (iv) u. 24. $4u/3, u/6; 7mu^2/12$.
26. $\frac{1}{3}(7 - 8e)u, \frac{1}{3}(7 + 16e)u$ in direction $\overrightarrow{AB}$; 1/8;
$3u/(1 + k), k > \frac{1}{2}$.
27. $e = 1, \frac{1}{3}, \frac{1}{5}, \frac{1}{7}, \ldots, \dfrac{1}{1 + 2n}, \ldots$

Exercise 9.1

1. (a) $n(n+2)$, (b) $\frac{1}{6}n(n-1)(2n+5)$, (c) $2^{n+1}-2$,
(d) $\frac{1}{12}n(n+1)(n+2)(3n+1)$, (e) $\frac{3}{2}(1-3^{-n})$,
(f) $\frac{1}{4}(n+4)^2(n+5)^2-100$.

2. (a) $\dfrac{n}{2(n+2)}$, (b) $\frac{3}{4}-\dfrac{1}{2(n+1)}-\dfrac{1}{2(n+2)}=\dfrac{n(3n+5)}{4(n+1)(n+2)}$,

(c) $\frac{1}{12}-\dfrac{1}{2(n+2)}+\dfrac{1}{2(n+3)}$.

3. (a) (i) 0, (ii) 1, (b) (i) $-n$, (ii) $n+1$. 4. $\frac{1}{3}n(n+1)(n+2)$
7. $\frac{1}{6}n(n+1)(2n+1)(3n^2+3n-1)$. 8. $(n+1)!-1$.
9. (a) $\frac{5}{2}n(n+1)$, (b) $\frac{1}{6}n(n+1)(38n+1)$.
10. (a) n^2, (b) $\frac{1}{3}n(2n-1)(2n+1)$.
11. (a) 352 800, (b) 56 953, (c) $\frac{1}{4}n^2(3n+1)(5n+3)$, (d) $-n^2(4n+3)$.
12. $r(r+3)$; $\frac{1}{3}n(n+1)(n+5)$.
13. $(3r-1)^2$; $\frac{1}{2}n(6n^2+3n-1)$.
14. $2r(2r-1)$; $\frac{1}{3}n(n+1)(4n-1)$.
15. $r(r+1)(r+4)$; $\frac{1}{12}n(n+1)(n+2)(3n+17)$.

16. $\dfrac{1}{r(r+3)}$; $\dfrac{11}{18}-\dfrac{1}{3}\left(\dfrac{1}{n+1}+\dfrac{1}{n+2}+\dfrac{1}{n+3}\right)$.

17. $\dfrac{1}{(2r-1)(2r+1)}$; $\dfrac{n}{2n+1}$.

18. $\dfrac{1}{(2r-1)(2r+1)(2r+3)}$; $\dfrac{1}{12}-\dfrac{1}{8(2n+1)}+\dfrac{1}{8(2n+3)}$.

19. $\dfrac{r}{(r+1)(r+2)(r+3)}$; $\dfrac{1}{4}+\dfrac{1}{2(n+2)}-\dfrac{3}{2(n+3)}$.

20. $\dfrac{x^{n+1}-1}{x-1}$; $\dfrac{nx^{n+1}-(n+1)x^n+1}{(x-1)^2}$.

21. $\dfrac{nx^{n+2}-(n+1)x^{n+1}+x}{(1-x)^2}$.

22. $\{1-a-(2n+3)(-a)^{n+1}+(2n+1)(-a)^{n+2}\}/(1+a)^2$.

Exercise 9.2

1. $1-\frac{1}{2}x^2+\frac{1}{24}x^4$. 2. $x+\frac{1}{3}x^3$. 3. $1+\frac{1}{2}x^2+\frac{5}{24}x^4$.
4. $-\frac{1}{2}x^2-\frac{1}{12}x^4$. 5. $x^2-\frac{1}{3}x^4$. 6. $x-\frac{1}{2}x^2+\frac{1}{6}x^3-\frac{1}{12}x^4$.
7. $x-\frac{1}{3}x^3$. 8. $1+x+\frac{1}{2}x^2-\frac{1}{8}x^4$. 9. $x+\frac{1}{6}x^3$.
10. $x+x^2+\frac{1}{3}x^3-\frac{1}{30}x^5-\frac{1}{90}x^6$.

Exercise 9.3

1. $1 - 3x + 6x^2 - 10x^3$. 2. $1 + x - \frac{1}{2}x^2 + \frac{1}{2}x^3$.
3. $\frac{1}{8} - \frac{3}{64}x + \frac{15}{1024}x^2 - \frac{35}{8192}x^3$. 4. $1 + 4x + 8x^2 + 16x^3$.
5. $\frac{1}{4} - \frac{1}{2}x + \frac{7}{16}x^2 - \frac{5}{16}x^3$. 6. $1 + 2x^2 + 8x^3$. 7. $1 - x + x^3$.
8. $1 - \frac{1}{2}x + \frac{11}{8}x^2 - \frac{29}{16}x^3$. 9. $1 + \frac{1}{3}x + \frac{8}{9}x^2 - \frac{49}{81}x^3$,
10. $1 - x + \frac{1}{2}x^2, |x| < 1$. 11. $2 + \frac{25}{12}x + \frac{1175}{288}x^2, |x| < \frac{1}{3}$.
12. $1 + \frac{1}{2}x^2, |x| < \frac{1}{5}$. 13. $\frac{1}{3} - \frac{2}{9}x + \frac{7}{27}x^2, |x| < 1$
14. $1 - 3x + 9x^2, |x| < \frac{1}{2}$. 15. $1 + \frac{5}{2}x + \frac{43}{8}x^2, |x| < \frac{1}{2}$.
16. $3 + x + 11x^2 + 25x^3 + 83x^4, |x| < \frac{1}{3}$.
17. $\frac{4}{3}x + \frac{8}{9}x^2 + \frac{28}{27}x^3 + \frac{80}{81}x^4 + \frac{244}{243}x^5, |x| < 1$.
18. $1 + 2x + x^2 + x^4 + 2x^5, |x| < 1$.
19. $4 - x + 12x^2 - 11x^3 + 38x^4, |x| < \frac{1}{2}$.
20. $2 - 2x + 2x^4 - 2x^5 + 2x^8, |x| < 1$.
21. $-1 + x + \frac{7}{4}x^2 + \frac{1}{2}x^3 - \frac{3}{16}x^4, |x| < \sqrt{2}$.
22. $3 + 3x + 9x^2; \{(-1)^r + 2^{r+1}\}x^r$.
23. $4x - 8x^2 + 28x^3; \{1 - (-3)^r\}x^r$.
24. $2 - \frac{3}{2}x + \frac{17}{4}x^2; \{(-2)^r + 2^{-r}\}x^r$.
25. $1 + \frac{1}{2}x - \frac{3}{8}x^2 + \frac{7}{16}x^3; |x| < \frac{1}{2}$; (a) 1·0885, (b) 1·019 428.
26. $(a + b) + 2(1 - a + b)x + (-4 - 2a + 3b)x^2 + 4(-1 - a + b)x^3$; 13, −8.
27. $n = 6, a = \frac{1}{2}$. 28. $a = 3, b = 1$; 70/13.

Exercise 9.4

1. (a) $1 - x + \frac{1}{2}x^2; \dfrac{(-1)^r x^r}{r!}$, (b) $1 + 3x + \frac{9}{2}x^2; \dfrac{3^r x^r}{r!}$,
(c) $1 + x^2 + \frac{1}{2}x^4; \dfrac{x^{2r}}{r!}$, (d) $e + ex + \frac{1}{2}ex^2; \dfrac{ex^r}{r!}$.
2. (a) $1 + 4x + 6x^2 + \frac{16}{3}x^3$, (b) $1 - 3x + \frac{7}{2}x^2 - \frac{13}{6}x^3$,
(c) $1 - x + \frac{3}{2}x^2 - \frac{7}{6}x^3$.
3. 2·7183; 0·3679.
4. (a) $3x - \frac{9}{2}x^2 + 9x^3; -\frac{1}{3} < x \leqslant \frac{1}{3}$. (b) $-\frac{1}{2}x - \frac{1}{8}x^2 - \frac{1}{24}x^3; -2 \leqslant x < 2$.
(c) $\ln 2 + \frac{1}{3}x - \frac{1}{18}x^2; -3 < x \leqslant 3$. (d) $1 - \dfrac{x}{e} - \dfrac{x^2}{2e^2}; -e \leqslant x < e$.
5. (a) $x - \frac{3}{2}x^2 + \frac{1}{3}x^3 + \frac{1}{4}x^4$, (b) $x - \frac{5}{2}x^2 + \frac{7}{3}x^3 - \frac{17}{4}x^4$,
(c) $-4x + 3x^2 - \frac{34}{3}x^3 + \frac{63}{2}x^4$. 6. 1·0986.
7. (a) $(1 - r)/r!$, (b) $(4 + 5r - 3r^2)2^{r-2}/r!$, (c) $(4r^2 + 1)/2^r r!$.
8. (a) $x + \frac{1}{2}x^2$, (b) $4 - 4x + 3x^2$, (c) $1 + \frac{15}{2}x + \frac{193}{8}x^2$, (d) $2x$,
(e) $1 + 2x + \frac{5}{2}x^2$, (f) $2 + x + \frac{11}{6}x^2$.
9. 0·6931. 10. $a = 2, n = -2$. 11. $a = \frac{1}{2}, b = 2$; 31/170.

Exercise 9.5

1. (a) 0·2955, (b) 0·9211, (c) −0·4161, (d) 0·8415.
2. (a) $3x - \frac{9}{2}x^3 + \frac{81}{40}x^5$, (b) $1 - 2x^2 + \frac{2}{3}x^4$, (c) $x - \frac{1}{2}x^2 + \frac{1}{6}x^3$,
   (d) $e - \dfrac{e}{2}x^2 + \dfrac{e}{6}x^4$, (e) $1 + \frac{1}{6}x^2 + \frac{7}{360}x^4$, (f) $x + \frac{1}{3}x^3 + \frac{2}{15}x^5$.
3. (a) $x + x^2 - \frac{2}{3}x^3 - \frac{1}{6}x^4$, (b) $2x - 2x^2 + \frac{4}{3}x^3$, (c) $\frac{4}{3}x^4$,
   (d) $1 + 2x + x^2 - \frac{2}{3}x^3 - x^4$.

Exercise 9.6

1. (a) $\frac{1}{3}n(n^2 - 1)$, (b) $\frac{1}{3}(n + 1)(4n^2 + 14n + 9)$, (c) $\dfrac{n+1}{n(2n+1)}$,
   (d) $\frac{3}{4} - \dfrac{1}{2n} - \dfrac{1}{2(n+1)}$.
2. (a) $\frac{1}{12}n(n + 1)(n + 2)(3n + 5)$, (b) $n(n + 1)(2n^2 + 2n - 1)$.
3. $\frac{1}{6}n^2(n + 1)^2(n - 1)$.
4. $\dfrac{1 - (n+1)x^{2n} + nx^{2n+2}}{(1 - x^2)^2}$
5. $(x + 1)(x + 2)(x + 3)$; $\dfrac{1}{x+1} + \dfrac{2}{x+2} - \dfrac{3}{x+3}$.
6. (a) $2^{n+1} - 2 + n^2$, (b) $a = -2, b = -1$; 73/74.
7. $1 + \dfrac{1}{2x^2} - \dfrac{1}{8x^4}$, $1 - \dfrac{1}{2x^2} - \dfrac{1}{8x^4}$.
8. $2 - \frac{1}{4}x - \frac{1}{64}x^2$; $2 + \frac{7}{4}x + \frac{175}{64}x^2$, $|x| < \frac{1}{2}$.
9. $2 + x - \frac{3}{2}x^3$.
10. (a) $3x - 3x^2 + 9x^3$, $\{1 - (-2)^r\}x^r$, $|x| < \frac{1}{2}$,
    (b) $\frac{1}{6}x + \frac{5}{36}x^2 + \frac{19}{216}x^3$, $\left(\dfrac{1}{2^r} - \dfrac{1}{3^r}\right)x^r$, $|x| < 2$,
    (c) $1 + 5x + 18x^2$, $\frac{1}{2}(5.3^r - 2^{r+1} - 1)x^r$, $|x| < \frac{1}{3}$.
11. (a) $1 - 2x - x^2 + 2x^3 + \frac{3}{2}x^4$; $|x| < 1/\sqrt{2}$,
    (b) $1 + 4x + 14x^2 + 40x^3 + 104x^4$; $|x| < \frac{1}{2}$,
    (c) $3 + 6x + 6x^2 + 12x^3 + 36x^4$; $|x| < \frac{1}{2}$.
12. (a) $x + \frac{5}{24}x^3 - \frac{1}{8}x^4$; for $|x| < 1$; $a = -\frac{1}{4}$,
    (b) $1 - \frac{1}{2}x + \frac{3}{8}x^2 - \frac{5}{16}x^3$, $\dfrac{1.3.5....(2r-1)}{2^r r!}(-1)^r x^r$; 0·099 504.
13. (a) (i) (0, 1), (ii) (0, 1), (iii) (1, 0), (iv) (0, 0);
    (b) $7x + \frac{1}{2}x^2 + \frac{43}{3}x^3 + \frac{49}{4}x^4$, $-\frac{1}{3} \leqslant x < \frac{1}{3}$; $\{(-2)^{n+1} + 3^n\}x^n/n$.
14. $1 + x + x^2 + \frac{1}{3}x^3$; 0·287.
15. (i) $x - \dfrac{x^2}{2} + \dfrac{x^3}{3} - \dfrac{x^4}{4} + \ldots + (-1)^{n+1}\dfrac{x^n}{n} + \ldots$; (ii) 0·6931; 1/3, 23/60.

16. $p^2 - 2, p^3 - 3p; -px - \frac{1}{2}(p^2 - 2)x^2 - \frac{1}{3}(p^3 - 3p)x^3$.
17. $1, \sqrt{3}, 2$.

Exercise 10.1

1. 0·697. 2. 5·731. 3. 45·992. 4. 4·154. 5. 0·144.
6. 0·825. 7. Greater. 8. Greater. 9. Greater.
10. Less. 11. Greater. 12. Less. 13. 0·8358. 14. 0·4783.
15. 0·6807. 16. 0·8945. 17. 0·2763. 18. 3·2413.
19. $1 + 12x^2 + 66x^4 + 220x^6 + 495x^8$; 1·7655; 1·9253.
20. 0·0711; 0·0715.
21. (a) 10·075, (b) 9·99. 22. (a) 7·175 $\mathrm{m\,s^{-1}}$, (b) 11·85 $\mathrm{m\,s^{-1}}$; 131 m.
23. (a) 422·5 J. (b) 423 J. 24. 0·7444. 25. 1·4241.
26. 1·68. 27. 0·524.

Exercise 10.2

1. 2·7, −0·7. 2. 0·7. 3. −2·1. 4. −0·6, 1, 1·6.
5. −1·8, −0·4, 1·2. 6. 0·8. 7. 0, ±1·9. 8. −3·6, −2·1, 1·3.
9. 1·1. 10. 1, 4·2. 11. One. 12. Two. 13. Three.
14. Five. 15. One; −3, −2. 16. One; 1, 2.
17. Three; −4, −3; 1, 2; 2, 3. 18. Three; −3, −2; −1, 0; 1, 2.
19. One; 1, 2. 20. One; 1, 2. 21. Two; −3, −2; 1, 2.
22. One; 2, 3. 23. 3·2. 24. −0·8, 1·6. 25. −3, 1; 1·1.
26. 2·2. 27. −5·2, −0·6, 2·9. 28. −0·2, 0·6, 2·0. 29. −4·1.
30. 1·8, 4·5. 31. −1·3, 1·9, 4·0.
32. (a) −1·75, 0·43, 1·32, (b) −1·6, 0·7, 1·1.
33. (a) −0·1, (b) 0·25, (c) 0·4, (d) 0·2.

Exercise 10.3

1. 1·20. 2. −1·11. 3. −1·52. 4. 2·39. 5. 1·56.
6. 0·79. 7. −1·90. 8. 1·17. 9. 2·803. 10. 1·650.
11. −0·830. 12. −1·834. 13. 1·278. 14. 2·219. 15. 0·453.
16. 3·6; 3·6913. 17. Two; 3·98, −1·06.
18. 2, $a := 2$, $h := 0{\cdot}01$, $g(x) := e^x - 2$; 1·68.

Exercise 10.4

1. (a) 2·24, (b) 1·26, (c) 0·79, 3·87. 2. (a) 1·414, (b) 3·162, (c) 3·606.
3. 1·71. 4. 0·769. 5. 1·821. 6. −3·794.
7. 0·1926; for $x_0 = -6$ the sequence diverges, for $x_0 = -5$ the sequence converges to the positive root.
8. 1; 1·260, 1·312. 9. 1·732.

10. 1·087 using $x_{n+1} = (4 - \frac{5}{2}x_n)^{1/3}$, $x_{n+1} = \dfrac{8}{2x_n^2 + 5}$ or Newton's method.

11. 2·618; decision box should ensure maximum error of 0·1. 12. 0·7.

Exercise 10.5

1. 3·69, 6·71. 2. 4, 3; (a) 4·826, (b) 4·8224. 3. 0·4778; 0·4783.
4. 0·410. 5. (i) 2·09, (ii) $-1{\cdot}53$, 0·26, 1·27; 0·5.
6. (i) two, (ii) $m < -\frac{1}{2}$. 7. $(0, 0)$, $(2, 4e^{-2})$. 8. $q < 0$.
9. 2·4; underestimate. 10. Three; 1·43, larger. 11. (ii) output "2, 3".
12. (a) 2·208, (b) 2·208. 13. 1, 3; -4, 0; $a < -4$ or $a > 0$; 4·1.
14. 3·43; 3·75; 15. 15. $(x - 2)(x - 3)(x^2 + x + 2)$; 3·065.
16. (a) 0·818, (b) 0·810. 17. 0·74. 18. 0·5, 0·73.
19. 1·2. 20. 0·77. 21. $x = (3x - 1)^{1/3}$; 1·53.

Exercise 11.1

3. (a) and (d). 4. 2, -1. 5. 1, 0, 2. 6. $\dfrac{dy}{dx} = 3x^2$.

7. $x\dfrac{dy}{dx} = y + x^2$. 8. $x\dfrac{dy}{dx} + y = 0$. 9. $\dfrac{dy}{dx} + y = 0$.

10. $x + y\dfrac{dy}{dx} = 0$. 11. $2x\dfrac{dy}{dx} = y$. 12. $x\dfrac{dy}{dx} = 2x - y$.

13. $x\dfrac{dy}{dx} = 2y + x^2$. 14. $2\dfrac{dy}{dx} + y\tan x = 0$.

15. $x\ln x\dfrac{dy}{dx} = y\ln y$. 16. $\dfrac{d^2y}{dx^2} + y = 0$.

17. $\dfrac{d^2y}{dx^2} - 2\dfrac{dy}{dx} - 3y = 0$. 18. $3\dfrac{dy}{dx}\left(\dfrac{d^2y}{dx^2}\right) + y\dfrac{d^3y}{dx^3} = 0$.

19. $\dfrac{d^3y}{dx^3} - 2\dfrac{d^2y}{dx^2} + \dfrac{dy}{dx} = 0$.

Exercise 11.2

1. $y = x^3 + x + c$. 2. $y = 2\sin\frac{1}{2}x + c$. 3. $y = \frac{1}{2}x^3 + \ln|x| + c$.
4. $y = x + \ln(x - 1)^2 + c$. 5. $y = \cos x + \frac{1}{3}x^3 + Ax + B$.

6. $y = e^{-x} - e^x + Ax + B$. 7. $y = \dfrac{1}{2x + k}$. 8. $y = \tan^{-1}(x + k)$.

9. $y = \ln|x + k|$. 10. $y = (c - x)^{-1/3}$. 11. $y = \frac{1}{3}(x + 1)^3 - 9$.
12. $y = 1 + \sqrt{(2x + 3)}$. 13. $y = x \sin x + \cos x + 1$.
14. $y = \ln \frac{1}{2}(1 + x^2)$. 15. $y = \frac{1}{4}(x - \sin 2x)$.
16. $y = 2 \ln x + x^2 - 4x + 3$. 17. $y = 3e^{-2x}$. 18. $y = 1 - e^x$.
19. $y = \tan(x + \frac{1}{4}\pi)$. 20. $y = (3 + e^x)/(3 - e^x)$. 21. $y = x/(1 + kx)$.
22. $y = kx + x \sin x$. 23. $y = kx^2 - x$. 24. $y = x - e^{1-x} + 2$.

25. $xe^y \dfrac{dy}{dx} + e^y;\ y = \ln\left(x - \dfrac{2}{x}\right)$.

Exercise 11.3

1. $y = (x^2 - x + c)^{1/3}$. 2. $y = 1/(c - 3x^2)$. 3. $y = -\ln(\cos x + c)$.
4. $y = \ln(e^x + c)$. 5. $y = \sin^{-1}(\frac{1}{2}x^2 + c)$. 6. $y = \tan^{-1}(3x + c)$.
7. $y = Ax$. 8. $y = A/(1 - x)$. 9. $y = A(x^2 + 1)^2$.
10. $y = A(x - 1)/(x + 1)$. 11. $y = x/(1 - 2x)$. 12. $y = 5e^{-x^2}$.
13. $y = 2 \sec x$. 14. $y = \frac{1}{2}\ln(x^2 + 1)$. 15. $y = 2e^{x^2} - 1$.
16. $y = xe^{x-1}$. 17. $y = \sin^{-1}(1 + x^3)$. 18. $y = \sin^{-1}(x - 1)$.
19. $y = 1/(1 - 2x^2)$. 20. $y = (x - 1)/(x + 1)$. 21. $y = Ae^{-x}/(1 - x)$.
22. $y = Ae^x(x - 1)/(x + 1)$. 23. $y = \tan(e^x + c)$. 24. $y = \ln(1 + Ae^{x^2})$.
25. $y = e^{-x}\sqrt{(2x + 1)}$. 26. $y = \sqrt{\{(\ln x)^2 + 9\}}$.
27. $x^2 - y^2 - 2xy + 2x + 2y = A$.
28. $x^2(x^2 - y^2) = A$.

Exercise 11.4

1. $x^2 + y^2 = 5$. 2. $y = 1/x^2$.

3. $\dfrac{dC}{dt} = kC;\ C = C_0 e^{kt};\ t = \dfrac{1}{k}\ln\frac{3}{2}$. 4. $\dfrac{d\theta}{dt} = -k\theta$.

5. $(1 + \sqrt{2})t$. 6. $\dfrac{dx}{dt} = -kx$; 93 days; 320 mg. 7. 40 min.

8. $x = 100(1 + e^{-t/10})$; 137 litres; 6 min 56 sec. 9. $30 \ln 2$ min., $100e^{-1}$ kg.

10. 2·99. 11. $t = \dfrac{\ln 3}{V\beta};\ m = \dfrac{V}{\alpha(1 + 3e^{-V\beta t})}$.

Exercise 11.5

1. $y = A \cos 4x + B \sin 4x$. 2. $y = A \cos \frac{\sqrt{7}}{2}x + B \sin \frac{\sqrt{7}}{2}x$.
3. $y = 2 + A \cos 3x + B \sin 3x$. 4. $y = 1 + A \cos \sqrt{10}x + B \sin \sqrt{10}x$.
5. $y = Ae^{6x} + Be^{-6x}$. 6. $y = Ae^{2\sqrt{2}x} + Be^{-2\sqrt{2}x}$.
7. $y = -\frac{1}{27} + Ae^{\sqrt{3}x} + Be^{-\sqrt{3}x}$. 8. $y = \frac{1}{4} + Ae^{2x} + Be^{-2x}$.
9. $y = 4 \cos 2x - 3 \sin 2x$. 10. $y = 3e^{4x} + 7e^{-4x}$. 11. $y = 6 \cos \frac{1}{6}x$.

12. $y = 4\sin\frac{4}{3}x - 5\cos\frac{4}{3}x$. 13. $y = -2e^{5x/2} - 7e^{-5x/2}$.
14. $y = 4e^{-2\sqrt{3}x}$. 15. $y = -\frac{4}{7} + 3e^{7x/3} + 6e^{-7x/3}$.
16. $y = \frac{8}{5} + 4\cos\frac{5}{2}x + 9\sin\frac{5}{2}x$. 17. $y = 2 + \sin\sqrt{3}x$.
18. $y = -3 + e^{9x/4} + e^{-9x/4}$. 19. $y = -3 + \sqrt{2}e^{3\sqrt{2}x}$.
20. $y = \frac{1}{2} - 4\sqrt{3}\cos 2\sqrt{3}x - 3\sqrt{3}\sin 2\sqrt{3}x$.

Exercise 11.6

1. $\left(\dfrac{dy}{dx}\right)^2 - 4x\dfrac{dy}{dx} + 4y = 0$.

2. $(2x+1)\dfrac{d^2y}{dx^2} - 2\dfrac{dy}{dx} - (2x+3)y = 0$.

3. $y = x^2$; $y = x^2 + 2x$.
4. (a) $\ln\frac{1}{3}(3 - 7u) = 7(t - 5)$, (b) $3(y + 18) = (x^2 + 9)^{3/2}$.
5. (a) $2x^3 + 11x^2 + 20x = 37 - 2t$, (b) $y = \{\frac{1}{2}(5e^{x+1} - 3)\}^{1/2}$.

6. $y = 36e^{-t/6} + 27$; 11 minutes; 27 degrees. 7. $y = \sin x + \dfrac{\cos x - 1}{1 + x}$.

8. (a) $x^2 - 4y^2 = 4$; $x - y\sqrt{3} = 1$.
9. (a) $\tan y = 1 - \cos x$, (b) $y = \ln\frac{1}{2}(1 + e^{2x})$.
10. (i) $y = e^{-x}(1 + x)^2$, (ii) $y^2 = 2x$.

12. $y = \dfrac{2x}{x^2 + cx + 1}$; (a) $y = \dfrac{2x}{x^2 + 1}$, (b) $y = \dfrac{2x}{x^2 + x + 1}$,

(c) $y = \dfrac{2x}{(x+1)^2}$.

13. $\dfrac{dy}{dx} = \dfrac{y}{x}$, $y = 2x$; $\dfrac{dx}{dt} = -2x^3$, $x = \dfrac{1}{\sqrt{(4t+1)}}$.

14. (a) $y\dfrac{d^2y}{dx^2} + \left(\dfrac{dy}{dx}\right)^2 = 0$.

15. (i) (a) $y = A + \ln[\ln(1/x)]$, (b) $y = A + \ln(\ln x)$, (iii) $y = x$, $y = 2 - x$.

16. 20°, $53\frac{3}{4}$°. 17. (a) $x = 2 + \dfrac{1}{12}\ln\left(\dfrac{5(3y-2)}{3y+2}\right)$, (b) $\dfrac{dp}{dt} = kp$.

18. (a) $y = \frac{1}{3}(5e^{3x-6} - 2)$, (b) $\dfrac{dm}{dt} = -km$; 75·98%.

19. $\dfrac{dP}{dt} = kP(N - P)$, $P = N/(1 + Ae^{-Nkt})$; $N = 400\,000$; N is the limiting size of P.

20. $\dfrac{dh}{dt} = -\lambda$, $\pi h\dfrac{dh}{dt} = -(Ck + \pi\lambda h)$.

21. (i) $\dfrac{dQ}{dt} = -kQ$; $Q = Q_0e^{-kt}$; $Q = 2{\cdot}5 \times 10^{-4}$, (ii) $y = 3e^{5x} + e^{-5x}$.
22. $y = 2 - 4\sin 3x$; -2, $\frac{2}{3}n\pi + \frac{1}{18}\pi$, $\frac{2}{3}n\pi + \frac{5}{18}\pi$.
23. (i) $(y + 1)e^{-y} = (1 - x)e^x$, (ii) $y = 4 - 2\sin 2x$; $\frac{1}{2}r\pi$, r an integer.
24. (i) $y = Ae^x/(1 + x)$, (ii) $y = \frac{1}{2}(1 - \cos 2x - \sin 2x)$.
25. $y = \frac{1}{2} + A\cos(2x^2) + B\sin(2x^2)$; $y = -\frac{1}{2} + Ae^{2x^2} + Be^{-2x^2}$.

Exercise 12.1

1. $\ln 3\,\text{m s}^{-1}$. 2. $\ln 5$ seconds. 3. $6\,\text{m s}^{-1}$. 4. $0{\cdot}585\,\text{m s}^{-1}$.
5. $2a$. 6. $\sqrt{(gR/3)}$. 7. (a) $2/x$, (b) $\sqrt{(4t + h^2)}$. 8. $-\frac{1}{2}e^x$.

Exercise 12.2

1. $\ln 3$. 2. $\frac{1}{2}\ln(1 + u^2)$. 3. $-\dfrac{1}{k}\{v + \ln(1 - v)\}$.

4. $\dfrac{a}{b^2}(2 - \ln 3)$. 5. $\dfrac{2u}{2 + su}$; $\dfrac{s^2}{4} + \dfrac{s}{u}$. 6. (i) 482 m, (ii) 24·1 s.

8. $\left[\dfrac{g}{k}(1 - e^{-2ks}) + u^2e^{-2ks}\right]^{1/2}$. 9. (iii) 11·1 s.

10. $\dfrac{dv}{dt} = \dfrac{k(U^2 - v^2)}{mv}$; $v\dfrac{dv}{dx} = \dfrac{k(U^2 - v^2)}{mv}$.

11. $\dfrac{m}{2k}\tan^{-1}(\frac{1}{2}u)$; $\dfrac{m}{2k}\ln(1 + \frac{1}{4}u^2)$. 12. $\dfrac{P^2}{k}\left(1 - \dfrac{1}{e}\right)$.

15. $\dfrac{m}{R^2}\left\{H\ln\left(\dfrac{H - Ru}{H - 2Ru}\right) - Ru\right\}$;

$\dfrac{m}{R^3}\left\{H^2\ln\left(\dfrac{H - Ru}{H - 2Ru}\right) - HRu - \frac{3}{2}R^2u^2\right\}$.

16. $\dfrac{mU^2}{2S}\ln\left(\dfrac{U^2}{U^2 - V^2}\right)$; $\dfrac{mU^2}{2S}\left\{U\ln\left(\dfrac{U + V}{U - V}\right) - 2V\right\}$.

Exercise 12.3

1. $5\,\text{m s}^{-1}$, $10\,\text{m s}^{-2}$. 2. (a) 2 m, (b) $\frac{2}{3}\pi$ s, (c) $3\sqrt{3}\,\text{m s}^{-1}$.
3. 6π s; 2·5 N. 4. (i) $270/\pi$, (ii) $\frac{2}{3}\sqrt{13}$ m, (iii) $540\sqrt{13}$ N.
5. $6\,\text{m s}^{-1}$, 2 m; $5{\cdot}9\,\text{m s}^{-1}$. 6. (i) π s, (ii) 2π s, (iii) $3\pi/2$ s.

7. (i) $\dfrac{5}{3\pi}$ m, (ii) $\frac{5}{2}\sqrt{3}\,\text{m s}^{-1}$, (iii) $\dfrac{1}{2\pi}$ m, (iv) $\frac{1}{18}$ s.

8. $2\sqrt{3}\,\text{m s}^{-1}$, 1/6 s; 8 J.
9. (a) $x = 3\cos 2t + 3\sin 2t$, (b) $-3, 6$, (c) $3\pi/8$.
10. (a) $x = 3\cos 4t + \sqrt{3}\sin 4t$, (b) $\pi/24$, (c) $2\sqrt{3}$.

Exercise 12.4

1. $5a/3$. 2. (i) $2l$, (ii) $\frac{1}{2}\pi\sqrt{(3l/g)}$, (iii) $15l/8$.
3. $\frac{1}{2}mg, \frac{1}{2}mg$; (i) $3l/2$, (ii) $\pi(2l/g)^{1/2}$. 4. 4.
5. $2\pi\sqrt{(2a/g)}$, a; $\frac{1}{2}a$ above 0. 6. $\sqrt{(2a\lambda/m)}$.
7. $5l/2$, $\pi(lm/\lambda)$, $\sqrt{(\lambda l/m)}$. 8. $2l$; $\pi\sqrt{\left(\frac{2l}{g}\right)}, \left(\frac{2u^2l + gl^2}{4g}\right)^{1/2}$.
9. $u\sqrt{\left(\frac{lm}{\lambda}\right)}; \frac{\pi}{2}\sqrt{\left(\frac{lm}{\lambda}\right)}; \pi\sqrt{\left(\frac{lm}{\lambda}\right)} + \frac{2l}{u}$. 10. $\frac{16}{9}a$; $\sqrt{(5ag)}$; $\frac{\pi}{6}\sqrt{\left(\frac{a}{g}\right)}$.
11. (a) 994 mm, (b) 991 mm, (c) 996 mm, (d) 164 mm.
12. (a) 143 cm, (b) 56 cm, (c) 36 cm, (d) 14 cm.

Exercise 12.5

1. $49\cos\theta - 24$, $58{\cdot}8\cos\theta - 19{\cdot}2$; 3·68, 25·8; 61°.
2. $6{\cdot}18\,\text{m s}^{-1}$, 19·35 N. 3. $\frac{1}{3}, \frac{1}{3}ag$.
5. $mg|3\cos\theta - 2|$; $\frac{2}{3} < \cos\theta < 1$. 6. $\sqrt{(2ag/3)}$.
7. $\frac{mu^2}{a} - 2mg + 3mg\cos\theta$; mg. 8. (ii) $\frac{7}{2}ag$, $\sqrt{(\frac{1}{2}ag)}$.
9. $\cos\alpha\sqrt{(2gl\sin\alpha)}$; no. 10. $\{2ag(1 + \sin\theta)\}^{1/2}$; (b) $\pi/6$, $(3ag)^{1/2}$; $(a/g)^{1/2}$.
11. $\{u^2 - 6ag(1 - \cos\theta)\}^{1/2}$; $\frac{mu^2}{2a} - 3mg(1 + \sin\phi)$.
12. $\frac{2}{3}mg(1 + \cos\theta)$, $\frac{1}{3}mg(5\sin\theta - 4\theta)$.

Exercise 12.6

1. Particle escapes from the earth.
2. $mv\frac{dv}{dx} = Ke^{-\alpha x} - \lambda m$; $\left\{\frac{2K}{\alpha m}\left(1 - e^{-\alpha x}\right) - 2\lambda x\right\}^{1/2}$.
3. $\frac{1}{2}(\mu/a)^{1/2} \leqslant V \leqslant \frac{3}{4}(\mu/a)^{1/2}$.
4. $5\cos 3t + 12\sin 3t + 2$; 15, 39; $t \approx 0{\cdot}967$. 5. $\frac{2k^2}{g}\left(1 - e^{-gt/k}\right) - kt$.
6. $\frac{1}{800}$ s; $\frac{dv}{dt} = -16 \times 10^3(2v)^{1/2}$, $\frac{1}{400}$ s; $v = Ae^{-kt}$, A and k positive constants.

7. (i) $2uT$, (ii) $T \ln 3$, (iii) $4mu^2$. 8. $v = gTe^{-x/gT^2}$

9. $v^2 = u^2 \tan\left(\frac{\pi}{4} - \frac{2gz}{u^2}\right)$. 10. (ii) $u\sqrt{(1 - e^{-2kx})}$.

13. (b) (i) $3\pi/8$ s, (ii) $26\frac{2}{3}$ m s^{-1}, (iii) $142\frac{2}{9}$ m s^{-2}, (iv) $\pi/16$ s.

14. $\frac{\pi}{4}$ m/s, $\frac{\pi^2}{8}$ m/s^2; $\frac{\pi\sqrt{3}}{8}$ m/s, $\frac{2}{3}$ s.

15. (i) 12 s, $24/\pi$ m, (ii) $t = 5$; $12\sqrt{2}$ s, $\sqrt{2}$ m s^{-1}.
16. $\ddot{x} = -\omega^2 x$, $2\pi/\omega$; $t = 12$ s, $t = 24$ s. 17. $4u/\pi$.

18. (i) 5·09 m, (ii) 3·60 m, (iii) $\frac{4}{\pi}\sin^{-1}\left(\frac{\pi}{16}\right)$; 2π N, $\mu \geqslant \pi/g$.

19. $\frac{400}{3}$ N; $\frac{2}{\sqrt{3}}$ m/s. 20. $2mg + \frac{2\lambda}{3l}x$. 21. $\sqrt{(\frac{1}{3}gl)}$.

23. $\frac{5\sqrt{2}}{24}l$. 25. $\{ag(2\cos\theta - 1)\}^{1/2}$, $2mg(3\cos\theta - 1)$; $8a/25$.

28. $\sqrt{(3ga/2)}$. 29. $\ddot{\mathbf{r}} = \mathbf{g} - k\dot{\mathbf{r}}$.

Exercise 13.1

1. (a) 1/7, (b) 1/8, (c) 21/13, (d) 7/4, (e) 15/8, (f) 3/4.
2. (a) (5, 0), (b) $(-2, -2)$, (c) $(2, -2)$, (d) $(-\frac{1}{2}, 1\frac{1}{4})$.
3. (a) $(12, -5)$, (b) $(6, -19)$, (c) $(-4, -1)$, (d) $(7, -8\frac{1}{2})$.
4. (a) 1, (b) 2, (c) $\sqrt{2}$, (d) $\sqrt{13}$, (e) 9, (f) $2\sqrt{10}$.
5. (a) 6, (b) 24, (c) 18, (d) $11\frac{1}{2}$, (e) 0, (f) $36\frac{1}{2}$.
6. (a) $2x + y - 1 = 0$, $x - 2y - 2 = 0$. (b) $x + y - 1 = 0$, $x - y - 1 = 0$.
   (c) $x = -\frac{1}{3}$, $y = \frac{1}{3}$. (d) $2x - y + 4 = 0$, $x + 2y - 13 = 0$.
7. 45°, 45°, 90°. 8. 45°, 26° 34′, 108° 26′. 9. $(-3/7, 12/7)$.
10. $(5\frac{5}{18}, 3\frac{1}{2})$. 11. 3. 12. 30° 58′. 13. (2, 3).
14. $(-\frac{1}{2}, -\frac{1}{2})$. 15. (i) (6, 7), (ii) 16/15, (iii) $8\sqrt{2}$, (iv) 80.

Exercise 13.2

4. $x^2 + y^2 - 6x - 2y - 10 = 0$, (5, 5); $x^2 + y^2 - 6x - 2y - 70 = 0$, $(-1, -7)$.
5. (11/5, 13/5).
6. $x^2 + y^2 - 2x - 2y - 23 = 0$, $x^2 + y^2 - 18x - 14y + 105 = 0$, $4x + 3y = 32$,
   $3x - 4y + 26 = 0$, $3x - 4y = 24$.
7. $x^2 + y^2 - 6x + 10y - 16 = 0$, $x^2 + y^2 + 12x - 14y - 40 = 0$;
   $3x - 4y = 4, 10$
8. $4x - 3y = 3$.
9. $x^2 + y^2 - 6x + 4y - 12 = 0$; $x^2 + y^2 + 14x - 6y - 42 = 0$.
12. (a) $2x - 2y - 1 = 0$, (b) $x^2 + y^2 - 2y - 1 = 0$.

13. (a) $x^2 + y^2 + x - 4y + 1 = 0$, (b) $x^2 + y^2 - 3y = 0$,
    (c) $x^2 + y^2 + 4x - 7y + 4 = 0$.
14. (a) $x^2 + y^2 + x - 4y - 9 = 0$, (b) $5x^2 + 5y^2 + 23x - 29y = 0$,
    (c) $2x^2 + 2y^2 - 7y - 23 = 0$.
15. $x + 1 = 0, x + 2\sqrt{6}y = 11 - 2\sqrt{6}, x - 2\sqrt{6}y = 11 + 2\sqrt{6}$.

Exercise 13.3

1. (a) $(3t^2, 6t)$, (b) $(3t, 3/t)$, (c) (t^2, t^3), (d) $(5t/2, 5/2t)$, (e) $(4t^2, 4t^3)$,
   (f) $(5t^2/2, 5t)$.
2. (a) $(3\cos\theta, 2\sin\theta)$, (b) $(3\cos\theta, 3\sin\theta)$, (c) $(\sqrt{5}\cos\theta, \sin\theta)$,
   (d) $(\cos\theta, \frac{2}{3}\sin\theta)$, (e) $(\cos\theta, 3\sin\theta)$, (f) $(5\cos\theta, 1 + 5\sin\theta)$.
3. $x + t^2y = 8t; ty = t^3x + 4 - 4t^4$. 4. $x + 4y = 4, 2y = 8x - 15$.
5. $x + 2\sqrt{3}y = 4, 4\sqrt{3}x - 2y = 3\sqrt{3}$.
7. $4x + y = 4c$ at $(\frac{1}{2}c, 2c)$, $4x + y + 4c = 0$ at $(-\frac{1}{2}c, -2c)$.
9. $x = t, y = \dfrac{3}{2t} - \dfrac{3t^3}{2}$.
10. $x + pqy = a(p + q), x + p^2y = 2ap; \left(\dfrac{2apq}{p+q}, \dfrac{2a}{p+q}\right)$. 14. $xy = 2c^2$.
15. $2y = 3tx - at^3; \dfrac{at^2}{4}, -\dfrac{at^3}{8}; t = 0$.
16. $(3\{p^2 + pq + q^2\}, \frac{9}{2}pq\{p + q\}); 9y^2 = 12x - 16$.
18. $\dfrac{x^2}{a^2} + \dfrac{y^2}{b^2} = 2$. 19. $x\sin\theta + y = 2\tan\theta; x^2 - y^2 = 4$.
20. (a) $d^2\sin^2\theta$, (b) $-1, 3/2$.

Exercise 13.4

1. (a) $4/5; (\pm 4, 0); x = \pm 25/4$, (b) $3/5; (\pm 3/2, 0); x = \pm 25/6$,
   (c) $\frac{1}{2}\sqrt{3}; (\pm\sqrt{3}, 0); x = \pm(4\sqrt{3})/3$, (d) $2/3; (\pm 2, 0); x = \pm 9/2$,
   (e) $\frac{1}{3}\sqrt{5}; (0, \pm\sqrt{5}); y = \pm(9\sqrt{5})/5$, (f) $\frac{1}{2}\sqrt{3}; (0, \pm\frac{1}{2}\sqrt{3}); y = \pm\frac{2}{3}\sqrt{3}$.
2. (a) $5/4; (\pm 5, 0); x = \pm 16/5; y = \pm\frac{3}{4}x$, (b) $\sqrt{2}; (\pm 2\sqrt{2}, 0); x = \pm\sqrt{2}; y = \pm x$,
   (c) $\frac{1}{2}\sqrt{5}; (\pm\sqrt{5}, 0); x = \pm 4/\sqrt{5}; y = \pm\frac{1}{2}x$.
3. (a) $\frac{1}{2}\sqrt{3}$, (b) $4/5$, (c) $\frac{1}{3}\sqrt{5}$, (d) $4/5$.
4. (a) $\dfrac{x^2}{16} + \dfrac{y^2}{12} = 1$, (b) $\dfrac{x^2}{225} + \dfrac{y^2}{144} = 1$.
5. $5x^2 + 9y^2 - 40x - 18y + 44 = 0$.
6. (a) $x = 3, y = 0$, (b) $2y = x + 4, 2x + y + 3 = 0$,
   (c) $9x + 4y = 20, 8x - 18y = 7$, (d) $2x + y + 12 = 0, 2y = x + 1$.
7. (a) $4xx_1 + 9yy_1 = 36$, (b) $\dfrac{xx_1}{a^2} + \dfrac{yy_1}{b^2} = 1$, (c) $\dfrac{xx_1}{3} - \dfrac{yy_1}{2} = 1$.

8. $2xy + 2kx + 2ky = 7k^2$.

11. (i) $2ax - 2by = a^2 - b^2$; (ii) $(0, -b)$, $\left(\dfrac{3\sqrt{3}}{5}b, \dfrac{4}{5}b\right)$.

12. $a^2 = 2c^2$.

Exercise 13.5

1. 7.
2. $x - 2y + 9 = 0$, $x + 3y = 16y$; $x^2 + y^2 - 2x - 10y + 1 = 0$, $x^2 + y^2 - 7x - 144 = 0$.
3. $\frac{1}{5}|a + 4|$; $a = 1$, $3x + 4y = 7$.
4. $(5, 5)$, $\sqrt{10}$; $3m^2 - 10m + 3 = 0$; $y = \frac{1}{3}x$, $y = 3x$; (i) $53° 8'$, (ii) $x^2 + y^2 + 2x - 14y + 40 = 0$.
5. $2x + y + 4 = 0$, $(2, -8)$, $(-4, 4)$; 108.
6. 2, 5; 3/10, $\pm 2/5$.
7. $(0, 8)$, $4\sqrt{2}$; $(9, -1)$, $5\sqrt{2}$; $(4, 4)$; 167°.
8. $-1, 5$; $3/4$; $0, -8/9$.
9. (i) $3x - 4y = 9$, (ii) $(-1, -3)$, $(\frac{23}{25}, -\frac{39}{25})$, $2x^2 + 2y^2 + x + 8y + 5 = 0$.
10. $(8, 2)$; 16/3.
12. $\left(ct + \dfrac{c}{t}, ct + \dfrac{c}{t}\right)$.
13. $-4c^2/a^2$; 1/2; 2:3.
14. $\left(\dfrac{8}{9}, \dfrac{16\sqrt{2}}{27}\right)$, $\left(\dfrac{2}{9}, -\dfrac{2\sqrt{2}}{27}\right)$.
15. $8\frac{2}{3}$.
16. $x\cos t + y\sin t = 2\sin t\cos t$; $3\pi/8$.
17. (a) $ay + bx = ab\sqrt{2}$, (b) $(ax - by)\sqrt{2} = a^2 - b^2$; $\frac{1}{4}\pi ab$, πab.
18. 0°, 60°; $(2, 0)$, $(1, 3/2)$.
19. $bx\sec\theta + ay\operatorname{cosec}\theta = ab$; $\dfrac{x^2}{a^2} + \dfrac{y^2}{b^2} = \left(\dfrac{ab}{a^2 - b^2}\right)^2$.
20. $t_1t_2 + t_2t_3 + t_3t_1 = 0$; $-2t$.

Exercise 14.1

1. (a) $5i$ (b) $i\sqrt{5}$, (c) $3 - 4i$, (d) $2 - 3\sqrt{2}i$.
2. (a) ± 2, (b) $\pm 2i$, (c) $\pm i\sqrt{2}$, (d) $\pm\frac{3}{2}i$, (e) $0, \pm i\sqrt{3}$, (f) $\pm 1, \pm i$.
3. (a) $1 \pm i$, (b) $3 \pm i$, (c) $2 \pm 3i$, (d) $\frac{1}{2} \pm i$, (e) $\frac{1}{4}(-3 \pm i\sqrt{7})$, (f) $\frac{1}{2}(1 \pm i\sqrt{3})$.
4. (a) $3 - 5i$, (b) $4 + 7i$, (c) $-3i$, (d) -4, (e) $-1 - i$.
5. (a) $7 + 9i$, $3 + 7i$, (b) $3 - i$, $1 + 7i$, (c) $8 - 4i$, $-2 + 2i$, (d) $2 + 7i$, $-4 - 3i$.
6. (a) $3 + 11i$, (b) $7 - 9i$, (c) $4 - 7i$, (d) 25, (e) $-21 - 20i$, (f) $-2 - 2i$
7. (a) $-i$, (b) i, (c) 1 (d) -1, (e) $-i$, (f) i.
8. (a) $-1 + 2i$, (b) $\frac{1}{5} + \frac{3}{5}i$, (c) $\frac{5}{2} + \frac{1}{2}i$, (d) $\frac{1}{2} - \dfrac{\sqrt{3}}{2}i$, (e) $-\frac{1}{2}i$, (f) $\frac{4}{25} - \frac{3}{25}i$.
9. (a) $\frac{4}{13}$, (b) $\frac{7}{25} + \frac{1}{25}i$, (c) 6.

10. (a) $a = -1, b = 1$, (b) $a = 2, b = \frac{1}{2}$.
11. (a) $\pm(2 - i)$, (b) $\pm(1 + i)$, (c) $\pm(3 + i)$, (d) $\pm(5 - 2i)$.
12. (a) $-4, 5$, (b) $2, 10$, (c) $0, 16$, (d) $-6, 34$.
13. (a) $(x + i)(x - i)$, (b) $(x + 5i)(x - 5i)$,
(c) $(3x + 2i)(3x - 2i)$, (d) $(x + 1 + 2i)(x + 1 - 2i)$,
(e) $(x + 3 + 2i)(x + 3 - 2i)$, (f) $(2x - 3 + 4i)(2x - 3 - 4i)$.
14. $0, 0; 1, 0; -\frac{1}{2}, \frac{1}{2}\sqrt{3}; -\frac{1}{2}, -\frac{1}{2}\sqrt{3}$. 15. $x = 4, y = -3$.
16. $2 - i, 2 + 3i$. 17. $32 + 47i, 32 - 47i; 53, 61$.
18. $-5 - 4i, 1 + 7i; -3, -1$.
19. (a) $0, \pm 1$, (b) $\pm(1 + \sqrt{2}), \pm(1 - \sqrt{2})$. 20. $\pm(3 + 2i), 2 + 4i, -1 + 2i$.

Exercise 14.2

5. -2. 6. $3 - i, -3$. 7. $1 - i, -2; 4$. 8. $2 + 3i, -1; -3, 9$.

9. $-i, \frac{1}{2}(-1 \pm \sqrt{5})$. 10. $-1 - i, 2 \pm i$. 11. $-\frac{1}{2} \pm \frac{\sqrt{3}}{2}i$.

12. $\alpha, \alpha\omega, \alpha\omega^2$; (a) $3, -\frac{3}{2} \pm \frac{3\sqrt{3}}{2}i$, (b) $-2, 1 \pm i\sqrt{3}$,

(c) $-i, \frac{\sqrt{3}}{2} + \frac{1}{2}, -\frac{\sqrt{3}}{2} + \frac{1}{2}i$.

13. (a) $2, -1 \pm i\sqrt{3}$, (b) $-1, \frac{1}{2} \pm i\frac{\sqrt{3}}{2}$, (c) $0, -\frac{3}{2} \pm i\frac{\sqrt{3}}{2}$.

14. (a) 0, (b) -1, (c) -1. 15. (a) $-\omega$, (b) $1 - \omega$, (c) $\omega - 1$.
16. $a^3 + b^3 + c^3 - 3abc$. 17. $9z^3 - 12z^2 + 6z - 1$. 18. $-1, -1$.

Exercise 14.3

1. (a) $3 + 2i$, (b) $-2 + 4i$, (c) $5i$, (d) $1 - 2i$, (e) -2.
2. (a) 5, (b) $\sqrt{29}$, (c) 3, (d) $\sqrt{2}$, (e) 4.
3. $-2 - i, 2 - i, -2 + i, -1 + 2i, 1 - 2i$. 4. $2i, -2 + 2i, \frac{1}{2} - \frac{1}{2}i$.
5. $6 + 8i, i$. 6. (a) $\pm i$, (b) $\pm(1 - i)$, (c) $\pm(2 + i)$, (d) $\pm(4 + i)$.
8. (a) $k(1 + 2i), k \geqslant 0$, (b) $k(1 + 2i), -1 \leqslant k \leqslant 0$, (c) $k(1 + 2i), k \leqslant -1$.

10. (a) 5, (b) $\sqrt{85}$, (c) $\sqrt{13}$. 12. $\frac{4}{3} + \frac{2\sqrt{2}}{3}i, \frac{2}{3} + \frac{4\sqrt{2}}{3}i$.

Exercise 14.4

1. (a) $2, \pi/6$, (b) $\sqrt{2}, -\pi/4$, (c) $4, \pi/2$, (d) $3, \pi$.
2. (a) $5, -0{\cdot}927$, (b) $\sqrt{5}, 2{\cdot}678$, (c) $\sqrt{10}, -1{\cdot}893$, (d) $\sqrt{34}, -0{\cdot}540$.
3. (a) $5, \pi/6$, (b) $1/2, -\pi/3$, (c) $1/\sqrt{2}, \pi/4$, (d) $\sqrt{2}, 3\pi/4$.
4. (a) $5, \pi/3$, (b) $1, -\pi/2$, (c) $1, -2\pi/3$, (d) $2, -3\pi/4$.

5. (a) $2\sqrt{3}\{\cos(-\pi/3) + i\sin(-\pi/3)\}$, (b) $8\{\cos 0 + i\sin 0\}$,
(c) $2\sqrt{2}\{\cos(-3\pi/4) + i\sin(-3\pi/4)\}$, (d) $1\{\cos(-\pi/2) + i\sin(-\pi/2)\}$
6. (a) $r, \theta - \pi$, (b) $r, \theta + \frac{1}{2}\pi$, (c) $r^2, 2\theta$, (d) $1, 2\theta$.
7. (a) -1, (b) $-\dfrac{\sqrt{3}}{2} + \frac{1}{2}i$, (c) $-\dfrac{1}{\sqrt{2}} - \dfrac{1}{\sqrt{2}}i$, (d) $-\frac{1}{2} + \dfrac{\sqrt{3}}{2}i$,
(e) $-i$, (f) $\frac{1}{2} + \dfrac{\sqrt{3}}{2}i$.
8. (a) $10, \pi/2; 2, \pi/3; 20, 5\pi/6; 5, \pi/6$,
(b) $4, 5\pi/6; \sqrt{2}, -\pi/4; 4\sqrt{2}, 7\pi/12; 2\sqrt{2}, -11\pi/12$.
9. $\sec\theta, \theta$.
10. $2\sin\frac{1}{2}\theta, \frac{1}{2}(\theta - \pi)$; $2\sin\frac{1}{2}\theta, \frac{1}{2}(\theta + \pi)$; $\frac{1}{2}\operatorname{cosec}\frac{1}{2}\theta, \frac{1}{2}(\pi - \theta)$; $\frac{1}{2}\operatorname{cosec}\frac{1}{2}\theta, \frac{1}{2}(\theta - \pi)$.
11. $2, \pi/3; 2, \pi/2$. 12. $\pm\pi/2$.
13. (a) $\frac{1}{13} + \frac{5}{13}i$, 1·37; 2·75, $\frac{1}{13}$, (b) $|z|^2 : 1$.

Exercise 14.5

12. $x^2 + y^2 + 4x = 0$. 13. $x^2 + y^2 - 9x - y + 16 = 0$. 14. $x = 2$.
15. $5x + 3y = 14$. 16. $2x^2 + 2y^2 + 25x + 75 = 0$.
17. $5x^2 + 5y^2 - 26x + 8y + 1 = 0$. 18. $x^2 + y^2 = 25$, excluding $(-5, 0)$.
19. $x^2 + y^2 - 2x + 2y = 0$, excluding $(2, 0)$. 20. $5x - 2y + 6 = 0$.
21. $xy = 1$. 22. $x^2 - y^2 = 1$. 23. $x(x^2 + y^2 - 1) = 0$, excluding $(0, 0)$.
24. $y(x^2 + y^2 - 9) = 0$, excluding $(0, 0)$. 36. (a) 9, (b) 3, (c) 3, (d) 4.
37. $5\sqrt{2} + 3, 5\sqrt{2} - 3$. 38. 24, 2. 39. $x^2 + y^2 + 10x + 16 = 0$.
40. (i) $1 \pm i\sqrt{5}$, (ii) $1 + i$.

Exercise 14.6

1. (a) 1, (b) $-i$, (c) $\frac{1}{2} - \dfrac{\sqrt{3}}{2}i$, (d) $-\frac{1}{2} + \dfrac{\sqrt{3}}{2}i$.
2. $2\left(\cos\dfrac{\pi}{6} + i\sin\dfrac{\pi}{6}\right)$; $512 - 512\sqrt{3}i$, $-\dfrac{\sqrt{3}}{256} + \frac{1}{256}i$.
3. $\sqrt{2}\left(\cos\dfrac{3\pi}{4} + i\sin\dfrac{3\pi}{4}\right)$; $256, -\frac{1}{8}i$. 4. (a) 1, (b) -1.
5. $4\cos^3\theta - 3\cos\theta, 3\sin\theta - 4\sin^3\theta, \dfrac{3\tan\theta - \tan^3\theta}{1 - 3\tan^2\theta}$.
6. $16\sin^5\theta - 20\sin^3\theta + 5\sin\theta, 1 - 12\sin^2\theta + 16\sin^4\theta$.
8. $32\cos^6\theta - 48\cos^4\theta + 18\cos^2\theta - 1, 32\cos^5\theta - 32\cos^3\theta + 6\cos\theta$,
$$\frac{6\tan\theta - 20\tan^3\theta + 6\tan^5\theta}{1 - 15\tan^2\theta + 15\tan^4\theta - \tan^6\theta}.$$

9. (a) $\frac{1}{16}(\cos 5\theta + 5\cos 3\theta + 10\cos\theta)$,
   (b) $\frac{1}{64}(\cos 7\theta + 7\cos 5\theta + 21\cos 3\theta + 35\cos\theta)$,
   (c) $\frac{1}{16}(2\cos\theta - \cos 3\theta - \cos 5\theta)$.
10. (a) $\frac{1}{4}(3\sin\theta - \sin 3\theta)$, (b) $\frac{1}{64}(35\sin\theta - 21\sin 3\theta + 7\sin 5\theta - \sin 7\theta)$,
   (c) $\frac{1}{64}(3\sin\theta + 3\sin 3\theta - \sin 5\theta - \sin 7\theta)$.
12. (a) $3\pi/8$, (b) $\pi/32$.

Exercise 14.7

1. $\dfrac{x}{x^2+y^2} - \dfrac{y}{x^2+y^2}i$. 2. $3, 1; -3, -1$. 3. $1 + 2i; 5\sqrt{2}; -2, -1$.
4. (i) $\pm(1 - 3i)$, (ii) $2 - 2i, 1 + i$.
5. Sum of squares of diagonals of a parallelogram = sum of squares of sides.
6. $\pm 1, \pm i, \pm\dfrac{1}{\sqrt{2}}(1 - i), -\dfrac{1}{\sqrt{2}}(1 + i)$.
7. (i) $-1 + i\sqrt{3}, 2, 2\pi/3; \dfrac{\sqrt{3}}{2} + \frac{1}{2}i, 1, \pi/6$,
   (ii) $1 \pm 2i$; (a) $p^2 = q - 4$, (b) $2p = q + 5$.
8. (a) $-3\sqrt{3} + 5i; 2\sqrt{13}, 2{\cdot}38$ rad, (b) $5, -20$.
9. (i) $-1 - i, 3\pi/4$, (ii) $2 - i, 2; -10$.
10. (b) $-1, \frac{1}{2} \pm \dfrac{\sqrt{3}}{2}i$, (c) $-1 - 3i, \frac{1}{2}(1 - 3\sqrt{3}) + \frac{1}{2}(3 + \sqrt{3})i$,
   $\frac{1}{2}(1 + 3\sqrt{3}) + \frac{1}{2}(3 - \sqrt{3})i$.
11. $-\frac{1}{2} \pm \dfrac{\sqrt{3}}{2}i; 1 \pm i\sqrt{3}, \frac{5}{2} \pm \dfrac{\sqrt{3}}{2}i; 28$.
12. (ii) $x^3 - 3abx + a^3 + b^3; -\sqrt[3]{2} - \sqrt[3]{4}, -\omega\sqrt[3]{2} - \omega^2\sqrt[3]{4}, -\omega^2\sqrt[3]{2} - \omega\sqrt[3]{4}$.
13. (i) $\pm(3 - 2i)$, (ii) $3 \pm i\sqrt{3}$. 14. (i) (a) $7/5, -4/5$; (b) $2 \pm i$.
15. (i) $1 - i, -1 \pm 2i$, (ii) $\frac{1}{2}(3 + i\sqrt{3}); \pi/6, 2\pi/3$.
16. (i) $2, 32; \pi/3, -\pi/3; -16\sqrt{3}$, (ii) $2 + 3i$. 17. (i) -1, (ii) $1 + 2i, 2\sqrt{5}$.
18. (i) $\sqrt{2}, -\pi/4; \frac{1}{2} + \frac{1}{2}i, 8i$. 20. (i) $2 - i, 3 - 4i; (2, -1)$.
21. (ii) $3x^2 + 3y^2 + 10x + 3 = 0$. 22. (i) $x = 1, y = 2$ or $x = -1, y = -2$.
23. (i) (a) $13, -23°$, (b) $1, 90°$; (ii) 12; (iii) $|z + 1 + i| \geqslant 4$.
24. $2\sin\theta; \frac{1}{2}\operatorname{cosec}\theta, -\theta; y = -\frac{1}{2}$. 25. $2\sqrt{2} - 2$.
26. $2\cos\frac{1}{2}\theta, \frac{1}{2}\theta; \left[2\cos\left(\dfrac{\theta}{2}\right)\right]^n\cos\left(\dfrac{n\theta}{2}\right) + i\left[2\cos\left(\dfrac{\theta}{2}\right)\right]^n\sin\left(\dfrac{n\theta}{2}\right)$.

Exercise 15.1

1. 10. 2. $3\frac{1}{2}$. 3. 14; 20 p. 4. (a) 1, (b) 9/8.
5. $p(2) = 2/15, p(3) = 8/15, p(4) = 1/3; 3{\cdot}2$. 6. 0·8 p loss.

7. $\frac{125}{216}, \frac{75}{216}, \frac{15}{216}, \frac{1}{216}$; 2 p loss. 8. (a) £122.50, (b) £112·672.
9. $p(x) = 1/6, x = 1, 2, 3, 4, 5; p(6) = 0; p(x) = 1/36, x = 7, 8, 9, 10, 11, 12; 4\frac{1}{12}; 6/17$.
10. (i) 24/625, (ii) 24/625, (iii) 2. 11. 28·5, 721·05. 12. 2·53, 0·305.
13. $1\frac{1}{2}, \frac{3}{4}; \frac{1}{20}, \frac{9}{20}, \frac{9}{20}, \frac{1}{20}$; 60%.

14.

r	0	1	2	3	4
$P(Y = r)$	0·09	0·24	0·34	0·24	0·09

r	0	1	2	3	4
$P(Z = r)$	0·447	0·232	0·222	0·072	0·027

$E(Z) = 1$, $\text{Var}(Z) = 1·2$.

15. $4\frac{1}{3}, 2\frac{2}{9}; 8\frac{2}{3}, 8\frac{8}{9}$. 16. $\frac{1}{2}(n + 1)$, 0·42.
17. $2/n(n + 1); \frac{1}{3}(2n + 1), \frac{1}{18}(n - 1)(n + 2)$.
18. $x = 1, 2, 3, \ldots, 9; p(x) = \frac{9}{45}, \frac{8}{45}, \frac{7}{45}, \ldots, \frac{1}{45}; \frac{11}{3}, \frac{2}{3}\sqrt{11}; 1;$

$$x = 1, 2, 3, \ldots; p(x) = \frac{10 - x}{45}.$$

19. (a) $1 + 2x + 3x^2 + \ldots + rx^{r-1} + \ldots$, (b) $p(1 - p)^{r-1}; 1/p$.
20. (i) $\frac{1}{3}$, (ii) $\frac{2}{9}$, (iii) $\frac{16}{243}; \frac{32}{243}$; 50 p; $\frac{16}{81}$; 0·177. 21. 3/16.

Exercise 15.2

1. (a) 0·279, (b) 0·457; 1·5, 1·25. 2. (a) 0·250, (b) 0·526; 2·5, 1·875.
3. 0·657; 7. 4. (a) 0·393, (*b*) 0·000 085 2; $n \geqslant 18$.
5. 24, 0·4; $(0·6 + 0·4t)^{24}$.
6. $(0·55 + 0·45t)^4$; (i) 4, (ii) $4(0·55)(0·45)^3$. 7. 0·228, 0·210.
8. 0·0464; 4, 2·4.
9. $\frac{4}{3}, \frac{356}{405}$;

No. of males	0	1	2	3	4
Expected no. of litters	160	320	240	80	10

10. 0·991, 0·977;

No. of ones	0	1	2	3	4	5	6
Expected no. of throws	335	402	201	54	8	1	0

11. (a) 0·315, (b) (i) 0·53, (ii) 5, 22, 37, 28, 8. 12. 26·4%, 62·4%.
13. (i) 0·201; (ii) (a) 0·0556, (b) 0·0367, (c) 0·6.
14. (i) 4·8, 0·98, (iii) 0·737, (iv) 0·388.
15. (a) $\frac{1}{27}$, (b) $\frac{2}{9}$, (c) $\frac{4}{9}$, (d) $\frac{8}{27}$, (e) 0·299, (f) 0·903, (g) 0·328.

16. (ii) 1/5. 17. $\dfrac{(n-r+1)p}{r(1-p)}$; (a) 8, (b) 2, 3.

18. (a) $\dfrac{135}{2048}$; (b) $\dfrac{459}{512}$; $\dbinom{n-1}{2}\left(\dfrac{1}{4}\right)^3\left(\dfrac{3}{4}\right)^{n-3}$; 8, 9.

19. (a) $(r+1)/2^{r+2}$, (b) 2, (c) 4. 20. $t/(6-5t)$; 6, 30.

Exercise 15.3

1. (a) 1/2, 1/12, (b) 3, 4/3, (c) $k/2$, $k^2/12$, (d) $(a+b)/2$, $(b-a)^2/12$.
2. 4/27; 9/5, 2, 3/5. 3. (a) 2·4, (b) 20; 1/3, 0·178.
4. (i) −3/16, (ii) 19/80, (iv) 2. 5. 4, 8/15, 11/225; 0·541.
6. 3/8, 0, 2/5; (i) 85/512, (ii) 85/152. 7. (i) 12, 1, (ii) 0·0523.
8. (i) 2/5, (ii) 13/5, (iii) 3/2.
9. 0·6; $F(x) = 0$ for $x \leqslant 0$, $F(x) = 4x^3 - 3x^4$ for $0 \leqslant x \leqslant 1$, $F(x) = 1$ for $x \geqslant 1$.
10. 28, 12, 6, 2, 1, 1.
11. (i) 2/3, (ii) 4/3, $(\sqrt{14})/6$, (iii) $F(x) = 0$ for $x \leqslant 0$, $F(x) = \frac{1}{3}x^2$ for $0 \leqslant x \leqslant 1$, $F(x) = \frac{1}{6}(-3 + 6x - x^2)$ for $1 \leqslant x \leqslant 3$, $F(x) = 1$ for $x \geqslant 3$, (iv) 13/24.
12. (iii) $F(x) = -\frac{1}{12}x^3$ for $x \leqslant -1$, $F(x) = \frac{1}{2} + \frac{1}{2}x - \frac{1}{12}x^3$ for $-1 \leqslant x \leqslant 1$, $F(x) = 1 - \frac{1}{12}x^3$ for $x \geqslant 1$, (iv) $E(X) = 0$, $\text{Var}(X) = 11/15$.
13. (ii) 1, 5/6, (iii) 1/4.
14. (i) $k = 2$, (ii) $f(x) = 2$ for $0 \leqslant x \leqslant 0{\cdot}5$, $f(x) = 0$ otherwise, (iii) 0·25, (iv) 0·144.
15. (i) 3/32, (ii) 11/16, (iii) $f(x) = 3(4 - x^2)/32$ for $-2 \leqslant x \leqslant 2$, $f(x) = 0$ otherwise, (iv) 0, 4/5.
16. $F(x) = 0$ for $x \leqslant 1$, $F(x) = \frac{1}{3}(x^2 - 1)$ for $1 \leqslant x \leqslant 2$, $F(x) = 1$ for $x \geqslant 2$; $G(y) = 0$ for $y \leqslant 1$, $G(y) = \frac{1}{3}(y^4 - 1)$ for $1 \leqslant y \leqslant \sqrt{2}$, $G(y) = 1$ for $y \geqslant \sqrt{2}$; $g(y) = 4y^3/3$ for $1 \leqslant y \leqslant \sqrt{2}$, $g(y) = 0$ otherwise.
17. $1/(2\sqrt{y})$; $(1/\sqrt{w}) - 1$.
18. $f(x) = 1/(b-a)$ for $a \leqslant x \leqslant b$, $f(x) = 0$ otherwise; $F(x) = 0$ for $x \leqslant a$, $F(x) = (x-a)/(b-a)$ for $a \leqslant x \leqslant b$, $F(x) = 1$ for $x \geqslant b$; $P(V \leqslant v) = 3 - 72/v$; (i) $f(v) = 72/v^2$, (ii) 0·4, (iii) 28·8 m.p.h.
19. (a) $a^2/18$, (b) $F(x) = 0$ for $x \leqslant a$, $F(x) = x^2/a^2$ for $0 \leqslant x \leqslant a$, $F(x) = 1$ for $x \geqslant a$, (c) (i) $1/(a+1)^2$, (ii) $2(1-y)$ for $0 \leqslant y \leqslant 1$, 0 otherwise.
20. $g(y) = 1$ for $0 \leqslant y \leqslant 1$, $g(y) = 0$ otherwise.

Exercise 15.4

1. (a) 0·6179, (b) 0·1151, (c) 0·97725, (d) 0·01287, (e) 0·1760, (f) 0·5558.
2. (a) 0·9378, (b) 0·1944, (c) 0·9715, (d) 0·02239, (e) 0·2065, (f) 0·2976.
3. (a) 0·9554, (b) 0·2119, (c) 0·6612.
4. (a) 0·3085, (b) 0·5793, (c) 0·0796.
5. (a) 0·9487, (b) 0·9750, (c) 0·04412
6. (a) 0·44, (b) 0·583, (c) −0·878.
7. (a) 0·524, (b) 2·326, (c) 3·291.

8. (a) 0·3%, (b) 6·6%, (c) 44%, (d) 1·2%.
9. 11·51%, 81%, 7·49%; £6528.64.
10. (i) 4·8%, (ii) 97·7%; 1111·5 to 1188·5.
11. (a) 37·82%, (b) 125·5 hrs to 194·5 hrs, (c) 0·405.
12. (a) 4 men, (b) 0·0446, (c) 1·746 m.
13. (a) 3, 62·3, (b) 15, 9·24, (c) 84·5, 111, (d) 8·31, 1·74.
14. 0·0122, 0·0102. 15. 104, 38·8.
16. (a) 0·205, 0·205, (b) 0·120, 0·119, (c) 0·0419, 0·0416, (d) 0·190, 0·179, (e) 0·0985, 0·0932, (f) 0·0493, 0·0474.
17. 0·0287. 18. 0·210. 19. (i) 0·026, (ii) 0·932, (iii) 0·923.
20. (a) 4320/15625; (b) 7120/15625; (a) 0·026, (b) 0·853.
21. (a) Negligible, (b) negligible, (c) 0·806.
22. (i) 0·648, (ii) 92·65%, (iii) 0·654.

Exercise 15.5

1. (a) 0·156, (b) 0·663. 2. (a) 0·353, (b) 0·285.
3. 0·181; $e^{-0\cdot02n}$; 5. 4. 0·337; 1536.

5.

r	0	1	2	3	4	5	6	7	8	9	10
(a)				0·001	0·005	0·015	0·037	0·074	0·120	0·160	0·176
(b)			0·002	0·008	0·019	0·038	0·063	0·090	0·113	0·125	0·125
(c)				0·001	0·005	0·015	0·037	0·073	0·119	0·160	0·177
(d)	0·001	0·002	0·005	0·011	0·021	0·036	0·057	0·080	0·103	0·120	0·126

r	11	12	13	14	15	16	17	18	19	20	21
(a)	0·160	0·120	0·074	0·037	0·015	0·005	0·001				
(b)	0·114	0·095	0·073	0·052	0·035	0·022	0·013	0·007	0·004	0·002	0·001
(c)	0·160	0·119	0·073	0·037	0·015	0·005	0·001				
(d)	0·120	0·103	0·080	0·057	0·036	0·021	0·011	0·005	0·002	0·001	

6.

r	0	1	2	3	4	5	6	7	8	9
(a)	0·122	0·270	0·285	0·190	0·090	0·032	0·009	0·002		
(b)	0·135	0·271	0·271	0·180	0·090	0·036	0·012	0·003	0·001	
(c)	0·132	0·223	0·291	0·223	0·101	0·027	0·004			
(d)	0·144	0·217	0·276	0·217	0·106	0·032	0·006	0·001		

7.

r	0	1	2	3	4	5	6	7	8	9
(a)	0·005	0·029	0·078	0·139	0·181	0·185	0·154	0·108	0·064	0·033
(b)	0·007	0·034	0·084	0·140	0·175	0·175	0·146	0·104	0·065	0·036
(c)	0·017	0·033	0·070	0·121	0·167	0·186	0·167	0·121	0·070	0·033
(d)	0·022	0·037	0·073	0·119	0·160	0·177	0·160	0·119	0·073	0·037

r	10	11	12	13
(a)	0·015	0·006	0·002	0·001
(b)	0·018	0·008	0·003	0·001
(c)	0·012	0·004	0·001	
(d)	0·015	0·005	0·001	

8.

r	0	1	2	3	4	5	6	7	8	9
(a)		0·002	0·009	0·025	0·054	0·090	0·123	0·144	0·146	0·129
(b)		0·003	0·011	0·029	0·057	0·092	0·122	0·140	0·140	0·124
(c)	0·003	0·005	0·013	0·027	0·050	0·080	0·112	0·137	0·146	0·137
(d)	0·004	0·007	0·015	0·030	0·052	0·080	0·110	0·132	0·140	0·132

r	10	11	12	13	14	15	16	17	18
(a)	0·102	0·073	0·047	0·028	0·015	0·007	0·003	0·001	0·001
(b)	0·099	0·072	0·048	0·030	0·017	0·009	0·005	0·002	0·001
(c)	0·112	0·080	0·050	0·027	0·013	0·005	0·002	0·001	
(d)	0·110	0·080	0·052	0·030	0·015	0·007	0·003	0·001	

9. (a) 0·368, (b)0·456. 10. (a) 0·258, (b) 0·567.
11. (a) Poisson approximation: 0·442,
 (b) Poisson approximation: 0·353,
 (c) Normal approximation: 0·119.
14. (i) 0·19, (ii) 0·21, (iii) 0·012.
15. (a) (i) 1·04, (ii) 53, 55, 29, 10, 3; (b) 0·121.
16. 0·647, 0·185. 17. (i) 0·55, (ii) 0·18.
18. £266C, £415C. 19. 0·460; 33.
20. 0·323; 172 to 228.

Exercise 15.6

1. 3·6, 14·5, 0·35, 28, 5·95. 2. 0·4, 1, 2·2; 0·84, 1·2.
3. $1\frac{2}{3}$, 3, $5\frac{4}{5}$; $20\frac{11}{15}$, $\frac{2}{9}$.
4. $\frac{1}{2}$, $\frac{3}{10}$, $\frac{24}{35}$; $-\frac{1}{5}$, $\frac{73}{2450} \approx 0{\cdot}0298$.
5. (a) 6, 2·5, (b) 4, 6, (c) 27, 40, (d) −3, 24,
 (e) 10, 8·5, (f) 2, 8·5, (g) 14, 28·5, (h) 24, 64.
6. (a) 31, 234, (b) 109, 50, 159. 7. 30 g, 0·3 g. 8. 200 g, 6·71 g.
9. (a) 2·1, 0·49; 0·8, 0·36, (b) no; independent,

(c)

xy	0	1	2	3	4	6
$p(xy)$	0·3	0·12	0·32	0·18	0·05	0·03

; yes,

(d)

$x + y$	1	2	3	4	5
$p(x + y)$	0·06	0·27	0·41	0·23	0·03

; yes.

10. (i) (a) 1·2, 0·56; 1·7, 0·81, (b) all values; not independent,

(c)

xy	0	1	2	3	4	6
$p(xy)$	0·27	0	0·45	0·1	0·1	0·08

; no,

(d)

$x+y$	0	1	2	3	4	5
$p(x+y)$	0·03	0·15	0·07	0·47	0·2	0·08

; no.

(ii) (a) 1, 0·6; 2, 0·6, (b) all values, not independent,

(c)

xy	0	1	2	3	4	6
$p(xy)$	0·3	0·2	0·05	0·2	0·2	0·05

; yes,

(d)

$x+y$	1	2	3	4	5
$p(x+y)$	0·05	0·4	0·1	0·4	0·05

; yes.

11. 1, 2, 4, 8. 12. £3.33.
13. $\frac{1}{210}, \frac{4}{35}, \frac{3}{7}, \frac{8}{21}, \frac{1}{14}$; 2·4, 0·8; 1·6, 0·8; 4, 1·13.
14. (i) 0·9, 1·2,

(ii)

z	-2	-1	0	1	2	3	4
$p(z)$	0·1	0·1	0·3	0·2	0·2	0·1	0

,

(iii) 0·2.

15. (i) 0·011, (ii) 0·036, (iii) 0·27. 16. 0·12 mm, 0·0583 mm; 1·98%
17. (a) 0·673, (b) 0·138. 18. (i) 0·089, (ii) 0·661.
19. (i) 0·841, (ii) 0·500, (iii) 0·421; 0·994.
20. (i) 0·823, (ii) 0·059; 1/4, 1/6. 21. $\frac{1}{2}r\pi, \frac{1}{2}r\pi; 4r/\pi$; 1/3.
22. (i) $\mu_x^2 + \sigma_x^2$; (ii) $\mu_x\mu_y; \mu_x^2\sigma_y^2 + \mu_y^2\sigma_x^2 + \sigma_x^2\sigma_y^2$; (iii) $\mu_y(\mu_x^2 + \sigma_x^2)$.
23. $\dfrac{\lambda^r\mu^{n-r}}{r!(n-r)!}e^{-(\lambda+\mu)}; \sum_{r=0}^{n}\dfrac{\lambda^r\mu^{n-r}}{r!(n-r)!}e^{-(\lambda+\mu)}$. 24. 0·121.
25. (b) $(1-p+pt)^m, (1-p+pt)^n; (1-p+pt)^{m+n}$.

Exercise 15.7

1. $(n+1)/2, (n^2-1)/12$; 1/2; (a) 2/105, (b) 16.
2. £16. 3. (a) 0·323, (b) $£1\frac{32}{65} \approx £1.49$.
4. 1/15; 2/3, 34/45; 1/75; 4/3, 68/45.
5. (i) $P(N=r) = \dfrac{1}{k}(1 \leqslant r \leqslant k), E(N) = \frac{1}{2}(k+1)$,

(ii) $P(N=r) = \dfrac{1}{k}\left(\dfrac{k-1}{k}\right)^{r-1} (r \geqslant 1), E(N) = k$; first strategy.

6. (i) 1/5, (ii) 8/9, (iii) 10, (iv) 8. 7. 0·0512.
8. (i) 0·168, (ii) 0·360, (iii) 0·309, (iv) 0·132, (v) 0·028, (vi) 0·002; 0·947.
9.

No. of defectives	0	1	2	3	4	5	6
Expected frequency	16	42	45	26	8	1	0

10. $\frac{1}{p}, \frac{1-p}{p^2}; \frac{2t^2}{(5-3t)(5-4t)}; \frac{194}{625} = 0{\cdot}3104.$
11. (i) 2, (ii) $1\frac{1}{2}$, (iii) $1\frac{5}{12}$,
(iv) $F(x) = 0$ for $x \leqslant 0$, $F(x) = x - \frac{1}{2}x^2$ for $0 \leqslant x \leqslant 1$, $F(x) = \frac{1}{2}$ for $1 \leqslant x \leqslant 2$,
$F(x) = \frac{1}{2}x^2 - 2x + \frac{5}{2}$ for $2 \leqslant x \leqslant 3$, $F(x) = 1$ for $x \geqslant 3$, (v) $\frac{3}{4}$.
12. (ii) $\sqrt{(3/5)}$. 13. 1, 1/6; 3/4; 5/6.
14. (a) 3/250, $6\frac{7}{8}$, (b) (1) £3.20, (ii) 0·792.
15. (i) 1/2, −1/4, (ii) 7/16,
(iii) $f(x) = \frac{1}{2}x$ for $0 \leqslant x \leqslant 1$, $f(x) = \frac{1}{2}$ for $1 \leqslant x \leqslant 2$,
$f(x) = \frac{3}{2} - \frac{1}{2}x$ for $2 \leqslant x \leqslant 3$, $f(x) = 0$ otherwise; 3/2; 5/12.
16. (a) $1 - e^{-\lambda x}$, (b) $\frac{1}{\lambda}, \frac{1}{\lambda^2}$, (c) $\frac{1}{\lambda}\ln 2$ (d) $\frac{1}{\lambda}\ln\left(\frac{20}{19}\right)$.
17. (i) 0·202, (ii) 0·410.
18. (i) 0·880, (ii) 9·69, 0·2339, (iii) 0·180.
19. £9360, 19·6%; (a) £20 000, £9360; (b) £19 800, £10 296.
20. (i) 0·60, (ii) 0·20, (iii) 0·95, (iv) 0·50.
21. (i) 0·227, (ii) 54·7, (iii) 78·7 cm^2, (iv) 0·579.
22. 1·514 mm, 0·051 mm; 3·028 mm, 0·072 mm; 0·895; 3·214 mm.
23. (i) 0·345, (ii) 0·692; 0·0033; 0·304.
24. (a) (i) 0·5, (ii)

No. of accidents	0	1	2	3	4	5
Expected no. of days	121	61	15	3	0	0

(b) Binomial model: 0·932, Poisson approximation: 0·929.
25. (i) 9, (ii) 26, (iii) 43; 8.
26. (a) (i) 0·0581, (ii) 0·0519; (b) $(0{\cdot}96)^n$, 74; (c) 0·955.
27. (i) 0·135, (ii) 0·271, (iii) 0·0290, (iv) 2, (v) 8·19.
28. (i) 0·419, (ii) 0·676.
29. (a) (i) 0·0952, (ii) 0·0905; (b) 0·853; (c) 0·632; 63·2, 4·82.
30. 1/12; 0, n/12; 16. 31. (b) 2·2.
32. (i) $F(x) = 0$ for $x \leqslant 1$, $F(x) = 1 - x^{-3}$ for $x \geqslant 1$;
(ii) 3/2, 3/4; p.d.f. of Y: $g(y) = 3y^2$ for $0 \leqslant y \leqslant 1$, $g(y) = 0$ otherwise.
33. (ii) 8/15,
(iii) $F(x) = 0$ for $x \leqslant 0$, $F(x) = x^2(2 - x^2)$ for $0 \leqslant x \leqslant 1$, $F(x) = 1$ for $x \geqslant 1$;
(iv) $g(y) = 8y^3(1 - y^4)$ for $0 \leqslant y \leqslant 1$, $g(y) = 0$ otherwise.
34. (a) (i) 5/2, (ii) 7,
(iii) $F(x) = 0$ for $x \leqslant 1$, $F(x) = \frac{1}{3}(x^2 - 1)$ for $1 \leqslant x \leqslant 2$, $F(x) = 1$ for $x \geqslant 2$;
(b) (i) 5/2, 3/4, (ii) interval from 1 to 4.

Exercise 16.1

1. 1, 0·8;

sample	0, 0	0, 1	0, 2	0, 3	1, 1	1, 2	1, 3	2, 2	2, 3	3, 3
probability	0·09	0·30	0·06	0·06	0·25	0·10	0·10	0·01	0·02	0·01
$\bar{X}$	0	0·5	1	1·5	1	1·5	2	2	2·5	3
V	0	0·25	1	2·25	0	0·25	1	0	0·25	0

$\bar{x}$	0	0·5	1	1·5	2	2·5	3
$p(\bar{x})$	0·09	0·30	0·31	0·16	0·11	0·02	0·01

v	0	0·25	1	2·25
$p(v)$	0·36	0·42	0·16	0·06

2.

$\bar{x}$	0	1/3	2/3	1	4/3	5/3	2	7/3	8/3	3
$p(\bar{x})$	0·027	0·135	0·252	0·242	0·174	0·108	0·040	0·018	0·003	0·001

v	0	2/9	2/3	8/9	14/9	2
$p(v)$	0·154	0·456	0·120	0·126	0·108	0·036

3. 2, 4·8;

sample	A, B	A, C	A, D	A, E	B, C	B, D	B, E	C, D	C, E	D, E
$\bar{X}$	1	1	4	2	0	3	1	3	1	4
V	1	1	4	0	0	9	1	9	1	4

$\bar{x}$	0	1	2	3	4
$p(\bar{x})$	0·1	0·4	0·1	0·2	0·2

v	0	1	4	9
$p(v)$	0·2	0·4	0·2	0·2

2, 1·8, 3; 2, 2·4, 2·4.

4.

$\bar{x}$	2/3	4/3	2	8/3	10/3
$p(\bar{x})$	0·2	0·2	0·1	0·4	0·1

v	8/9	32/9	56/9	8
$p(v)$	0·4	0·1	0·4	0·1

2, 0·8, 4; 2, 1·6, 3·2.

5. (a) 0·096, (b) 0·096.
6. 0·623, 0·789.
7. 0·332, 0·0587; 0·0092.
8. (a) (i) 1·75, 0·01, (ii) 1·65, 0·0225; (b) 0·1, 0·180.
9. $N(-2, 1{\cdot}5)$; (a) 0·949, (b) 0·200.
10. (a) 28, (b) 822.
11. $1{\cdot}645/\sqrt{n}$; 215.
12. (a) 0·463 using binomial distribution,
    (b) 0·336 using normal approximation with continuity correction,
    (c) 0·185 without continuity correction (0·174 with correction),
    (d) 0·0023 without continuity correction.
13. (a) 0·608, (b) 0·834, (c) 0·996.

Exercise 16.2

1. (a) 4·1, (b) 2·66, (c) 0·75. 2. (a) 43·2, (b) 68·0, (c) 0·375.
3. 2·41, 2·34; 2·41, 2·32. 4. 59·7, 249; 59·7, 225.
5. 2·07, 1·83; 0·35. 6. $\theta/2$, $\theta^2/12$; 6·2, 39·2.
7. $3\theta/4$, $3\theta^2/80$; 9·6, 122.
8. (a) I: 44·5, 373; II: 42, 233·25; 43·25, 346; (b) 43·25, 325.
9. (a) I: 9·25, 12·7; II: 9, 10·7; 9·17, 13·5; (b) 9·17, 12·7.
10.

$\{X_1, X_2\}$	$\{0,0\}$	$\{0,1\}$	$\{0,2\}$	$\{1,1\}$	$\{1,2\}$	$\{2,2\}$
Probability	$\frac{1}{4}\theta^2$	$\theta - \theta^2$	$\frac{1}{2}\theta^2$	$(1-\theta)^2$	$\theta - \theta^2$	$\frac{1}{4}\theta^2$

$(X_1 - X_2)^2$	0	1	4
Probability	$1 - 2\theta + \frac{3}{2}\theta^2$	$2\theta - 2\theta^2$	$\frac{1}{2}\theta^2$

;

$\frac{1}{2}\theta + \frac{1}{2}\theta^2$; Z, $\mathrm{Var}(Z) = \frac{1}{2}\theta - \frac{1}{2}\theta^2 < \mathrm{Var}(Y)$.

11. $\dfrac{r_1}{n_1}, \sqrt{\left\{\dfrac{p(1-p)}{n_1}\right\}}; \sqrt{\left\{\dfrac{p(1-p)(n_1+n_2)}{4n_1n_2}\right\}}; \dfrac{1}{3} < \dfrac{n_1}{n_2} < 3.$

Exercise 16.3

1. (a) 836·1 g, 847·9 g, (b) 840·1 g, 845·9 g. 2. (a) 3·10, 3·42, (b) 3·05, 3·47.
3. (a) 439 g, 497 g, (b) 451·3 g, 469·7 g.
4. (a) 0·083, 0·177, (b) 0·075, 0·185.
5. (a) 3·81 m, 4·09 m, (b) 3·78 m, 4·12 m, (c) 0·059, 0·128, (d) 0·050, 0·137.
6. 0·774, 0·959; 178. 7. (a) 36, (b) 50.
8. (a) 235·8 ml, 246·2 ml, (b) 236·1 ml, 249·9 ml, (c) −10·2 ml, 14·2 ml.
9. (a) 0·741 kg, 0·879 kg, (b) 0·835 kg, 0·945 kg, (c) 0·0391 kg^2, (d) 0·014 kg, 0·146 kg.
10. (i) 34%, (ii) 29·4%, 38·6%, (iii) 2156. 11. 26·9, 8·65; 25·7, 28·1.
12. [−294 g, 294 g].

Exercise 16.4

1. 0·045; (a) no, (b) yes.
2. Reject claim if number of passes is less than (a) 6, (b) 5, (c) 4; true significance level (a) 3·81%, (b) 0·50%, (c) 0·04%.
3. Reject hypothesis if cricketer scores more than 30 runs in 0, 6 or 7 innings; 0·047.
4. Yes. 5. (a) No, (b) Yes.
6. Approximately normal with mean 1·65 m, s.d. 0·007 m; yes, significant at 5% level.
7. No, not significant at 5% level.

8. (a) Reject hypothesis, (b) accept hypothesis.
9. Strong evidence against supposition (probability of more than 290 successes by guesswork is less than 0·002).
10. (a) No, (b) No.
11. (a) Males: $\mu_m = 1{\cdot}65$, variance $= 0{\cdot}0025$; females: $\mu_f = 1{\cdot}55$, variance $= 0{\cdot}00375$;
    (b) 0·103;
    (c) H_0: $\mu_m - \mu_f = 0{\cdot}08$, H_1: $\mu_m - \mu_f > 0{\cdot}08$; may reject H_0 at 5% level and at 1% level.
12. 4·41; [9·87, 10·73]; 3·61; H_0: $\mu_x - \mu_y = 0$, H_1: $\mu_x > \mu_y$; evidence not significant at 3% level.
13. [2·60, 3·12]; 0·567; yes, results significant.
14. [0·270, 0·355]; 0·25; accept hypothesis that proportion of serious injuries is greater amongst drivers not wearing seat belts.
15. H_0: $\pi_A - \pi_B = 0$, H_1: $\pi_A - \pi_B > 0$; accept theory of experts; 0·569, 0·711.
16. H_0: $\pi = 0{\cdot}25$, H_1: $\pi > 0{\cdot}25$, where π is probability of correctly identifying the suit of a card; 5·08%; 0·181.
17. Accept H_0 if fewer than 59 germinate; 0·380.

Exercise 16.5

1. $N(\mu, \sigma^2/n)$; 0·106; 1240 units.
2. (a) Approximately normal with mean 283 g, s.d. 2·3 g; (b) 1, 0·0558.
3. $N(54, 64)$; 0·023, 15·7; $N(5, 144)$; 0·662.
4. 46·8;

m	46	$46\frac{1}{3}$	$46\frac{2}{3}$	47	$47\frac{1}{3}$
$p(m)$	1/120	3/20	2/5	19/60	1/8

; 0·3055.

5. 0·9723, 0·9877. 6. 105, 78·1; [84·8, 125·2].
7. [1·550, 1·650] using unbiased estimate of variance (or [1·551, 1·649] using sample variance); [0·102, 0·298] in both cases.
8. (i) 0·833,
   (ii) 0·180 using binomial distribution (0·181 using normal approximation),
   (iii) H_0: $p = 0{\cdot}5$, H_1: $p \neq 0{\cdot}5$; conclusion: $p \neq 0{\cdot}5$.
9. (a) 10·56%, 6·68%, (b) £98.88, (c) yes, using one-tailed test.
10. Accept claim, since probability of 7 or more successes is less than 5%; stronger evidence to support claim, since probability of 13 or more successes is approximately 1%.
11. 11·6, 1·89; would not reject at 5% level; H_0: mean $= 12$, H_1: mean > 12; $12 + 4{\cdot}652/\sqrt{n}$.
12. (i) 80, 16, (ii) 0·8192, (iii) H_0: cure rate $= 80\%$, H_1: cure rate $> 80\%$; 6·92%; 0·608.
13. 8·00667, 0·000175; accept alternative hypothesis; accept alternative hypothesis.
14. (a) [341·07, 341·33]; (b) not significantly different.
15. (a) Yes, (b) Yes. 16. (ii) [0·476, 0·524].
17. 0·590; (i) no, (ii) $r \leqslant 2$.
18. 35, 3·028. 19. 64, 452·84. 20. 1·9, 0·837.

Exercise 17.1

1. High positive correlation. 2. Negative correlation.
3. Probably uncorrelated. 4. Fairly low positive correlation.
5. No linear correlation, but possibly some non-linear relationship.

Exercise 17.2

1. 0·994. 2. −0·816. 3. 0·050. 4. 0·661. 5. −0·032.
6. (a) −7·78, (b) −0·659.
7. (a) 14, 1·02; 6·05, 1·07, (b) −0·75, (c) −0·683. 8. 0·878.
10. (a) $r(x, z) = 1$, (b) $r(x, z) = 1$, (c) $-1 \leqslant r(x, z) \leqslant 1$.
13. $\left(\sum u^2\right)\lambda^2 + 2\left(\sum uv\right)\lambda + \sum v^2$ 14. 0, 0. 15. 0·04, 0·059. 16. 0, 0.

Exercise 17.3

1. (a) 0·429, (b) 0·543, (c) −0·314. 2. (a) 0·333, (b) 0·467, (c) −0·2.
3. P, Q: 0·75; Q, R: 0·567; P, R: 0·317.
4. P, Q: 0·556; Q, R: 0·444; P, R: 0·222.
5. 1, 1. 6. −0·858, −0·711. 7. −0·15, −0·111.
8. 0·530, 0·444. 9. 0·317, 0·167. 10. −0·049, 0, 0·048.
11. −0·781; −1, −1. 12. −0·922, −0·8.

Exercise 17.4

1. $y = 114{\cdot}38 - 1{\cdot}45x$; (a) no, (b) yes: 61.
2. $y = 32{\cdot}05 + 0{\cdot}73x$; first case: 78.
3. (ii) (22, 380), (iv) $y = 22 + 16x$; £422.
4. $y = -5{\cdot}35 + 0{\cdot}47x$; $x = 11{\cdot}81 + 2{\cdot}10y$.
5. $y = 18{\cdot}63 - 0{\cdot}76x$; $x = 19{\cdot}93 - 0{\cdot}88y$.
6. $y = 240{\cdot}82 + 0{\cdot}051x$; $x = 248{\cdot}84 + 0{\cdot}049y$.
7. $y = 0{\cdot}798 + 0{\cdot}020x$; $x = -9{\cdot}35 + 22{\cdot}22y$.
8. $y = 46576 - 0{\cdot}78x$; $x = 1841{\cdot}5 - 0{\cdot}0013y$.
9. (a) $r\sqrt{\left(\frac{S_{yy}}{S_{xx}}\right)}$, (b) $\frac{1}{r}\sqrt{\left(\frac{S_{yy}}{S_{xx}}\right)}$; "$y$ on x" line: $y = \bar{y}$; "x on y" line: $x = \bar{x}$.

Exercise 17.5

1. (a) −0·930, (b) $r_S = 0{\cdot}143$, $r_K = 0$.
2. (a) $r_S = 0{\cdot}143$, $r_K = 0{\cdot}071$, (b) 0·152.
3. (a) $r_S = 0{\cdot}6$, $r_K = 0{\cdot}4$, (b) $x = -1{\cdot}17 + 1{\cdot}23y$.
4. $y = 2{\cdot}90 + 0{\cdot}56x$; 0·792. 5. 0·603; $W = -75{\cdot}90 + 0{\cdot}89h$.

6. (a) 0·3, 1·65, (c) $y = 1{\cdot}257 + 1{\cdot}31x$, (e) 3·42 tonnes. 7. $-0{\cdot}036$.
8. (a) 2·5, 2·45, (b) 0·85, 0·4475, (c) 0·425, (d) $y = 1{\cdot}2 + 0{\cdot}5x$, (e) 2·95.
9. 0·309. 10. (ii) $y = 48{\cdot}35 + 2{\cdot}75x$, (iii) 0·787.
11. (i) $-0{\cdot}374$, (ii) $-0{\cdot}256$, (iii) scatter diagram suggests non-linear relationship.
12. (i) $y = -2{\cdot}40 + 0{\cdot}49x$, (ii) 0·997, (iii) no.
13. (i) $y = 5680 + 1{\cdot}33x$, (ii) 28300.
14. (a) Invalid, (b) invalid, (c) valid, (d) invalid.
15. (i) $y = 16 + 0{\cdot}4x$, (ii) 0·330 to 0·470; 6·60 to 9·40.

Exercise 18.1

1. (a) $(3, 0, 2), (0, 2, 0)$, (b) $(2, -1, -1), (0, 2, 0)$, (c) $(5, -10, 0), (2, 5, -4)$.
2. (a) $\begin{pmatrix} 1 & 0 & 0 \\ 0 & 1 & 0 \\ 0 & 0 & -1 \end{pmatrix}$, (b) $\begin{pmatrix} 5 & 0 & 0 \\ 0 & 5 & 0 \\ 0 & 0 & 5 \end{pmatrix}$,

   (c) $\begin{pmatrix} -1 & 0 & 0 \\ 0 & 1 & 0 \\ 0 & 0 & -1 \end{pmatrix}$, (d) $\begin{pmatrix} 1 & 0 & 0 \\ 0 & 1 & 0 \\ 0 & 0 & 2 \end{pmatrix}$,

   (e) $\begin{pmatrix} 1 & 0 & 0 \\ 0 & 0 & -1 \\ 0 & 1 & 0 \end{pmatrix}$, $\begin{pmatrix} 1 & 0 & 0 \\ 0 & 0 & 1 \\ 0 & -1 & 0 \end{pmatrix}$.

3. (a) 90° rotation about z-axis, $(0, 1, 0) \to (1, 0, 0)$. (b) Central inversion.
   (c) Orthogonal projection onto x, z plane. (d) Reflection in x, z plane.
   (e) 120° rotation about line $x = y = z$, $(1, 0, 0) \to (0, 0, 1)$.
   (f) Two-way stretch, $\times$ 3, in both x- and y-directions.
4. (a) Rotation through θ about z-axis, from x-axis towards y-axis.
   (b) Reflection in plane $y = z$.
   (c) Shear with y, z plane invariant, $(1, 0, 0) \to (1, 0, 1)$.
   (d) 120° rotation about line $x = -y = -z$, $(0, 1, 0) \to (0, 0, 1)$.
   (e) 180° rotation about line $x = z$, $y = 0$.
   (f) Mapping onto plane $x = y$.
5. (a) $\begin{pmatrix} 0 & -1 & 0 \\ -1 & 0 & 0 \\ 0 & 0 & 1 \end{pmatrix}$, (b) $\begin{pmatrix} \frac{1}{2} & \frac{1}{2} & 0 \\ \frac{1}{2} & \frac{1}{2} & 0 \\ 0 & 0 & 1 \end{pmatrix}$,

   (c) $\begin{pmatrix} \cos\theta & 0 & -\sin\theta \\ 0 & 1 & 0 \\ \sin\theta & 0 & \cos\theta \end{pmatrix}$ or $\begin{pmatrix} \cos\theta & 0 & \sin\theta \\ 0 & 1 & 0 \\ -\sin\theta & 0 & \cos\theta \end{pmatrix}$,

   (d) $\begin{pmatrix} 1 & 0 & 1 \\ 0 & 1 & 1 \\ 0 & 0 & 1 \end{pmatrix}$, (e) $\begin{pmatrix} 0 & -1 & 0 \\ -1 & 0 & 0 \\ 0 & 0 & 1 \end{pmatrix}$.

6. $\begin{pmatrix}-1 & 0 & 0\\ 0 & 1 & 0\\ 0 & 0 & -1\end{pmatrix}, \begin{pmatrix}1 & 0 & 0\\ 0 & 1 & 0\\ 0 & 0 & 1\end{pmatrix}$; **M**: 90° rotation about y-axis, $(1,0,0) \to (0,0,1)$, $\mathbf{M}^2$: 180° rotation about y-axis, $\mathbf{M}^4$: identity transformation.

7. $\begin{pmatrix}0 & 0 & 1\\ 0 & 1 & 0\\ 1 & 0 & 0\end{pmatrix}, \begin{pmatrix}0 & 0 & -1\\ 0 & 1 & 0\\ -1 & 0 & 0\end{pmatrix}$; **P**: reflection in y, z plane, **Q**: 90° rotation about y-axis, $(1,0,0) \to (0,0,-1)$,

QP: reflection in plane $x = z$, **PQ**: reflection in plane $x + z = 0$.

8. (a) $\begin{pmatrix}1 & 0 & 0\\ 0 & 0 & -1\\ 0 & 1 & 0\end{pmatrix}$, (b) $\begin{pmatrix}-1 & 0 & 0\\ 0 & -1 & 0\\ 0 & 0 & -1\end{pmatrix}$, (c) no inverse,

(d) $\begin{pmatrix}1 & 0 & 0\\ 0 & -1 & 0\\ 0 & 0 & 1\end{pmatrix}$, (e) $\begin{pmatrix}0 & 0 & 1\\ 1 & 0 & 0\\ 0 & 1 & 0\end{pmatrix}$, (f) $\begin{pmatrix}\frac{1}{3} & 0 & 0\\ 0 & \frac{1}{3} & 0\\ 0 & 0 & 1\end{pmatrix}$.

9. (a) $\begin{pmatrix}\cos\theta & \sin\theta & 0\\ -\sin\theta & \cos\theta & 0\\ 0 & 0 & 1\end{pmatrix}$, (b) $\begin{pmatrix}1 & 0 & 0\\ 0 & 0 & 1\\ 0 & 1 & 0\end{pmatrix}$, (c) $\begin{pmatrix}1 & 0 & 0\\ 0 & 1 & 0\\ -1 & 0 & 1\end{pmatrix}$,

(d) $\begin{pmatrix}0 & -1 & 0\\ 0 & 0 & 1\\ -1 & 0 & 0\end{pmatrix}$, (e) $\begin{pmatrix}0 & 0 & 1\\ 0 & -1 & 0\\ 1 & 0 & 0\end{pmatrix}$, (f) no inverse.

10. $\mathbf{r} = \lambda(\mathbf{i} - \mathbf{j} + 2\mathbf{k})$; the plane $2x - 5y + 3z = 0$.
11. $\mathbf{r} = \lambda(\mathbf{i} - 3\mathbf{j} + 4\mathbf{k}) + \mu(3\mathbf{i} + 5\mathbf{j} - 2\mathbf{k})$; $x - y - z = 0$.

12. $\begin{pmatrix}-2 & 2 & -2\\ -4 & 4 & -4\\ -2 & 2 & -2\end{pmatrix}, \begin{pmatrix}0 & 0 & 0\\ 0 & 0 & 0\\ 0 & 0 & 0\end{pmatrix}$; **M**: mapping onto plane $x - y + z = 0$, $\mathbf{M}^2$: mapping onto line $\mathbf{r} = \lambda(\mathbf{i} + 2\mathbf{j} + \mathbf{k})$. $\mathbf{M}^3$: mapping onto origin.

Exercise 18.2

1. (a) $\begin{pmatrix}3 & 1 & -3\\ -2 & -3 & -2\\ 5 & -1 & 1\end{pmatrix}$, (b) $\begin{pmatrix}1 & 2 & -11\\ 6 & -6 & 11\\ 10 & 3 & -8\end{pmatrix}$, (c) $\begin{pmatrix}0 & -4 & 9\\ -6 & 1 & -11\\ -5 & -1 & -3\end{pmatrix}$,

(d) $\begin{pmatrix}-7 & -1 & 5\\ 19 & 2 & -11\\ -10 & -3 & 3\end{pmatrix}$, (e) $\begin{pmatrix}1 & -1 & 3\\ -2 & 3 & -8\\ 2 & -2 & 7\end{pmatrix}$.

2. $\begin{pmatrix}16 & 0 & 0\\ 0 & 16 & 0\\ 0 & 0 & 16\end{pmatrix}$; $\frac{1}{16}\begin{pmatrix}1 & 3 & 0\\ 2 & -1 & 3\\ -1 & 0 & 1\end{pmatrix}$, $\frac{1}{16}\begin{pmatrix}1 & 3 & -9\\ 5 & -1 & 3\\ 1 & 3 & 7\end{pmatrix}$.

3. $\begin{pmatrix} 0 & 0 & 1 \\ -1 & 0 & 0 \\ 0 & -1 & 0 \end{pmatrix}$, $\begin{pmatrix} 1 & 0 & 0 \\ 0 & 1 & 0 \\ 0 & 0 & 1 \end{pmatrix}$; $\begin{pmatrix} 0 & 0 & 1 \\ -1 & 0 & 0 \\ 0 & -1 & 0 \end{pmatrix}$.

4. $\begin{pmatrix} -3 & 0 & 0 \\ 0 & -3 & 0 \\ 0 & 0 & -3 \end{pmatrix}$; $\frac{1}{3}\begin{pmatrix} -4 & -2 & 1 \\ 1 & -1 & -1 \\ -1 & -2 & -2 \end{pmatrix}$.

5. $\begin{pmatrix} 1 & -3 & 2 \\ -2 & 0 & 1 \\ 0 & 2 & 3 \end{pmatrix}$, $\begin{pmatrix} 3 & 0 & -2 \\ -1 & 4 & 1 \\ 2 & -1 & 0 \end{pmatrix}$

6. $\begin{pmatrix} 9 & 0 & 0 \\ 0 & 9 & 0 \\ 0 & 0 & 9 \end{pmatrix}$; $\frac{1}{9}\mathbf{P}^T$.

7. (a) -5, (b) 11, (c) -11, (d) 0, (e) 142, (f) -35.
8. (a) 0, (b) -354, (c) 0, (d) 840, (e) $-20\,160$, (f) 0.
9. (a) $-a-b+2c$, (b) $13a+5b-c$, (c) $a+2b-5c$.
10. (a) 5, (b) 28, (c) -9. 11. 6, 6, -2.

Exercise 18.3

1. $\begin{pmatrix} 2 & 1 & 1 \\ -9 & -4 & -6 \\ 6 & 3 & 4 \end{pmatrix}$, $\begin{pmatrix} 2 & 1 & 1 \\ -9 & -4 & -6 \\ 6 & 3 & 4 \end{pmatrix}$.

2. $\begin{pmatrix} -5 & -3 & -2 \\ 15 & 8 & 6 \\ 7 & 4 & 3 \end{pmatrix}$, $\begin{pmatrix} 5 & 3 & 2 \\ -15 & -8 & -6 \\ -7 & -4 & -3 \end{pmatrix}$.

3. $\begin{pmatrix} 4 & 8 & -4 \\ 14 & 28 & -14 \\ 6 & 12 & -6 \end{pmatrix}$, no inverse.

4. $\begin{pmatrix} -4 & -1 & 8 \\ 5 & -1 & -1 \\ 3 & 3 & -6 \end{pmatrix}$, $\frac{1}{9}\begin{pmatrix} -4 & -1 & 8 \\ 5 & -1 & -1 \\ 3 & 3 & -6 \end{pmatrix}$.

5. $\begin{pmatrix} -8 & -9 & 5 \\ 6 & 3 & 0 \\ -13 & -9 & 10 \end{pmatrix}$, $\frac{1}{15}\begin{pmatrix} 8 & 9 & -5 \\ -6 & -3 & 0 \\ 13 & 9 & -10 \end{pmatrix}$.

6. $\begin{pmatrix} 6 & -2 & 7 \\ -2 & -1 & -4 \\ -3 & 1 & -6 \end{pmatrix}$, $\frac{1}{5}\begin{pmatrix} 6 & -2 & 7 \\ -2 & -1 & -4 \\ -3 & 1 & -6 \end{pmatrix}$.

7. $\begin{pmatrix} 0 & 0 & 0 \\ 0 & 0 & 0 \\ 0 & 0 & 0 \end{pmatrix}$, no inverse.

8. $\begin{pmatrix} -9 & 9 & 0 \\ -7 & 4 & 2 \\ 17 & -11 & -1 \end{pmatrix}$, $\frac{1}{9}\begin{pmatrix} -9 & 9 & 0 \\ -7 & 4 & 2 \\ 17 & -11 & -1 \end{pmatrix}$.

9. $\begin{pmatrix} -11 & -3 & -1 \\ -1 & -6 & -2 \\ -8 & -6 & 5 \end{pmatrix}$, $\frac{1}{21}\begin{pmatrix} 11 & 3 & 1 \\ 1 & 6 & 2 \\ 8 & 6 & -5 \end{pmatrix}$.

10. $\begin{pmatrix} -2 & 12 & -2 \\ -2 & 0 & 2 \\ 2 & -4 & 2 \end{pmatrix}$, $\frac{1}{4}\begin{pmatrix} -1 & 6 & -1 \\ -1 & 0 & 1 \\ 1 & -2 & 1 \end{pmatrix}$.

11. $\begin{pmatrix} 14 & 0 & 7 \\ 56 & 0 & 28 \\ 14 & 0 & 7 \end{pmatrix}$, no inverse.

12. $\begin{pmatrix} -25 & -32 & -10 \\ -10 & -13 & -4 \\ -18 & -23 & -7 \end{pmatrix}$, $\begin{pmatrix} 25 & 32 & 10 \\ 10 & 13 & 4 \\ 18 & 23 & 7 \end{pmatrix}$.

13. $\begin{pmatrix} 0 & -1 & 1 \\ -1 & 0 & 1 \\ 0 & 0 & 1 \end{pmatrix}$, $\frac{1}{2}\begin{pmatrix} 5 & 6 & -15 \\ -7 & -8 & 21 \\ -1 & -2 & 5 \end{pmatrix}$.

Exercise 18.4

1. $\begin{pmatrix} -1 & -1 & 1 \\ 2 & 3 & -1 \\ -2 & -2 & 1 \end{pmatrix}$; (a) $-1, 4, -3$, (b) $2, -2, 1$, (c) $0, 0, 0$.

2. $\frac{1}{10}\begin{pmatrix} 5 & 3 & -1 \\ -5 & 1 & 3 \\ -5 & -5 & 5 \end{pmatrix}$; (a) $1, -1, -2$, (b) $1, 0, -2$, (c) $\frac{1}{5}, \frac{7}{5}, 0$.

3. (a) $2, 1, -10$, (b) $-3, 2, 3/2$, (c) $1, -1/3, -2/3$, (d) $1/4, 3/4, -7/4$, (e) $-1, 0, 2$, (f) $-2/7, 5/7, 13/7$.
4. (a) No solution, (b) $\lambda, 3 - 2\lambda, 1 + 3\lambda$.
5. (a) No solution, (b) $\lambda, 4\lambda - 7, 2 + 5\lambda$.
6. (a) $\lambda, \mu, -5 - 4\lambda + 3\mu$, (b) no solution
7. (a) $\lambda, -\frac{1}{3}(4 + \lambda), \lambda$, (b) $\lambda, \mu, \frac{1}{2}(\lambda - 3\mu - 4)$.
8. $\lambda, 7\lambda - 10, 5\lambda - 7$.
9. (a) $k = 0$; $2, \lambda, \lambda - 1$, (b) $k = -3/2$; $\lambda, \frac{1}{4}(7 - 10\lambda), \frac{1}{4}(3 + 2\lambda)$.
10. $4 - 4k$; $k = 1$; $\lambda, 2\lambda - 5, \lambda - 2$.

11. (i) Three planes meeting in a single point;
    (ii) each plane parallel to the line of intersection of the other two;
    (iii) three planes with common line of intersection.
12. $1, 2, -1; -5; -5, -11; 3\lambda - 5, \lambda, 5\lambda - 11$.

Exercise 18.5

1. (a) $\begin{pmatrix} 1 & 0 & 0 \\ 1 & 2 & -1 \\ -1 & -1 & 1 \end{pmatrix}$, (b) $\begin{pmatrix} 2 & 1 & 0 \\ 1 & 1 & 1 \\ 0 & 0 & 1 \end{pmatrix}$,

   (c) $\begin{pmatrix} 1 & 2 & 3 \\ 0 & 1 & 2 \\ 0 & 0 & 1 \end{pmatrix}$, (d) $\begin{pmatrix} 1 & 0 & 0 \\ 3 & 1 & 0 \\ 1 & 2 & 1 \end{pmatrix}$.

2. (a) $\begin{pmatrix} 20 & 5 & -2 \\ 11 & 3 & -1 \\ -8 & -2 & 1 \end{pmatrix}$, (b) $\begin{pmatrix} -10 & -7 & 2 \\ -6 & -4 & 1 \\ -7 & -5 & 1 \end{pmatrix}$,

   (c) $\begin{pmatrix} 0 & 1 & 0 \\ 1 & 0 & 1 \\ 0 & 1 & 1 \end{pmatrix}$, (d) $\begin{pmatrix} 7 & 4 & 2 \\ 3 & 2 & 1 \\ -5 & -3 & -1 \end{pmatrix}$.

3. (a) $\begin{pmatrix} -3 & 2 & 1 \\ 9 & -5 & -2 \\ -5 & 3 & 1 \end{pmatrix}$, (b) $\begin{pmatrix} 1 & 0 & 2 \\ 0 & 0 & 1 \\ -3 & 1 & -7 \end{pmatrix}$,

   (c) no inverse, (d) $\begin{pmatrix} 0 & 0 & 1 \\ -3 & 1 & -11 \\ 1 & 0 & 3 \end{pmatrix}$.

4. (a) $\frac{1}{2}\begin{pmatrix} -2 & 0 & 1 \\ -4 & 2 & -1 \\ 0 & 0 & 1 \end{pmatrix}$, (b) no inverse,

   (c) $\frac{1}{6}\begin{pmatrix} 2 & -4 & 7 \\ 0 & 6 & -12 \\ 0 & 0 & 3 \end{pmatrix}$, (d) $\frac{1}{2}\begin{pmatrix} 1 & -2 & -1 \\ -4 & 22 & 8 \\ 2 & -12 & -4 \end{pmatrix}$.

5. (a) $\begin{pmatrix} 0 & -1 & 2 \\ -1 & 2 & -1 \\ 1 & -1 & 0 \end{pmatrix}$, (b) $\frac{1}{2}\begin{pmatrix} 1 & -3 & 0 \\ -1 & 3 & 1 \\ 0 & -2 & 0 \end{pmatrix}$,

   (c) $\frac{1}{3}\begin{pmatrix} -4 & 0 & 1 \\ 3 & 0 & 0 \\ 14 & 3 & -5 \end{pmatrix}$, (d) no inverse

6. (a) $\begin{pmatrix} 0 & 1 & 0 \\ 0 & 0 & 1 \\ 1 & 0 & 0 \end{pmatrix}$, (b) $\begin{pmatrix} 1 & 0 & 0 \\ 0 & \frac{1}{2} & 0 \\ 0 & 0 & \frac{1}{3} \end{pmatrix}$, (c) $\begin{pmatrix} 0 & 0 & \frac{1}{3} \\ 0 & \frac{1}{2} & 0 \\ 1 & 0 & 0 \end{pmatrix}$, (d) $\begin{pmatrix} 1 & 0 & 0 \\ 0 & 0 & -1 \\ 0 & -1 & 0 \end{pmatrix}$.

7. $$\mathbf{A}^{-1} = \begin{pmatrix} 1 & 0 & 0 \\ -4 & 1 & 0 \\ -5 & 2 & 1 \end{pmatrix}, \mathbf{B}^{-1} = \frac{1}{4}\begin{pmatrix} 2 & 2 & 10 \\ 1 & -1 & -5 \\ -1 & 1 & 1 \end{pmatrix}.$$

8. $$\begin{pmatrix} -16 & 9 & 4 \\ -5 & 3 & 1 \\ 4 & -2 & -1 \end{pmatrix}, \begin{pmatrix} 1 & -1 & 0 \\ 0 & 1 & -1 \\ 0 & 0 & 1 \end{pmatrix};$$
$$\begin{pmatrix} -11 & 6 & 3 \\ -9 & 5 & 2 \\ 4 & -2 & -1 \end{pmatrix}, \begin{pmatrix} -16 & 41 & -30 \\ -5 & 13 & -10 \\ 4 & -10 & 7 \end{pmatrix}.$$

Exercise 18.6

1. (a) Reflection in the plane $y + z = 0$.
   (b) 180° rotation about z-axis and enlargement × 2.
   (c) 30° rotation about y-axis, from x-axis to z-axis.

2. (a) $\begin{pmatrix} \cos 2\theta & \sin 2\theta & 0 \\ \sin 2\theta & -\cos 2\theta & 0 \\ 0 & 0 & 1 \end{pmatrix}$, (b) $\begin{pmatrix} \cos 2\theta & \sin 2\theta & 0 \\ \sin 2\theta & -\cos 2\theta & 0 \\ 0 & 0 & -1 \end{pmatrix}$.

3. $(x, y, -z), (-y, x, z), (y, x, -z)$;
$$\begin{pmatrix} 1 & 0 & 0 \\ 0 & 1 & 0 \\ 0 & 0 & -1 \end{pmatrix}, \begin{pmatrix} 0 & -1 & 0 \\ 1 & 0 & 0 \\ 0 & 0 & 1 \end{pmatrix}, \begin{pmatrix} 0 & 1 & 0 \\ 1 & 0 & 0 \\ 0 & 0 & -1 \end{pmatrix};$$
$\begin{pmatrix} 1 & 0 & 0 \\ 0 & -1 & 0 \\ 0 & 0 & -1 \end{pmatrix}$ 180° rotation about x-axis,

$\begin{pmatrix} -1 & 0 & 0 \\ 0 & 1 & 0 \\ 0 & 0 & -1 \end{pmatrix}$ 180° rotation about y-axis;

reflection in x, z plane, reflection in y, z plane.

4. $3x - 2y + z = 0$; the line $\mathbf{r} = \lambda(9\mathbf{i} + 3\mathbf{j} - 2\mathbf{k})$.
5. The line $x = 1$, the line $y = -x$; anti-clockwise rotation through 90° about $(1, -1)$.

6. (a) $\begin{pmatrix} 1 & 0 & 0 \\ 0 & 1 & 0 \\ 0 & 0 & 1 \end{pmatrix}$, (b) $\begin{pmatrix} 1 & -1 & -1 \\ -1 & 3 & -1 \\ 0 & 1 & 3 \end{pmatrix}$.

7. (a) $\mathbf{I} - \mathbf{P}$, (c) $\mathbf{I} - 2\mathbf{P}$.

9. (a) $\begin{pmatrix} 9 & -10 & 7 \\ -4 & 5 & -3 \\ 1 & -1 & 1 \end{pmatrix}$, (b) $\begin{pmatrix} 1 & 0 & 0 \\ 0 & 1 & 0 \\ 0 & 0 & 1 \end{pmatrix}$, $\begin{pmatrix} -2 & 1 & -1 \\ 1 & 2 & 1 \\ 3 & 2 & 2 \end{pmatrix}$.

10. (i) $\begin{pmatrix} -2 & -1 \\ 1\frac{1}{2} & \frac{1}{2} \end{pmatrix}$, (ii) no inverse, $\Delta = 0$, (iii) no inverse, $\Delta = 0$,

(iv) no inverse (not square), (v) $\begin{pmatrix} 1 & -1 & 1 \\ 0 & 1 & -1 \\ 0 & 0 & 1 \end{pmatrix}$.

11. $\frac{1}{5}\begin{pmatrix} -1 & 9 & -3 \\ -2 & 3 & -1 \\ 2 & -13 & 6 \end{pmatrix}$; $\frac{1}{5}(-a + 9b - 3c)$, $\frac{1}{5}(-2a + 3b - c)$, $\frac{1}{5}(2a - 13b + 6c)$.

12. $\begin{pmatrix} 18 & -5 & 2 \\ -10 & 3 & -1 \\ 7 & -2 & 1 \end{pmatrix}$; 7, -3, 3; $\begin{pmatrix} 5 & -1 & 1 \\ 4 & -1 & 0 \\ -3 & 1 & 0 \end{pmatrix}$.

13. $\begin{pmatrix} -1 & 1 & 1 \\ -7 & 5 & 4 \\ -4 & 3 & 2 \end{pmatrix}$; 1, 0, -1; $\begin{pmatrix} 3 & -4 & 6 \\ 11 & -21 & 33 \\ 6 & -12 & 19 \end{pmatrix}$.

14. 2, $-7/3$; $\alpha = 2$: λ, $4\lambda - 1$, $1 - 3\lambda$; $\alpha = -7/3$: λ, $4\lambda + 10/3$, $-3\lambda - 17/9$.

15. (i) $\frac{1}{3}$, $-\frac{1}{3}$, $\frac{2}{3}$, (ii) 5, 3; $\mathbf{K}$ does not satisfy $R_3 = 5R_1 + 3R_2$.

16. $\begin{pmatrix} 11 & -18 & 0 \\ 6 & -10 & 0 \\ 6 & -9 & -1 \end{pmatrix}$.

17. 5; λ, $3 - 2\lambda$, λ; line mapped onto point (3, 6, 9).

18. $\begin{pmatrix} 3 & 2 & 1 \\ 2 & 1 & 0 \\ 1 & 0 & 0 \end{pmatrix}$, $\begin{pmatrix} 0 & 0 & 1 \\ 0 & 1 & -2 \\ 1 & -2 & 1 \end{pmatrix}$.

19. (i) $-4/3$, $3/2$, (ii) 0, 1, $3/2$; $a = -4/3$, no solutions; $a = 3/2$, solutions exist.

Exercise 19.1

1. Group. 2. No inverses. 3. 0 has no inverse.
4. Not closed. 5. Group. 6. Group. 7. No inverses.
8. Group 9. Not associative. 10. No identity or inverses.
11. No identity or inverses. 12. Group. 13. Not closed, no inverses.
14. Not associative, no identity or inverses. 15. Group. 16. Group.
17. 1 has no inverse. 20. 1, -1, $\frac{1}{2}(-1 + i\sqrt{3})$, $\frac{1}{2}(-1 - i\sqrt{3})$, $\frac{1}{2}(1 - i\sqrt{3})$.

21. (a) Not closed, not associative, s has no inverse.
    (b) Not associative, no identity or inverses.
    (c) Not associative, q has no inverse. (d) Group.
22. (a) $\{q\}, \varnothing$, (b) $\{p\}, \{p\}$, (c) $\{s\}, \{p,q\}$, (d) $\{r\}, \{q\}$.
24. (a) B, (b) I.

$*$	I	A	B	C
I	I	A	B	C
A	A	B	C	I
B	B	C	I	A
C	C	I	A	B

15. (i) $\frac{1}{2}$; $1/4x$; (ii) $\begin{pmatrix} 1 & 0 \\ 0 & 1 \end{pmatrix}, \begin{pmatrix} -1 & 0 \\ 0 & -1 \end{pmatrix}$.

Exercise 19.2

1. No. 3. (a) 4, (b) 2, (c) 1, (d) 3, (e) 4, (f) 3.
4. (a) 6, (b) 6, (c) 5, (d) 3, (e) 5, (f) 3.
5. $\{1\}, \{5\}, \{6\}, \{1,9\}, \{6,4\}, \{6,2,4,8\}, \{1,3,7,9\}$.
6. $\{1,5\}, \{1,7\}, \{1,11\}, \{4,8\}, \{9,3\}$.
7. $\{1,2,4,8\}, \{1,4,11,14\}, \{1,4,7,13\}, \{6,3,9,12\}$.

Exercise 19.3

1. (a) B, (b) R_2, (c) R_1, (d) R_1.
3.

	I	R_1	R_2	R_3	X	Y	A	B
I	I	R_1	R_2	R_3	X	Y	A	B
R_1	R_1	R_2	R_3	I	A	B	Y	X
R_2	R_2	R_3	I	R_1	Y	X	B	A
R_3	R_3	I	R_1	R_2	B	A	X	Y
X	X	B	Y	A	I	R_2	R_3	R_1
Y	Y	A	X	B	R_2	I	R_1	R_3
A	A	X	B	Y	R_1	R_3	I	R_2
B	B	Y	A	X	R_3	R_1	R_2	I

4. (a) 2, (b) 4, (c) 10, (d) infinite. 5. (a) 8, (b) 4, (c) 8, (d) 12.
6. (a) False; parallel mirror lines produce a translation.
   (b) True.
   (c) False: a translation perpendicular to mirror line produces a reflection.
   (d) False: equal and opposite rotations produce a translation.
   (e) False: only true if centre of rotation is on mirror line, otherwise result is a glide-reflection.

7. (a) $\begin{pmatrix}1 & 2 & 3 & 4\\ 2 & 3 & 4 & 1\end{pmatrix}$, (b) $\begin{pmatrix}1 & 2 & 3 & 4\\ 4 & 3 & 2 & 1\end{pmatrix}$,

   (c) $\begin{pmatrix}1 & 2 & 3 & 4\\ 1 & 2 & 3 & 4\end{pmatrix}$, (d) $\begin{pmatrix}1 & 2 & 3 & 4\\ 1 & 3 & 2 & 4\end{pmatrix}$.

8. $\begin{pmatrix}1 & 2 & 3\\ 1 & 2 & 3\end{pmatrix}$, $\begin{pmatrix}1 & 2 & 3\\ 1 & 3 & 2\end{pmatrix}$, $\begin{pmatrix}1 & 2 & 3\\ 2 & 1 & 3\end{pmatrix}$, $\begin{pmatrix}1 & 2 & 3\\ 3 & 2 & 1\end{pmatrix}$,

   $\begin{pmatrix}1 & 2 & 3\\ 3 & 1 & 2\end{pmatrix}$, $\begin{pmatrix}1 & 2 & 3\\ 2 & 3 & 1\end{pmatrix}$.

9. (a) 3, (b) 2, (c) 5, (d) 6, (e) 8.
10. (a) Rotation through 180°. (b) Reflection in BD.
    (c) Reflection in line through mid-points of AD and BC.
    No.

Exercise 19.4

3. $\{\varnothing, X\}, \{\varnothing, Y\}, \{\varnothing, Z\}$. 4. $\{e, g\}$.
5. $\{\mathbf{I}, \mathbf{A}, \mathbf{B}\}, \{\mathbf{I}, \mathbf{C}\}, \{\mathbf{I}, \mathbf{D}\}, \{\mathbf{I}, \mathbf{E}\}$. 6. $\{I, R_1, R_2\}, \{I, A\}, \{I, B\}, \{I, C\}$.
7. $\{10, 8\}, \{10, 4, 16\}$. 8. $\{1, -1\}, \{1, -1, i, -i\}$.
9. $\{I, B\}, \{I, D\}, \{I, E\}, \{I, F\}, \{I, G\}, \{I, A, B, C\}$.
10. $\{I, H_1\}$, $\{I, H_2\}$, $\{I, H_3\}$, $\{I, X\}$, $\{I, Y\}$, $\{I, Z\}$, $\{I, C\}$, $\{I, H_1, H_2, H_3\}$, $\{I, H_1, X, C\}$, $\{I, H_1, Y, Z\}$, $\{I, H_2, Y, C\}$, $\{I, H_2, X, Z\}$, $\{I, H_3, Z, C\}$, $\{I, H_3, X, Y\}$.
11. (a) Yes, (b) No, (c) No, (d) Yes.
12. (a) No, (b) Yes, (c) No, (d) Yes. 13. No. 14. No.
15. 1, 2, 3, 4. 16. $i, -i$. 17. No. 18. Rotations of 60° or 300°.
19. 3, 5. 20. No.
21. x, x^3, x^5, x^7; e: 1, x^2: 4, x^4: 2, x^6: 4.
22. All except e which has order 1.
23. I: 1, A: 4, B: 2, C: 4, D: 4, E: 4, F: 4, G: 4.
24. (i) I: 1, A: 4, B: 2, C: 4, D: 2, E: 2, F: 2, G: 2.
    (ii) $\{I, B\}, \{I, D\}, \{I, E\}, \{I, F\}, \{I, G\}$.
    (iii) $\{I, A, B, C\}$; $B = A^2, C = A^3, I = A^4$.
    (iv) $\{I, B, D, F\}$ or $\{I, B, E, G\}$, no element of order 4.
    (v) I, B.
25. 4, 4.

Exercise 19.5

1. (a), (b), (c) and (e); (d) and (f).
2. (a), (d) and (f); (b), (c) and (e).
3. Rectangle or rhombus; given groups each have element of order 4, suggested symmetry groups do not.
4. The set does not contain the identity and is not closed.
8. a, d.

Exercise 19.6

2. Identity is 0; true.
3. e, e, b, d; closed, identity, inverses; $\{e\}$, $\{e, a\}$, $\{e, b\}$, $\{e, c\}$, $\{e, d\}$.
7. 3, 4, 5, 9. 8. 24.
9. p_1: 1, p_2: 3, p_3: 3, p_4: 2, p_5: 2, p_6: 2; $\{p_1, p_2, p_3\}$; p_2 or p_3; $\{p_1, p_4\}$, $\{p_1, p_5\}$, $\{p_1, p_6\}$.
10. (i) Identity 8, inverse elements 2 and 4, 10 and 12, 6 is self-inverse; 7 has no inverse.
    (ii) Rectangle or rhombus; square.
11. (a) (i) $a = b = 0$, $\begin{pmatrix} \exp(-a) & -b \\ 0 & \exp(a) \end{pmatrix}$, (ii) Two possibilities are given by $a \in \mathbb{R}$, $b = 0$ and $a \in \mathbb{Z}$, $b \in \mathbb{R}$.
13. Top right $\begin{matrix} \mathbf{B} & \mathbf{BA} & \mathbf{BA}^2 \\ \mathbf{BA}^3 & \mathbf{B} & \mathbf{BA} \\ \mathbf{BA}^2 & \mathbf{BA}^3 & \mathbf{B} \end{matrix}$ Bottom left $\begin{matrix} \mathbf{BA}^2 & \mathbf{BA}^3 & \mathbf{B} \\ \mathbf{BA}^3 & \mathbf{B} & \mathbf{BA} \\ \mathbf{B} & \mathbf{BA} & \mathbf{BA}^2 \end{matrix}$
14. (i) Yes, (ii) no, set does not contain inverses.
15. $$A = \begin{pmatrix} 0 & 1 & 0 \\ 1 & 0 & 0 \\ 0 & 0 & 1 \end{pmatrix};$$
$$I, A, \begin{pmatrix} 1 & 0 & 0 \\ 0 & 0 & 1 \\ 0 & 1 & 0 \end{pmatrix}, \begin{pmatrix} 0 & 0 & 1 \\ 0 & 1 & 0 \\ 1 & 0 & 0 \end{pmatrix}, \begin{pmatrix} 0 & 1 & 0 \\ 0 & 0 & 1 \\ 1 & 0 & 0 \end{pmatrix}, \begin{pmatrix} 0 & 0 & 1 \\ 1 & 0 & 0 \\ 0 & 1 & 0 \end{pmatrix};$$
$$\left\{\begin{pmatrix} a & b & c \\ a & b & c \end{pmatrix}, \begin{pmatrix} a & b & c \\ b & a & c \end{pmatrix}\right\}, \left\{\begin{pmatrix} a & b & c \\ a & b & c \end{pmatrix}, \begin{pmatrix} a & b & c \\ a & c & b \end{pmatrix}\right\},$$
$$\left\{\begin{pmatrix} a & b & c \\ a & b & c \end{pmatrix}, \begin{pmatrix} a & b & c \\ c & b & a \end{pmatrix}\right\}; \left\{\begin{pmatrix} a & b & c \\ a & b & c \end{pmatrix}, \begin{pmatrix} a & b & c \\ b & c & a \end{pmatrix}, \begin{pmatrix} a & b & c \\ c & a & b \end{pmatrix}\right\}.$$
16. $\begin{matrix} \mathbf{M}_3 & \mathbf{M}_1 & \mathbf{M}_6 \\ \mathbf{M}_1 & \mathbf{M}_2 & \mathbf{M}_5 \\ \mathbf{M}_5 & \mathbf{M}_6 & \mathbf{M}_1 \end{matrix}$ $\{\mathbf{M}_1, \mathbf{M}_2, \mathbf{M}_3\}$, $\mathrm{F} \leftrightarrow \mathbf{M}_1$, $\mathrm{G} \leftrightarrow \mathbf{M}_2$, $\mathrm{H} \leftrightarrow \mathbf{M}_3$ or $\mathrm{F} \leftrightarrow \mathbf{M}_1$, $\mathrm{G} \leftrightarrow \mathbf{M}_3$, $\mathrm{H} \leftrightarrow \mathbf{M}_2$.
17. Two possibilities are: $(\mathbb{R}, +)$, $\begin{pmatrix} 1 & \ln a \\ 0 & 1 \end{pmatrix} \leftrightarrow \ln a$; $(\mathbb{R}^+, \times)$, $\begin{pmatrix} 1 & \ln a \\ 0 & 1 \end{pmatrix} \leftrightarrow a$.
18. Yes. 20. Yes.

Index